D0558878

MUSCLE AS A FOOD, 2

Proceedings of an international symposium
sponsored by the University of Wisconsin
1969
With the support of United States Public Health
Service Research Grant 7D–00158 and a
special grant from the American Meat Institute
Foundation

# *The Physiology and Biochemistry of Muscle as a Food, 2*

EDITED BY

E. J. Briskey
R. G. Cassens
B. B. Marsh

THE UNIVERSITY OF WISCONSIN PRESS
*Madison, Milwaukee, and London, 1970*

Published by
The University of Wisconsin Press
Box 1379, Madison, Wisconsin 53701
The University of Wisconsin Press, Ltd.
27–29 Whitfield Street, London, W.1



Printed in the United States of America by
Kingsport Press, Inc., Kingsport, Tennessee

ISBN 0–299–05680–5
LC 66–22849

*To the memory of*

L. L. KASTENSCHMIDT

*1939–1970*

# Contents

# *Contributors*

*R. H. Adrian*
Physiological Laboratory
University of Cambridge
Cambridge, England

*N. Arakawa*
Department of Biochemistry, Biophysics and Animal Science
Iowa State University
Ames, Iowa 50010

*C. H. Beatty*
Oregon Regional Primate Research Center
Beaverton, Oregon 97005
and
University of Oregon Medical School
Portland, Oregon

*R. Bischoff*
Department of Anatomy
Washington University
St. Louis, Missouri 63110

*R. M. Bocek*
Oregon Regional Primate Research Center
Beaverton, Oregon 97005
and
University of Oregon Medical School
Portland, Oregon

*E. J. Briskey*
Muscle Biology Laboratory
University of Wisconsin
Madison, Wisconsin 53706

*M. Brooke*
University of Colorado
School of Medicine
Denver, Colorado 80220

*W. A. Busch*
Department of Biochemistry, Biophysics and Animal Science
Iowa State University
Ames, Iowa 50010

*R. G. Cassens*
Muscle Biology Laboratory
University of Wisconsin
Madison, Wisconsin 53706

*E. Cosmos*
Institute for Muscle Diseases, Inc.
515 East 71st St.
New York, New York 10021

*V. Dubowitz*
Children's Hospital
University of Sheffield
Sheffield, England

*S. Ebashi*
Department of Pharmacology
University of Tokyo
Tokyo, Japan

*J. A. Faulkner*
Department of Physiology
University of Michigan
Ann Arbor, Michigan

*T. Fukazawa*
Faculty of Agriculture
Hokkaido University
Sapporo, Japan

*G. F. Gauthier*
Department of Biological Sciences
Wellesley College
Wellesley, Massachusetts 02181

*J. Gergely*
Department of Muscle Research
Retina Foundation
Harvard Medical School
Boston, Massachusetts 02114

*G. Goldspink*
Muscle Research Laboratory
University of Hull
Hull, England

*D. E. Goll*
Department of Biochemistry, Biophysics and Animal Science
Iowa State University
Ames, Iowa 50010

*E. E. Gordon*
Research Director
Department of Physical Medicine
Jefferson Medical College
Philadelphia, Pennsylvania 19107

*D. Green*
Institute for Enzyme Research
University of Wisconsin
Madison, Wisconsin 53706

*R. A. Harris*
Department of Biochemistry
Indiana University Medical Center
Indianapolis, Indiana 46203

*R. J. Havel*
Department of Medicine and Cardiovascular Research Institute
University of California Medical Center
San Francisco, California

*S. Heywood*
Department of Genetics and Cell Biology
University of Connecticut
Storrs, Connecticut 06268

*H. Holtzer*
Department of Anatomy
University of Pennsylvania
Philadelphia, Pennsylvania 19104

*B. Issekutz, Jr.*
Department of Physiology and Biophysics
Dalhousie University
Halifax, Nova Scotia, Canada

*L. L. Kastenschmidt*
Muscle Biology Laboratory
University of Wisconsin
Madison, Wisconsin 53706

*B. R. Landau*
Department of Medicine
Case Western Reserve University
Cleveland, Ohio 44106

*H. A. Lardy*
Institute for Enzyme Research
University of Wisconsin
Madison, Wisconsin 53706

*R. A. Lawrie*
Department of Applied Biochemistry and Nutrition
University of Nottingham
Loughborough, Leics,
England

*D. Lister*
British Meat Research Institute
Langford, Bristol
England

*B. B. Marsh*
Meat Industry Research Institute of New Zealand
Hamilton, New Zealand

*K. Maruyama*
Biological Institute
University of Tokyo
Komaba-Cho Meguro-Ku
Tokyo, Japan

*W. F. H. M. Mommaerts*
Department of Physiology
University of California
Los Angeles, California 90024

*L. D. Peachey*
Department of Biochemistry and Biophysics
University of Pennsylvania
Philadelphia, Pennsylvania 19104

*S. V. Perry*
Department of Biochemistry
University of Birmingham
Birmingham, England

*R. M. Robson*
Department of Animal Science
University of Illinois
Urbana, Illinois 61801

*R. K. Scopes*
British Meat Research Institute
Langford, Bristol
England

*D. B. Slautterback*
Department of Anatomy
University of Wisconsin
Madison, Wisconsin 53706

*M. H. Stromer*
Department of Biochemistry, Biophysics and Animal Science
Iowa State University
Ames, Iowa 50010

*A. Veis*
Department of Biochemistry
Northwestern University
Evanston, Illinois 60611

*A. Weber*
Department of Biochemistry
School of Medicine
St. Louis University
St. Louis, Missouri 63103

*J. R. Whitaker*
Department of Food Science
University of California
Davis, California

*E. Widdowson*
Department of Infant Nutrition
University of Cambridge
Cambridge, England

# *Preface*

This book epitomizes the concept that has given rise to the development of an Institute of Muscle Biology on the University of Wisconsin campus. It represents the proceedings of the second Symposium on the Physiology and Biochemistry of Muscle as a Food, held on the University of Wisconsin campus, July 14–16, 1969. The first symposium of this series, held in 1965, served as a foundation for the present symposium, which consisted of eight sessions probing such topics as: muscle cell development, red and white muscle, muscle membrane systems, myofibrillar proteins, stromal and sarcoplasmic proteins, muscle adaptation, muscle metabolism, and the biology of muscle as a food. Supplementary chapters were contributed by E. Cosmos, G. F. Gauthier, and J. A. Faulkner. Discussion sessions, chaired by distinguished scientists, were directed toward the clarification of significant questions in basic and applied aspects of muscle biology, with many direct applications to the use of muscle as a food.

The planning and editorial committee wishes to thank all contributors, session chairmen, and panel review members for their efforts and their enthusiastic cooperation. We also express our appreciation to the symposium's advisory committee, consisting of T. Fukuzawa, D. E. Goll, R. Hamm, R. A. Lawrie, B. B. Marsh, R. B. Sleeth, A. D. Stevens, and G. F. Stewart. Dr. Stewart, as Secretary-General of the International Committee of Food Science and Technology, cosponsored the symposium and offered evaluative and helpful suggestions for future programs. For the interest and cooperation of our colleagues and graduate students in muscle biology, the planning committee is grateful. Finally, we would like to thank Mrs. Jean Zwaska and Mrs. M. Fitzsimmons for their ready assistance and secretarial help.

Special appreciation is due to the United States Public Health Service, for research grant FD–00158, and to the American Meat Institute Foundation for a special symposium grant. Supplemental support was provided by: the American Meat Science Association; the Graduate School, the

College of Agricultural and Life Sciences, and the Meat and Animal Science Department, University of Wisconsin, Madison; the Wisconsin Alumni Research Institute, Madison; Jones Dairy, Fort Atkinson, Wisconsin; Oscar Mayer & Co., Madison, Wisconsin. We are especially grateful to Mr. Dean Griffith, Executive Vice President, and Dr. Louis Sair, Vice President, of the Griffith Laboratories, for their action in making additional funds available for the publication of these proceedings.

We wish to thank Dean G. S. Pound and Associate Dean R. W. Bray, College of Agricultural and Life Sciences, for making it possible for Dr. B. B. Marsh to hold a Visiting Professorship at the University of Wisconsin for a six-month period during the editing of these proceedings.

E. J. B.
R. G. C.
B. B. M.

*Madison, Wisconsin*
*January, 1970*

# MUSCLE AS A FOOD, 2

# Chapter 1

# Muscle as a Food

B. B. MARSH

"In the past two decades . . . fundamental knowledge of the physiological and biochemical properties and behaviour of muscle has increased out of all recognition. Perhaps because of the bewildering rate of growth of this fundamental knowledge and the constantly changing conception of muscle which has resulted, there has not been . . . any striking application of the principles of modern biochemistry to the technology of handling of meat animals and meat."

These words are neither recent nor my own; they have been quoted from the first page of the first paper in Volume 1 of *Advances in Food Research,* which appeared just 21 years ago. The paper, by E. C. Bate-Smith, was called "The physiology and chemistry of rigor mortis, with special reference to the aging of beef," a title very similar to that of the present symposium, with its clearly implied recognition of the substantial contribution which physiology and biochemistry can make to meat science. But one's attention is drawn more particularly to his thoughts on "the bewildering growth of fundamental knowledge," on "the constantly changing conception of muscle," and on the absence of "any striking application of these principles" to meat technology. These views are even more relevant today than they were in 1948; for although our knowledge of meat and its qualities has grown enormously in the intervening years, knowledge of muscle and its behavior has increased even more. The gap between discovery and application is still widening. The output of new information, the development of new concepts, the postulation of new theories—these are accelerating at such a rate that we may liken muscle biology to an expanding universe and ourselves to the particles within it, fast receding from each other and suffering the inevitable consequence of greater and greater isolation. Compounding the problem, new branches of science splinter from the old, and studies which a few years ago were clearly together within one particular discipline are now so altered that they are scarcely recognizable as relatives. Expansion and specialization, vital though they are, demand a high fee, and we in muscle biology are paying the price by learning more about less with every forward step.

This symposium gives us the opportunity to demonstrate that isolation and fragmentation have not yet separated us entirely from our erstwhile close neighbors. We shall be able to establish that we are not merely prepared but eager to share both our knowledge and our problems, and to prove that depth of study and breadth of vision are not necessarily related inversely. If we are successful in attaining these general objectives, we cannot fail to attain two which are more specific: first, to make the meat scientist not only *aware* of the current situation in muscle science but thoroughly *familiar* with it, so that meat and muscle become truly synonymous in his projective thinking; second (and equally important), to broaden the outlook of the basic muscle biologist to the point where he is conscious not only of the contributions he can make to meat science but also of the benefits he can receive from it.

First and foremost, muscle is a machine for translating chemical energy into physical movement. Like almost all machines it has a supporting structure, an energy store, a fuel entry, and a waste exhaust system. Like most machines it possesses a set of controls to make it function quickly or slowly or not at all. Like some machines it is fitted with a communication system permitting its operation from a distance, and a set of gears to accommodate both large and small loadings. It comes in a wide variety of shapes and sizes (to suit a great range of tasks and situations), and in a limited selection of colors, based on functional rather than esthetic requirements; it also has a built-in alternative power source immediately available for short periods of crisis. Unlike most machines it is able to build, maintain, and repair itself from relatively simple materials, has various compensatory mechanisms to allow adjustment to new conditions, and quite literally is constructed to last a lifetime. Finally, unique among machines, at the end of its working life it is both edible and nutritious, and becomes in fact an extremely valuable item of commerce.

It should not surprise us, then, to find that muscle is such a complex tissue; it could be nothing else in view of the number and variety of functions it must perform. To those whose interest in it to the present has been confined to the end product, this very complexity may be something of a comfort, for it adequately explains why meat has proven an almost impossibly difficult material to control, to standardize, or to improve.

Long before the machine becomes a foodstuff—indeed, before it is even performing fully its mechanical function—it must be assembled from its constituent parts. The study of growth and development of muscle is of great significance in relation to the health of the animal, the efficiency of its musculature in life, and the properties and amount of its meat in death. It is necessarily a topic of vast complexity and wide scope: among its many aspects we may note the synthesis of proteins, the structural organization and synchronization of component parts, and the involved pattern of differentiation dictated by a variety of functional requirements. If the normal characteristics of muscle are to be modified, or if its abnormal properties or behavior are to be corrected, then the myogenetic and developmental pattern of the tissue must first be understood.

As an example of the spectacular results already achieved in another field of protein production, I draw your attention to the recent finding that wool yield is at least trebled if a gram or two of sulfur-containing amino acid is introduced daily into the fourth stomach of a sheep. A trebling of animal size might present embarrassing problems in farm management; nevertheless the possibility of a greatly increased rate of growth is an exciting one, in view of the pathetically low efficiency of the animal as a converter of plant protein to muscle.

During growth and development the composition and character of muscle alter very considerably, even when all controllable conditions are held constant, for we are unable to control the changes associated with age. Quite apart from these age-induced alterations, however, the tissue adjusts to the conditions imposed on the organism by its environment. Changes in the pattern of animal husbandry, introduced either by accident or design, greatly influence the properties of muscle both as a contractile machine and as a foodstuff. The effects of feeding level and of activity are of special significance, for they are readily controllable by farm management. Although the results of various treatments have long been recognized, the complex mechanisms by which they exert their effects are still being unraveled; we must appreciate that very much more is involved than a mere increase in pigment with exercise, or a decline in fat content with low feed intake.

Let us also remember that overactivity and undernutrition are but two of several imposed states to which muscle must adapt itself. Major quantitative and qualitative changes may occur in the musculature following any of a variety of stresses, whether their application has been intentional or inadvertent, of long or short duration, or by physical, physiological, or psychological means.

In providing for the performance of work, nature has gone to some trouble to ensure that, for each muscle, the store of energy is big enough and the rate of its replenishment fast enough to meet the likely demand, yet at the same time to avoid oversupplying it to the point of wastefulness. But muscles differ in the amount of work they do, in the speed at which they do it, and in the duration for which it is maintained. To place a standard power unit in all of them would be as uneconomic and inefficient as to fit every road vehicle with an engine of the same size and type.

This problem of differing energy demand has been partly overcome by a differentiation during development into red and white categories. The color-coding of a muscle provides a clue to the metabolism—glycolytic or respiratory—which predominates during normal activity, and can be partly related to the nature of the energy requirement: for large or small effort, fast or slow movement, brief or sustained action.

If differentiation involved no more than a difference in myoglobin and respiratory enzyme content, it could be ignored as a factor in the post mortem behavior of muscle; since death is followed swiftly by a cessation of aerobic activity, post mortem metabolism is almost entirely anaerobic in both red and white muscles. But the differing level of respiratory activity is associated in life with a varying pattern of metabolites, and as a consequence red and white muscles may enter the prerigor phase stocked to quite different extents with certain substrates. The course of rigor onset is thus influenced by tissue respiration in life; the properties of meat are affected by a process virtually ceasing at slaughter. Looked at in this way, the color of meat is seen to be far more important in studies of meat quality than merely as another aspect of consumer appeal.

Now to consider the two purposes for which muscle exists: actively, in life, to move; passively, in death, to nourish. In view of the prominence given to these features of muscle in the symposium program, it is justifiable to deal with them in some detail; and because of their equal importance to us, it is desirable to introduce them together. This is done most conveniently by considering the tissue during the early post mortem period, while it retains many of the properties of the living material yet is fast approaching a static and lifeless condition. This prerigor phase, being but a step away from the live muscle on the one hand and the dead meat on the other, seems an eminently suitable position from which to look in both directions—if not at the same time, then at least in quick succession while the afterimage of the previous view still persists clearly.

The states of life and death may be regarded as two overlapping circles, a sort of biological Venn diagram from the New Math syllabus, and progress from one to the other—from muscle to meat—necessarily involves a journey through the area common to both. The basic muscle scientist is interested primarily in the living organism but it is to the tissue in the early post mortem phase that he turns for his experimental material, the information he derives being extrapolated back to the living condition. The meat scientist has as his principal concern the quality of the end product, but he too is realizing (perhaps rather belatedly) the great importance of the prerigor state. It is becoming increasingly obvious to him that changes provoked during this period may affect very appreciably the properties of the ultimate material, and the information he obtains is extended forward to the inert condition. The common ground between life and death, the post mortem but prerigor phase, is therefore as vital to the one area of investigation as to the other.

Movement requires a source of energy and a means of converting it to physical action. Muscle enters the prerigor period with both and leaves it with neither, for during rigor onset the energy supply is depleted and the contractile machine locks into a stable and immobile configuration. The rundown from a dynamic to a static condition is not to be regarded solely as a degenerative phase, however, for some of the processes accompanying it are very similar to those which occur during contraction in the living tissue. Thus prerigor muscle is easily extended by a relatively light load, and relaxed muscle in the living organism is readily stretched by the action of a small external force; in both, the lack of resistance is due to the ease with which actin and myosin filaments can slide smoothly past each other. Similarly, muscle in rigor is almost inextensible (as shown by the resistance met when we attempt to move a limb of a "set" carcass), and activated living muscle is equally so (for what purpose would be served by a contracting muscle which extended easily under load?). The inability to stretch is due in both cases to the cross links or bridges which form between actin and myosin, their presence effectively locking the filaments together and preventing any possible sliding movement.

Turning from changes in extensibility to changes in length, however, we encounter major differences in character between the shortening which may accompany rigor onset and that which occurs in living stimulated muscle—differences of speed, of magnitude, and of strength—and we must heed the warning that shortening in rigor is not to be regarded as a slow and irreversible contraction. Nevertheless there are other all-too-readily produced post mortem effects preceding or accompanying rigor onset which are associated on the one hand with the most devastating havoc among the desirable attributes of meat and on the other with a shortening which is faster, stronger, and greater than that seen in normal rigor. Even if these phenomena of thaw rigor and cold shortening are, like normal rigor, not to be looked upon as true contractions, it is abundantly clear that the detrimental effects they cause are due to the same actin-myosin interaction as that which produces shortening in the living tissue. The concept of bridging between filaments of the myofibrillar proteins is thus as vital to meat quality investigators as it is to students of muscular contraction. Indeed it becomes a strong link directly joining the most basic of muscle studies with the most applied aspects of meat acceptability.

From our vantage point in the early post mortem period, therefore, we are very conscious of the major significance of the actin-myosin combination. If we travel back in time to the living tissue, it is found at the center of the contracting and working machine. If we stay where we are, it is encountered as the cause of increasing rigidity, post mortem shortening, and declining water-retention. If we move forward to the ultimate product, it is met as the principal supplier of protein, the chief source of desirable texture, and by far the major ingredient of excessive toughness.

The fuel which our machine will use is ATP, and undoubtedly its most spectacular and interesting function is to drive the engine. Indeed, the fascination of moving parts may lessen our awareness of the host of other vital tasks it performs, and of the several routes by which its supply and immediate availability are ensured.

Our knowledge of the enzymes which determine the level of ATP, though sufficient to give a consistent picture of the general features of muscle behavior, is still quite inadequate to account for many specific aspects. We are aware of its plasticizing and contracting roles in the living tissue, of the ionic equilibria which strongly influence its coming and going, and of the physical consequences of its gradual disappearance from early post mortem muscle. But there remains very much to learn of the factors which determine its breakdown, and of the effects which its breakdown determines. Granted, in post mortem tissue we have an excellent outline of the rigor process and its energy relations which broadly fits all mammalian muscle, and sufficient knowledge to relate variations in its pattern within a species to changes in the imposed conditions before or after death. But the day when this matrix could be profitably fitted to other species, or even to other muscles within a species, is long past; it is the individual feature rather than the broad outline which must now receive attention. In our study of post mortem muscle we must consider more and more the factors which determine energy production and utilization in the living tissue: the purpose which the muscle serves, the speed at which it moves, the duration of its periods of continuous use, the age, species, and degree of domestication of the animal, the stresses (both long- and short-term) to which it has been exposed. Only when the facts of life in muscle are known can we expect to influence its qualities in death.

Finally, before we start our machine a mechanism must be fitted to ensure that energy may be released at a controllable rate. Since the first demonstration 30 years ago that the contractile component itself sparks the fuel, the problem has been not to release the energy but to contain it. Enormous effort has been expended in many laboratories to explain the obvious ability of muscle to relax as well as contract, and the picture, still emerging, is indeed an involved one. The central role of the calcium ion as an activator of contraction has been recognized for a long time, and its removal has long been suspected as the function of the relaxing mechanism; but only recently have the steps in the opposing processes become clear.

Despite the complexity of this subject of control mechanisms, it is certainly not a topic of academic or fundamental interest only. The control of post mortem chemical and physical change is every bit as important to the meat scientist as is that of the corresponding processes in life to the muscle physiologist. Those who have set up a simple thaw rigor experiment will know the disastrous effect that a lifting of the rigor control mechanism can have on the appearance, the fluid retention, and the tenderness of meat. The condition of watery or pale, soft, exudative (PSE) pork, a problem of considerable economic importance in major pig-producing countries, is clearly related to the very rapid rate of glycolysis found in the affected muscles soon after death, and appears to be a direct consequence of a partial loss of control over the normal post mortem glycolytic mechanism. In New Zealand, which each year exports a weight of frozen meat equal to that of four times its human population, much attention is being given to the massive shortening displayed by prerigor muscle when it is exposed to freezing temperatures, for the great toughening due to shortened actomyosin appears to be a direct result of the failure of a control mechanism. In meat quality studies, then, as in research on other aspects of muscle biology, it is not enough to explain why the tissue functions; we must also explain why it does not function.

This, then, is our picture of muscle, and from its vague outline we may assess the magnitude of our task. We are to assemble the tissue from its parts, develop it to a size and strength and speed optimal for the work it will perform, and adapt it to meet the stresses it may encounter. It is to be differentiated to increase its efficiency for the specific function it will undertake, and provided with fuel and the means to use it. It must be given a sensitive control mechanism allowing the most delicate adjustment of its energy usage and power output. In its final form it is to possess not only the vital properties of a highly efficient machine but also the desirable qualities of a highly acceptable foodstuff.

Of necessity, our approach to this task will be by a process of fragmentation. We shall tear the tissue apart and study it piece by piece and function by function, the better to understand the nature and purpose of every component. But when at the end we put the fragments together again, let us build from them but a single structure and not the two tissues, separately labeled muscle and meat, with which we started. This is the objective of the Madison group whose concept of muscle as a food is the basis of the present symposium.

# PART 1

## Muscle Cell Development

# Chapter 2

# Biosynthesis of Muscle Proteins

S. M. HEYWOOD

Muscle is a tissue uniquely suited for the study of the synthesis of structural proteins and their ultimate organization into a complex structural arrangement. The proteins comprising the contractile apparatus have been studied in some detail and a great deal is known concerning their chemistry, their morphology, and their role in muscle contraction (Szent-Györgyi, 1960; Perry, 1967). However, relatively little is known concerning their biosynthesis. In order to understand the processes involved in muscle building in developing systems, as well as in the adult organism, it is necessary to examine the manner by which the proteins of the myofibril are synthesized, the various controls exerted on their synthesis, the temporal relationship of their synthesis to each other, and the manner by which newly synthesized proteins or polypeptide subunits polymerize to form the sarcomere.

Embryonic chick muscle offers a valuable material for studying the synthesis of muscle proteins. It has a low level of free ribonuclease and the ribosomes are not attached to membranous material (Heywood, Dowben, and Rich, 1968). It is very active in protein synthesis and accumulates over a relatively short period of time a large amount of cell-specific proteins. Furthermore, a great deal is known concerning the morphological changes occurring during muscle development (Fischman, 1967; Allen and Pepe, 1965; Przybylski, 1966). An analysis of the proteins synthesized as well as the rates of their synthesis can, therefore, be correlated to the morphological changes occurring during muscle development.

As a first approach to this problem an attempt was made to prepare an active, cellfree, polyribosomal system from embryonic chick skeletal muscle. Other investigators, using low ionic strength, have previously concluded that muscle generally contains relatively few ribosomes and that

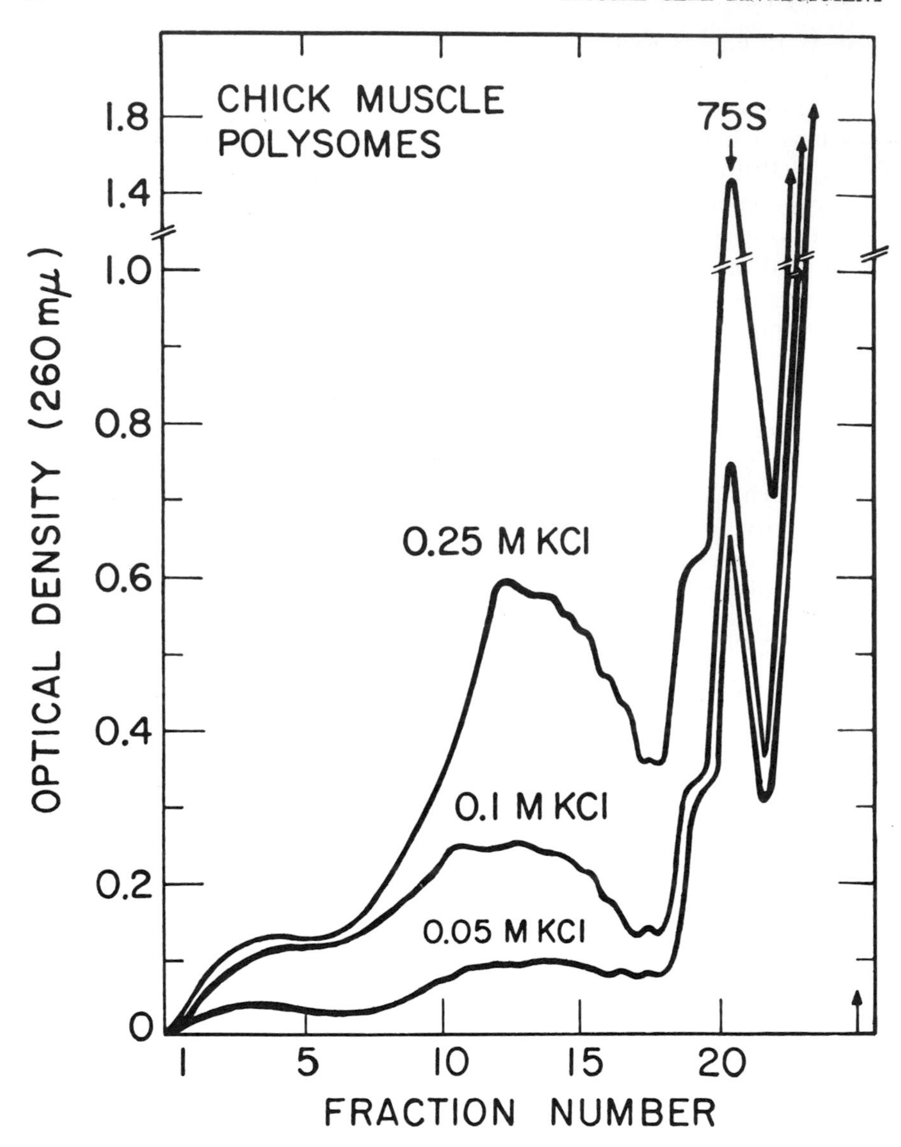


*Fig. 2.1.* Sucrose density-gradient patterns of embryonic chick muscle polysomes prepared with extraction buffers of various concentrations of KCl plus 0.01 M $MgCl_2$ and 0.01 M Tris (*p*H 7.4). From Heywood, Dowben, and Rich (1967).

these are inactive in comparison with ribosomes of other tissues (Earl and Korner, 1965; Florini and Breuer, 1965).

As shown in Fig. 2.1, the amount of polysome material recovered from chick muscle homogenate is related to the ionic strength of the buffer used. With 0.25 M KCl, the buffer system has an ionic strength of near 0.3. Higher concentrations of KCl in the buffer did not result in a greater increase in the yield of polysomes. The similarity in ionic conditions for obtaining a high yield of polysomes and for extracting myosin (Szent-Györgyi, 1960) suggested the possibility that a coprecipitation of myosin with polysomes resulted under low ionic strength conditions. That this was the case was demonstrated by the fact that polysomes from other tissues and cells could be precipitated in the presence of myosin (Heywood, Dowben, and Rich, 1968). Furthermore, myosin was found to coprecipitate with both species of ribosomal RNA as well as with synthetic polyribonucleotides. These results confirm those of Zak, Rabinowitz, and Platt (1967) who reported that the main RNA components of myosin preparations were the ribosomal RNAs.

Proteins are made largely on polyribosomes and not on single ribosomes (Rich, Warner, and Goodman, 1963). Nascent polypeptide chain radioactivity on single ribosomes is normally a result of degradation of polysomal material during isolation. In order to determine the extent of degradation of polysomal material during the isolation procedure, nascent polypeptide chains were labeled by injecting embryos with $^{14}$C-labeled amino acids. As shown in Fig. 2.2, there is a minimum of radioactivity per ribosome associated with the single ribosomes (75 S). The radioactivity per ribosome associated with the major class of polysomes

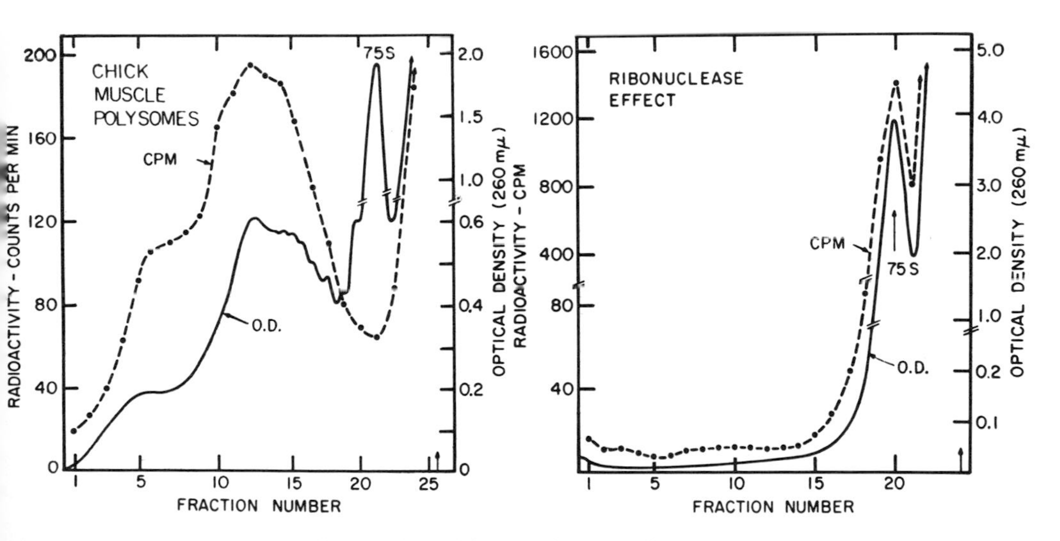


*Fig. 2.2.* Nascent polypeptide-chain radioactivity associated with embryonic chick muscle polysomes (*left*). An equal aliquot of cytoplasmic extract heated with 2 mg/ml ribonuclease for 15 min at 2° C prior to sedimentation, resulting in the disappearance of optical density and radioactivity in the polysome region (*right*). From Heywood, Dowben, and Rich (1968).

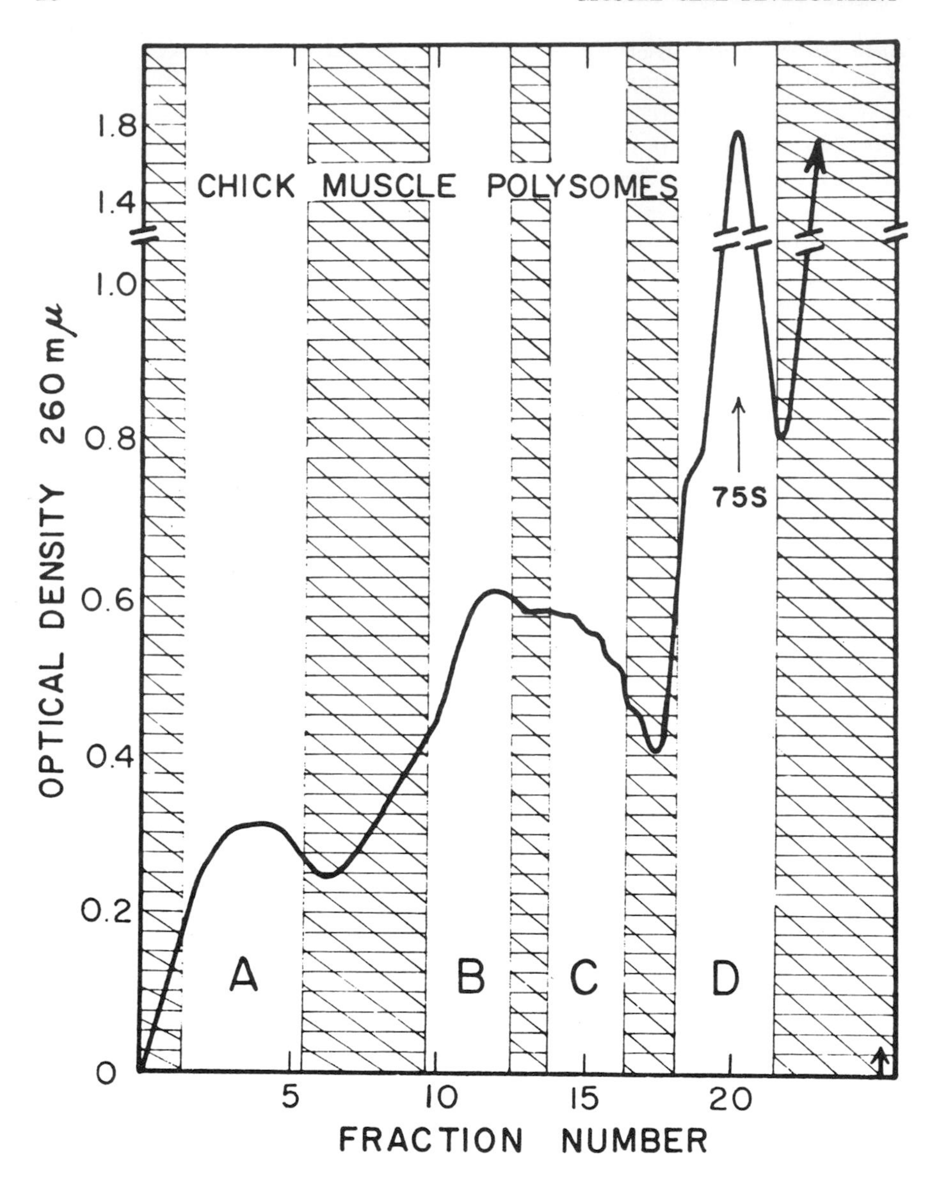


*Fig. 2.3.* Sucrose gradient analysis of embryonic chick muscle cytoplasmic extract. Fractions A–D were prepared for cellfree protein synthesis. From Heywood and Rich (1968).

(fraction 12–16) is one-half of that associated with the more rapidly sedimenting polysomes (fraction 4–7). This suggests that the latter contain longer polypeptide chains than the smaller polysomal material. Upon examination in the electron microscope, these rapidly sedimenting polysomes were observed to contain 55–65 ribosomes extended in a somewhat coiled configuration (Heywood, Dowben, and Rich, 1967). Large polysomes of this size have been observed previously in tissue sections (Allen and Pepe, 1965; Przybylski, 1966) and in muscle cell cultures (Fischman, 1967).

The three major structural proteins of the myofibril are myosin, actin, and tropomyosin. The molecular weights of these proteins (or subunit molecular weight) are 200,000, 60,000, and 35,000 respectively (Kielley and Harrington, 1960; Dreizen, Hartshorne, and Stracker, 1966; Frederiksen and Holtzer, 1968; Hanson and Lowy, 1963; Lewis, Maruyama, Carroll, Kominz, and Laki, 1963; Woods, 1966). It would be expected, therefore, that the polypeptides comprising these proteins would be made on polysomes of different size classes, the exception being if they were synthesized on polycistronic messenger ribonucleic acids (mRNAs). The polysomes obtained from the chick muscle homogenate were collected in four separate fractions as shown in Fig. 2.3. These fractions, consisting of different size classes of polysomes, were incubated in a cellfree amino acid incorporating system. The radioactive products synthesized by the polysomes were layered on acrylamide gels and electrophoresed. A control gel containing myosin, actin, and tropomyosin was run in parallel with the sample gels. The results of such an analysis are shown in Fig. 2.4. It can be seen that the A polysomes, containing 55–65 ribosomes, synthesize mainly myosin while B polysomes, containing 15–25 ribosomes, synthesize actin in addition to a number of unidentified polypeptides. C polysomes, containing 5–8 ribosomes, synthesize tropomyosin; and fraction D, consisting of single ribosomes, was found to be relatively inactive and no specific polypeptide could be identified with it. These analyses have been carried out further by chemically isolating and purifying the radioactive products to constant specific activity (Heywood and Rich, 1968). These results support the conclusion that the three different size classes of polysomes are involved with separate synthesis of myosin, actin, and tropomyosin. Furthermore, it suggests that the synthesized proteins assume the characteristics of the native molecules. The relationship of the size of the polysome to the size of the polypeptide chain agrees with results of similar analysis of polysomes from other cell types (Warner, Rich, and Hall, 1962; Kiko and Rich, 1964). The fact that myosin, actin, and tropomyosin are synthesized on different size classes of polysomes suggests that their mRNAs are not polycistronic. In order to verify these conclusions, however, a peptide analysis of the polypeptides synthesized *in vitro* will be required.

If, in fact, the mRNAs for myosin, actin, and tropomyosin are not polycistronic, regulation of their synthesis may involve differential gene

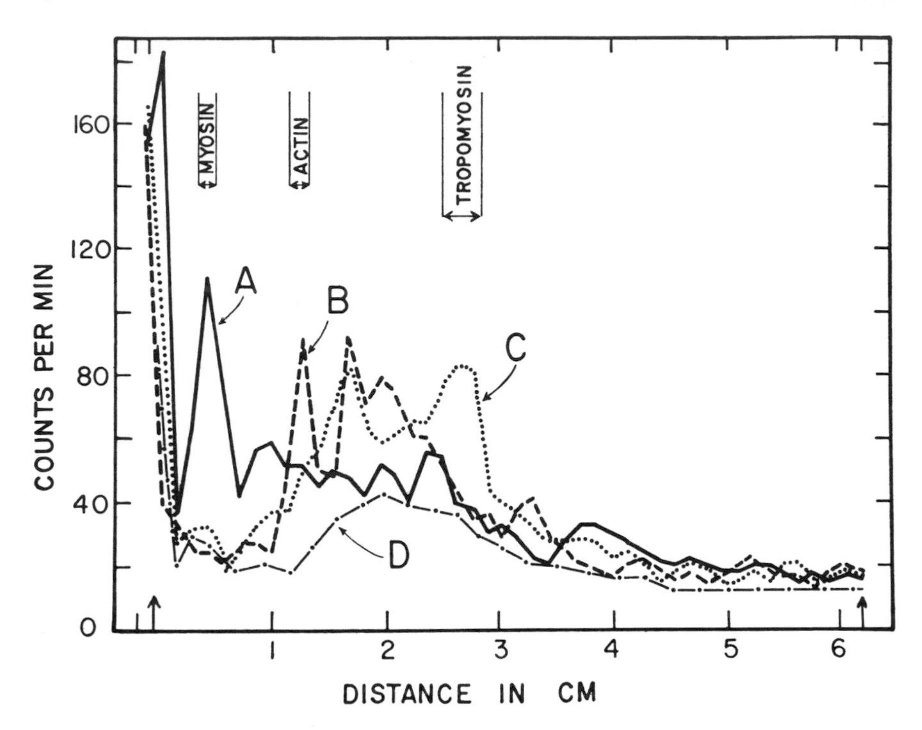


*Fig. 2.4.* Distribution of radioactive proteins synthesized from fractions A–D, Fig. 2.3, on acrylamide gel electrophoresis. From Heywood and Rich (1968).

activity. One then may expect a temporal sequence in the appearance of these components of the myofibril during muscle development. Other modes of regulation, evoking a similar response, cannot, however, be ruled out. That a sequence exists in the appearance of these proteins is suggested by studies involving the formation of myofilaments in developing chick skeletal muscle (Fischman, 1967; Obinata, Yamamoto, and Maruyama, 1966; Przybylski, 1966). Two varieties of filaments directly involved with the formation of sarcomeres have been seen: actin filaments 60–70 Å in diameter, and thicker myosin filaments 140–160 Å in diameter. At earlier stages of development there appears to be a predominance of actin-like filaments over the myosin filaments (Fischman, 1967). The changes observed in the polysome pattern at different ages of embryonic development appear to reflect the sequence of events observed by the electron microscope (Fig. 2.5). In the 10-day embryo there is a sharp peak in the polysome region associated with actin synthesis, while the peak of myosin-synthesizing polysomes is relatively small. By 14 days there is a significant increase in the number of myosin polysomes with

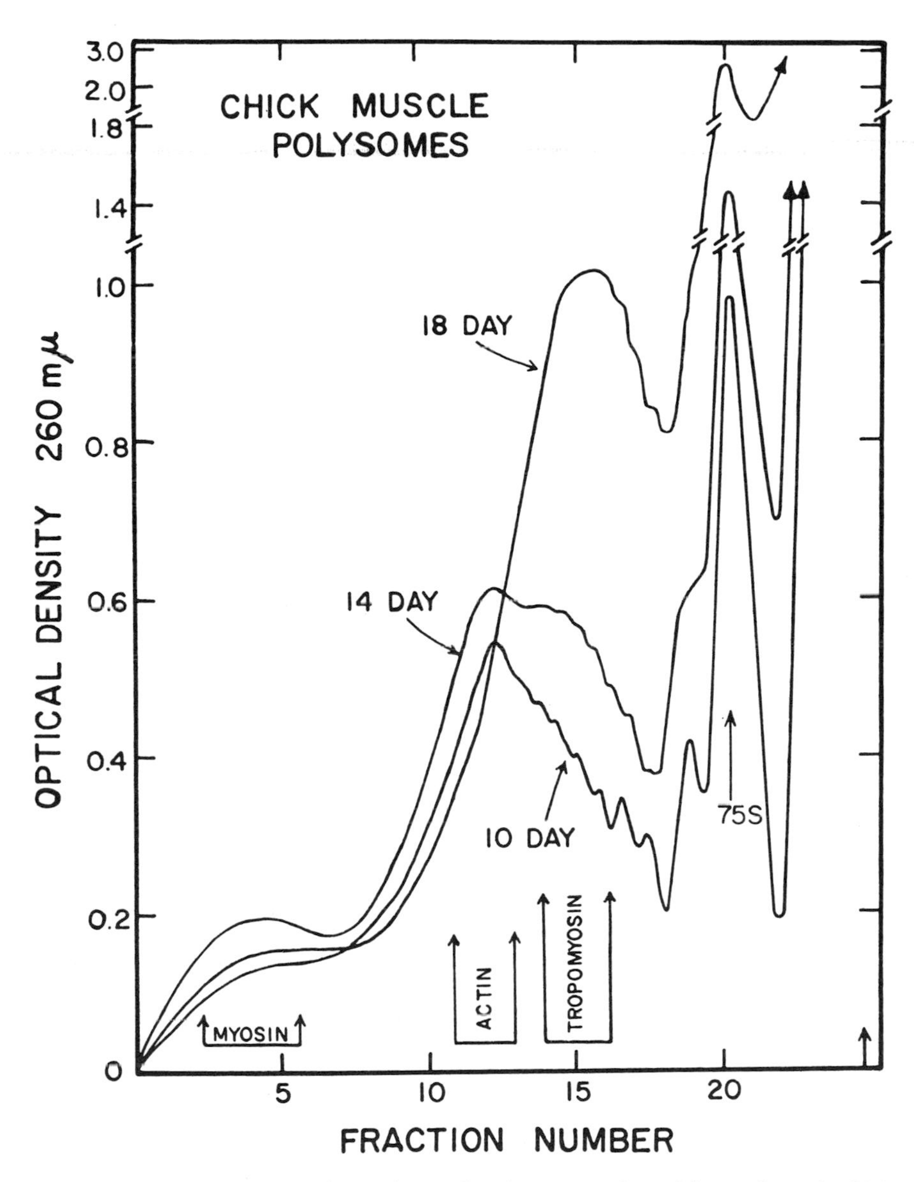


*Fig. 2.5.* Sucrose gradient analyses of cytoplasmic extracts from 0.7 g embryonic chick leg muscle at different ages of embryological development. The positions labeled with myosin, actin, and tropomyosin correspond to fractions A–C in Fig. 2.3 from 14-day chick muscle polysomes. From Heywood and Rich (1968).

very little change in the number of polysomes in the actin-synthesizing region. The percentage of myosin-synthesizing polysomes (MSP) to total polysomes has been found to increase from 7% to 20% during this interval of time. At 14 days an increase in the size class of polysomes associated with tropomyosin synthesis is observed. This increase continues up to the 18th day of development. This may represent an increase in tropomyosin synthesis. The fact that the major peak of polysomes is derived from a variety of cell types (myotubes, myoblasts, fibroblasts) and is synthesizing a large number of proteins makes any conclusion regarding the association of changes in actin and tropomyosin polysome peaks with the appearance of actin and tropomyosin structures very tenuous. On the other hand, the MSP appear to be solely involved with myosin synthesis, and a direct correlation between the appearance of thick filaments, fusion of myoblasts, and the increase in MSP may be possible.

Unlike chick skeletal muscle, embryonic heart has been shown to proliferate and simultaneously synthesize the proteins making up the contractile apparatus (Manasek, 1968). If this is the case, a change in polysome pattern, reflecting a sequential synthesis of proteins, may not be observed. As shown in Fig. 2.6, very little difference exists between the polysome patterns of heart from 8–18 days of development either quantitatively or qualitatively. A general correspondence is noted between the polysome profile of 14–day skeletal muscle and heart. Both cell types contain large polysomes sedimenting near the bottom of the gradient.

The ability to isolate a single class of polysomes involved in the synthesis of myosin led to the possibility of isolating and characterizing myosin mRNA. Total RNA extracted from MSP has been found to stimulate the synthesis of a protein which cochromatographs on DEAE cellulose with myosin and which migrates during acrylamide gel electrophoresis identically to myosin (Heywood and Nwagwu, 1968). If the RNA associated with the MSP is analyzed on sucrose density gradients after labeling with $^{32}P$, a distinct species of RNA sedimenting at approximately 26 S is observed. This species of RNA is not found to be associated with other size classes of polysomes or single ribosomes. Utilizing a cellfree amino acid incorporating system derived from muscle tissue, the 26 S RNA was tested for messenger activity. This species of RNA was judged to be the presumptive mRNA for the large 200,000-molecular-weight subunit of myosin by the following criteria: (1) It causes an increase in amino acid incorporation and the formation of polysomes when added to a cellfree system. (2) It stimulates the synthesis of a protein which precipitates with antiserum prepared against myosin and which cochromatographs with myosin on DEAE cellulose. (3) It specifi-

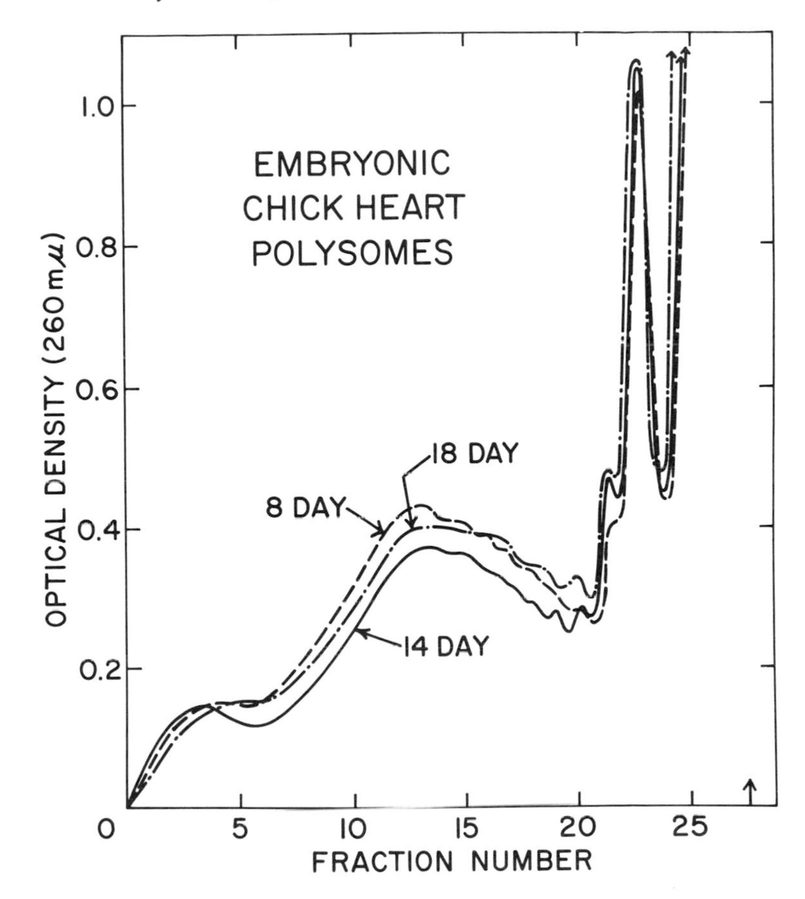


*Fig. 2.6.* Sucrose density analysis of cytoplasmic extracts from 0.5 g embryonic chick heart at different ages of embryological development. The polysomes were prepared as described in Fig. 2.2.

cally stimulates the synthesis of only the 200,000-molecular-weight subunit of myosin identified by acrylamide gel electrophoresis. Other RNAs from either the MSP or non-MSP from muscle tissue were not found to fill these criteria for myosin mRNA (Heywood and Nwagwu, 1969).

The presumption that the 26 S RNA isolated from MSP is the mRNA coding for the large molecular subunit of myosin has found further support by testing the messenger activity of the RNA in a heterologous cellfree system. This system, utilizing ribosomes from chicken reticulocytes, excludes the possibility that the 26 S RNA is specifically stimulating endogenous mRNA. It was found that addition of 26 S RNA from

MSP to such a heterologous cellfree system resulted in the synthesis of a protein which both chromatographed and migrated on acrylamide gels with myosin (Heywood, 1969). Nevertheless, final characterization of 26 S RNA as myosin mRNA will require peptide analysis of the synthesized product.

An mRNA coding for a 200,000-molecular-weight polypeptide would have about 6,000 nucleotides, thereby having a molecular weight of approximately $1.7 \times 10^6$. A polynucleotide of this molecular weight would be expected to sediment close to 28 S ribosomal RNA which has a molecular weight of $1.6 \times 10^6$ (Hamilton, 1967). The fact that myosin mRNA sediments more slowly may be due to the fact that it has a more extended configuration than ribosomal RNA. Nevertheless, a general agreement exists between the size of the messenger RNA and the polypeptide for which it codes.

Similar experiments are under way, utilizing RNA fractions from smaller polysomes, to synthesize actin, tropomyosin, and troponin. In this manner the synthesis and ultimate laying down of the myofibril will be studied in a well-defined system. This should reveal, in part, the interrelationship between the synthesis of the subunit components of the molecules and the polymerization of these components into mature molecules interacting with each other to build the contractile apparatus.

With the progress of the work on the synthesis of cell-specific proteins in cellfree systems, it became pertinent to ask to what extent the increase of cell-specific proteins, which occurs during muscle development in the intact embryo, can be interpreted quantitatively in terms of translational and transcriptional mechanisms. This aspect of the problem has been studied in collaboration with H. Herrmann at the University of Connecticut. Some aspects of muscle development in the limb of the chick embryo which have to be considered in such an analysis are here briefly reviewed.

The initial condensation of rapidly proliferating myogenic cells occurs in the limb of the 5-day embryo (Kitiyakara, 1959) and a high rate of proliferation of these cells continues to the 7th day of development with 70% of all myogenic cells still dividing at that time (Marchok and Herrmann, 1967). A sharp decline in the proliferative rate follows, with the number of proliferating cells dropping to 38%, 15%, and 2.5% of the total cell population on the 9th and 18th days prehatching and the 8th day posthatching respectively.

Although on the 11th day about 70% of all the myogenic cells in the limb have ceased proliferating, only 12% of all cells have fused and formed myotubes, the histological precursors of the mature muscle fiber. Histologically the rapid appearance of muscle fibers during the prehatching period can be readily followed (Marchok and Herrmann, 1967) but quantitative data about the progress of fiber formation from single myoblasts are not available as yet. During myoblast fusion and fiber formation, critical changes take place in the protein-forming system of the muscle cell. Stockdale and Holtzer (1961) have shown that the myoblast of skeletal muscle does not synthesize myosin before fusion but that the presence of myosin can be detected a short time after fusion has taken place. This suggests that before fusion, protein synthesis is directed mainly toward production of non-cell-specific proteins, e.g., cell proteins required for cell division and maintenance of cell metabolism. After myoblast fusion, a part of the activity of the protein-forming system is diverted toward synthesis of cell-specific proteins. The acceleration in the increase of myosin levels in muscle tissue has been described (Baril and Herrmann, 1967).

As pointed out earlier, the phase of chick muscle development which is being studied at the present time is just this period of transition. Parallel to the readjustments in the activities of the protein-forming system, changes in RNA synthesis have been found to occur (Marchok and Wolff, 1968). Evidence for the hormonal controls of this transition has also been reported (De la Haba, Cooper, and Elting, 1966).

A perspective of the magnitudes of the changes in protein-synthetic activities during chick leg muscle development can be obtained from the data in Table 2.1. In this table we have compiled the measured amounts of myosin and the estimated quantities of total cell-specific proteins as well as the levels of nonspecific proteins on the 9th and 17th day prehatching and the 29th day and the 6th month posthatching. It can be seen that during the transition period (9th day prehatching to 8th day

TABLE 2.1

*Quantities of protein in the total leg musculature of the developing chick (mg)*

| Age | Myosin* | Total cell-specific protein † | Total non-cell-specific protein ‡ |
|---|---|---|---|
| 9E | 0.02 | 0.03 | 1.1 |
| 17E | 1.5 | 2.2 | 28 |
| 8H | 16 | 24 | 156 |
| 180H | 5000 | 7500 | 7500 |

* Calculated from direct determinations of myosin/g tissue and weight of total tissue.

† Calculated from the myosin values assuming a 2/1 ratio of myosin/actin + tropomyosin.

‡ Total protein − cell-specific protein.

posthatching) the cell-specific and non-cell-specific proteins increase 8,000-fold and 141-fold respectively, and for the entire developmental period 250,000-fold and 7,500-fold respectively. It can be recognized that the cell-specific proteins increase much more rapidly during this transition period. However, the absolute quantities of the cell-specific proteins, relative to the non-cell-specific proteins, remain small until after 28 days posthatching. Thus up to the 8th day posthatching, the amount of myosin formed is about 1/10 of the amount of non-cell-specific protein. Since the quantity of MSP increases during this period from 7% to 20% of the total polysomes, the efficiency of myosin production per polysome unit is not greater than that of the polysomes synthesizing other proteins. Therefore, the large relative increase in myosin seems to be a result of an increase in the quantity of MSP relative to the increase in other polysomes and is not due to a higher efficiency of the MSP.

It should be pointed out that it has not been determined so far to what extent the slower increase in the amounts of nonmyosin proteins is due to a declining rate of synthesis or an increased breakdown of some non-cell-specific proteins. It is known, however, that after cessation of cell proliferation, the kinases involved in DNA synthesis disappear, presumably at different developmental periods (Scholl, Herrmann, and Roth, 1968; Stockdale, 1970). Also, glucose-6-phosphate dehydrogenase, essential for the formation of the pentose portion of the nucleic acids, was found to decline during this transition period of muscle development (Love, Stoddard, and Grasso, 1970).

In attempting an analysis of the mechanism of these changes in protein synthesis in the intact embryo, we have followed the transfer of labeled amino acids from the blood plasma into the cell pool and the polysome-bound peptides. Using a mixture of 14 amino acids for labeling and expressing the specific activities of amino acids in the three compartments as cpm/m$\mu$m amino acid, it became apparent that 10–20 min after injection of label into the embryonic circulation, the nascent peptides attained levels of specific activity which were 4–5 times higher than the corresponding values in the cell pool. At this time, when the specific activity of the polysome-bound peptides attained a maximum, it reached the rapidly declining specific activity of the blood plasma. This labeling pattern suggested that the amino acids in the cell pool are not the direct precursors of the amino acids in the polysome-bound peptides. Instead, the blood plasma amino acids, after their transport through the cell membranes, were used for peptide synthesis without significant dilution by the cell pool.

The interpretation of a direct transport of amino acids from blood to polysomes was corroborated by double labeling experiments with diaphragm muscle (Hider, Fern, and London, 1969) and by the kinetics of amino acid uptake into myosin in the intact embryo (Rourke and Herrmann, unpublished observations). However, polysome labeling after injection of certain individual amino acids gave results which did not correspond to those obtained with the amino acid mixture. Therefore, the assumption of a direct transport of amino acids can be used only hypothetically for further calculations of the quantities of myosin synthesized by the developing muscle tissue. The results of these calculations could be compared with the directly observed figures for myosin accumulation in the developing muscle. As a first approximation, the time required for maximal labeling of the polysome-bound peptides was assumed to represent the synthesis time for a myosin subunit of 200,000 molecular weight. Maximal labeling was observed at about 20 min. From this and from the quantity of MSP, the amount of myosin synthesized per day per g of muscle tissue for the 14-day embryo was calculated to be about 520 μg/g tissue per day. The direct determinations of the increase in myosin content of developing muscle tissue give values of 350 μg/g tissue per day. It can be seen that these values are within the same range and hence support the assumptions which underlie these calculations. In another series of measurements, myosin synthesis was calculated from the incorporation of phenylalanine into purified myosin. This calculation gave a value of about 200 μg/g tissue per day, which is again within the range of the directly determined myosin accumulation. Although these results are as yet tentative, they show that the analysis of the activities of the component steps of protein synthesis in muscle of the intact embryo is possible and that a correlation of results obtained with the cellfree system and in the intact embryo should eventually yield an unusually complete picture of the molecular mechanisms of protein synthesis during muscle development.

## ACKNOWLEDGMENTS

Contribution No. 176 from the Institute of Cellular Biology, University of Connecticut, Storrs, Connecticut. Supported by grant HD–03316 from the National Institute of Child Health and Human Development (NIH).

## *References*

Allen, E. R., and F. A. Pepe. 1965. Ultrastructure of developing muscle cells in the chick embryo. *Amer. J. Anatomy 116*:115.

Baril, E., and H. Herrmann. 1967. Studies of muscle development II. Immunological and enzymatic properties and accumulation of chromatographically homogeneous myosin of the leg musculature of the developing chick. *Dev. Biol. 15*:318.

De la Haba, G., D. W. Cooper, and V. Elting. 1966. Hormonal requirements for myogenesis of striated muscle *in vitro*. *Proc. Nat. Acad. Sci. U.S. 56*:1719.

Dreizen, P., D. J. Hartshorne, and A. Stracker. 1966. The subunit structure of myosin. *J. Biol. Chem. 241*:443.

Earl, C. N., and A. Korner. 1965. The isolation and properties of cardiac ribosomes and polysomes. *Biochem. J. 94*:721.

Fischman, D. A. 1967. An electron microscope study of myofibril formation in embryonic chick skeletal muscle. *J. Cell Biol. 32*:557.

Florini, J. R., and C. B. Breuer. 1965. Amino acid incorporation into protein by cell-free preparations from rat skeletal muscle. *Biochem. 4*:253.

Frederiksen, D. W., and A. Holtzer. 1968. The substructure of the myosin molecule: production and properties of the alkali subunits. *Biochem. 7*:3935.

Hamilton, M. G. 1967. The molecular weight of the 30 s RNA of Jenson Sarcoma ribosomes as determined by equilibrium centrifugation. *Biochim. Biophys. Acta 134*:473.

Hanson, J., and J. Lowy. 1963. The structure of F-actin and of actin filaments isolated from muscle. *J. Mol. Biol. 6*:46.

Herrmann, H., A. Marchok, and E. Baril. 1967. Growth rates and differential function of cells. *Nat. Cancer Inst. Monogr. 26*:303.

Heywood, S. M. 1969. Synthesis of myosin on heterologous ribosomes. *Cold Spring Harbor Symposia on Quantitative Biology*. (In press.)

Heywood, S. M., R. M. Dowben, and A. Rich. 1967. The identification of polyribosomes synthesizing myosin. *Proc. Nat. Acad. Sci. U.S. 57*:1002.

———. 1968. A study of muscle polyribosomes and the coprecipitation of polyribosomes with myosin. *Biochem. 7*:3289.

Heywood, S. M., and M. Nwagwu. 1968. *De novo* synthesis of myosin in a cell-free system. *Proc. Nat. Acad. Sci. U.S. 60*:229.

———. 1969. Partial characterization of presumptive myosin messenger RNA. *Biochemistry 8*:3839.

Heywood, S. M., and A. Rich. 1968. *In vitro* synthesis of native myosin, actin, and tropomyosin from embryonic chick polyribosomes. *Proc. Nat. Acad. Sci. U.S. 59*:590.

Hider, R. C., E. B. Fern, and D. R. London. 1969. Relationship between intracellular amino acids and protein synthesis in the extensor digitorum longus muscle of rats. *Biochem. J. 114*:171.

Kielley, W. W., and W. F. Harrington, 1960. A model for the myosin molecule. *Biochim. Biophys. Acta 41*:401.

Kiko, Y., and A. Rich. 1964. Induced enzyme formed on bacterial polyribosomes. *Proc. Nat. Acad. Sci. U.S. 51*:111.

Kitiyakara, A. 1959. The development of non-myotomic muscles of the chick embryo. *Anat. Rec. 133*:35.

Lewis, M. S., K. Maruyama, W. R. Carroll, D. Kominz, and K. Laki. 1963. Physical properties and polymerization reactions of native and inactivated G-actin. *Biochem. 2*:34.

Love, D. S., F. J. Stoddard, and J. A. Grasso. 1970. Endocrine regulation of embryonic muscle development: hormonal control of DNA accumulation, pentose cycle activity and myoblast proliferation. *Dev. Biol.* (In press.)

Manasek, F. J. 1968. Mitosis in developing cardiac muscle. *J. Cell Biol. 37*:191.

Marchok, A. C., and H. Herrmann. 1967. Studies of muscle development I. Changes in cell proliferation. *Dev. Biol. 15*:129.

Marchok, A. C., and J. A. Wolff. 1968. Studies of muscle development IV. Some characteristics of RNA polymerase activity in isolated nuclei from developing chick muscle. *Biochim. Biophys. Acta 155*:378.

Obinata, T. M., M. Yamamoto, and K. Maruyama. 1966. The identification of randomly formed thin filaments in differentiating muscle cells of the chick embryo. *Dev. Biol. 14*:192.

Perry, S. V. 1967. The structure and interactions of myosin. *In Progr. Biophys. Mol. Biol. 17*:325.

Przybylski, R. Z. 1966. Ultrastructural aspects of myogenesis in the chick. *Lab. Invest. 15*:836.

Rich, A., J. R. Warner, and H. M. Goodman. 1963. The structure and function of polysomes. *In Cold Spring Harbor Symposia on Quantitative Biology 28*:269.

Scholl, A., H. Herrmann, and J. Roth. 1968. Studies of muscle development III. An evaluation of thymidylate kinase activity with respect to cell proliferation in developing chick leg muscle. *Life Sci. 7*:91.

Stockdale, F. E. 1970. Changing levels of DNA polymerase activity during the development of skeletal muscle tissue *in vivo*. *Dev. Biol.* (In press.)

Stockdale, F., and H. Holtzer. 1961. DNA synthesis and myogenesis. *Exp. Cell Res. 24*:508.

Szent-Györgyi, A. G. 1960. Proteins of the myofibril, p. 1. *In Structure and Function of Muscle,* Vol. 2. New York, Academic Press.

Warner, J. R., A. Rich, and C. Hall. 1962. Electron microscope studies of ribosomal clusters synthesizing hemoglobin. *Science 138*:1399.

Woods, E. F. 1966. The dissociation of tropomyosin by urea. *J. Mol. Biol. 16*:581.

Zak, R., M. Rabinowitz, and C. Platt. 1967. Ribonucleic acids associated with myofibrils. *Biochem. 6*:2493.

# Chapter 3

# *Mitosis and Myogenesis*

H. HOLTZER AND R. BISCHOFF

## INTRODUCTION

Elsewhere the theme has been developed that on a molecular level there are no undifferentiated, naive, virginal, or uncommitted cells (Holtzer, 1968, 1970; Holtzer and Abbott, 1968).* The biosynthetic programs of all cells—zygote, blastula or gastrula cells, or terminally differentiated nerve, liver, or pancreatic cells—are governed by cytoplasmic-nuclear interactions. Cells without nuclear messages are, according to current dogma, inconceivable, although the unique synthetic activities of zygote, morula, or blastula cells are unrecognized at present. The kinds of molecules a given cell synthesizes at any time are determined in part by factors inherited from its mother cell and in part by factors impinging on the cell since its inception at metaphase. In early development cell lineages are established. The subsets of covertly differentiated precursor cells that constitute these lineages are the obligatory antecedents of the terminally differentiated cells described in textbooks of histology. The core problem of differentiation is not how the synthesis of molecules like myosin and myoglobin is augmented or dampened, but characterizing the unique biosynthetic activities in the succession of transitory phenotypes that lead to the myoblast, with its machinery for translating the terminal luxury molecules (Holtzer, 1967) of mature muscle. We would predict, for example, that no exogenous molecule will directly transform a zygote or morula or blastula cell into a viable cell synthesizing myosin and myoglobin. On the other hand, *after* a series of mitoses some of the progeny of a particular morula or blastula cell do acquire *for the first time* the machinery to produce molecules characteris-

* Abbreviations used: BU, 5-bromouracil; BUdR, 5-bromodeoxyuridine; FUdR, 5-fluorodeoxyuridine; TdR, thymidine; DNP, dinitrophenol; S, period of DNA synthesis; $G_1$, presynthetic period; $G_2$, postsynthetic period; M, mitosis.

tic of muscle. We postulate that the changes required to transform a zygote or morula or blastula cell into a cell synthesizing myosin occur only as the consequence of a series of step-wise events, and that these discrete increments are coupled to a sequence of "quantal" cell cycles (Holtzer and Abbott, 1968; Holtzer, Bischoff, and Chacko, 1969; Holtzer, 1970). Interspersed with these quantal cell cycles are proliferative mitoses associated with simple increase in cell number. According to this scheme there are several distinctly different generations or phenotypic classes of myogenic precursor cells. The shift from one class to the next is dependent upon a quantal cell cycle. The increase in number of cells within each class is a function of proliferative cell cycles.

In this paper we will review some experiments which indicate that the conversion of replicating presumptive myoblasts, which do not fuse or synthesize myosin, into nonreplicating myoblasts, which do fuse and synthesize myosin, is coupled to and paced by the mitotic history of the cell.

## FUSION, MYOSIN SYNTHESIS, AND THE MITOTIC CYCLE

It is approximately a decade since my collaborators and I presented first indirect, and then direct, evidence that in embryogenesis and in regeneration, multinucleated myotubes are the result of the fusion of mononucleated myogenic cells (Holtzer, Marshall, and Finck, 1957; Holtzer, 1958; Holtzer, Abbott, and Lash, 1958; Holtzer, 1961). In spite of continued efforts in many laboratories (Konigsberg, 1965; Betz, Firket, and Reznik, 1966; Allen and Pepe, 1965; Przybylski and Blumberg, 1966; Fischman, 1967; Coleman and Coleman, 1968), neither the biological nor biochemical events associated with and responsible for fusion are understood. Much will be learned about the properties of the plasmalemma when we understand the molecular basis of surface restructuring during fusion. Direct efforts to follow fusion with the electron microscope have not thus far been particularly informative (Ishikawa, Bischoff, and Holtzer, 1968). Cinematography of fusion *in vitro* confirms our early report that mononucleated cells tend to line up and, after a succession of tentative pseudopodial contacts, abruptly fuse.

A few straightforward questions relate to the discriminatory properties of fusing myogenic cells. Do myogenic cells fuse with nonmyogenic cells? Do skeletal muscle cells fuse with cardiac or smooth muscle cells? To answer these questions, fibroblasts, kidney and liver cells, dedifferentiated chondryocytes (Chacko, Abbott, Holtzer, and Holtzer, 1969), smooth muscle cells from the gizzard, and cardiac muscle cells were labeled with $^{3}$H-TdR. These labeled cells were added to cultures in which myotubes were forming. Invariably the labeled cells were excluded from the forming myotubes (Okazaki and Holtzer, 1965). Clearly the surface-surface interaction initiated when competent myogenic cells contact and recognize each other is different from that following collisions with heterotypic cells. It has been claimed that myoblasts from species as different as mouse and chick can form hybrid muscle syncitia (Wilde, 1959; Yaffe and Feldman, 1965; Maslow, 1969), but critical inspection of their published data is not, in our opinion, convincing. Whether the behavior of the cell surface of fusing cells has more in common with events responsible for phagocytosis (Cohen and Ehrenreich, 1969) or with events responsible for fusion of egg and sperm (Colwin and Colwin, 1963; Metz, 1967) is not known. Likewise, whether fusion induced by either living or UV-irradiated virus (Harris, Watkins, Ford, and Schoefl, 1966) or the fusion observed in neoplastic cells (Ephrussi, Scaletta, Stenchever, and Yoshida, 1964) has much in common with fusion of myogenic cells is still to be determined.

Earlier work utilizing phase contrast microscopy, $^{3}$H-TdR and fluorescein-labeled antibodies against myosin suggested that fusion could occur between the following kinds of myogenic cells: (1) young myotubes synthesizing myosin and actin; (2) young myotubes synthesizing contractile proteins and mononucleated cells not yet synthesizing myosin; (3) two mononucleated cells of which one, both, or neither were synthesizing contractile proteins. In brief, there is no obligatory sequence between the synthesis of myosin and fusion. In contrast, there is a rigid and invariable correlation between DNA synthesis and fusion and between DNA synthesis and myosin synthesis. Fused cells normally never enter S. Cells synthesizing myosin normally never enter S (Stockdale and Holtzer, 1961; Okazaki and Holtzer, 1965).

As evidence is accumulating from a variety of differentiating systems that some cell divisions are more crucial than others with respect to overt differentiation, we have been analyzing in greater detail the relationships between the mitotic cycle and the formation of muscle.

It was found that normal cells in metaphase, or cells arrested in metaphase with mitotic inhibitors, do not fuse (Bischoff and Holtzer, 1968*a;* Ishikawa, Bischoff, and Holtzer, 1968). It is likely that the cell surface of myogenic cells in metaphase is significantly altered with respect to the cell surface in other parts of the mitotic cycle. The following experiments provide direct measurement of the point in the mitotic cycle at which myogenic cells acquire surface properties compatible with fusion. Cultures consisting of dividing mononucleated cells and young myotubes were pulse-labeled with $^{3}$H-TdR and fixed at intervals to determine the minimum time for a labeled mononucleated cell to become incorporated into a myotube (Bischoff and Holtzer, 1969*a*). The earliest fusion of labeled cells took place 8–10 hours from the end of S. By 16–20 hours many labeled nuclei had become incorporated into myotubes (Fig. 3.1). Correlation of this interval with the length of the cell cycle phases measured in the same cultures led to the following conclusions (see Fig. 3.2): (1) Fusion cannot occur in the S, $G_2$, or M periods of the presumptive myoblast. (2) One or both daughter cells arising from a particular division become competent to fuse only after spending a minimum of 5 hours in the $G_1$ or diploid phase. (3) Cells probably withdraw from the division cycle prior to fusion.

From this and related data we suggest that in normal development there is a quantal division which produces one or two daughter cells which, for the first time in that cell lineage, acquire the competence to fuse. It is unlikely that such competent myogenic cells undergo additional proliferative mitoses during which their ability to fuse waxes and wanes as they progress through successive mitotic cycles. It is an open question whether the crucial event in this quantal cell cycle is an event associated with the parental DNA synthetic period or the $G_2$ or M period or the $G_1$ of the myoblast itself (for discussion see Bischoff and Holtzer [1969*a*], Bischoff, Holtzer, and Chacko [1969], and Holtzer [1970]).

A critical test of the quantal division concept would be to demonstrate that the terminal cell cycle separating the transformation of a presumptive myoblast into a myoblast is an obligatory event. In other words, could a *presumptive myoblast* in $G_1$ fuse, given the opportunity, or must it undergo another division? To hold cells in $G_1$ we utilized the *in vitro* phenomenon of density-dependent inhibition of division. Under these conditions cells come to rest in the $G_1$ or diploid state (Defendi and Manson, 1963). Cultures were set up at either high or low density. In the sparse situation cells were separated by several diameters, while in the dense cultures the cells were confluent. The cultures were grown for 20 hours to permit division. Both series were then dissociated and replated at high density to test for the frequency of fusible myoblasts. During the following 3 days, under identical conditions, the cells grown at high density for the initial 20 hours formed very sparse myotubes, while the cells grown at low density formed many large hypernucleated myotubes. Control experiments were carried out to show that presumptive myoblasts had not been selectively eliminated during their sojourn at high density. To test if there was, in fact, inhibition of replication in crowded cultures, cells were grown at high and low densities and the frequency of proliferating cells was determined with both Colcemid and $^3$H-TdR for successive 4-hour intervals throughout the 20-hour period. From 16 to 20 hours there was a 60% inhibition in the frequency of mitotic cells and a 75% inhibition in the number of cells in S in the crowded cultures.

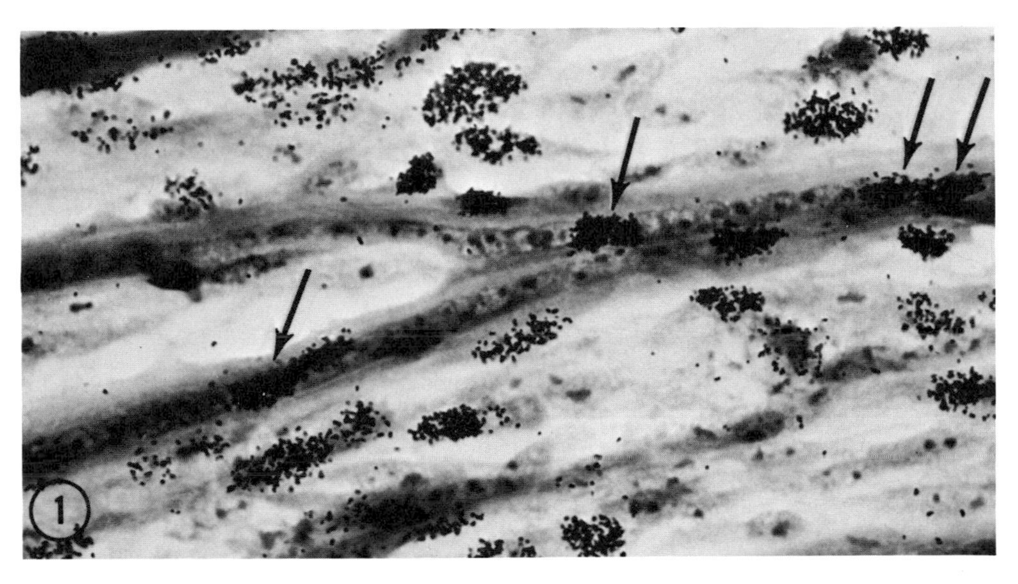

*Fig. 3.1.* Radioautograph of 2-day culture fixed 16 hours after the mononucleated cells were labeled with a 30-min exposure to $^{3}$H-TdR (0.5 $\mu$C/ml medium). Labeled nuclei begin to appear in myotubes at 8 hours postlabeling. This field shows a bifurcating myotube containing several labeled nuclei (*arrows*). $\times$ 380.

Comparison of the inhibition of mitosis and DNA synthesis suggests that cells in the crowded cultures are able to complete the cycle they are in at the time of plating but are blocked in $G_1$ from entering another cycle (see Nameroff and Holtzer, 1969). These results are consistent with the proposal that some of the myogenic cells in the initial inoculum are obligated to undergo at least one more division—a quantal division—before fusing (Bischoff and Holtzer, 1969*b*).

Fusion is affected by factors other than mitosis. Lack of $Ca^{++}$, the presence of DNP, azide, antimycin-A, inappropriate substrates, changes in *p*H, or prolonged treatment with cyclohexamide all block fusion. The blocking effect of antimycin-A and azide on fusion is relatively irreversible; *p*H shifts, exposure to protein synthesis inhibitors, and $Ca^{++}$ depletion are relatively reversible. Competent myoblasts will not fuse in medium depleted of $Ca^{++}$ over a 48-hour period. Fusion takes place, however, within 10 hours after restoring $Ca^{++}$. This fusion occurs without the myoblasts undergoing an intervening round of DNA synthesis providing that the cells underwent terminal quantal mitosis either before being cultured or while *in vitro*. Without further analysis it is difficult to determine if inhibitory agents block fusion *per se* or by preventing the accumulation of competent cells by cell division (e.g., colchicine

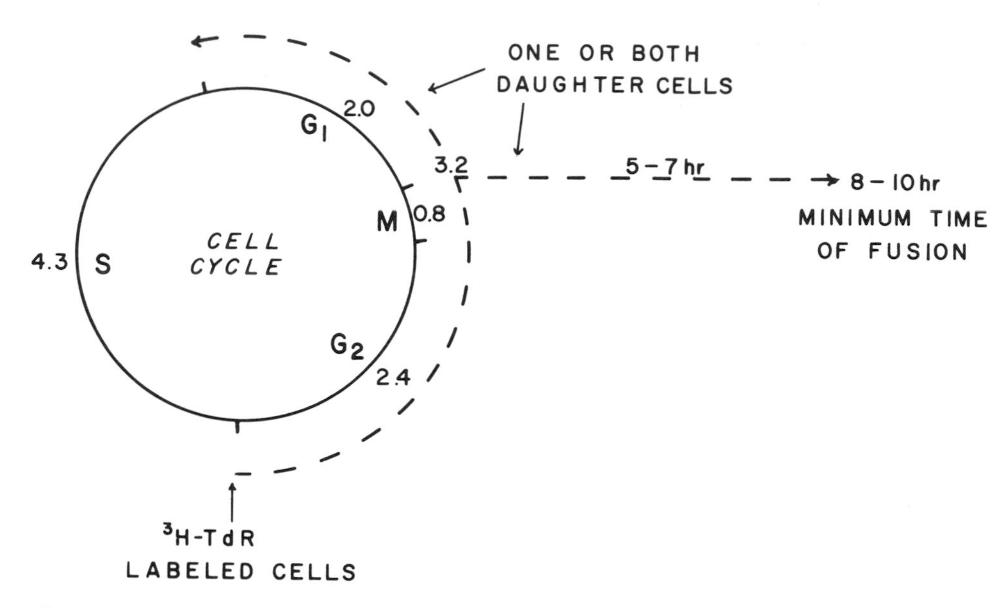


*Fig. 3.2.* Diagram of the relation between the cell cycle and myoblast fusion. The fate of proliferating cells was followed by marking cells in S with a 30-min pulse of ${}^3$H-TdR. Those cells labeled at the end of S (O time) should be the first to become incorporated into myotubes. About 3.2 hours elapse before labeled daughter cells enter $G_1$. Those daughter cells destined to fuse spend a minimum of 5–7 hours in $G_1$ ($G_0$ ?), while cells that continue to proliferate enter S after only 2 hours in $G_1$.

blocks fusion by preventing cells from entering $G_1$, but has no effect on the fusion of competent myoblasts [Bischoff and Holtzer, 1968*a*]).

More recently we have been studying blocks to fusion under more physiological conditions (Bischoff and Holtzer, 1969*a*). Labeled competent myogenic cells have been added to successively older cultures in which myotube formation is proceeding. Labeled cells fuse with myotubes up to 2 or 3 days old. They will not fuse with older or more mature myotubes. Whether the block to fusion resides in the myotube membrane itself or in some extracellular deposit such as the basement membrane or surface coat is yet unknown. This barrier is germane to the origin and mode of segregation of the satellite cells in mature muscle (Mauro, 1961; Ishikawa, 1966; Church, Noronka, and Allbrook, 1966).

A problem related to overt differentiation and satellite cell formation is that of why replicating myogenic stem cells cease to divide (Bischoff and Holtzer, 1969*a*) and what induces them to reenter the mitotic cycle during muscle regeneration in mammals or during regeneration of the salamander limb or tail (Holtzer, Avery, and Holtzer, 1954; Holtzer,

1958). There are now some experimental data suggesting that the extracellular matrix produced by several different kinds of tissues dampens the multiplication of myogenic cells. When myogenic cells are added to nonreplicating but viable cellular substrates of muscle, liver, or kidney, the myogenic cells neither replicate nor migrate (Nameroff and Holtzer, 1969). This is not due to simple depletion of the ambient medium. Pending further analysis it is proposed that extracellular mucopolysaccharides secreted by complex, essentially postmitotic tissues check the multiplication and migration of myogenic cells. It will be of interest to learn if this inhibitory influence is due to secretory products of the various epithelial cells and myotubes, or whether it is due to the products of fibroblasts present in the three tissues.

## EFFECTS OF BUdR ON MYOGENESIS

Over a range of concentrations BUdR has a remarkable effect on the differentiation of many types of cells. Replicating chondrocytes and amnion cells are selectively inhibited from synthesizing chondroitin sulfate and hyaluronic acid respectively (Abbott and Holtzer, 1968; Holtzer and Abbott, 1968; Bischoff and Holtzer, 1968*b;* Schulte-Holthausen, Chacko, Davidson, and Holtzer, 1969; Chacko, Holtzer, and Holtzer, 1969; Lasher and Cahn, 1969). In addition BUdR reversibly suppresses the outgrowth of neurites by replicating presumptive neuroblasts and melanin synthesis by replicating pigment cells (Biehle and Holtzer, unpublished data); BUdR also suppresses the synthesis of collagen and hyaluronic acid by replicating fibroblasts (Marker, Warren, Holtzer, and Holtzer, unpublished data). In other systems, BUdR has been found to prevent antibody synthesis by immunologically competent cells (O'Brien and Coons, 1963; Dutton, Dutton, and Vaughn, 1960), the appearance of zymogen granules in pancreatic cells (Wessells, 1964), and the formation of echinochrome by sea urchin embryos (Gontcharoff and Mazia, 1967). BUdR preferentially blocks the synthesis of luxury molecules, namely those molecules synthesized by cells that are *not* required for the viability of the cell synthesizing them (Holtzer, 1968; Holtzer, Bischoff, and Chacko, 1969). The effect of the thymine analog on the synthesis of essential molecules, namely those molecules present in most cells and essential for the viability of the cells synthesizing them, is much less obvious.

Earlier reports to the effect that BUdR suppresses fusion and contractile protein synthesis, with a much lesser effect on viability and reproductive activity (Stockdale, Okazaki, Nameroff, and Holtzer, 1964; Okazaki and Holtzer, 1965), have been confirmed (Coleman and Coleman, 1968;

Coleman, Coleman, and Hartline, 1969). More recently (Bischoff and Holtzer, 1969*b*), four questions have been posed: (1) Will BUdR-suppressed, but replicating, presumptive myoblasts suffer impairment of their genetic commitment to myogenesis after several generations in the suppressed state? (2) How many rounds of DNA synthesis in the presence of BUdR are required to suppress fusion and myosin synthesis? (3) What is the effect of BUdR on reproductive integrity and the synthesis of essential molecules? (4) By varying the concentration of BUdR will fusion be suppressed but not myosin synthesis?

*1. Reversibility of BUdR suppression*

The descendants of cells reared in BUdR for 5 days form myotubes when subcultured and replated in normal medium (Okazaki and Holtzer, 1965). Experiments were performed to test the capacity of BUdR-suppressed cells to express their terminal myogenic phenotype after protracted growth in BUdR. Freshly isolated cells were grown in BUdR for a total of 15 days. At 5-day intervals during this period the cells were subcultured and replated at low density to maintain a high rate of cell division. A few attenuated myotubes formed in the BUdR primary cultures, but no myotubes or elongated cells appeared in the two subcultures (Fig. 3.3). After 15 days the suppressed cells were replated in normal medium. No multinucleated cells were observed for the next 10

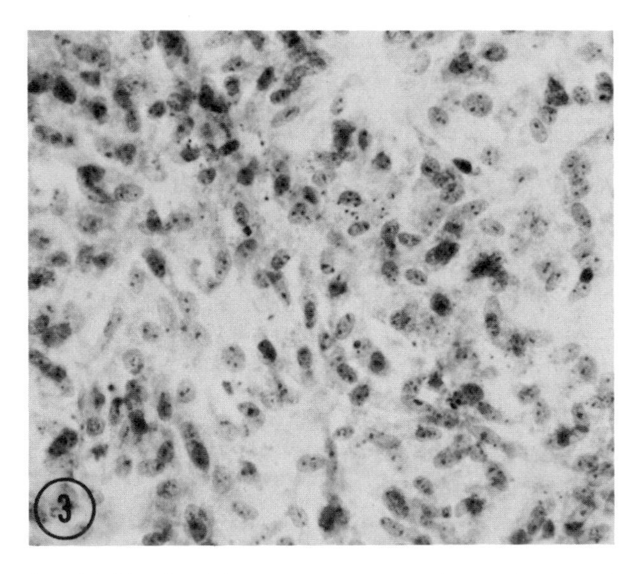


***Fig. 3.3.*** **Five-day-old second generation culture after a total of 10 days in 50 $\mu$g/ml BUdR. Virtually all cells show the BUdR-suppressed syndrome. $\times$ 320.**

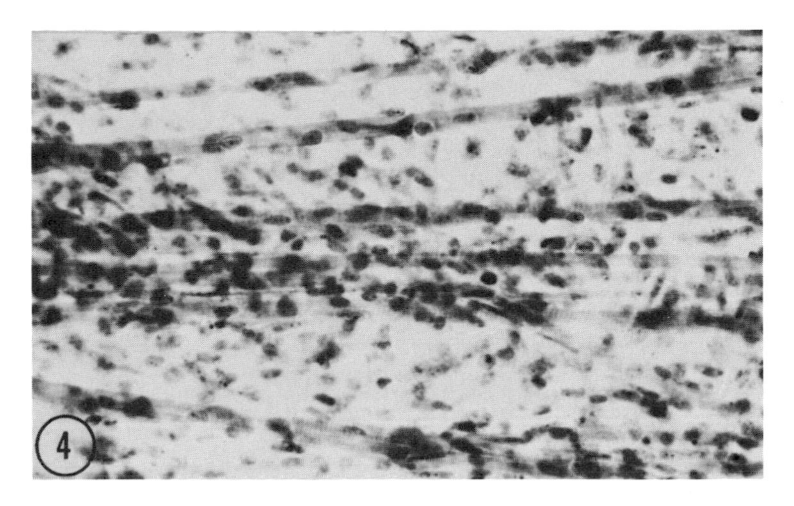


*Fig. 3.4.* Seven-day-old culture showing myotubes formed by the progeny of cells shown in Fig. 3.3. These cells were grown in BUdR for 15 days, then in normal medium for 17 days according to the subculture protocol described in the text. The proportion of fused nuclei is approximately the same as that found in normal primary cultures. × 300.

days of continuous culture. The cells were again subcultured in normal medium. Multinucleated myotubes appeared in 3 days and increased in frequency with continued culture (Fig. 3.4). Thus, many BUdR-suppressed myogenic cells reproduce themselves with high efficiency for a total of 25 days. Under these conditions there is no "overgrowth" of myogenic cells by fibroblasts (see also Chacko, Abbott, Holtzer, and Holtzer, 1969). These results are in contrast to the loss of presumptive myoblasts that occurs in less than a week when myogenic cells are allowed to fuse in normal medium (Bischoff and Holtzer, 1969*a*).

It is a challenge to propose a molecular model for BUdR-suppressed myogenic cells. For example, such cells do not revert to blank, "undifferentiated" cells. They do not deposit chondroitin sulfate, or synthesize albumin or melanin, etc.; this indicates that BUdR suppression does not disturb most of the genetic controls that characterize myogenic precursor cells. If BUdR "freezes" the level of differentiation, and the relief of that inhibition by new DNA synthesis allows the cells to jump into the next phenotypic "orbit," then progressive differentiation on a genetic level involves, primarily, derepression rather than repression. Myogenic cells do not synthesize sets of luxury molecules characteristic of cartilage, liver, or pigment cells. This view is in keeping with earlier speculations (Holtzer, 1961), but is in sharp contrast to those of Zwilling (1968) and Lash (1968).

*2. Effect of one round of DNA synthesis in the presence of BUdR*

That BUdR is incorporated into the DNA of myogenic cells has been demonstrated by gradient centrifugation and radioautography using $^3$H-BUdR (Stockdale, Okazaki, Nameroff, and Holtzer, 1964; Okazaki and Holtzer, 1965). The following experiments indicate that incorporation of the analog for one S period suffices to block fusion and myosin synthesis during the $G_1$ of the daughter cells. The design of the experiment was to permit incorporation of BUdR into the DNA made during one S period, and then to challenge the cells to display their capacity to fuse or synthesize contractile proteins over the next 4 days. Cells were grown continuously for 4 days in $10^{-6}$M FUdR. At this concentration of FUdR multiplication is promptly inhibited. After 12 hours in FUdR, in one series BUdR was added, in another series BUdR plus TdR was added, and in the third series the cells simply remained in FUdR. The concentration of TdR used in the second series prevented the suppressive effects of BUdR. The cultures in the three series were incubated for 7 hours (the S period of these cells is about 5 hours), the cells washed and suspended with trypsin and plated in fresh dishes to remove residual BUdR. In effect, the cells were allowed to synthesize DNA for one 7-hour period either in the presence of BUdR alone or in BUdR plus TdR. As shown in Fig. 3.5, cells in the FUdR + BUdR series exhibited the typical BUdR-suppressed syndrome: excessively flattened cells which did not fuse. Cells in the FUdR + BUdR + TdR series exhibited many myotubes with normal cross-striated myofibrils (Fig. 3.6). FUdR alone did not prevent fusion of the modest number of cells in these cultures that do not have to divide again before fusing. To avoid objections inherent in the use of FUdR, these results were confirmed in a different type of experiment. It has been shown that in normal 2-day cultures many cells incorporate $^3$H-TdR, divide, and fuse without going through another S period (Bischoff and Holtzer, 1969*a*). When labeled BUdR is used instead of TdR, those cells that incorporate the label are prevented from fusing. $^3$H-BUdR-labeled nuclei are found only in mononucleated cells after 24 hours; they are not found in multinucleated myotubes (Bischoff and Holtzer, 1970).

In summary, unifilar substitution of BU for thymine prevents fusion and synthesis of myofibrils in the progeny of the affected cells. Incorporation during one S period is effective even when thymidylate synthesis is not blocked by FUdR.

*3. Effect of BUdR on growth*

In view of the frequent reports of the inhibitory action of BUdR on cell proliferation (Littlefield and Gould, 1960; Puck and Kao, 1967; Berkovitz, Simon, and Toliver, 1968; Toliver and Simon, 1967), the experiment summarized in Fig. 3.7 was performed. Cells were grown for 5

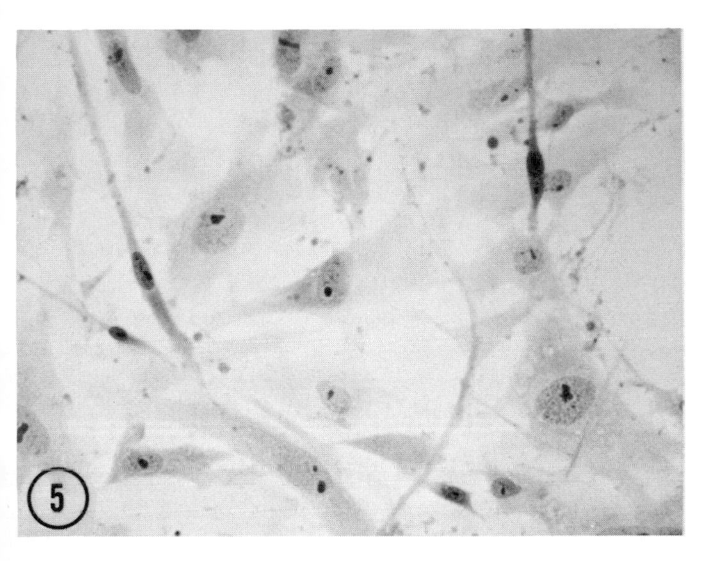


*Fig. 3.5.* Four-day-old culture grown continuously in $10^{-6}$ M FUdR except for a 7-hour period on day 1 during which 50 μg/ml BUdR was added. Cells are excessively enlarged and show the typical BUdR-suppressed syndrome. Myotubes are not present in these cultures. × 280.

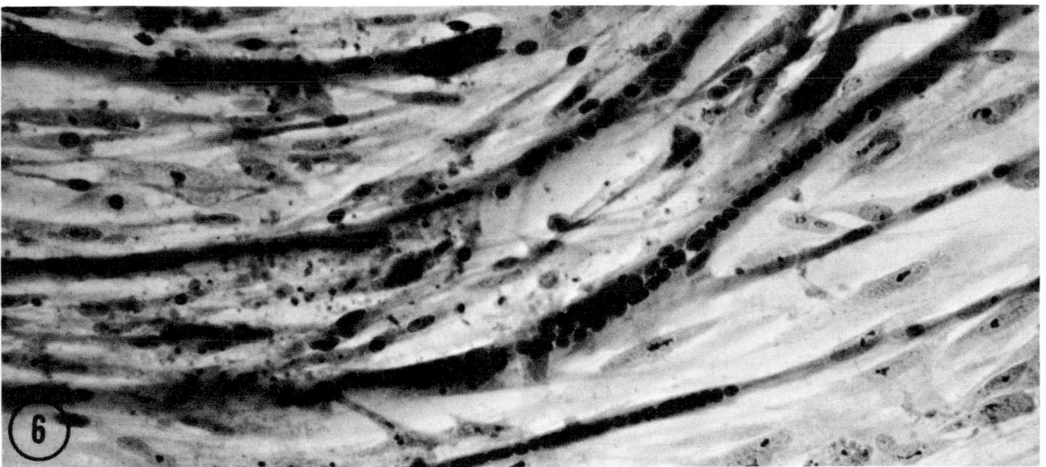


*Fig. 3.6.* Same protocol as in Fig. 3.5, except the block to DNA synthesis was circumvented for 7 hours with BUdR plus excess TdR. Cell morphology and myotube formation are normal. × 280.

days in 1, 5, or 50 μg/ml BUdR and the increase in total number of nuclei was determined. Cells grown in 50 μg/ml BUdR failed to keep pace with the controls after the first 24 hours and by 4 days produced 75% fewer nuclei than did the normal myotube-containing cultures. Despite the reduction in the cumulative number of cells, there was no significant difference in the mitotic index at 2, 3, or 4 days. Obviously, more cells died in the BUdR cultures than in the control cultures. In contrast, cultures in 5 μg/ml BUdR produced the same or slightly more cells than did the control cultures. The inhibition of growth by BUdR is

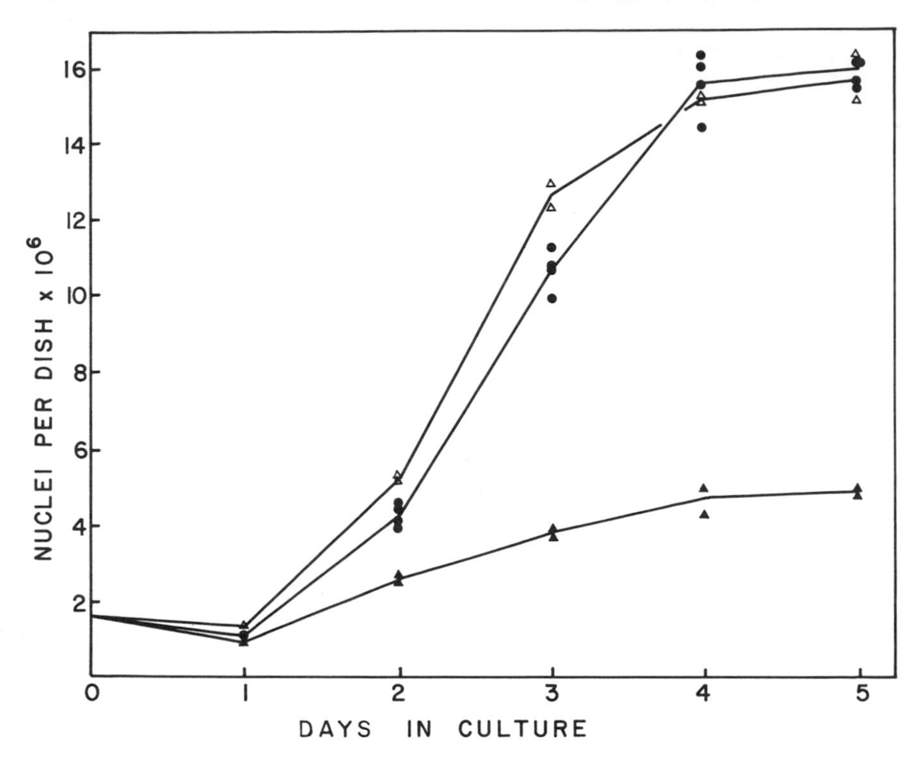


*Fig. 3.7.* Effect of BUdR on cell production. Cultures were initiated with $1.5 \times 10^6$ cells per 60 mm dish and the total number of nuclei was counted with a hemocytometer each day. The cytoplasm was digested with 0.1 M citric acid before counting nuclei. *Closed circles,* no BUdR added; *open triangles,* 5 μg/ml BUdR; *closed triangles,* 50 μg/ml BUdR.

concentration-dependent (Fig. 3.8). At low concentrations of BUdR there is actually a stimulation of cell production as compared with controls. Alteration of cell morphology parallels the concentration of the analog—the greatest degree of flattening and membrane expansion occurs at 50 μg/ml, while cells grown in 1 μg/ml are elongate and exhibit no excessive spreading. Intermediate concentrations produce intermediate effects on morphology. In view of the increase in cell number observed with lower analog concentrations, it might be profitable to regard the action of BUdR as forcing the presumptive myoblasts to stay in the mitotic cycle. According to this view, myogenic cells prevented from leaving the division cycle could not activate the synthetic activities required for fusion and contractile protein synthesis.

The modest effects on cell reproduction that we observe are in contrast to those found with HeLa and L-cells. This may in part be due to the

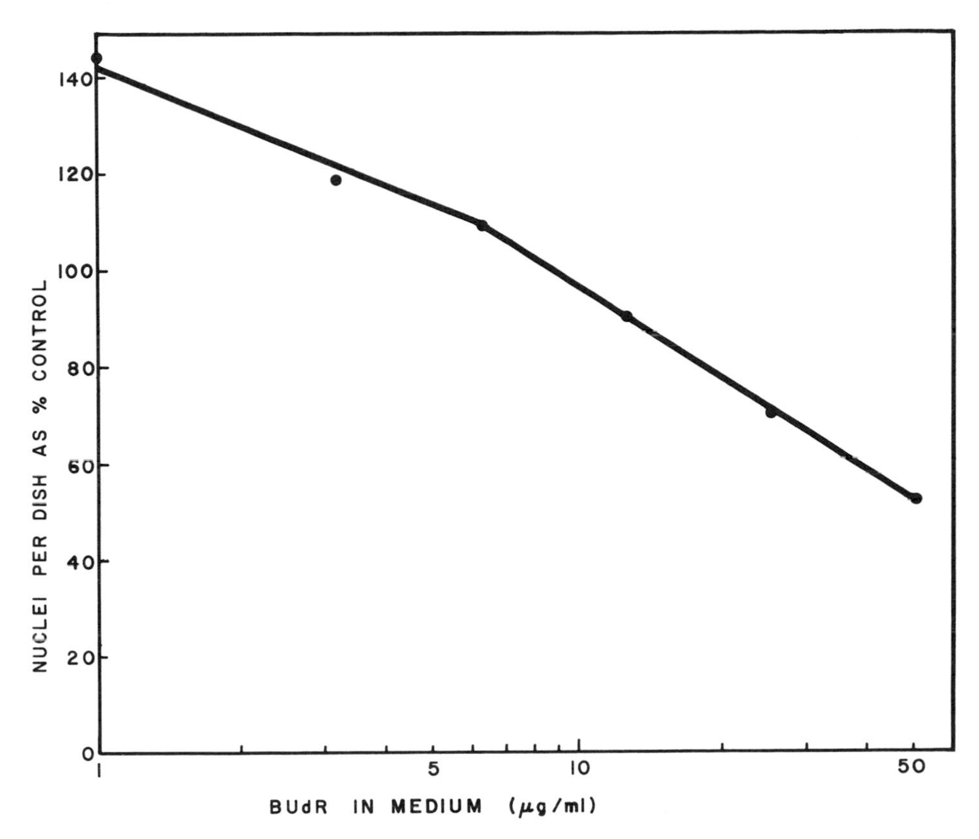


*Fig. 3.8.* Effect of BUdR on cell production as a function of concentration. $1.5 \times 10^6$ cells were placed in 60 mm dishes and cultured for 3 days in various concentrations of BUdR. Nuclei were counted in citric acid and plotted as a percentage of the untreated control value. Each point is the average of duplicate dishes. Myoblast fusion was blocked in all concentrations of BUdR.

problems associated with using "plating efficiency" as a criterion for viability (see also Holtzer and Abbott, 1968; Chacko, Holtzer, and Holtzer, 1969) and in part to the intracellular efficiency of thymidylate synthetase in the cells being analyzed. BUdR is used to select mutants (cells lacking thymidine kinase) for many kinds of experiments with long-term cultured cells (Littlefield, 1966); similar usage with normal cells is open to question.

*4. Uncoupling of fusion and contractile protein synthesis*

Myoblast fusion is blocked throughout the entire range of BUdR concentrations used (see Fig. 3.8). Despite the normal appearance of cells in 1 μg/ml BUdR, fusion does not take place. In older cultures grown in this concentration many cells become highly elongated and aligned in closely packed swirls. Electron microscopy of such cultures reveals that many of these mononucleated cells form thick and thin filaments.

If BUdR is acting directly on genes, then the genes regulating the production of molecules responsible for fusion may be more sensitive to the disruption caused by BU-DNA than those responsible for the production of myosin and actin.

*Effect of BUdR on protein synthesis*

A thesis worth proving or disproving is that the genetic control of luxury molecules is more prone to minor perturbations than is the machinery regulating the production of essential molecules (Holtzer and Abbott, 1968). Fig. 3.9 demonstrates that total protein synthesis in replicating cells is not greatly affected by high or low concentrations of BUdR. Replicating normal cells and replicating myogenic cells suppressed by BUdR cannot be distinguished at the level of *total* protein synthesis.

As an example of an essential molecule, the lactate dehydrogenase (LDH) activity of BUdR-suppressed cultures has been followed (Carlson, Murison, and Holtzer, unpublished observations). In terms of total activity per DNA and in terms of isozyme patterns in acrylamide gel, there are no great differences between myogenic cells in normal medium and those suppressed by BUdR.

If the effects of BUdR are due solely to its replacement of thymine in newly formed DNA, then the paradox of its minor action on essential molecules might be due to: (1) gene redundancy for the coding of essential molecules; (2) a higher thymine content per codon for luxury molecules; (3) a difference in the efficiency of BUdR competition with thymidylate for insertion into the genes for luxury molecules. Alternatively, although the primary site of BUdR action appears to be DNA (see Bischoff and Holtzer, 1970), its selectivity need not imply a direct effect on a list of specific genes for luxury molecules. BUdR may act on the

*Fig. 3.9.* Incorporation of $^{3}$H-leucine by normal and BUdR-suppressed cells. Cells were grown in 0, 5, or 50 $\mu$g/ml BUdR for 5 days, subcultured and continued in the analog. The second generation cultures were labeled for 4 hours on day 1, 3, or 5 with 5 $\mu$C/ml $^{3}$H-leucine, fixed, and the radioactivity was determined in a scintillation counter. Each point represents the average of 3 dishes. Second generation cultures prepared from 5-day normal primary cultures do not form myotubes and thus provided a source of mononucleated cells for comparison with the BUdR-suppressed cells. Nuclei were counted in citric acid. *Closed circles,* untreated control cells; *open circles,* 50 $\mu$g/ml BUdR; *triangles,* 5 $\mu$g/ml BUdR.

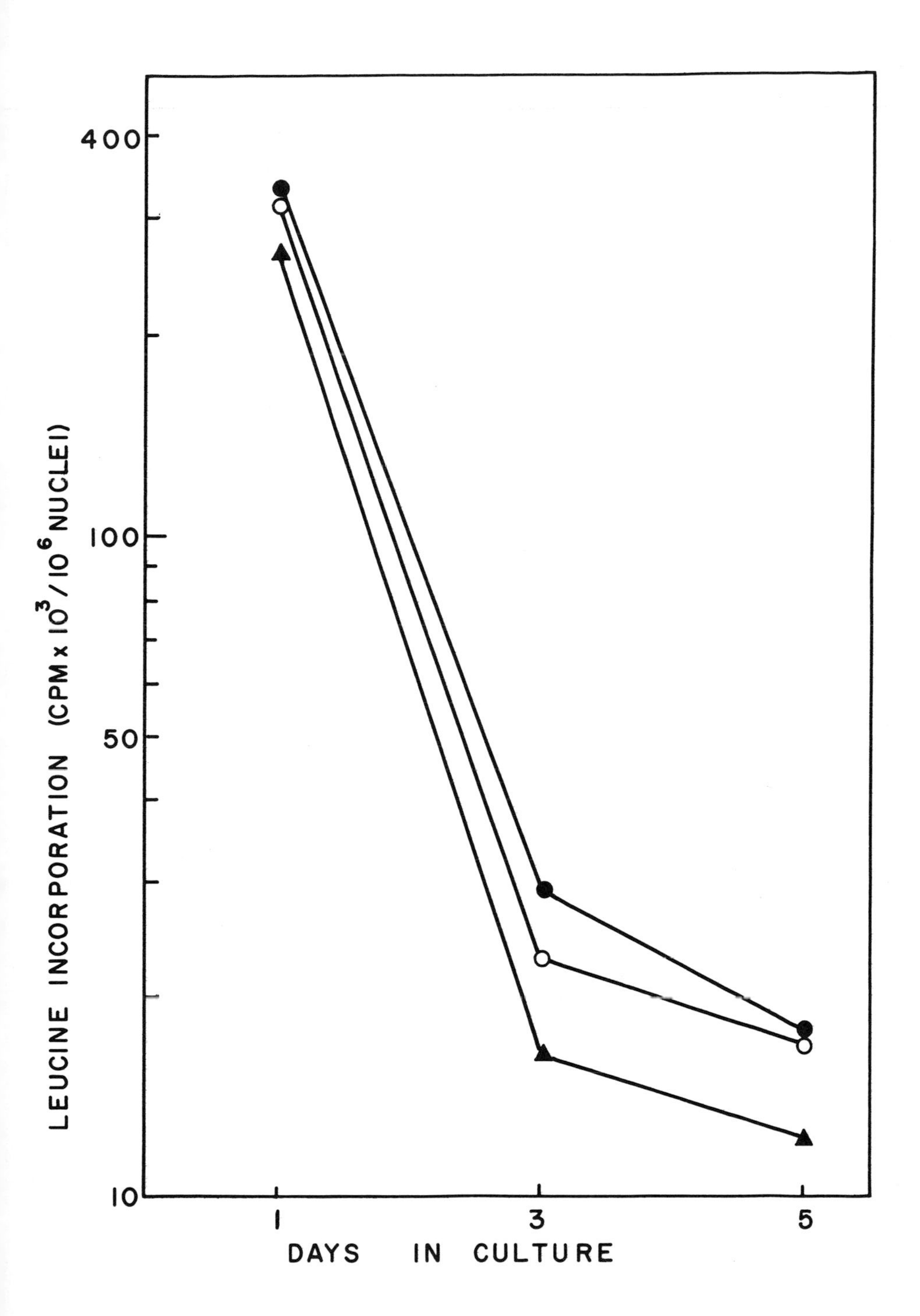

400
100
50
10
1
3
5
DAYS IN CULTURE
LEUCINE INCORPORATION (CPM x 10$^3$/10$^6$ NUCLEI)

same molecule in all cells, for example, on a glycoprotein destined for the cell membrane; and the suppressive effect on luxury molecules in a variety of cells may indicate only that the synthesis of such molecules requires a normal cell surface (Abbott and Holtzer, 1968; Holtzer and Abbott, 1968; Holtzer, 1970).

## DISCUSSION

### *Model for myogenesis*

Cell differentiation is an historical process. It is concerned with generation gaps—daughter cells synthesizing molecules not synthesized by the mother cell. When both daughter cells display phenotypic properties—cryptic or apparent—displayed by the mother cell, they are the products of a proliferative mitosis (Fig. 3.10). Such mitoses are responsible for growth: the increase in number of equivalent cells. Proliferative mitoses do not lead to diversification or progressive differentiation. As schematized in Fig. 3.10, the kinds of mitoses that result in new phenotypes have been termed quantal mitoses (Holtzer, 1967; Holtzer and Abbott, 1968; Holtzer, Bischoff, and Chacko, 1969; Holtzer, 1970). This category of mitosis figured prominently in the biological literature three and four decades ago (e.g., Wilson, 1925; Korschelt, 1882), but with a few notable exceptions (e.g., Stebbins, 1967; Kühn, 1965) has been ignored in the recent literature on differentiation.

Fig. 3.11 represents a scheme of step-wise differentiation, from covert to overt expression, centering on a set number of quantal cell cycles, alternating with variable numbers of proliferative mitoses. This scheme accounts for all data pertaining to myogenesis, and, with but minor qualifications, for the differentiation of all cell types. The assumption underlying this scheme is that zygotes, or morula or blastula cells, as differentiated cells, are limited in their capacity to respond to environmental changes. Such cells do not and cannot commence synthesizing terminal luxury molecules characteristic of other differentiated cells unless restructured by mitotic divisions. "X" number of quantal mitoses (the mitosis that shifts cells from one vertical column in Fig. 3.11 to the next vertical column) are required to convert the various generations of precursor cells into the terminal phenotype of a given lineage. Only a succession of quantal mitoses boosts a myogenic $\alpha$ cell to a myogenic $\beta$ cell, to a presumptive myoblast, to a myoblast. It is worth stressing that variable numbers of proliferative mitoses (or mitoses within the vertical columns in Fig. 3.11) can intervene between the quantal mitoses. Failure in the past to distinguish between proliferative and quantal mitosis resulted in the misleading distinction between systems exhibiting determinative cleavage, or mosaic systems, and those exhibiting indetermina-

PROLIFERATIVE VS. QUANTAL MITOSIS

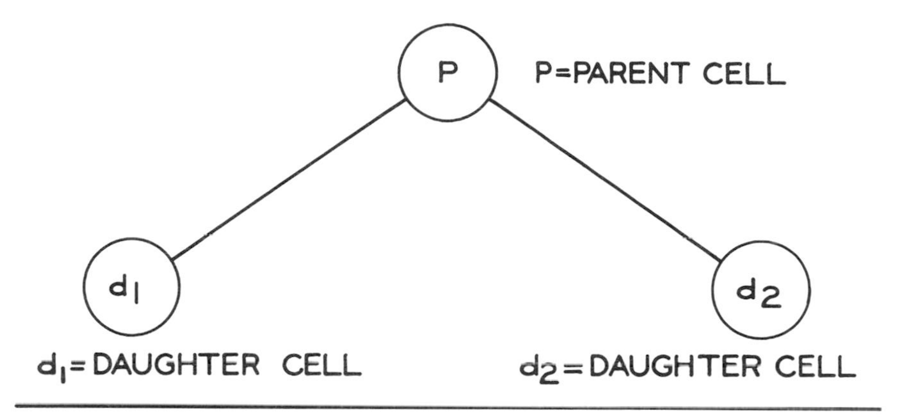


I. PROLIFERATIVE MITOSIS $d_1 = d_2 = P$

EXAMPLES: REGENERATING LIVER, CORNEA, FIBROBLASTS, HELA CELLS, ETC.

II. QUANTAL MITOSIS

$d_1 = d_2 \neq P$

$d_1 \neq d_2 = P$

$d_1 \neq d_2 \neq P$

EXAMPLES: STRATUM GERMINATIVUM, HEMATOCYTOBLASTS; PRESUMPTIVE MYOBLASTS, NEUROBLASTS, CHONDROBLASTS, LIVER AND PANCREATIC CELLS, ETC.

*Fig. 3.10.* Scheme of phenotypic relation between parent and daughter cells in proliferative vs. quantal mitosis.

tive cleavage, or regulative systems. All differentiation in all systems is dependent upon determinative or quantal divisions; these are the divisions that lead to the "determined" state of classical embryologists.

The concept of quantal divisions does not eliminate the importance of environmental factors impinging upon the genome in guiding differentiation. Indeed, extracellular cues acting during the course of a *quantal*

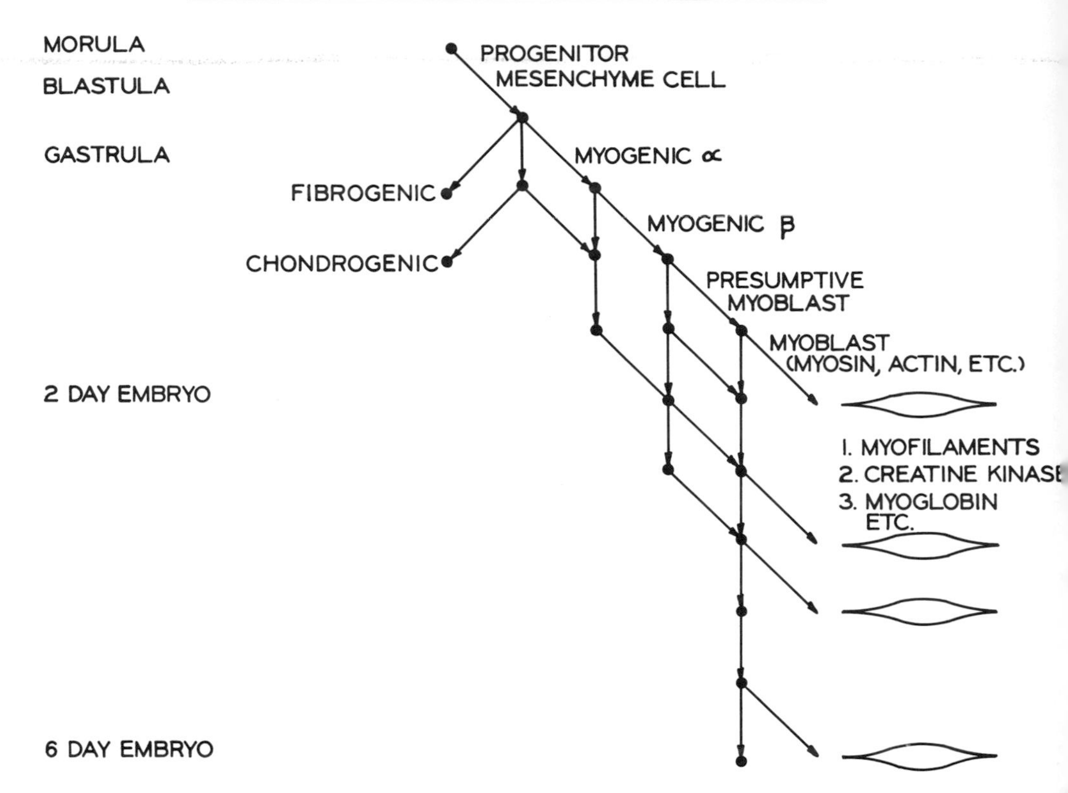


*Fig. 3.11.* Model for myogenesis based on the concept of a modest but rigid number of quantal mitoses alternating with variable numbers of proliferative mitoses. From progenitor mesenchyme cell there are 4 successive quantal cell cycles leading to 4 distinctly different transitory myogenic phenotypes before a determined myogenic cell translates terminal luxury molecules. These terminal luxury molecules include those required for fusion, as well as myosin, myoglobin, and tropomyosin. It is postulated that myogenic cells ancestral to the myoblast are constitutively incapable of fusing or synthesizing terminal luxury molecules without undergoing the requisite numbers of additional quantal mitoses. Each quantal mitosis would segregate a cytoplasmic factor and/or derepress a portion of the genome; this factor or derepressed state would be transmitted to that cell's progeny. It is also postulated that myogenic α cells, myogenic β cells, and presumptive myoblasts translate unique luxury molecules which are not myosin, myoglobin, or tropomyosin. A myoblast emerging from its last mitosis on day 2 or day 6 is in all major respects similar to one emerging from its terminal mitosis in a post hatched chick.

division are essential for the reprogramming of the daughter cells, whereas the same environment acting during a *proliferative* division would not result in transformation of the daughter cells.

This model accounts for compensatory development or regulation following extirpation in developing embryos (Weiss, 1939). Removal of many, but not all, presumptive myoblasts or any class of myogenic precursor cells would be compensated provided the remaining cells undergo a series of proliferative mitoses. Radical extirpation of all presumptive myoblasts or all classes of myogenic precursor cells from a local region should, and does, lead to the total absence of muscle (Holtzer and Detwiler, 1953; Holtzer, 1958). Similar observations have been made on numerous other developing systems: limb buds (Harrison, 1918), nervous system (Wenger, 1951; Holtzer, 1951) and vertebral cartilage (Holtzer and Detwiler, 1953) to mention but a few. The strictest and clearest dependency of differentiation on a precise succession of obligatory quantal mitoses occurs in mosaic systems where, at times, a single division ushers in two new phenotypes.

Likewise, the determining element in growth control in tissues may be the balance between quantal and proliferative divisions. As skeletal muscle tissue matures *in vitro,* presumptive myoblasts decline drastically in relation to postmitotic myoblasts (Bischoff and Holtzer, 1969*a*). We interpret this to mean that most divisions in the older cultures are of the quantal type that lead to nondividing daughter cells, thereby restricting increase in cell number in the tissue.

What kinds of luxury molecules are synthesized by presumptive myoblasts, or myogenic $\alpha$, $\beta$, or $\gamma$ cells? Such cells must have inherited or synthesized molecules not present in other cells. Yet currently these whole classes of molecules which are responsible for the covertly differentiated or "determined" state of myogenic precursor cells are unknown. Developing heart cells synthesize a hyaluronic-acid-like molecule (Holtzer and Matheson, 1970). It is not unlikely that the various kinds of myogenic precursor cells not only synthesize unique mRNAs and proteins, but unique kinds of polysaccharides as well. The single most elusive problem in cell differentiation is to catalogue those luxury molecules that are responsible for the "determined" state of precursor cells.

Two issues relating to quantal cell cycles invite further probing: (1) Are quantal divisions determined before the actual mitosis in the $G_1$, S, or $G_2$ of the mother cell, or does a particular mitosis become a quantal mitosis only in retrospect because of events occurring in the early $G_1$ of covertly differentiated daughter cells? (2) What are the critical phases of a quantal division cycle which permit divergence of phenotypes in the progeny? For example, if the critical event of derepression is tied directly to DNA synthesis in the mother cell, the expression of this event in either transcription or translation might still require the $G_1$ cytoplasm of the daughter cell.

ACKNOWLEDGMENTS

This work was supported by grants HD00189, 5T01 HD00030–03 from the National Institute of Child Health and Human Development (NIH) and GB5047X from the National Science Foundation. Dr. Holtzer is the recipient of Research Career Development Award 5K3 HD2970 (NIH).

## *References*

Abbott, J., and H. Holtzer. 1968. The loss of phenotypic traits by differentiated cells V. The effect of 5-bromodeoxyuridine on cloned chondrocytes. *Proc. Nat. Acad. Sci. 59:*1144.

Allen, E., and F. Pepe. 1965. Ultrastructure of developing muscle cells in the chick embryo. *Amer. J. Anat. 116:*115.

Berkovitz, A., E. H. Simon, and A. Toliver. 1968. DNA replication in the absence of cell division in BUdR-FUdR treated HeLa cells. *Exp. Cell Res. 53:*497.

Betz, E. H., H. Firket, and M. Reznik. 1966. Some aspects of muscle regeneration. *Int. Rev. Cytol. 19:*203.

Bischoff, R., and H. Holtzer. 1968*a*. The effect of mitotic inhibitors on myogenesis in vitro. *J. Cell Biol. 36:*111.

———. 1968*b*. Inhibition of hyaluronic acid synthesis by BUdR in cultures of chick amnion cells. *Anat. Rec. 160:*317.

———. 1969*a*. Mitosis and the processes of differentiation of myogenic cells in vitro. *J. Cell Biol. 41:*188.

———. 1969*b*. A quantal mitosis and fusion. *J. Cell Biol. 43:*13a.

———. 1970. Inhibition of myoblast fusion after one round of DNA synthesis in 5-bromodeoxyuridine. *J. Cell Biol. 44:*134.

Chacko, S., J. Abbott, S. Holtzer, and H. Holtzer. 1969. The loss of phenotypic traits by differentiated cells VI. Behavior of the progeny of a single chondrocyte. *J. Exp. Med. 130:*417.

Chacko, S., S. Holtzer, and H. Holtzer. 1969. Suppression of chondrogenic expression in mixtures of normal chondrocytes and BUdR-altered chondrocytes grown in vitro. *Biochem. Biophys. Res. Comm. 34:*183.

Church, J. C. T., R. F. X. Noronka, and D. B. Allbrook. 1966. Satellite cells and skeletal muscle regeneration. *Brit. J. Surg. 53:*638.

Cohen, Z. A., and B. A. Ehrenreich. 1969. The uptake, storage, and intracellular hydrolysis of carbohydrates by macrophages. *J. Exp. Med. 129:*201.

Coleman, J. R., and A. W. Coleman. 1968. Muscle differentiation and macromolecular synthesis. *J. Cell. Comp. Physiol. 72* (suppl. 1) :19.

Coleman, J. R., A. W. Coleman, and E. J. H. Hartline. 1969. A clonal study of the reversible inhibition of muscle differentiation by the halogenated thymidine analog 5-bromodeoxyuridine. *Dev. Biol. 19*:527.

Colwin, L. H., and A. L. Colwin. 1963. Role of the gamete membranes in fertilization in *Saccoglossus kowalevskii* (*Enteropneusta*) I. The acrosomal region and its changes in early stages of fertilization. *J. Cell Biol. 19*:477.

Defendi, V., and L. A. Manson. 1963. Analysis of the life-cycle in mammalian cells. *Nature 198*:359.

Dutton, R. W., A. W. Dutton, and J. H. Vaughn. 1960. The effect of 5-bromouracil deoxyribonucleoside on the synthesis of antibody in vitro. *Biochem. J. 75*:230.

Ephrussi, B., L. J. Scaletta, M. A. Stenchever, and M. C. Yoshida. 1964. Hybridization of somatic cells in vitro. *Symp. Int. Soc. Cell Biol. 3*:13.

Fischman, D. A. 1967. An electron microscope study of myofibril formation in embryonic chick skeletal muscle. *J. Cell Biol. 32*:557.

Gontcharoff, M., and D. Mazia. 1967. Developmental consequences of introduction of bromouracil into the DNA of sea urchin embryos during early division stages. *Exp. Cell Res. 46*:315.

Harris, H., J. F. Watkins, C. E. Ford, and G. I. Schoefl. 1966. Artificial heterokaryons of animal cells from different species. *J. Cell Sci. 1*:1.

Harrison, R. G. 1918. Experiments on the development of the fore limb of *Amblystoma,* a self-differentiating equipotential system. *J. Exp. Zool. 25*:413.

Holtzer, H. 1951. Reconstitution of the urodele spinal cord following unilateral ablation. *J. Exp. Zool. 117*:523.

———. 1958. The development of mesodermal structures in regeneration and embryogenesis. *In* C. Thornton (ed.), *Regeneration in vertebrates.* Univ. of Chi.

———. 1961. Aspects of chondrogenesis and myogenesis, p. 35. *In* D. Rudnick (ed.), *Synthesis of Molecular and Cellular Structure.* Ronald Press, New York.

———. 1967. Mutually exclusive activities during myogenesis, p. 57. *In* A. T. Milhorat (ed.), *Exploratory Concepts in Muscular Dystrophy and Related Disorders.* Excerpta Medica Foundation, Amsterdam.

———. 1968. Induction of chondrogenesis, a concept in quest of mechanisms, p. 152. *In* R. Fleischmajer (ed.), *Epithelial-Mesenchymal Interactions.* Williams and Wilkins, Baltimore.

———. 1970. Quantal and proliferative mitoses in the differentiation of muscle, cartilage and red blood cells. *In* H. Padykula (ed.), *Gene Expression in Somatic Cells.* Academic Press, New York.

Holtzer, H., and J. Abbott. 1968. Oscillations of the chondrogenic phenotype in vitro, p. 1. *In* H. Ursprung (ed.), *Stability of the Differentiated State.* Springer-Verlag, Berlin.

Holtzer, H., J. Abbott, and J. Lash. 1958. On the formation of multinucleated myotubes. *Anat. Rec. 131*:567.

Holtzer, H., G. Avery, and S. Holtzer. 1954. Some properties of the regenerating limb blastema cells of salamanders. *Biol. Bull. 107*:313.

Holtzer, H., R. Bischoff, and S. Chacko. 1969. Activities of the cell surface during myogenesis and chondrogenesis. *In* R. T. Smith and R. A. Good (eds.), *Cell Recognition*. Appleton-Century-Crofts, New York.

Holtzer, H., and S. Detwiler. 1953. Induction of skeletogenous cells. *J. Exp. Zool. 123*:335.

Holtzer, H., J. Marshall, and H. Finck. 1957. An analysis of myogenesis by the use of fluorescent antimyosin. *J. Cell Biol. 3*:705.

Holtzer, H., and D. Matheson. 1970. Induction of chondrogenesis in the embryo, p. 1214. *In* A. Balczs (ed.), *Chemistry and Molecular Biology of the Extracellular Matrix*. Academic Press, London.

Ishikawa, H. 1966. Electron microscopic observations of satellite cells with special reference to the development of mammalian skeletal muscle. *Z. Anat. Entwicklungsgesch. 125*:45.

Ishikawa, H., R. Bischoff, and H. Holtzer. 1968. Mitosis and intermediate-sized filaments in developing skeletal muscle. *J. Cell Biol. 38*:538.

Konigsberg, I. R. 1965. Aspects of cytodifferentiation of skeletal muscle, p. 337. *In* R. DeHaan and H. Ursprung (eds.), *Organogenesis*. Holt, Rinehart & Winston, New York.

Korschelt, E. 1882. Uber Bau und Entwicklung des *Dinophilus apatris*. *Z. Wiss. Zool. 37*:315.

Kühn, A. 1965. *Vorlesung en uber Entwicklungsphysiologie* (2d ed.). Springer-Verlag, Berlin.

Lash, J. 1968. Chondrogenesis: genotypic and phenotypic expression. *J. Cell. Comp. Physiol. 72* (suppl.):35.

Lasher, R., and R. D. Cahn. 1969. The effects of 5-bromodeoxyuridine on the differentiation of chondrocytes in vitro. *Dev. Biol. 19*:415.

Littlefield, J. W. 1966. The use of drug-resistant markers to study the hybridization of mouse fibroblasts. *Exp. Cell Res. 41*:190.

Littlefield, J. W., and E. A. Gould. 1960. The toxic effect of 5-bromodeoxyuridine on cultured epithelial cells. *J. Biol. Chem. 235*:1129.

Maslow, D. 1969. Cell specificity in the formation of multinucleated striated muscle. *Exp. Cell Res. 54*:381.

Mauro, A. 1961. Satellite cells of skeletal muscle fibers. *J. Biophys. Biochem. Cytol. 9*:493.

Metz, C. B. 1967. Gamete surface components and their role in fertilization, p. 163. *In* C. B. Metz and A. Monroy (eds.), *Fertilization,* Vol. 1. Academic Press, New York.

Nameroff, M., and H. Holtzer. 1969. Interference with myogenesis. *Dev. Biol. 19*:380.

O'Brien, T. F., and A. H. Coons. 1963. Studies on antibody production VII. The effect of 5-bromodeoxyuridine on the in vitro anamnestic antibody response. *J. Exp. Med. 117*:1063.

Okazaki, K., and H. Holtzer. 1965. An analysis of myogenesis using fluorescein-labeled amtimyosin. *J. Histochem. Cytochem. 13*:726.

Przybylski, R. J., and J. M. Blumberg. 1966. Ultrastructural aspects of myogenesis in the chick. *Lab. Invest. 15*:863.

Puck, T. T., and F. T. Kao. 1967. Genetics of somatic mammalian cells V. Treatment with 5-bromodeoxyuridine and visible light for isolation of nutritionally deficient mutants. *Proc. Nat. Acad. Sci. 58*:1227.

Schulte-Holthausen, H., S. Chacko, E. Davidson, and H. Holtzer. 1969. Enzyme changes in BUdR-suppressed chondrocytes. *Proc. Nat. Acad. Sci. 64*:650.

Stebbins, G. L. 1967. Gene action, mitotic frequency and morphogenesis in higher plants. *Dev. Biol. 1* (suppl.) :113.

Stockdale, F., and H. Holtzer. 1961. DNA synthesis and myogenesis. *Exp. Cell Res. 24*:508.

Stockdale, F., K. Okazaki, M. Nameroff, and H. Holtzer. 1964. 5-bromodeoxyuridine: effect on myogenesis in vitro. *Science 146*:533.

Toliver, A., and E. H. Simon. 1967. DNA synthesis in 5-bromouracil tolerant HeLa cells. *Exp. Cell Res. 45*:603.

Weiss, P. 1939. *Principles of Development.* Henry Holt, New York.

Wenger, B. S. 1951. Determination of structural patterns in the spinal cord of the chick embryo studied by transplantation between brachial and adjacent levels. *J. Exp. Zool. 116*:123.

Wessells, N. K. 1964. DNA synthesis, mitosis and differentiation in pancreatic acinar cells in vitro. *J. Cell Biol. 20*:415.

Wilde, C. 1959. Differentiation in response to the biochemical environment, p. 3. *In* D. Rudnick (ed.), *Cell, Organism and Milieu.* Ronald Press, New York.

Wilson, E. B. 1925. *The Cell in Development and Heredity* (3rd ed.). Macmillan, New York.

Yaffe, D., and M. Feldman. 1965. The formation of hybrid multinucleated muscle fibers from myoblasts of different genetic origin. *Dev. Biol. 11*:300.

Zwilling, E. 1968. Morphogenetic phases in development, p. 184. *In* M. Locke (ed.), *The Emergence of Order in Developing Systems.* Academic Press, New York.

# Chapter 4

# *The Role of the Innervation of the Functional Differentiation of Muscle*

W. F. H. M. MOMMAERTS

One of the recurrent subjects in these symposia has been the theme of differences between various types of muscle. On the one hand, we are dealing with distinctions observed when comparing different animal sources, on the other hand, with often equally striking differences among the muscles of one animal. When we discuss these characteristics, we must keep in mind that they represent the chemical and structural counterparts of two profound biological principles: function, and the genetically determined potential of organic differentiation. These principles interact constantly along a line of causal and teleological connections: from among a spread of genetically allowed possibilities, some are repressed and some expressed, thus giving cause to a pattern of differentiation by which the functional and behavioral characteristics of the animal emerge, its individual activities, and its survival as a species. I would like to discuss this problem in a broad fashion, concentrating upon some of our recent findings which promise to open up the investigation of connections between genetic information, molecular organization, and functional expression.

At the macroscopic level, the functional unit of the musculature would seem to be the individual muscle, but there is a prominent organizational entity at the level just below this, which is the motor unit. I am limiting the discussion to the important category of monosynaptically innervated voluntary striated muscles, which make up the bulk of the musculature in the human, in meat-producing animals, and in mammals in general. As schematically indicated in Fig. 4.1, a motor unit consists of a motoneurone in the ventral horn of the spinal cord, with its long axon, the terminal axon branches with their neuromyonal end plates, and the muscle fibers of myones. From the muscle's own standpoint, the motor

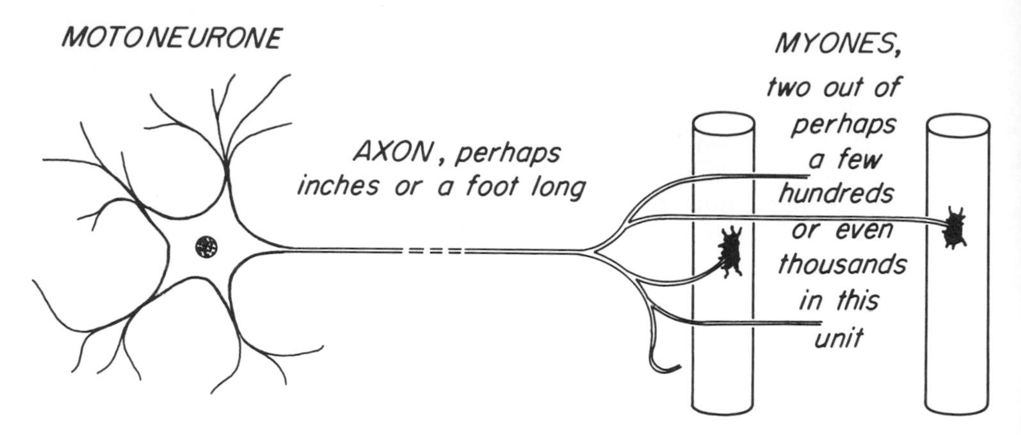


***Fig.*** *4.1.* Schematic drawing of a motor unit.

unit is a group of muscle fibers innervated collectively by one motor cell. The number of motor units in a muscle varies, but in larger muscles is likely to be of the order of a hundred or so. The number of myones in a unit likewise varies: it can be a few or a few dozen, or it can be thousands. The functional significance of the motor unit is primarily that it responds as a whole to the excitation of one motor cell, however elicited, so that the nervous system can effect a fractional activation of parts of the muscle selectively.

The motor control of this unitary activity has been the subject of several classical investigations. Most recently, a series of papers from Henneman's laboratory (Henneman and Olson, 1965; Henneman, Somjen, and Carpenter, 1965*a, b*) dealt with the problem. Among the findings were that, in the cat, muscle fibers seem to fall within three categories with respect to contraction velocity, fatigability, and demonstrable or presumable metabolic characteristics; that motor units appear to consist of only one type of muscle fibers; and that motor units of different characteristics are physiologically activated in certain sequences and by certain rules as to the size and characteristics of the unit and its motoneurone. What concerns us at the moment is that one motor axon, innervating perhaps hundreds or even thousands of muscle fibers, associates exclusively with only one among the several myonal types, even if the muscle contains all three. This could be explained in one of two ways. Either the characteristics of the muscle fibers are ontogenetically fixed, and specialized axons wander toward the type of myone with which they are destined to associate; or all the myones are genetically equal, and their specialization becomes determined by the neurones to which they become attached. The information just reported does not allow a decision, but lines of evidence to which we shall turn will suggest that the latter explanation applies.

The main experiment starts with what it is convenient to call the Buller and Eccles experiment (Buller, 1965; Buller, Eccles, and Eccles, 1960*a, b;* Eccles, 1967). In this experiment, one investigates one muscle each of the predominantly fast and slow types, F and S muscles. In the cat, these could be the flexor hallucis or digitorum longus, and the soleus, respectively. These differ three- to fourfold in contraction velocity. They also differ in some other characteristics which we shall discuss subsequently. In the experiment, one effects a surgical crossing of the nerves leading to these muscles, as shown schematically in Fig. 4.2. The nerves are sectioned, and interchanged before reconnecting. The nerve fibers in the distal ends, disconnected from their cell bodies, degenerate; the voids they leave are gradually replaced by the axons from the other nerve, which make new connections upon the muscle fibers. After 6 months or longer, the results are tested. What was an F muscle has now become an S muscle, nearly indistinguishable with respect to its contraction characteristics. The S muscle in turn has become somewhat faster, without however reaching the velocity of the F muscle in this instance (it does so in certain muscles of the rat [Close, 1965]). This experiment shows very clearly that the nerve determines the characteristics of the muscle, at least to the limits of the potentialities remaining in an adult cat.

Contraction velocity may be a major physiological characteristic, but it is not a property which by itself allows an explanation as to how the nerve might execute its formative influence, and it is to this question that we have addressed ourselves (Mommaerts, 1968; Buller and Mommaerts, 1969; Buller, Mommaerts, and Seraydarian, 1969; Mommaerts, Buller, and Seraydarian, 1969). Our starting point will be the assumption that the splitting of ATP is the rate-limiting step in the process whereby chemical energy leads to the contractile phenomenon.

After the discovery by Engelhardt and Ljubimowa (1939) of the relation between myosin and adenosine triphosphatase, and especially after the striking discoveries by Szent-Györgyi (1942) on actomyosin contractility *in vitro,* it seemed that myosin–adenosine triphosphatase was at the foundation of the chemodynamic transduction process. But a difficulty soon arose (Mommaerts, 1946; Mommaerts and Seraydarian, 1947): muscle is rich in $Mg^{++}$, and this ion is essential to the contraction process. Yet at the same time, in the presence of other salt, it inhibits myosin–adenosine triphosphatase so strongly that its residual activity seems to account for only about 1% of the natural rate of ATP utilization that one has to accept in order to account for known phenomena according to several estimates (Mommaerts, 1950), or according to what

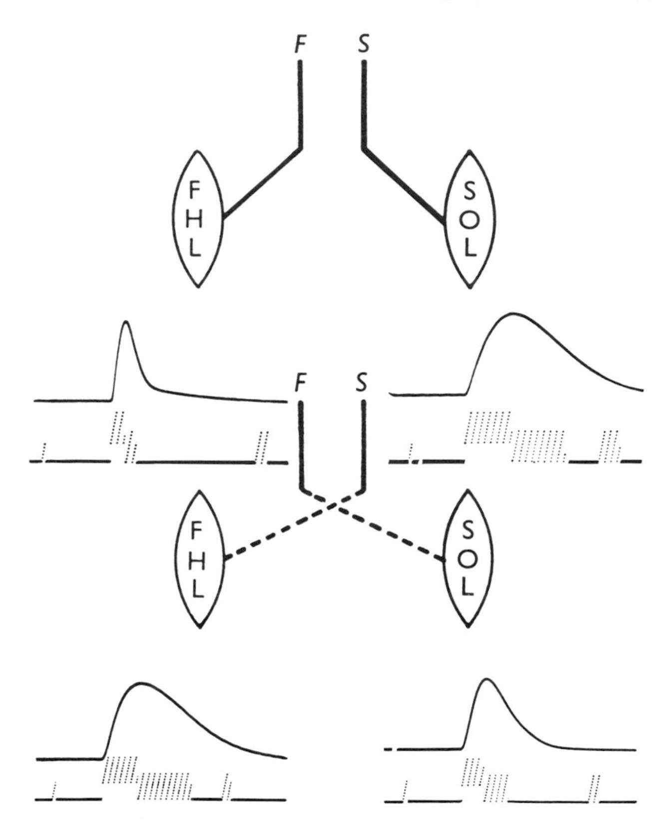


*Fig. 4.2.* Schematic representation of nerve crossing experiment, with records showing resulting changes in contraction velocity and twitch duration. From Buller and Lewis (1965).

can now be measured by direct experimentation (Mommaerts, 1969). Thus, to paraphrase our conclusion of the late 1940's, myosin–adenosine triphosphatase cannot play the role in contraction that was attributed to it. Perry (1951) and Perry and Grey (1956) have found, however, that actomyosin and particularly myofibrils can be quite active in the presence of $Mg^{++}$, and so the difficulty appears solved. Not completely, perhaps, and certainly not with a comfortable margin, for when we take the actual rates into account, we must conclude that myofibrillar adenosine triphosphatase can just barely account for the breakdown rates *in vivo.* Contrasting this with an enzyme such as that which transfers phosphoryl groups from ATP to creatine or vice versa, and which seems to be present in a hundredfold excess, we have a first indication that the splitting of ATP may be the rate-limiting reaction in the chain of transformations. A second argument was provided recently in a comparative investigation by

Bárány (1967), who determined the myosin– and actomyosin–adenosine-triphosphatase rates on proteins obtained from widely different muscles. For a number of these, data were available on their contraction velocities as well and a plot of these (Fig. 4.3A) shows a remarkably close correlation between contraction velocity and adenosine triphosphatase over a wide range.

Thus, the assertion that the adenosine triphosphatase step is rate-limiting is quite well founded. We must, however, analyze somewhat further its fundamental meaning. First of all, is the catalytic activity primarily a property of myosin or of actin, or of both, and how does it relate to the events proposed in Huxley's (1957) guiding theory of the contractile mechanism? I have dealt with this question elsewhere and need not repeat the full argument (Mommaerts, 1969). It suffices for now to state that it is the myosin partner which determines the adenosine triphosphatase velocity, as shown by Bárány's findings (Fig. 4.3B) that there is a perfect correlation between myosin– and actomyosin–adenosine triphosphatase; indeed, in most of his experiments only the myosins were prepared individually and were combined with the same actin, because the choice of the latter did not seem to matter. For the first approximation, therefore, the following statement suggests itself: physiological ATP splitting is a cooperative event between actin and myosin occurring during the translocational process at the cross bridges. The participation of actin, whether allosterically or through its direct involvement in a reaction step, causes $Mg^{++}$ to be an activator for the adenosine triphosphatase reaction, as it is for the translocation process at large. But the specific properties of the myosin moiety determine at what speed both reactions take place.

Upon further consideration it may yet emerge that this statement is still rather derivative. In the living fibril, bridge connections are made and broken, and it is likely that in each effective cycle one molecule of ATP is split. It could be that in isolated fibrils and even in actomyosin something similar happens; in fact, these entities contract or show syneresis during the first moments of incubation in the enzymic assay and, once shortened, may well continue to execute abortive contraction cycles. Then, it might be that the real rate-limiting step is the formation of the active myosin-actin combination, upon which the ATP splitting follows. This outcome might not be unwelcome, because Huxley's theory assumes the complex formation, not the adenosine triphosphatase, to be rate-limiting. We are designing an experimental approach to these questions. The result of this line of thought might be that the specific properties of myosin determine primarily the velocity of its repetitive combinations with actin, and thus determine the actomyosin–adenosine-triphosphatase rate indirectly, perhaps as the result of a temporarily ensuing transcon-

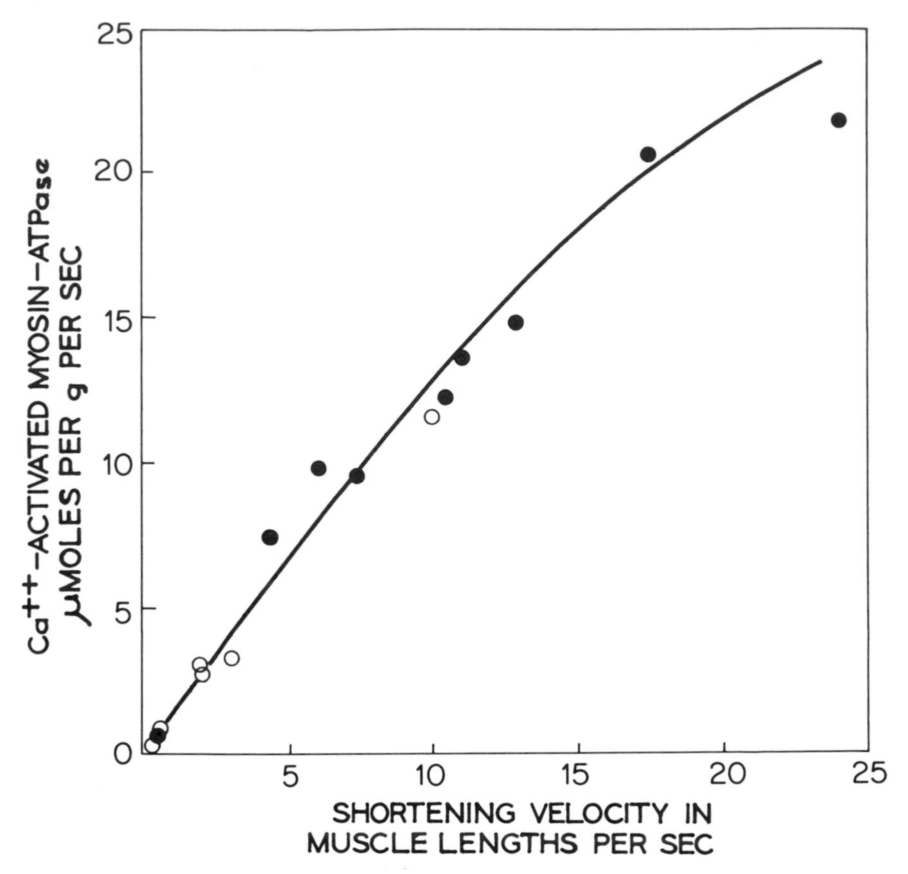


*Fig. 4.3. Above,* relation between contraction velocity and myosin–adenosine triphosphatase activity. *Right,* relation between myosin– and actomyosin–adenosine triphosphatase activity. Both sets of data refer to muscles of different origins, which span a wide range of velocities. From Bárány (1967).

formation of the myosin into an active adenosine triphosphatase once in each cyclic event. With these possibilities on the horizon, it still remains, therefore, that myosin– and fibrillar adenosine triphosphatase rates measured *in vitro* in appropriate media retain a firm connection with the problems at issue.

With this background, let us return to the mechanism of the Buller and Eccles transformation, with respect to myosin and myofibrillar protein. Our result was clear: in the crossing F → S, in which the contraction velocity becomes reduced three- to fourfold, the myosin– or myofibrillar adenosine triphosphatase diminishes in the same proportion. In the crossing S → F, when the contraction velocity increases little, the adenosine triphosphatase increases little too. These findings, therefore, indi-

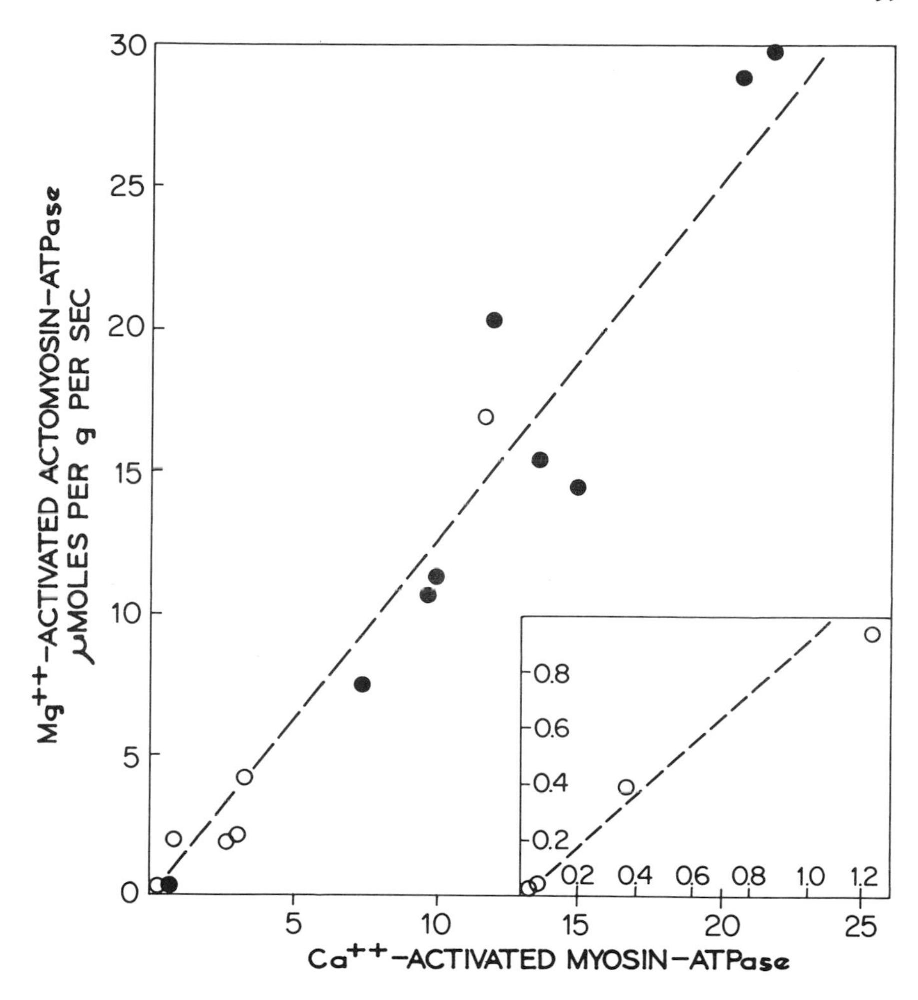


cate that muscles differ in contraction velocity because they have different myosins; and that the innervation determines the kind of myosin the muscles synthesize.

Contraction velocity is not, however, the only difference between F and S muscles. They differ strikingly in the duration of the contraction cycle. This property is likewise altered by cross innervation, in a more nearly symmetrical fashion for the two opposite crosses (Buller and Lewis, 1965). We do not have a good single hypothesis to explain the mechanism of this, as we do for contraction velocity. We know (Ebashi and Endo, 1969) that the excitatory process leads to a liberation of $Ca^{++}$ which, for reasons not concerning us now, activates the contraction process by suppressing an inhibitory interaction; and that the active period is terminated by a resequestering of the $Ca^{++}$ by the sarcoreticular membrane system. A more active $Ca^{++}$ transport into this system would surely terminate the activity cycle sooner, as in the F muscles. We do not know, however, whether this is (1) a shorter duration of the excitatory state during which ions are released concerning the nature of which nothing is known; or a more rapid onset of the reuptake by either (2) a greater extent of development of the sarcoreticular system, or (3) a greater activity with respect to $Ca^{++}$ uptake per unit surface or volume. We have found that the $Ca^{++}$-uptake activity of isolated vesicles shows a difference between F and S muscles, and that this difference changes reciprocally upon cross innervation. But the situation is ambiguous for several technical reasons, and we cannot really claim to have explained the differences in duration of the contraction cycle and their reversal in the Buller and Eccles experiment.

The third property we have examined has a very different physiological context again. Among the skeletal muscles there are some which work steadily at moderate power, as does also the heart, and which are so designed with respect to their myonal organization (Mommaerts, 1963) that they can, by oxidative phosphorylation, resynthesize ATP at about the rate it is being used. These, therefore, will normally function in a steady state, and for teaching purposes I have called these "pay-as-you-go muscles." The S muscles, which seem to be mostly used in posture and standing, and hence function steadily, would be representatives of this. The other type of muscles, our present F muscles among them, may have bouts of intense activity, but they are not designed to resynthesize ATP at the corresponding high rate during those bursts; hence they accrue a metabolic debt, physiologically manifest as an oxygen debt, which necessitates a subsequent period of rest and restitution by delayed oxidation. For this reason, I have called these the "twitch-now-pay-later muscles." The metabolic events during the deficit period are various, a diminution of phosphoryl-creatine being among them; but prominent is the occurrence of an intense glycolysis by which some of the exhaustion of high-energy phosphates is counteracted.

In view of this distinction, we have examined the F and S muscles with respect to lactic dehydrogenase, the last enzyme to act in the anaerobic conversion of glycogen. This enzyme occurs in the F muscles in vastly greater quantity. But there is more to it than this difference in amount. In intense glycolytic metabolism, lactic dehydrogenase serves to reduce pyruvate to lactate, in order to balance the oxidative step at the phosphotriose level, NAD acting as the carrier. In the oxidative muscle type, the enzyme serves instead to oxidize lactate to pyruvate; and, inasmuch as lactate is merely one among the possible circulating fuels, not much of the enzyme is needed. Lactic dehydrogenase can occur in different isozyme forms, with kinetic preferences favoring the reduction of pyruvate or the oxidation of lactate, respectively. Such isozymes in this case are separable by electrophoresis, and one has to discover multiple forms (e.g., Markert and Moller, 1959; Wieland and Pfleiderer, 1957) of which the opposing extremes are known as the muscle or M type (reductive) and the heart or H type (oxidative), while intermediary types result from the permutation of M and H subunits in a tetrameric enzyme. As will be expected after this buildup, we found the F muscle to contain the M type and the S muscle the H type enzyme, usually quite purely. After nerve crossing, the following changes occur. In the F → S cross, the amount of lactic dehydrogenase diminishes almost though not completely to the S level; in the S → F cross, it increases but little; with respect to isozyme pattern, both crosses lead to an intermediary mixture.

There are other biochemical distinctions between the muscles, subjected to neural control; these have been reviewed by Guth (1968), and I pass over them since they, too, will fit into the set of distinctions indicated above—e.g., a number of them would relate to the oxidative aspect of the S type of metabolism. It remains, however, to mention another example of a neural influence, one which refers to a functionally distinct system, the excitable and conductive cell membrane itself. Ordinarily, this membrane is stimulated through the motor nerve, by a cholinergic mechanism (cf. Katz, 1967), and the region of the surface membrane near the end plate is highly sensitive to acetylcholine, while the remainder of the membrane is not. Upon denervation in the frog, the entire membrane becomes acetylcholine-sensitive (Axelsson and Thesleff, 1959; Miledi, 1960). This shows that the presence of a functional innervation suppresses a potential property of the cell membrane—curiously, only at a distance from the junction, not at the junction itself. Clearly, this important finding also belongs to the set of phenomena we are discussing. So far, it has not yet lent itself to a further specification of its nature, in part perhaps because we do not know just what kind of property the findings reflect.

These phenomena suggest a great deal, in terms of the question of how the functional architecture of the neuro-effector system expresses itself by selecting and repressing features from among the broader genetic potentialities.

First of all, by what means does the innervation exert its specifying influence? Two possibilities suggest themselves at once. It might be the efferent impulse pattern, or it might be a matter of formative substances wandering down the nerve. A case might be made for either, and this is a topic for lively discussions which, at this moment, are likely to end in a draw. I would tend to agree with a conclusion by Eccles (1967), who considers it unlikely that "the information provided by the frequency of discharges of slow and fast motoneurones would be sufficiently discriminative to serve as the basis for the differentiation of speed of muscle contractions. The discharges of any particular motoneurone are far too diversified in frequency and not of some standard frequency as in the test experiments." Still, keeping in mind that physiologically, over the long range, there may be a consistent difference between regular discharges at some low frequency, as opposed to more widely spaced periods of high-frequency discharge, we cannot dismiss the possibility out of hand. Both this mechanism and trophic substances might well play a role, with one or the other perhaps figuring predominantly with respect to specific functional characteristics. Another intermediary standpoint could be that the impulse pattern influences the rate of migration of certain materials, or the muscle's permeability for, or active uptake of, limiting substrates. Surely, substances do migrate centrifugally along axons—a topic now receiving renewed attention—and when all is said and done, it is likely that some of these play a role in the mechanism of the phenomena we are discussing.

Whichever the mechanism, one general conclusion is obvious. While, according to classical tenets, the genetic potential is encoded in DNA, and embryonic and postnatal differentiation is a matter of selective repression and realization of features, our findings now specify that the nervous system plays a role in these selections.

Could one defend the thesis that the findings are a result of denervation rather than of cross innervation? In our work, at least, the most striking changes are those in the crossing F → S, and grossly speaking the result of this change is not unlike that in denervation. This hypothesis would break down when some of the findings are considered in detail, and especially when Close's findings (1965) are drawn into the comparison. But let us follow up the case where it looks strongest: in the cross F → S, the fibrillar adenosine triphosphatase becomes slowed, as it does in denervation. There are two replies. First, the muscle does not remain denervated; the reestablishment of a functional innervation can be shown by stimulation of the newly connected nerve. But, one might say, only the neuromyonal impulse conduction becomes functional, not the formative influence. A bit contrived, perhaps, but possible. The second answer is, however, that even the effect of denervation by itself shows that there is a formative influence; and, since in the F → S cross there is functional activity, the change cannot be due to total disuse. This is about as far as the argument can be carried at present. Further work will clarify the matter.

At this point, we must reiterate that nerve crossing effects a changeover of some properties but not of all. Why not all? Apparently, at any rate at the fairly late stages of development we have examined, the developmental potential has already become restricted. It is remarkable that an adult cat, upon permutation of its nervous connections, can modify its effector properties considerably. It is not astonishing that it cannot do so completely, for nothing in the order of things dictates the necessity of making provisions for the possibility that somebody will cross-connect the nerves. The cross-innervated muscle does not become modified completely; only some characteristics which are still in a plastic state do so. Thus, the F muscle connected to an S nerve becomes slowly contracting because it changes to a different myosin, lengthens its cycle duration because of some changes in the sarcoreticulum, and loses its glycolytic capacity; and, as indicated by the findings of others (Prewitt and Salafsky, 1967; Bücher and Pette, 1965) it may gain in oxidative capacity and hence, presumably, in persistent nonfatigable activity. The S muscle, reversely connected, becomes but little faster; it shortens its active state cycle and thus reduces its tension output in a twitch which may not, however, be a large handicap if the muscle works with tetani; it does not gain a significant glycolytic capability as far as indicated by lactic dehydrogenase, and will therefore fatigue even faster, inasmuch as its metabolism would effectively be like that in McArdle's disease (Mommaerts, Illingworth, Pearson, Guillory, and Seraydarian, 1959; Larner and Villar-Palasi, 1959; Schmid, Robbins, and Traut, 1959; Pearson, Rimer, and Mommaerts, 1961). The cats survive in a sheltered laboratory setting, but might be handicapped in the competitive life of the jungle or the alley. We must remember that the experiment is artificial, does not relate to events in natural ecosystems, and if it did would merely lead to the elimination of the unfittest. But at the laboratory level, the phenomena supply a link between the problems of genetic coding and functional expression which we hope to develop in further work.

Returning to Fig. 4.1, it is worth emphasizing that the key to the specifications lies in the motoneurone soma in the spinal cord, though possibly in turn this may be modified by influences upon it from other sources. This neurone, perhaps 50 $\mu$ in diameter, acting over a distance of many inches or even feet, branches and connects to hundreds or thousands of myones, and determines essential features of the differentiation of a mass of muscle that may weigh a gram. Sketching the situation this way, one is reminded of that occasionally rediscovered showcase of phys-

*Fig. 4.4.* A cell of the parasol alga, *Acetabularia.* The nucleus shown in the base of the cell determines the specific differentiation of the parasol.

iological genetics, *Acetabularia* (Hämmerling, 1963), where the nucleus in the lower end of the cellular stalk determines the specific properties of the parasol perhaps an inch distant (Fig. 4.4). The motor unit is no less impressive an example of developmental control exerted over a distance. Thus, we have placed the phenomenon in a general biological context.

## ACKNOWLEDGMENTS

This work was supported in part by grant HE-11351, NIH, and the Muscular Dystrophy Associations of America, Inc.

## *References*

Axelsson, J., and S. Thesleff. 1959. A study of supersensitivity in denervated mammalian skeletal muscle. *J. Physiol. 147:*178.

Bárány, M. 1967. Adenosine triphosphatase activity of myosin correlated with speed of muscle shortening. *J. Gen. Physiol. 50:*197.

Bücher, Th., and D. Pette. 1965. Ueber die Enzymaktivitätsmuster in Bezug zur Differenzierung der Skeletmuskulatur. *Verhandl. Deutsch. Ges. inn. Med. 71,* Kongress 104.

Buller, A. J. 1965. Mammalian slow and fast skeletal muscle. *In* D. R. Curtis and A. K. McIntyre (eds.), *Studies in Physiology.* Springer-Verlag, Berlin.

Buller, A. J., J. C. Eccles, and R. M. Eccles. 1960*a,* Differentiation of fast and slow muscles in the cat hind limb. *J. Physiol. 150:*399.

———. 1960*b*. Interactions between motoneurones and muscles in respect of the characteristic speeds of their responses. *J. Physiol. 150:* 417.

Buller, A. J., and D. M. Lewis. 1965. Further observations on mammalian cross-innervated muscle. *J. Physiol. 178:*343.

Buller, A. J., and W. F. H. M. Mommaerts. 1969. Myofibrillar ATPase as a determining factor for contraction velocity, and its changes upon experimental cross-innervation. *J. Physiol. 201:*46.

Buller, A. J., W. F. H. M. Mommaerts, and K. Seraydarian. 1969. Enzymic properties of myosin in fast and slow twitch muscles of the cat following cross-innervation. *J. Physiol. 205:*581.

Close, R. 1965. Effects of cross-union of motor nerves to fast and slow skeletal muscles. *Nature 206:*831.

Ebashi, S., and M. Endo. 1969. Calcium ion and muscle contraction. *Progr. Biophys. Mol. Biol. 18:*123.

Eccles, J. C. 1967. The effects of nerve cross-union on muscle contraction, p. 151. *In* A. T. Milhorat (ed.), *Exploratory Concepts in Muscular Dystrophy and Related Disorders.* Excerpta Medica Foundation, Amsterdam.

Engelhardt, V. A., and M. N. Ljubimowa. 1939. Myosin and adenosine triphosphatase. *Nature 144:*668.

Guth, L. 1968. Trophic influences of nerve on muscle. *Physiol. Rev. 48:*645.

Hämmerling, J. 1963. Nucleo-cytoplasmic interactions in Acetabularia and other cells. *Ann. Rev. Plant Physiol. 14:*65.

Henneman, E., and C. B. Olson. 1965. Relations between structure and function in the design of skeletal muscles. *J. Neurophysiol. 28:*581.

Henneman, E., G. Somjen, and D. O. Carpenter. 1965*a*. Functional significance of cell size in spinal motoneurons. *J. Neurophysiol. 28:*560.

———. 1965*b*. Excitability and inhibitability of motoneurons of different sizes. *J. Neurophysiol. 28:*599.

Huxley, A. F. 1957. Muscle structure and theories of contraction. *Progr. Biophys. Biophys. Chem. 7:*257.

Katz, B. 1967. *Nerve, Muscle and Synapse.* McGraw-Hill Publ. Co., New York.

Larner, J., and C. Villar-Palasi. 1959. Enzymes in glycogen storage myopathy. *Proc. Nat. Acad. Sci. 45:*1239.

Markert, C. L., and F. Moller. 1959. Multiple forms of enzymes: tissue, ontogenetic and specific species patterns. *Proc. Nat. Acad. Sci. 45:*753.

Miledi, R. 1960. The acetylcholine sensitivity of frog muscle fibers after complete or partial denervation. *J. Physiol. 151:*1.

Mommaerts, W. F. H. M. 1946. Myosin and adenosinetriphosphatase in muscular activity. *Science 104:*605.

———. 1950. *Muscular Contraction, a Topic in Molecular Physiology.* Interscience, New York.

———. 1963. The muscle cell and its functional architecture. *Amer. J. Med. 35:*606.

———. 1968. Muscle energetics: biochemical differences between muscles as determined by the innervation. *Proc. Int. Union Physiol. Sci. 6:*1161.

———. 1969. Energetics of muscular contraction. *Physiol. Rev. 49*:427.

Mommaerts, W. F. H. M., A. J. Buller, and K. Seraydarian. 1969. The modification of some biochemical properties of muscle by cross-innervation. *Proc. Nat. Acad. Sci. 64*:128.

Mommaerts, W. F. H. M., B. Illingworth, C. M. Pearson, R. J. Guillory, and K. Seraydarian. 1959. A functional disorder of muscle associated with the absence of phosphorylase. *Proc. Nat. Acad. Sci. 45*:791.

Mommaerts, W. F. H. M., and K. Seraydarian. 1947. A study of the adenosine triphosphatase activity of myosin and actomyosin. *J. Gen. Physiol. 30*:401.

Pearson, C. M., D. G. Rimer, and W. F. H. M. Mommaerts. 1961. A metabolic myopathy due to absence of muscle phosphorylase. *Amer. J. Med. 30*:502.

Perry, S. V. 1951. The adenosine triphosphatase activity of myofibrils isolated from skeletal muscle. *Biochem. J. 48*:257.

Perry, S. V., and T. C. Grey. 1956. A study of the effects of substrate concentration and certain relaxing factors on the magnesium-activated myofibrillar adenosine triphosphatase. *Biochem. J. 64*:184.

Prewitt, M. A., and B. Salafsky. 1967. Effect of cross-innervation on biochemical characteristics of skeletal muscles. *Amer. J. Physiol. 213*:295.

Schmid, R., P. W. Robbins, and R. R. Traut. 1959. Glycogen synthesis in muscles lacking phosphorylase. *Proc. Nat. Acad. Sci. 45*:791.

Szent-Györgyi, A. 1942. *Myosin and Muscular Contraction* (by I. Banga, T. Erdös, M. Gerendás, W. F. H. M. Mommaerts, F. B. Straub, and A. Szent-Györgyi). S. Karger, Basel and New York.

Wieland, T., and G. Pfleiderer. 1957. Nachweis der Heterogenität von Milchsäure-Dehydrogenasen verschiedenen Ursprungs durch Trägerelektrophorese. *Biochem. Z. 329*:112.

# Chapter 5

# *Summary and Discussion of Part 1*

PANEL MEMBERS: D. E. GOLL, *Chairman*
G. BEECHER
M. REEDY
R. E. DAVIES
L. LORAND
G. BORISY

*Goll:* It seems appropriate to begin a symposium on muscle biology with a section on the origin, development, and synthesis of muscle tissue. Part 1 has examined muscle cell development at three levels of complexity —the molecular level, the cellular level, and the interaction between innervation and the functional development of muscle in the mature organism. I think it is obvious that our understanding of muscle development at any of these levels of organization is still very much in its infancy, but it is equally obvious from the chapters in Part 1 that muscle development is currently a very active and exciting field, and that we may expect significant developments in this area in the next several years.

At the molecular level, Dr. Heywood described in Chapter 2 his very elegant work on isolation of the mRNA directing myosin synthesis. Further progress in this area may lead to discovery of some of the reactions that control rate of myosin synthesis. Extension of these efforts to the other myofibrillar proteins would provide information about control of the rate of myofibrillar protein biosynthesis, and thereby invite attempts to increase feed-to-muscle conversion ratios in our domestic animals by suppressing or negating these controls that limit the rate of myofibrillar protein synthesis. Moreover, by use of muscle biopsy samples and biochemical assays of those enzymes or reactions that are rate-limiting in myofibrillar protein synthesis, it may be possible to obtain a rapid and accurate estimate of an animal's meat-producing potential while it is still quite young.

Dr. Holtzer, in Chapter 3, has described the problems in attempting to discover how an undifferentiated, embryonic cell is transformed into a muscle cell, and why it becomes a muscle cell rather than a liver or a nerve cell. The mechanism of cell differentiation remains a very complex and intriguing problem whose solution is essential to understanding muscle development. Dr. Holtzer proposes that undifferentiated blastula or morula cells are not transformed into muscle cells by the action of some "inducer" molecule, but rather that undifferentiated cells are converted into differentiated myoblasts by a series of ordered, step-wise events, and these events are coupled to particular kinds of mitoses. These Dr. Holtzer terms "quantal mitoses," and they are distinguished from ordinary mitoses that lead simply to an increase in cell members ("proliferative mitoses"). Understanding how undifferentiated cells are transformed into muscle cells that synthesize myofibrillar proteins also has important implications for the use of muscle as a food. Such understanding, for example, may eventually make it possible to use tissue culture centers to produce quantities of muscle large enough to serve as a food supply. Such tissue culture production of meat would have several inherent advantages over our present system of using animals to produce meat: (1) the supplied nutrients would be used entirely in muscle production, and inedible items such as bone, hides, hooves, horns, hair, and viscera would not be produced; (2) one would have convenient and flexible control over the "kind" of muscle produced, i.e., the ratio of "T-bone muscle" to "round-steak muscle" could be controlled and varied at will; (3) reproductive and management problems associated with animal husbandry would not exist. Other advantages could also be listed. Although the production of meat in tissue culture centers is still only a future possibility, such a method of meat production clearly has important social and scientific implications.

In Chapter 4, Professor Mommaerts describes some evidence, recently obtained in his laboratory, showing that muscle differentiation, at least in terms of functional properties in the mature muscle cell, is influenced by innervation. Thus, the kind of protein synthesized in a muscle cell can be changed from that characteristic of "white" muscle cells to that characteristic of "red" muscle cells, simply by changing the innervation of the cell. The extent to which the protein synthetic mechanism in a differentiated muscle cell can be altered by changing innervation is not yet clear, but it seems unlikely that drastic alterations, such as converting a muscle cell into an epithelial cell, can be effected by altering innervation. Rather, it appears that the quantal mitoses already experienced by the time a cell can be distinguished as a muscle cell limit the potentialities of the cell to further change. Since quantal mitoses in the embryo occur before innervation, but differentiation into "red" and "white" muscle cells occurs only after innervation of the cells, it may be postulated that innervation controls functional differentiation of a muscle cell once the cell has differentiated into a muscle cell. Further evidence, however, is needed to determine clearly the role of innervation in development of the functional properties of muscle.

Professor Mommaerts' finding that cross innervation can cause "white" muscles to change to "red" muscles and "red" muscles to change, at least partially, to "white" muscles, also has some clear implications in the use of muscle as a food. Ashmore and coworkers have recently reported that muscles from "double-muscled" bovine animals contain a much higher proportion of "white" cells than do muscles from normal bovine animals, but that very few other physiological or biochemical differences exist between muscles from these two types of animals. Since "double-muscled" animals exhibit a tremendous propensity for muscle or meat production, these findings suggest that conversion of "red" muscle cells into "white" muscle cells, either by changing innervation or by any other means, may enhance the value of an animal for meat production.

Clearly, the information presented in the three papers in Part 1 suggests some completely new approaches to the problem of meat production, and also indicates some potentially fruitful areas for future research in the related areas of muscle biology and meat science.

DISCUSSION

*Lawrie:* Have your experiments led you to suspect that the classification of muscle as "fast" or "slow" (or intermediate) is insufficient to account for all types of muscle action?

*Mommaerts:* The classification is, of course, a very global one and certainly does not suffice to characterize all muscles. But it is often useful and has served us well in the present work.

*Lorand:* Does cross innervation cause a change in myoglobin concentration?

*Mommaerts:* We have not examined myoglobin content. However, Bach (1948) performed a tendon transfer and observed that red muscle turns paler and pale muscle turns redder. It is still a mystery how tendon transfer causes this result.

*Davies:* What happens to myosin in muscle that is responding to cross innervation? Is it a modification of the myosin that is present or has it all turned over completely and been replaced by new myosin? We know, for instance, that the lifetime of myosin from mouse muscle is about 3 weeks, but is anything known about turnover of myosin in cat muscle?

*Mommaerts:* I am convinced that we are dealing with the synthesis of new myosin molecules, but I must admit that we have no explicit evidence on this point. Six months are allowed for the cross-innervation experiments. The distal nerve fibers probably degenerate during the 1st month, new fibers begin to move in, and new end plates are subsequently formed. That still leaves time for the synthesis of new myosin. It is, of course, difficult to learn more about this, but we have started efforts to study the conditions of protein synthesis in adult muscle. Unlike the results with chicken embryo, the incorporation rates we get are exceedingly low and we are encountering considerable difficulty. Suppose there is some incorporation. Then all one knows is that there has been some incorporation in myosin, but the type of myosin is not known. We have therefore concentrated our efforts on lactate dehydrogenase, as described in Chapter 4, because the isozyme separations are easily made and the results would be meaningful.

*Sréter:* We have investigated lactate dehydrogenase in denervated muscle of rabbit and rat. The decrease in enzyme activity is very remarkable in fast muscle, after a long period of denervation, but there is almost no change in slow muscle.

*Peter:* We demonstrated in animals and later with denervating diseases in humans that the fast-type isoenzymes of lactate dehydrogenase do have a tendency to disappear as the amount of the slow type of isoenzymes increases.

*Davies:* You said that denervation causes the muscle to go from fast to slow type. How much of the major effect can be accounted for by denervation or how much has to be associated with the possibility that a new nerve may cause a change in the muscle type?

*Mommaerts:* First of all we do not have a denervated muscle, because innervation reestablishes itself. At worst, one could say the nerves are there, they conduct impulses, but they do nothing else. Even if they were denervated or even if some other experiments were thrown into the argument, then it still shows an influence of the nerve. I admit that the case would be clearer if we would show the change from fast to slow and slow to fast, because then there would be no question regarding this matter. I see, of course, the weakness you are pointing at; we can go from fast to slow but we cannot go from slow to fast. According to a paper by Close (1965), however, such a transformation can be accomplished. I think that the lactate dehydrogenase experiment does supply part of the answer and I hope that we will soon have more information.

*Perry:* Would Dr. Mommaerts accept the view that the varying enzymic properties of myosin preparations from different muscles of a given species or after cross innervation are due to the myosin preparation from the various muscles being composed of different proportions of, say, two isozymes which have different specific adenosine triphosphatase activities?

*Mommaerts:* Yes, definitely. It is a moot question whether we would be helped by speaking of isozymes when regarding the wide range of muscles existing throughout the animal kingdom. But if in one animal there were to be two major kinds of myosin, then the proposed terminology could be very helpful in analogy with other well-studied cases. And the viewpoint that these two could occur in varying proportions would help us understand the occurrence of different intermediate cases as well.

*Sréter:* We have described some properties of myosin from red and white muscle. Susceptibility to trypsin digestion and alkali lability are of interest. The tryptic digestion of red muscle myosin resembles that for cardiac myosin, since red myosin is more resistant than white myosin. Are these properties changed in cross-innervation experiments?

*Mommaerts:* We have not investigated the difference with respect to tryptic digestion. We are touching upon alkali lability, but have no data so far, after nerve crossing.

*Sréter:* Did you make negatively stained preparations of the fragmented sarcoplasmic reticulum from the two types and also from the two types after cross innervation?

*Mommaerts:* No. Beside all other objections to that part of the work, it is also technically very difficult to make a good preparation from the amount of muscle available (under 100 mg). We do not take this as a defeat, but are trying to devise a scheme to isolate the sarcoplasmic reticulum system in its original integrity in order to have a better measure of its $Ca^{++}$-uptake activity and possibly also to find the system from which one could force the release of $Ca^{++}$.

*Gergely:* How can one relate the changes in $Ca^{++}$ uptake, observed on cross innervation, to the changes in contraction time? Are they in the right direction?

*Mommaerts:* This, as I have pointed out, is the most tentative part of our work. Both the extent of the development of the $Ca^{++}$-sequestering system and the amount of $Ca^{++}$ released may differ. But the differences in uptake velocity of isolated vesicles do run in the right direction, although I have mentioned that this could be an experimental artifact based upon differences in resistance to disintegration during the isolation procedure.

*Lorand:* What do the ordinates represent on the $Ca^{++}$-uptake curve: rate or extent of $Ca^{++}$ uptake?

*Mommaerts:* They represent $Ca^{++}$ uptake after 8 or 12 min under standardized conditions.

*Lorand:* Do the ordinates in your myosin adenosine triphosphatase curves (Fig. 4.3) represent maximum velocities or do they pertain to a single arbitrary case only?

*Mommaerts:* In some cases they are initial velocities determined from a time curve, in other instances they are single points during the initial part of the ATP splitting.

*Bailey:* What happens to the population of mitochondria in muscle tissue which has changed from fast to slow or slow to fast? Do the number and activities of mitochondria change when you change the type of innervation?

*Mommaerts:* I would refer, in this connection, to the analyses of oxidative enzymes that have been performed by several authors (reviewed by Guth, 1968).

*Brooke:* I would like to comment on the errant points of Dr. Mommaerts' graph. We have been interested in muscle transformation from one type to another. A reinnervated muscle at the end of some period of time often appears functionally normal, but it may consist of a few huge fibers and extensive fibrous tissue. I think the fibrous tissue may restrict the use of the muscle and thereby limit the adaptation that can occur in it, regardless of the signals from its new nerve source. This fibrous tissue may certainly affect contraction speed.

*Mommaerts:* Yes, that is certainly so. The myosin was, of course, assayed after isolation, but your point is that one may very well also get some deviation in the mechanical analysis of the crossing. In other words, it would be the abscissa value that is at fault. That is quite correct, and is again in line with the fact that the chemical correlations among themselves are much better.

*Reedy:* What is known about the distinguishing peculiarities of the two families of muscle nuclei, of RNAs, and of spinal motoneurones, associated with these fast and slow muscles, especially during the F-S crossing process you have described?

*Mommaerts:* I have no knowledge to contribute to these important questions. With respect to RNA, I would venture that distinction might not easily be found by chemical analyses, but rather by biochemical assays in some appropriate test system.

*Reedy:* At what stages in the life of a cat has the F-S transformation in contractile physiology been shown to occur? Are presenile cats as adept at it as halfgrown kittens?

*Mommaerts:* It can occur in both stages mentioned, the kitten and the adult. It is possible that properties which do not (or only slightly) change over in the adult are altered more effectively if the crosses are performed in kittens. A change in the architecture and responsivity of the vascular system might be two cases in point.

*Reedy:* Carlson (1967, 1968) has recently confirmed Studitsky's remarkable regeneration experiments. Regeneration can occur after a muscle has been excised, minced, and replaced in the original fascial compartment. Within about 30 days a competent whole muscle, complete with tendon and nerve connections, has regenerated (albeit with some loss of bulk and some increase in fibrous connective tissue). Do you see any appeal in the idea of basing the F-S crossover experiment on this; i.e., in grinding up the two muscles and then packing the mince from the fast muscle back into the fascial compartment of the slow muscle, and vice versa, while leaving the cut nerves alone so far as possible?

*Mommaerts:* If the experiment is repeatable, this would then indeed be a valuable additional route for investigation.

*Beecher:* Are the events that occur in this regeneration (described above by Reedy) similar to the events that occur in normal neonatal differentiation and development of muscle cells?

*Holtzer:* I know of no difference between what you term neonatal differentiation and that occurring during normal regeneration or when a muscle mince is replaced within a scooped-out muscle, a type of experiment the Russians have been performing for years.

*Beecher:* In accidents where the innervations of several fingers have been severed, the nerve ends may not always be rejoined with their original partners during repair and reconnection. However, the higher centers of the central nervous system eventually overcome this mismatch so that under voluntary control the muscles of the desired finger are innervated. Using the same analogy in the studies with cross innervations, do you feel that the transition of F myosin to S myosin might eventually revert back to F myosin?

*Mommaerts:* This is not impossible, but would be a difficult subject for experimentation.

*Dhalla:* Are there any differences in the chemical compositions of the motoneurones to slow and fast fibers? If there are no differences, would you speculate on what regulates the development of these muscles?

*Mommaerts:* There is probably no chemical difference that would reveal itself in the gross analysis of groups of chemicals. The problem is that in the mixed nerve one does not know which fibers go where. There is a certain correlation between fiber diameter and the type of muscle fiber it goes to. Generally the larger nerve fibers go to the larger motor units. If one fiber has to nourish 5,000 muscle fibers—not in a nutritional sense, but in some kind of a formative sense—then it should be larger than if it only has to nourish a hundred. There is, however, only a 20% difference in fiber diameter, and a much greater difference in size of the motor unit. So, in other words, one would have to go to the utmost final branches in order to know that a certain branch goes only to the F muscle and another branch goes only to the S muscle. Even then one would have difficulty because there are also sensory fibers and the sensory fibers are likely to differ. I would say that various sensory fibers are likely to amount to 30% in the mixture. This is less than the majority, but nevertheless not negligible. Now, what should one assay for? Certainly routine analyses will not show anything of interest because such differences as there may be are likely to be subtle. It will certainly be a part of a future program to find out the difference between a branch to a soleus nerve and a branch to one of the others.

*Beecher:* How does the muscle cell translate the electrical impulses received from its innervation into biochemical information in terms, say, of synthesizing new and specific proteins or modifying existing proteins in the muscle cell?

*Mommaerts:* This is the crux of the matter, assuming that it is the impulse pattern which is responsible. As is pointed out in Chapter 4, however, this is not the only possibility. It also remains to be considered that the impulse pattern acts indirectly, by determining the summated amount of activity and its distribution over time.

*Beecher:* Does the information that a muscle spindle sends over the afferent network have any influence on the type of metabolic activity (oxidative versus glycolytic) that is predominant in the effector muscle cell?

*Mommaerts:* Very likely there are differences between the muscle spindles of different types of muscles, and it would be an important question for future research to see whether they become modified in cross innervation.

*Davies:* It is stated that F muscles "are not designed to resynthesize ATP at the corresponding high rate during those bursts"; I presume you mean by oxidative phosphorylation, since ATP levels do not fall.

*Mommaerts:* Yes, I was referring to oxidative, or for that matter glycolytic, resynthesis; mainly to the former since only that, in general, can support a steady state activity.

*Forrest:* Hník, Jermanovà, Vyklický, and Zelená (1967) observed no change in either ultrastructure or contraction velocity of red or white muscle as a result of cross innervation in the chick. They reasoned that their results were due to the fact that they were working with muscles more homogeneous for the red and white types than most investigators. Do you think this explanation is plausible?

*Mommaerts:* I have no explanation.

*Holtzer:* I do not know of the work cited. But I know of nothing in the cross-innervation experiments that rules out the differential degeneration of one set of fibers, leaving the other set to survive.

*Eason:* Several biochemical reactions occur during the activation-contraction cycle of the muscle tissue used in the nerve crossover type experiments. What do you consider to be the rate-limiting step that takes place in this series of reactions?

*Mommaerts:* There is reason to believe that (a) the rate-limiting step is in the chemomechanical transduction process itself, and (b) this correlates with the adenosine triphosphatase activity of isolated myosin or myofibrils. The quoted paper by Bárány suggests this very explicitly, but my current review (Mommaerts, 1969) discusses this more broadly on page 496.

*Marsh:* Is it conceivable that information gained from crossover experiments could lead to modification of muscle for its use as food?

*Mommaerts:* In principle, yes. Of course, it will not be practical to do this by nerve crossing, but the understanding gained eventually will not fail to be helpful in the judgment of other possible treatments, e.g., the amount and nature of exercise, that might lead to the goal.

*Bock:* When the synthesis of muscle fibrils is under control at a translational level, it is likely that the three-dimensional structure plays a part in this control. Is there an interaction of the nascent proteins on the polysomes with the myofibril that is specifically a function of those nascent proteins? That is to say, at slightly lower salt concentrations could precipitation of the polysomes be specific for muscle polysomes as distinguished from other polysomes? Could actin, myosin, or tropomyosin, or some combination of these, replace the initiation factor and stimulate other tissue to make better use of the message?

*Heywood:* We have found no evidence for the combination of actin with nascent myosin even at ionic strength of about 0.15. It is possible that we could miss this because you would expect that the heavy meromyosin end of the molecule would have to be the free end of the nascent polypeptide. It is not clear yet which end of the molecule contains the N-terminal amino acid. If the light meromyosin is at the N terminal, it is unlikely that actin could combine with myosin in a physiological manner. We do not have information on tropomyosin in this regard. The initiation factors appear to be washed off purified ribosomes that have been released of nascent chains. The factors appear to be very specific. The reticulocyte factors apparently initiate hemoglobin synthesis. The muscle factors seem to initiate, as far as we know, tropomyosin and myosin synthesis. There is no evidence at all that there is any actin, tropomyosin, or myosin obtained with these initiation factors. These initiation factors are involved with protein initiation in the sense that they are binding the mRNA to the ribosomes, and thereby aid in the formation of the initiation complex. They bind the mRNA specifically to the 40 S subunit. Therefore, it seems unlikely that our "initiation factors" are related to these other proteins. They are very much like initiation factors in *E. coli* systems, known as A, B, and C factors. Our general objective has been an examination of translation controls. It would be very interesting to see if the nascent polypeptide chains actually could polymerize with the other molecules making up the contractile apparatus. It seems more likely that nascent myosin will interact with myosin. It seems quite likely to me that if the light meromyosin end of the chain projects from the polysome it would bind with other myosin molecules, and in this way build a myosin filament. In fact, we have seen some electron micrographs that have actually revealed a thin filament coming off one end of the polysome, or a series of them of increasing length.

*Perry:* If I understood Dr. Heywood correctly, he got incorporation into the heavy chains of myosin but not the light. Although I have doubted the relation of the light components to the heavy for some years, I have become convinced more recently that they are, in fact, real components of myosin. Therefore, if myosin is being synthesized, I think both components should be radioactive. I imagine that these two polypeptide chains come together at a fairly early stage. Also, is there evidence that the large component of myosin is in fact moving on the acrylamide gel? In our experience, this is a rather difficult thing to do unless the gel has a very low cross linking.

*Heywood:* We have been making myosin, but that is not really the correct terminology. The question is, are the myosin isozymes really turning over, or are these small low-molecular-weight components of myosin allosteric affectors that change the properties of the basic enzymatic binding properties of the myosin molecule? Our evidence suggests that we are making the large, 200,000-molecular-weight subunit of myosin. I have provided evidence that this was a single mRNA, that this group of polysomes that sediments at the bottom of the gradient is involved in the synthesis of this subunit of myosin and not the total myosin molecule. The low-molecular-weight components have probably been synthesized on other smaller polysomes that sediment more to the top of the gradient. Therefore, it is not a polycistronic message, and it is not the isolated mRNA synthesizing myosin, but rather a component of the myosin molecule of 200,000 molecular weight.

The acrylamide gels are low cross linkage gels, and about 95–98% of the material we add to the gel enters the gel. It seems inconceivable that most of the protein put on the gel is being left behind as the large-molecular-weight component. Also, we have separated the 200,000-molecular-weight subunit by guanidine hydrochloride and Sephadex G-200 column chromatography. When this preparation is concentrated and put on acrylamide gels, the isolated large subunit of myosin moves as a single band and the small isolated components from the same column will move ahead.

*Gergely:* Is there evidence that small subunits incorporate radioactive amino acids and that these subunits become part of the newly formed myosin?

*Heywood:* When the isolated 26 S mRNA is utilized, only the large 200,000-molecular-weight subunit of myosin is synthesized in the cellfree system. The size of the mRNA (approximately 6,000 nucleotides) compares with the size of the large myosin subunit (approximately 2,000 amino acids). It is assumed that the smaller subunits of myosin are synthesized separately on smaller polysomes.

*Gergely:* I presume that Dr. Heywood synthesizes whole myosin, but he has been describing the chain isolated from heavy polysomes. Do you have experiments where you add cold myosin, then reisolate everything and finally determine the radioactivity in the various chains?

*Heywood:* Incorporation only occurred in the large-molecular-weight chain when we used isolated mRNA. We were not making total myosin.

*Lorand:* The work of Dr. Heywood and his colleagues really shows the complexities one encounters when protein synthesis is studied in cells other than *E. coli*. This difficulty is amplified by the large size of protein like myosin. It seems to me, however, that Dr. Heywood put this latter property of myosin to a rather good advantage.

As Dr. Heywood points out repeatedly in Chapter 2, one of the keys to this kind of work is the establishment of identity between the *in-vitro*-synthesized proteins and the naturally isolated ones. Clearly, one needs to move towards fingerprinting. With actin and tropomyosin this should be an attainable goal. For the time being, would it be reasonable to hope for an electron microscopic identification of nascent myosin attached to the polysomal structure? Could, for instance, the adenosine triphosphatase in the head position of this protein be visualized by histochemical tests? I mean, of course, with the addition of the light subunit which—as Stracher's group showed—would be a necessary component for the enzyme.

Finally, could Dr. Heywood comment, on the basis of his biosynthetic studies, on the possibility or even the likelihood that the long heavy chain of myosin is constituted of repeating block polypeptide sequences rather than a single unique one?

*Heywood:* We are presently attempting to identify nascent myosin chains by use of the electron microscope. It is hoped that by utilizing the techniques of Lowey, Slayter, Weeds, and Baker (1969) we will be able to visualize the interaction of the nascent myosin chains with the light subunits as well as with actin in the presence of the light subunits. Our evidence as well as that of Lowey suggests that repeating blocks of polypeptides are unlikely in the makeup of the long heavy chain of myosin.

*Kornguth:* Does antimyosin antibody precipitate the newly synthesized myosin or the ribosome-bound myosin?

*Heywood:* We have not tested the antimyosin antibody against nascent myosin. In the experiments in which we tested for the *in vitro* synthesis of myosin, only the 150,000 $\times g$ supernatants of the reaction mixtures were used. Allen and Terrence (1969), however, have shown that antimyosin will react with nascent myosin chains on the polysomes.

*Eason:* Have you tested any of your synthetized protein for enzyme activity (i.e., myosin adenosine triphosphatase)?

*Heywood:* Yes, we attempted this. The background adenosine triphosphatase activity was too high to draw any conclusions, however. By using isolated polysomes from muscle tissue, we obtained some evidence that the nascent myosin molecules exhibited their characteristic adenosine triphosphatase activity.

*Borisy:* There is biochemical evidence from chick striated muscle which indicates that the muscle proteins are synthesized in particular temporal order; first actin, then myosin, then tropomyosin. Dr. Heywood's evidence from cardiac muscle indicates that perhaps they are synthesized concurrently. Might the different organization of cardiac and striated skeletal muscle be related to the difference in sequence in the synthesis of the muscle proteins?

*Holtzer:* I know of no evidence to indicate seriously that there is any significant temporal lapse in the synthesis of one protein vs. another. There have been reports in the literature of excess amounts of actin. I do not believe they have ever been verified. I would be most pleased to learn about any critical evidence to indicate that great amounts of myosin are synthesized before or after great amounts of actin.

*Whitaker:* Would you comment on the control mechanism or mechanisms which turn on the biosynthesis of a protein at a specific time in cell differentiation?

*Holtzer:* From data based on microbial studies one could propose all sorts of schemes. However, from data based on tissue cells, as far as I am concerned, there is virtually nothing that can be said. Our ignorance in this area, in spite of hundreds of papers in the literature, is total.

*Reedy:* Evidently tissue culture cells can synthesize myosin, even filaments and sarcomeres, without waiting for direction from the nervous system as to whether fast or slow myosin is appropriate. Do you suppose that they would be producing some immature or embryonic myosin, analogous to the fetal hemoglobin which precedes adult hemoglobins? Is the muscle more or less innervated before you excise it and culture it?

*Holtzer:* I am dubious about the existence of immature or embryonic myosin. It may well exist but I submit that the evidence thus far presented is equally well explained by assuming the embryonic myosin is contaminated; myosin extracted from embryos is notoriously difficult to separate from DNA, all sorts of RNA, other proteins, and mucopolysaccharides.

Even if there is an embryonic myosin it would not be analogous to fetal hemoglobin. The two different kinds of hemoglobin are produced in two very different kinds of red blood cells. In the chick, for example, the cells producing the fetal hemoglobin are morphologically very different from the definitive red blood cells. Presumably embryonic myosin would be made in the same muscle fibers as mature myosin. This would mean a change in the kind of myosin a given cell synthesizes. It is also intriguing to ask how a thick embryonic myosin filament surrounded by six actin filaments might be replaced by a thick mature myosin filament. In brief, until the objections of trace contamination of other molecules, or even of allosteric changes associated with the extraction procedures, are eliminated, I am sceptical of the claims of embryonic myosin.

Muscle cultures have been set up with tissue before and after innervation *in vivo*. Whatever role innervation has with respect to monitoring muscle, it can categorically be stated that it plays no role in the basic events of myogenesis, such as proliferation of presumptive myoblasts, fusion, synthesis of contractile proteins, initiation of transverse tubule system, etc. Both in salamander embryos and insect embryos there are reports that the nervous system may play some role for myogenic cells earlier than presumptive myoblasts.

*Heywood:* I agree with Dr. Holtzer. It impresses me that the chick embryo synthesizes tremendous amounts of cell-specific proteins in a very few days. The fetal chick is very active at 14 days. Just a few days later it is going to escape from the egg. So, if it is in fact a fetal myosin, it must be a very active one. We have not been able to find any electrophoretic differences between adult myosin and embryonic myosin from chicken.

*Reedy:* Dr. Heywood has found that intact embryos use external or plasma amino acids in addition to cell pool amino acids to synthesize myosin. Is there any corresponding finding for myosin-synthesizing cells in culture?

*Holtzer:* We have just begun experiments that eventually might help answer this question. At the moment all I can say is that they certainly incorporate exogenous amino acids.

*Borisy:* Dr. Holtzer, you indicated that cells can undergo either a quantal mitosis or a proliferative mitosis. How many divisions or how many filial ancestors does a muscle cell have? Does a cell ever undergo what you would call proliferative mitosis in normal development?

*Holtzer:* A definite answer to the latter question is yes. I said that in tissue culture you can get something for nothing, particularly with BUdR. I hope that someone interested in muscle as a food understands that he can put one myogenic cell in tissue culture and recover literally hundreds if not thousands. With BUdR one can keep the cells proliferating for weeks. If the BUdR is removed, and provided the cells undergo normal mitosis in normal medium, thus yielding cells with normal RNA, they start making myosin. I believe that the quantal mitosis machinery is the important aspect and that there are "X" number of quantal mitoses that are necessary to produce the cell that suddenly has the machinery to make myosin.

*Beecher:* What events predict and control the extent of mitosis of muscle cells?

*Holtzer:* No more is known of why myogenic cells cease replicating than of why a nerve, pancreas, liver, or any other normal cell stops dividing. One could pose the question of whether a myogenic cell ceases replicating in order to translate for myosin. I suspect that is the case, though it is clear that in the case of cardiac myoblasts, one division or at the most two might be possible, after a cell has translated for myosin. *In vivo,* before there is any significant muscle activity, there are big and small muscles with different numbers of nuclei, and one must invoke genetic factors controlling the mitotic activity of individual muscles.

*Beecher:* What factors control the number of nuclei a muscle cell contains? Is the number of nuclei that a muscle cell contains related to the function or metabolic activity of the muscle?

*Holtzer:* In part I commented on this in answer to the previous question. As far as I know, there is no documentation that use hypertrophy involves an increase in number of nuclei.

*Kornguth:* Is there any indication that different surface antigens are present at different phases of the cell cycle in muscle cells?

*Holtzer:* This is an intriguing possibility that is currently being investigated. However, as you well know this is not an easy problem and to my knowledge cell-specific surface antigens, though postulated, have never been convincingly demonstrated.

*Kastenschmidt:* What is known about the physical and chemical modifications of the cell surface that presumably occur during the various stages of the cell cycle?

*Holtzer:* Essentially nothing.

*Davies:* When the mononucleated cells fuse, do the cell membranes become the sarcoplasmic reticulum of the newly formed fiber?

*Holtzer:* There is no evidence to that effect.

*Aberle:* What is the nature of the membrane of the myotube and is it analogous to the sarcolemma?

*Holtzer:* As far as is known, prior to growth the membranes of the myotube are made up by the membranes of the individual mononucleated cells. Subsequently additional membrane must be synthesized by the enlarging myotube. This membrane by definition is the sarcolemma. Carbohydrate and collagen are added to the older (about 5 days) myotubes in culture.

*Dubowitz:* How do you know that the labeled cell in metaphase is tightly applied to the myotube and not just overlying it, i.e., superimposed on the culture?

*Holtzer:* When is something "lying against" rather than "tightly applied"? Under the electron microscope the distance between the two is minimal, and essentially that describes the relationship between the cell membrane of the muscle fiber and the cell membrane of the satellite cell.

*M. Seraydarian:* Is there a hypothesized role for the filaments that form prior to the myotubules? Do they persist in the cell?

*Holtzer:* Every cell that has been looked at, to my knowledge, contains the 100-Å filaments. The significance of this at the moment invites much speculation, but warrants little serious discussion.

*Davies:* The important question seems to be, how embryonic is embryonic? The chicken can escape from the egg with its own power, but it is quite clearly a different matter with mice. Dr. Goldspink found that the sarcomere lengths in fetal mice are not as long as those in the adult, but Dr. Holtzer noticed that the sarcomeres that are laid out are the same length in embryonic chicken as in the adult. Is it true that sarcomeres never become longer from the time that you first observe them?

*Holtzer:* If there is any difference between a chicken and a mouse in terms of something as fundamental as laying down the basic architecture of a skeletal cell, I will be very surprised. In our earlier work we stressed the sarcomere. I stand corrected because we should have stressed the A band. In point of fact, you have the same evolution of a change in the sarcomere length in a chicken as you have in a mouse. What you observe when you first can see a respectable myofibril is that there is virtually no I band. But the A band, the length of the definitive myosin filament, never changes. I do not really know how to define embryonic myosin vs. adult myosin because in a 2-day chick embryo one sees the beating heart and in a 3-day chick embryo skeletal myofibrils are visible. Material that we begin with, be it a 13-day or 15-day chick embryo, is no less embryonic than that which came out of a 3-day chick embryo. I know of no evidence to indicate that myosin that will be synthesized by a satellite cell from a 40-year-old man during regeneration is in any way different from the myosin in an embryo.

*Davies:* The immediate answer to the question then is that sarcomere lengths are shorter when the fibers are first visible, but the A band is constant.

## *References*

Allen, E. R. and C. F. Terrence. 1969. Immunochemical and ultrastructural studies of myosin synthesis. *Proc. Nat. Acad. Sci. 60*:1209.

Bach, L. M. N. 1948. Conversion of red muscle to pale muscle. *Proc. Soc. Expl. Biol. Med. 67*:268.

Carlson, B. M. 1967. An investigation into a method for the stimulation of regeneration of skeletal muscle. *Anat. Rec. 157*:225.

———. 1968. Regeneration research in the Soviet Union. *Anat. Rec. 160*:665.

Close, R. 1965. Effects of cross-union of motor nerves to fast and slow skeletal muscles. *Nature 206*:831.

Guth, L. 1968. Trophic influences of nerve on muscle. *Physiol. Rev. 48*:645.

Hník, P., I. Jirmanovà, L. Vyklický and J. Zelená. 1967. Fast and slow muscles of the chick after nerve cross-union. *J. Physiol. 193*:309.

Lowey, S., H. S. Slayter, A. G. Weeds and H. Baker. 1969. Substructure of the myosin molecule I. Subfragments of myosin by enzymic degradation. *J. Mol. Biol. 42*:1.

Mommaerts, W. F. H. M. 1969. Energetics of muscular contraction. *Physiol. Rev. 49*:427.

# PART 2

## Red and White Muscle

# Chapter 6

# *Differentiation of Fiber Types in Skeletal Muscle*

V. DUBOWITZ

Enzyme histochemistry has added a new dimension to the study of skeletal muscle. The early anatomists recognized differences in the color and structure of different animal muscles and were able to divide these into "red" and "white" or "granular" and "nongranular" muscles. Subsequently, the physiologists tried to correlate variations in speed of contraction (fast and slow) of different muscles with these variations in color or granularity, and the biochemists were able to demonstrate differences in the enzyme content of the muscles.

With the advent of modern histochemical techniques, it became possible to localize enzyme systems and other chemical constituents at a cellular level and this opened the way for a direct correlation of the functional activity of individual fibers with their morphology.

## FIBER TYPES

On the basis of the reciprocal activity of phosphorylase and various dehydrogenases in individual muscle fibers, we suggested the subdivision into two fiber types—type I fibers rich in dehydrogenases and poor in phosphorylase, and type II fibers rich in phosphorylase and poor in the dehydrogenases (Dubowitz and Pearse, 1960*a, b*). We also recognized the presence of fibers of intermediate activity with any particular enzyme reaction. Subsequent workers have tried to categorize these "intermediate" fibers as specific types.

Stein and Padykula (1962) defined three fiber types (A, B, and C) in the rat gastrocnemius, based on the succinic dehydrogenase reaction. In serial section they correlated the distribution of glycogen, two forms of adenosine triphosphatase, and esterase with the distribution of succinic dehydrogenase and drew up a histochemical profile for each fiber type. They correlated the A fiber with the classical "white" fiber (our type II fiber), while the B and C represented two forms of "red" fiber (our type

I). In the gastrocnemius all three types were present, but the soleus contained only B and C.

Romanul (1964) added to the complexity. He studied complete cross sections of the rat calf with a wide variety of enzyme systems including cytochrome oxidase, the diaphorases, various dehydrogenases, phosphorylase, and esterase. By correlating the relative activities of all these enzymes in individual fibers, he was able to recognize eight possible fiber types. The gastrocnemius and plantaris contained all eight, while the soleus had only three. These eight fiber "types" formed a spectrum which could be broadly divided into three groups. At one end were the fibers with a high capacity to utilize glycogen, a low lipid metabolism, low oxidative metabolism, and a low myoglobin content. At the other end of the spectrum were the fibers with a low capacity for glycogen breakdown, a very high lipid metabolism, high oxidative metabolism, and high myoglobin content. In between were a group with moderate ability to utilize glycogen, moderate lipid metabolism, high oxidative metabolism, and a high myoglobin content. These three groups are substantially similar to the three fiber types of Stein and Padykula (1962).

Argument about the number and the characteristics of these fiber types has continued. In particular, differences have been observed in specific muscles in various animals. Thus in the soleus of the rat and monkey, a "slow," "red" muscle, Bocek and Beatty (1966) observed that the variation in succinic dehydrogenase activity noted by Stein and Padykula (1962) was irregular, and that the more deeply staining fibers were also high in phosphorylase. It was thus difficult to correlate these fibers in the "red" soleus muscle with any of the three types in "white" muscles.

More recently, Nyström (1968) has described his very detailed studies on the muscles of the cat. Like Stein and Padykula (1962), he defined three types, A, B, and C, on the basis of the succinic dehydrogenase reaction (pale, intermediate, and dark staining respectively). He then assessed $NADH_2$-tetrazolium reductase, lipids, myofibrillar adenosine triphosphatase, phosphorylase, and glycogen within these fiber types in serial section. He found that in the "fast-white" muscle, such as the gastrocnemius, the three types were readily recognizable.

The C fiber stained darkly for succinic dehydrogenase, $NADH_2$-tetrazolium reductase, and lipids, but gave only a faint reaction for phosphorylase and glycogen, and a very weak or negative reaction for myofibrillar adenosine triphosphatase.

The A fiber showed a weak reaction of succinic dehydrogenase, $NADH_2$-tetrazolium reductase, and lipids, and a strong reaction when stained for phosphorylase, glycogen, and myofibrillar adenosine triphosphatase. The B fiber, clearly recognizable as intermediate in sections stained for succinic dehydrogenase, $NADH_2$-tetrazolium reductase, and lipids, showed a staining intensity similar to that of the A type for phosphorylase, glycogen, and adenosine triphosphatase.

In "slow-red" muscles such as the crureus and soleus, Nyström found that the majority of fibers stained homogeneously for succinic dehydrogenase, $NADH_2$-tetrazolium reductase, and myofibrillar adenosine triphosphatase. The staining intensity of adenosine triphosphatase was higher than that of the C type in "fast-white" muscles, but lower than that of the B type. The staining for phosphorylase was similar to that of the C type. The reaction for succinic dehydrogenase and $NADH_2$-tetrazolium reductase was similar in intensity to the C type, but the formazan granules seemed less tightly packed. Nyström concluded that the general fiber type in "slow-red" muscles, such as soleus and crureus, was not directly comparable with any of the three types in "fast-white" muscles, and suggested that this be looked upon as a separate type, S.

In addition, Nyström also observed in the soleus and crureus a few aberrant fibers which gave a much more intense reaction for adenosine triphosphatase. These fibers also stained more darkly for succinic dehydrogenase but were not particularly striking in sections stained for $NADH_2$-tetrazolium reductase.

In mammalian muscle the general consensus is that there are three fiber types (A, B, and C), which can be recognized in "fast-white" muscles (Fig. 6.1). "Slow-red" muscles tend to have one type of fiber, which approximates to type C, but differs slightly from it (Fig. 6.2). In addition, there may be present a small number of atypical fibers, which do not conform to any of the other types.

In human muscle the distribution is less complex than in animals. All skeletal muscles are histochemically mixed in fiber type (Fig. 6.3). With the enzyme systems discussed above, the fiber population falls more clearly into two main fiber types with less tendency to intermediate activity. One circumstance, however, under which one can consistently identify three subgroups of fibers dependent on staining activity is with the periodic acid Schiff reaction for glycogen (using cryostat frozen sections). In the study of normal and diseased human muscle the subdivision into type I and type II fibers has been retained (Engel, 1962; Dubowitz, 1968). The fibers correlate fairly closely with the C and A fibers respectively in animal muscle.

### DEVELOPING MUSCLE

The histochemical techniques which have helped to define these fiber types have also made possible the study of their development and differentiation. In some animals, such as the guinea pig, and also in man, the

*Fig. 6.1.* Rabbit flexor digitorum longus (FDL). Note 3 fiber types: strong, weak, and intermediate staining. $NADH_2$-tetrazolium reductase (TR). × 110.

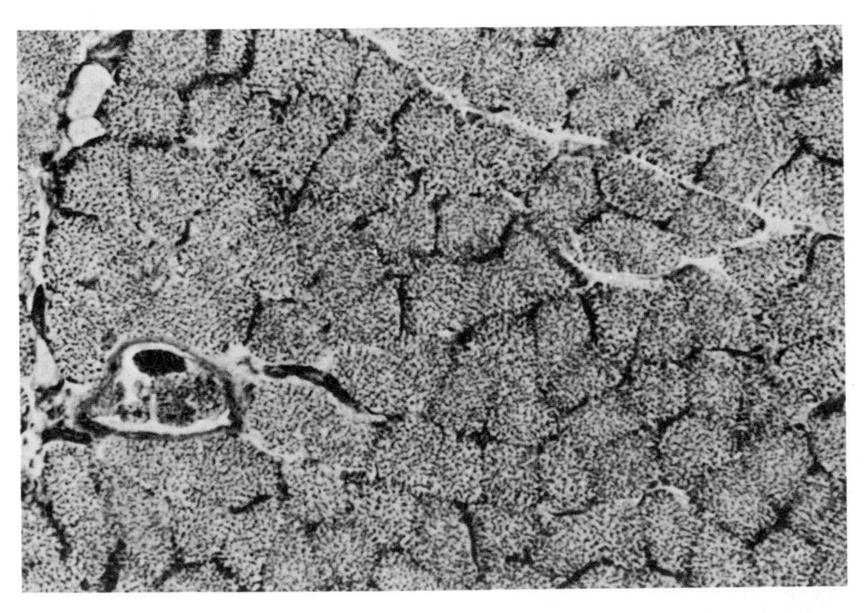

*Fig. 6.2.* Rabbit soleus. Note uniformity of enzyme reaction. Intrafusal fibers of muscle spindle show variation. $NADH_2$-TR. × 175.

*Fig. 6.3.* Human deltoid. Shows 2 fiber types. $NADH_2$-TR. × 175.

skeletal muscle is fully differentiated into its fiber types at birth. In other animals such as the mouse, rat, and rabbit, differentiation only occurs postnatally (Dubowitz, 1963, 1965*a, b,* 1968). While it is difficult to define the exact fiber type in the undifferentiated state in the developing animal, it does approximate more to the type I (or C) fiber.

In addition to these differences between animals, there are also variations from one muscle to another in the same animal. Thus, some muscles may show histochemical differentiation before others; and there may also be earlier differentiation with some enzymes than with others (Dubowitz, 1965*a,* 1968; Nyström, 1968). This may account for some of the apparently divergent observations in the literature.

In contrast to my observation that the skeletal muscle of the rat and mouse were undifferentiated at birth, Wirsén and Larsson (1964) found differentiation in the thoracic muscles of the fetal and newborn mouse. These conclusions were, however, based entirely on their observations with the phosphorylase reaction. At 16 days they could identify primary and secondary fibers and by the 19th day tertiary fibers as well. With the succinic dehydrogenase reaction and with lipid stain there was no difference between fibers before birth, so they did not use these methods for further study.

I think it may be important to distinguish between a variation in the strength of reaction of a single enzyme in different fibers of a muscle, and the differentiation of muscle into fiber types. For the recognition of the fiber types as seen in mature adult muscle, it is probably necessary for the muscle fiber to show the characteristic profile as judged by at least two enzyme systems. A useful selection for screening in this way are succinic dehydrogenase, phosphorylase, and myofibrillar adenosine triphosphatase.

Individual muscles may show characteristic patterns of change. Thus, in the chick both the pectoralis and the gastrocnemius initially contain only one fiber type, which conforms to type I (Cosmos, 1966). The pectoralis changes postnatally into predominantly type II fibers, while the gastrocnemius differentiates into three types. This work was based on the phosphorylase and succinic dehydrogenase activities.

With the succinic dehydrogenase reaction, Nyström (1966) observed that in the gastrocnemius of the newborn cat, the fibers were initially all of one type and subsequently differentiated into two and then three types. The soleus also consisted of one type initially but remained uniform in fiber type, except for a transitory period of differentiation into two types between 2 and 7 weeks.

Karpati and Engel (1967) recorded an initial period of differentiation in the soleus of the newborn guinea pig. This was based mainly on the myofibrillar adenosine triphosphatase reaction, but also on other reactions, details of which were not presented.

In a study of development of the soleus of the cat and rabbit, I observed marked differences with different enzymes. Thus the adenosine triphosphatase and, to a lesser extent, the phosphorylase reaction might show marked differentiation at a time when the succinic dehydrogenase and $NADH_2$-tetrazolium reductase reactions were uniform (Dubowitz, 1968) (Figs. 6.4–6.8).

Recently Nyström (1968) has reported on his experience with a number of enzymes in the postnatal differentiation of several muscles of the cat. In newborn kittens, the fibers within "slow-red" soleus and "fast-white" gastrocnemius muscles appeared fairly uniform, except for the adenosine triphosphatase reaction. With time, different fiber types appeared in both muscles. This was more striking in the gastrocnemius, where it progressed to the mature three-fiber pattern. In contrast, the early differentiation in the soleus was soon suppressed. In the gastrocnemius lightly and darkly staining fibers were visible with succinic dehydrogenase by 10 days and distinct by 15 days. The B fiber, however, did not appear until about 6 weeks of age. Other hindleg muscles followed a similar sequence, but in the foreleg the differentiation started earlier.

In the newborn kitten, light and dark staining fibers could be distinguished at times with the phosphorylase reaction in the soleus and the gastrocnemius, but with the adenosine triphosphatase reaction the presence of light and dark staining fibers was much more marked. These fibers, however, were uniformly rich in oxidative enzymes, lipids, and

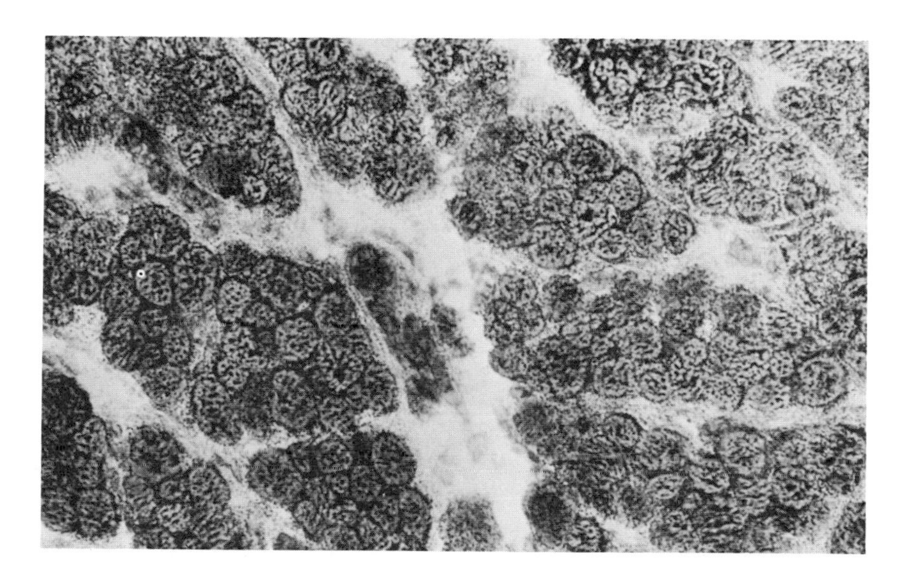

*Fig. 6.4.* Rabbit (age 2 days). FDL. Uniformity of fiber activity. $NADH_2$-TR. × 120.

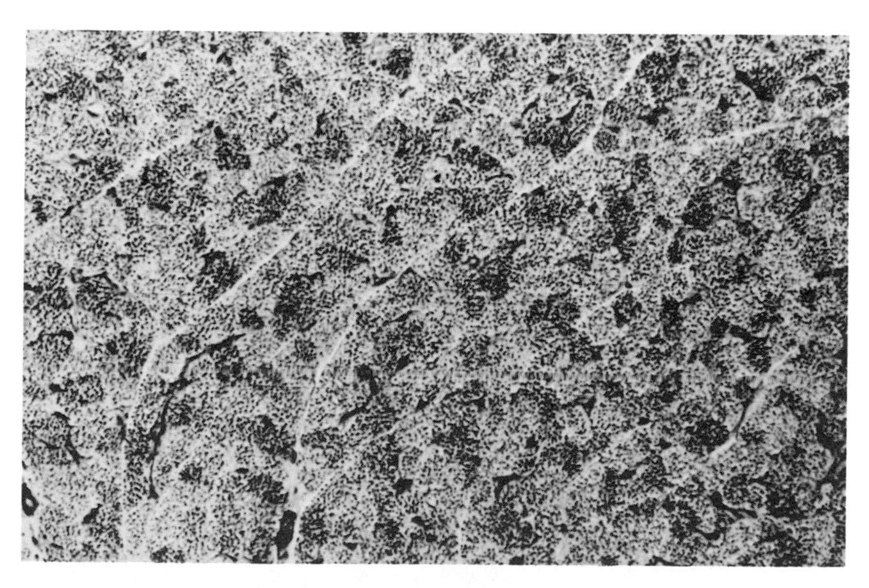

*Fig. 6.5.* Rabbit (age 13 days). FDL. Shows clearcut division into fiber types. $NADH_2$-TR. × 100.

glycogen, and Nyström was unable to correlate them with the A, B, and C fibers of adult cat muscle.

One may thus conclude that in animals all muscle starts in an undifferentiated state. Differentiation into adult fiber types proceeds at varying rates and some enzymes show differentiation before others.

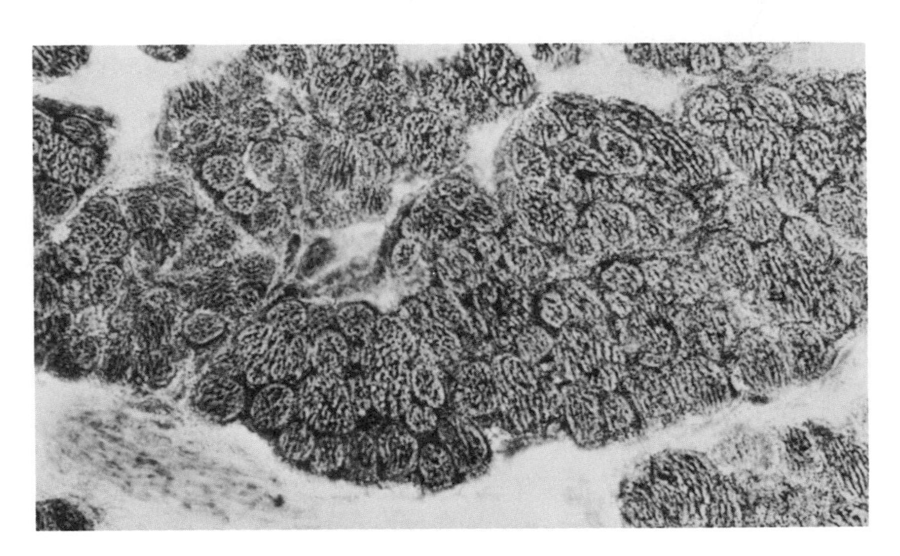

*Fig. 6.6.* Rabbit (age 2 days). Soleus. Uniform activity. $NADH_2$-TR. × 120.

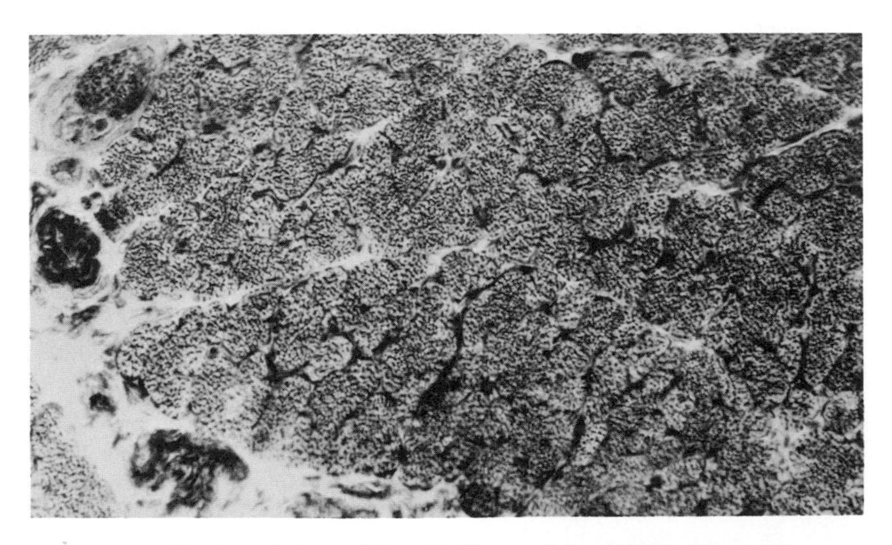

*Fig. 6.7.* Rabbit (age 13 days). Soleus. Uniform activity. $NADH_2$-TR. × 100.

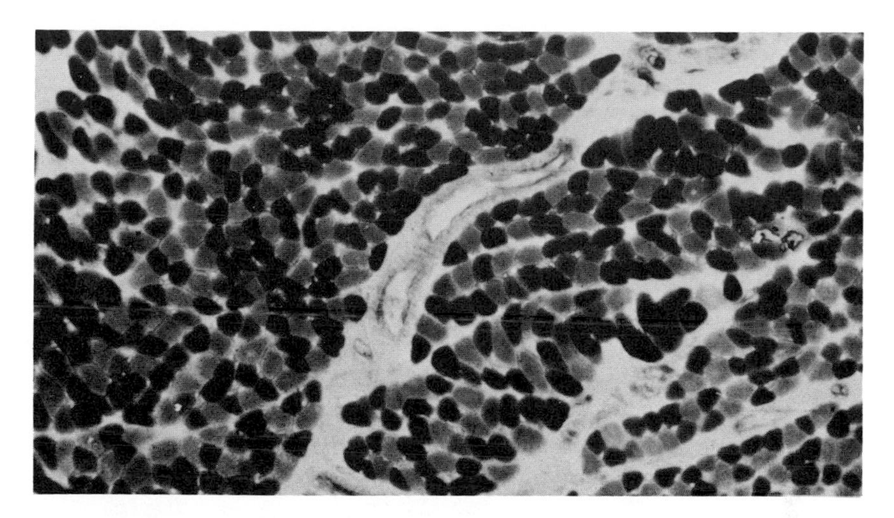

*Fig. 6.8.* Rabbit (age 13 days). Soleus. Clearcut variation in fiber activity. Adenosine triphosphatase. × 100.

In developing *human muscle* the histochemical pattern differs from that seen in animals (Dubowitz, 1965*b*). Up to about 18 weeks' gestation, one cannot recognize the fiber types of mature muscle. As in developing animals there may be some variation in enzyme activity with individual enzymes such as adenosine triphosphatase (Fig. 6.9). After about 20 weeks the muscle shows full differentiation into type I and type II fibers with similar enzyme profiles to mature muscle (Fig. 6.10). However, the type I fibers only comprise about 5–10% of the total. After about 30 weeks' gestation, all skeletal muscle is composed of approximately equal numbers of type I and type II fibers, as in the adult.

## CONTRACTILE PROPERTIES

While a broad correlation has proved possible between histochemical fiber types and the contractile properties of the muscle as a whole, there are still a number of aspects needing explanation. Thus, slow muscles in animals, such as soleus, crureus, and anconeus, contain predominantly one type of fiber approximating to type I or C fibers, rich in oxidative enzymes and thus adapted for sustained activity. On the other hand, fast muscles, such as flexor digitorum longus or flexor hallucis longus, are not composed of purely type II or A fibers as might have been anticipated, but are always a mixture of two or three fiber types. This is difficult to correlate with the findings of the physiologists that slow or fast muscle appears to be homogeneous and does not show a mixed pattern of activity (Buller, Eccles, and Eccles, 1960*a*). Evidence is accumulating, however, that mammalian muscle may be composed of motor units with different contraction times (Gordon and Phillips, 1953; Bessou, Emonet-Denand, and Laporte, 1963; Anderson and Sears, 1964).

Henneman and Olson (1965) tried to correlate histochemical with motor unit properties. Using the staining technique for mitochondrial

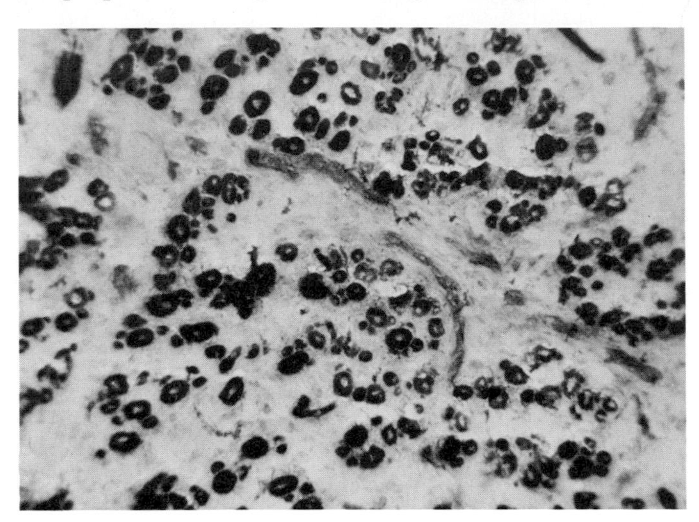

*Fig. 6.9.* Human fetus (12 weeks' gestation). Deltoid. Note overall uniform activity in myotubes. Adenosine triphosphatase. × 65.

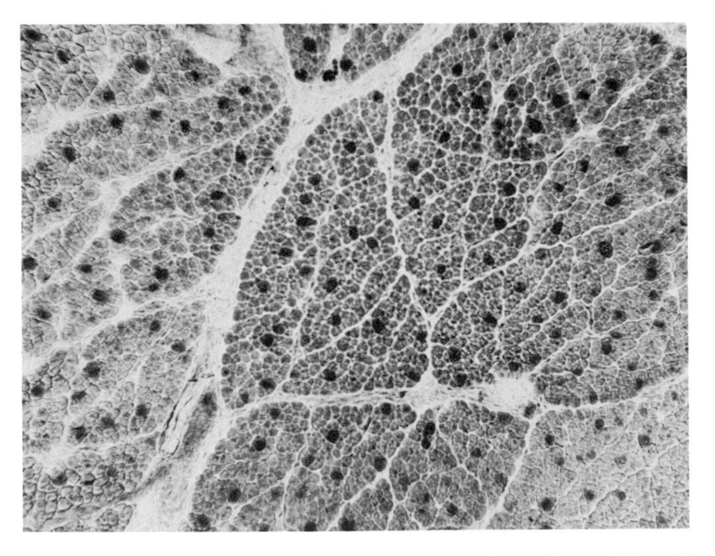

*Fig. 6.10.* Human fetus (22 weeks' gestation). Triceps. Note 2 fiber-type patterns. $NADH_2$-TR. × 75.

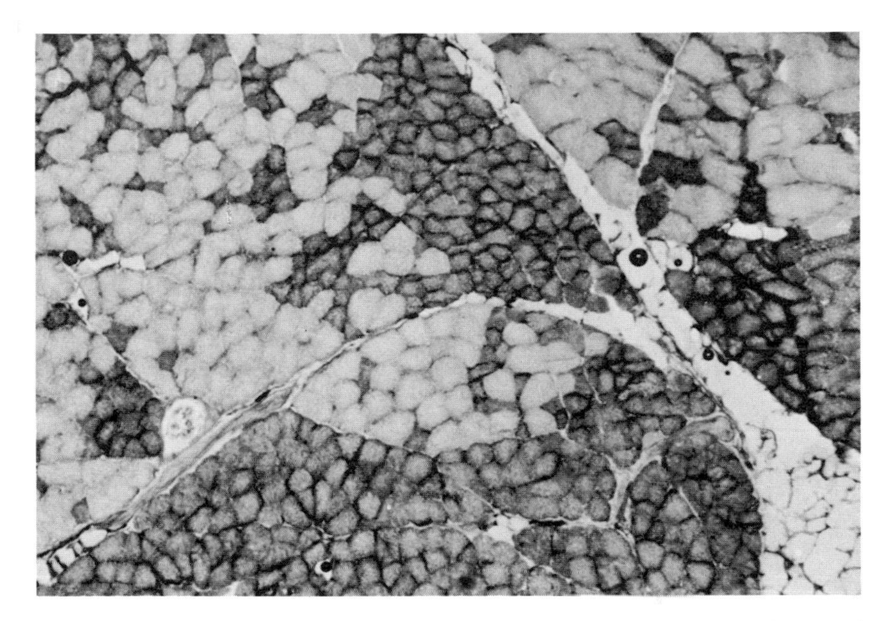

*Fig. 6.11.* Rabbit FDL cross-innervated by nerve to soleus. Note large clusters of type I (dark) fibers, and presence of type II fibers. No obvious intermediate fibers. $NADH_2$-TR. × 45.

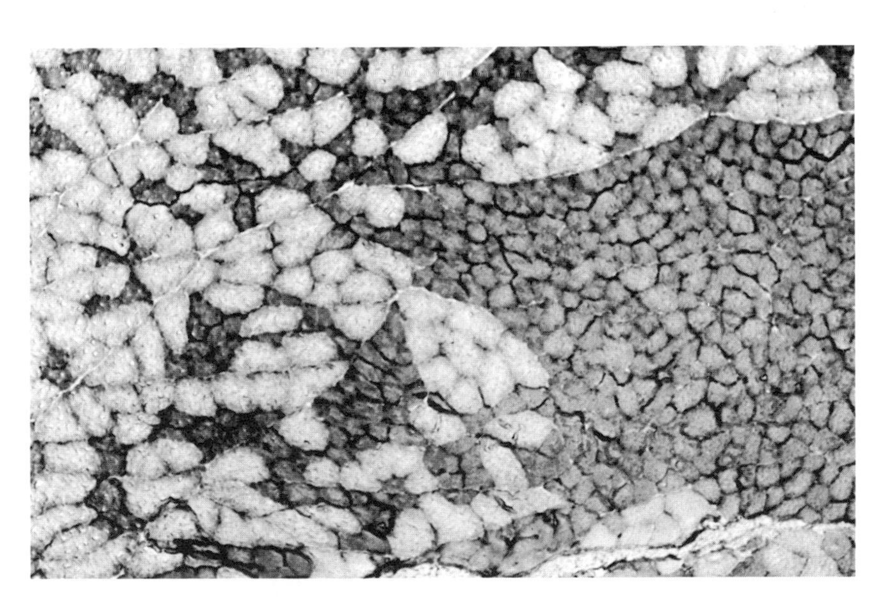

*Fig. 6.12.* Cat flexor hallucis longus cross-innervated by nerve to soleus. Note large bundle of uniform fibers (type I) resembling normal soleus. Rest of muscle composed of two fiber types with no intermediate fibers. $NADH_2$-TR. × 45.

adenosine triphosphatase, they concluded that the one fiber type in the soleus and three fiber types in the gastrocnemius correlated respectively with one and three types of motor unit. Wuerker, McPhedran, and Henneman (1965) and McPhedran, Wuerker, and Henneman (1965) classified the three types of motor unit on the basis of their size, their twitch contraction time, and their conduction velocity. The fastest contracting muscle fibers, which correlated with type II or A fibers, were innervated by the fastest conducting nerve fibers. It is somewhat surprising, though, that although classifying soleus as a one-fiber type muscle, McPhedran, Wuerker, and Henneman (1965) reported a wider variation in the contraction times of the soleus motor units than in the gastrocnemius.

In an interesting series of experiments Kugelberg and Edström (1968) and Edström and Kugelberg (1968) recently showed that repetitive stimulation of the sciatic nerve had a differential influence on the phosphorylase activity and glycogen content of the A, B, and C fibers of the anterior tibial muscle. On this basis they were subsequently able to map out the individual fibers making up a motor unit, by repetitive stimulation of single motor nerve fibers.

In developing muscle in the cat, Buller, Eccles, and Eccles (1960*a*) showed that all muscles were initially slow in their contraction time. This correlates well with the initial uniformity of histochemical pattern and approximation to type I or C fibers. With development, the potentially fast muscles gradually became faster, to reach their adult speed by about 4 weeks. Slow muscles, such as the soleus, became slightly faster but then reverted to the slow speed compatible with the adult state.

## FACTORS INFLUENCING FIBER DIFFERENTIATION

Buller, Eccles, and Eccles (1960*b*) showed that by crossing the nerves of the slow soleus and fast flexor hallucis muscles of the cat, one could change the contractile properties of these muscles. They postulated a chemical influence from the nervous system passing down the axon to the muscle and directly influencing its contractile properties. They further suggested that the same mechanism might be responsible for the changes in contractile properties of developing muscle. As an alternative to the chemical mediation of the neural influence on the muscle, Vrbová (1963) has suggested that the frequency and character of the impulses down the nerve might be the controlling factor.

By the application of histochemical techniques to this interesting experimental model, we were able to show, both in the cat and the rabbit, that cross innervation changes not only the contractile properties of the muscle but also the histochemical enzyme profile, and presumably the whole metabolic makeup of the muscle (Dubowitz, 1967). We noted that the changes in a fast muscle such as flexor hallucis were much more striking than those in the slow soleus. The flexor hallucis or flexor digitorum, which normally contains a predominance of large type II fibers with small clusters of type I fibers of smaller diameter (Fig. 6.1), showed after cross innervation by the nerve of soleus large areas of uniform type I fibers indistinguishable in pattern from normal soleus (Figs. 6.11–6.12).

There is one other observation of interest, and possibly of significance. Whereas in the normal flexor hallucis longus one can readily recognize three fiber types with the $NADH_2$-tetrazolium reductase reaction (Fig. 6.1), in the cross-innervated muscle there only appear to be 2 fiber types (Figs. 6.11–6.12). This supports the view that the nerve in the first place only determines two basic fiber types and that the intermediate fibers may be variants of these fiber types, perhaps induced by other factors.

Michael Bárány (unpublished observations, 1966) estimated the $Ca^{++}$-, actin- and EDTA-activated adenosine triphosphatase levels in these cross innervated muscles. Although he found some changes—in the right direction—these were slight and he considered that they were not marked enough to be outside the range of experimental error of the methods. When one looks at the focal nature of the histochemical changes (Figs. 6.11–6.12), one can understand why the biochemical changes assessed in the muscle as a whole may be less striking. However, as anticipated, significant changes in various enzymes such as pyruvic kinase, aldolase, malic dehydrogenase, and isocitric dehydrogenase have been demonstrated biochemically (Prewitt and Salafsky, 1967).

The influence of the nervous system on the determination of fiber type in muscle is now well established. It would be of interest if this could in due course be correlated with other influences of the nerve on muscle, such as the trophic effect. This in turn could give us a bridge in our understanding of the trophic influence of many other factors on muscle. For instance, nutrition and exercise have a marked influence; disuse leads to atrophy; castration can cause atrophy of specific muscles such as the levator ani of the rat; and hormones, such as cortisone, can produce marked atrophy. Some of these influences may well be selective on specific fiber types in muscle. A better understanding of many of these processes may throw further light on our understanding of the differentiation of muscle into its fiber types.

## ACKNOWLEDGMENTS

This work has been supported by grants from The Muscular Dystrophy Group of Great Britain, The Medical Research Council, and The Sheffield University Research Fund. I am grateful to Mrs. J. Binns, Mrs. B. Lunn, and Miss C. Heinzmann for technical assistance and Mr. A. K. Tunstill for photography.

## *References*


Anderson, P., and T. A. Sears. 1964. The mechanical properties and innervation of fast and slow motor units in the intercostal muscles of the cat. *J. Physiol. 173*:114.

Bessou, P., F. Emonet-Denand, and Y. Laporte. 1963. Relation entre la vitesse de conduction des fibres nerveuses motrices et le temps de contraction de leurs unites motrices. *C. R. Acad. Sci. 256*:5625.

Bocek, R. M., and C. H. Beatty. 1966. Glycogen synthetase and phosphorylase in red and white muscle of rat and rhesus monkey. *J. Histochem. Cytochem. 14*:549.

Buller, A. J., J. C. Eccles, and R. M. Eccles. 1960*a*. Differentiation of fast and slow muscles in the cat hind limb. *J. Physiol. 150*:399.

———. 1960*b*. Interactions between motoneurones and muscles in respect of the characteristic speeds of their responses. *J. Physiol. 150*:417.

Cosmos, E. 1966. Enzymatic activity of differentiating muscle fibers I. Development of phosphorylase in muscles of the domestic fowl. *Dev. Biol. 13*:163.

Dubowitz, V. 1963. Enzymatic maturation of skeletal muscle. *Nature 197*:1215.

———. 1965*a*. Enzyme histochemistry of skeletal muscle I. Developing animal muscle. *J. Neurol. Neurosurg. Psychiat. 28*:516.

———. 1965*b*. Enzyme histochemistry of skeletal muscle II. Developing human muscle. *J. Neurol. Neurosurg. Psychiat. 28*:519.

———. 1967. Cross-innervated mammalian skeletal muscle: histochemical, physiological and biochemical observations. *J. Physiol. 193*:481.

———. 1968. *Developing and Diseased Muscle. A Histochemical Study*. Spastics International Medical Publications in association with William Heinemann Medical Books Ltd., London.

Dubowitz, V., and A. G. E. Pearse. 1960*a*. Reciprocal relationship of phosphorylase and oxidative enzymes in skeletal muscle. *Nature 185*:701.

———. 1960*b*. A comparative histochemical study of oxidative enzyme and phosphorylase activity in skeletal muscle. *Histochemie 2*:105.

Edström, L., and E. Kugelberg. 1968. Histochemical composition, distribution of fibres and fatiguability of single motor units. Anterior tibial muscle of the rat. *J. Neurol. Neurosurg. Psychiat. 31*:424.

Engel, W. K. 1962. The essentiality of histo- and cytochemical studies of skeletal muscle in the investigation of neuromuscular disease. *Neurology 12*:778.

Gordon, G., and C. G. Phillips. 1953. Slow and rapid components in a flexor muscle. *Quart. J. Exp. Physiol. 38*:35.

Henneman, E., and C. B. Olson. 1965. Relations between structure and function in the design of skeletal muscles. *J. Neurophysiol. 28*:581.

Karpati, G., and W. K. Engel. 1967. Neuronal trophic function. A new aspect demonstrated histochemically in developing soleus muscle. *Arch. Neurol. 17*:542.

Kugelberg, E., and L. Edström. 1968. Differential histochemical effects of muscle contractions on phosphorylase and glycogen in various types of fibres: relation to fatigue. *J. Neurol. Neurosurg. Psychiat. 31*:415.

McPhedran, A. M., R. B. Wuerker, and E. Henneman. 1965. Properties of motor units in a homogeneous red muscle (soleus) of the cat. *J. Neurophysiol. 28:*71.

Nyström, B. 1966. Succinic dehydrogenase in developing cat leg muscles. *Nature 212:*954.

———. 1968. Histochemistry of developing cat muscles. *Acta Neurol. Scand. 44:*405.

Prewitt, M. A., and Salafsky, B. 1967. Effect of cross innervation on biochemical characteristics of skeletal muscles. *Amer. J. Physiol. 213:*295.

Romanul, F. C. A. 1964. Enzymes in muscle I. Histochemical studies of enzymes in individual muscle fibers. *Arch. Neurol. 11:*355.

Romanul, F. C. A., and J. P. Van Der Meulen. 1967. Slow and fast muscles after cross innervation. Enzymatic and physiological changes. *Arch. Neurol. 17:*387.

Stein, J. M., and H. A. Padykula. 1962. Histochemical classification of individual skeletal muscle fibers of the rat. *Amer. J. Anat. 110:*103.

Vrbová, G. 1963. The effect of motoneurone activity on the speed of contraction of striated muscle. *J. Physiol. 169:*513.

Wirsén, C., and K. S. Larsson. 1964. Histochemical differentiation of skeletal muscle in fetal and newborn mice. *J. Embryol. Exp. Morph. 12:*759.

Wuerker, R. B., A. M. McPhedran, and E. Henneman. 1965. Properties of motor units in a heterogeneous pale muscle (gastrocnemius) of the cat. *J. Neurophysiol. 28:*85.

# Chapter 7

# *The Ultrastructure of Three Fiber Types in Mammalian Skeletal Muscle*

G. F. GAUTHIER

## INTRODUCTION

It has long been known that the skeletal muscles of an individual mammal differ in color (Lorenzini, 1678; see Ciaccio, 1898), and that the fibers composing skeletal muscles differ in their microscopic appearance (Ranvier, 1874). Even within a given muscle, the fibers appear to differ from one another (Grützner, 1884; Knoll, 1891), and as many as three types of fibers were described by Bullard in 1919. Histochemical procedures for localizing enzymic activity have confirmed these early observations and have extended them by revealing additional differences among muscle fibers (Padykula, 1952; Ogata, 1958; Nachmias and Padykula, 1958; Dubowitz and Pearse, 1960). The histochemical approach has, furthermore, underlined the functional significance of fiber types, because it is possible to relate differences in enzymic activity to metabolic differences. It has recently been demonstrated, using autoradiographic procedures, that even fibers which appear to be identical by other histochemical procedures may exhibit difference when metabolic turnover is taken into consideration (Coimbra, 1969). Finally, analysis of mammalian skeletal muscle using the electron microscope has demonstrated that fibers can be further distinguished on the basis of ultrastructural differences (Pellegrino and Franzini, 1963; Padykula and Gauthier, 1963, 1967*b;* Forssmann and Matter, 1966; Gauthier and Padykula, 1966; Shafiq, Gorycki, Goldstone, and Milhorat, 1966; Shafiq, Gorycki, and Milhorat, 1969; Tice and Engel, 1967; Gauthier, 1969).

Despite the striking manifestations of heterogeneity, however, differences among fibers within a particular muscle have not always been fully considered (Pellegrino and Franzini, 1963; Gustafsson, Tata, Lindberg, and Ernster, 1965; Tice and Engel, 1967; Miledi and Slater, 1968). Cytological characteristics of a particular fiber type have, because of this, been incorrectly attributed to some form of experimental or pathological alteration. It is obviously important that differences among fibers be recognized when analyzing changes that arise from experimental treatment or when interpreting physiological measurements on whole muscle; yet the functional significance of heterogeneity cannot be adequately appreciated unless the ultrastructural features of the individual fibers have been clearly established. Once they are established, the task of relating ultrastructure to physiological data remains a difficult one, since the physiological measurements usually reflect, at least in the mammal, characteristics of the whole muscle. Recent studies indicate, however, that three physiologically distinct types of motor units exist in skeletal muscles of the rat (Close, 1967*a, b*). The presence of three morphologically distinct types of muscle fibers and of their corresponding neuromuscular junctions in rat skeletal muscle has been demonstrated in our laboratory, and the ultrastructural features of these three fibers are described below.

### THE THREE FIBER TYPES AS DISTINCT ENTITIES

We have been able to establish criteria for the identification of three distinct types of fibers in skeletal muscles of the albino rat. Because the pattern of distribution of fibers is closely related to the color of the whole muscle, we have chosen to adopt the terms "red," "white," and "intermediate" as designations for these fibers. The validity of this terminology will be discussed in a separate section of this report. The three fiber types have been identified consistently by using a variety of histochemical procedures. In a group of serial sections of diaphragm muscle (Fig. 7.1), for example, the same fibers were identified using three different procedures. In these three histochemical preparations, there are obvious differences among fibers. Fibers 2 and 3, for example, were repeatedly identified as red fibers. Fiber 12 is a white fiber, and 5 an intermediate fiber. It was calculated, furthermore, in all three preparations, that red fibers constitute 60% of the total population, while white and intermediate fibers constitute only 20% each.

Mitochondrial content is a particularly suitable cytological criterion for the identification of fiber types. When procedures known to demonstrate mitochondrial enzymes or the phospholipid components of mitochondria are used, a characteristic cytological pattern is observed, and reactive sites are occupied by mitochondria in equivalent sections examined with the electron microscope. Mitochondrial adenosine triphosphatase (Fig. 7.1B) has the same pattern of distribution among fibers of the rat diaphragm as sites stained with Sudan black (Figs. 7.1C, 7.18D). Although triglyceride droplets are also stained in the latter, they can be distinguished from phospholipid components of the fibers by their circular profiles and intense blue-black color. A similar pattern of distribution is exhibited by the mitochondrial enzyme, succinic dehydrogenase, and this is illustrated in the semitendinosus muscle of the rat (Fig. 7.10).

In general, mitochondrial content is inversely related to the diameter of the fibers (Figs. 7.1B, 7.1C, and 7.10; Tables 7.1–2). Fibers with a small diameter are rich in mitochondria, especially along the periphery of the fibers. These are equivalent to classical "red" fibers. Fibers with the largest diameter have the lowest mitochondrial content, and these are equivalent to classical "white" fibers. The third type of fiber is intermediate in size, and is similar in many respects to the red fiber except that mitochondria are less abundant, especially at the periphery of the fibers.

TABLE 7.1
*Comparative dimensions of fiber types in the rat diaphragm*

| Fiber type | Fiber diameter (average) | Width of Z line | |
|---|---|---|---|
| | | Average | SD |
| Red | 27 μ | 634 Å | 31.2 |
| Intermediate | 34 μ | 433 Å | 39.4 |
| White | 44 μ | 339 Å | 30.5 |

From Padykula and Gauthier (1967*a*), and unpublished observations.

The form and intracellular distribution of mitochondria are most clearly demonstrated, at the light microscopic level, in Epon sections (2μ) stained with toluidine blue. In red fibers (Figs. 7.2, 7.3, 7.12) circular and filamentous profiles are present in the interior of the fibers. These represent interfibrillar rows of mitochondria and paired mitochondria at the I bands, respectively (Figs. 7.4, 7.6). Large spherical mitochondria form conspicuous aggregations at the periphery of the fibers (Figs. 7.2, 7.3, 7.12). In contrast, the mitochondria of white fibers (Figs. 7.3, 7.13) consist almost entirely of filamentous profiles which reflect paired mitochondria at the I bands (Figs. 7.4, 7.5, 7.8). Large spherical mitochondria are sparse or absent in these fibers, and peripheral regions are noticeably free of mitochondrial aggregations (Figs. 7.3–5). Intermediate fibers have many of the features of red fibers, but subsarcolem-

TABLE 7.2
*Analysis of fiber population in the rat semitendinosus*

| Region of muscle | Type of fiber | Fiber diameter (average) | % of fiber types | Area contributed by each fiber type (% of total area) |
|---|---|---|---|---|
| Red band | Red | 45 μ | 52 | 44 |
| | Intermediate | 52 μ | 40 | 44 |
| | White | 59 μ | 9 | 13 |
| White band | Red | 47 μ | 4 | 2 |
| | Intermediate | 47 μ | 14 | 7 |
| | White | 69 μ | 82 | 91 |

From G. F. Gauthier (1969).

mal aggregations and interfibrillar rows of mitochondria are less conspicuous.

The same three fiber types in both the diaphragm and semitendinosus can be identified with the electron microscope. The distribution of mitochondria, in particular, confirms observations made at the light microscopic level, and the ultrastructural appearance reveals additional distinctive features that permit identification of these fibers. In the red fiber (Figs. 7.6, 7.14), subsarcolemmal aggregations consist of large, closely packed mitochondria. Their abundant cristae consist primarily of parallel sheets. Similar mitochondria are arranged in longitudinal rows between the myofibrils, and triglyceride droplets are frequently associated with them (Fig. 7.6). In addition, paired mitochondria encircle the myofibrils at the I bands. These "bracelet-like" mitochondria are seen as small elliptical profiles in most longitudinal sections of the muscle fiber. When the plane of section is tangential to the surface of the myofibril, however, the mitochondria are sectioned longitudinally, and elongated profiles extend transversely across the myofibrils (Fig. 7.6). The intermediate fiber (Figs. 7.7, 7.15) is, in many respects, similar to the red fiber except that mitochondria are somewhat smaller and their cristae are less closely packed. The Z lines are noticeably thinner in the intermediate than in the red fiber, and this is an important feature which distinguishes these two fibers. When both fibers are included in a single section (Fig. 7.15), the contrasting appearance is readily evident. In the white fiber (Figs. 7.8, 7.16), mitochondria are sparse, even in the subsarcolemmal region. Usually only the paired I-band mitochondria characteristic of all three fibers are present. The Z lines are narrowest in this fiber. Actual measurements in diaphragm fibers (Table 7.1) show that the red fiber has the smallest fiber diameter and the widest Z lines. The white fiber, in contrast, has the largest diameter, but the Z lines are about half as wide as in the red fiber. In the intermediate fiber, both these dimensions are between those of the red and white fibers.

Differences exist, also, in the sarcoplasmic reticulum. The distribution of this membrane system is the same in all three fibers and is typical of that seen in other mammalian skeletal muscles. Triads occur at the A-I junctions and are closely associated with paired mitochondria at the I bands. Longitudinal tubules extend from the terminal cisternae toward

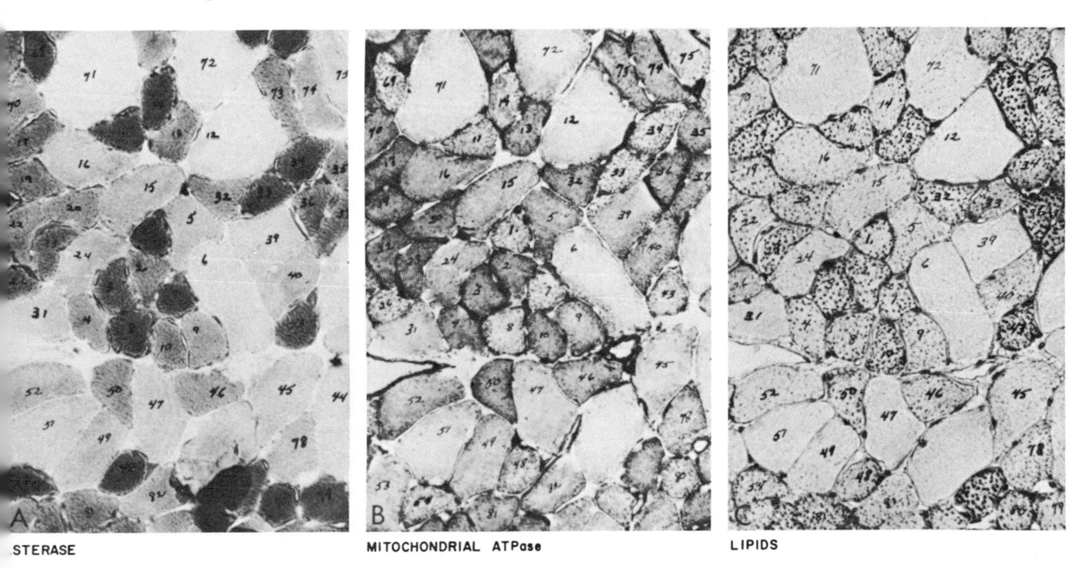


**SERIAL CRYOSTAT SECTIONS OF DIAPHRAGM - ALBINO RAT**

| Classification of Fibers Based on Cytochemical Comparisons | | |
|---|---|---|
| Type of Fiber | Percentage | Typical Examples in the Numbered Photographs Above |
| RED | 60 | 1, 2, 3, 7, 8, 43 |
| INTERMEDIATE | 20 | 5, 15 |
| WHITE | 20 | 12, 71 |

ig. 7.1. Diaphragm, serial cryostat sections. *A,* nonspecific esterase activity, using α-naphthyl acetate Pearse, 1960). *B,* adenosine triphosphatase, *p*H 7.2, in the presence of p-hydroxymercuribenzoate (Padyula and Gauthier, 1963; Gauthier, 1967). *C,* Sudan black B, demonstrating both triglyceride droplets nd phospholipid components of mitochondria. In all three preparations, the degree of enzymic activity nd the amount of lipid is inversely related to fiber diameter. From H. A. Padykula and G. F. Gauthier unpublished).

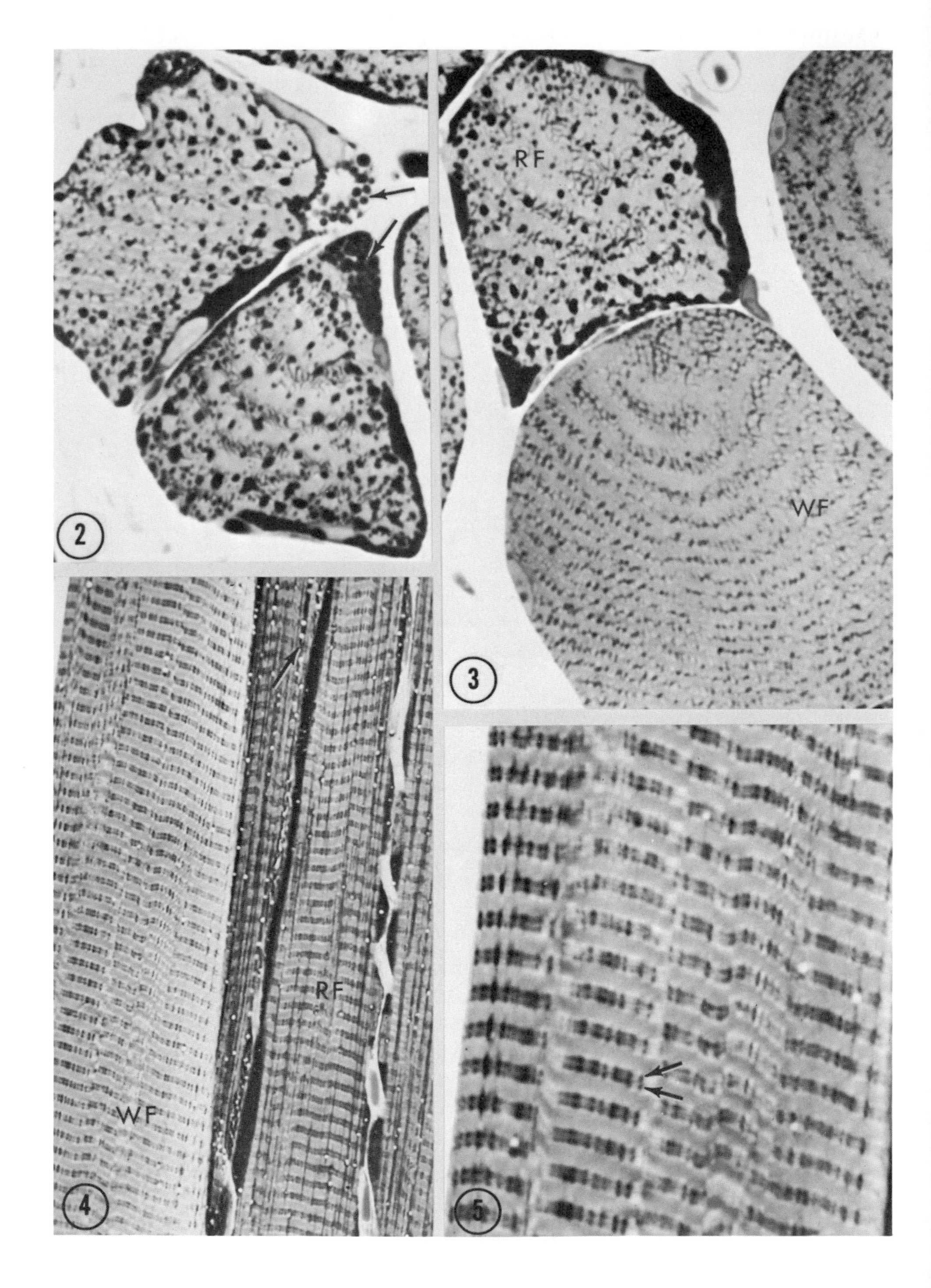
2
RF
WF
3
RF
WF
4
5

*Fig. 7.2.* Diaphragm, toluidine blue. Transverse sections of two red fibers. Large spherical mitochondria form conspicuous aggregations at the periphery (*arrows*). Circular and filamentous profiles in the interior of the fibers represent longitudinal interfibrillar rows and paired mitochondria encircling the myofibrils at the I bands, respectively. × 1,200.

*Fig. 7.3.* Diaphragm, toluidine blue. Transverse section showing a red fiber (*RF*) rich in large mitochondria, adjacent to a white fiber (*WF*), where mitochondria consist almost entirely of filamentous profiles which reflect pairs at the I bands (Fig. 7.5). The peripheral regions of the white fiber are noticeably free of mitochondrial aggregations. × 1,200.

*Figs. 7.4–5.* Diaphragm, toluidine blue. Longitudinal sections. Portions of three red fibers (*RF*) and one white fiber (*WF*) are included in *Fig. 7.4* (× 720). Large interfibrillar mitochondria are aligned with lipid droplets (negative images) to form longitudinal rows (*arrow*) in the red fiber. Paired mitochondria occur at the I bands in both types of fibers. Their alignment over the I band creates the appearance of a dark band at this position. In *Fig. 7.5* (× 1,600) the paired mitochondria associated with the I bands are seen in a white fiber. Elliptical profiles (arrows) on either side of the Z lines represent the major form of mitochondria in this fiber. Large peripheral mitochondria are absent. From L. A. Bunting (1969).

the centers of the A bands to form a more or less transverse network in the region of the H band. A difference is apparent in the form of the system at the H band. In the red fiber, an elaborate open network of narrow tubules extends across the myofibrils in this region. In the same region of the white fiber, on the other hand, there is a more compact arrangement, which consists of broad parallel tubules or flattened sacs. The contrasting appearance of this portion of the sarcoplasmic reticulum has been established in red and white fibers of both the diaphragm and semitendinosus muscles of the rat (Gauthier, 1969).

There are, therefore, a number of reliable criteria for identifying three fiber types, and the validity of these criteria has been confirmed by the ultrastructural analysis of two different skeletal muscles of the rat, namely the diaphragm (Gauthier and Padykula, 1966; Padykula and Gauthier, 1967*b*), and a hindlimb muscle, the semitendinosus (Gauthier, 1969). All three of these fibers represent distinct entities, and their special features permit identification repeatedly at both the light and electron microscopic level. The individuality of skeletal muscle fiber types is indicated also by histophysiological experiments. It has been demonstrated, using histochemical criteria, that each of three fiber types in the anterior tibial muscle of the rat responds in a characteristic manner to electrical stimulation (Kugelberg and Edström, 1968).

It has been suggested (Goldspink, 1962*a;* Goldspink and Rowe, 1968; Rowe and Goldspink, 1969) that differences among the component fibers of a skeletal muscle represent merely a phase of transformation from one type to another, and that the red fiber, in particular, is a relatively primitive form, capable of differentiating into a white fiber. However,

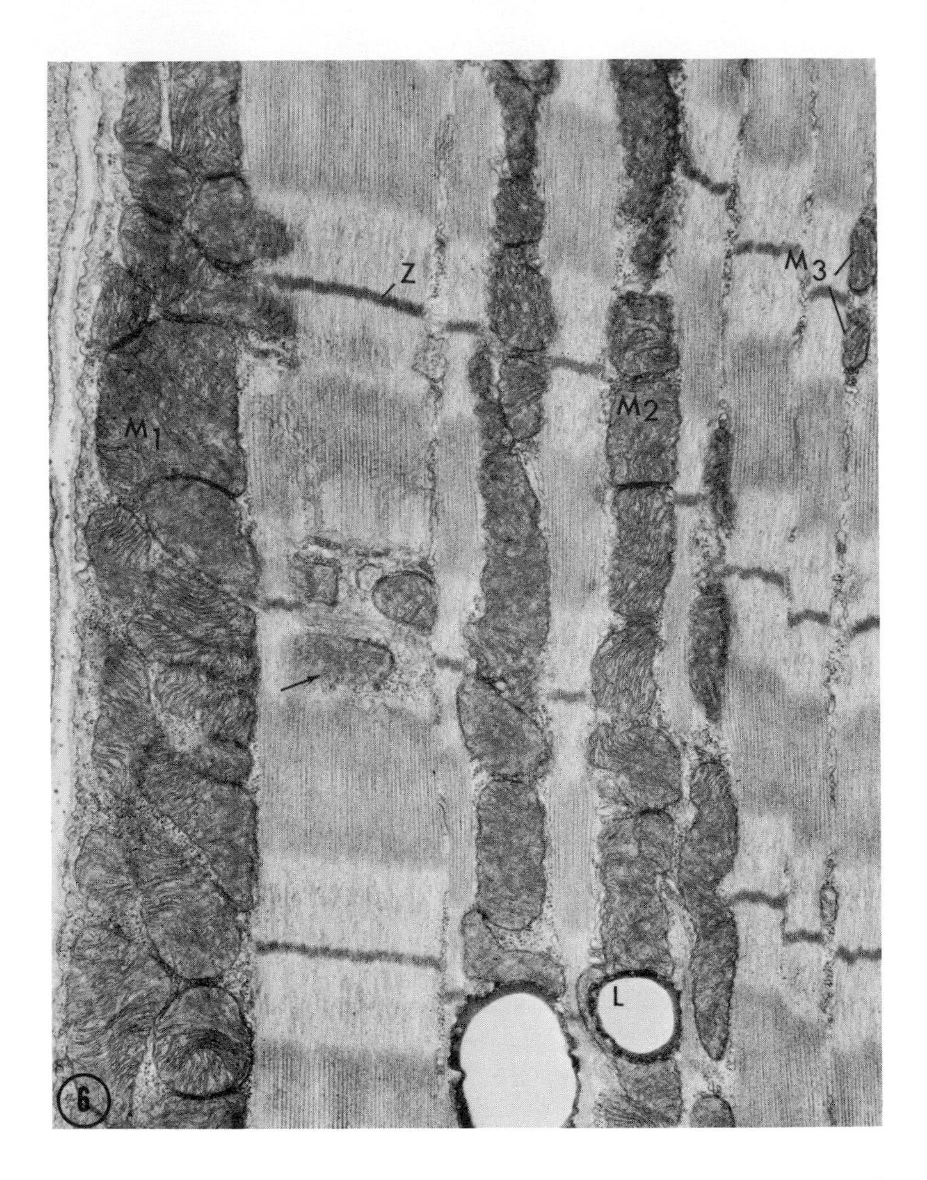


*Fig. 7.6.* Diaphragm, red fiber. This electron micrograph shows the subsarcolemmal region in longitudinal section. Large mitochondria with closely packed cristae form conspicuous aggregations ($M_1$) just beneath the sarcolemma and longitudinal rows ($M_2$) between myofibrils. Lipid droplets (*L*) are closely aligned with these large mitochondria. Paired mitochondria ($M_3$) are located at the I bands. These elliptical profiles are sections through filamentous mitochondria which encircle the myofibrils. The myofibril at the left is sectioned close to the surface so that a portion of one of these paired mitochondria is viewed as it traverses the myofibril (*arrow*). Z lines (*Z*) are wider in the red than in either the intermediate (Fig. 7.7) or the white fiber (Fig. 7.8). × 13,500.

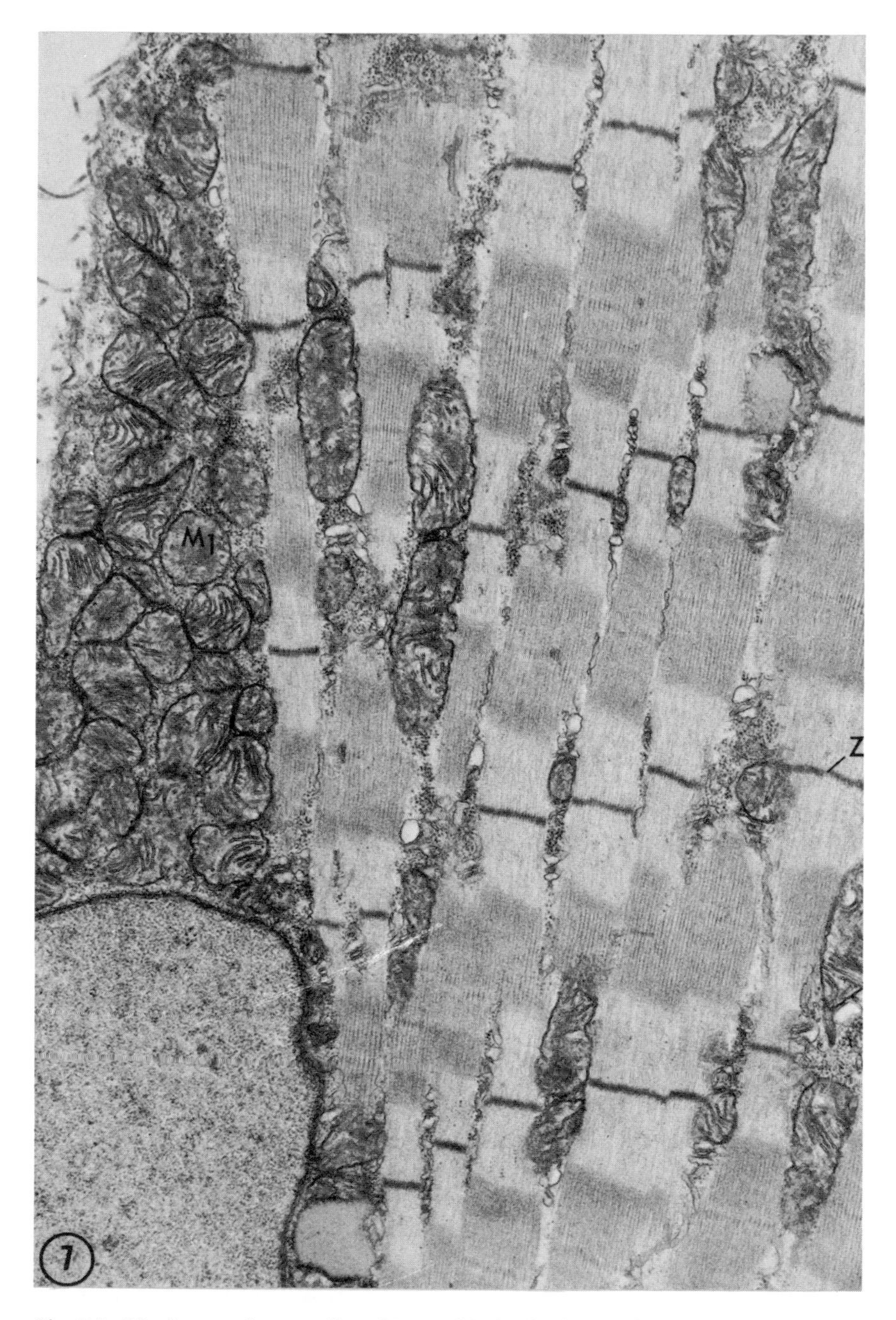


*Fig. 7.7.* Diaphragm, intermediate fiber. This is similar to the red fiber **(Fig. 7.6)** except that mitochondria ($M_1$) tend to be somewhat smaller and their cristae less closely packed than in the red fiber. Z lines (*Z*) are narrower in the intermediate than in the red fiber. × 13,500.

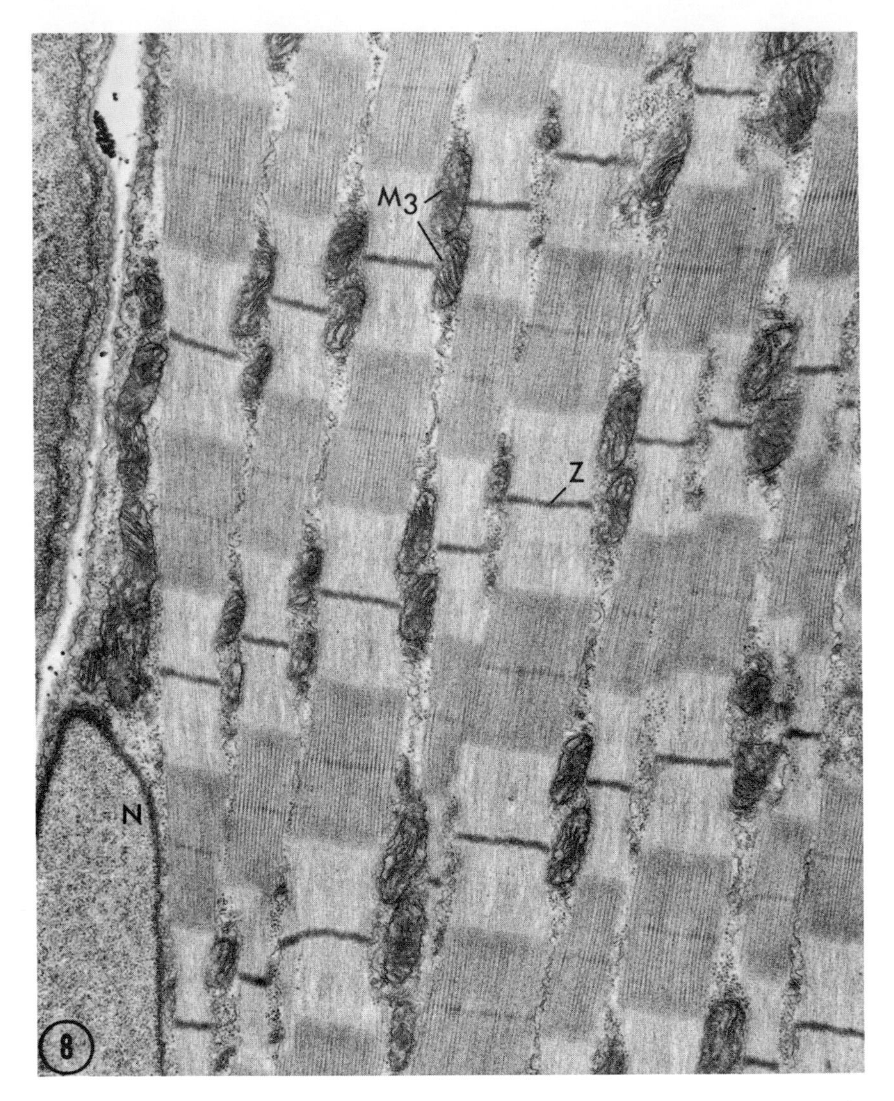


*Fig. 7.8.* Diaphragm, white fiber. Subsarcolemmal mitochondria are sparse, even in the nuclear region (*N*). Mitochondria consist almost entirely of the paired filamentous form at the I bands ($M_3$). Z lines (*Z*) are narrowest in this fiber. Compare with Figs. 7.6–7. × 13,500.

during postnatal development, the muscle fibers of the rat diaphragm are not equivalent to adult red fibers (Bunting, 1969). These differentiating fibers are alike, and they resemble superficially the adult red fiber, but their ultrastructure is distinctive. Even at 8 days, the diameter of the fibers is smaller than that of the red fiber, and the mitochondrial distri-

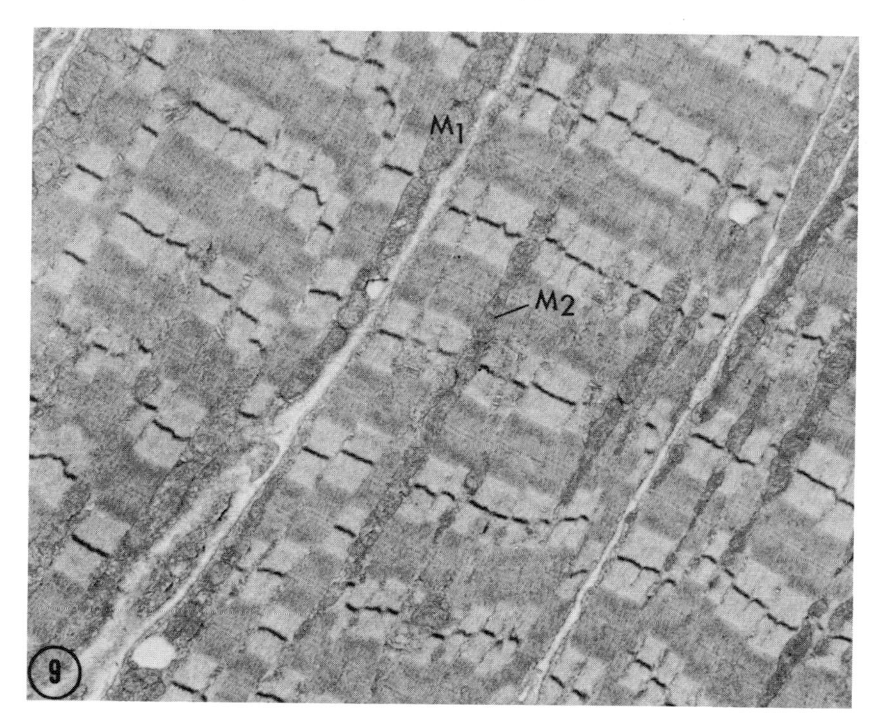


*Fig. 7.9.* Diaphragm, 8 days postnatal. These fibers resemble adult red fibers superficially, but their diameter is smaller than that of the red fiber and their ultrastructure is distinctive. Interfibrillar rows of mitochondria ($M_2$) are present as in the red fiber, but subsarcolemmal mitochondria ($M_1$) are less abundant. Paired mitochondria at the I bands are, for the most part, absent. × 6,500. From L. A. Bunting (1969).

bution is different (Fig. 7.9). Subsarcolemmal aggregations are less conspicuous than in the red fiber, and paired mitochondria at the I bands are sparse or absent. The sarcoplasmic reticulum differs from that of the red fiber also. The entire membrane system consists of an elaborate network extending along the full length of the sarcomere. It is similar to the membrane system described in other developing muscles (Ezerman and Ishikawa, 1967; Walker, Schrodt, and Bingham, 1968; Schiaffino and Margreth, 1969). While these observations do not eliminate the possibility that the red fiber may represent a developing form, they do demonstrate that, when the ultrastructural features of typical developing fibers are analyzed, they cannot be equated to adult red fibers. Furthermore, during later stages of postnatal development, fibers whose mitochondrial content resembles that of red and white fibers appear simultaneously; that is, white fibers do not "follow" red fibers. Also, these cytological

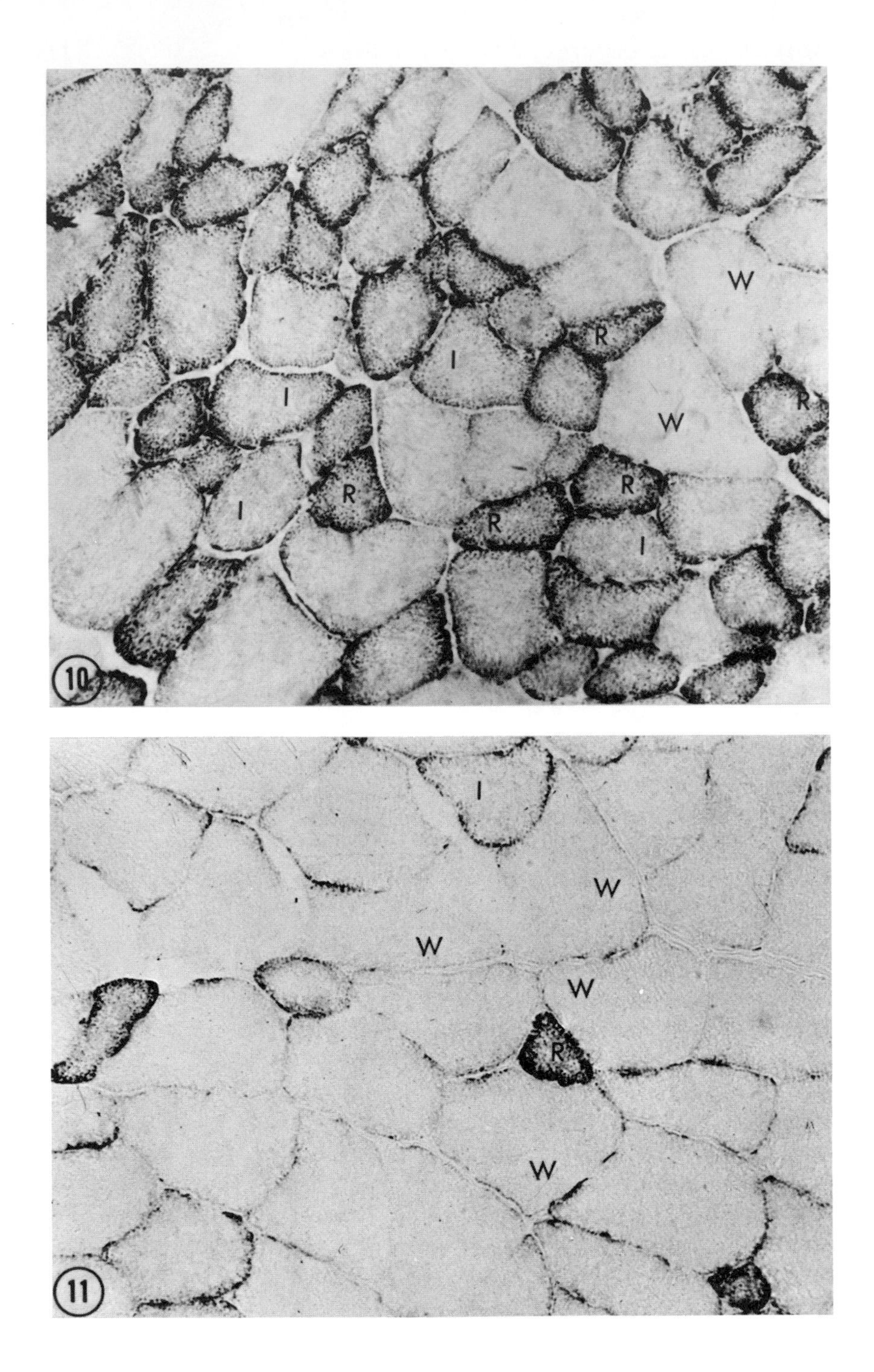
10
W
R
I
I
R
W
R
I
R
R
I
11
I
W
W
W
R
W

*Fig. 7.10.* Semitendinosus, succinic dehydrogenase. Red band. This region of the muscle consists primarily of fibers rich in mitochondrial enzymic activity, especially at the periphery of the fibers. There are numerous small red (*R*) and intermediate fibers (*I*) and a few white fibers (*W*) . × 225.

*Fig. 7.11.* Semitendinosus, succinic dehydrogenase. White band. This region consists primarily of fibers with a low mitochondrial content (white fibers) . Only a few red and intermediate fibers are present. Fibers are similar to corresponding fiber types in the red band (*R, I, W*) , except that white fibers tend to be even larger in the white than in the red region. × 225.

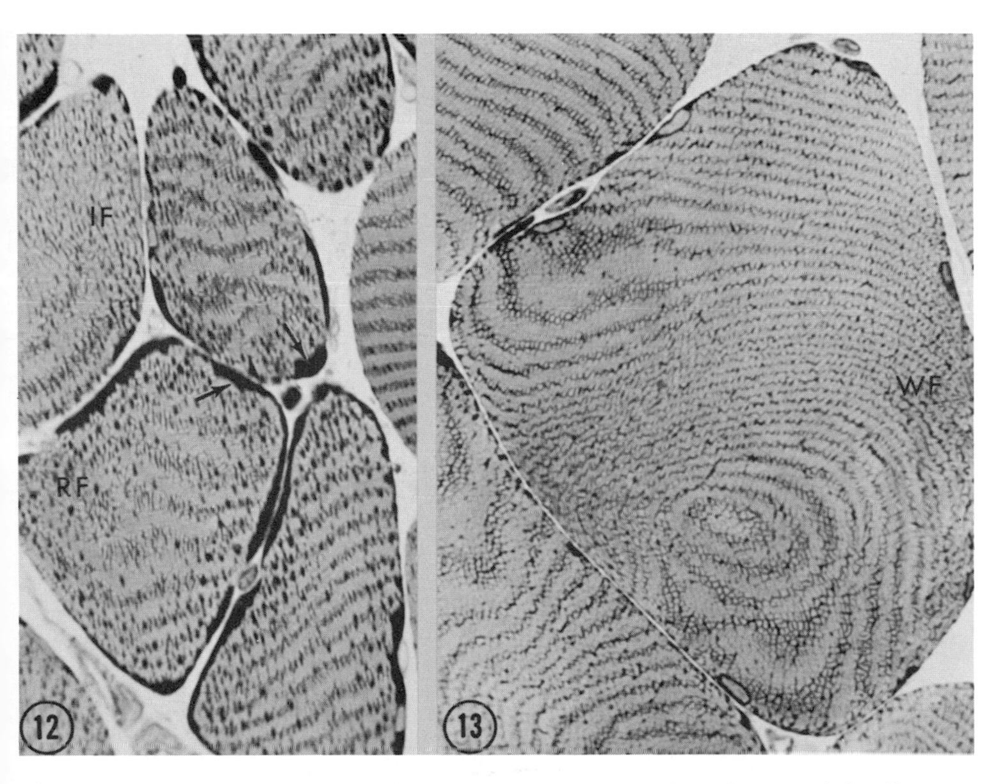


*Fig. 7.12.* Semitendinosus, toluidine blue. Transverse section. Red band. Most of the fibers are relatively small and rich in mitochondria (*RF, IF*) , particularly in the subsarcolemmal regions (*arrows*) . × 760. From G. F. Gauthier (1969) .

*Fig. 7.13.* Semitendinosus, toluidine blue. Transverse section. White band. A single white fiber (*WF*) is representative of the predominant type in this region. Mitochondria are less abundant even in the subsarcolemmal region. Filamentous profiles of paired I-band mitochondria are the major form. × 760. From G. F. Gauthier (1969) .

differences occur before there is any difference in fiber diameter. These findings do not, consequently, support the notion that the mitochondrial content of the white fiber reflects a "dilution" (Goldspink, 1962*b*) of a preexisting mitochondrial population in a red fiber. They do, however, support the concept that there are three distinctive fiber types in the skeletal muscles of mammals.

14
M1
M2
M3
Z

*Fig. 7.14.* Semitendinosus, red fiber. In this electron micrograph, the appearance is similar to that of the red fiber in the diaphragm (Fig. 7.6). Subsarcolemmal aggregations of mitochondria ($M_1$) consist of large, closely packed mitochondria with abundant cristae. Similar mitochondria are arranged in longitudinal rows ($M_2$) between myofibrils. Paired mitochondria encircle myofibrils at the I bands. These are usually seen as elliptical profiles ($M_3$) on opposite sides of the Z line. As in the diaphragm, the Z line ($Z$) is widest in the red fiber. × 17,200.

## THE RELATIONSHIP TO COLOR

A relationship can be demonstrated between the ultrastructural and cytochemical features of the three fiber types and the color of a muscle (Gauthier, 1969). The semitendinosus muscle of the rat is divided into a red band and a white band which can be seen with the naked eye. By taking advantage of this clear separation, it is possible to analyze the cytological composition of red and white regions within a single muscle. Using the histochemical demonstration of succinic dehydrogenase activity as a measure of mitochondrial content, it can be demonstrated that the red portion of the muscle (Fig. 7.10) consists primarily of fibers rich in mitochondria (red and intermediate fibers), while the white portion (Fig. 7.11) consists almost entirely of fibers with a low mitochondrial content (white fibers). Cytological features (Figs. 7.12–13) are similar to those of equivalent fibers in the diaphragm.

In general, the fibers of the semitendinosus are larger in cross section (Table 7.2) than those of the diaphragm (Table 7.1), but in each case the red fibers are the smallest and the white fibers the largest. Diameters of corresponding fiber types in the two regions of the semitendinosus are close except that the white fiber is even larger in the white region than in the red. Red and intermediate fibers are numerous in the red band, and together they constitute 92% of the population. In the white band, however, there is a total of only 18% of these two types. Because the most numerous fibers in the red region also have the smallest diameters, their contribution to the total population of that region is more accurately represented by a measurement of the percentage of the cross-sectional area occupied. Thus 88% of the red band is occupied by red and intermediate fibers, and only 9% of the white band is occupied by these two types (Table 7.2). Red and intermediate fibers contribute 44% each to the cross-sectional area of the red band. Redness in the semitendinosus, therefore, is reflected in an equal contribution of both red and intermediate fibers.

The ultrastructural appearance of the three fibers composing this muscle is similar to that observed in the three corresponding fibers of the

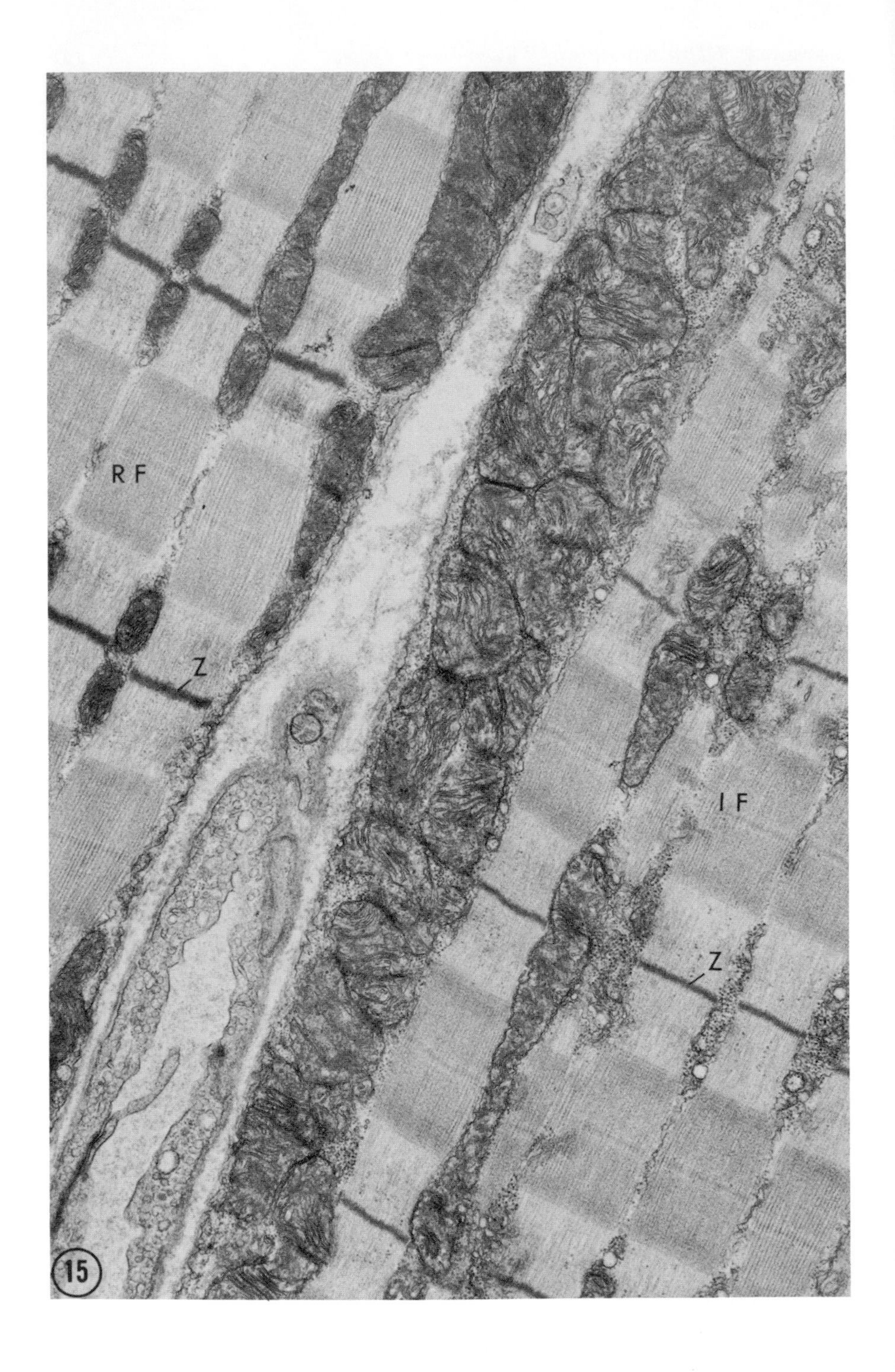
RF
Z
IF
Z
15

*Fig. 7.15.* Semitendinosus, intermediate fiber adjacent to a red fiber included in the same section. The fibers are similar except that mitochondrial cristae are less closely packed in the intermediate (*IF*) than in the red fiber (*RF*). Z lines (*Z*) are noticeably thinner in the intermediate than in the red fiber. × 17,200.

diaphragm, which suggests that they are equivalent. In the red fiber (Fig. 7.14), large mitochondria with closely packed cristae are abundant beneath the sarcolemma and in longitudinal interfibrillar rows. Additional paired mitochondria are present at the I bands. The intermediate fiber (Fig. 7.15) is similar except that subsarcolemmal and interfibrillar mitochondria tend to be smaller, and they have fewer cristae. Z lines are clearly thinner in the intermediate than in the red fiber. In the white fiber (Fig. 7.16), mitochondria are sparse; usually there are no subsarcolemmal aggregations or interfibrillar rows; and the paired mitochondria at the I bands are the predominant type. Z lines are narrowest in the white fiber. Differences, therefore, in mitochondrial content, width of the Z line, and in the sarcoplasmic reticulum again make it possible to recognize three fiber types. The relationship between these features and the color of the muscle supports the use of the terms red, white, and intermediate as simple designations for the three fiber types in mammalian skeletal muscles. It is important, furthermore, that both red and intermediate fibers can contribute to the redness of a muscle.

Mitochondrial content, in particular, is an appropriate criterion for the classification of fiber types. Cytochemical methods used to demonstrate mitochondria are relatively easy to perform; the results are reproducible and readily interpreted; and reactive sites correspond to specific structures seen with the electron microscope. Fibers with abundant mitochondria correspond, furthermore, to the "granular" or "protoplasm-rich" fibers of the older literature (e.g., Grützner, 1884; Knoll, 1891). The identification of the red fiber, therefore, is based for the most part on light microscopic criteria that are similar to those used in earlier studies. Chemical measurements likewise indicate that oxidative enzymes are higher in red than in white muscles (Green, 1951), and that the amount of succinic dehydrogenase, in particular, varies directly with the percentage of red fibers present in the muscle (Beatty, Basinger, Dully, and Bocek, 1966; Bocek and Beatty, 1966). In their comparative biochemical and histochemical analyses, Bocek and Beatty (pers. comm.) have designated as "red" fibers those which, according to our criteria, might be designated "red" or "intermediate"; thus chemical as well as histochemical data suggest that both red and intermediate fibers contrib-

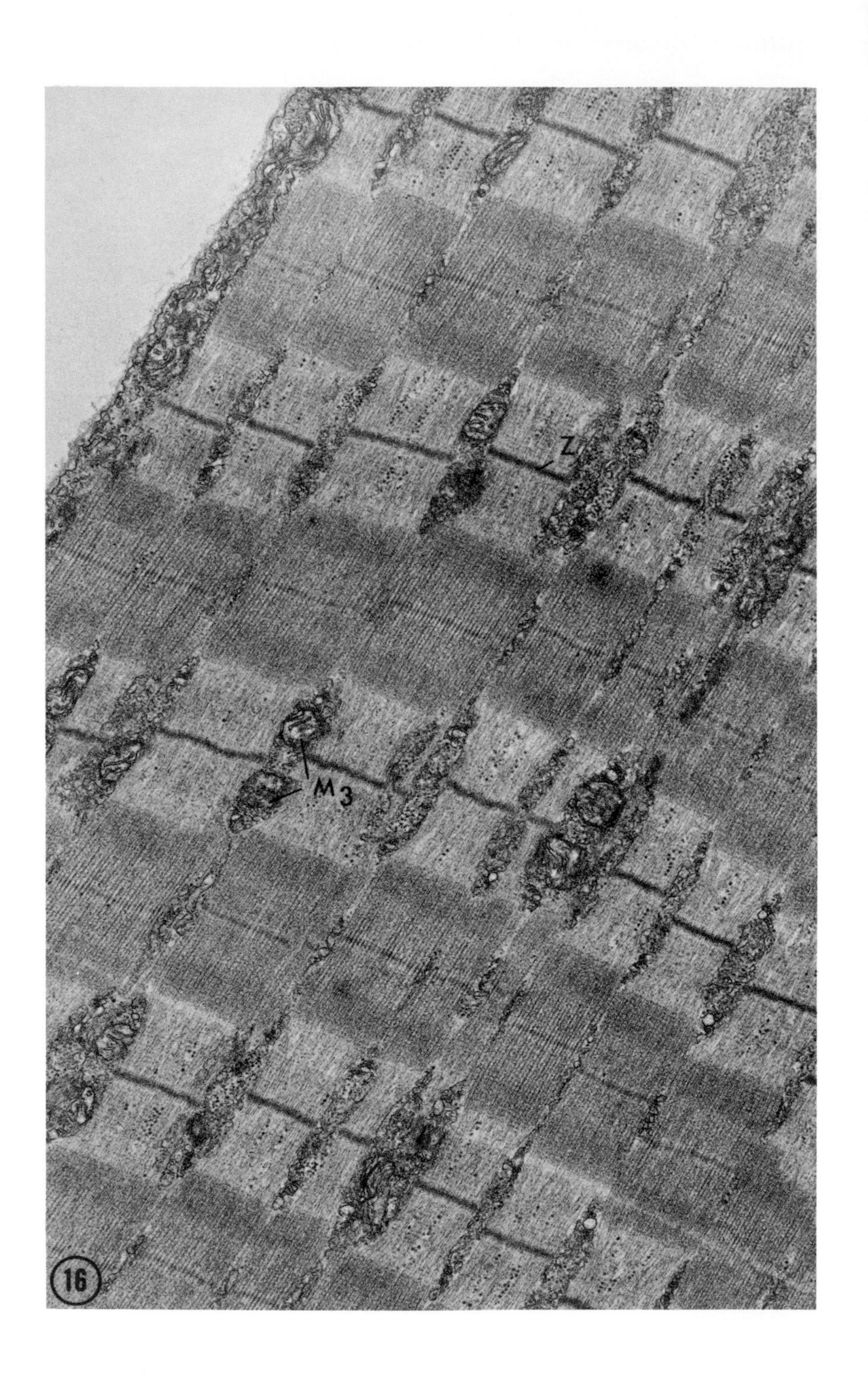
Z
M3
16

*Fig. 7.16.* Semitendinosus, white fiber. Subsarcolemmal mitochondria as well as their cristae are sparse. Paired mitochondria ($M_3$) at the I bands predominate. Z lines (Z) are narrower than in either the intermediate or red fiber (Figs. 7.14 and 7.15). × 17,200.

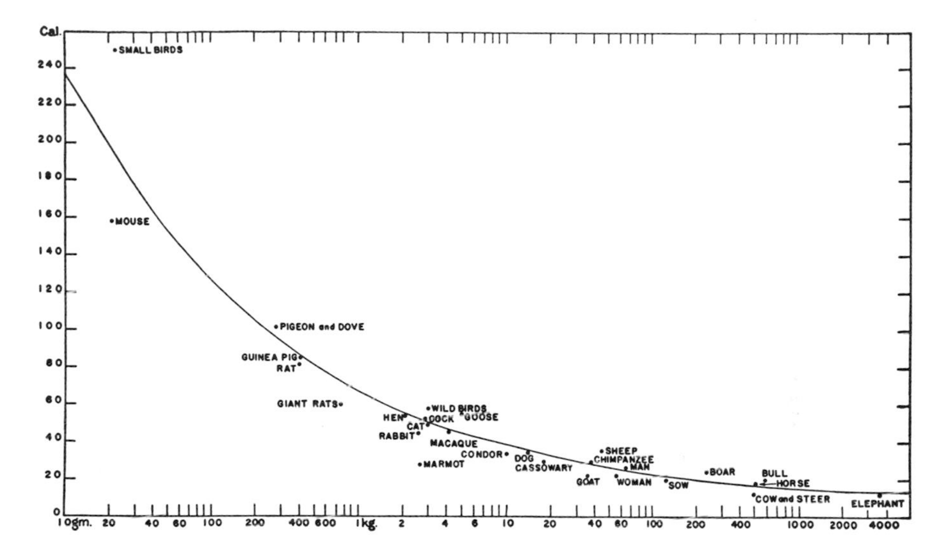


*Fig. 7.17.* Semilogarithmic graph showing the inverse relationship between heat production per kg of body weight and body size in mammals and birds. From F. G. Benedict (1938).

ute to the redness of a muscle, and, in both cases, mitochondrial content is a measure of redness.

Other classifications of muscle fiber types include numerical and alphabetical designations (e.g., Dubowitz and Pearse, 1960; Engel, 1962; Stein and Padykula, 1962), but they are somewhat confusing. Use of the designations I and II, for example, does not provide for an intermediate fiber, and classification as A, B, or C can be confused with similar neurological terminology. An older classification (Krüger, 1929) is based on the configuration of myofibrils (Fibrillenstruktur and Felderstruktur). Though perhaps the most descriptive, this nomenclature involves difficulties of interpretation arising from plane of section. The limitations of the terms "red" and "white" to designate fiber types have been pointed out (Miller, 1967; Schmalbruch, 1967; James, 1968). Nevertheless, this terminology is both descriptive and simple, and it has, as its basis, a relationship to the color of the whole muscle.

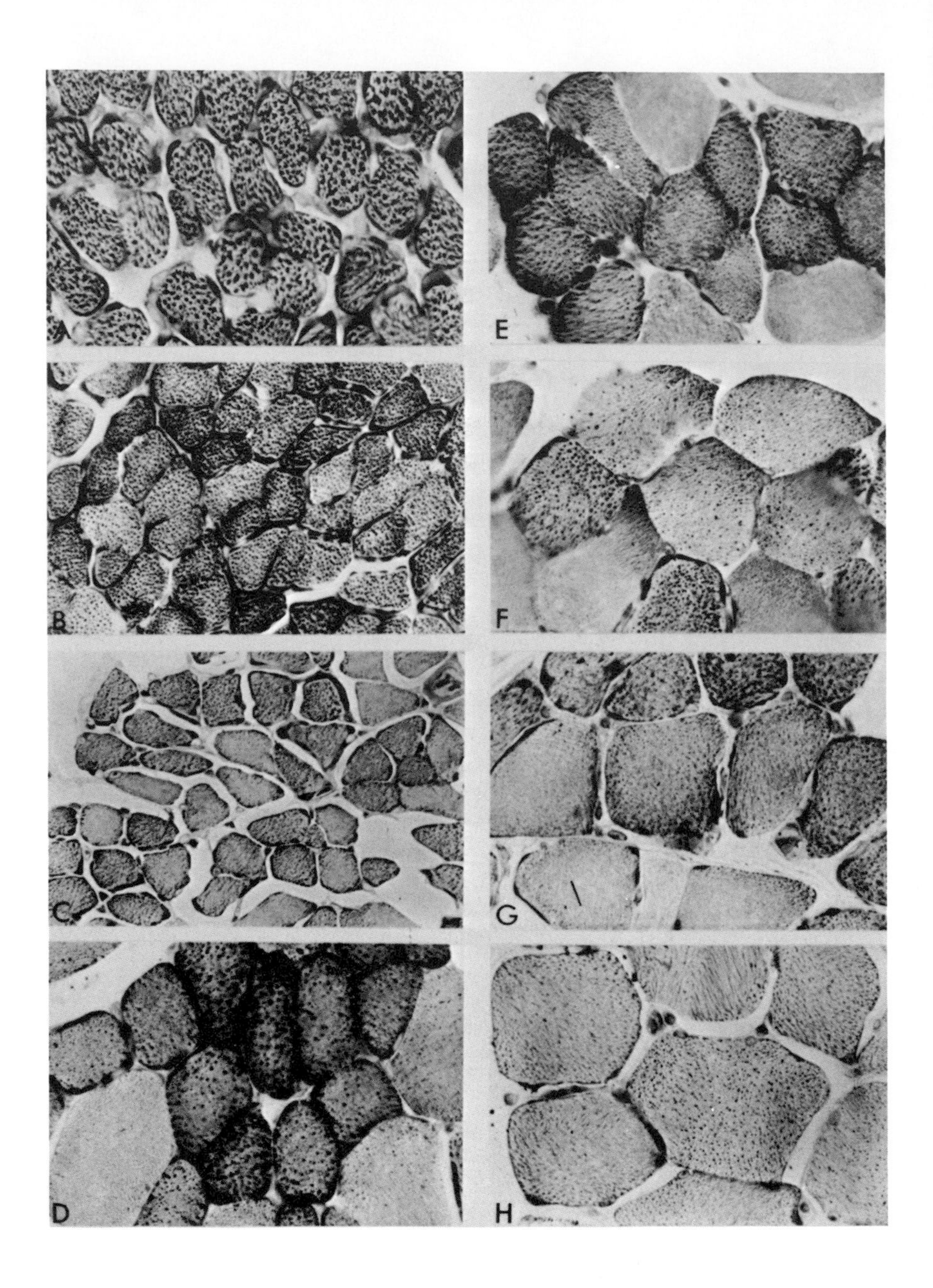


*Fig. 7.18.* Diaphragm, Sudan black. Transverse sections photographed at the same magnification and arranged in order of increasing body size. *A,* harvest mouse. *B,* shrew. *C,* albino mouse. *D,* albino rat. *E,* squirrel. *F,* cat. *G,* man. *H,* cow. In the smallest mammals, fibers are uniformly small and rich in mitochondria. As body size increases, fiber diameter increases, and the fiber population becomes heterogeneous. In the largest mammal, all fibers are large and have a low mitochondrial content. × 320. From G. F. Gauthier and H. A. Padykula (1966).

THE RELATIONSHIP TO FUNCTION

Most measurements of physiological properties of skeletal muscles are made on whole muscles, and in the mammal at least it is difficult to relate these measurements to functional differences in individual muscle fibers. We have approached this problem indirectly by studying the diaphragm, a muscle which varies quantitatively according to body size of the animal. Metabolic rate is inversely related to body size in mammals (Fig. 7.17). As body size increases, heat production decreases (Benedict, 1938). Likewise, breathing rate is inversely related to body size (Crosfill and Widdicombe, 1961). It was suggested (Krebs, 1950) that the skeletal musculature is most likely responsible for the reciprocal relationship between metabolic rate and body size; and a direct relationship exists between the respiratory rate of diaphragm muscle *in vitro* and body size of the animal (Bertalanffy and Pirozynski, 1953). It therefore seemed possible that these quantitative differences might be reflected in the microscopic appearance of the diaphragm. Accordingly, the diaphragms of 36 different mammalian species were examined cytochemically (Gauthier and Padykula, 1966), and it was demonstrated that the fiber composition of the diaphragm is similarly related to body size (Fig. 7.18). The diaphragm of a small mammal is composed of small fibers rich in mitochondria (red fibers), and the ultrastructural features of these fibers resemble closely those of the red fiber of the rat diaphragm described earlier. The diaphragm of a large mammal, on the other hand, consists almost entirely of large white fibers, which have ultrastructural features similar to those of the white fiber of the rat diaphragm. Mammals of intermediate body size have diaphragms which contain mixtures of fiber types. The red fiber, therefore, reflects a higher metabolic rate and faster breathing rate than does the white fiber; and this perhaps reflects also a greater speed of contraction of the diaphragm muscle.

These data appear to contradict physiological data, which tend to relate the "redness" of a muscle to "slowness" of contraction. The conclusion that the red fiber is a slow fiber must be challenged, however, for there are other examples of relatively "fast" muscles which have the ultrastructural (e.g., Revel, 1962) and cytochemical (e.g., Hall-Craggs, 1968) characteristics of typical red fibers. It is important, moreover, to note that the soleus, a muscle commonly used in physiological analyses, is composed, as judged by light microscopic criteria, primarily of fibers which resemble intermediate fibers in the rat (Stein and Padykula, 1962) and entirely of this fiber type in the cat (Henneman and Olson, 1965; Nyström, 1968). Hence, in this particular muscle, "slowness" is reflected in a high content of intermediate, not red, fibers. The relationship between fiber type and speed of contraction is, therefore, not a simple

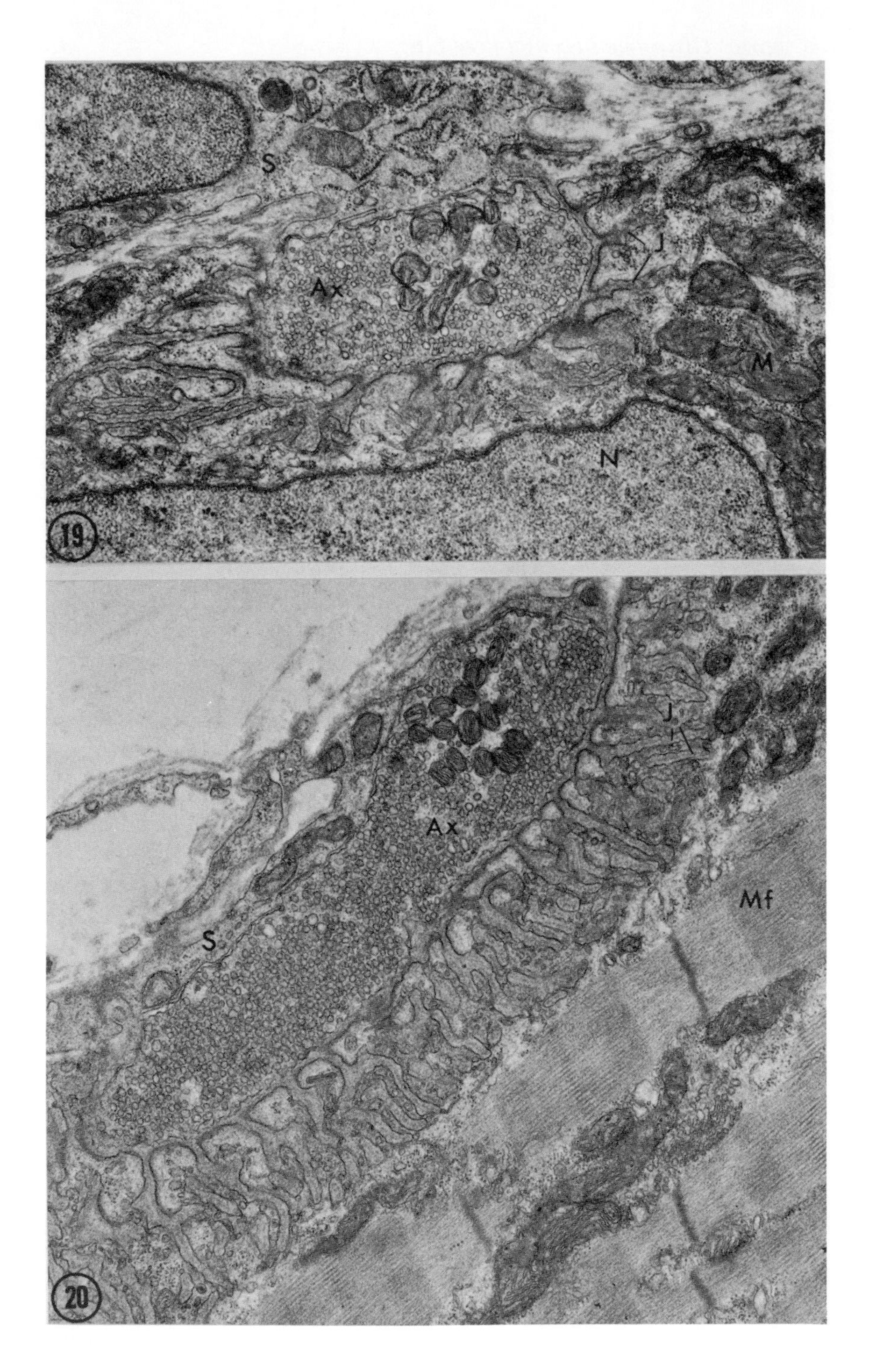
S
J
Ax
M
N
19
J
Ax
Mf
S
20

*Fig. 7.19.* Diaphragm. Neuromuscular junction, red fiber. The axonal ending (*Ax*) is more or less elliptical and contains moderate numbers of vesicles. The sarcoplasmic surface is invaginated to form a few relatively short junctional folds (*J*). Part of a Schwann cell (*S*), with its nucleus, is present in the upper left, and caps the axonal ending. A muscle nucleus (*N*) and numerous subsarcolemmal mitochondria (*M*) are located just below the axonal ending. × 17,500.

*Fig. 7.20.* Diaphragm, neuromuscular junction, white fiber. The axonal ending (*Ax*) is flat and more elongated than in the red fiber (Fig. 7.19). Axonal vesicles are closely packed, and juctional folds (*J*) are longer and more numerous than in the red fiber. A portion of a Schwann cell (*S*) is visible above the axonal ending, and myofibrils (*Mf*) are located below. × 17,500.

one, and other factors, such as duration of contraction, muscle tension, and metabolic rate of the animal must be considered when making such an analysis. Clearly, however, the red fiber is "equipped" for a high level of activity. Oxidative machinery (mitochondria) is concentrated at the cell surface and is intimately associated with the contractile apparatus and with an available source of metabolic fuel (triglyceride droplets). The diameter of the fiber is small, and thus an ample surface-to-volume ratio facilitates metabolic exchange. Surface area of the sarcoplasmic reticulum is enhanced by an elaborate configuration of tubules in the region of the H band, and this feature, in particular, is most likely related to speed of contraction.

The relationship of the nervous system to the properties of a whole muscle and the distribution of fiber types is most dramatically demonstrated in experiments with cross innervation. Physiological properties of fast and slow muscles are, to a great extent, reversed by cross innervation, as demonstrated by Buller, Eccles, and Eccles (1960) in the cat and by Close (1965) in the rat. The biochemical composition (Prewitt and Salafsky, 1967; Buller, Mommaerts, and Seraydarian, 1969; Mommaerts, Buller, and Seraydarian, 1969; Mommaerts, 1970) and the histological pattern of fiber types (Dubowitz, 1967; Romanul and Van Der Meulen, 1967; Yellin, 1967) are similarly altered. Evidence for a direct relationship between the motoneurone and a specific type of muscle fiber is found in recent physiological studies of motor units. The distribution of motor units suggests a relationship to the pattern of fiber types in cat muscles (Henneman and Olson, 1965). In fact, three distinct types of motor units have been demonstrated in hindlimb muscles of the rat (Close, 1967*a, b*). It is probable, moreover, that each motor unit is composed of one of three types of muscle fibers, and this is supported by a correlated physiological and histochemical analysis, in which single motor units were stimulated. Stimulation of a given motor nerve brought about a selective histochemical response in one type of muscle fiber only (Edström and Kugelberg, 1968). It is not surprising, therefore, that red, white, and intermediate fibers exhibit ultrastructural differences at their neuromuscular junctions (Padykula and Gauthier, 1970). In the red fiber (Fig. 7.19), axonal endings are elliptical, and they contain only moderate numbers of vesicles. The sarcoplasmic surface is invaginated to form a few relatively short junctional folds. In the white fiber (Fig. 7.20), axonal endings are flat and elongated, axonal vesicles are closely packed, and junctional folds are longer and more numerous than in the red fiber. The neuromuscular junction of the intermediate fiber has characteristics which resemble each of the other two fiber types. The axonal ending is wider than that of the red fiber, however, and perhaps even longer than that of the white fiber. Junctional folds are less closely spaced than in the red fiber, yet even longer than in the white fiber. The appearance of the neuromuscular junction is thus another important criterion which distinguishes the three fiber types at the ultrastructural level. The role of the neuromuscular junction in the pattern of fiber types is suggested by preliminary studies which indicate that, during postnatal development of the diaphragm, neuromuscular junctions exhibit differences (Fitts, 1969) before distinct types of muscle fibers appear (Bunting, 1969).

## CONCLUSIONS

Three distinct types of skeletal muscle fibers can be identified consistently on the basis of mitochondrial content, width of the Z line, and configuration of the sarcoplasmic reticulum. The relationship of these features to the color of the whole muscle justifies the simple and descriptive terms "red," "white," and "intermediate" to designate the fibers. Ultrastructural differences in the neuromuscular junctions further distinguish the three fiber types and show, moreover, that structural differences exist, not only in the muscle fibers themselves but also in the motoneurones serving them. The individuality of the three muscle fibers and of their neuromuscular relationships is consistent with physiological evidence for three separate types of motor units.

## ACKNOWLEDGMENTS

The author is grateful to Dr. Helen A. Padykula for encouragement and criticism during the course of the studies represented in this report.

These investigations were supported by grants from the United States Public Health Service (HD–01026) and the Muscular Dystrophy Associations of America, Inc.

## *References*

Beatty, C. H., G. M. Basinger, C. C. Dully, and R. M. Bocek. 1966. Comparison of red and white voluntary skeletal muscles of several species of primates. *J. Histochem. Cytochem. 14*:590.

Benedict, F. G. 1938. *Vital energetics. A study in comparative basal metabolism.* Carnegie Inst. Washington Publ. No. 503.

Bertalanffy, I. von, and W. J. Pirozynski. 1953. Tissue respiration, growth, and basal metabolism. *Biol. Bull. 105*:240.

Bocek, R. M., and C. H. Beatty. 1966. Glycogen synthetase and phosphorylase in red and white muscle of rat and rhesus monkey. *J. Histochem. Cytochem. 14*:549.

Bullard, H. H. 1919. Histological as related to physiological and chemical differences in certain muscles of the cat. *The Johns Hopkins Hospital Rep. 18*:323.

Buller, A. J., J. C. Eccles, and R. M. Eccles. 1960. Interactions between motoneurones and muscles in respect of the characteristic speeds of their responses. *J. Physiol. 150*:417.

Buller, A. J., W. F. H. M. Mommaerts, and K. Seraydarian. 1969. Enzymic properties of myosin in fast and slow twitch muscles of the cat following cross-innervation. *J. Physiol. 205*:581.

Bunting, L. A. 1969. Cytological features of striated muscle fibers during postnatal differentiation. Honors thesis, Department of Biological Sciences, Wellesley College.

Ciaccio, V. 1898. La découverte des muscles blancs et des muscles rouges, chez le lapin, revendiquée en taveur de S. Lorenzini. *Arch. Ital. Biol. 30*:287.

Close, R. 1965. Effects of cross-union of motor nerves to fast and slow skeletal muscles. *Nature 206*:831.

———. 1967*a*. Dynamic properties of fast and slow skeletal muscles of mammals, p. 142. *In* A. T. Milhorat (ed.), *Exploratory Concepts in Muscular Dystrophy and Related Disorders.* Excerpta Medica Foundation, Amsterdam.

———. 1967*b*. Properties of motor units in fast and slow skeletal muscles of the rat. *J. Physiol. 193*:45.

Coimbra, A. 1969. Radioautographic studies of glycogen synthesis in the striated muscle of rat tongue. *Amer. J. Anat. 124*:377.

Crosfill, M. L., and J. G. Widdicombe. 1961. Physical characteristics of the chest and lungs and the work of breathing in different mammalian species. *J. Physiol. 158*:1.

Dubowitz, V. 1967. Cross-innervated mammalian skeletal muscle: histochemical, physiological and biochemical observations. *J. Physiol. 193*:481.

Dubowitz, V., and A. G. E. Pearse, 1960. A comparative histochemical study of oxidative enzyme and phosphorylase activity in skeletal muscle. *Histochemie 2*:105.

Edström, L., and E. Kugelberg. 1968. Histochemical composition, distribution of fibres and fatiguability of single motor units. Anterior tibial muscle of the rat. *J. Neurol. Neurosurg. Psychiat. 31*:424.

Engel, W. K. 1962. The essentiality of histo- and cytochemical studies of skeletal muscles in the investigation of neuromuscular disease. *Neurology 12*:778.

Ezerman, E. B., and H. Ishikawa. 1967. Differentiation of the sarcoplasmic reticulum and T system in developing chick skeletal muscle *in vitro*. *J. Cell Biol. 35*:405.

Fitts, R. N. 1969. Cytological studies of the differentiation of motor endplates and of the phrenic nerve in the diaphragm of the albino rat. Honors thesis, Department of Biological Sciences, Wellesley College.

Forssmann, W. G., and A. Matter. 1966. Ultrastruktureller Nachweis von zwei Myofibrillentypen in den Muskelfasern der Rattenzwerchfells. *Experientia 22*:816.

Gauthier, G. F. 1967. On the localization of sarcotubular ATPase activity in mammalian skeletal muscle. *Histochemie 11*:97.

———. 1969. On the relationship of ultrastructural and cytochemical features to color in mammalian skeletal muscle. *Z. Zellforsch. 95*:462.

Gauthier, G. F., and H. A. Padykula. 1966. Cytological studies of fiber types in skeletal muscle. A comparative study of the mammalian diaphragm. *J. Cell Biol. 28*:333.

Goldspink, G. 1962*a*. Studies of postembryonic growth and development of skeletal muscle. *Proc. Roy. Irish Acad. 62*:135.

———. 1962*b*. Biochemical and physiological changes associated with the postnatal development of the biceps brachii. *Comp. Biochem. Physiol. 7*:157.

Goldspink, G., and R. W. D. Rowe. 1968. Studies on postembryonic growth and development of skeletal muscle II. Some physiological and structural changes that are associated with the growth and development of skeletal muscle fibres. *Proc. Roy. Irish Acad. 66*:85.

Greene, D. E. 1951. The cyclophorase system, p. 15. *In* J. T. Edsall (ed.), *Enzymes and Enzyme Systems*. Harvard Univ.

Grützner, P. 1884. Zur Anatomie und Physiologie der quergestreiften Muskeln. *Rec. Zool. Suisse 1*:665.

Gustafsson, R., J. R. Tata, O. Lindberg, and L. Ernster. 1965. The relationship between the structure and activity of rat skeletal muscle mitochondria after thyroidectomy and thyroid hormone treatment. *J. Cell Biol. 26*:555.

Hall-Craggs, E. C. B. 1968. The contraction times and enzyme activity of two rabbit laryngeal muscles. *J. Anat. 102*:241.

Henneman, E., and C. B. Olson. 1965. Relations between structure and function in the design of skeletal muscles. *J. Neurophysiol. 28*:581.

James, N. T. 1968. Histochemical demonstration of myoglobin in skeletal muscle fibres and muscle spindles. *Nature 219*:1174.

Knoll, P. 1891. Über protoplasmaarme und protoplasmareiche Muskulatur. *Denkschr. Akad. Wiss. Wien 58*:633.

Krebs, H. A. 1950. Body size and tissue respiration. *Biochim. Biophys. Acta 4*:249.

Krüger, P. 1929. Über einen möglichen Zusammenhang zwischen Struktur Funktion, and chemischer Beschaffenheit der Muskeln. *Biol. Zbl. 49*:616.

Kugelberg, E., and L. Edström, 1968. Differential histochemical effects of muscle contractions on phosphorylase and glycogen in various types of fibres: relation to fatigue. *J. Neurol. Neurosurg. Psychiat. 31*:415.

Miledi, R., and C. R. Slater. 1968. Some mitochondrial changes in denervated muscle. *J. Cell Sci. 3*:49.

Miller, J. E. 1967. Cellular organization of rheusus extraocular muscle. *Invest. Ophthal. 6*:18.

Mommaerts, W. F. H. M. 1970. Role of the innervation of the functional development of muscle. *This volume,* p. 53.

Mommaerts, W. F. H. M., A. J. Buller, and K. Seraydarian. 1969. The modification of some biochemical properties of muscle by cross-innervation. *Proc. Nat. Acad. Sci. 64*:128.

Nachmias, V. T., and H. A. Padykula. 1958. A histochemical study of normal and denervated red and white muscles of the rat. *J. Biophys. Biochem. Cytol. 4*:47.

Nyström, B. 1968. Histochemistry of developing cat muscles. *Acta Neurol. Scand. 44*:405.

Ogata, T. 1958. A histochemical study of the red and white muscle fibers I. Activity of the succinoxydase system in muscle fibers. *Acta Med. Okayama 12*:216.

Padykula, H. A. 1952. The localization of succinic dehydrogenase in tissue sections of the rat. *Amer. J. Anat. 91*:107.

Padykula, H. A., and G. F. Gauthier. 1963. Cytochemical studies of adenosine triphosphatases in skeletal muscle fibers. *J. Cell Biol. 18*:87.

———. 1967*a*. Ultrastructural features of three fiber types in the rat diaphragm. (Abstr.) *Anat. Rec. 157*:296.

———. 1967*b*. Morphological and cytochemical characteristics of fiber types in normal mammalian skeletal muscle, p. 117. *In* A. T. Milhorat (ed.), *Exploratory Concepts in Muscular Dystrophy and Related Disorders.* Excerpta Medica Foundation, Amsterdam.

———. 1970. The ultrastructure of the neuromuscular junctions of mammalian red, white, and intermediate skeletal muscle fibers. *J. Cell Biol.* (In press.)

Pearse, A. G. E. 1960. *Histochemistry. Theoretical and Applied* (2d ed.). Little, Brown, Boston.

Pellegrino, C., and C. Franzini. 1963. An electron microscope study of denervation atrophy in red and white skeletal muscle fibers. *J. Cell Biol. 17*:327.

Prewitt, M. A., and B. Salafsky. 1967. Effect of cross innervation on biochemical characteristics of skeletal muscles. *Amer. J. Physiol. 213*:295.

Ranvier, L. 1874. De quelques faits relatifs à l'histologie et à la physiologie des muscles striés. *Arc. Physiol. Norm. et Path.* 2, Sér. I:5.

Revel, J. P. 1962. The sarcoplasmic reticulum of the bat cricothyroid muscle. *J. Cell Biol. 12*:571.

Romanul, F. C. A., and J. P. Van Der Meulen. 1967. Slow and fast muscles after cross innervation. Enzymatic and physiological changes. *Arch. Neurol. 17*:387.

Rowe, R. W. D., and G. Goldspink. 1969. Muscle fibre growth in five different muscles in both sexes of mice I. Normal mice. *J. Anat. 104*:519.

Schiaffino, S., and A. Margreth. 1969. Coordinated development of the sarcoplasmic reticulum and T system during postnatal differentiation of rat skeletal muscle. *J. Cell Biol. 41*:855.

Schmalbruch, H. 1967. Fasertypen in der Unterschenkelmuskulatur der Maus. *Z. Zellforsch. 79*:64.

Shafiq, S. A., M. A. Gorycki, L. Goldstone, and A. T. Milhorat. 1966. Fine structure of fiber types in normal human muscle. *Anat. Rec. 156*:283.

Shafiq, S. A., M. A. Gorycki, and A. T. Milhorat. 1969. An electron microscope study of fibre types in normal and dystrophic muscles of the mouse. *J. Anat. 105*:281.

Stein, J. M., and H. A. Padykula. 1962. Histochemical classification of individual skeletal muscle fibers of the rat. *Amer. J. Anat. 110*:103.

Tice, L. W., and A. G. Engel. 1967. The effects of glucocorticoids on red and white muscles in the rat. *Amer. J. Path. 50*:311.

Walker, S. M., R. Schrodt, and M. Bingham. 1968. Electron microscope study of the sarcoplasmic reticulum at the Z line level in skeletal muscle fibers of fetal and newborn rats. *J. Cell Biol. 39*:469.

Yellin, H. 1967. Neural regulation of enzymes in muscle fibers of red and white muscle. *Exp. Neurol. 19*:92.

*Chapter 8*

# *Some Comments on Neural Influence on the Two Histochemical Types of Muscle Fibers*

M. H. BROOKE

That different types of muscle fiber exist is a phenomenon which has been known for a considerable time. The differentiation may be made on the grounds of appearance, red muscle being contrasted with white, or on the basis of physiological properties when slowly contracting muscle may be contrasted with fast-contracting muscle. We have chosen the histochemical techniques to separate the two fiber types. Dubowitz (1960) pointed out that there is an inverse relationship between the oxidative enzyme activity in the muscle fiber and phosphorylase activity, as demonstrated histochemically. Those fibers which are rich in oxidative activity and poorly demonstrated with the phosphorylase stain are described as type I fibers. Type II fibers possess the reverse property, namely that the histochemical reaction for oxidative enzyme activity is poorly developed, whereas that for phosphorylase is marked (Figs. 8.1–3). We have preferred to use the calcium method for the demonstration of adenosine triphosphatase (Padykula and Herman, 1955) in order to identify fiber types (Fig. 8.4) because, under pathological conditions, the correlation between oxidative enzyme histochemistry and phosphorylase reactions may be lost. Thus, for example, atrophic fibers in denervation may all appear heavily stained with the oxidative reactions but with the standard adenosine triphosphatase reaction they are differentiated into two distinct fiber types, type I (lightly stained) and type II (heavily stained) (Figs. 8.5 and 8.6). This should not be construed as meaning that the adenosine triphosphatase activity of type II fibers is absolutely higher than that of type I. It is probably a reflection of the *p*H at which the reaction is carried out. The standard histochemical reaction takes place at *p*H 9.4 and the adenosine triphosphatase system responsible for the development of the stain in type I fibers is alkali-labile. Preincubation at *p*H 10.3 prevents development of the reaction in the type I fibers (Figs.

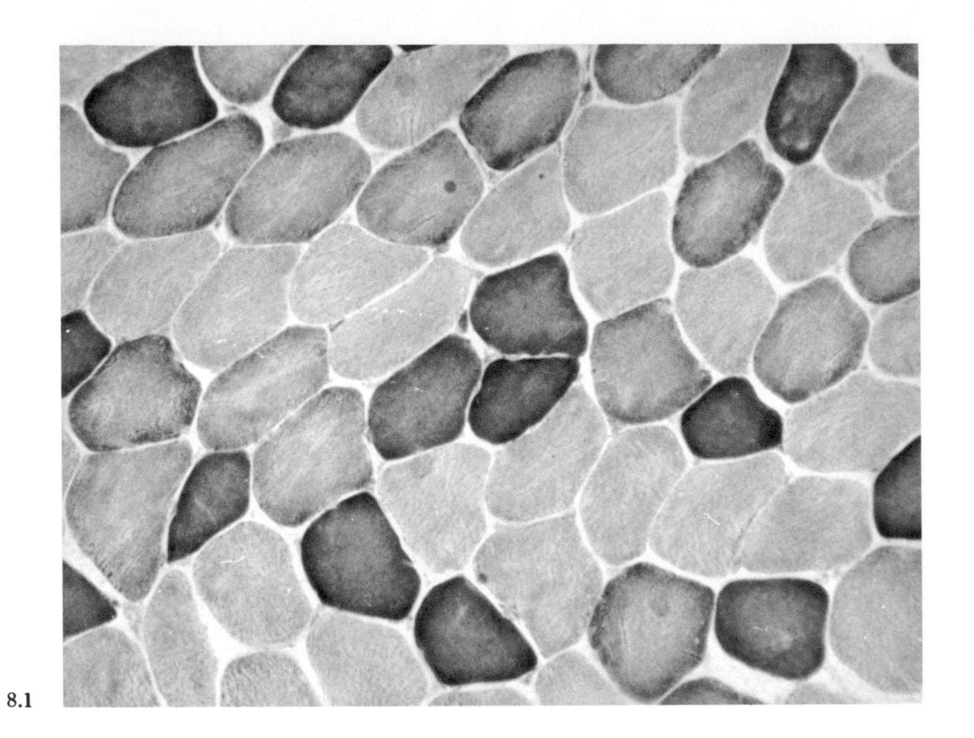

8.1

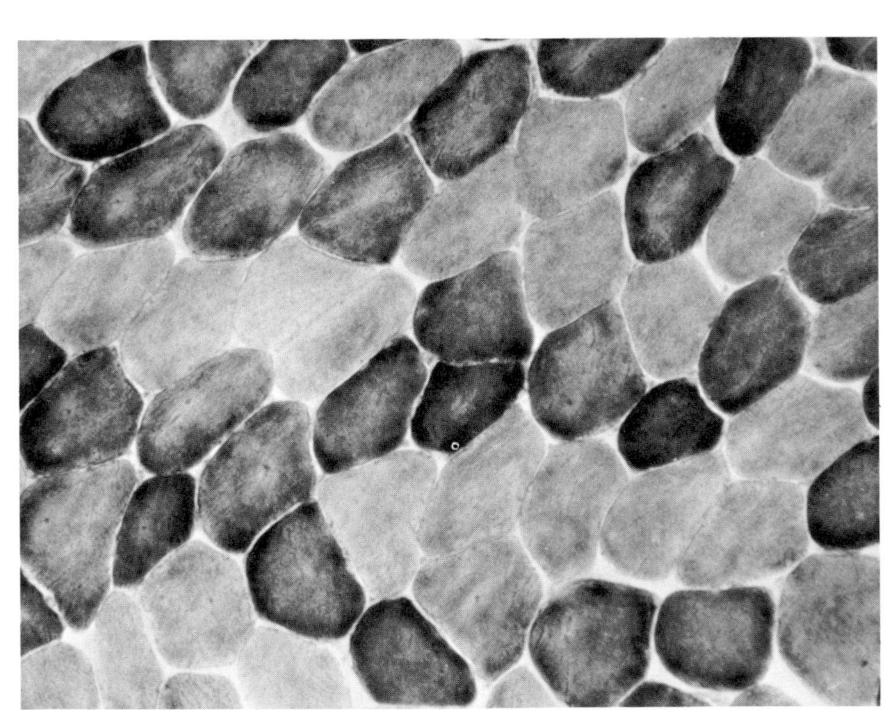

8.2

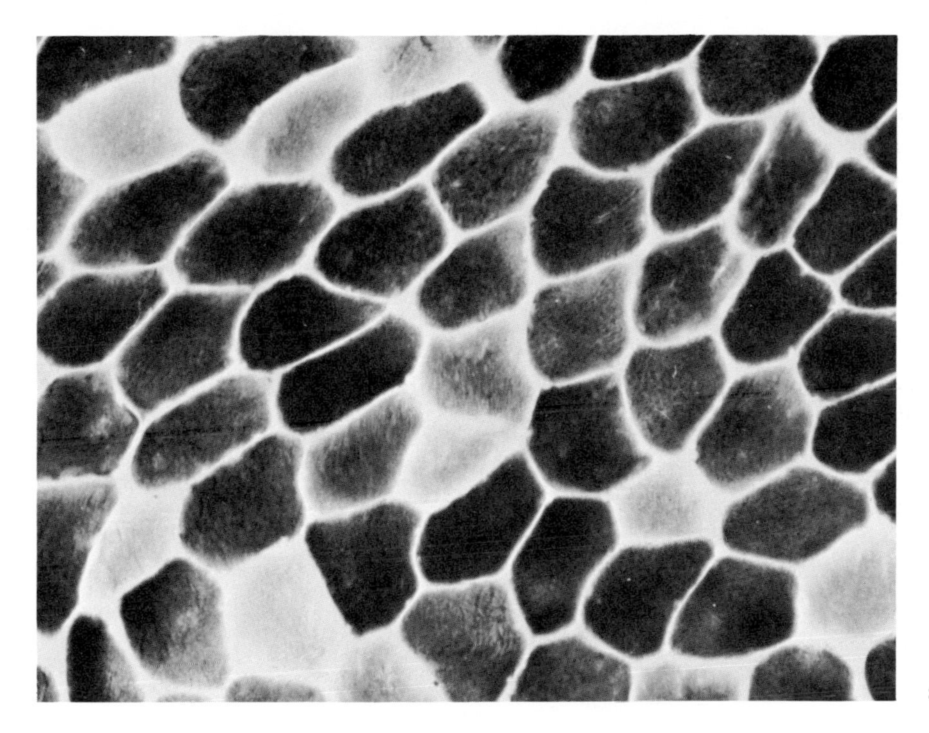

8.3

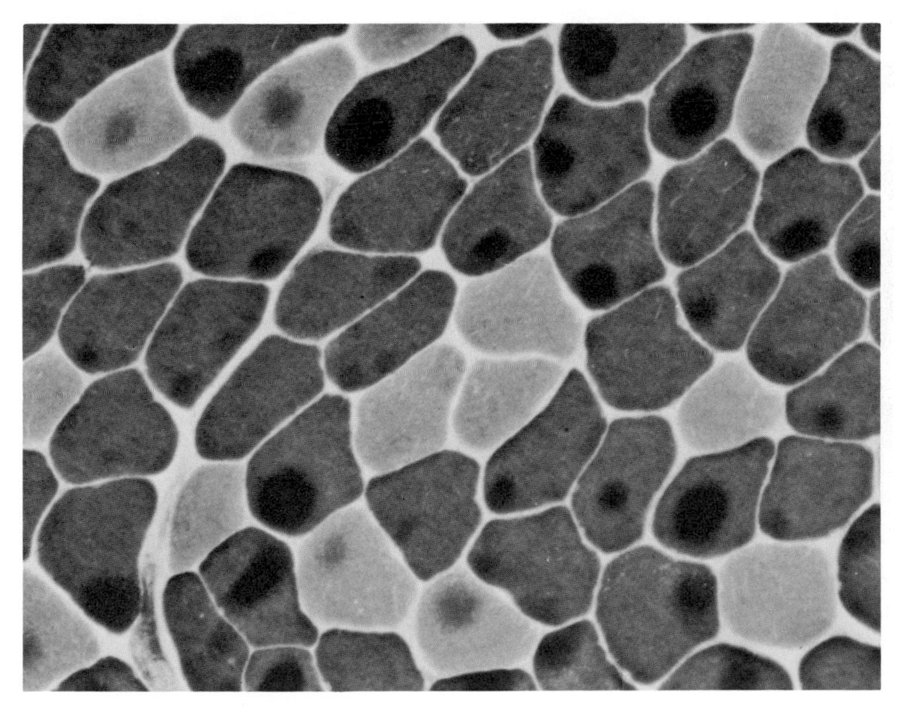

8.4

*Figs. 8.1–4*. Serial sections of a male human muscle biopsy, identifying the same muscle fibers in cross section with the four different histochemical reactions. *Fig. 8.1* shows the reaction for NADH-tetrazolium reductase. *Fig. 8.2* shows the succinic dehydrogenase reaction. With these two reactions, the darkest fibers are type I and the lightest fibers are type II, but there are fibers intermediate between the two. *Fig. 8.3* shows the phosphorylase reaction. In general, the fibers which are light with the reaction for oxidative enzymes are dark with the phosphorylase reactions and vice versa. This is not, however, an absolute correlation as can be seen by examining the three fibers towards the center of the field that are very dark with the NADH diaphorase reaction. Only one of these fibers is light with the phosphorylase stain, the other two being intermediate in staining properties. *Fig. 8.4* shows the routine adenosine triphosphatase reaction at *p*H 9.4. In this reaction only two fiber types are demonstrable in human muscle and the differentiation is always very clear. Consequently, we have used this reaction to type fibers. When type I and type II fibers are mentioned it should be stressed that we are referring to routine adenosine triphosphatase type I fibers, which are light, and routine adenosine triphosphatase type II fibers, which are dark. These generally correlate with type I and II fibers seen in the oxidative enzyme reactions and in the phosphorylase reactions but this is not necessarily true all the time.

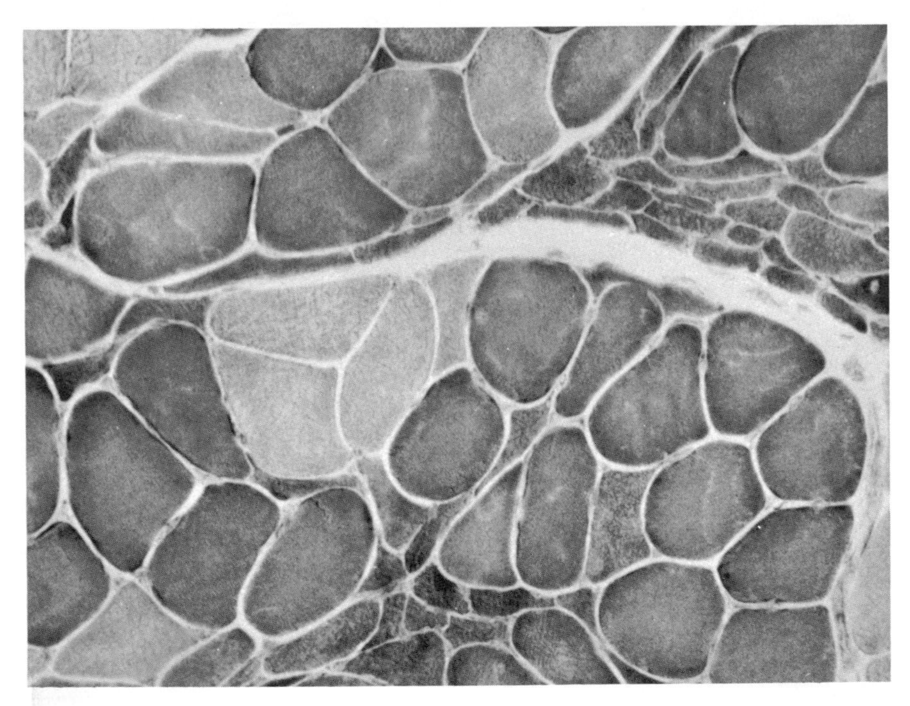

*Fig. 8.5*. NADH-tetrazolium reductase reaction in a denervated muscle. The small fibers are all either dark or intermediate, and two fiber types cannot easily be distinguished (cf. Fig. 8.6.) .

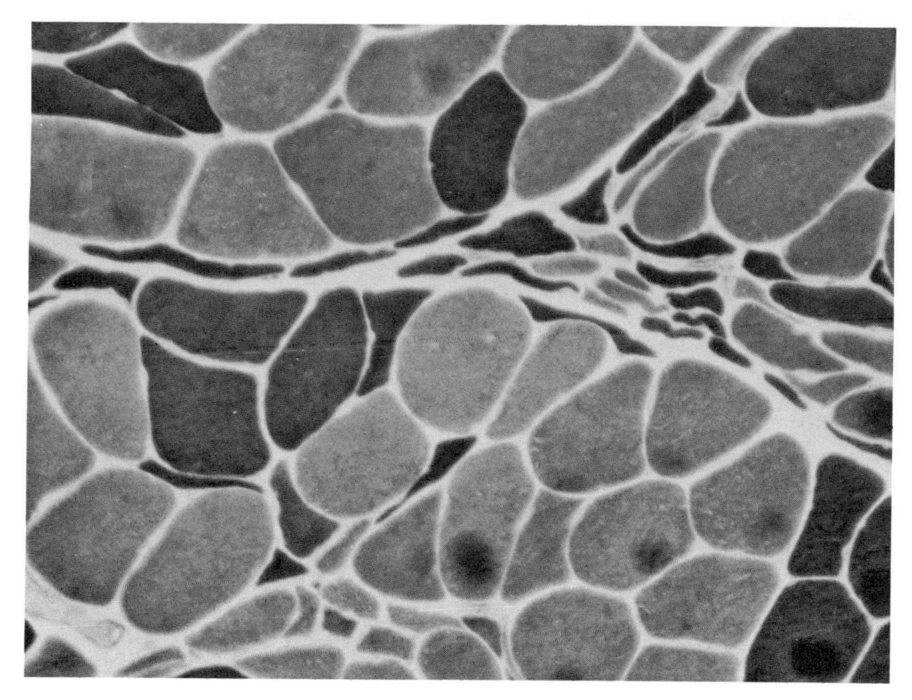

*Fig. 8.6.* A routine adenosine triphosphatase reaction on a serial section to that in Fig. 8.5, with two fiber types clearly differentiated among the small fibers (cf. Fig. 8.5).

8.7–8), but not in the type II fibers, whose adenosine triphosphatase system is acid-labile. Preincubation of the tissue section below *p*H 4.5 inhibits reaction in the type II fibers, and accentuates it in the type I muscle fibers, leading to an apparent reversal in the adenosine triphosphatase pattern (Fig. 8.9) (Brooke and Kaiser, 1969). Brooke and Kaiser (1970) have recently characterized three types of fibers, but here only types I and II will be discussed.

It is worth noting at this point that the correlation between the various characteristics of the fibers—anatomic, physiological, biochemical, and histochemical—is not firmly established. A muscle such as the soleus, however, is in most mammals always a red muscle, always slowly contracting, and containing a high proportion of adenosine triphosphatase type I fibers determined histochemically. The gastrocnemius, on the other hand, is in most mammals a white muscle, physiologically a fast muscle and containing mostly histochemically type II fibers.

It is now clear that the typing of a muscle fiber is not an immutable characteristic. The innervation of the muscle has been shown to be of

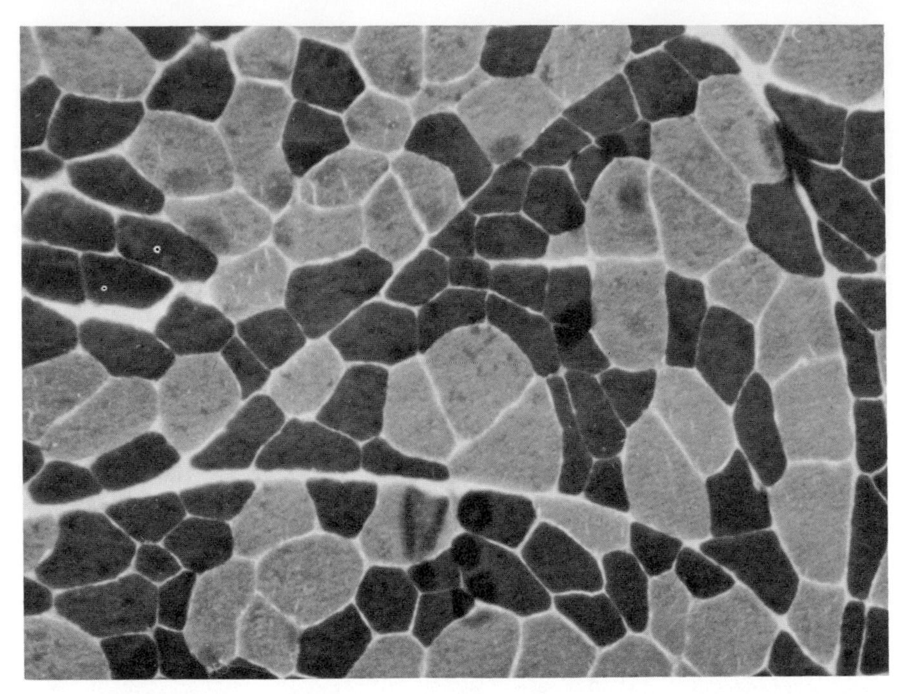
8.7

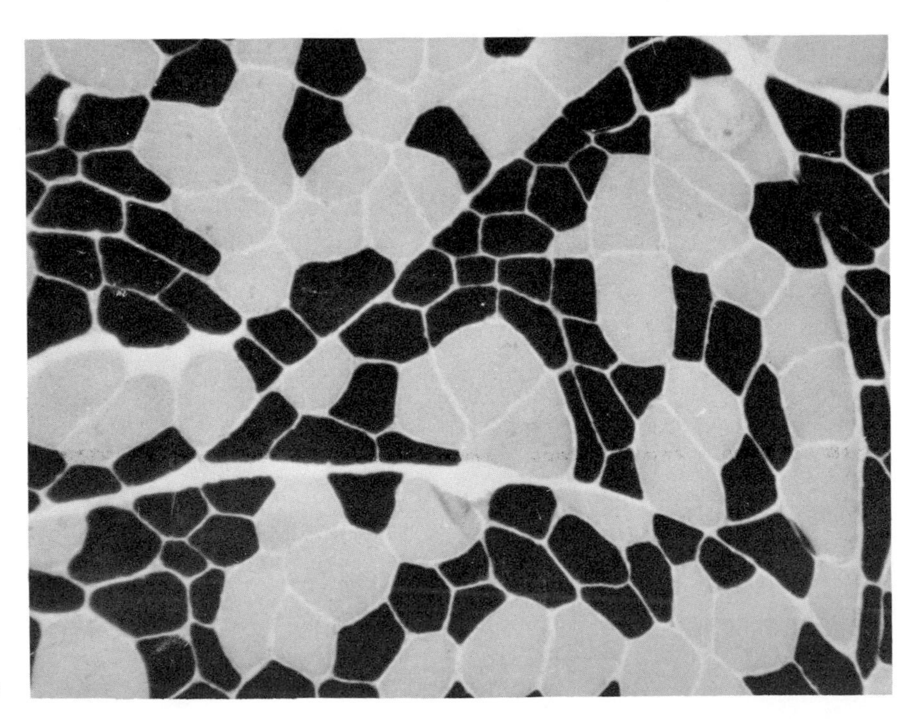
8.8

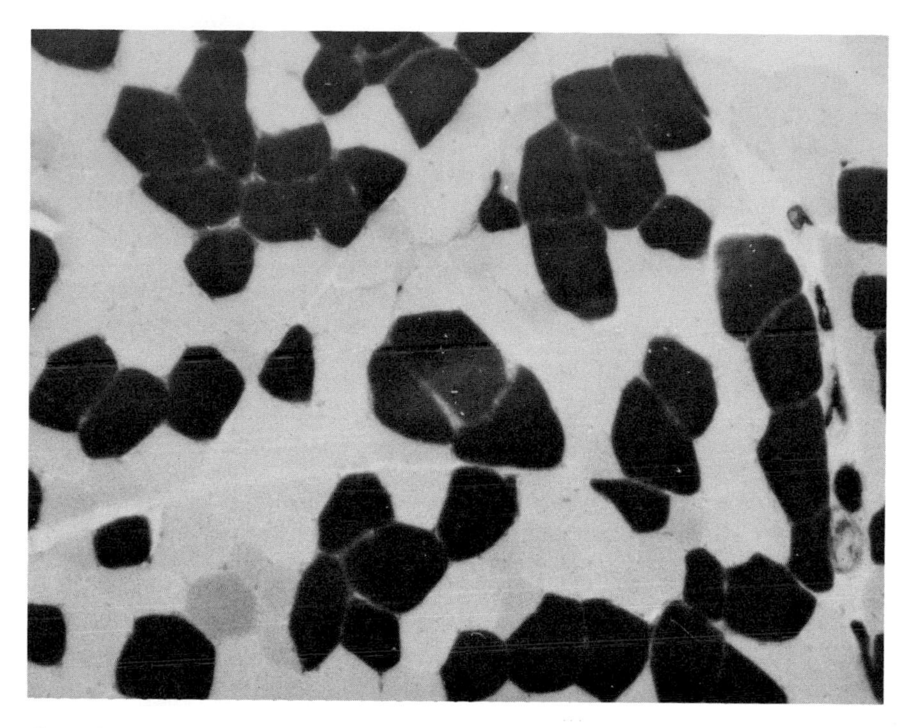

8.9

*Figs. 8.7–9.* Three different adenosine triphosphatase reactions on serial sections of a female human muscle biopsy. *Fig. 8.7* is the routine reaction, with the two fiber types (light, type I; dark, type II) clearly differentiated. *Fig. 8.8,* the same reaction after preincubation at *p*H 10.3. The reaction in the type I fibers is inhibited and these fibers are even lighter than with the routine reaction. *Fig. 8.9,* the same reaction after preincubation at *p*H 4.3. The reaction in type II fibers has been inhibited, that in type I fibers is accentuated, leading to a "reversal" of the reaction.

great importance in determining the fiber types. Reversing the innervation between the soleus and gastrocnemius changes the histochemical and physiological characteristics of the muscle (Dubowitz, 1967; Guth, Watson, and Brown, 1968; Romanul and Van Der Meulen, 1967; Robbins, Karpati, and Engel, 1969). Thus, a soleus muscle which receives a new innervation from the nerve formerly supplying the gastrocnemius becomes a faster muscle physiologically with a higher proportion of type II muscle fibers within it. In some animals at birth, the soleus is a mixed muscle consisting of both type I and type II fibers, but during early development, the number of the type II fibers diminishes until the muscle is predominantly composed of type I fibers. If, however, the soleus is denervated before such differentiation takes place, further differentiation is prevented and the muscle remains a mixed muscle (Karpati and Engel, 1967). These experiments suggest that the histochemical properties as well as the physiological properties of the muscle fibers are influenced by the nerve supply to the muscle. This neural influence could be due to either a direct trophic action of the nerve on the muscle or an indirect action, due to the nature of the contraction pattern induced by the pattern of firing of the nerve. Whichever it may be, there is no question but that the fibers have the capability of changing their histochemical and physiological types.

The influence of the innervation on the typing of fibers also brings up for consideration the concept of the motor unit. The nerve supply to muscle arises from cell bodies in the anterior horn of the spinal cord. An axon from a single anterior horn cell branches peripherally to supply many muscle fibers. The collection of muscle fibers supplied by one anterior horn cell is called the motor unit. The number of muscle fibers in a given motor unit varies from a few dozen fibers to several hundred depending upon the muscle in question. If the histochemical muscle fiber type is related to the nerves supplying these muscle fibers, then all the muscle fibers of a given unit should be of the same type. Until recently this was not certain, but the work of Doyle and Mayer (1969) has suggested that this is indeed so. Thus, the duality of neuromuscular function extends back at least as far as the anterior horn cell.

A similar and even more complex duality is seen in the sensory apparatus of striated muscle (Boyd and Davey, 1966; Brown and Matthews, 1966). The muscle spindle is a sensory organ consisting of several small "intrafusal" fibers within a connective tissue sheath. Two varieties of sensory endings, called primary annulo-spiral and secondary flower spray endings, are associated with these intrafusal fibers. Stretching of the muscle spindle causes the sensory ending of the intrafusal fibers to discharge, and the impulse is conducted to the spinal cord. At this site, it reaches the anterior horn cell and an impulse is then conducted down the motoneurone to cause the extrafusal muscle fiber to contract. This system can, therefore, act as a servomechanism whereby stretch on a muscle causes contraction of the muscle to resist the stretching movement. The motor supply of the intrafusal fibers is derived via the gamma efferent system. Two kinds of neuromuscular junctions are seen. The first is the so-called gamma trail ending and the second the gamma plate ending. There are also two kinds of intrafusal muscle fibers. The larger fiber with a central region occupied by a collection of nuclei is known as the nuclear bag fiber, characterized histochemically as adenosine triphosphatase type I (Fig. 8.10). The second type of fiber is more numerous, smaller, and is termed a nuclear chain fiber. Its histochemical characteristics are type II.

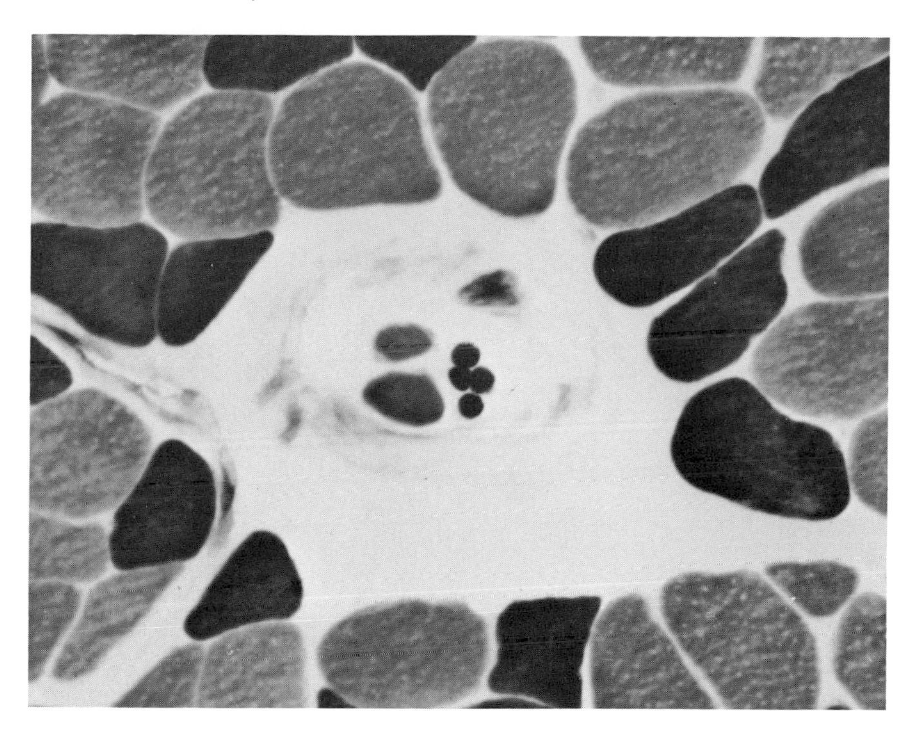

*Fig. 8.10.* A human muscle spindle demonstrated with the adenosine triphosphatase reaction. Large fibers in the center are type I, the small round fibers associated with them are type II. The capsule surrounding the intrafusal fibers can barely be seen with this reaction.

There is considerable dispute as to the exact way in which the muscle spindle functions. The following is an oversimplification, perhaps, of one particular view. Basically, the muscle spindle is a sensory "servo" organ which responds to stretch. Its bias is set by the system of gamma efferent fibers, which fall into two categories, static and dynamic. Stimulation of the dynamic system of gamma efferent fibers alters the bias in such a way that the rate of firing from the sensory receptors of the muscle spindle is increased during the dynamic phase of stretching. Stimulation of the static efferent system does not increase the rate of firing during the dynamic phase of stretch, but increases the rate of firing of the sensory organ at any given length. One view (Boyd and Davey, 1966) is that the static fibers of the gamma efferent system mainly supply the nuclear bag or type I intrafusal fibers, whereas the dynamic efferent fibers mainly supply the nuclear chain, or histochemical type II fibers. Functionally, the static gamma efferent fibers could be responsible for signalling the "demanded length" of the muscle. The static gamma fibers will be able to set the tone of the muscle spindle fibers so that the constant and reflex discharge of the annulo-spiral ending will cause sufficient contraction in the extrafusal fiber to set the muscle at a given length. The dynamic gamma efferent system, however, will be more likely to act as a method of smoothing out movements of muscle as they are being made.

Another property of muscle which has been clearly established is that when the muscle is subjected to repeated work, it will hypertrophy in the sense that its gross volume will increase. When the muscle is rendered immobile, it will atrophy and the gross volume will decrease (Helander, 1966; Siebert, 1928). These changes in size are of considerable interest, since they suggest that it might be possible to obtain information about the higher neural control of muscle fibers by studying their response in terms of such changes. Various pathological and physiological situations have been investigated and will be summarized below (Brooke and Engel, 1969*a, b, c, d*). We have used the histogram to determine the size of muscle fibers. A photograph is taken at 100 × magnification of a 10-$\mu$ section made transversely through the muscle. The histochemical procedure used to demonstrate the muscle fibers in this section is the adenosine triphosphatase reaction. Different methods for measuring the fiber size have been recommended, and have ranged from a measurement of surface area to measuring the circumference of the fiber. We have chosen a method which combines simplicity with reasonable accuracy, that of measuring the "lesser fiber diameter," defined as the maximum diameter across the lesser aspect of the muscle fiber (for an illustration, see Fig. 8.11). This measurement is designed to overcome the distortion which occurs when a muscle is cut obliquely (Fig. 8.12), producing an oval appearance in the fiber. Unless the lesser diameter is measured, an erroneously large measurement will result, as Fig. 8.12 clearly shows. A histogram is constructed by measuring the lesser diameter of 200 fibers and plotting the number of fibers against the size of the fibers, measured to the nearest 10 $\mu$. In order to obtain better statistical correlations, histograms from uniform groups of patients are summated to produce a "megahistogram." For example, biopsies were taken from the vastus muscle of a series of normal men and the individual histograms summated to form a megahistogram representing the normal male vastus (Fig. 8.13). The details of this study are given elsewhere (Brooke and Engel, 1969*a*). Similarly, a megahistogram was constructed from the vastus of normal women.

There is an interesting difference between these two histograms. The sizes of the type I fibers in the two sexes are comparable. The type II fibers, on the other hand, are smaller than the type I fibers in the female

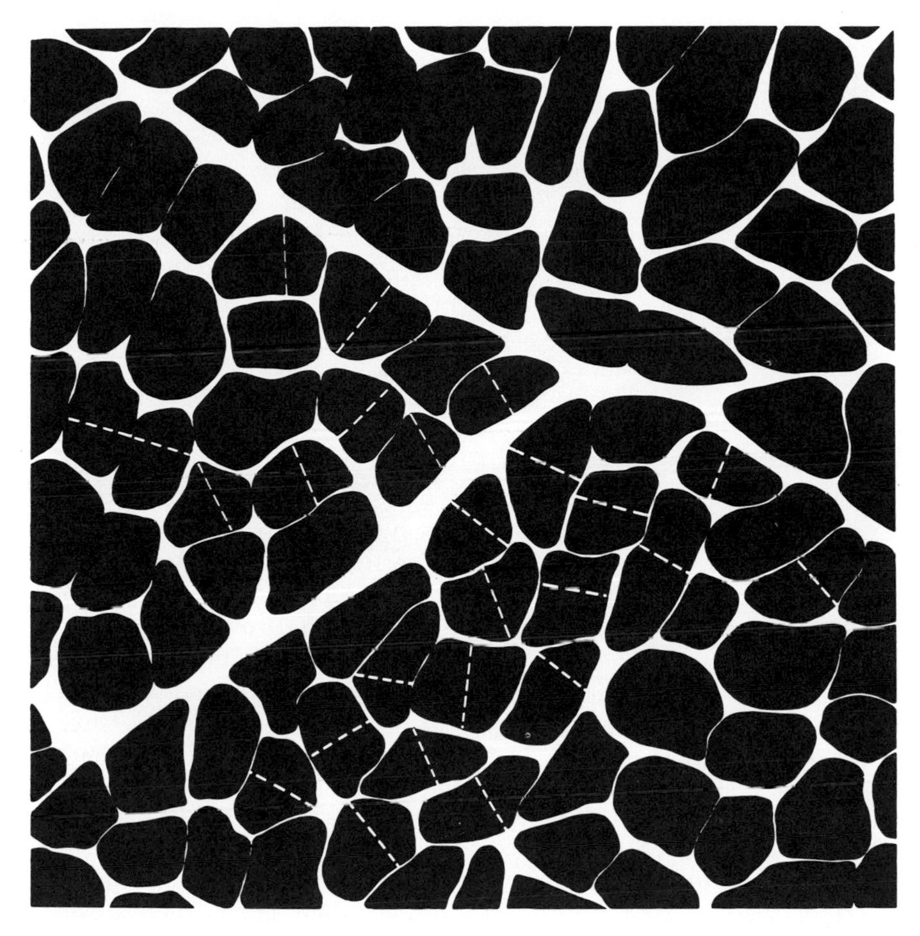

*Fig. 8.11.* The measurements taken in order to construct histograms of the lesser fiber diameters.

and larger in the male. The same difference between the sexes is seen in the biceps muscle. Examples of male and female biopsies are seen in Figs. 8.14–15. It is tempting to speculate that this difference represents the functional difference between male and female muscle. The more powerful muscle of the male may, in some way, be associated with the larger type II fibers, but the question arises as to whether these fibers are a primary or a secondary phenomenon. They may become larger because the male is called upon to exert greater efforts than the female (secondary hypertrophy), or they may be a feature of the male muscle that enables the male to exert more powerful effort than the female. This would be a primary phenomenon. Speculation on this point is uncertain.

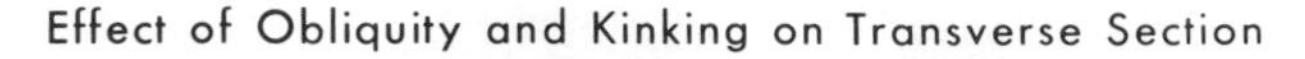


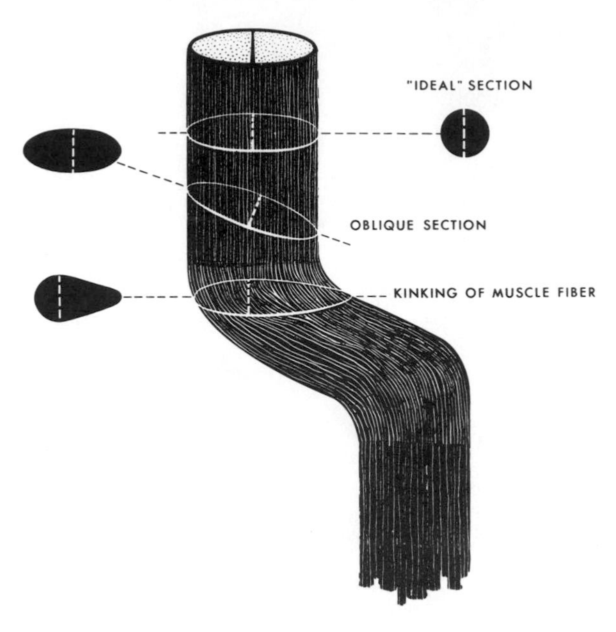


*Fig. 8.12.* Effect of changes in the apparent size of the fiber produced by either an oblique section or by kinking of the fibers.

There is, however, an interesting exception to the group of female biopsies. Certain of the women, unaffected by any disease themselves, were mothers of children with various neuromuscular diseases whose handicap was so severe as to make the child unable to walk without assistance. In the biceps of these women, the type II fibers were significantly larger than the type II fibers of the other female groups (Fig. 8.13). It is not certain that mothers of severely affected children do any more physical work than their counterparts without such children, but a great physical effort is required in caring for a child who can neither stand nor walk without assistance. Thus, we feel that the large type II fibers in this group of women represent a secondary phenomenon associated with the performance of hard physical work.

The association of work with large type II fibers would suggest the association of inactivity with atrophy of the fibers. This is the case in humans, in whom many forms of inactivity occur, whether of a physiological nature, such as that due to bedrest, or whether due to diseases causing impairment of muscle functions, such as periodic paralysis, myas-

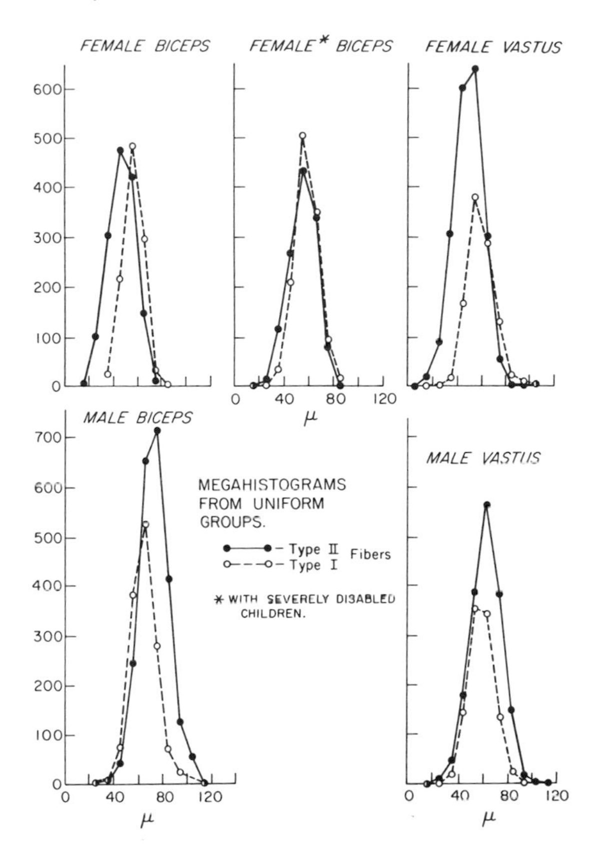


*Fig. 8.13.* Megahistograms of normal human muscle biopsies, illustrated here, for clarity, as frequency polygons. In the male muscles the type II fibers (*solid line*) are slightly larger than type I fibers (*dashed line*). In the female, the reverse is true.

thenia gravis, and arthritis. In all of these examples, type II fiber atrophy is seen.

Thus far, we have discussed the occurrence of changes in the size of the type II fibers. In the conditions listed above, no change occurs in the size of type I fibers. It is necessary, therefore, to consider the conditions in which changes of the type I fibers occur. Human disease of the motoneurones such as occurs in the entity amyotrophic lateral sclerosis provides an example of atrophy of both fiber types. In this illness, there is a progressive loss of motor nerve cells, involving the upper motoneurone or pyramidal system in addition to the lower motoneurone or anterior horn

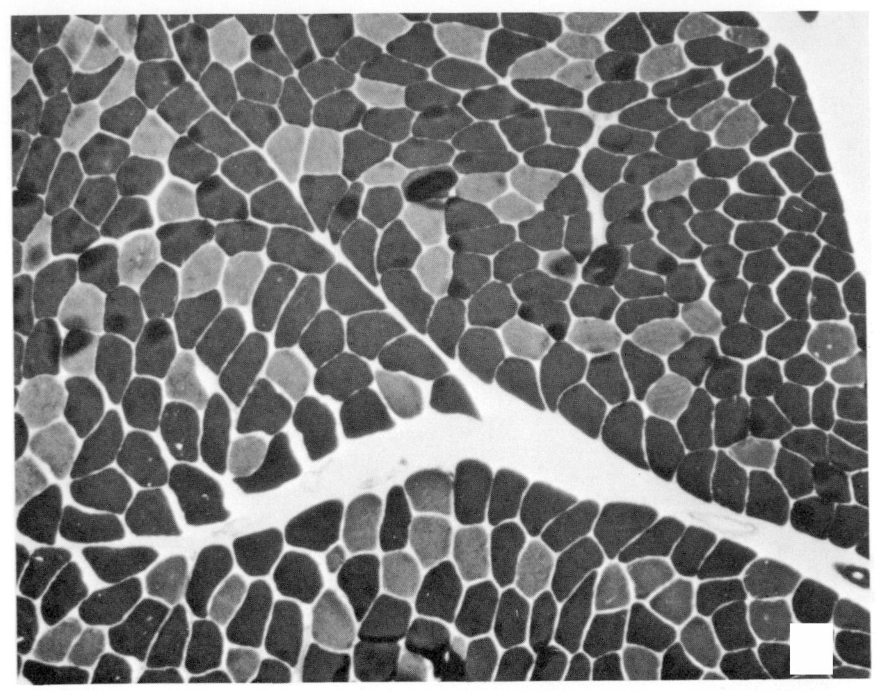

8.14

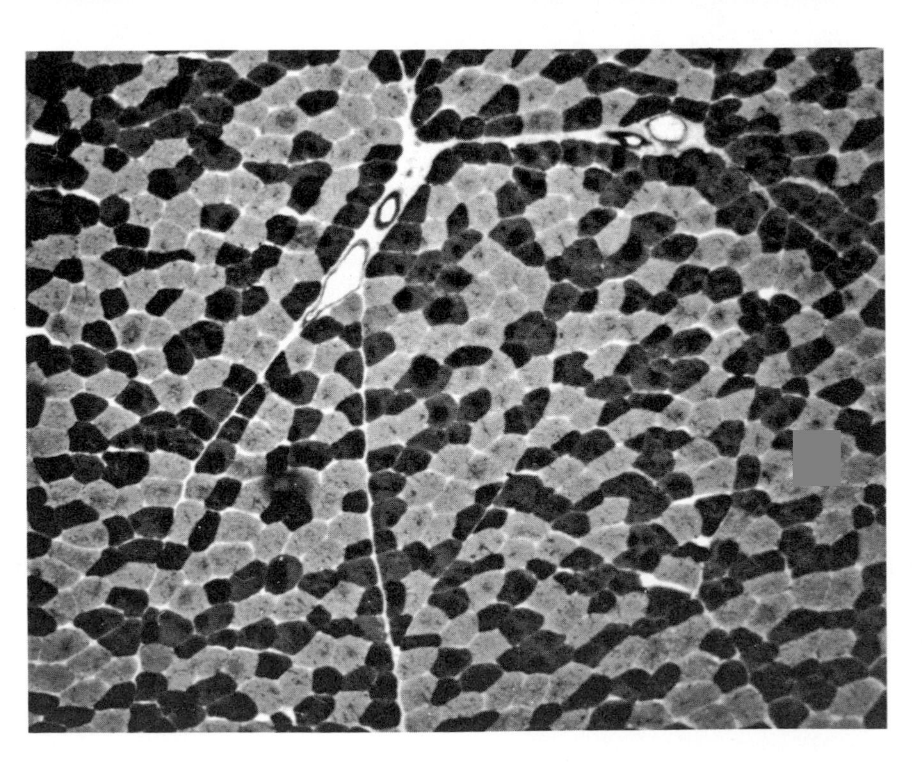

8.15

*Fig. 8.14.* Male human muscle, adenosine triphosphatase reaction. The type II (dark) fibers are slightly larger than the type I fibers.

*Fig. 8.15.* Female human muscle, adenosine triphosphatase reaction. Type II (dark) fibers in general smaller than type I fibers.

cell. Clinically, it is possible to differentiate the three varieties: (1) those patients in whom there is a predominant involvement of the lower motoneurone; (2) those in whom there is equal involvement of the upper and lower motoneurones; and (3) those in whom the disease involves predominantly the upper motoneurone. In the patients in the first group the megahistograms show that both fiber types exhibit a population of fibers that are abnormally small (Fig. 8.16a). This would be anticipated because those fibers which were receiving their innervation from degenerating anterior horn cells will atrophy. The small fibers of both fiber types are an indication of an involvement of motor nerve cells of both types. Additionally, these patients also demonstrate a population of abnormally large type II fibers. Large fibers or hypertrophied fibers are usually associated with increased work. It is likely that the population of large type II fibers is associated with the efforts of the patient to overcome the weakness caused by the disease. Thus, the fibers which still retain their innervation will be used to effect this additional effort. Again, there is the association between type II fiber hypertrophy and strong voluntary effort. An example of such a biopsy is seen in Fig. 8.17.

The patients from group (2) demonstrate a slightly different megahistogram. The population of small atrophic fibers is as in the previous group (Fig. 8.16b), but the large type II fibers are lacking. It seems, therefore, that integrity of the pyramidal tracts or upper motoneurone is necessary in order for hypertrophy of type II fibers to occur. One of the possible arguments against this point is that patients with involvement of both upper and lower motoneurones tend to be more severely involved clinically than those with only involvement of the lower motoneurone. It is, therefore, possible that the lack of type II fiber hypertrophy might represent only the severity of the disease. However, the third group of patients, in whom involvement of the upper motoneurone was the predominant finding, were, if anything, less severely handicapped than those with involvement of only the lower motoneurone. In group (3) patients, type II atrophy was marked. Involvement of the upper motoneurone or pyramidal tracts occurs in diseases other than amyotrophic lateral sclerosis and the atrophy of type II fibers is seen in miscellaneous pyramidal

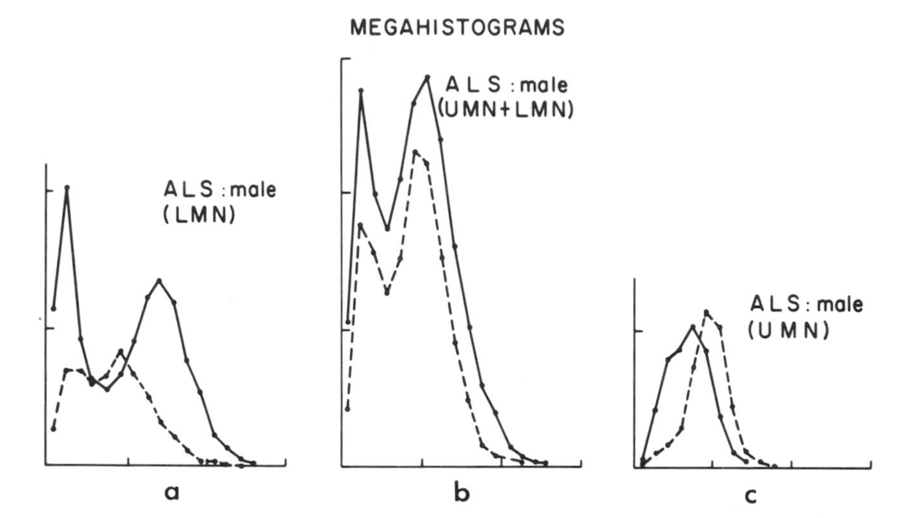


*Fig. 8.16.* Megahistograms (to same scale) from three groups of patients with motoneurone disease. The first calibration mark on the abscissa represents fibers with a lesser diameter of 60 $\mu$, roughly the normal size of type I human muscle fibers. (a) Patients with motoneurone disease involving predominantly the lower motoneurone; (b) patients in whom both upper and lower motoneurones are involved; (c) patients in whom predominantly the upper motoneurone is involved. In (b) there are two peaks in the megahistograms of both fiber types. The additional peak represents a population of small fibers.

The difference between (a) and (b) lies in the presence of very large type II fibers (*solid line*) in (a). In (c) the type II fibers (*solid line*) are smaller than the type I fibers (*dashed line*).

tract lesions. There is no atrophy of type I fibers, however, in this last group of patients. It would seem possible to reproduce the findings of type II atrophy by sectioning the pyramidal tracts as, for example, in cordotomy. However, in most cordotomized animals, the hindlimbs become fixed in a position of extreme plantar flexion at the ankle and extension at the knee. The changes produced in this situation can be duplicated by passively pinning the joints so that the limb is fixed in this position. Thus, the effect of experimental cord section is still unclear.

To summarize, type II fibers become larger in situations associated with increased demands for "powerful" activity and atrophy in situations where the muscle is not used. This disuse may be physiological or pathological due to disease which limits activity. Further, the integrity of the pyramidal motor system is necessary for the maintenance of type II

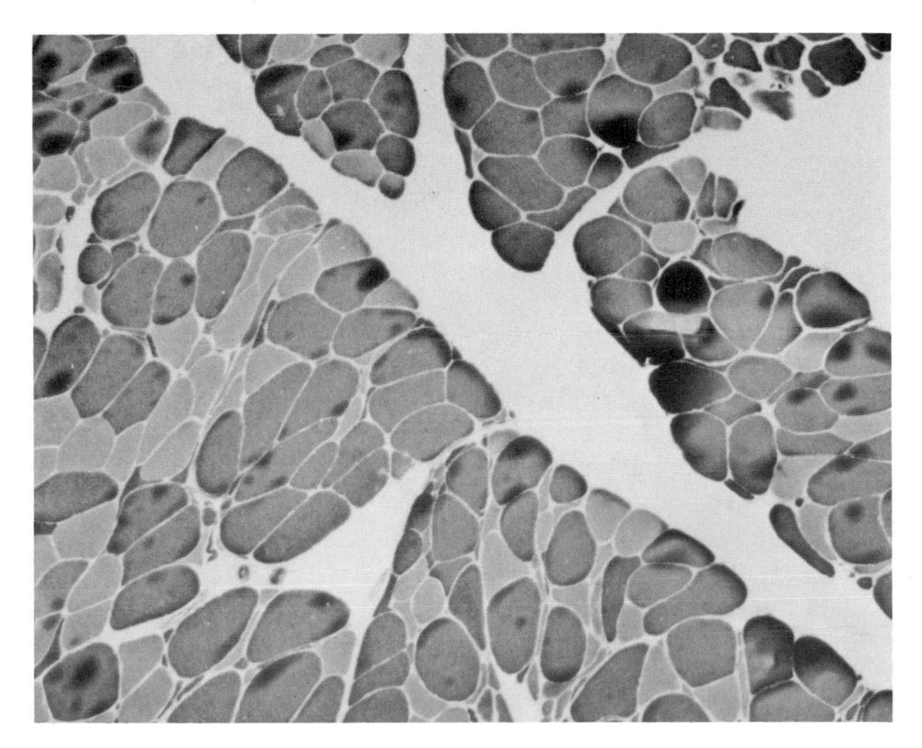

*Fig. 8.17.* Human muscle biopsy from a patient with motoneurone disease of the predominantly lower motoneurone type, adenosine triphosphatase reaction, showing very large type II (dark) fibers.

fiber size. Under these circumstances, the size of type I fibers is not altered.

The independence in the behavior of the two fiber types could be explained by innate metabolic differences: i.e., type I fibers may be incapable of atrophy and hypertrophy even though subjected to the same stimuli—activity or inactivity—which cause changes in the size of the type II fibers. We have seen, however, that in human motoneurone disease in which nerve fibers supplying both fiber types are affected both type I fibers and type II fibers atrophy. An alternative explanation is that the nervous circuits controlling type II fiber activity are independent of those circuits supplying type I fiber activity. This would suggest that type II fiber activity is associated with powerful voluntary activity which is mediated in association with the pyramidal system before reaching the anterior horn cells. It also suggests that type I fibers are not involved in such activity.

It is necessary to turn to experimental situations in animals for information about changes in type I fiber size, although comparing results obtained from animal experimentation with those obtained from humans in health and disease is not without danger. If the soleus and gastrocnemius in the cat are tenotomized, within 2 to 4 weeks the type I fibers atrophy and the type II fibers hypertrophy (Fig. 8.18) (Engel, Brooke, and Nelson, 1965). This would suggest that in the situation of tenotomy, type I fibers are inactive and type II fibers are overactive. There are no anatomic lesions in the nerve supply to the muscles. Thus, the inactivity of type I fibers must be due to some other cause of a decrease in nerve potentials reaching the muscle. Following tenotomy the muscle spindle within the muscle will be entirely relaxed, and, since there is no attachment for the tendon, it will be very difficult to produce any tension on the spindle. Its sensory endings that respond to stretch will be less active and there will be a marked reduction in the impulses passing back to the extrafusal muscle fibers via the usual monosynaptic reflex arc. Thus, one source of efferent impulses to the extrafusal muscle fibers will be removed or diminished. This could be one explanation for the presence of atrophied type I fibers, if it is assumed that they are

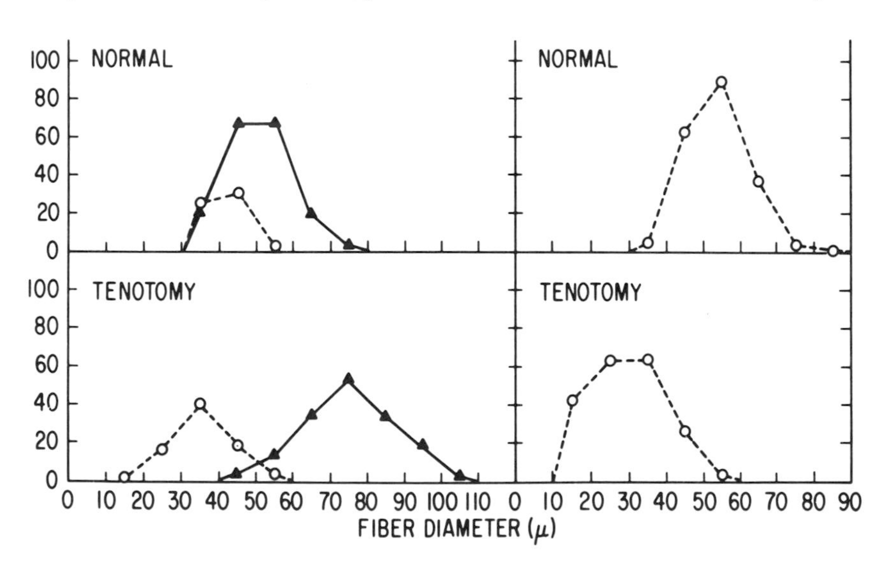


*Fig. 8.18.* Histograms constructed from the gastrocnemius (*left*) and soleus (*right*) muscles from a normal cat and from one which has undergone an Achilles tenotomy. *Dashed line,* type I fibers, *solid line,* type II fibers.

activated reflexly by a system of neurones associated with the muscle spindles. The second aspect to consider is the type II fiber hypertrophy. If enlargement of type II fibers is associated with strong voluntary activity, then the type II fiber hypertrophy seen in the tenotomized muscle should be due to the same thing. All the proprioceptive information available to the animal will suggest that the tenotomized muscle is weak, i.e., that the foot cannot be plantar flexed. It is possible that the animal will try to overcome this "weakness" by strong voluntary contraction. To test the hypothesis that the hypertrophy of type II fibers was indeed being mediated by contraction of the muscle, some animals were subjected to tenotomy and denervation at the same time (Fig. 8.18). This procedure prevented the hypertrophy of type II fibers, showing that the factor which causes such enlargement is dependent upon the integrity of the nerve, and suggesting that it is due to contraction of the muscle. A further experiment to try to evaluate the influence of voluntary contraction upon the type II hypertrophy involved pinning the hindlimb of an animal in extreme plantar flexion. This produced shortening of the muscle in the same fashion as tenotomy. Atrophy of both type I and type II fibers occurred. The lack of type II fiber hypertrophy might be explained by the fact that the proprioceptive information available to the animal, in this case, would not be that of inability to plantar flex the foot but rather of extreme plantar flexion which might discourage any voluntary attempt at further plantar flexion. Thus, the soleus and gastrocnemius muscle would not be contracted.

To return again to the example of simple tenotomy, the fact that the type I fibers do not hypertrophy suggests that they do not receive any innervation from the voluntary, pyramidal-tract-associated nerve fibers.

The hypothesis that we are presenting for consideration is that most of the powerful or voluntary activity associated with pyramidal system function is mediated by a contraction of type II muscle fibers. The type I muscle fibers are activated by a system which is associated with reflex, spindle-associated impulses, suggesting that the fibers are also under the higher control of centers that influence spindle activity such as the cerebellum and the brain stem centers. There is no direct proof of the validity of this hypothesis, although there exists supporting evidence that there is some independence in the control of type I and type II fibers.

In the skipjack tuna, red and white muscle fibers do not exist together within the same muscle but are separated into muscles which are homogeneously of one type or the other (Rayner and Keenan, 1967). Electrophysiological studies have shown that the red muscles, which would be the equivalent of type I or slow muscles, are used for basal swimming

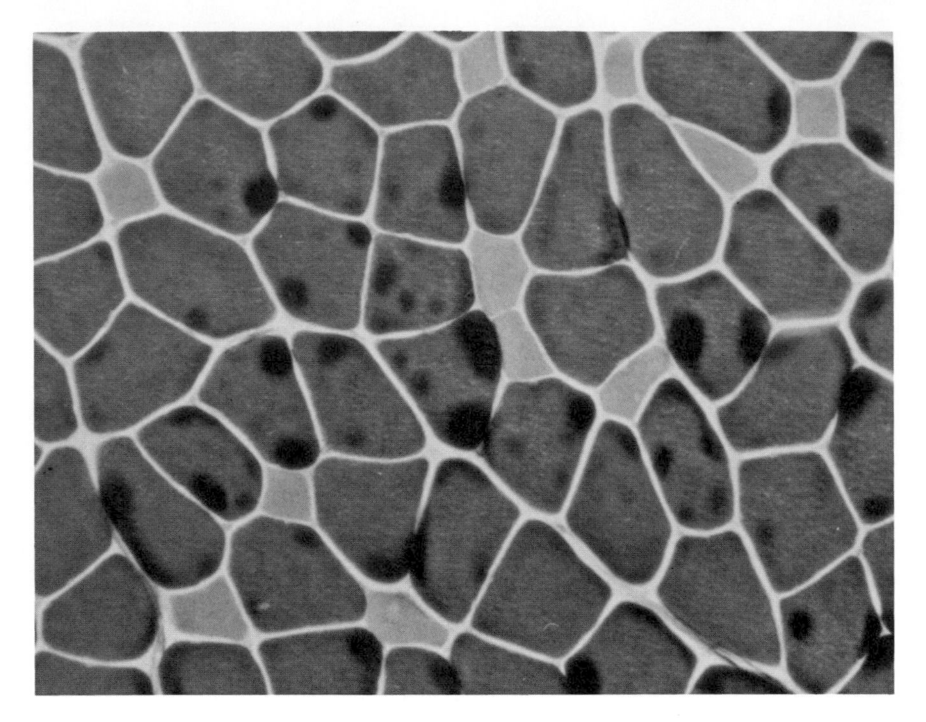

*Fig. 8.19.* Human muscle biopsy from a patient with myotonic dystrophy, adenosine triphosphatase reaction, showing effect of disease. Type I fibers are markedly smaller than type II.

motions. The white muscles are used for short bursts of high activity such as feeding or escape behavior.

Merton (1953) proposed, on physiological grounds, that muscle function was controlled by two pathways. Ordinary movements such as were necessary for maintaining posture were thought to be mediated via a system associated with the muscle spindle which acted as a servomechanism for adjusting the length of the muscle. The second mechanism was thought to be a direct action of higher motor centers upon the anterior horn cells and it was suggested that this mechanism was responsible for urgent sudden movements. This theory forms a parallel to the hypothesis suggested in the present paper.

It was noted by O'Leary, Heinbecker, and Bishop (1932) that in monkeys experimentally infected with the poliomyelitis virus, paralysis of the muscles resulted from damage of anterior horn cells. Stimulation of the cord above the level of the anterior horn cells supplying the paralyzed muscles produced no response. Stimulation of the nerve roots, however, produced a contraction of muscle, indicating a lesion between the suprasegmental supply and the nerve root. It was postulated that this was due to the disease of the anterior horn cell since some days later the motor roots did not respond to stimulation. Brooks (1953), however, found patients with poliomyelitis of sufficient severity to cause complete voluntary paralysis of the muscle. Nevertheless, stimulation of the peripheral nerve produced a contraction of the muscle, and the sensitivity to stimulation was retained for a long period of time. This might suggest that the problem did not arise from progressive degeneration of the motor nerve secondary to anterior horn cell disease but that there were muscle fibers whose nerve supply was not under voluntary control. The classical example of type I fiber atrophy in adults occurs in the biceps muscle of patients with myotonic dystrophy (Fig. 8.19). The nature of this disease is unclear but Daniel and Strich (1964) have suggested that there is a morphological abnormality of the muscle spindle.

The muscle spindle and gamma efferent system is thought to be under some degree of cerebellar control, and it might be expected that in cerebellar disease there would be changes in the type I fibers. We do not have enough evidence in adults to be certain of this. It is of interest that in children's biopsies there is an association between cerebellar damage and type I fiber atrophy and between pyramidal tract disease and type II fiber atrophy (Brooke and Engel, 1969*d*).

In conclusion, we have presented evidence that the two histochemical muscle fiber types behave differently and independently with regard to changes in the fiber size. It may be that such differences are an innate property of the muscle fibers themselves, reflecting biochemical differences between the two fiber types. It is difficult, however, to form a theory which would explain the behavior of fiber types on these grounds. There is a simpler hypothesis: that voluntary powerful or sudden activity, in which the pyramidal motor system is involved, is mediated by type II fibers, and that spindle-associated reflex activity is mediated via the histochemical type I fibers of muscle. It seems likely that the type I fibers are not under the same voluntary control as the type II fibers. There is no evidence, however, that the type II fibers have no reflex control. Indeed, it is possible that the double system of muscle activity involves not only the muscle fibers and the motoneurones but also extends to the spindles. Thus, for example, extrafusal fibers of each type may be modified by impulses from intrafusal fibers of the same type. So marked is the difference between the behavior of type I and type II fibers, that this hypothesis demands further investigation.

## *References*

Boyd, I. A., and M. R. Davey. 1966. The distribution of two types of small motor nerve fiber to different muscles in the hind limb of the cat, p. 59. *In* R. Granit (ed.), *Muscular Afferents and Motor Control.* Wiley, New York.

Brooke, M. H., and W. K. Engel. 1969*a*. The histographic analysis of human muscle biopsies with regard to fiber types 1. Adult male and female. *Neurology 19*:221.

———. 1969*b*. The histographic analysis of human muscle biopsies with regard to fiber types 2. Diseases of the upper and lower motor neurons. *Neurology 19*:378.

———. 1969*c*. The histographic analysis of human muscle biopsies with regard to fiber types 3. Myotonias, myasthenia gravis and periodic paralysis. *Neurology 19*:469.

———. 1969*d*. The histographic analysis of human muscle biopsies with regard to fiber types 4. Children's biopsies. *Neurology 19*:591.

Brooke, M. H., and K. K. Kaiser. 1969. Some comments on the histochemical characterization of muscle adenosine triphosphatase. *J. Histochem. Cytochem. 17*:431.

———. 1970. Three myosin ATPase systems. *J. Histochem. Cytochem.* (In press.)

Brooks, D. 1953. Nerve conduction in poliomyelitis, p. 35. *In* G. E. W. Wolstenholme (ed.), *The Spinal Cord.* Little, Brown, Boston.

Brown, M. C., and P. B. C. Matthews. 1966. On the subdivision of the efferent fibers to muscle spindles into static and dynamic fusimotor fibers. *In* B. L. Andrew (ed.), *Control and Innervation of Skeletal Muscle.* E. and S. Livingstone, London.

Daniel, P. M., and S. J. Strich. 1964. Abnormalities in the muscle spindles in dystrophia myotonica. *Neurology 14*:310.

Doyle, A. M., and R. F. Mayer. 1969. Histochemistry and topology of the motor unit in the cat. (Abstr.) *Neurology 19*:296.

Dubowitz, V. 1960. Reciprocal relationship of phosphorylase and oxidative enzyme activity in muscle. *Nature 185*:701.

———. 1967. Cross innervated mammalian skeletal muscle: histochemical, physiological and biochemical observation. *J. Physiol. 193*:481.

Engel, W. K., M. H. Brooke, and P. G. Nelson. 1965. Histochemical studies of denervated or tenotomised cat muscle. *Ann. N.Y. Acad. Sci. 138* (1):160.

Guth, L., P. K. Watson, and W. C. Brown. 1968. Effects of cross reinnervation on some chemical properties of red and white muscles in the rat and cat. *Exp. Neurol. 20*:52.

Helander, E. 1966. General considerations of muscle development, p. 19. *In* E. J. Briskey, R. G. Cassens, and J. C. Trautman (eds.), *The Physiology and Biochemistry of Muscle as a Food.* Univ. of Wis.

Karpati, G., and W. K. Engel. 1967. Neuronal trophic function. A new aspect demonstrated histochemically in developing soleus muscle. *Arch. Neurol. 17*:542.

Merton, P. A. 1953. Speculations on the servocontrol of movement, p. 63. *In* G. E. W. Wolstenholme (ed.), *The Spinal Cord*. Little, Brown, Boston.

O'Leary, J. L., D. Heinbecker, and G. H. Bishop. 1932. Nerve degeneration in poliomyelitis. *Arch. Neurol. Psych. 28*:272.

Padykula, H. A., and E. Herman. 1955. The specificity of the histochemical method for adenosine triphosphatase. *J. Histochem. Cytochem. 3*:170.

Rayner, M. D., and M. J. Keenan. 1967. The role of red and white muscles in the swimming of the skipjack tuna. *Nature 214*:392.

Robbins, N., G. Karpati, and W. K. Engel. 1969. Histochemical and contractile properties in the cross-innervated guinea pig soleus muscle. *Arch. Neurol. 20*:318.

Romanul, F. C. A., and F. P. Van Der Meulen. 1967. Slow and fast muscles after cross innervation. Enzymatic and physiological changes. *Arch. Neurol. 17*:387.

Siebert, W. W. 1929. Untersuchungen über Hypertrophie des Skelettmuskels. *Z. Klin. Med. 109*:350.

*Chapter 9*

# *Biochemistry of the Red and White Muscle*

C. H. BEATTY AND R. M. BOCEK

The terms "red" and "white" when applied to muscle denote a compendium of morphologic and physiologic differences observed by many investigators in numerous vertebrate species. These differences in muscle have interested investigators for over a century (Nachmias and Padykula, 1958), and the pioneering investigations on the biochemical as well as structural and functional differences in the two types of muscle were summarized by Needham (1926). The classic postulate regarding the metabolic and functional differences between red and white muscle states that white muscle, capable of rapid but brief contractions, primarily utilizes glycolysis for energy production whereas red muscle, which can contract for prolonged periods, relies chiefly on oxidative mechanisms (Szent-Györgyi, 1953). Although these broad biochemical differences between red and white muscle have been accepted for some time, more precise information on the metabolism of the two types of muscles is needed. Recent studies have indicated that red and white muscle fibers respond differently to denervation and to certain disease processes (Bajusz, 1965; Pellegrino and Franzini, 1963; Fahimi and Roy, 1966). The slow muscle in the dystrophic mouse was less affected than the fast muscle (Brust, 1956) and the pale, soft, exudative condition in pig muscle has been correlated with a high percentage of intermediate fibers in the affected muscles (Cooper, Cassens, and Briskey, 1969). The biochemical bases for these different responses are not clear.

It is well known that red and white muscles are not homogeneous with respect to fiber type but are composed of varying ratios of red,* white, and intermediate fibers. An exception is the rabbit psoas, which is reported to contain only white fibers (Arangio and Hagstrom, 1969).

* Synonyms for red fibers found in the literature include: dark, slow, small, type I, C, tonic and slow-twitch fibers. White fibers are designated light, fast, large, type II, A, phasic, tetanic, and fast-twitch.

Soleus muscle, often considered to be a uniformly red-fibered muscle, apparently consists of two types of red fibers (Stein and Padykula, 1962; Bocek and Beatty, 1966). It is interesting to note that in two species of prosimians, the slow loris and the potto, which move with characteristic deliberateness, all muscles examined, including the gastrocnemius, brachioradialis, and sartorius, are red muscles, uniformly composed of soleus-type fibers (Bocek and Beatty, unpublished observations). Most mammalian muscles, however, contain at least three types of fibers (Ogata and Mori, 1964; Stein and Padykula, 1962; Moody and Cassens, 1968). Red fibers are usually smaller in diameter than white fibers, have a higher mitochondrial content, and consequently are rich in oxidative enzyme activities (Dubowitz and Pearse, 1960); they contain more myoglobin (Chinoy, 1963; James, 1968), lipase activity (Piantelli and Rebollo, 1967), and triglycerides (Adams, Denny-Brown, and Pearson, 1962). White fibers have a higher glycogen content and higher myofibrillar adenosine triphosphatase, phosphorylase, and glycolytic enzyme activities than do red fibers (Engel, 1962; Dubowitz and Pearse, 1960). Fibers intermediate in activity also have been observed (Ogata and Mori, 1964; Stein and Padykula, 1962); and subtypes within the classifications of red, white, and intermediate have been described on the basis of histochemical evidence (Romanul, 1964) and quantitative assay of enzyme activity (Dawson and Romanul, 1964). Ultrastructural differences between red, white, and intermediate fibers have also been described (Gauthier, 1969). Numerous additional instances of structural and histochemical differences between red and white muscle fibers are recorded in the literature, but a listing of these differences is not within the scope of this paper.

Since red and white skeletal muscles are not homogeneous in their fiber types, metabolic and functional differences between intact muscles are a reflection of the predominant fiber type present. Furthermore, the proportion of fiber types present may vary widely in a single muscle, as, for instance, in the medial and lateral heads of the rat gastrocnemius (Romanul, 1964) and in the superficial and deep portions of the rhesus monkey brachioradialis (Beatty, Basinger, Dully, and Bocek, 1966). Characterization of muscle according to fiber content is necessary for an adequate interpretation of experimental results. Too often studies, especially on man, are carried out with little knowledge of muscle composition. This fact is important since we have found that the metabolism of red and white fiber groups, for instance, differs as much as or more than the effects of imposed experimental conditions. When interpreting metabolic data based on histochemical evidence, one should remember that histochemical techniques give only limited information. Factors such as diffusion of enzymes from intracellular sites, the translocation of metabolites during incubation of tissue sections, and the absence of activating metabolites or of normal electron carrier molecules in dehydrogenase reactions, can give false results (Van Wijhe, Blanchaer, and Jacyk, 1963*a;* Fahimi and Karnovsky, 1966). The effects of freezing on enzyme activity and correlation with actual cytolocalization also should be considered. Ideally, investigations with individual or small groups of exclusively red or white muscle fibers yield the most precise information on localization of enzymes and metabolites (Van Wijhe, Blanchaer, and Stubbs, 1964) and on the physiological response of the fiber types (Sexton and Gersten, 1967). Whenever possible, it is preferable to correlate histochemical results with quantitative enzyme assays and with *in vitro* and *in vivo* data.

ENZYME ACTIVITIES

Table 9.1 lists data from quantitative enzyme assays for various red and white muscles from several species of mammals. Although it is impossible, even in a single species, to correlate the activities for a given enzyme as determined by different investigators, relative activities for red and white muscle correlate well with histochemical data. Glycolytic enzyme activities are higher in white muscle, with the exception of hexokinase. This enzyme is not exclusively a glycolytic enzyme since glucose-6-phosphate is a branch point leading to the glycogen-metabolizing enzyme system as well as to the pentose cycle reactions. Phosphorylase activity is higher in white muscle, but glycogen synthetase activity is higher in red. In assays of a single enzyme representing amino acid metabolism (glutamic oxaloacetic transaminase) and of pentose cycle enzymes, activities were greater in red than in white muscle. Fatty acid metabolism, indicated by $\beta$-hydroxy acyl Co-A dehydrogenase activity, was also higher in red muscle. Studies on avian muscle lipase concur with these data (George and Talesara, 1961). Enzymes of the citric acid cycle and cytochrome oxidase are also higher in red muscle. The significance of these enzyme activities in the metabolism of red and white muscle will be discussed later. As might be expected from the fact that white muscle requires more rapid energy production, adenosine triphosphatase activities are higher in white than in red muscle (Ogata, 1960; Seidel, Sréter, Thompson, and Gergely, 1964; Bárány, Bárány, Reckard, and Volpe, 1965; Sréter, Seidel, and Gergely, 1966; Maddox and Perry, 1966). Both the presence of a more highly developed sarcoplasmic reticulum (SR) (Pellegrino and Franzini, 1963; Franzini-Armstrong, 1964) and the observation that $Ca^{++}$ uptake was greater by white muscle grana (fragmented SR) than by red (Sréter, 1964) are compatible with the faster

TABLE 9.1

*Quantitative differences in enzyme activities of mammalian red and white muscle*

| Enzyme and species | | Muscle | Enzyme units | Reference |
|---|---|---|---|---|
| | | GLYCOGEN METABOLISM | | |
| Phosphorylase | | | | |
| guinea pig | (W) | adductor | 10.4 ± 2.3[1] | Stubbs and Blanchaer |
| | (R) | quadratus femoris | 19.2 ± 2.9 | (1965) |
| rabbit | (R) | soleus | 135[2] | Pette (1966*a*) |
| | (W) | adductor magnus | 2800 | |
| rhesus | (R) | sartorius | 14 ± 2.4[1] | Bocek and Beatty |
| monkey | (W) | brachioradialis | 23 ± 2.8 | (1966) |
| rat | (R) | semimembranosus | 26 ± 8.7[1] | Bocek and Beatty |
| | (W) | semimembranosus | 86 ± 8.6 | (1966) |
| rabbit | (R) | soleus | 13.3 ± 6.7 | Burleigh and Schimke |
| | (W) | adductor magnus | 110 ± 15 | (1968) |
| rat | (R) | soleus | 7.5[1] | Dawson and Romanul |
| | (W) | gastrocnemius | 32 | (1964) |
| Glycogen synthetase | | | | |
| guinea pig | (R) | adductor | 1.5 ± 0.14 | Stubbs and Blanchaer |
| | (W) | quadratus femoris | 0.71 ± 0.09 | (1965) |
| rhesus | (R) | sartorius | 28.4 ± 2.5[7] | Bocek and Beatty |
| monkey | (W) | brachioradialis | 14.7 ± 2.8 | (1966) |
| rat | (R) | semimembranosus | 34.1 ± 1.8[7] | Bocek and Beatty |
| | (W) | semimembranosus | 22.6 ± 1.1 | (1966) |
| guinea pig | (R) | vastus lateralis | 116 ± 4.4[3] | Jeffress, Peter, and |
| | (W) | vastus lateralis | 91 ± 5.1 | Lamb (1968) |
| | | GLYCOLYTIC ENZYMES | | |
| Hexokinase | | | | |
| rabbit | (R) | soleus | 93[2] | Pette (1966*a*) |
| | (W) | adductor magnus | 12 | |
| rabbit | (R) | soleus | 129 ± 22[2] | Burleigh and Schimke |
| | (W) | adductor magnus | 27 ± 4 | (1968) |
| guinea pig | (R) | vastus lateralis | 11.3 ± 1.0[2] | Peter, Jeffress, and |
| | (W) | vastus lateralis | 8.1 ± 0.7 | Lamb (1968) |
| Phosphohexose isomerase | | | | |
| rabbit | (R) | soleus | 120 ± 5.5[1] | Burleigh and Schimke |
| | (W) | adductor magnus | 830 ± 118 | (1968) |
| Phosphofructokinase | | | | |
| rabbit | (R) | pectineus | 729[2] | Opie and Newsholme |
| | (W) | gastrocnemius | 1,710 | (1967) |
| Fructose 1,6-diphosphatase | | | | |
| rabbit | (R) | pectineus | 27[2] | Opie and Newsholme |
| | (W) | gastrocnemius | 50 | (1967) |

| Enzyme and species | Muscle | Enzyme units | Reference |
|---|---|---|---|
| Aldolase | | | |
| rabbit | (R) soleus | 25 ± 5[1] | Burleigh and Schimke (1968) |
| | (W) adductor magnus | 178 ± 38 | |
| cat | (R) soleus | 250[5] | Prewitt and Salafsky (1967) |
| | (W) flexor digitorum longus (FDL) | 78 | |
| rat | (R) soleus | 32[4] | Dawson and Romanul (1964) |
| | (W) gastrocnemius | 320 | |
| Glyceraldehyde phosphate dehydrogenase | | | |
| rabbit | (R) soleus | 10,850[2] | Pette (1966*a*) |
| | (W) adductor magnus | 87,260 | |
| Pyruvate kinase | | | |
| cat | (R) soleus | 280[5] | Prewitt and Salafsky (1967) |
| | (W) FDL | 1,675 | |
| rat | (R) soleus | 182[4] | Dawson and Romanul (1964) |
| | (W) gastrocnemius | 850 | |
| Lactate dehydrogenase | | | |
| guinea pig | (R) soleus | 114 ± 51[1] | Blanchaer, Van Wijhe, and Mozersky (1963) |
| | (W) quadratus femoris | 313 ± 35 | |
| rat | (R) soleus | 750[4] | Dawson and Romanul (1964) |
| | (W) gastrocnemius | 2,625[4] | |
| pig | (R) serratus ventralis | 6.3 ± 0.6[4] | Beecher, Kastenschmidt, Hoekstra, Cassens, and Briskey (1969) |
| | (W) longissimus dorsi | 15.0 ± 0.7 | |
| cat | (R) vastus intermedius | 100[9] | Guth, Watson, and Brown (1968) |
| | (W) rectus femoris | 192 | |
| rabbit | (R) soleus | 1.4 ± 0.06[5] (heart form) | Garcia-Buñuel, Garcia-Buñuel, Green, and Subin (1966) |
| | | 0.4 ± 0.04 (muscle form) | |
| | (W) gastrocnemius | 0.4 ± 0.04 (H) | |
| | | 6.0 ± 0.3 (M) | |
| rabbit | (R) pectineus | 1,474[2] (H) | Opie and Newsholme (1967) |
| | | 1,030 (M) | |
| | (W) gastrocnemius | 1,930 (H) | |
| | | 3,142 (M) | |
| Phospho enolpyruvatecarboxykinase | | | |
| rabbit | (R) pectineus | 23[2] | Opie and Newsholme (1967) |
| | (W) leg muscle | 120 | |

| Enzyme and species | Muscle | Enzyme units | Reference |
|---|---|---|---|
| α-glycerophosphate dehydrogenase | | | |
| rat | (R) soleus | 22[4] | Dawson and Romanul (1964) |
| | (W) gastrocnemius | 350 | |
| guinea pig | (R) soleus | 15.9 ± 6.3[1] | Blanchaer, Van Wijhe, and Mozersky (1963) |
| | (W) quadratus femoris | 31 ± 8.5 | |
| FATTY ACID METABOLISM | | | |
| β-OH acyl CoA-dehydrogenase | | | |
| rabbit | (R) soleus | 250[2] | Pette (1966*a*) |
| | (W) adductor magnus | 62 | |
| CITRIC ACID CYCLE ENZYMES | | | |
| Citrate synthetase | | | |
| rabbit | (R) soleus | 340[2] | Pette (1966*a*) |
| | (W) adductor magnus | 86 | |
| Isocitric dehydrogenase | | | |
| cat | (R) soleus | 42[5] | Prewitt and Salafsky (1967) |
| | (W) FDL | 17 | |
| rat | (R) soleus | 41[4] | Dawson and Romanul (1964) |
| | (W) gastrocnemius | 5 | |
| Malic dehydrogenase | | | |
| cat | (R) soleus | 2,080[5] | Prewitt and Salafsky (1967) |
| | (W) FDL | 1,385 | |
| rat | (R) soleus | 1,360[4] | Dawson and Romanul (1964) |
| | (W) gastrocnemius | 590 | |
| Succinic dehydrogenase | | | |
| rhesus | (R) soleus | 28.7 ± 0.8[6] | Beatty, Basinger, Dully, and Bocek (1966) |
| monkey | (W) brachioradialis | 5.8 ± 0.2 | |
| rat | (R) semimembranosus | 58.8 ± 3[6] | Beatty, Basinger, Dully, and Bocek (1966) |
| | (W) semimembranosus | 21.6 ± 0.2 | |
| pig | (R) serratus ventralis | 2.6 ± 0.4[8] | Beecher, Kastenschmidt, Hoekstra, Cassens, and Briskey (1969) |
| | (W) longissimus dorsi | 1.2 ± 0.2 | |
| Cytochrome oxidase | | | |
| rat | (R) soleus | 4.2[4] | Dawson and Romanul (1964) |
| | (W) gastrocnemius | 2.0 | |

| Enzyme and species | Muscle | Enzyme units | Reference |
|---|---|---|---|
| | AMINO ACID METABOLISM | | |
| Glutamicoxala-cetic transa-minase | (R) pectineus | 840[2] | Opie and Newsholme |
| rabbit | (W) leg muscle | 458 | (1967) |
| rat | (R) soleus | 94[4] | Dawson and Romanul |
| | (W) gastrocnemius | 27 | (1964) |
| | PENTOSE CYCLE ENZYMES | | |
| Glucose-6-$PO_4$ dehydrogenase | | | |
| rat | (R) soleus | 0.42[4] | Dawson and Romanul |
| | (W) gastrocnemius | 0.20 | (1964) |
| 6-phosphogluconic acid dehydro-genase | | | |
| rat | (R) soleus | 100[9] | |
| | (W) gastrocnemius | 80 | |
| cat | (R) vastus intermedius | 100[9] | |
| | (W) rectus femoris | 73 | |

Values are single determinations or means ± SE. When a single muscle is listed as both red and white, samples from the red and white areas were analyzed.

[1] μmoles substrate utilized or product formed/min per g muscle wet wt.
[2] μmoles substrate utilized or product formed/hour per g muscle wet wt.
[3] nanomoles product formed/min per mg protein.
[4] Units/g muscle.
[5] Units/min per g noncollagenous protein.
[6] μg formazan/15 min per mg nitrogen.
[7] μmoles product/min per 100 mg N.
[8] μg formazan/15 min per g wet wt.
[9] Arbitrary units.

contraction-relaxation time of white muscle since $Ca^{++}$ uptake and release by the SR have been associated with the contractile process.

It is also known that certain exceptions to the white-fast and red-slow muscle relationships exist, as, for example, the mammalian diaphragm (Padykula and Gauthier, 1967). Certain extraocular muscles and the tensor tympani and stapedius muscles of the ear are also morphologically different from skeletal muscle containing slow "felderstructur" muscle fibers similar to those found in the frog muscle (Fernand and Hess,

1969). The metabolism of these two highly specialized muscles has not been investigated.

## BLOOD FLOW

The contrasting physiological and biochemical properties of predominantly red and white muscle have generally been considered appropriate adaptations to the different functions of these muscles. Several laboratories have reported that the blood supply to red is greater than that to white muscle fibers, as might be expected from the greater dependence of red muscle on blood-borne substrates (Romanul, 1965; Reis, Wooten, and Hollenberg, 1967). In studies on anesthetized as compared with unanesthetized animals, the preservation of a threefold higher blood flow through the all red soleus than through the gastrocnemius indicates that this differential blood flow is not due to the greater activity of the soleus in maintaining posture. The higher blood flow of red muscle, therefore, is appropriately matched to its different function and metabolism. There is no way to measure the effect of blood flow on the uptake by individual fibers *in vivo,* and the effect of this differential flow is impossible to assess. In a few instances, the *in vitro* metabolism of red and white muscle has been checked by *in vivo* experiments, and the differences appeared to be similar to those *in vitro* (Bocek, Peterson, and Beatty, 1966; Goldberg, 1967).

## INTERMEDIARY METABOLISM—INTRODUCTION

In contrast to the abundant literature on histochemical differences and differences in the levels of metabolic constituents and enzyme activities between predominantly red and white muscle, few *in vitro* or *in vivo* data on the metabolism of the two muscle types have been published. This paucity is due partly to the lack of a suitable skeletal muscle fiber preparation. Most work on muscle metabolism has been done on isolated rat diaphragm or perfused rat heart, preparations that are neither typical of striated skeletal muscle nor suitable for comparison of red vs. white muscle metabolism. Diaphragm muscle is a specialized muscle that cannot be characterized in the usual manner. In contrast to voluntary skeletal muscle in which the SR is less abundant in red than in white fibers, the red fibers of the diaphragm have more SR (Padykula and Gauthier, 1967). The fiber composition of the diaphragm varies as a function of the size and metabolic rate of the animal, the rapidly contracting shrew diaphragm being uniformly composed of red fibers and the more slowly contracting cow diaphragm of white fibers. Furthermore, histological examination of the diaphragm reveals that it is covered on both sides by a complex peritoneal and pleural mesothelial membrane.

Mesothelial tissue from the peritoneum is composed of cells containing mitochondria and endoplasmic reticulum, and has numerous small microvilli projecting into the peritoneal cavity. The mesothelial covering of the diaphragm has a separate metabolism that converts hypoxanthine to uric acid, a reaction that could not be demonstrated in the underlying muscle tissue (Peterson, Beatty, and Bocek, 1961). Thus, the metabolism of diaphragm represents not only the metabolism of diaphragm muscle but also the separate metabolism of its mesothelium. Cascarano, Rubin, Chick, and Zweifach (1964) have investigated permeability changes in diaphragm and have concluded that metabolic phenomena studied in this preparation require careful evaluation regarding the limitations imposed by mesothelial barriers.

Our laboratory, therefore, extended *in vitro* studies to include voluntary skeletal muscle, and a technique was developed for preparing viable fiber bundles from the adductor muscles of the hindlimbs of rats. Predominantly red and white areas of the muscles were separated visually, and this visual differentiation was substantiated by analytical determinations (Beatty, Basinger, Dully, and Bocek, 1966). When rhesus monkeys were used, biopsies were obtained under light Sernylan anesthesia [1-(1-phenylcyclohexyl) piperidine HCl]. The major part of the sartorius (red) muscle or the superficial area of the brachioradialis (white) was removed. Removal of these superficial muscles is a simple rapid procedure that does not incapacitate the monkey. The histochemical characterization of the sartorius for succinic dehydrogenase and phosphorylase activities demonstrated that it is composed of 60–70% red fibers and that the ratio of red to white fibers is uniform throughout the muscle. The superficial part of the brachioradialis is composed of about 25% red fibers. By careful dissection along fasciculi, fibers are easily separated into groups about 50 to 60 mm long $\times$ 1 $mm^2$. The majority of muscle fibers in each group are continuous from end to end of the biopsy, and the presence of many intact cell membranes is assured since the muscle fiber group preparation is as capable of concentrating $^{14}C$-labeled $\alpha$-aminoisobutyric acid as is the isolated diaphragm (Peterson and Beatty, 1961). Incubation of fiber groups with insulin almost doubled the ratio of the concentration of $^{14}C$ in tissue water over that in the incubation medium for skeletal muscle fiber groups and diaphragm, and anoxic conditions abolished the ability of both preparations to accumulate $\alpha$-aminoisobutyric acid-$^{14}C$.

## GLYCOGEN METABOLISM

The present theory of a cyclic glycogen metabolism in muscle assumes that the rate-limiting catabolic and anabolic enzymes are phosphorylase and glycogen synthetase, respectively. Histochemical studies on resting muscle (mixed muscles) have usually revealed a reciprocal relationship between these two enzymes in individual fibers, the red fibers staining deeply for synthetase, the white for phosphorylase (Hess and Pearse, 1961; Bocek and Beatty, 1966). Unlike the staining pattern in mixed muscles, succinic dehydrogenase, phosphorylase, and glycogen synthetase were highly active in the same fibers of the all-red muscles, soleus and quadratus femoris (Bocek and Beatty, 1966). These data do not appear to be consistent since one assumes that glycogenesis as well as glycogenolysis is higher in white muscles. These histochemical results have, however, been confirmed by enzyme assays and extended to include data on the active forms of these enzymes (Stubbs and Blanchaer, 1965; Bocek and Beatty, 1966); both histochemical and analytical data indicate that synthetase is more active in predominantly red muscle than in white, whereas the converse is true of phosphorylase.

Results of an *in vitro* study of glycogen metabolism are in agreement with those of Domonkos (1961), who showed that the glycogen level of white muscle fiber groups decreases more precipitously than that of red during incubation (Beatty, Peterson, and Bocek, 1963; Bocek, Peterson, and Beatty, 1966) (Table 9.2). Bär and Blanchaer (1965), however,

TABLE 9.2

*Total glycogen counts/min per g tissue in red and white muscle of the rat*

| Duration of incub. (min) | mg/g tissue | | | $10^3$ cpm per g tissue | |
|---|---|---|---|---|---|
| | Red | White | P* | Red | White† |
| Control‡ | 2.9 ± 0.2 | 3.4 ± 0.4 | <0.02 | | |
| 10 | 3.1 ± 0.2 | 3.3 ± 0.3 | NS | 33 ± 2 | 3 ± 1 |
| 20 | 3.2 ± 0.1 | 3.0 ± 0.3 | NS | 59 ± 13 | 6 ± 2 |
| 30 | 3.2 ± 0.3 | 3.1 ± 0.2 | NS | 105 ± 20 | 10 ± 3 |
| 45 | 3.0 ± 0.1 | 2.6 ± 0.3 | NS | 113 ± 22 | 11 ± 1 |
| 60 | 3.2 ± 0.2 | 3.2 ± 0.2 | NS | 149 ± 21 | 31 ± 9 |
| 90 | 2.4 ± 0.2 | 1.9 ± 0.1 | <0.05 | 123 ± 10 | 20 ± 4 |
| 120 | 2.2 ± 0.2 | 1.9 ± 0.2 | <0.01 | 110 ± 14 | 36 ± 9 |

Red and white adductor muscle fiber groups from hind limbs of rats incubated in Krebs bicarbonate buffered medium, *p*H 7.4, plus 150 mg glucose/100 ml and 2.5 μc glucose-U-$^{14}$C. Specific activity of glucose in the medium was 425 × $10^3$ counts/min per mg glucose. Values are means ± SE, wet wt of tissue, 6 experiments. P = probability of the difference occurring by chance. NS = not significant. Glycogen expressed as glucose equivalents.

* Statistical analysis on basis of paired observations.

† P for the difference red vs. white <0.01 in all instances.

‡ End of equilibration period.

From Bocek, Peterson, and Beatty (1966).

found an increase in glycogen in both types of rat muscle during incubation. This discrepancy can be explained by differences in the initial glycogen concentrations of the tissues; those reported by Bär and Blanchaer were markedly lower than those found by Bocek, Peterson, and Beatty (1966). Furthermore, Bär and Blanchaer used rat diaphragm as a typical red muscle. This tissue is specialized for constant motion and has creatine phosphate and glycogen levels over 50% lower than those found in predominantly red rat adductor muscles (Beatty, Peterson, Bocek, and West, 1959). When we repeated the experiments of Bär and Blanchaer, our results agreed with theirs when the muscle was incubated in phosphate-buffered medium. Bär and Blanchaer suggested that this inverse relationship between initial glycogen values and net glycogen change can be explained by the stimulation of synthetase I formation with low glycogen levels.

When incorporation of glucose-$^{14}$C into glycogen of rat skeletal muscle fiber groups was measured at short intervals over a 2-hour incubation period, the incorporation of label from glucose-$^{14}$C into glycogen was much greater in red than in white muscle throughout the entire experi-

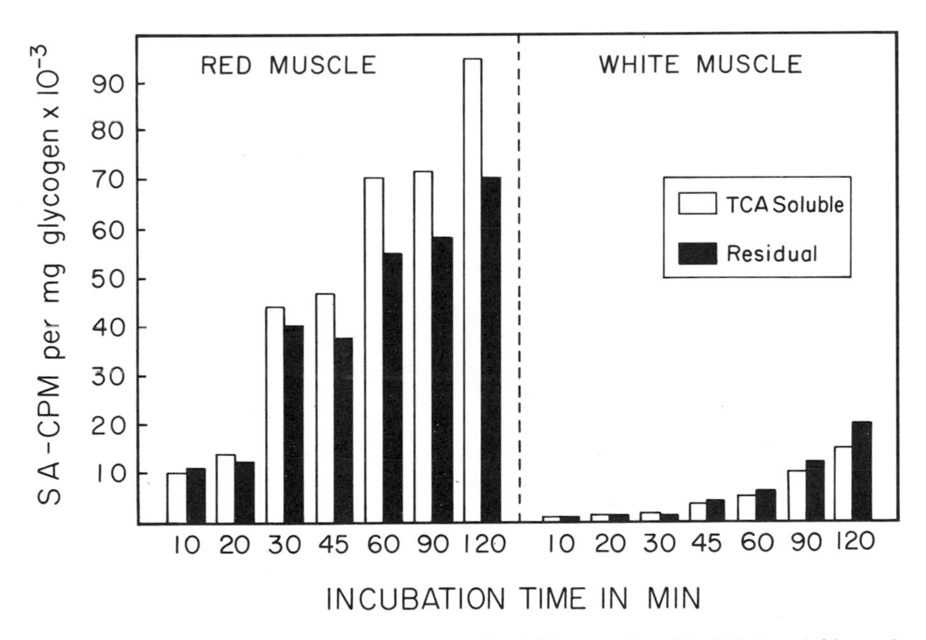


*Fig. 9.1.* Changes in the specific activities of trichloroacetic-acid (TCA)-soluble and residual glycogen in red and white adductor muscles of the rat over a 2-hour incubation, typical experiment. (See also Table 9.2.) From Bocek, Peterson, and Beatty (1966).

ment, even during the period when the glycogen levels were constant (Table 9.2). Changes in the specific activities of glycogen showed the same differences between the two types of muscle (Fig. 9.1). Bär and Blanchaer (1965) also reported greater *in vitro* incorporation of glucose-$^{14}C$ into glycogen of red muscle (diaphragm) compared with that of white muscle (external oblique). Insulin increased the incorporation of label from glucose-$^{14}C$ into glycogen 2.5- and 2.8-fold for red and white muscle fiber groups respectively, although the absolute increase was much greater in red muscle (Dully, Bocek, and Beatty, 1969). These data suggest the presence of two or more pools of glucose-6-$PO_4$ in both types of muscle. Only one of these pools was insulin-sensitive and glycogen was formed primarily from the insulin-sensitive pool. Analysis of muscle glycogen after i.v. injection of glucose-$^{14}C$ demonstrated again a higher $^{14}C$ activity in red than in white muscle glycogen (Table 9.3). In this instance, the larger blood flow through the red muscle *in vivo* favors a higher incorporation by this muscle. The ATP levels were the same in both types of muscle at the end of a 2-hour incubation period, and the creatine phosphate was about 30% higher in white muscle fiber groups (Beatty, Peterson, and Bocek, 1963). The higher glycolytic activity of the white muscle fiber preparation may have contributed to maintenance of creatine phosphate.

In contrast to the results with rats, the glycogen level of white muscle from the rhesus monkey was lower than that of the red (Table 9.4). The glycogen content of excised rhesus muscle was relatively stable, and samples, frozen in liquid nitrogen within 12 sec of biopsy or analyzed after 5 min at room temperature, showed no difference in concentrations. Although muscle biopsies were obtained under various conditions in order to avoid depletion of glycogen, the level in white muscle was always

TABLE 9.3

*Specific activity of glycogen* (in vivo *experiments*)

| Muscle | No. of samples | SA ($10^3$ cpm/mg glycogen) min after glucose-$^{14}C$ | |
|---|---|---|---|
| | | 15–30 min | 90 min |
| Red | 17 | 1.29 ± 0.27 | 1.74 ± 0.54 |
| White | 17 | 0.48 ± 0.11 | 0.92 ± 0.29 |
| P red vs. white | | <0.05 | NS |

Unpaired red and white muscle samples removed after i.v. injection of approximately 50 μc glucose-U-$^{14}C$ into rats. Values are means ± SE.

From Bocek, Peterson, and Beatty (1966).

TABLE 9.4

*Comparison of total glycogen levels in red and white muscle of the rhesus monkey*

| Muscle* | N | Glycogen, mg/g tissue, wet wt |
|---|---|---|
| Soleus | 6 | 7.28 ± 0.66 |
| Sartorius | 15 | 6.91 ± 0.26 |
| Brachioradialis | 16 | 5.62 ± 0.33 |
| P (sol. vs. br.) | | <0.001 |
| P (sol. vs. sart.) | | NS |
| P (sart. vs. br.) | | <0.001 |

Values are means ± SE. N = no. of experiments.

* Soleus (red), sartorius (predominantly red), and brachioradialis (predominantly white).

From Dully, Bocek, and Beatty (unpublished data).

lower than in red. The rhesus monkey is highly excitable, and it is known that stress causes a greater decrease in glycogen levels of white than of red muscle. Apparently the stress of being confined to a restraining chair even though the animals had adapted to the situation for a week, or the excitement of having people near even though anesthesia was induced without handling the monkey, was sufficient to cause glycogen breakdown. Hultman (1967) has reported that in man the low muscle glycogen levels found immediately after exercise to complete muscular exhaustion did not return to resting values in 20 hours. Since we waited only 4 hours after induction of anesthesia, muscle glycogen may not have been reconstituted.

### GLYCOLYSIS

The glucose uptake of predominantly red muscle fiber groups (Table 9.5) was greater than that of white, a finding that agrees with reports of a higher hexokinase activity in red muscle. However, glycogenolysis plus glycolysis as measured by chemical lactate and pyruvate productions was at least twofold higher in white than in red muscle (Bocek, Basinger, and Beatty, 1966; Domonkos, 1961). Glycolysis, as measured by lactate-$^{14}$C plus pyruvate-$^{14}$C production from glucose-$^{14}$C and by calculation of the percentage of label from glucose uptake appearing in lactate and pyruvate, was higher in white than in red muscle, although glucose uptake was higher in red muscle (Table 9.5). These data agree with reports that glycolytic enzymes, with the exception of hexokinase, are higher in white muscle (Table 9.1). It should also be recognized that in rapidly metabo-

TABLE 9.5

*Glucose uptake and percentage of metabolized glucose-¹⁴C appearing in various metabolic fractions of red and white muscle of rats*

| Muscle | Glucose uptake, mg/g per hr, wet wt | % incorporation* | | | |
|---|---|---|---|---|---|
| | | Glycogen | Lactate | Pyruvate | $CO_2$ |
| Red | 1.90 ± 0.09 | 22.5 ± 1.2 | 32.7 ± 1.6 | 5.0 ± 0.2 | 9.7 ± 0.7 |
| White | 1.65 ± 0.04 | 9.2 ± 0.1 | 44.0 ± 2.7 | 8.1 ± 0.4 | 6.4 ± 0.4 |
| P | <0.025 | <0.001 | <0.005 | <0.001 | <0.001 |

Values are means ± SE, 10 experiments.

* Percent incorporation defined as total $^{14}C$ activity in the metabolite divided by the counts per min removed from the medium (glucose-$^{14}C$ uptake) corrected for lactate-$^{14}C$ and pyruvate-$^{14}C$ production into the medium during incubation × 100. Experimental conditions as in Table 9.2.

From Bocek, Basinger, and Beatty (1966).

lizing systems, rates of metabolic activity cannot be estimated adequately by determining levels of constituents. In an overall calculation, the amount of lactate and pyruvate found in the medium is a balance between production and utilization. Therefore, lactic and pyruvic acid accumulations, for example, indicate only minimal rates of glycolysis, since the concentrations of lactate and pyruvate are influenced by oxidation of these constituents via the citric acid cycle, to which we now turn.

TABLE 9.6

*Succinic dehydrogenase activity in homogenates of red and white muscle from rhesus monkey*

| | Quadratus femoris* | Soleus † | Sartorius ‡ | Brachioradialis |
|---|---|---|---|---|
| μg formazan/mg N | | | | |
| Female | 41.0 ± 2.0(7) | 31.3 ± 2.8(6) | 17.9 ± 1.2(8) | 5.9 ± 0.6(8) |
| Male | — | — | 23.8 ± 2.1(6) | 5.4 ± 0.7(5) |
| P, male vs. female | | | <0.02 | NS |
| Histochemical characterization § | 4+ | 4+ | 3+ | 1+ |

Values are means ± SE; number of animals given in parentheses.

* P, Q.f. vs. sol. <.025.

† P, sol. vs. sart. <.005.

‡ P, sart. vs. Br. <.005.

§ Histochemical classification: 1+, less than 25% red fibers; 2+, 25–50% red fibers; 3+, 50–74% red fibers; 4+, more than 75% red fibers or 100% red fibers (two types of fibers).

From Beatty, Basinger, Dully, and Bocek (1966).

TABLE 9.7

*Succinic dehydrogenase activity in homogenates of predominantly red areas of muscle as compared with predominantly white areas of same muscle*

| | μg formazan/mg nitrogen | | |
|---|---|---|---|
| Muscle | Red areas* | White areas† | P‡ |
| Rat | | | |
| Semimembranosus & semitendinosus | 58.8 ± 3.0 | 21.6 ± 2.0(18) | <0.005 |
| Rhesus monkey | | | |
| Semimembranosus & semitendinosus | 20.6 ± 1.1 | 8.1 ± 1.0(6) | <0.005 |
| Tibialis anterior | 29.1 ± 3.1 | 21.2 ± 1.2(6) | <0.025 |
| Gastrocnemius | 29.1 ± 2.0 | 10.6 ± 0.9(5) | <0.01 |
| Brachioradialis | 17.2 ± 1.0§ | 7.0 ± 0.1(6) | <0.02 |

Values are means ± SE; number of animals given in parentheses.
* Histochemical analysis 50% red fibers, 3+.
† Histochemical analysis 25% red fibers, 1+.
‡ Statistical analysis on basis of paired observations, red vs. white muscle.
§ The red area of this muscle is small and lies close to the bone.
From Beatty, Basinger, Dully, and Bocek (1966).

## CITRIC ACID CYCLE

The evidence that citric acid cycle (CAC) activity is higher in red than in white muscle is overwhelming, and we shall make no attempt to review all the data. Numerous workers have used either histochemical (Dubowitz and Pearse, 1960; Engel, 1962) or analytical techniques (Table 9.1), and we have demonstrated a direct correlation between the qualitative histochemical classification and the quantitative measurements of succinic dehydrogenase in red and white muscles (Table 9.6). It should be noted that although the enzyme levels are higher in the rat, the maximum relative differences in activity between red and white areas of the same muscle are similar in the rat (2.7 ×) and rhesus monkey (2.5 to 2.8 ×) (Table 9.7). In general, the histochemical and analytical evidence has been confirmed by *in vitro* experiments.

In our *in vitro* experiments, a larger percentage of the glucose uptake appeared in the $^{14}CO_2$ fraction of red than of white muscle (Table 9.5). Since the glucose uptake of red muscle was also greater, more glucose-$^{14}C$ was oxidized by the CAC and more $^{14}CO_2$ was produced (Table 9.8). If the oxidation of lactate and pyruvate is greater in red than in white muscle, lactate and pyruvate accumulations likewise decrease in red muscle in the absence of any difference in rate of glycogenolysis and glycolysis between the two muscle types. When the total ATP formed

TABLE 9.8

$^{14}CO_2$ *production from glucose-*$^{14}C$ *by predominantly red and white muscle fiber groups of the rat*

| Duration of incub. (min) | $10^3$ counts/min per g tissue, wet wt | | | |
|---|---|---|---|---|
| | Red muscle | White muscle | Diff.* | P† |
| 10 | 5.2 ± 1.2 | 3.5 ± 0.7 | − 1.7 ± 0.7 | <0.06 |
| 20 | 10.8 ± 2.3 | 8.2 ± 1.4 | − 2.2 ± 1.2 | NS |
| 30 | 16.2 ± 2.9 | 10.5 ± 2.0 | − 5.7 ± 1.2 | <0.01 |
| 45 | 25.5 ± 4.3 | 19.6 ± 4.7 | − 8.0 ± 1.8 | <0.025 |
| 60 | 45.5 ± 9.9 | 27.0 ± 4.7 | −18.5 ± 5.8 | <0.025 |
| 90 | 80.6 ± 17.7 | 36.7 ± 5.3 | −42.8 ± 7.2 | <0.025 |
| 120 | 103.0 ± 21.0 | 58.3 ± 8.9 | −44.3 ± 18.0 | <0.05 |

Experimental conditions as in Table 9.2. Values are means ± SE, 6 experiments.

* Difference ± SE of difference.

† Statistical analysis on basis of paired observations.

From Bocek, Peterson, and Beatty (1966).

from the conversion of glucose to lactate, pyruvate, and $CO_2$ (Table 9.5) in the two preparations is calculated, it is found that red muscle has a gain of 59 μmoles of ATP/g per hour and white muscle 54 μmoles of ATP/g per hour. Oxidation via the pentose cycle is so low that this pathway is disregarded in the calculation (Beatty, Peterson, Basinger, and Bocek, 1966).

Although glucose uptake and oxygen consumption are higher in red than in white muscle, the percentage of oxygen uptake accounted for by carbohydrate oxidation is also similar in the two types of muscle (resting state). Based on glucose uptakes of 10.6 and 9.2 μmoles/g per hour and oxygen consumptions of 32 and 22 μmoles/g per hour for red and white muscle, respectively, the percentage of oxygen uptake accounted for by glucose-$^{14}C$ oxidation was 19% for red and 16% for white muscle. Glycogen of a relatively low specific activity was also being degraded, however, and these glucosyl units oxidized during the experimental period. The magnitude of the unlabeled carbohydrate fraction metabolized can be estimated from dilution of pyruvate-$^{14}C$ or lactate-$^{14}C$. From the specific activity of glucose in the incubation medium ($425 \times 10^3$ cpm/mg) and of pyruvate (302 and $247 \times 10^3$ cpm/mg), it can be calculated (SA pyruvate/SA glucose × 100) that 71% and 58% of the pyruvate produced by red and white muscle respectively originated from glucose-$^{14}C$. After correction for this dilution of the glucose-$^{14}C$, the percentage of oxygen uptake accounted for by oxidation of carbohydrate was 27% for red and

28% for white muscle. In agreement with these data, the respiratory quotients of red and white muscle fiber groups are similar (0.82 ± 0.05 SE and 0.79 ± 0.05 SE respectively) and typical of values for the oxidation of a mixture of substrates (Cassens, Bocek, and Beatty, 1969). Presumably, voluntary skeletal muscle has a supply of endogenous lipids available as a substrate for respiration *in vitro*. While the data on lactate and pyruvate productions appear to confirm the thesis that white muscle depends more on glycogenolysis and glycolysis for energy production than does red, one must remember that carbohydrate does not seem to be the major energy source in either predominantly red or white muscle under the conditions of these experiments. This conclusion is not in agreement with many workers who regard glycolysis as the predominant energy pathway in white muscle (Citoler, Benítez, and Maurer, 1967; Reis, Wooten, and Hollenberg, 1967). Opie and Newsholme (1967) comment that "the substrate for oxidation in white muscle is probably glucose or glycogen rather than lipid, which is more important in red muscle." *In vitro* experiments with stimulated muscle fiber groups might yield different results, however.

Several groups of workers have demonstrated higher respiratory rates in red muscle homogenates and mitochondria, muscle slices, and muscle fiber groups (Ogata, 1960; Bär and Blanchaer, 1965; Domonkos and Latzkovits, 1961; Beatty, Peterson, and Bocek, 1963; Cassens, Bocek, and Beatty, 1969). Ogata (1960) demonstrated a greater stimulation of respiration for red than for white muscle with pyruvate and succinate as substrates. Carbon dioxide production from lactate-1-$^{14}C$ was greater in mitochondrial preparations from red than from white muscle (Bär and Blanchaer, 1965); however, lactate uptakes were not measured in these experiments. Pyruvate, acetoacetate, and $\alpha$-ketoglutarate were utilized to a greater extent by red than by white muscle (Domonkos, 1961; Beatty, Peterson, and Bocek, 1963). Likewise, $^{14}CO_2$ production from acetoacetate-3-$^{14}C$ was higher by red than by white muscle fiber groups. Increased uptake and $^{14}CO_2$ production does not, however, necessarily prove increased CAC activity. When both acetoacetate-3-$^{14}C$ uptake and oxidation to $^{14}CO_2$ were measured in the same preparation, the percentage increase in uptake and in cpm/mg of tissue appearing in $^{14}CO_2$ was the same for red and white muscle (Table 9.9). In this case the greater $^{14}CO_2$ production of red muscle could be due to a greater acetoacetate uptake that was secondary to a higher level of activating enzyme rather than to a higher CAC activity. In agreement with histochemical data (Dubowitz and Pearse, 1960), more acetoacetic acid was converted to $\beta$-hydroxybutyrate by red muscle fiber groups (Beatty, Peterson, and Bocek, 1963). This again is evidence of a higher oxidative activity in red muscle since

TABLE 9.9

*Acetoacetic acid uptake and utilization by red and white muscle fiber groups from the rat*

| | N | Red | White | P* |
|---|---|---|---|---|
| Acetoacetic acid uptake μmoles/g per hr, wet wt | 6 | 5.3 | 3.4 | <0.01 |
| $^{14}CO_2$ production cpm/mg per hr, wet wt | 6 | 106 ± 10 | 74 ± 7 | <0.01 |
| % uptake in $CO_2$ | 6 | 72 ± 4 | 69 ± 4 | NS |

Values are means ± SE. Krebs glycylglycine buffered medium, 0.7 mM acetoacetate (3.2 μc/flask acetoacetate-3-$^{14}$C) and 8.5 mM glucose. N = no. of experiments.

* Statistical analysis on basis of paired observations.

From Beatty, Peterson, and Bocek (1963).

β-hydroxybutyrate dehydrogenase is reported to be attached to the respiratory enzyme assembly in the mitochondrial membrane.

### LACTATE AND α-GLYCEROPHOSPHATE METABOLISM

Using histochemical techniques, Van Wijhe, Blanchaer, and Jacyk (1963*b*) reported about equal amounts of α-glycerophosphate dehydrogenase (αGPDH) in red and white skeletal muscle fibers in the presence of phenazine methosulfate (PMS) and a higher αGPDH activity in white

TABLE 9.10

*α-Glycerophosphate concentrations of muscle from rats following 2-hour incubation under aerobic or hypoxic conditions*

| | N | αGP concentration (μmoles/g wet wt) Aerobic 95% $O_2$ + 5% $CO_2$ | Hypoxic 95% $N_2$ + 5% $CO_2$ | P* aerobic vs. hypoxic |
|---|---|---|---|---|
| Red muscle groups | 6 | 0.12 ± 0.02 | 0.53 ± 0.05 | <.005 |
| White muscle groups | 6 | 0.06 ± 0.02 | 0.22 ± 0.04 | <.005 |
| P red vs. white muscle | | <.02 | <.01 | |

Values are means ± SE. Krebs bicarbonate buffered medium, *p*H 7.4 plus 150 mg glucose per 100 ml. N = no. of experiments.

* Statistical analysis on basis of paired observations.

From Peterson, Gaudin, Bocek, and Beatty (1964).

muscle by a quantitative assay, also with PMS (Blanchaer, Van Wijhe, and Mozersky, 1963). According to these authors, a low diaphorase level in this preparation of white muscle accounted for the previous erroneous impression that αGPDH was low in white muscle (histochemical data). Pette (1966*b*) has also reported a higher level of αGPDH in white than in red muscle. The oxygen consumption of white mitochondria was higher than that of red with α-glycerophosphate (αGP) as a substrate, and white muscle mitochondria had a higher rate of oxygen consumption with αGP than with lactate, succinate, or $NADH_2$ (Blanchaer, 1964). When lactate or succinate was the substrate, the $QO_2$ of red muscle mitochondria was higher than with αGP. Blanchaer concluded from these data that an αGP shuttle coupling the reactions generating $NADH_2$ in the cytoplasm of muscle with the mitochondrial respiratory chain may be more important in white than in red muscle, whereas direct oxidation of $NADH_2$ may be more important in red muscle mitochondria. *In vitro* experiments in our laboratory demonstrated that after 2 hours' incubation in medium containing 100 mg% glucose, there was more αGP in red than in white muscle fiber groups and that incubation under hypoxic conditions increased the concentration of αGP about fourfold in both types of muscle (Table 9.10). Possibly these higher αGP levels in red muscle fiber groups after incubation are a reflection of more active utilization of this compound by white muscle mitochondria than by red, as suggested by Blanchaer's data. Turnover experiments are required to solve this problem.

In the presence of PMS, lactate dehydrogenase activity was at least as high in white as in red muscle fibers (histochemical data), and quantitative assays demonstrated higher activity in both the whole homogenate and the supernatant of white muscle (Blanchaer, Van Wijhe, and Mozersky, 1963; Van Wijhe, Blanchaer, and Jacyk, 1963*a*). Not only are lactate dehydrogenase activities higher in white muscle (Table 9.1) but the distribution of the isozymes is specific in red and white fibers (Van Wijhe, Blanchaer, and Stubbs, 1964). In general, the isozyme distribution of individual fibers agreed with previous data on predominantly red and white muscle homogenates. In single red muscle fibers isozymes I, II, and III were found to be moderate to strong in activity; however, these fibers also contained isozymes IV and V. Isozyme V was strongly represented in single white fibers; the remaining isozymes were much less active. These isozyme patterns of red and white fibers can be correlated with the kinetic properties of lactate dehydrogenase subunits and the different physiological roles of these subunits (Dawson, Goodfriend, and Kaplan, 1964).

TABLE 9.11

*Succinic dehydrogenase activity in homogenates of soleus (Sol) and brachioradialis (BR) muscles of fetal, neonatal, and adult rhesus monkeys as measured by formazan production*

| | μg formazan/10 mg wet wt | | | μg formazan/mg N | | |
|---|---|---|---|---|---|---|
| Series | Sol | BR | P* | Sol | BR | P* |
| Fetal † | | | | | | |
| 90-day | 1.6 ± 0.1 | 1.3 ± 0.1 | <0.02 | 11.9 ± 0.4 | 9.8 ± 0.8 | <0.02 |
| 120-day | 3.3 ± 0.2 | 1.8 ± 0.2 | <0.005 | 14.2 ± 0.6 | 8.1 ± 1.0 | <0.005 |
| P 90- vs. 120-day | <.005 | NS | | <.05 | NS | |
| 150-day | 3.4 ± 0.4 | 1.8 ± 0.2 | <0.005 | 13.5 ± 0.9 | 6.6 ± 1.0 | <0.005 |
| P 120- vs. 150-day | NS | NS | | NS | NS | |
| Infant ‡ | | | | | | |
| 1–4-day (neonate) | 3.3, 3.2 | 1.6, 1.8 | | 11.5, 11.4 | 5.3, 7.0 | |
| 2–6-week (infant) | 3.5, 4.7 | 1.9, 2.4 | | 11.5, 15.3 | 6.4, 7.6 | |
| 2-Year | 10.6, 11.1 | 2.7, 2.5 | | 32.0, 34.0 | 8.1, 7.9 | |
| Adult § ‖ | 10.8 ± 0.6 | 2.3 ± 0.3 | <0.005 | 33.1 ± 2.8 | 6.3 ± 0.6 | <0.005 |
| P adult vs. 150-day | <0.005 | NS | | <0.005 | NS | |

Values are means ± SE.

* Statistical analysis on basis of paired observations, sol vs. BR.

† Each mean represents duplicate determinations on five fetuses.

‡ Each value represents duplicate or triplicate determinations on different monkeys.

§ Single determinations on nine monkeys.

‖ Samples of BR obtained from superficial portion (2 mm deep) of entire length of muscle.

From Beatty, Basinger, and Bocek (1967).

## PENTOSE CYCLE ACTIVITY

Red muscle has higher glucose-6-phosphate dehydrogenase and 6-phosphogluconic acid dehydrogenase activities than does white (Table 9.1), a difference to be expected since red muscle has a higher rate of protein synthesis and therefore needs a larger supply of ribose for RNA synthesis. However, the overall dehydrogenase values reported were low for both types of muscle. When pentose cycle (PC) activity of predominantly red muscle fiber groups *in vitro* was calculated from the specific yields (percentage of label from the uptake appearing in a metabolic product × $10^{-2}$) of $^{14}CO_2$ from glucose-1- and glucose-6-$^{14}C$, less than 0.5% of the glucose uptake was metabolized via this pathway even under conditions which should stimulate PC activity (Beatty, Peterson, Basinger, and Bocek, 1966). Because of the low activity of the PC, quantitative differences between red and white muscle could not be measured in this preparation.

## FETAL MUSCLE METABOLISM

One of the major areas of investigation at the Oregon Regional Primate Research Center is that of fetal and neonatal physiology. Our group has been interested in the metabolism of rapidly growing muscle tissue from the rhesus monkey. Before investigating the intermediary metabolism of red and white fetal muscle, we had to ascertain when muscles differentiate in this species. Succinic dehydrogenase activity was lower in the soleus muscle from the rhesus fetus than in that from the adult (Table 9.11). By 90 days (about 50% of gestation), however, enzyme activity was higher in red than in white fetal muscle. By 120 days (73% of gestation), the fetal muscle showed a pattern of differentiation similar to that of the adult (Fig. 9.2); these analytical results agree with the histochemical data. Since the dry weight and nitrogen content are lower and the percentage of total protein (nitrogen) represented by collagenous

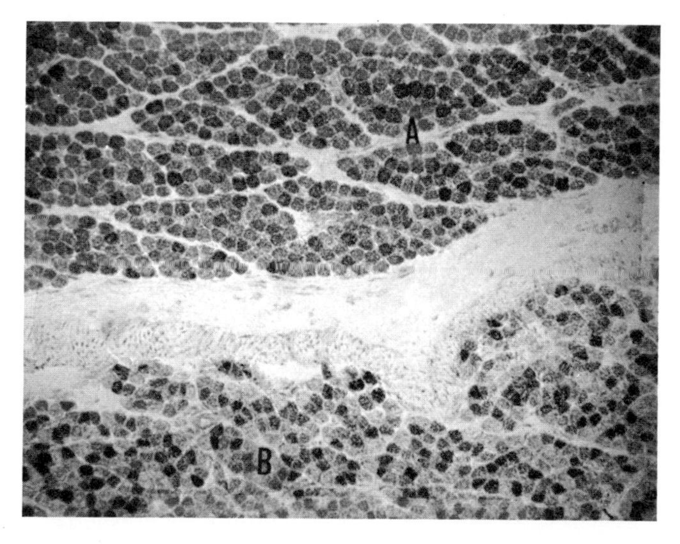


*Fig. 9.2. Area A,* soleus of a 120-day rhesus fetus; all red muscle with two types of red fibers. *Area B,* plantaris; mixed fiber types. Stained for succinic dehydrogenase activity. × 100.

TABLE 9.12

*Hydroxyproline, collagen, total nitrogen, and percentage of dry weight in predominantly red and predominantly white muscle of the rhesus monkey*

| Series | Hydroxyproline (mg/g wet wt) | Collagen* (wet wt) (%) | Hydroxyproline (mg/g N) | Nitrogen (wet wt) (g/100 g) | Dry wt (%) |
|---|---|---|---|---|---|
| Fetal | | | | | |
| 90-day red† (5) | 127 ± 5 | 0.95 | 99 ± 4 | 1.29 ± 0.03 | 11.9 ± 0.4 |
| white‡ (4) | 169 ± 14§ | 1.26 | 132 ± 11§ | 1.27 ± 0.03 | 11.8 ± 0.4 |
| 120-day red (5) | 240 ± 11 | 1.79 | 107 ± 5 | 2.27 ± 0.07 | 18.2 ± 0.2 |
| white (5) | 322 ± 36‖ | 2.40 | 146 ± 16‖ | 2.21 ± 0.08 | — |
| 150-day red (7) | 245 ± 22 | 1.83 | 93 ± 8 | 2.63 ± 0.07 | 20.2 ± 0.4 |
| white (7) | 418 ± 17** | 3.12 | 159 ± 10** | 2.62 ± 0.08 | — |
| Infant | | | | | |
| 1–5 day red (3) | 292 ± 28 | 2.18 | 101 ± 10 | 2.88 ± 0.06 | 20.6 ± 0.4 |
| white (3) | 386 ± 15‖ | 2.87 | 134 ± 5‖ | — | — |
| 2–6 week red (5) | 305 ± 9 | 2.28 | 98 ± 3 | 3.17 ± 0.07 | 21.9 ± 0.3 |
| white | — | — | — | — | — |
| Adult | | | | | |
| Red†† (15) | 182 ± 15 | 1.36 | 56 ± 5 | 3.24 ± 0.05 | 24.3 ± 0.2 |
| White‡ (15) | 353 ± 29** | 2.63 | 108 ± 9** | 3.27 ± 0.06 | 24.8 ± 0.2 |

Values are means ± SE. P values for red vs. white muscle were calculated on the basis of paired observations. Numbers in parentheses represent number of animals (duplicate or triplicate samples on each sample).

* Assuming that hydroxyproline represents 13.4% of collagenous protein.

† Mixed thigh muscle, predominantly red.

‡ Brachioradialis.

§ P for red versus white muscle <0.02.

‖ P for red versus white muscle <0.05.

** P for red versus white muscle <0.01.

†† Sartorius.

From Beatty, Basinger, and Bocek (1967).

protein is higher in fetal than in adult muscle in the species previously studied (Dickerson and Widdowson, 1960), we needed to determine these parameters for the red and white muscle of the rhesus fetus as a base of reference for calculating metabolic data. In agreement with the histochemical data, the hydroxyproline concentration (collagen) on the basis of wet weight was lower in red than in white muscle as early as 90 days fetal age (Table 9.12). When the hydroxyproline values were calculated on the basis of nitrogen content of muscle, the concentrations were also lower in red than in white muscle and were constant for each type of muscle from 90 days fetal age through the 2- to 6-week period. The differences in collagen content of the adult and fetal series were greater for red than for white muscle. In rapidly growing red muscle, hydroxyproline (collagen) levels were almost 100% higher than in the adult, and in rapidly growing white muscle this difference was about 40%. Unfortunately, the amount of white muscle available in the rhesus fetus was not sufficient to prepare satisfactory muscle fiber groups. However, when the oxygen consumption and $CO_2$ production of predominantly red muscle from fetuses of various ages and from adults were compared, the $QO_2$ values and the $CO_2$ productions were higher at 90 days, similar at 125 days, and lower at 155 days gestational age than those of adult muscle (Table 9.13). These data were calculated on the basis of noncollagenous protein nitrogen (NCN) to correct for variations in the content of water and collagenous protein (Beatty, Basinger, and Bocek, 1968). The glucose uptake of this preparation of fetal muscle was higher at 90 and 125 days gestational age than that of adult muscle (Table 9.14). According to Guth (1968), red fibers are the more primitive form of muscle fibers.

TABLE 9.13

*Oxygen consumption and $CO_2$ production of muscle fiber groups from fetal, infant, and adult rhesus monkeys*

| | Fetal-days | | | Infant | |
|---|---|---|---|---|---|
| | 89–90 | 120–129 | 150–160 | (2–6 wk) | Adult |
| $QO_2$ µmoles/mg NCN hr | | | | | |
| First hour | 1.20 ± 0.04* | 0.75 ± 0.06 | 0.63 ± 0.03† | 0.67 ± 0.06 | 0.78 ± 0.05 |
| Second hour | 1.03 ± 0.03* | 0.61 ± 0.02 | 0.51 ± 0.01* | 0.62 ± 0.05 | 0.65 ± 0.03 |
| | (4) | (6) | (6) | (3) | (14) |
| $CO_2$ µmoles/mg NCN | | | | | |
| | 1.68 ± 0.14* | 1.04 ± 0.07 | 0.91 ± 0.05† | — | 1.09 ± 0.05 |
| | (3) | (5) | (5) | | (7) |
| RQ‡ | 0.76 ± 0.06 | 0.78 ± 0.06 | 0.80 ± 0.06 | — | 0.78 ± 0.02 |
| | (3) | (5) | (5) | | (7) |

Muscle incubated for 2 hours in Krebs glycylglycine buffered medium, *p*H 7.4 at 37°C plus 100 mg of glucose per 100 ml; the gas phase $O_2$. Values are means ± SE. Numbers in parentheses denote number of monkeys, duplicate, triplicate, or quadruplicate flasks in each experiment. NCN = noncollagenous protein nitrogen.

* P for fetal vs. adult series <0.01.

† P for fetal vs. adult series <0.05.

‡ Calculated for 2-hour period.

From Beatty, Basinger, and Bocek (1968).

TABLE 9.14

*Glucose uptakes of muscle fiber groups from fetal, infant, and adult rhesus monkeys*

| | Fetal-days | | | Infant | | |
|---|---|---|---|---|---|---|
| | 89–90 | 120–129 | 150–160 | 1–5 days | 2–6 wk | Adult |
| Glucose uptake μg/mg NCN*/hr | 78 ± 6† (5) | 57 ± 2† (8) | 47 ± 1 (7) | 44 ± 4 (5) | 48 ± 3 (6) | 42 ± 2 (7) |

Muscle incubated for 2 hours in Krebs bicarbonate buffered medium plus 100 mg glucose per 100 ml medium; 1.2 μc glucose-$^{14}$C per ml was added at the end of the 15-min equilibration period. Values are means ± SE. Numbers in parentheses denote the number of monkeys (duplicate flasks in each experiment).

* Noncollagenous protein nitrogen.

† P for fetal vs. adult <.005.

From Beatty, Basinger, and Bocek (1968).

TABLE 9.15

*Comparison of palmitate-1-$^{14}$C uptake and distribution by red and white muscle fiber groups of the rat*

| Muscle | Uptake μmoles/ g per 2 hr | % label from uptake in lipid | % lipid label in triglycerides | % lipid in muscle |
|---|---|---|---|---|
| Red | 1.12 ± 0.10 | 84 ± 6 | 67 ± 1 | 1.42 ± 0.08 |
| White | 1.10 ± 0.07 | 68 ± 3 | 49 ± 2 | 0.87 ± 0.04 |
| P* red vs. white | NS | <0.005 | <0.005 | <0.005 |

Values are means ± SE of data from 4 rats, duplicate flasks. Red and white muscle fiber groups were incubated in Krebs glycylglycine buffered medium plus 8.3 μmoles glucose and 0.5 μmoles palmitate per ml, 0.85 μc of palmitate-1-$^{14}$C was added at the end of a 15-min equilibration period and the incubation continued for 2 hours.

* P values calculated on the basis of paired observations.

Since the major portion of the histochemical differentiation (succinic dehydrogenase) in rhesus fetuses occurs between 90 and 120 days fetal age, the conversion of red to white fibers in this species correlates with the *in vitro* oxygen uptake, that is, the undifferentiated red 90-day fetal muscle consumes more oxygen than the more differentiated, although predominantly red, 120-day fetal muscle.

### LIPID METABOLISM

Many laboratories have demonstrated, histologically and analytically, a sharp difference in the distribution of lipids of red and white muscle (Fröberg, 1967; Masoro, Rowell, and McDonald, 1964), red muscle having a greater supply. George and Talesara (1961) have reported that the lipase activity in red pigeon pectoral muscle is almost three times greater than that in white muscle. Concurring with these observations, Wirsén (1965) showed by autoradiography that after i.v. injection of albumin-bound palmitate-$^{14}C$, red pigeon breast muscle took up more activity than did white. It has also been demonstrated that trout red muscle is able to oxidize more $^{14}C$-labeled hexanoate, octanoate, and myristic acid to $^{14}CO_2$. Although the uptake of albumin-bound palmitate-$^{14}C$ by red and white rat muscle did not differ, the percentage of uptake appearing in the lipid fraction was greater in red muscle and so was the incorporation (Table 9.15). The distribution of the label in the lipid fraction was also different in the two series although 50% or more of the label appeared in the triglyceride fraction of both types of muscle.

### GLUCOSE FATTY-ACID CYCLE

The effect of free fatty acids (FFA) on the metabolism of cardiac and diaphragm muscle has been extensively investigated in the perfused rat heart and isolated rat diaphragm (Ruderman, Toews, and Shafrir, 1969). The investigations with perfused rat heart indicate that high levels of FFA increase FFA utilization and decrease glucose uptake. This work on the interrelationships between glucose and fatty acid oxidation in cardiac muscle supports the glucose fatty-acid cycle theory proposed by Randle, Garland, Hales, and Newsholme (1963), which states that increased availability of fatty acids for oxidation is responsible for the abnormalities in carbohydrate metabolism of muscle in diabetes, starvation, and carbohydrate deprivation. The results with rat diaphragm *in vitro,* however, are conflicting (Ruderman, Toews, and Shafrir, 1969). Furthermore, heart and diaphragm are constantly active muscles that morphologically and perhaps biochemically are not typical of striated skeletal muscle; they comprise a relatively small proportion of the total muscle mass. In comparison with the amount of data on cardiac and diaphragm muscle, the paucity of information on the effect of FFA on voluntary skeletal muscle is amazing. Levari and Kornbleuth (1966) incubated muscle slices with and without low levels of palmitate (0.3 mM) and found no differences in the conversion of glucose-U-$^{14}C$ to lactate-$^{14}C$, glycogen-$^{14}C$, or $^{14}CO_2$. Cassens, Bocek, and Beatty (1969) incubated predominantly red and white skeletal muscle fiber groups with and without octanoate (0.8 and 4.0 mM) and observed no difference in glucose uptake (Table 9.16) even though 30–40% of the total $CO_2$ originated from octanoate. Octanoate, however, increased oxygen consumption in both types of muscle, a fact that suggested that this substance uncoupled oxidative phosphorylation. Despite the differences in

TABLE 9.16

*Effect of 4.0 mM octanoate on carbohydrate metabolism in red and white muscle of the rhesus monkey*

| | Red muscle | | White muscle | | P red vs. white muscle | |
|---|---|---|---|---|---|---|
| | −octanoate | +octanoate | −octanoate | +octanoate | −octanoate | +octanoate |
| Glucose uptake* | | | | | | |
| mg/g per hr | 0.83 ± 0.03 | 0.87 ± 0.03 | 0.98 ± 0.05 | 1.03 ± 0.05 | <.02 | <.01 |
| P for octanoate† | | NS | | NS | | |
| Oxygen consumption | | | | | | |
| μmoles/g per 2 hrs | 27.0 ± 0.8 | 43.8 ± 2.2 | 23.2 ± 0.8 | 38.3 ± 1.4 | <.05 | <.01 |
| P for octanoate† | | <.001 | | <.001 | | |
| $CO_2$ production | | | | | | |
| μmoles/g per 2 hrs | 22.0 ± 1.7 | 37.0 ± 1.7 | 17.6 ± 1.3 | 31.5 ± 2.2 | <.025‡ | NS |
| P for octanoate† | | <.001 | | <.001 | | |
| RQ | 0.82 ± 0.05 | 0.84 ± 0.03 | 0.76 ± 0.05 | 0.82 ± 0.05 | NS | NS |
| P for octanoate† | | NS | | NS | | |

Red and white fiber groups were incubated in Krebs glycylglycine buffered medium, *p*H 7.4, plus 1 mg glucose and 1.33 μc glucose-U-$^{14}C$/ml. Values are means ± SE, 15 to 18 pairs of flasks each for red and white muscle (wet wt) from 9 monkeys.

* Calculated from $^{14}C$-glucose uptake values corrected for $^{14}C$-lactate production in the media during incubation.

† P for the addition of 4.0 mM octanoate.

‡ Paired observations.

From Cassens, Bocek, and Beatty (1969).

metabolism between predominantly red and white muscle, octanoate affected both types of muscle in a similar manner. Incubation of predominantly red muscle fiber groups with and without 1.5 mM palmitate (molar ratio FFA/albumin = 3) also had no effect on glucose-$^{14}C$ uptake ($-0.2 \pm 0.4$ SE $\mu$moles glucose per g tissue, wet wt per hour) but decreased the percentage of label from the uptake appearing in $^{14}CO_2$ ($-2.2 \pm 0.1$ SE) Beatty and Bocek, unpublished observations). These results suggest that fatty acid inhibition of glucose uptake is important in heart and perhaps in diaphragm muscle but not in red and white voluntary skeletal muscle.

### PROTEIN METABOLISM

It appears reasonable to assume that higher levels of RNA correlate with higher rates of metabolism since the RNA level is higher in red than in white muscle (Margreth and Novello, 1964; Goldberg, 1967). Recently, Opie and Newsholme (1967) have reported that glutamic-oxaloacetic transaminase is higher in red muscle. There appears to be less glycine in red than in white muscle, both in the resting and in the contracted state (Krieg and Kirsten, 1961). Results from the few *in vivo* and *in vitro* experiments on protein metabolism in the two types of muscle generally agree with the above data. Beatty, Peterson, Bocek, Craig, and Weleber (1963) reported that glycine-1-$^{14}C$ incorporation was higher in predominantly red than in white muscle fiber groups after a 2-hour incubation in Krebs-bicarbonate buffered medium (Table 9.17).

TABLE 9.17

*Effect of glucagon on the incorporation of glycine-$^{14}C$ into protein from thigh adductor muscle fiber groups of rats*

| | Controls | +glucagon (16 $\mu$g/ml) | | |
|---|---|---|---|---|
| | Protein* (cpm/mg) | Protein* (cpm/mg) | % decrease | P† for % decrease |
| White muscle | 43.0 ± 3.2(12) | 33.4 ± 2.8(11) | 22.0 ± 5.0(11) | <0.005 |
| Red muscle | 59.3 ± 4.2(12) | 42.5 ± 3.5(12) | 28.7 ± 4.8(12) | <0.005 |
| P red vs. white muscle | <0.01 | <0.02 | NS | |

Krebs bicarbonate buffered medium, *p*H 7.4, plus glycine-1-$^{14}C$, 0.31 $\mu$c/ml (0.13 mg/100 ml); 2 hours incubation. Values are means ± SE. Numbers of flasks given in parentheses.

* Biuret method.

† Statistical analysis on basis of paired observations.

From Beatty, Peterson, Bocek, Craig, and Weleber (1963).

Glucagon added to the medium caused a similar percentage decrease in the incorporation of $^{14}C$ from glycine into the protein of both types of muscle, providing evidence that glucagon has a similar and direct effect on protein metabolism in both types of muscle. It is postulated that 3′, 5′-cyclic adenylic acid (cyclic AMP) is involved in the action of glucagon and other hormones. Addition of 1 mM cyclic AMP to the incubation medium decreased incorporation of label from glycine-$^{14}C$ into the protein of white but not of red muscle fiber groups from adult rats, whereas such incorporation decreased with the addition of cyclic AMP in both types of fiber preparations from younger rats (Table 9.18) (Peterson, 1967, and pers. comm.). Short (1969) has demonstrated that methionine-$^{14}C$ incorporation into protein is more rapid in red than in white muscle and is stimulated by insulin in both fiber types. Other workers using *in vivo* techniques have reported that red muscle incorporates more amino acids into both the sarcoplasmic and myofibrillar fractions than does white (Dreyfus and Vibert, 1967; Goldberg, 1967). Citoler, Benítez, and Maurer (1967) measured the incorporation of $^{3}H$-labeled amino acids into muscle protein by autoradiography and found that all the amino acids tested except alanine were more rapidly incorporated into red muscle protein.

## CROSS INNERVATION

The work of Buller, Eccles, and Eccles (1960), which demonstrated that the contractile properties of fast and slow muscles can be reversed by cross innervation, has stimulated a number of studies on the biochemical properties of such muscles. As might be expected in biological systems, the results are somewhat ambiguous. Reversal of contraction speed and histochemical enzymatic characteristics of the fibers in cross-innervated fast and slow muscle were demonstrated by Romanul and Van Der Meulen (1967). Histochemical techniques also showed conversion of fast to slow fibers in the kitten, adult cat, and young rabbits; but significant changes from slow to fast with cross innervation were found only in the rabbit (Dubowitz, 1967). Quantitative enzyme assays with rat muscle indicated that phosphorylase and lactic dehydrogenase increased in slow muscle innervated by fast nerve with no changes in cross-innervated fast muscle (Guth, Watson, and Brown, 1968). Prewitt and Salafsky (1967), however, found quantitative decreases of oxidative and glycolytic enzymes to be the most striking changes in cross-innervated slow and fast muscle respectively; small but significant increases in glycolytic enzymes were found in the cross-innervated (slow) soleus. Drahota and Gutmann (1963) found that whereas the $K^+$ concentration did not change in phasic muscle innervated with the soleus nerve supply, both $K^+$ and

TABLE 9.18

*The effect of cyclic-3′,5′-AMP on incorporation of glycine-1-$^{14}C$ into protein of adductor muscle fiber groups from rats*

| | | cpm/mg protein N | | | | | |
|---|---|---|---|---|---|---|---|
| | | Adults | | | Young | | |
| | N | Red | White | P | Red | White | P |
| Control | 29 | 932 ± 35 | 632 ± 26 | <0.005 | 1117 ± 53 | 871 ± 58 | <0.05 |
| 3′,5′-AMP | 7 | 847 ± 61 | 448 ± 18 | <0.005 | 939 ± 69 | 796 ± 51 | >0.10 |
| P for 3′,5′-AMP | | NS | <0.01 | | <0.01 | <0.02 | |

Adductor muscle fiber groups were incubated in Krebs phosphate buffered medium plus 1.5 mg glucose and 1 × $10^{-3}$mmoles cyclic AMP/ml in appropriate flasks. Values are means ± SE. N = no. of rats.

Statistical analysis on the basis of paired observations, red vs. white muscle.

From Peterson (1967) and pers. comm.

glycogen content increased in the soleus when innervated by the fast extensor muscle nerve. Incomplete conversion of slow to fast muscle on cross innervation, as indicated by contractile properties, has also been reported (Buller, Ranatunga, and Smith, 1968). As Guth, Watson, and Brown (1968) pointed out, the discrepancies in these results probably cannot be ascribed to technical errors or even to species differences. Rather, variations in the fiber type of the muscles, the physiological state of the muscles, the viability of the regenerated nerves, and the interrelationships between nervous activity and the amount of work done by the muscle especially after reinnervation—all may affect the metabolic and functional characteristics of the muscles.

Protein constituents are also influenced by innervation, and Guth and Watson (1967) showed that when the soleus was cross-innervated with nerves to a white muscle, the muscle protein electrophoretic patterns changed to resemble those of fast muscle. Protein patterns of mixed fiber areas of the gastrocnemius were converted to those of the soleus on cross innervation, but the pattern of a homogenous fiber area of the gastrocnemius did not change (Guth, Watson, and Brown, 1968). Since changes in enzyme activities and in soluble protein patterns with cross innervation were not comparable, it was suggested that different neural mechanisms are involved. An increase in the ratio of the content of hemoglobin in fast:slow muscle was demonstrated with cross innervation (McPherson and Tokunaga, 1967). An increase in the capillary network surrounding the fibers of a fast muscle cross-innervated with the soleus nerve has also been reported (Romanul and Van Der Meulen, 1967). Adenosine triphosphatase activity increased in the soleus (Karpati and Engel, 1967) and decreased in fast muscle (Dubowitz, 1967) on cross innervation.

The data from these cross-innervation studies indicate that the predominant metabolic pathways as well as the structure and contractile responses of red and white muscle are controlled by neural influences, although the relationship is not clearly defined.

## EXERCISE

Most investigations on the metabolism of red and white muscle have been made on muscles in a resting state, although it is reasonable to assume that some metabolic pathways are stimulated when the muscles are contracting and performing work. Recent studies of Hultman (1967) have shown that glycogen stores in muscle are related to work performance; however, the relative importance of this observation in red and white muscle was not determined in their studies on man. Short, Cobb, Kawabori, and Goodner (1969) studied the effect of moderate training on carbohydrate metabolism and found that in untrained rats glycogen concentration and *in vitro* glucose uptake were similar in red and white muscle. In trained rats, glycogen concentration was higher in red than in white muscle and the glucose uptake was lower by white muscle than it was by red. Except for an apparent reversal of glycogen content in the two types of muscle, these authors found no evidence that glucose metabolism is affected differently by exercise training other than an apparent exaggeration of the normal, small differences between red and white muscle. Significant increases in both total glycogen synthetase and the glucose-6-$PO_4$ independent form of the enzyme were found in both types of muscle with prolonged training; the increase in activity apparently is greater in red than in white muscle (Jeffress, Peter, and Lamb, 1968). In a histochemical study on the effect of contraction induced by low frequency stimulation, Kugelberg and Edström (1968) found that changes in phosphorylase and glycogen content were most pronounced in A (white) fibers, less in the B (intermediate) fibers, and least in the C (red) fibers of the anterior tibialis muscle of the rat. Soleus fibers showed no changes. These authors concluded that fatigue in the anterior tibialis was mainly the result of exhaustion of the glycolytic machinery of the A fibers in which glycogenolysis unbalanced resynthesis at low stimulation frequencies. Hexokinase activity increased almost twofold in both red and white muscle with exercise (Peter, Jeffress, and Lamb, 1968). These observations support the view that red muscle depends on glycolytic as well as on oxidative mechanisms for energy of contraction.

Prolonged training increased the cytochrome and succinate oxidase activities of mitochondrial preparations from both soleus and gastrocnemius, but the increases were not strikingly different in the two muscles (Holloszy, 1967). The part of the gastrocnemius analyzed was not defined, however, and may have been predominantly red.

Sréter (1963) found that the intracellular $Na^+$ and $K^+$ distributions in red and white muscles were altered by prolonged stimulations (up to 6 hours). $K^+$ in white muscle declined steadily on stimulation whereas in red muscle the depletion diminished with time. $Na^+$ influx in white muscle increased tenfold with stimulation but only a 50% increase was found in red muscle. The sum of the intracellular $Na^+$ accumulation did not balance the $K^+$ depletion.

## CONCLUDING REMARKS

Voluntary skeletal muscle accounts for about 40% of the body weight and, by virtue of its mass, the metabolism of this tissue is of major importance to the total body economy. In most mammals, the majority of muscles appear to be of mixed fiber types (25 to 75% red fibers); there are no adequate data on the relative amounts of all-red, predominantly red (>75% red fibers), and predominantly white (<25% red fibers) muscle present. In the rhesus monkey, we have estimated that all-red muscle represents only about 4% of the muscle mass. We did not attempt to determine the amount of white muscle as the white areas constitute the most superficial part of many muscles and differentiation and separation of the predominantly white muscle from the underlying intermediate or red area is technically a difficult procedure. Predominantly red and white muscles have been shown to differ in almost every parameter investigated with histological and histochemical techniques and quantitative enzyme assays. Investigations of the two types of muscle *in vitro* and *in vivo* have also demonstrated differences in carbohydrate, protein, and lipid biochemistry. It appears imprudent to correlate function with histochemical and biochemical results except in general terms; the weight of evidence, however, strongly supports the original basic theory that red and predominantly red muscles, which have higher rates of oxidative metabolism, are adapted for sustained activity and prolonged energy production and that predominantly white muscle, with a higher glycolytic activity, is adapted for sudden bursts of activity. From the data accumulated in our laboratory we must also conclude that glycogenolysis and glycolysis are important sources of energy in red muscle and that energy sources other than glucose and glycogen, presumably lipids, are utilized to a large extent by predominantly white muscle. Most metabolic investigations have been on resting muscle, however, and it is reasonable to assume that studies of working muscle may present quite different metabolic pictures.

Biochemical control mechanisms for glycolysis have been studied intensively in heart and to a lesser extent in diaphragm muscle. Almost nothing has been reported on these mechanisms for red and white voluntary skeletal muscle. An understanding of the factors that control or limit the flux through metabolic pathways is necessary in order to comprehend problems relating to the quantitative aspects of these pathways; we are beginning an investigation of control mechanisms in red and white muscle both in the resting and working states. Our studies have indicated that the free fatty acids, octanoate and palmitate, which inhibit glucose uptake in heart, have no effect on uptake in resting voluntary skeletal muscle, but this investigation is in its early stages.

## ACKNOWLEDGMENTS

Publication No. 292 of the Oregon Regional Primate Research Center, supported in part by grant FR–00163 (NIH). The investigations by C. H. Beatty and R. M. Bocek reported in this paper were supported by grants from the National Institute of Arthritis and Metabolic Diseases and the National Institute of Child Health and Human Development (NIH) and from the Muscular Dystrophy Associations of America, Inc.

## *References*

Adams, R., D. Denny-Brown, and C. M. Pearson. 1962. *Diseases of Muscle* (2d ed.), p. 42. Paul B. Hoeber, New York.

Arangio, G. A., and J. W. C. Hagstrom. 1969. The histochemical classification of rabbit hindlimb striated muscle. *J. Histochem. Cytochem. 17*:127.

Bajusz, E. 1965. Comparative enzyme histochemistry of myopathies. Similarities and differences between dystrophic and denervated muscle, p. 555. *In* W. M. Paul, E. E. Daniel, C. M. Kay, and G. Monckton (eds.), *Muscle*. Pergamon Press, New York.

Bär, U., and M. C. Blanchaer. 1965. Glycogen and $CO_2$ production from glucose and lactate by red and white skeletal muscle. *Amer. J. Physiol. 209*:905.

Bárány, M., K. Bárány, T. Reckard, and A. Volpe. 1965. Myosin of fast and slow muscles of the rabbit. *Arch. Biochem. Biophys. 109*:185.

Beatty, C. H., G. M. Basinger, and R. M. Bocek. 1967. Differentiation of red and white fibers in muscle from fetal, neonatal and infant rhesus monkeys. *J. Histochem. Cytochem. 15*:93.

———. 1968. Oxygen consumption and glycolysis in fetal, neonatal, and infant muscle of the rhesus monkey. *Pediatrics 42*:5.

Beatty, C. H., G. M. Basinger, C. C. Dully, and R. M. Bocek. 1966. Comparison of red and white voluntary skeletal muscles of several species of primates. *J. Histochem. Cytochem. 14*:590.

Beatty, C. H., R. D. Peterson, G. M. Basinger, and R. M. Bocek. 1966. Major metabolic pathways for carbohydrate metabolism of voluntary skeletal muscle. *Amer. J. Physiol. 210:*404.

Beatty, C. H., R. D. Peterson, and R. M. Bocek. 1963. Metabolism of red and white muscle fiber groups. *Amer. J. Physiol. 204:*939.

Beatty, C. H., R. D. Peterson, R. M. Bocek, N. C. Craig, and R. Weleber. 1963. Effect of glucagon on incorporation of glycine-$C^{14}$ into protein of voluntary skeletal muscle. *Endocrinology 73:*721.

Beatty, C. H., R. D. Peterson, R. M. Bocek, and E. S. West. 1959. Acetoacetate and glucose uptake by diaphragm and skeletal muscle from control and diabetic rats. *J. Biol. Chem. 234:*11.

Beecher, G. R., L. L. Kastenschmidt, W. G. Hoekstra, R. G. Cassens, and E. J. Briskey. 1969. Energy metabolites in red and white striated muscles of the pig. *J. Agr. Food Chem. 17:*29.

Blanchaer, M. C. 1964. Respiration of mitochondria of red and white skeletal muscle. *Amer. J. Physiol. 206:*1015.

Blanchaer, M. C., M. Van Wijhe, and D. Mozersky. 1963. Oxidation of lactate and α-glycerophosphate by red and white skeletal muscles I. Quantitative studies. *J. Histochem. Cytochem. 11:*500.

Bocek, R. M., G. M. Basinger, and C. H. Beatty. 1966. Comparison of glucose uptake and carbohydrate utilization in red and white muscle. *Amer. J. Physiol. 210:*1108.

Bocek, R. M., and C. H. Beatty. 1966. Glycogen synthetase and phosphorylase in red and white muscle of rat and rhesus monkey. *J. Histochem. Cytochem. 14:*549.

Bocek, R. M., R. D. Peterson, and C. H. Beatty. 1966. Glycogen metabolism in red and white muscle. *Amer. J. Physiol. 210:*1101.

Brust, M. 1966. Relative resistance to dystrophy of slow skeletal muscle of the mouse. *Amer. J. Physiol. 210:*445.

Buller, A. J., J. C. Eccles, and R. M. Eccles. 1960. Interactions between motoneurones and muscles in respect of the characteristic speeds of their responses. *J. Physiol. 150:*417.

Buller, A. J., K. W. Ranatunga, and J. Smith. 1968. Influence of temperature on the isometric myograms of cross innervated mammalian fast twitch and slow twitch skeletal muscles. *Nature 218:*877.

Burleigh, I. G., and R. T. Schimke. 1968. On the activities of some enzymes concerned with glycolysis and glycogenolysis in extracts of rabbit skeletal muscles. *Biochem. Biophys. Res. Comm. 31:*831.

Cascarano, J., A. D. Rubin, W. L. Chick, and B. J. Zweifach. 1964. Metabolically induced permeability changes across mesothelium and endothelium. *Amer. J. Physiol. 206:*373.

Cassens, R. G., R. M. Bocek, and C. H. Beatty. 1969. The effect of octanoate on carbohydrate metabolism in red and white muscle of the rhesus monkey. *Amer. J. Physiol. 217:*715.

Chinoy, N. J. 1963. Histochemical localization of myoglobin in the pigeon breast muscle. *J. Anim. Morphol. Physiol. 10:*74.

Citoler, P., L. Benítez, and W. Maurer. 1967. Autoradiographic studies of protein synthesis rates in red and white muscle fibers. *Exp. Cell Res. 45*:195.

Cooper, C. C., R. G. Cassens, and E. J. Briskey. 1969. Capillary distribution and fiber characteristics in skeletal muscle of stress-susceptible animals. *J. Food Sci. 34*:299.

Dawson, D. M., T. L. Goodfriend, and N. O. Kaplan. 1964. Lactate dehydrogenases: functions of the two types. *Science 143*:929.

Dawson, D. M., and F. C. A. Romanul. 1964. Enzymes in muscles II. Histochemical and quantitative studies. *Arch. Neurol. 11*:369.

Dickerson, J. W. T. and E. M. Widdowson. 1960. Chemical changes in skeletal muscle during development. *Biochem. J. 74*:247.

Domonkos, J. 1961. The metabolism of the tonic and tetanic muscles I. Glycolytic metabolism. *Arch. Biochem. Biophys. 95*:138.

Domonkos, J., and L. Latzkovits. 1961. Metabolism of the tonic and tetanic muscles II. Oxidative metabolism. *Arch. Biochem. Biophys. 95*:144.

Drahota, Z., and E. Gutmann. 1963. Long-term regulatory influences of the nervous system on some metabolic differences in muscles of different function, p. 143. *In* E. Gutmann and P. Hník (eds.), *The Effect of Use and Disuse on Neuromuscular Functions*. Elsevier, Amsterdam.

Dreyfus, J. C., and M. Vibert. 1967. Metabolic differences between the red and white muscles of the normal rabbit. *Rev. Franç. Etud. Clin. Biol. 12*:343.

Dubowitz, V. 1967. Cross-innervated mammalian skeletal muscle: histochemical, physiological and biochemical observations. *J. Physiol. 193*:481.

Dubowitz, V., and A. G. E. Pearse. 1960. A comparative histochemical study of oxidative enzyme and phosphorylase activity in skeletal muscle. *Histochemie 2*:105.

Dully, C. C., R. M. Bocek, and C. H. Beatty. 1969. The presence of two or more glucose-6-phosphate pools in voluntary skeletal muscle and their sensitivity to insulin. *Endocrinology 84*:855.

Engel, W. K. 1962. The essentiality of histo- and cytochemical studies of skeletal muscle in the investigation of neuromuscular disease. *Neurology 12*:778.

Fahimi, H. D., and M. J. Karnovsky. 1966. Cytochemical localization of two glycolytic dehydrogenases in white skeletal muscle. *J. Cell Biol. 29*:113.

Fahimi, H. D., and P. Roy. 1966. Cytochemical localization of lactate dehydrogenase in muscular dystrophy of the mouse. *Science 152*:1761.

Fernand, V. S. V., and A. Hess. 1969. The occurrence, structure and innervation of slow and twitch muscle fibres in the tensor tympani and stapedius of the cat. *J. Physiol. 200*:547.

Franzini-Armstrong, C. 1964. Fine structure of sarcoplasmic reticulum and transverse tubular system in muscle fibers. *Fed. Proc. 23*:887.

Fröberg, S. O. 1967. Determination of muscle lipids. *Biochim. Biophys. Acta 144*:83.

Garcia-Buñuel, L., V. M. Garcia-Buñuel, L. Green, and D. K. Subin. 1966. Lactate dehydrogenase forms in denervation and disuse atrophy of red and white muscle. *Neurology 16*:491.

Gauthier, G. F. 1969. On the relationship of ultrastructural and cytochemical features to color in mammalian skeletal muscle. *Z. Zellforsch. 95*:462.

George, J. C., and C. L. Talesara. 1961. Quantitative study of the distribution pattern of certain oxidizing enzymes and a lipase in red and white fibers of the pigeon breast muscle. *J. Cell. Comp. Physiol. 58*:253.

Goldberg, A. L. 1967. Protein synthesis in tonic and phasic skeletal muscles. *Nature 216*:1219.

Guth, L. 1968. Trophic influences of nerve on muscle. *Physiol. Rev. 48*:645.

Guth, L., and P. K. Watson. 1967. The influence of innervation on the soluble proteins of slow and fast muscles of the rat. *Exp. Neurol. 17*:107.

Guth, L., P. K. Watson, and W. C. Brown. 1968. Effects of cross-reinnervation on some chemical properties of red and white muscles in the rat and cat. *Exp. Neurol. 20*:52.

Hess, R., and A. G. E. Pearse. 1961. Dissociation of uridine diphosphate glucose-glycogen transglucosylase from phosphorylase activity in individual muscle fibers. *Proc. Soc. Exp. Biol. Med. 107*:569.

Holloszy, J. O. 1967. Biochemical adaptations in muscle—effects of exercise on mitochondrial oxygen uptake and respiratory enzyme activity in skeletal muscle. *J. Biol. Chem. 242*:2278.

Hultman, E. 1967. Physiological role of muscle glycogen in man, with special reference to exercise. *Circulation Res. 20*:I–99.

James, N. T. 1968. Histochemical demonstration of myoglobin in skeletal muscle fibres and muscle spindles. *Nature 219*:1174.

Jeffress, R. N., J. B. Peter, and D. R. Lamb. 1968. Effects of exercise on glycogen synthetase in red and white skeletal muscle. *Life Sci. 7*:957.

Karpati, G., and W. K. Engel. 1967. Transformation of the histochemical profile of skeletal muscle by "foreign" innervation. *Nature 215*:1509.

Krieg, U., and E. Kirsten. 1965. Freie Aminosäuren im roten und weissen Muskel vom Kaninchen. *Biochem. Z. 341*:543.

Kugelberg, E., and L. Edström. 1968. Differential histochemical effects on muscle contractions and on phosphorylase and glycogen in various types of fibres: relation to fatigue. *J. Neurol. Neurosurg. Psychiat. 31*:415.

Levari, R., and W. Kornblueth. 1966. Metabolic characteristics of extraocular muscles of rabbits: comparison of heart and latissimus dorsi. *Israel J. Med. Sci. 3*:513.

McPherson, A., and J. Tokunaga. 1967. The effects of cross-innervation on the myoglobin concentration of tonic and phasic muscles. *J. Physiol. 188*:121.

Maddox, C. E. R., and S. V. Perry. 1966. Differences in the myosins of the red and white muscles of the pigeon. *Biochem. J. 99*:8P.

Margreth, A., and F. Novello. 1964. Observations on the chemical determination and distribution of ribonucleic acid in several striated muscles. *Exp. Cell Res. 35*:38.

Masoro, E. J., L. B. Rowell, and R. M. McDonald. 1964. Skeletal muscle lipids—analytical method and composition of monkey gastrocnemius and soleus muscles. *Biochim. Biophys. Acta 84*:493.

Moody, W. G., and R. G. Cassens. 1968. Histochemical differentiation of red and white muscle fibers. *J. Anim. Sci. 27*:961.

Nachmias, V. T., and H. A. Padykula. 1958. A histochemical study of normal and denervated red and white muscles of the rat. *J. Biophys. Biochem. Cytol. 4*:47.

Needham, D. M. 1926. Red and white muscle. *Physiol. Rev. 6*:1.

Ogata, T. 1960. Differences in some labile constituents and some enzymatic activities between red and white muscle. *J. Biochem. (Tokyo) 47*:726.

Ogata, T., and M. Mori. 1964. Histochemical study of oxidative enzymes in vertebrate muscle. *J. Histochem. Cytochem. 12*:171.

Opie, L. H., and E. A. Newsholme. 1967. Activities of fructose, 1,6-diphosphatase, phosphofructokinase and phosphoenolpyruvate carboxykinase in white muscle and red muscle. *Biochem. J. 103*:391.

Padykula, H. A., and G. F. Gauthier. 1967. Discussion, p. 130. *In* A. T. Milhorat (ed.), *Exploratory Concepts in Muscular Dystrophy and Related Disorders.* Excerpta Medica Foundation, Amsterdam.

Pellegrino, C., and C. Franzini. 1963. An electron microscope study of denervation atrophy in red and white skeletal muscle fibers. *J. Cell Biol. 17*:327.

Peter, J. B., R. N. Jeffress, and D. R. Lamb. 1968. Exercise: effects on hexokinase activity in red and white skeletal muscle. *Science 160*:200.

Peterson, R. D. 1967. Effects of cyclic-3′,5′-AMP on incorporation of glycine-1-$^{14}$C into protein of muscle fibers from adult and young rats. *Fed. Proc. 26*:821.

Peterson, R. D., and C. H. Beatty. 1961. Metabolic studies on rat muscle fibers. *Fed. Proc. 20*:302.

Peterson, R. D., C. H. Beatty, and R. M. Bocek. 1961. High energy phosphate compounds of rat diaphragm and skeletal muscle. *Amer. J. Physiol. 200*:182.

Peterson, R. D., D. Gaudin, R. M. Bocek, and C. H. Beatty. 1964. α-Glycerophosphate metabolism in muscle under aerobic and hypoxic conditions. *Amer. J. Physiol. 206*:599.

Pette, D. 1966*a*. Mitochondrial enzyme activities, p. 28. *In* J. M. Tager, S. Papa, E. Quagliariello, and E. C. Slater (eds.), *Regulation of Metabolic Processes in Mitochondria.* Elsevier, Amsterdam.

———. 1966*b*. Energieliefernder Stoffwechsel des Muskels unter Zellphysiologischem Aspekt, p. 492. *In* E. Kuhn (ed.), *Progressive Muskeldystrophie, Myotonie, Myasthenie.* Springer-Verlag, Berlin.

Piantelli, A., and M. A. Rebollo. 1967. The lipase in the adult skeletal muscular tissue and during development. *Acta Histochem. 26*:1.

Prewitt, M. A., and B. Salafsky. 1967. Effect of cross innervation on biochemical characteristics of skeletal muscles. *Amer. J. Physiol. 213*:295.

Randle, P. J., P. B. Garland, C. N. Hales, and E. A. Newsholme. 1963. The glucose fatty-acid cycle: its role in insulin sensitivity and the metabolic disturbances of diabetes mellitus. *Lancet 1*:785.

Reis, D. J., F. Wooten, and M. Hollenberg. 1967. Differences in blood flow of red and white skeletal muscle in the cat. *Amer. J. Physiol. 213*:592.

Romanul, F. C. A. 1964. Enzymes in muscle I. Histochemical studies of enzymes in individual muscle fibers. *Arch. Neurol. 11*:355.

———. 1965. Capillary supply and metabolism of muscle fibers. *Arch. Neurol. 12*:497.

Romanul, F. C. A., and J. P. Van Der Meulen. 1967. Slow and fast muscles after cross innervation—enzymatic and physiological changes. *Arch. Neurol. 17*:387.

Ruderman, N. B., C. J. Toews, and E. Shafrir. 1969. Role of free fatty acids in glucose homeostasis. *Arch. Int. Med. 123*:299.

Seidel, J. C., F. A. Sréter, M. M. Thompson, and J. Gergely. 1964. Comparative studies on myofibrils, myosin, and actomyosin from red and white rabbit skeletal muscle. *Biochem. Biophys. Res. Comm. 17*:662.

Sexton, A. W., and J. W. Gersten. 1967. Isometric tension differences in fibers of red and white muscles. *Science 157*:199.

Short, F. A. 1969. Protein synthesis by red and white muscle in vitro: effect of insulin and animal age. *Amer. J. Physiol. 217*:327.

Short, F. A., L. A. Cobb, I. Kawabori, and C. J. Goodner. 1969. Influence of exercise training on red and white rat skeletal muscle. *Amer. J. Physiol. 217*:327.

Sréter, F. A. 1963. Cell water, sodium, and potassium in stimulated red and white mammalian muscles. *Amer. J. Physiol. 205*:1295.

———. 1964. Comparative studies on white and red muscle fractions. *Fed. Proc. 23*:930.

Sréter, F. A., J. C. Seidel, and J. Gergely. 1966. Studies on myosin from red and white skeletal muscles of the rabbit I. Adenosine triphosphatase activity. *J. Biol. Chem. 241*:5772.

Stein, J. M., and H. A. Padykula. 1962. Histochemical classification of individual skeletal muscle fibers of the rat. *Amer. J. Anat. 110*:103.

Stubbs, S. St. George-, and M. C. Blanchaer. 1965. Glycogen phosphorylase and glycogen synthetase activity in red and white skeletal muscle of the guinea pig. *Canad. J. Biochem. 43*:463.

Szent-Györgyi, A. 1953. *Chemical Physiology of Contraction in Body and Heart Muscle*, p. 105. Academic Press, New York.

Van Wijhe, M., M. C. Blanchaer, and W. R. Jacyk. 1963*a*. The oxidation of lactate and α-glycerophosphate by red and white skeletal muscle I. Quantitative studies. *J. Histochem. Cytochem. 11*:500.

———.1963*b*. The oxidation of lactate and α-glycerophosphate by red and white skeletal muscle II. Histochemical studies. *J. Histochem. Cytochem. 11*:505.

Van Wijhe, M., M. C. Blanchaer, and S. St. George-Stubbs. 1964. The distribution of lactate dehydrogenase isozymes in human skeletal muscle fiber. *J. Histochem. Cytochem. 12*:608.

Wirsén, C. 1965. Autoradiography of injected albumin bound 1-$C^{14}$-palmitate in pigeon pectoralis muscle. *Acta Physiol. Scand. 65*:120.

# Chapter 10

# Ontogeny of Red and White Muscles: The Enzymic Profile and Lipid Distribution of Immature and Mature Muscles of Normal and Dystrophic Chickens

E. COSMOS

Morphogenic and metabolic analyses emphasize important ontogenetic differences between major muscle types of normal chickens and of chickens with hereditary muscular dystrophy. In both strains, red muscle attains matured characteristics by the time of hatching whereas white muscle demonstrates major metabolic changes during *ex ovo* maturation. The inability of dystrophic muscle to complete this chemical conversion successfully is noted early *ex ovo;* with further development, there is an impairment of function and an eventual deterioration of the white muscle (Cosmos, 1966; Cosmos and Butler, 1967).

Gross examination of adult fowl shows that the breast muscle is pale compared to the deep red color of leg muscles. Chemically, the pale pectoralis responds strongly to enzymes of the glycolytic pathway; in contrast, the red soleus responds strongly to tests for mitochondrial enzymes. This relationship between pale or dark color and response to enzymes of anaerobic or aerobic metabolism has been verified in a variety of muscles tested (gastrocnemius, flexor digitorum longus, anterior and posterior latissimus dorsi, and pectoralis minor) in adult chickens. In the late embryo, however, the muscles are pale in color even though they demonstrate strong reactions to mitochondrial enzymes. With *ex ovo* maturation muscles destined to be white increase in anaerobic enzymic activity but remain pale; red muscles retain aerobic enzymic activity but increase in redness, a reflection of a change in myoglobin content. As a result of these observations our use of the terms "red" and "white" muscle is based on analyses and observations of the matured condition: red refers to muscles with strong mitochondrial or aerobic enzymic activity, and white to muscles with strong glycolytic or anaerobic enzymic activity.

## ENZYMIC DISTRIBUTION OF RED AND WHITE MUSCLE

To identify fiber types of adult birds, the mitochondrial enzyme succinic dehydrogenase (SDH) and the glycolytic enzyme phosphorylase were used, since these enzymes lend themselves both to histochemical distribution studies of tissue sections and to quantitative analyses of muscle homogenates. Further, the histochemical test for phosphorylase* permits easy identification of fiber types since the iodine-polysaccharide complex formed results in a red color (short-chain polysaccharide) for weak activity, a lavender to purple for intermediate, and a brilliant blue (long-chain polysaccharide) response for strong enzymic activity. These color reactions can be verified colorimetrically with homogenates (Cosmos, 1966).

Histochemical analyses with frozen sections of adult muscle indicate that the soleus is composed of two fiber types, red and intermediate, but that the pectoralis is a nearly homogeneous white muscle. Similar examination of muscles from adult birds with hereditary muscular dystrophy shows that the soleus is essentially similar to that of the normal chicken; however, a striking deviation from the normal response is noted in the dystrophic pectoralis. In contrast to the nearly homogeneous reaction of normal muscle, the dystrophic muscle shows a mixed enzymic response to both aerobic and anaerobic enzymes, a variability in fiber diameter, and a more loosely packed structure. To interpret these observations in dystrophic muscle, it was necessary to determine whether the dystrophic process was a regression from the normal adult pattern or whether it was an arrestment of differentiation early in development. Evidence for the latter interpretation could be obtained only if ontogenetic studies reveal a stage early in development when the normal pectoralis is a heterogeneous tissue showing a mixed fiber pattern similar to that of the adult dystrophic pectoralis.

Histochemical distribution studies of the normal pectoralis show the short-chain polysaccharide of weak phosphorylase (red) reactions in the

* Phosphorylase incubation media for frozen sections: 40 ml 0.2 M sodium acetate buffer, *p*H 5.6; 10 ml 0.2 M sodium fluoride, *p*H 6.0; 50 ml distilled $H_2O$; 400 mg glucose 1-phosphate dipotassium · 2 $H_2O$; 80 mg adenosine 5′-monophosphate; 8 mg glycogen (shellfish); 1.2 ml insulin (40 units/ml); 7.2 g polyvinylpyrrolidone (m wt 10,000); final *p*H 5.7. Frozen sections are incubated directly at 32° C for periods up to 60 min, rinsed in several washes of distilled $H_2O$, and differentiated in an iodine-iodide stock solution (3 g potassium iodide and 1.3 g iodine to 100 ml $H_2O$) diluted 1:20 v/v with distilled $H_2O$. Mounting medium is 1 part iodine-iodide stock solution to 5 parts glycerine. Slides stored at 4° C in total darkness to preserve reaction.

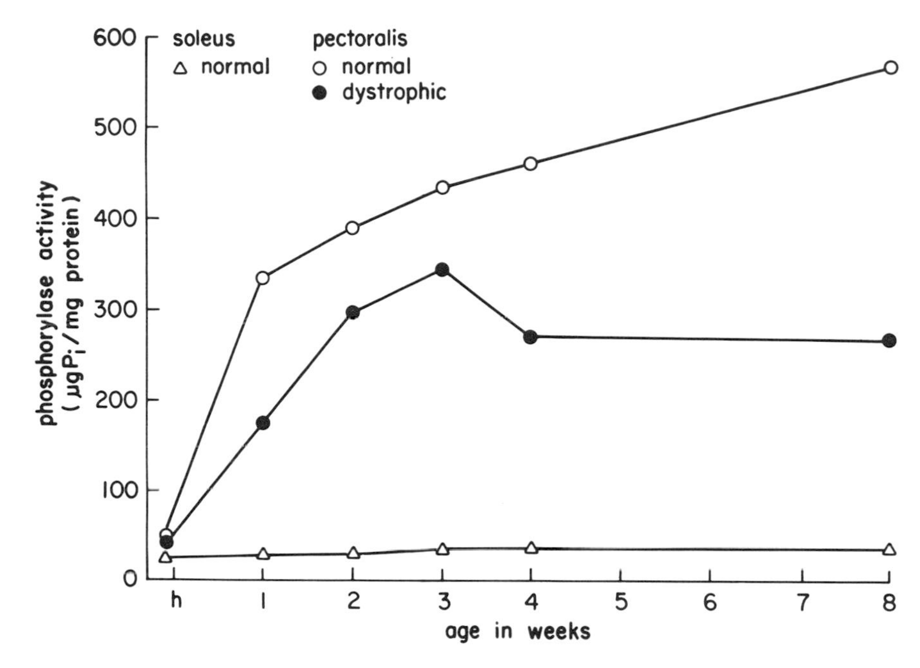


*Fig. 10.1.* Analyses of phosphorylase activity of homogenates (Cosmos, 1966) of the normal soleus and of the normal and dystrophic pectoral muscles from the late embryo to 8 weeks *ex ovo*. Activity is expressed as μg inorganic phosphate/mg of noncollagenous protein released after a 15-min incubation. *h,* hatching.

embryonic tissue, a mixture of polysaccharide responses in the immature muscle, and the long-chain polysaccharide (blue) reaction characteristic of strong phosphorylase activity in the mature fiber. The dystrophic muscle gives a response similar to that of the normal during the embryonic stage; however, *ex ovo,* the mixed fiber condition characteristic of the immature pattern of normal muscle is maintained (Cosmos, 1966), as evidenced also by histochemical reactions of SDH activity (Cosmos and Butler, 1967).

## ENZYMIC ANALYSES OF RED AND WHITE MUSCLE

In Fig. 10.1 are shown enzymic analyses of homogenates of the normal soleus and of the normal and dystrophic pectoral muscles. The period analyzed is from the late embryo through 8 weeks *ex ovo* since at this time the normal muscle reaches chemical maturity and the dystrophic tissue shows enzymic and functional changes without great loss of muscle tissue. In normal birds phosphorylase activity of the pectoralis shows a 14-fold increase during *ex ovo* maturation whereas the soleus shows only a 1.4-fold increase for the same period. In contrast, measurements with the mitochondrial enzyme SDH again show little change in activity during differentiation of red muscle but a decrease in activity during the same period for white muscle (Fig. 10.2). These analyses support the view that red muscle attains its mature enzymic profile before hatching but that white muscle exhibits an alteration in its metabolic behavior from aerobic to anaerobic during *ex ovo* ontogeny. This ability of white muscle to alter its metabolic characteristics during the differentiating period seems to be lacking in the dystrophic pectoralis. Deviations from normal are noted early *ex ovo* as an inability both to maintain an increment in phosphorylase activity (Fig. 10.1) and to demonstrate the normal decrease in SDH activity (Fig. 10.2). Histochemical preparations show that the elevated SDH activity maintained in the abnormal muscle is restricted to certain cells only; other fibers show weaker than normal activity. As the muscle continues to mature, cells characterized by high SDH activity increase in number and in intensity of reaction. These cells are not maintained, however, but eventually decrease in number and are often absent in aged muscle. Analyses of the red muscle in the dystrophic animal (Cosmos, unpublished observations) show an enzymic profile

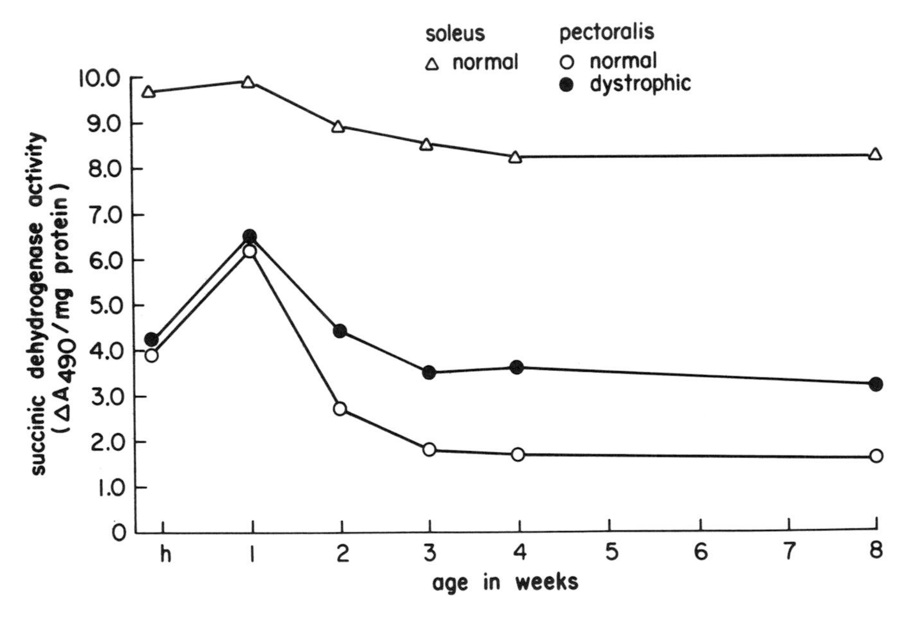


*Fig. 10.2.* Analyses of the succinic dehydrogenase activity of homogenates (Cosmos and Butler, 1967) of the same muscles shown in Fig. 10.1. Activity is expressed as the change in absorbency (ΔA) at 490 mμ/mg of noncollagenous protein at the end of a 15-min incubation.

essentially similar to that of the normal soleus. Thus the muscle with the delayed maturation is uniquely selected as a target for the condition of dystrophy during the early ontogenetic periods.

By 2 months *ex ovo* the normal pectoralis stabilizes to a mature chemical composition but the dystrophic muscle remains in a constant state of flux both structurally and functionally. During the immature period the abnormal tissue seems capable of replenishing lost protein. With the onset of sexual maturity, however, this capability is greatly diminished. Fig. 10.3 shows the noncollagenous protein analysis of the normal and dystrophic pectoral muscles. A maximum protein content is attained in the normal muscle at about 2 to 3 weeks *ex ovo* (Cosmos and Butler, 1967) and is maintained at approximately 200 mg/g for 24 months. The protein content of the dystrophic muscle reaches a maximum at about 3 to 4 weeks *ex ovo* and is maintained at a fairly constant

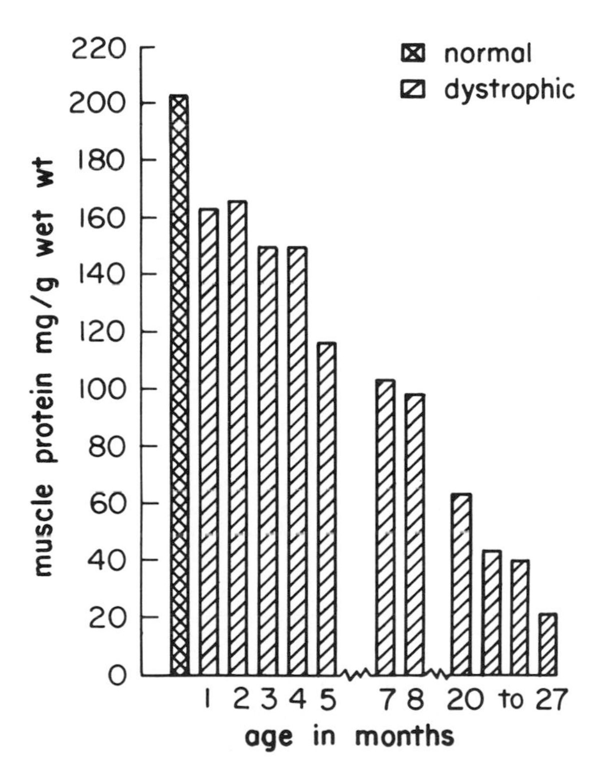


*Fig. 10.3.* The noncollagenous protein content (Cosmos and Butler, 1967) expressed as mg/g wet wt of normal and dystrophic pectoral muscles.

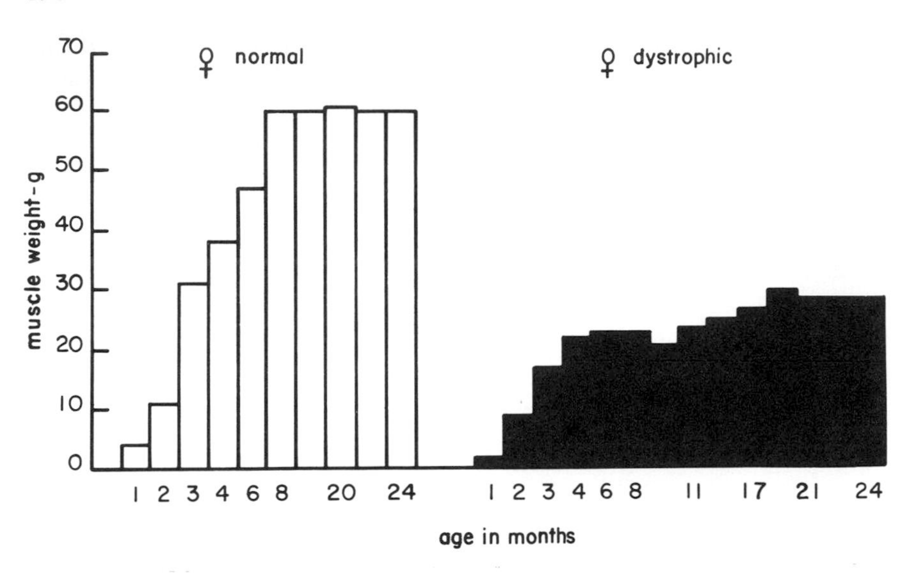


*Fig. 10.4.* Correlation of muscle weight (g) with age. The normal pectoralis continues to increase up to 8 or 9 months but the dystrophic, only up to sexual maturity (about 5 months). Weights are for the right pectoral muscles of female chickens.

level to sexual maturity; following this period there is a continual loss of muscle protein. A correlation of weight with age (Fig. 10.4) indicates that after sexual maturity only the normal pectoralis continues to increase in size. For the same period, the weight of the dystrophic muscle remains relatively constant even though the total protein content has declined to 10% of its original value (cf. Figs. 10.3 and 10.4). Microscopic examination of the abnormal tissue reveals the presence of lipid in those areas depleted of muscle protein.

LIPID DISTRIBUTION IN RED AND WHITE MUSCLE

In normal muscle of adult chickens lipid is noted extracellularly around blood vessels and nerves and intracellularly as discrete droplets closely associated with mitochondria. Furthermore, cells with low mitochondrial enzymic activity show weak lipid reactions and those with high aerobic enzymic activity demonstrate strong lipid reactions (Nachmias and Padykula, 1958). Thus white muscle gives weak lipid responses and red muscle, characterized by a rich vascular and neural supply and by a high level of mitochondrial enzymes, responds strongly to lipid tests. The latter also amasses increasing amounts of lipid with age but only between fascicles at sites of blood vessels and nerves (Cosmos, unpublished observations). However, the dystrophic pectoralis shows a progressive increase in lipids intracellularly as well as accumulations around cells and fascicles. Eventually, as the process of dystrophy progresses, large areas of muscle are totally replaced by adipose cells (Butler, Kehinde, Cosmos, and Milhorat, 1969).

Figs. 10.5–12b show the distribution of lipid in sections of the normal and dystrophic pectoralis from the early embryo (12 days) to 21 months *ex ovo*. In the embryonic tissue large masses of lipid are noted in regions of active growth (Fig. 10.5). As development progresses, lipid grains concentrate in the cells and around blood vessels and nerves. Fibers from the pectoralis of a newly hatched chicken demonstrate a uniformly strong lipid reaction (Fig. 10.6). By 7 days *ex ovo,* intracellular lipid is greatly

*Figs. 10.5–12.* Photomicrographs of the pectoral muscles of normal (Figs. 10.5–8) and dystrophic (Figs. 10.9–12b) chickens, showing lipid distribution by the Oil red O and Sudan black B methods. Both techniques demonstrate neutral fats; in addition, the Sudan method stains phospholipids both of myelin sheath and of mitochondria. A hematoxylin counterstain is used with the Oil Red O technique only. All tissues were frozen in isopentane at liquid nitrogen temperature, mounted on chucks, cut in a cryostat (−27° C) at 14 μ, fixed in formol-calcium, and stained for lipids.

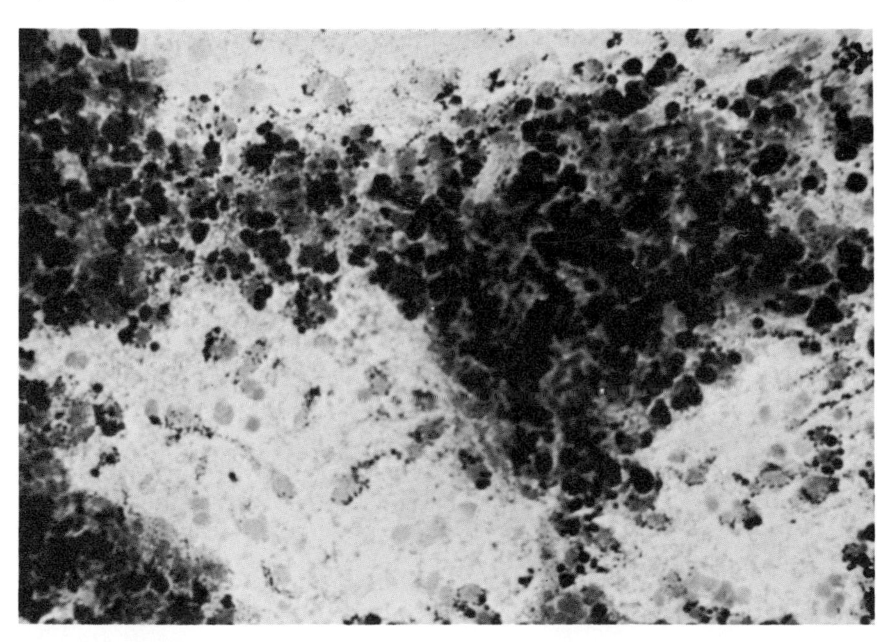

*Fig. 10.5.* Normal pectoralis, 12-day embryo: large masses of lipid (black areas) coincide with sites of active cell growth. In lighter areas lipid grains are seen in the cytoplasm of isolated cells. Oil Red O. × 352.

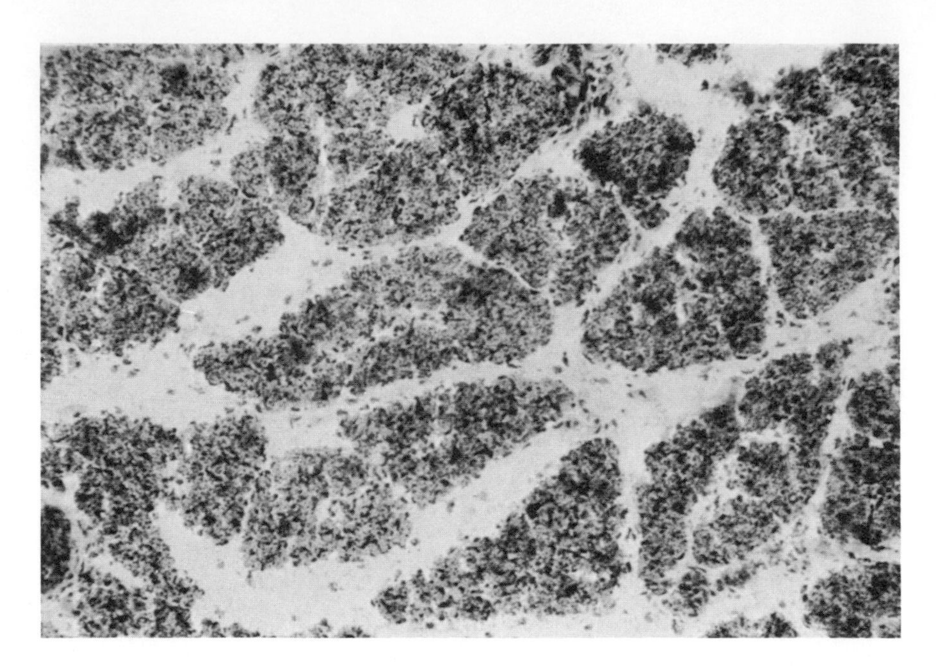

*Fig. 10.6.* Normal pectoralis, 1 day *ex ovo:* the strong lipid reaction is localized over the cells (black dots). The more heavily stained areas between cells but within fascicles correspond to lipids associated with blood vessels and nerves. The counterstain used in this preparation stains nuclei of isolated cells located between fascicles. Oil Red O.×142.

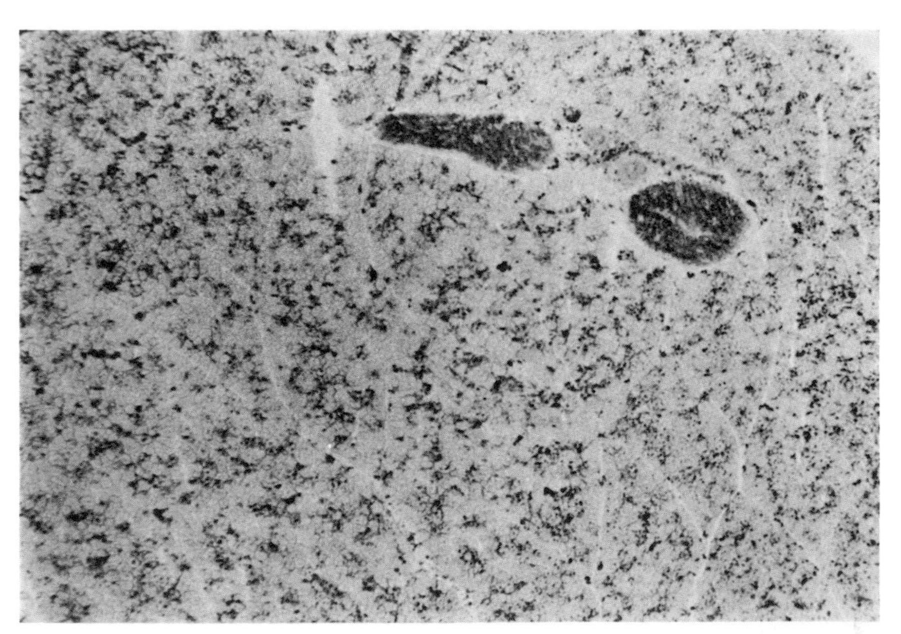

*Fig. 10.7.* Normal pectoralis, 7 days *ex ovo:* mixed fiber response, indicating cells at various stages of differentiation. A heavy reaction is associated with the high phospholipid content of the myelin sheath of the two large nerves. Sudan black B. × 142.

*Fig. 10.8.* Mature normal pectoralis, 2 months *ex ovo:* all fibers demonstrate a weak intracellular lipid response. Sudan black B. × 28.

diminished in some fibers but not in others (Fig. 10.7). This transition from a homogeneous to a heterogeneous response suggests that all fibers do not differentiate at equal rates. Cells with the stronger lipid stain are either newly formed fibers or ones demonstrating a delayed maturation. White muscle which is fully matured gives a lipid reaction which is uniformly weak (Fig. 10.8). During early ontogeny the dystrophic muscle is similar to the normal; however, abnormalities are detected at the mixed fiber stage as early as 2 weeks *ex ovo* (Fig. 10.9). Here a variability in the number of reactive areas in different fibers is noted. By 2 months *ex ovo* the variability of the intracellular lipid response is greatly accentuated (Fig. 10.10a); further, cells strong in mitochondrial enzymic activity are also strong in lipids (Fig. 10.10b). In the mature muscle (Fig. 10.11) lipid begins to accumulate between cells and fascicles; eventually, large areas of tissue become infiltrated with adipocytes (Figs. 10.12a–b).

In the immature state many dystrophic cells retain a high concentration of intracellular lipid, and in the mature state there is in addition a stimulation of extracellular adipogenesis. The origin of adipose cells in

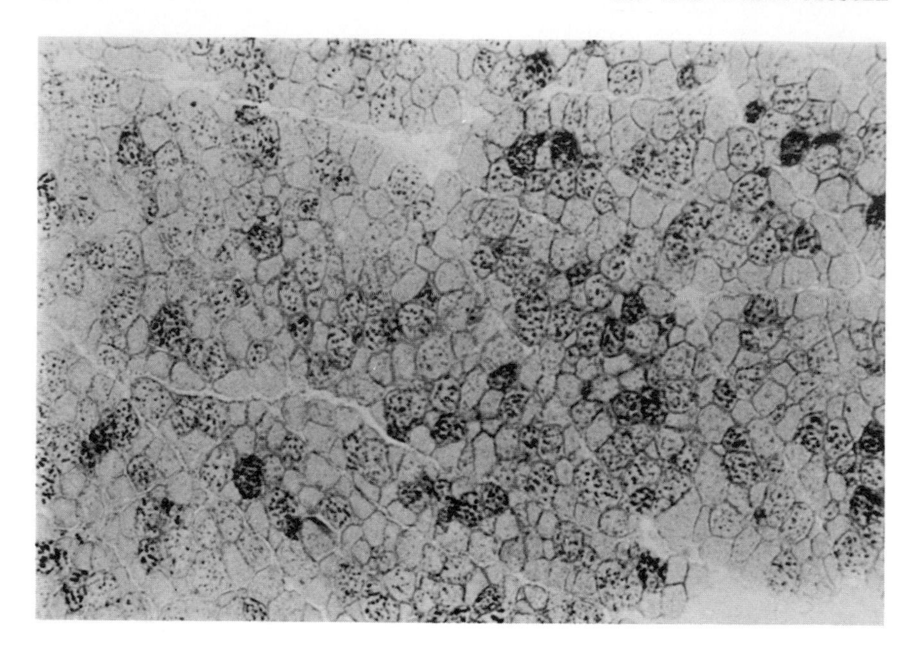

*Fig. 10.9.* Dystrophic pectoralis, 2 weeks *ex ovo:* black grains are heavier in some cells than in others. Sudan black B. × 227.

areas left vacant by degenerated fibers is difficult to explain; one theory is that these are not true adipose cells but may represent phagocytic mesenchymal cells of local origin capable of storing fat (Wassermann, 1964). Alternatively, these may be true adipocytes which had been stimulated to develop from dormant precursors once protein synthesis had been depressed. Various stages of adipogenesis (Simon, 1965) have not been noted in the mature dystrophic muscle; instead, fully matured adipocytes are present. It is tempting to speculate that the adult adipocyte actually has as a precursor the lipid-laden cell of the immature muscle. These cells which are filled with small lipid droplets and give strong SDH reactions (cf. Fig. 10.10a, b) are not easily demonstrated in aged tissue. It is possible that the metabolism of fat in these cells fails increasingly and that the lipid droplets eventually fuse and fill the entire cell. Hyperactive mitochondria showing heavy lipid accumulation have been noted in human myopathies (Coleman, Nienhuis, Brown, Munsat, and Pearson, 1967) and have been associated with an impairment of lipid utilization. An inability of mitochondria to oxidize fatty acids and an increase in fat synthesis have also been reported in mouse dystrophies (Lin, Hudson, and Strickland, 1969).

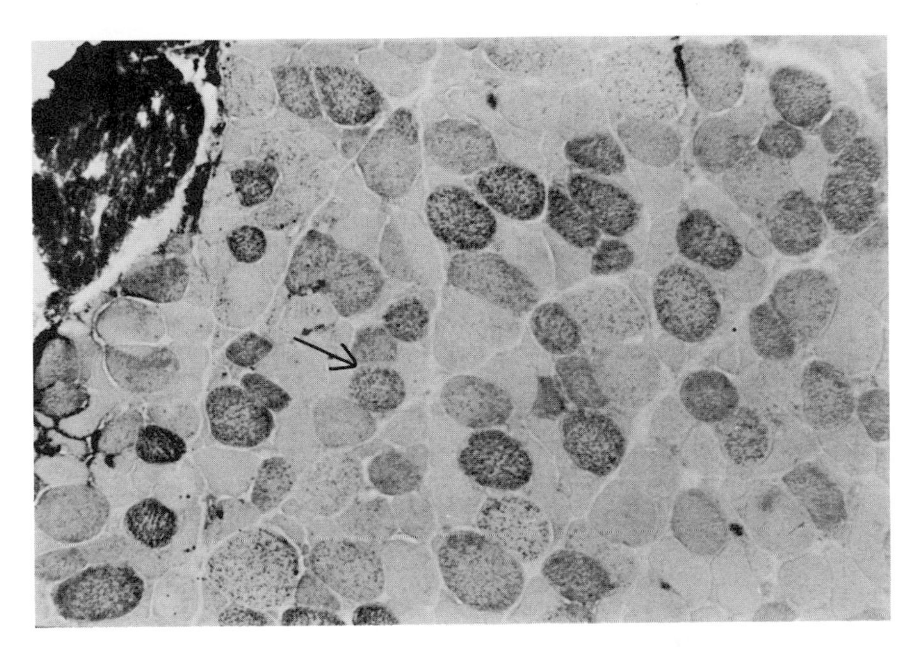

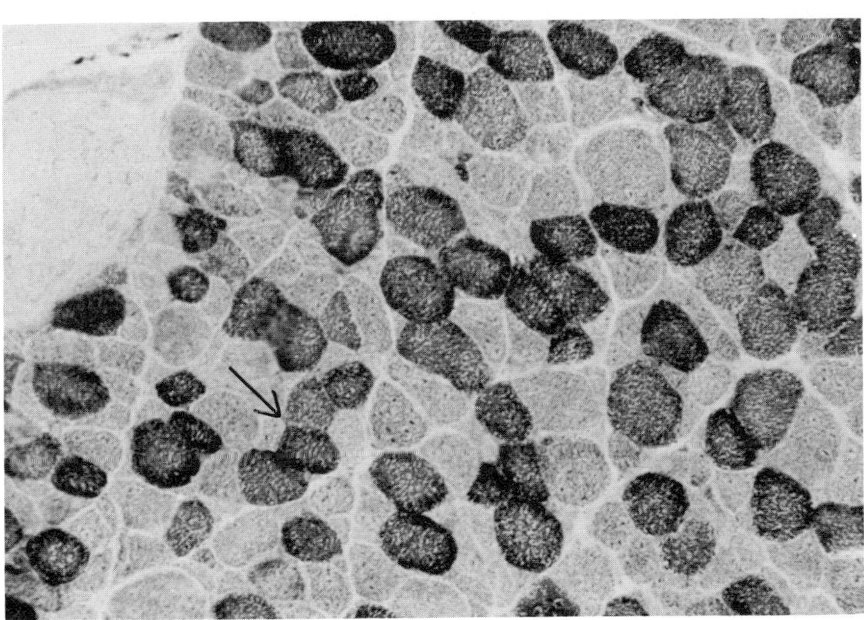

*Fig. 10.10.* (*a, above*) Dystrophic pectoralis, 2 months *ex ovo:* variable lipid response is still confined intracellularly. Sudan black B. (*b, below*) Section adjacent to (a) shows SDH reaction. Arrows point to similar cell groups; note that cells strong in the SDH reaction also give strong lipid responses. Nerve at upper left shows no response in (b) . × 142.

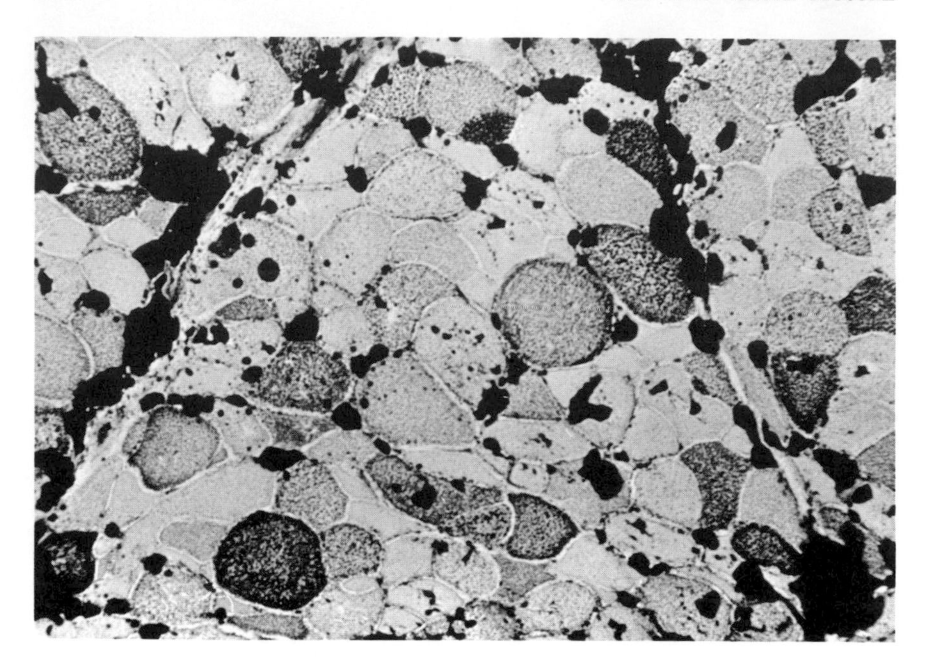

*Fig. 10.11.* Mature dystrophic pectoralis: note variability of reaction in different cells and accumulation of adipose tissue between cells and between fascicles. Sudan black B. × 142.

GENERAL COMMENTS

There seem to be two distinct periods in the development of the dystrophic muscle: (1) the immature stage, when the mechanism which converts the muscle from aerobic to anaerobic metabolism is faulty; and (2) the sexually mature period, when the ability to synthesize protein is greatly depressed and muscle structure is no longer maintained.

The factors controlling differentiation of fiber types have received ample consideration in the last decade from both physiological and histochemical approaches in fibers of various species. Recently the diverse development of neurones innervating fiber types has also been examined structurally and physiologically (Nyström, 1968). Results from cross-union experiments strongly suggest that neurones dictate both the physiological and metabolic patterns of red and white fibers during the period of early differentiation. However, conversion of one fiber type to the other has never been complete, and red fibers are less easily altered than are white ones (Dubowitz, 1967). This is of interest because white fibers show an alteration in their physiology from slow at birth to fast in postnatal life (Buller, Eccles, and Eccles, 1960), in their metabolism from aerobic to anaerobic for the same periods (Cosmos, 1966; Cosmos

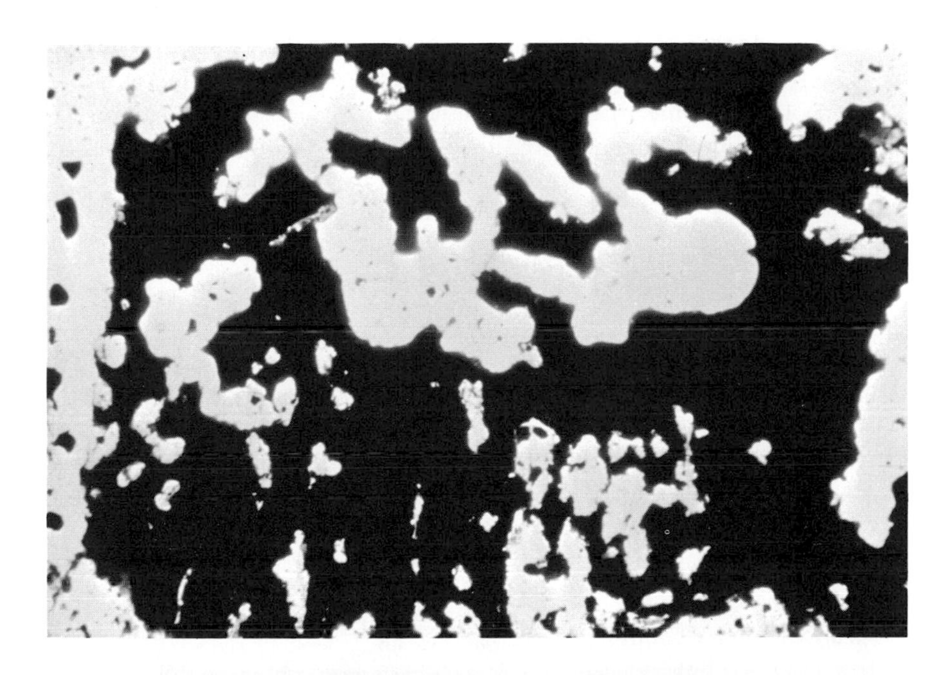

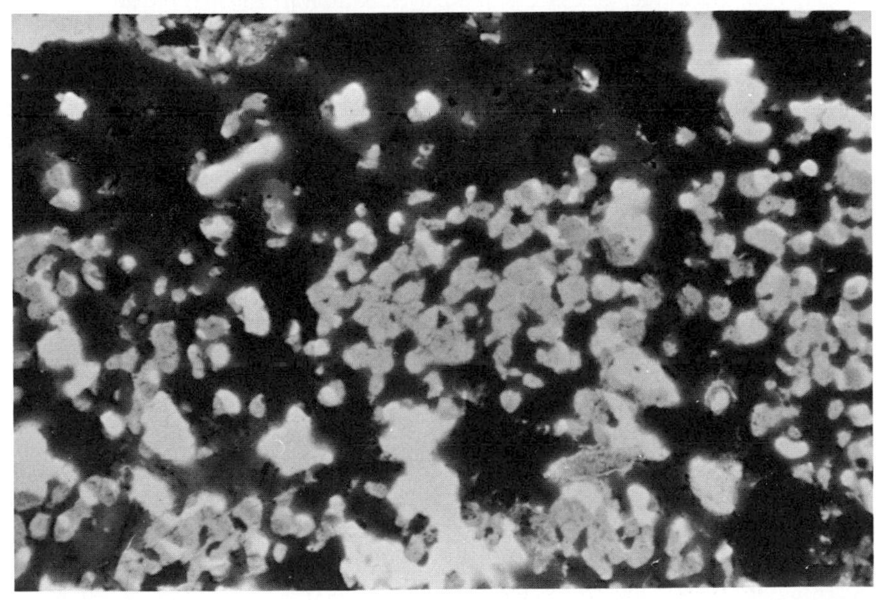

*Fig. 10.12.* (*a, above*) Dystrophic pectoralis, 21 months: massive replacement of muscle fibers by lipid. Sudan black B. (*b, below*) Section of an adult pectoralis from a region demonstrating lipid replacement and some intact cells. Oil Red O. × 28.

and Butler, 1967), and in both physiological function and histochemical profile after cross innervation (Dubowitz, 1967; Romanul and Van Der Meulen, 1967). Thus the white muscle, which is more amenable to alterations in function and in structure, is the one more easily affected by metabolic alterations of dystrophy.

## ACKNOWLEDGMENTS

I wish to express my sincere appreciation and gratitude to Jane Butler for her advice and assistance and to Dr. Ade T. Milhorat for many stimulating discussions and valuable suggestions during the preparation of this manuscript.

This investigation was supported by a grant from the Muscular Dystrophy Associations of America, Inc., and by USPHS award NB–06942–03.

## *References*

Buller, A. J., J. C. Eccles, and R. M. Eccles. 1960. Differentiation of fast and slow muscles in the cat hind limb. *J. Physiol. 150*:399.

Butler, J., O. Kehinde, E. Cosmos, and A. T. Milhorat. 1969. Lipid distribution in developing muscle of normal and dystrophic chickens. *Fed. Proc. 28*:782.

Coleman, R. F., A. W. Nienhuis, W. J. Brown, T. L. Munsat, and C. M. Pearson. 1967. New myopathy with mitochondrial enzyme hyperactivity. *J. Amer. Med. Ass. 199*:118.

Cosmos, E. 1966. Enzymatic activity of differentiating muscle fibers I. Development of phosphorylase in muscles of the domestic fowl. *Dev. Biol. 13*:163.

Cosmos, E., and J. Butler. 1967. Differentiation of fiber types in muscle of normal and dystrophic chickens. A quantitative and histochemical study of the ontogeny of muscle enzymes, p. 197. *In* A. T. Milhorat (ed.), *Exploratory Concepts in Muscular Dystrophy and Related Disorders.* Excerpta Medica Foundation, Amsterdam.

Dubowitz, V. 1967. Cross-innervation of fast and slow muscle. Histochemical, physiological and biochemical studies, p. 164. *In* A. T. Milhorat (ed.), *Exploratory Concepts in Muscular Dystrophy and Related Disorders.* Excerpta Medica Foundation, Amsterdam.

Lin, C. H., A. J. Hudson, and K. P. Strickland. 1969. Fatty acid metabolism in dystrophic muscle *in vitro. Life Sci. 8*:21.

Nachmias, V., and H. Padykula. 1958. A histochemical study of normal and denervated red and white muscles of the rat. *J. Biophys. Biochem. Cytol. 4*:47.

Nyström, B. 1968. Fiber diameter increase in nerves to "slow-red" and "fast-white" cat muscles during post-natal development. *Acta Neurol. Scand. 44*:265.

Romanul, F. C., and J. P. Van Der Meulen. 1967. Slow and fast muscles after cross innervation. *Arch. Neurol. 17*:387.

Simon, G. 1965. Histogenesis, p. 101. *In* A. Renold and G. Cahill (eds.), *Handbook of Physiology. Section 5: Adipose Tissue.* Amer. Physiol. Soc., Washington.

Wassermann, F. 1964. The concept of the "fat organ," p. 22. *In* K. Rodahl and B. Issekutz, Jr. (eds.), *Fat as a Tissue.* McGraw-Hill, New York.

# Chapter 11

# *Summary and Discussion of Part 2*

PANEL MEMBERS: R. G. CASSENS, *Chairman*
R. M. BOCEK
D. S. DAHL
R. HAMM
S. MORITA
W. MOODY

*Cassens:* The distinction between "red" and "white" muscle has been recognized for some time, and the past 10 years has seen an amazing increase in our knowledge about it—particularly about the characteristics and properties of the individual fibers. With this growth in knowledge has come a growth in awareness of the importance of the simple terms red and white to both past and future work in the field. This concern has been shared by a broad spectrum of people—physiologists, biochemists, neurologists, athletes, and food scientists, to name a few. A number of chapters in this symposium speak directly or indirectly of red and white muscle, but the five chapters that form Part 2 represent the views of five experts who have made great contributions to specific areas of research on red and white muscle.

Dr. Dubowitz began in Chapter 6 with a general discussion of the problems associated with histochemical typing or categorization of fiber types. Fast white mammalian muscle generally has three types—A (representing white fibers), B, and C (corresponding to two forms of red fibers). Slow red muscle from mammals has one fiber type that somewhat resembles C. There are two main types in human muscle, with less tendency toward intermediate types. These are referred to as types I and II; they correlate rather well with types C and A above.

The main thrust of Dr. Dubowitz's chapter is directed at the problem of fiber-type differentiation during growth and development. In guinea pig and man, muscles are fully differentiated at birth; in mouse, rat, and rabbit, differentiation occurs postnatally. Although it is difficult to define fiber type in undifferentiated muscle, Dr. Dubowitz believes that the undifferentiated fibers approximate type I or C fibers. Two problems associated with the investigation of fiber type should be recognized. Various muscles show differentiation at different times, and a difference in rate of differentiation is also found with different histochemical enzyme techniques. Chemical mediation of neural influence and frequency and character of impulses are the two possibilities most favored as causative agents of differentiation. Dr. Dubowitz believes that the influence of the neural system in determining fiber types is well established, and that studies in this area might form the bridge for elucidating the influence of certain other factors such as nutrition, exercise, and hormones on muscle. Some influences may well act selectively on specific fiber types in muscle.

Dr. Gauthier leaves no doubt, in Chapter 7, that there are clear, consistent, ultrastructural differences amongst red, white, and intermediate fibers. These three types were identified initially by histochemical methods; then the electron microscope was used to study the distribution and characteristics of the mitochondria, the form of the sarcoplasmic reticulum and the width of the Z line. The Z line, for example, is narrowest in white fibers and is clearly thinner in intermediate than red fibers. Readily apparent differences also exist for the other properties examined in view of fiber type. Intermediate fibers contribute, together with red fibers, to the apparent redness of the whole muscle. Dr. Gauthier defends her terminology of red, white, and intermediate as being simple and having as its basis a relationship to the color of the whole muscle. The work was extended to the examination of the developing fibers because of suggestions that the red fiber is a primitive form capable of differentiating into a white fiber. The differentiating fibers of the rat diaphragm are alike, and they resemble superficially the adult red fiber, but their ultrastructure is distinctive. The ultrastructural features of developing fibers cannot be equated to adult red fibers, and during later stages of postnatal development, fibers whose mitochondrial content resembles that of red and white fibers appear simultaneously—white fibers do not therefore "follow" red fibers. Also described in the chapter is work on diaphragm that attempts to relate structural findings to function.

M. H. Brooke's Chapter 8, although oriented toward some complicated neural influences, offers an interesting basis for some very useful experimentation on meat animals. His general theme is that by a rather simple technique of combined fiber-type–size approach, much information can be gained about higher nervous control. The chapter leads the reader through a review of the neural influence, the motor unit, and the concept of work hypertrophy and immobility atrophy. Dr. Brooke's work leads to the conclusion that the two histochemical muscle fiber types behave differently and independently with regard to changes in fiber size. He concedes that such differences may be an innate property of the muscle fibers themselves, reflecting a biochemical difference. But he prefers the hypothesis that voluntary powerful or sudden activity is mediated by type II fibers while spindle-associated reflex activity is mediated via the histochemical type I fibers.

The forte of white muscle has been recognized to be rapid but brief contractions, that rely primarily on glycolysis for energy production. This contrasts with red muscle, that contracts for prolonged periods of time and relies chiefly on oxidative metabolism. These biochemical characteristics of red and white muscle have been dealt with extensively and expertly by C. H. Beatty and R. M. Bocek in Chapter 9. They emphasize that almost without exception muscle is nonhomogeneous for fiber type, and that the resultant metabolism is dependent on the predominant fiber type. Differences in metabolism between red and white fiber groups may be as great as, or greater than, those due to imposed experimental conditions. Diaphragm and perfused heart have been two favored preparations for metabolic studies, but in fact neither of these is representative of striated skeletal muscle. The authors have therefore developed a technique for preparation of viable fiber bundles of predominant fiber type, and in this chapter they illustrate their approach with work from their own laboratory. They conclude that the weight of evidence supports the theory that predominantly red muscles have higher rates of oxidative metabolism, and are adapted for sustained activity and prolonged energy production, and that predominantly white muscle, with a higher glycolytic activity, is adapted for sudden bursts of activity. They also conclude, however, that glycogenolysis and glycolysis are important sources of energy in red muscle and that energy sources other than glucose and glycogen, presumably lipids, are utilized to a large extent by predominantly white muscle.

A study of the ontogeny of red and white muscle has been used by E. Cosmos in Chapter 10 to obtain information about muscular dystrophy. Red muscle is mature at hatching in chicken while white muscle shows major changes in metabolism during *ex ovo* maturation. The work of Dr. Cosmos was designed to answer the following question: Does the diseased tissue show a normal course of development followed by deterioration to the abnormal state, or is dystrophy a result of arrest of differentiation of white muscle? The results revealed that muscle with delayed maturation was uniquely selected as a target for the condition of dystrophy during early ontogenetic periods. The normal process of differentiation appears to be faulty during the early stage of conversion from aerobic to anaerobic metabolism. Then, the muscle structure is no longer maintained in the mature period. Literature is cited to show that white muscle is more amenable to alterations in structure and function than red. Cosmos then points out the interesting finding that white muscle is also more easily affected by the metabolic alterations of dystrophy. Lipid distribution in normal and dystrophic muscle is also discussed.

Much is known about fiber types in the muscles of small laboratory animals, but very little about them in muscles of bovine, porcine, and ovine animals. Muscle growth and quantity is important—could it be due to type II hypertrophy, and can this be controlled? Or, does fiber-type composition influence metabolism in post mortem muscle and therefore eventually the quality of muscle as a food? These are simple questions; perhaps the chapters in Part 2 will help to answer them.

## DISCUSSION

*Dahl:* Are there any absolute differences between red and white muscle fibers or do their constituents differ only relatively?

*Beatty:* For our work, I would say there are no absolute differences. Van Wijhe, Blanchaer, and Stubbs (1964) did more specific work on individual red and white muscle fibers; they found that white fibers contained almost exclusively the lactate dehydrogenase isozymes IV and V. In our work with predominantly red or predominantly white muscle, however, we have found as much as a 16-fold difference in specific metabolic measurements. Reference is also made to the work of Van Wijhe, Blanchaer, and Jacyk, 1963*a, b*.

*Sréter:* Recently we have done some experiments on the gastrocnemius muscle of the rabbit that bear on the question of the heterogeneity of fibers within a single muscle. They also show that there may be a considerable discrepancy between the appearance of a muscle—or a portion of a muscle—and its biochemical characteristics. The superficial layers of the gastrocnemius muscle consist almost entirely of white fibers while the deep layers appear to contain a considerable number of red fibers. We compared the activities of eight glycolytic enzymes, the myosin adenosine triphosphatase activity, and the $Ca^{++}$ uptake of fragmented sarcoplasmic reticulum in the two layers. The differences we found were rather small, differing by a factor of two or less. But if we took the semitendinosus or soleus muscle and compared it with the white adductor muscle there was a much greater difference. The glycolytic enzyme activities, for instance, were almost one order of magnitude higher in the white muscles. Do you think that if you had chosen a predominately red muscle (soleus, for example), the differences that you showed would be much greater?

*Beatty:* No, not in the rhesus monkey. We evaluated glycogen synthetase, phosphorylase, and succinic dehydrogenase stains, and succinic dehydrogenase homogenate activities in our search for the reddest and whitest muscles in the rhesus monkey. It is amazing how well these activities correlate with the visual assessment of the muscles. The soleus, which had a 5-fold greater activity in succinic dehydrogenase than the brachioradialis and a 1.7-fold greater activity than the sartorius was, metabolically speaking, similar to the sartorius in regard to its *in vitro* metabolism. Perhaps stimulated muscles might show greater differences.

*Dahl:* I noted that many of the experiments quoted in your paper made use of mixed muscle even though one type was in predominance. Would such studies on rhesus sartorius, for example, be significantly altered by the presence of 30–40% white fibers when you were interested in red fibers?

*Beatty:* Certainly the 30% of white fibers present in the sartorius affect our results. Now that the soleus appears to be a mixture of red and intermediate fibers, however, we know of no all white or all red mammalian muscles, certainly not in the primate. Actually, we have also obtained almost an order of magnitude difference in succinic dehydrogenase enzyme activity in some of the rhesus muscles, such as the quadratus femoris vs. the brachioradialis (7-fold; Table 9.6). Unfortunately, only certain muscles are suitable for the preparation of muscle fiber groups and the quadratus is not one of these. The fibers should be longitudinally arranged in the muscle and one must be able to tease groups of them apart with minimum trauma. The point we wish to emphasize is that the same percentage of the oxygen consumption was accounted for by the oxidation of glucose + glycogen in our predominantly red as in our predominantly white muscle. It would be most surprising to us if the 25% of red fibers in the white muscle preparation accounted for about the same amount of lipid oxidation as the 70% of red fibers in the red muscle. We think the white fibers must be utilizing a considerable amount of lipid.

*Dhalla:* What class of endogenous lipids is utilized during skeletal muscle contraction?

*Beatty:* Our laboratory has no data as yet on contracting muscle, and we know of no data on endogenous lipid utilization in red and white voluntary skeletal muscle.

*Moody:* Many biochemical and histochemical mechanisms have been studied in heart and diaphragm muscle, yet relatively little has been reported on these mechanisms for red and white skeletal muscle. Why? May one expect similar results or something quite different:

*Beatty:* Many of the same basic metabolic pathways are present in diaphragm, voluntary skeletal muscle, liver, brain, etc., but the relative importance of these metabolic pathways varies enormously from tissue to tissue and their sensitivity to hormones also varies. For instance, we were able to demonstrate a catabolic effect of glucagon on protein metabolism of voluntary skeletal muscle but not diaphragm (Peterson, Beatty, and Bocek, 1963; Beatty, Peterson, Bocek, Craig, and Weleber, 1963).

*Hamm:* You pointed out that diaphragm is not suitable for comparison of red vs. white muscle. In our studies on muscle enzymes we used diaphragm muscle as representative of "red muscle," but your objections against it are quite convincing. Is your statement also valid for studies on enzymes of muscle mitochondria and of the matrix? With a series of muscles with different myoglobin content including different skeletal muscles and diaphragm, we found that, concerning the correlation between mitochondrial enzymes or lactic dehydrogenase (LDH) and myoglobin content, the diaphragm muscle fits into the general pattern very well. Does this observation disagree with your findings?

*Beatty:* We found that rat diaphragm, at least, had a high succinic dehydrogenase activity, about the same as the soleus. This might not be true for other species since Gauthier and Padykula (1966) have shown that bovine diaphragm, for example, is composed primarily of white fibers.

*Hamm:* You mentioned that LDH V was found in large amounts in white fibers. We find the same in bovine and porcine muscles. LDH isozymes rich in subunit M were highly and negatively correlated with the myoglobin content of muscles. This subunit seems to be related to the relatively high glycolytic activity of white muscle fibers. The LDH subunit H seems to be more or less bound by subcellular particles. The LDH isozymes, rich in subunit B, appeared, however, to be equally distributed in white and red bovine and porcine muscles. Do you know something about the metabolic role of LDH I and LDH II in skeletal muscles?

*Beatty:* According to Dawson, Goodfriend, and Kaplan (1964), the H-type subunit (heart) or enzyme $H_4$ is predominantly active in aerobic metabolism where energy is continuously required. In these tissues pyruvate does not usually accumulate and when it does its inhibiting effect on the activity of this enzyme causes conversion of pyruvate to glycogen or glucose rather than lactate. The M subunit (muscle) or enzyme $M_4$ is primarily concerned with the conversion of pyruvate to lactate to stimulate glycolysis and meet the energy requirements of a short burst of activity.

*Hamm:* You mentioned that red muscle shows a higher glutamic-oxaloacetic transaminase activity (GOT) than white muscle. We have found the same in bovine and porcine muscles. A highly significant correlation between GOT activity and myoglobin content was found for different porcine muscles (the same for glutamic-pyruvic transaminase). It is interesting that this high correlation was found not only for the mitochondrial isozyme $GOT_M$ but also for the sarcoplasmic isozyme $GOT_S$. Can you explain the reason for this apparent coupling between the sarcoplasmic and the mitochondrial transaminase activity with regard to the protein metabolism?

*Beatty:* I cannot offer an explanation.

*Hamm:* According to Table 9.1, red muscles show a higher activity of mitochondrial enzymes and a lower activity of glycolytic enzymes than white muscles. We plotted the activity of mitochondrial and glycolytic enzymes against the myoglobin content of muscle and found that the increase in the mitochondrial activity and the decrease in the glycolytic activity with rising myoglobin content is greater in porcine muscles than in bovine muscles. At low myoglobin content, for instance, the activity of succinic dehydrogenase and aconitase in bovine and porcine muscle was not much different. At high myoglobin content, however, the activity of these enzymes was much higher in porcine muscle than in bovine muscle. I am somewhat surprised that in the red muscle of a nonruminant the mitochondrial enzyme activity was higher and the LDH V activity was lower than in the red muscle of a ruminant, because the muscles of ruminants should use more lipids and less glycogen for energy production than muscles of nonruminants. Do you have an explanation for this observation?

*Beatty:* No, and I know of no *in vitro* work on the uptake and utilization of lipid and carbohydrate by the skeletal muscle of either ruminants or nonruminants.

*Price:* Why should glycogen stores be greater in predominantly white muscle than in predominantly red muscle?

Does muscle function (i.e., historical exercise pattern) and blood supply influence expected and observed glycogen levels in both muscle types within any animal at any specific time?

*Beatty:* The answer is that these factors would be expected to influence glycogen levels. It is generally accepted, however, that in the resting state white muscle has higher glycogen levels than red, and these higher and more labile stores are consistent with the ability of white muscle to contract for limited periods anaerobically.

*Dahl:* You comment that the rhesus monkey is highly excitable; could not this comment apply to many other experimental animals in reference to stress and glycogen depletion? It has been my experience that anesthesia or slaughter of animals is stressful to them; would curarization better preserve the muscle from at least motor activity (e.g., agonal convulsions)? Would either the levels of epinephrine or adenyl cyclase give some measure of glycogen breakdown from stress in these preparations?

*Beatty:* The catching of the monkey and the sight of the experimenters are the stressful factors which we cannot seem to eliminate. I do not think that curarization would help solve this problem.

*Moody:* If muscles or portions of muscles are excised by biopsy techniques or removed at the time the animal is exsanguinated, may one expect valid glycogen activity using the PAS or other histochemical techniques? What effect does excision have upon the glycogen levels of the muscle, and how is one to know if all muscles react similarly?

*Beatty:* We have also frozen muscle *in situ* by immersing the leg of an anesthetized rat in liquid nitrogen. In this instance, the glycogen levels were the same as those obtained by biopsy.

*Briskey:* We have observed that the muscle of the pig, unlike that of the rat, responds dramatically to the mere act of excision. Lactic acid may increase by 15–20 $\mu$m/g and 10–12 $\mu$m/g of creatine phosphate may be broken down in the muscle of the stress-susceptible pig (Lister, Sair, Will, Schmidt, Cassens, Hoekstra, and Briskey, 1970) as a result of excision anoxia. The cryoprobe of Teeter, Carr, Tsai, and Briskey (1969) may be used to obtain tissue that more nearly approximates the *in situ* state of the muscle during life.

*Moody:* What is there about the PAS reaction for glycogen that consistently reveals three subgroups of fibers in skeletal muscle, and can one rely on this technique when studying predominantly white or red muscle?

*Brooke:* I think one can use the PAS reaction for studying muscle, but the information derived from such a study should not be assumed automatically to be true for other properties of muscle such as adenosine triphosphatase or the specific categorization of fibers into types I and II or fast and slow.

*George:* You have shown that the red skeletal muscle has a higher level of glycogen than the white. It is known that red fibers have a higher level of glycogen synthetase activity than white. What is the role of the glycogen store in the red fibers, which presumably utilize lipids for energy? Could this glycogen store serve as a source of oxaloacetate for fatty acid oxidation?

*Beatty:* The higher glycogen content in red than in white muscle has been demonstrated, as far as we know, only in rhesus muscle and we feel that this relationship, which is the reverse of that found in many mammalian species, may not necessarily be true in an unstressed monkey. As demonstrated by Hultman (1967), energy release from glycogen breakdown appears to be indispensable for heavy work, and in all probability does serve as a source of oxaloacetate.

*Briskey:* Beecher, Cassens, Hoekstra, and Briskey (1966) also observed that the red muscles of the pig had higher glycogen levels than white muscles, if the animals were heat-stressed before excision of the samples —however, the red muscles had lower glycogen levels than white muscles when compared in anesthetized control animals.

*Dahl:* What do you believe is the significance of the higher glycogen synthetase activity for red fibers than for white in mixed muscle?

*Beatty:* We really have no satisfactory explanation. The fact that we observed a 10-fold higher rate of glucose-$^{14}$C incorporation into glycogen in red muscle in the early periods of a 2-hour incubation indicates that under specific circumstances the enzyme assays reflect *in vitro* activities. Why red muscle would need a greater capacity to regenerate glycogen in the face of an apparent slower mechanism for glycogenolysis is not clear. An estimation of glycogen turnover rates and a correlation with the activities of the active and inactive forms of synthetase and phosphorylase may provide some clues to the answer. We are also planning to investigate stimulated muscle.

*Dahl:* Do you have any information on gluconeogenesis or other catabolic activities of protein in muscle, in reference to atrophy from disease or disuse?

*Beatty:* We have done no work in this field.

*Beecher:* Would you comment or speculate on the glyconeogenic activity of red and white fiber groups and the role glyconeogenesis plays in maintaining or replenishing glycogen concentrations in those two fiber groups?

*Beatty:* Bär and Blanchaer (1965) demonstrated the incorporation of lactate-1-$^{14}$C into glycogen. Under the conditions of this experiment there was no difference in the glyconeogenic effect of lactate in red and white muscle. The glycogen levels at the beginning of their experiment were low. The external oblique and diaphragm were used as white and red muscle.

*Green:* The best one can do at the present stage of our knowledge is to contrast red and white muscle. But is there perhaps a more fundamental difference that would only be recognizable at what we may call the ultrastructural level? I would like to suggest that there could be an ultrastructural level at which, for example, only mitochondria were concerned in the energizing process; in another type of fiber the glycolytic system would be involved. Could there be, at the ultrastructural level, one type of fiber in which actomyosin is not associated with any mitochondria and another type in which it is associated entirely with mitochondrial systems but perhaps is lacking in the glycolytic systems? Since you are dealing with population mixtures you really cannot see more than a trend and there is great difficulty in preparing a homogeneous system.

*Beatty:* There might be a clue in the fact that the specific isozymes of lactate dehydrogenase are present in individual muscle fibers, both red and white. Also, we have been able to show, if you accept the soleus as a red muscle, that is, as a muscle with no white fibers, that the soleus has about the same pattern of carbohydrate metabolism in the resting state (*in vitro*) as the sartorius, which is predominantly red (25–30% red fibers).

*Green:* You are still dealing at the gross level with a mixture of red and white, even if you call it the most unambiguous red muscle. It is just more red than it is white. There may be a level, an ultrastructural level, where you really have not so much a quantitative as a qualitative difference, but nobody has yet found it.

*Beatty:* Do you mean specifically with the mitochondria?

*Green:* At a more molecular level you might, in fact, find the pure types, and the difference at that level would be enormous.

*Beatty:* Blanchaer (1964) has done some work with red and white muscle mitochondria. Of course, they collected them under the same circumstances from predominantly red and white muscle. The next step would be to isolate individual red and white fibers but that also presents problems because there is a range of histochemical reaction intensity for individual fibers.

*Goldspink:* If red muscle fibers are smaller than white muscle fibers, as they almost invariably seem to be, then your red muscle samples would contain more fibers than the equivalent weight samples of white muscle (perhaps as many as 3 or 4 times more). Could this not, therefore, explain why the red muscle samples were in general more biochemically active than the white muscle samples?

*Beatty:* Certainly the oxidative metabolism of red muscle is higher than that of white and this higher metabolic rate is a function of the number of red fibers present because of their higher mitochondrial density. Perhaps a better way of looking at this would be to consider metabolic rates not as a function of the number of fibers but in relation to the total cross-sectional area occupied by the predominant fiber type present.

*Goldspink:* We have studied the relationship between succinic dehydrogenase and fiber size. We did this by measuring the amount of stain deposited in individual muscle fibers, and in fact we get a straight line relationship between the size of the muscle fibers and the total amount of succinic dehydrogenase. The concentration within the fiber differs of course, but the total amount within a fiber is the same. The reason is simply this: the mitochondria are diluted during the growth process by additional myofibril material. There are several important consequences to this. One is that a sample of white muscle will not give the same number of muscle fibers in that sample as would red muscle because of the difference in size. Therefore you would certainly expect some quantitative differences in the biochemistry of the sample. So it seems to be very important that we should know not only the concentration but the total amount within a fiber. We believe that the fiber may not have enough mitochondria to provide sufficient energy, by aerobic means, and has to rely more on glycolysis. One gets an increase in phosphorylase and the other enzymes, so it is not always necessary to look for neurogenic factors to explain fiber types; the growth processes can explain why there are different histochemical types within the same muscle.

*Dubowitz:* Would Dr. Goldspink explain why the non-NAD linked mitochondrial enzyme α-glycerophosphate dehydrogenase is not diluted in the same way in the white fibers as estimated histochemically?

*Goldspink:* We have not studied this enzyme, but there are certain variations between muscle fibers and perhaps muscle fibers just vary in their content of α-glycerophosphate dehydrogenase.

*Dubowitz:* I suggest that if you studied the α-glycerophosphate dehydrogenase you would get a linear relation in the opposite direction because it is high in the larger type II fiber.

*Goldspink:* I do not dispute the fact that phosphorylase is greater in the type II fibers, but I suggest this is a consequence of the further growth that the fibers undergo. Perhaps it may be that when a small fiber grows into a large fiber, the change in its metabolism also necessitates more α-glycerophosphate dehydrogenase and therefore more of this enzyme is synthesized by the large fiber.

*Peachey:* There are some ultrastructural studies which clearly show that muscle fibers within a given animal or in a given muscle have greatly differing numbers of mitochondria. Sometimes the larger fibers have a higher density of mitochondria as well as larger mitochondria.

*Goldspink:* It is necessary to carry out quantitative studies on the mitochondrial content of many individual muscle fibers before any conclusions can be drawn. Size of fiber and growth of the fiber determine the histochemical properties of the fiber.

*Peachey:* It seems to be possible for a fiber to get bigger and as it does so, to increase the number of mitochondria in even greater proportion than the increase in size.

*Goldspink:* Certain kinds of exercise will cause an increase in mitochondria. According to our findings from electron microscopic work and cytochemistry, however, this does not occur during "normal" growth.

*Brooke:* Seligman, Veno, Morizono, Wasserkrug, Katzoff, and Hanker (1967) have claimed that there are, in fact, two kinds of mitochondria as evaluated by succinic dehydrogenase at the electron microscopic level and that they both exist within the same fiber. Also, one must be extremely careful when talking about the histochemical reactions for oxidative enzymes, because the formazan binds so tightly to the lipid membranes that it is very difficult to be sure of localization. Where the formazan does not bind, it is impossible to demonstrate any enzyme; where it does bind it will demonstrate enzyme if there is any in the vicinity. I do not know if vicinity means micron or millimicron. It is possible not only to reverse lactate dehydrogenase to make the other fibers stain more intensely, but it is also possible to reverse the adenosine triphosphatase staining pattern. I have great difficulty sometimes in being convinced of the histochemical validity of some of these stains. Although they are a very useful technique, I think to correlate biochemistry and histochemistry is fraught with danger.

*Morita:* Does myoglobin content correlate well with fiber-type composition of muscle? For example, beef longissimus muscle contains about 25 times more myoglobin than rabbit muscle. Does it depend on fiber-type composition?

*Dubowitz:* Myoglobin does correlate with fiber type. Type I fibers have a high myoglobin content and type II a low. Varying gradations of intermediate fibers may also be present. James (1968) recognized five gradations of staining. Differences in total myoglobin content in different muscles may be determined by the proportion of type I and type II fibers and probably also by the absolute content of myoglobin per type I or other fiber, which may well vary from one species to another. It is interesting that some of the cross-innervation work done by McPherson and Tokunaga (1967) on myoglobin content showed that there was a reversal or a tendency to reversal of the two types.

*Morita:* Our histochemical work (Morita, Cassens, and Briskey, 1969) has shown a strong myoglobin reaction in red fibers but an essentially negative reaction in white fibers of porcine muscle. However, even the white fibers of bovine muscle have some reaction for myoglobin.

*Peachey:* I would be rather surprised if myoglobin content kept an absolute correlation with mitochondrial content in a muscle if one begins to compare different species. For example, in diving mammals the purpose of myoglobin is to allow the animals to stay under water for an extensive length of time and as far as I know all the muscles of seals and whales are red. These animals do not have as severe a need for tonic muscles or postural muscles, being floating animals, as do animals such as the monkey. So I would think that the kind of correlation which was brought up a minute ago might occur within one animal. One could probably find a violation of this correlation by looking at some of the diving mammals, however.

*Allin:* It is true that myoglobin and mitochondrial content are in very different ratios in different species, but it need not follow that myoglobin and mitochondria vary independently within the individual fibers of a given muscle in a given organism.

*Hamm:* In our laboratory we use the biochemical determination of the activity of a particular enzyme, such as succinic dehydrogenase, and the content of myoglobin for identification of the nature of the muscle. There is a positive correlation between myoglobin content and the activity of mitochondrial enzymes; there is a negative correlation with glycolytic enzymes. Do you think that this method is sufficient for identification of the overall metabolic activity of muscle or would it be necessary also to differentiate the fiber types histochemically?

*Dubowitz:* Biochemical studies of enzymes or myoglobin can only tell you the activity of the muscle as a whole. In order to know the composition of the muscle in terms of different fiber types, parallel histochemical studies are essential.

*Dalrymple:* Why do the mixed muscles of some species (e.g., bovine) show a uniform checkerboard distribution of red and white fibers while some other species (e.g., porcine) show histochemical patterns whereby red fibers are found in groups of several fibers completely surrounded by white fibers?

*Dubowitz:* The difference in fiber-type distribution is a known fact. Probably all muscles, if one could just work it out carefully enough, would have a specific profile, and one should be able to examine it under the microscope and identify it from the pattern. Certainly in some muscles like the pigeon breast muscle the composition is almost entirely type I; there are a few type II fibers, but these are almost always scattered around the periphery. Certainly some muscles like the gastrocnemius in the rat and the mouse consistently have areas with more of one fiber type than the other. I think one also has to be very careful about the use of the term "red areas and white areas" in order not to cause confusion—whether one is talking purely of a visual appearance or whether one is talking of the basis of histochemical observations. If the fiber type is dependent on the innervation, then clustering of one fiber type, as in porcine muscle, may be related to the way in which single motoneurones branch to supply adjacent fibers. Then one must try to explain why there is a difference in the pattern of innervation. Perhaps this is determined by the function or activity of the particular muscle. The fiber-type distribution is a known fact, but the significance is quite unknown.

*Bocek:* Dr. Dubowitz, you stated that in the human adult, all muscles are composed of about equal numbers of two types of fiber. How extensively have human muscles been surveyed? In most mammals, including the rhesus monkey, the soleus muscle is composed of predominantly red fibers. It seems unusual that the human soleus, histochemically, would be different from that of other species.

*Dubowitz:* Unlike the mouse, where one can readily study a complete cross section of a muscle, the human is a difficult subject to assess. Biopsy material is limited in quantity, and autopsy material not very good for enzyme assessment. I have done (1968, unpubl. data) counts of the proportion of fiber types in a large number of different muscles obtained during surgical procedures and have been unable to demonstrate any consistent differences between proximal and distal muscles or flexors and extensors and muscles with different functions. Certainly the soleus in the human appears to be as mixed histochemically as the gastrocnemius.

*Dahl:* What is the significance of intermediate (type B) fibers? Why are they not present in humans where all muscles are mixed, compared to lower forms? Do the intermediate fibers have an intermediate twitch profile?

*Dubowitz:* Intermediate fibers are probably a distinct group in animals. Intermediate fibers do occur in humans but seem to be less numerous and less clearcut than in animal muscles. Physiologists are just beginning to recognize different motor units within apparently "pure" slow or fast muscle. Future studies will probably reveal similar gradations between the physiological extremes of motor unit type as in histochemical studies. Cross-innervation studies support the view that there are really only two basic fiber types and I think the intermediate fibers are variations of these.

*Moody:* What is the significance of the variation in muscle fiber size between red and white muscle fibers within a single muscle bundle? Is there more variation in size among red fibers than white? At what stage during differentiation is size-difference most apparent?

*Brooke:* The amount of variability in the size of fibers depends upon the species being examined. It is slight in the human and marked in the pig, for example. I do not know the significance of this.

*Blumer:* Have identical twins been used to compare muscle fiber types? If so, is the pattern of red and white fibers more nearly alike than with full sibs?

*Dubowitz:* I think that, in terms of the fairly crude comparison of the same muscle in different subjects, there is so little difference that it would be difficult to distinguish identical twins from unrelated subjects, let alone sibs. Perhaps with more refined mapping, differences from one individual to another may be apparent. The same pattern of similarity for a given muscle is noted in animals as well, both in littermates and in unrelated animals of the same species.

*Hamm:* You mentioned adenosine triphosphatase activity several times, but you did not comment on the type of adenosine triphosphatase. It would be important to know, with regard to the metabolic type of fiber, whether the adenosine triphosphatase is localized in the mitochondria, the myofibrils, or the sarcotubular system. Is an exact histochemical differentiation between these different forms of adenosine triphosphatase possible and, if so, how are their activities correlated with the particular types of muscle fiber?

*Dubowitz:* My remarks were all on myofibrillar adenosine triphosphatase demonstrated at *p*H 9.4 by the method of Padykula and Herman (1955). There is a method for showing mitochondrial adenosine triphosphatase (Wachstein and Meisel, 1955) and this has a stronger reaction in type I fibers.

*Dhalla:* On the basis of staining techniques, you have shown some differences in phosphorylase enzymes in the red and white muscles. Could you tell us what type of phosphorylase you are dealing with and what is the significance of this phosphorylase physiologically, particularly when there is a great deal of high energy phosphate compound readily available for muscle contraction?

*Dubowitz:* Since AMP is added to the incubation medium, we are probably assessing total phosphorylase by the histochemical techniques used. Methods are available for differential study of phosphorylase *a* and *b*. We presume the phosphorylase activity reflects the ability to utilize glycogen particularly for the most rapid unsustained activity of the muscle fiber.

*Moody:* Is there a reliable histochemical procedure for determining lipase activity of skeletal muscle, and if so is this a good approach to use in studying lipid metabolism of muscle types?

*Dubowitz:* There are methods available for lipase activity. J. C. George has studied this extensively (see George and Ambadkar, 1963).

*Moody:* What effect does rapid freezing have on the enzyme activity of muscle and at what conditions of time and temperature should one store frozen sections?

*Dubowitz:* From a biochemical point of view the more rapidly a muscle is frozen the less morphological artifact is produced and the better are the enzyme reactions, particularly those for mitochondrial enzymes. Frozen specimens should be stored at as low a temperature as possible. If they are kept at −10° C there will be a loss of much activity of sarcoplasmic enzymes within a month. Tissues also tend to dry with storage and should be kept in sealed containers. Freezedrying techniques are the answer to prolonged storage. As regards the frozen sections, some reactions such as adenosine triphosphatase are permanent, others such as SDH mounted in aqueous medium fade with time and fat, if present, tends to spread over the section.

*Weber:* Since it seems that most muscles are mixed and have red and white fibers, are there two sets of motoneurones? Has anybody done the anatomy and double innervation?

*Dubowitz:* I think the anatomy of two sets of motoneurones has now been reasonably worked out. Kugelberg and Edström (1968) found that by stimulation of the sciatic nerve, after dissecting it down to physiologically individual nerves, they got a differential loss of phosphorylase and glycogen from the different fiber types. They could map out the whole territory of individual motor units. This seemed to correlate with a single nerve supplying single fiber types. This should be considered in addition to the work of Henneman and Olson (1965). I think one can now assume that one anterior horn cell of one nerve supplies uniform fibers.

*Dahl:* You have given several examples of diseases associated with type II fiber atrophy and associated the atrophy with the imposed inactivity. Why do you not see selective type II atrophy in the dystrophies which are diseases with similar inactivity?

*Brooke:* In dystrophies one does see type II atrophy in severely affected cases, but this process is masked by the enormous variation in the size of all fibers which is the hallmark of the disease.

*Dahl:* Can you tell the type II atrophy of disease states from that of normal females? Are there any other criteria, such as configurational changes (i.e., "roungulation" and "angulation") that will serve to distinguish the pathological from the physiological?

*Brooke:* This is discussed elsewhere (Brooke and Engel, 1969) ; such differentiation can be made by the use of histograms and "atrophy-hypertrophy" factors.

*Bocek:* It has been suggested that in the cat the red muscle fibers of the soleus differ histochemically from red fibers that are found in the mixed muscles (Nyström, 1968) . Is there any evidence that the diseases associated with the upper or lower motoneurones affect the characteristics of the red fibers of the soleus any differently from those of the red fibers of the mixed muscles?

*Brooke:* I do not really have any information on that: we have not been able to biopsy the soleus muscle from human patients, nor do we have any autopsy material. In terms of animal experimentation, there is a difference histochemically between the type I muscle fiber of the soleus and the type I muscle fiber of the gastrocnemius, but they do not seem to respond very differently in terms of their response to pathological situations.

*Dahl:* Is there any evidence that the red and white fibers have different functions even in a mixed muscle (e.g., do we stand on our red fibers and run on our white) ?

*Dubowitz:* I think this is probably correct. Studies on the influence of environmental factors such as exercise on the fiber type of a given muscle will probably answer this in due course.

*Morita:* Can muscle fibers be changed from one type to another by such factors as stress?

*Dubowitz:* We do not have the answer to that one yet, but there certainly is a fair amount of work accumulating on the selective effect of stress, such as exercise, on specific fiber types. Perhaps in due course we shall see a complete changeover in fiber type under influences other than the innervation.

*Hamm:* Inactivity should be associated with atrophy of type II fibers. Does physical work (training with an ergometer) result in a hypertrophy of type II fibers?

*Brooke:* I think so, at least as judged by some of our preliminary results using experimental animals. The same is suggested by examining type II fibers in human subjects, as discussed in Chapter 8.

*Dahl:* Can the type II "hypertrophy" seen in the tenotomized cat be due to the shortening of the muscle, making the fibers appear to be more plump? What is the effect of extension vs. flexion on fiber size?

*Brooke:* The "tenotomized-denervated" muscle is also shortened, as is the muscle fixed in a fully shortened position; in neither situation is type II hypertrophy observed.

*Dahl:* Have the cat tenotomy experiments been tried on any other animal, since there are differences between species?

*Brooke:* I do not know of other histochemical studies.

*Bocek:* What happens to blood circulation in tenotomized muscle? Since red muscle is more dependent on blood-borne substrates, would a possible diminished blood supply be involved in atrophy of the red fiber?

*Brooke:* I have no information on this question.

*Dahl:* Can a muscle fiber-type pattern be altered by changing its function, for instance by altering the muscle insertion so that it is a flexor rather than an extensor?

*Dubowitz:* There are no conclusive data on this yet, but I think the answer will prove to be yes. It is an aspect I am currently studying.

*Morita:* Do you have any information on factors, other than nerve, that influence muscle function—for example, hormones?

*Beatty:* I am sure that many factors such as hormones influence muscle biochemistry. The only hormones we have investigated are insulin and glucagon, and we could not distinguish any difference in the effect of these hormones on red and white muscle.

*Morita:* What happens to the mitochondria and sarcoplasmic reticulum following cross innervation?

*Beatty:* Dr. Mommaerts reported that after cross innervation there is a reversal in the rate of $Ca^{++}$ release by sarcoplasmic reticulum vesicles from fast and slow muscles (Chapter 4). We know of no electron microscopic studies of muscle after cross innervation although morphological differences between red, white, and intermediate fibers at the electron microscopic level have been reported (Padykula and Gauthier, 1967; Gauthier, 1969).

*Bocek:* In view of the results of cross-innervation studies and the eight or so fiber types described, Guth, Watson, and Brown (1968) have suggested that the typing of fibers on a rigid categorical basis ignores the fact that control of fiber types by "trophic" nerve function may be in a dynamic state; that is, the intermediate fiber may well represent fiber in transition from one type to another. Would you speculate on this comment in terms of Nyström's (1968) work that shows the late appearance of B fiber in the developing cat gastrocnemius?

*Dubowitz:* I find it difficult to visualize eight different nerve types with eight different dynamic states of neurotrophic influence. I think if Romanul used twice the number of enzyme stains he might have defined 16 fiber types instead of eight. I wonder whether, in fact, the nerve does not determine only the two extreme types, and that intermediate fiber types are determined by other factors such as activity?

*Edgerton:* As a general response to some of Dr. Brooke's comments, I would like to point out that much of the inconsistency in muscle histochemistry lies in the misinterpretation by the investigator, not the technique itself. For example, Stein and Padykula (1962) described three fiber types in rat skeletal muscle, using adenosine triphosphatase, SDH, and esterase. They stated, and others have confirmed (Edgerton, Gerchman, and Carrow, 1969), that the rat soleus contains intermediate and red fibers. The intermediate fibers demonstrated moderate SDH and low adenosine triphosphatase activity. Red fibers had high SDH and high adenosine triphosphatase activity. White fibers, which are common to most muscles, have low SDH and high adenosine triphosphatase activity. Romanul (1964), using a series of glycolytic and oxidative enzymes, described eight fiber types but reduced the classification into three groups (A, B, and C). On the basis of Romanul's metabolic description of these three groups, his classification of A, B, and C corresponds to Stein and Padykula's (1962) white, red, and intermediate fibers. Metabolically, these fibers show relatively high glycolytic, high glycolytic and oxidative, and high oxidative activity respectively. Further confusion has evolved from the use of another classification (type I and II), which has been paralleled to red and white fibers, respectively (Engel, 1962). Both intermediate and red fibers are usually considered to be type I in this two-category classification. Typing fibers as I or II was initially done with adenosine triphosphatase, which divides the fibers into two distinct populations. Since it was reported that type I fibers (light adenosine triphosphatase activity), with some exceptions, had high oxidative enzyme activity (Engel, 1962) and type II fibers (high adenosine triphosphatase activity) had low oxidative enzyme activity, muscle fibers are now classified by some investigators as type I or II using only oxidative enzymes, not adenosine triphosphatase, as the criterion. This has been another source of confusion, since a table in Stein and Padykula (1962) clearly shows that adenosine triphosphatase was low in intermediate fibers and high in white and red fibers. Work in my laboratory (Edgerton, Gerchman, and Carrow, 1969) and photographs from a recent publication

(Yellin, 1969) confirm Stein and Padykula's and Romanul's original observation. I think we can say with certainty that type I and II fibers do not correspond to red and white fibers, respectively. Type I fibers more closely parallel intermediate (oxidative) fibers, and type II parallel white (glycolytic) or red (glycolytic and oxidative) fibers.

Since contractile properties of the various fiber types are a point of frequent speculation, it is appropriate to mention another observation usually overlooked. Close (1967), in a study on the contractile properties of individual motor units of the rat soleus, revealed that three of 30 motor units were relatively fast (CT = 18 msec) and 27 of 30 units were relatively slow (CT = 38 msec). Since the proportion of intermediate to red fibers histochemically corresponded approximately to the proportion of slow to fast motor units, Close hypothesized that the intermediate fibers, not the red, are likely to be the slower ones in the rat soleus. Adenosine triphosphatase, as demonstrated histochemically at a *p*H of 9.4, is in agreement with the contraction times reported by Close. Although it cannot be specifically stated which chemical property of adenosine triphosphatase is being histochemically demonstrated, it does grossly correlate with contraction times. The major question arises when one sees a continuous distribution of contraction times (Henneman and Olson, 1965) for motor units in the soleus or gastrocnemius of the cat, but two distinct adenosine triphosphatase reactions (relatively low and high) are seen histochemically. A comparison of contractile properties and a hypothesis to explain the two populations of fibers that are demonstrated with adenosine triphosphatase have been reported (Robbins, Karpati, and Engel, 1969). Two populations of fibers with adenosine triphosphatase activity are actually what one might expect after seeing the data of Bárány, Bárány, Reckard, and Volpe (1965).

The validity of the assumption that red fibers are slow was questioned further by Hall-Craggs (1968) when he found that the thyroarytenoid was an extremely fast-contracting muscle (CT = 6.5 msec), but all of the fibers were red in the rabbit. The cricothyroid was histochemically mixed, resembling other fast muscles, and had contractile properties similar to such fast muscles as the extensor digitorum longus or lateral gastrocnemius of the leg.

In summary, I think we can say that skeletal muscle histochemistry has been the victim of gross misinterpretations and faulty assumptions which have concealed much significant information.

*Allin:* Dr. Edgerton attempts to relate histochemical typology to contractile speed. This is very unsafe. Close (1967) has presented evidence that the rat extensor digitorum longus has uniformly fast motor units only, yet this muscle is histochemically mixed. It is also unsafe to consider the "intermediate" fibers of rat limb muscles homologous to those of other mammals. In cats it is the "red" fibers which are low in myofibrillar adenosine triphosphatase activity, not the intermediates.

*Dubowitz:* I think the terms "red" and "white" should be restricted to the color of a muscle as a whole and perhaps macroscopic parts of the muscle, but not applied to individual fibers. Stein and Padykula (1962) defined their fiber types, on the basis of the SDH reaction, as strong-, weak-, and intermediate-reacting. In terms of physiological properties "redder" or "whiter" are thus relative and arbitrary. I think the only useful correlation is between contractile properties on the one hand, and either fiber types, based on enzyme histochemistry, or total enzymic content of the muscle or parts of the muscle assessed biochemically. It is of interest that Close (1967) was able to demonstrate fast and slow units in the slow soleus, but not in any fast muscles such as flexor hallucis longus. Might I also correct Dr. Edgerton—the typing of fibers as type I and II was not initially based on the adenosine triphosphatase reaction, as he suggests, but on the oxidative enzymes and phosphorylase (Dubowitz and Pearse, 1960).

*Barnard:* In view of all the evidence (histochemical and physiological) that there are more than two types of fibers, why do you still use a simple type I and II classification?

What is the evidence that different types of exercise produce different effects on type I and II fibers and what is the evidence that electrical stimulation produces different results as you stated?

*Brooke:* As is stressed in Chapter 8, the classification of fiber types depends on the property used. With many properties there are three or more different categories of fibers. We have used the routine adenosine triphosphatase reaction at *p*H 9.4 to classify muscle fibers. With this reaction, in human muscle and in the animal muscles discussed, there are two main fiber types. I am not denying the existence of more than two fiber types when other criteria are used.

The evidence for the different effects of different forms of muscle activity is based on some studies in which our laboratory is presently involved.

I would like to clear up one thing: I do not mean to imply that histochemistry is not a valid investigative technique. It is, however, a technique for which you should be aware that there are disadvantages. One must take into account the binding of end products to muscle in many of the stains before one can interpret the results.

When one talks about exercise one has to differentiate the kind of exercise because, for example, the changes in the plantaris, not only in the rat but also in the cat and guinea pig, depend upon the type of exercise that one provokes in the animal. Running on a treadmill gives information which is very difficult to interpret. One needs to know whether it is uphill, whether the animal has to make sudden jumps, etc. Full knowledge of the experimental situation is extremely important.

*Dahl:* Does the onset of differentiation of the fetal muscle correlate with the time of contact with the anterior horn cell during development?

*Dubowitz:* The differentiation probably correlates in some way with the maturation of the lower motoneurone (the peripheral nerve part of it). In the human the nerve makes contact with the muscle fiber very early, probably around 8–10 weeks gestation. Clearcut differentiation into recognizable fiber types occurs after the 20th week, and coincides with the development of mature motor end plates. Further correlative studies in this sphere should prove useful.

*Moody:* Is the differentiation of muscle fibers into various types a result of the adaptation of enzymes to activity, age, or growth of the muscle? Does this vary from one species to another?

*Dubowitz:* In my early comparative studies on various laboratory animals (Dubowitz, 1963) there seemed to be a correlation between the presence or absence of differentiation of the muscle at birth, the activity of the newborn animal, and its general development, for instance, fur development. Animals of long gestation, such as the guinea pig, show more maturity and muscle differentiation than those of short gestation such as the mouse and rat, but the correlation is not absolute for all animals. I think nutrition may also influence the speed of differentiation. Rats from small litters grow faster and show earlier subdivisions of their muscle into fiber types.

*Moody:* Are type I fibers observed in fetal muscles the same as type I fibers seen in older, more mature muscle?

*Dubowitz:* Although the undifferentiated fetal muscle is more like the type I than type II fibers, I think that quantitative studies on individual fibers will probably show some differences between the immature and the mature muscles.

*Hamm:* You said, "It may be important to distinguish between a variation in the strength of reaction of a single enzyme in different fibers of a muscle, and the differentiation of muscle into fiber types." If I understood this statement correctly, differences in the enzyme activities of different fiber types might not be necessarily due to differences in the metabolic type of the muscle fiber. I would like to know the biochemical reason for this statement.

*Dubowitz:* I was referring particularly to developing muscle, where with one enzyme reaction all fibers may still be uniform in activity while with another reaction there may be variation. I think, therefore, that it can be misleading to speak of fiber types on the basis of a single enzyme reaction, e.g., adenosine triphosphatase. I do not think fibers suddenly differentiate during development into mature enzyme types, but that some enzymes may well appear and reach differential levels before others. A specific fiber type should be based on at least two or preferably three different enzymes, in other words, a complete "enzymic profile."

*Hamm:* You demonstrated, by staining the fiber for certain enzymes, that in some animals the differentiation of muscle fiber types occurs postnatally. These enzymes are localized in particular compartments or organelles of the cell, e.g., SDH in the mitochondria, phosphorylase in the matrix, adenosine triphosphatase in the myofibril. How is the change in the histochemical activity of enzymes related to the development of the subcellular compartment in which this enzyme is localized? Does, for instance, an increase in mitochondrial activity during development mean an increase in number or size of mitochondria and vice versa?

*Dubowitz:* I do not think we have analyzed sufficient quantitative data to answer that. I am not aware, for example, of any studies on or changes in mitochondrial counts per fiber or levels of enzyme per fiber or per mitochondria during development.

*Moody:* Would you explain in more detail your thinking regarding the statement that, embryologically, all muscle consists initially of one fiber type?

*Dubowitz:* Muscle initially is undifferentiated by enzyme histochemical methods and therefore appears uniform. Although there may be slight variations of intensity of reaction with a particular enzyme, one cannot define fiber types within the muscle at this stage. In a particular newborn animal (e.g., cat or rabbit) the undifferentiated slow muscle such as the soleus is indistinguishable from the undifferentiated flexor digitorum longus while 2 weeks later there is a striking difference between the two muscles.

*Morita:* Which is the primitive fiber—red or white?

*Beatty:* The statement is often made in the literature that red fibers are more primitive than white. This assumption is based partly on the observation that the contraction times of muscles from newborn cats are slow and resemble those of adult red muscle. However, the red fibers in differentiated muscle probably differ from the immature fibers present in fetal muscle in many respects. Both red and white fibers could have differentiated from immature fibers to a similar extent, and until we have information on morphological and biochemical changes during

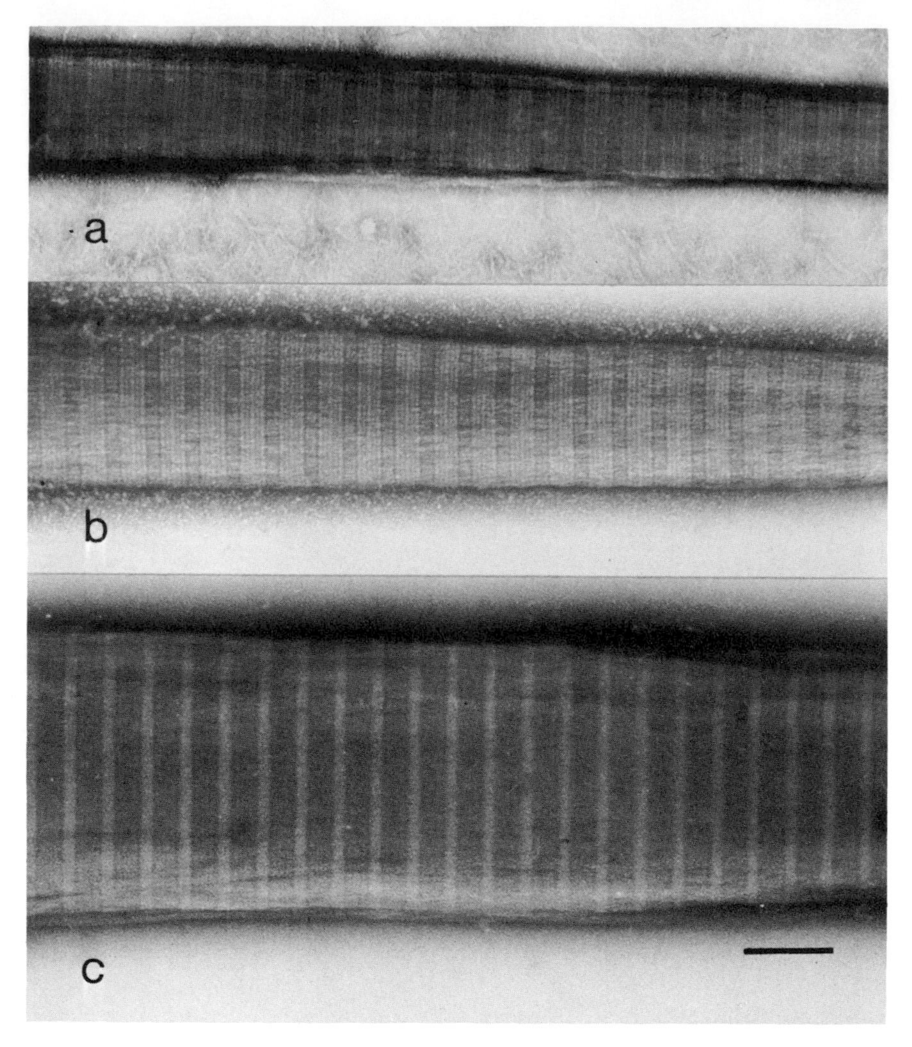


*Fig. 11.1.* Comparison of band patterns of tryptic LMM paracrystals. (*a*) cardiac muscle; (*b*) red muscle; (*c*) white muscle. Negative staining with 1% uranyl acetate. The bar indicates 1,000 Å.

differentiation of the two fiber types it may be premature to call the red fiber the more primitive fiber.

*Gergely:* In connection with the discussion concerning differences between white and red muscles, I should like to discuss briefly some recent observations which we (Nakamura, Sréter, and Gergely, 1969) have made. They show that myosins from the two types of muscle differ not only in their adenosine triphosphatase activity and in their stability at *p*H 9 or more (Sréter, Seidel, and Gergely, 1966) but also in their α-helical light meromyosin (LMM) portion. Tryptic LMM aggregates were examined with the electron microscope after negative staining with uranyl acetate. Both types of aggregates show a 430 Å periodicity (cf. Philpott and Szent-Györgyi, 1954; Huxley, 1963). There are, however, clearcut differences between the two types of aggregates with respect to the pattern within the repeating units. Aggregates—or as they are usually referred to, paracrystals—of white muscle LMM show lightly stained 100 Å bands and a heavily stained 330 Å band. In contrast, red muscle LMM paracrystals exhibit a wider—250–280 Å—lightly stained band and a narrower 150–180 Å dark band. Within the lightly stained band two doublets of 25–30 Å width are seen. It is interesting that the pattern shown by cardiac LMM aggregates is the same as that exhibited by skeletal red muscle LMM paracrystals (Fig. 11.1).

These findings make it possible to examine the electron microscopic fingerprint of different myosins and should be helpful in the study of the transformation of the character of muscles under a variety of conditions, including denervation, cross innervation, or increased activity.

## *References*

Bär, U. and M. C. Blanchaer. 1965. Glycogen and $CO_2$ production from glucose and lactate by red and white skeletal muscle. *Amer. J. Physiol. 209:*905.

Bárány, M., K. Bárány, T. Reckard, and A. Volpe. 1965. Myosin of fast and slow muscles of the rabbit. *Arch. Biochem. Biophys. 109:*185.

Beatty, C. H., R. D. Peterson, R. M. Bocek, N. C. Craig and R. Weleber. 1963. Effect of glucagon on incorporation of glycine-$C^{14}$ into protein of voluntary skeletal muscle. *Endocrinology 73:*721.

Beecher, G. R., R. G. Cassens, W. G. Hoekstra and E. J. Briskey. 1966. Red and white fiber content and associated post-mortem properties of seven porcine muscles. *J. Food Sci. 30:*6, 969.

Blanchaer, M. C. 1964. Respiration of mitochondria of red and white skeletal muscle. *Amer. J. Physiol. 206:*1015.

Brooke, M. H. and W. K. Engel. 1969. The histographic analysis of human muscle biopsies with regard to fiber types 2. Diseases of the upper and lower motor neurons. *Neurology 19:*378.

Close, R. 1967. Properties of motor units in fast and slow skeletal muscles of the rat. *J. Physiol. 193:*45.

Dawson, D. M., T. L. Goodfriend, and N. O. Kaplan. 1964. Lactic dehydrogenases: functions of the two types. *Science 143:*929.

Dubowitz, V. 1963. Enzymatic maturation of skeletal muscle. *Nature 197*:1215.

Dubowitz, V., and A. G. E. Pearse. 1960. A comparative histochemical study of oxidative enzyme and phosphorylase activity in skeletal muscle. *Histochemie 2*:105.

Edgerton, V. R., L. Gerchman, and R. E. Carrow. 1969. Histochemical changes in rat skeletal muscle after exercise. *Exp. Neurol. 24*:110, 123.

Engel, W. K. 1962. The essentiality of histo- and cytochemical studies of skeletal muscle in the investigation of neuromuscular disease. *Neurology 12*:778.

Gauthier, G. F. 1969. On the relationship of ultrastructural and cytochemical features to color in mammalian skeletal muscle. *Z. Zellforsch. 95*:462.

Gauthier, G. F. and H. Padykula. 1966. Cytological studies of fiber types in skeletal muscle. A comparative study of the mammalian diaphragm. *J. Cell Biol. 28*:333.

George, J. C. and P. M. Ambadkar. 1963. Histochemical demonstration of lipids and lipase activity in rat testis. *J. Histochem. Cytochem. 11*:420.

Guth, L., P. K. Watson and W. C. Brown. 1968. Effects of cross-reinnervation on some chemical properties of red and white muscles of rat and cat. *Exp. Neurology 20*:52.

Hall-Craggs, E. G. B. 1968. The contraction times and enzyme activity of two rabbit laryngeal muscles. *J. Anat. 102*:241.

Henneman, E. and C. B. Olson. 1965. Relations between structure and function in the design of skeletal muscles. *J. Neurophysiol. 28*:581.

Hultman, E. 1967. Physiological role of muscle glycogen in man with special reference to exercise. *Circulation Res. 20* (no. 3) (supp. 1) :1.

Huxley, H. E. 1963. Electron microscope studies on the structure of natural and synthetic protein filaments from striated muscle. *J. Mol. Biol. 7*:281.

James, N. T. 1968. Histochemical demonstration of myoglobin in skeletal muscle fibres and muscle spindles. *Nature 219*:1174.

Kugelberg, E. and L. Edström. 1968. Differential histochemical effects of muscle contractions on phosphorylase and glycogen in various types of fibres: relation to fatigue. *J. Neurol. Neurosurg. Psychiat. 31*:415.

Lister, D., R. A. Sair, J. A. Will, G. R. Schmidt, R. G. Cassens, W. G. Hoekstra, and E. J. Briskey. 1970. Metabolism of striated muscle of "stress-susceptible" pigs breathing oxygen or nitrogen. *Amer. J. Physiol. 218*:102.

McPherson, A. and J. Tokunaga. 1967. The effects of cross-innervation on the myoglobin concentration of tonic and phasic muscles. *J. Physiol. 188*:121.

Morita, S., R. G. Cassens and E. J. Briskey. 1969. Localization of myoglobin in striated muscle of the domestic pig: benzidine and $NADH_2$-TR reactions. *Stain Tech. 44*:283.

Nakamura, A., F. A. Sréter and J. Gergely. 1969. Electron microscope studies on aggregates of cardiac and red muscle light meromyosin. *Biophys. J. 9*:A-6.

Nyström, B. 1968. Histochemistry of developing cat muscles. *Acta Neurol. Scand. 44*:405.

Padykula, H. A. and G. F. Gauthier. 1967. Ultrastructural features of three fiber types in the rat diaphragm. (Abstr.) *Anat. Rec. 157*:296.

Padykula, H. A. and E. Herman. 1955. The specificity of the histochemical method for adenosine triphosphatase. *J. Histochem. Cytochem. 3*:170.

Peterson, R. D., C. H. Beatty, and R. M. Bocek. 1963. Effects of insulin and glucagon on carbohydrate and protein metabolism of adductor muscle and diaphragm. *Endocrinology 72*:71.

Philpott, D. E. and A. G. Szent-Györgyi. 1954. The structure of light meromyosin: an electron microscopic study. *Biochim. Biophys. Acta 15*:165.

Robbins, N., G. Karpati, and W. K. Engel. 1969. Histochemical and contractile properties in the cross-innervated guinea pig soleus muscle. *Arch. Neurol. 20*:318.

Romanul, F. C. A. 1964. Enzymes in muscle I. Histochemical studies of enzymes in individual muscle fibers. *Arch. Neurol. 11*:355.

Seligman, A. M., H. Veno, Y. Morizono, H. Wasserkrug, L. Katzoff, and J. Hanker. 1967. Electron microscopic demonstration of dehydrogenase activity with a new osmiophilic ditetrazolium salt (TC-NBT). *J. Histochem. Cytochem. 15*:1.

Sréter, F. A., J. C. Seidel, and J. Gergely. 1966. Studies on myosin from red and white skeletal muscles of the rabbit I. Adenosine triphosphatase activity. *J. Biol. Chem. 41*:24, 5772.

Stein, J. M. and H. A. Padykula. 1962. Histochemical classification of individual skeletal muscle fibers of the rat. *Amer. J. Anat. 110*:103.

Teeter, C., S. C. Carr, R. Tsai, and E. J. Briskey. 1969. A cryobiopsy technique for assessing metabolite levels in skeletal muscle. *Proc. Soc. Exp. Biol. Med. 131*:5.

Van Wijhe, M., M. C. Blanchaer, and W. R. Jacyk. 1963*a*. The oxidation of lactate and α-glycerophosphate by red and white skeletal muscle I. Quantitative studies. *J. Histochem. Cytochem. 11*:500.

———. 1963*b*. The oxidation of lactate and α-glycerophosphate by red and white skeletal muscle II. Histochemical studies. *J. Histochem. Cytochem. 11*:505.

Van Wijhe, M., M. C. Blanchaer, and S. St. George-Stubbs. 1964. The distribution of lactate dehydrogenase isozymes in human skeletal muscle fiber. *J. Histochem. Cytochem. 12*:608.

Wachstein, M. and E. Meisel. 1955. The distribution of demonstrable succinic dehydrogenase and of mitochondria in tongue and skeletal muscle. *J. Biophys. Biochem. Cytol. 1*:483.

Yellin, H. 1969. A histochemical study of muscle spindles and their relationship to extrafusal fiber types in the rat. *Amer. J. Anat. 125*:31.

# PART 3

## Muscle Membrane Systems

# Chapter 12

# *The Conformational Basis of the Energizing Process in the Mitochondrion*

D. E. GREEN AND R. A. HARRIS

In 1967 we proposed the hypothesis that intrinsic to the coupling of electron transfer to synthesis of ATP in the mitochondrion is the conservation of conformational energy (Penniston, Harris, Asai, and Green, 1968; Harris, Penniston, Asai, and Green, 1968; Green, Asai, Harris and Penniston, 1968). It is now possible to adduce a larger body of experimental evidence in support of this hypothesis, and moreover, to specify more precisely the mechanisms by which conformational energy may be generated and conserved. The unit of energy transduction in the mitochondrion is the tripartite repeating unit of the cristal membrane (Fig. 12.1). The basepiece of this unit contains any one of the four complexes of the electron transfer chain (Green, Allmann, Bachmann, Baum, Kopaczyk, Korman, Lipton, MacLennan, McConnell, Perdue, Rieske, and Tzagoloff, 1967); the headpiece contains the complex of proteins which catalyzes the hydrolysis of ATP (Kagawa and Racker, 1966; Kopaczyk, Asai, and Green, 1968). We may consider the tripartite repeating unit as a machine which can be activated at either end—by electron transfer in the basepiece or by hydrolysis of ATP in the headpiece. Either process can set in motion the same sequence of events in the repeating unit. In a general way we may say that intrinsic to our thesis is the notion that the tripartite repeating unit can exist in one of three different conformations (nonenergized, energized, and energized-twisted). Addition of substrate (which leads to electron transfer) or addition of ATP (which leads to hydrolysis of ATP) will induce the generation of the energized conformation from the nonenergized conformation. The same additions in presence of inorganic phosphate will induce the generation of the energized-twisted conformation from the nonenergized conformation via the intermediary step of the energized conformation. The fact that the same final states can be induced by two independent reactions which proceed

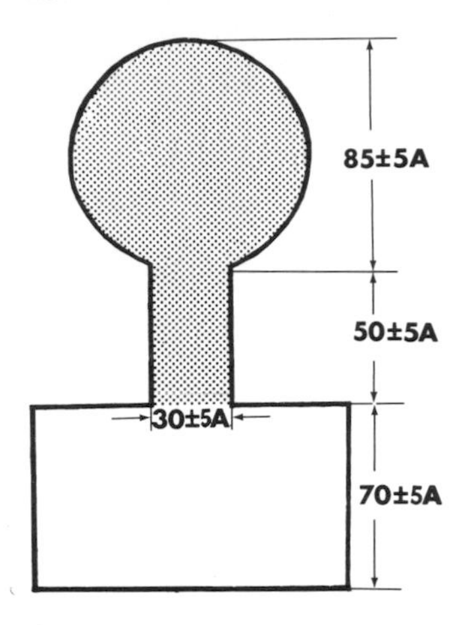


*Fig. 12.1.* The dimensions of the macrotripartite repeating unit. (Measurements and drawing provided by J. Asai and M. A. Asbell.)

*Fig. 12.2 (facing page).* The three basic configurational states of beef heart mitochondria. The incubation medium contained 1 mg of mitochondrial protein per ml and was 0.25 M in sucrose and 5 mM in Tris-Cl, *p*H 7.4: *A,* nonenergized configuration obtaining in the presence of rotenone (2 μg/mg of protein); *B,* energized configuration obtaining in the presence of rotenone plus succinate (5 mM): *C,* energized-twisted configuration obtaining in the presence of rotenone, succinate, and $P_i$, *p*H 7.4, (10 mM); and *D,* nonenergized configuration obtaining in the presence of rotenone, succinate, $P_i$, and m-ClCCP ($10^{-6}$).

in two different parts of the same repeating unit means that there must be a mechanism by which a perturbation initiated in the basepiece is transmitted to the headpiece and vice versa. We are suggesting that the role of the stalk is to transmit this perturbation either from basepiece to headpiece or from headpiece to basepiece.

Electron transfer can induce a conformational change in the repeating unit. We are proposing that the repeating unit in the energized conformation is at a higher energy level than the repeating unit in the nonenergized conformation and that the free energy released during electron transfer is conserved as conformational energy. In other words, redox energy becomes transduced to conformational energy and eventually conformational energy has to be transduced into the bond energy of the pyrophosphate bond of ATP. The order of transductions can be reversed. Hydrolysis of ATP would also induce the conformational change of the repeating unit and the conformationally perturbed repeating unit could then compel reversal of the electron transfer process, e.g., reduction of $NAD^+$ by succinate. This in broad outline is the essence of the conformational hypothesis. Now, let us consider the foundation of experimental evidence on which the conformational hypothesis rests.

## THE CONFIGURATIONAL CYCLE

Isolated beef heart mitochondria show very dramatically configurational changes in the cristal membranes when exposed to energizing

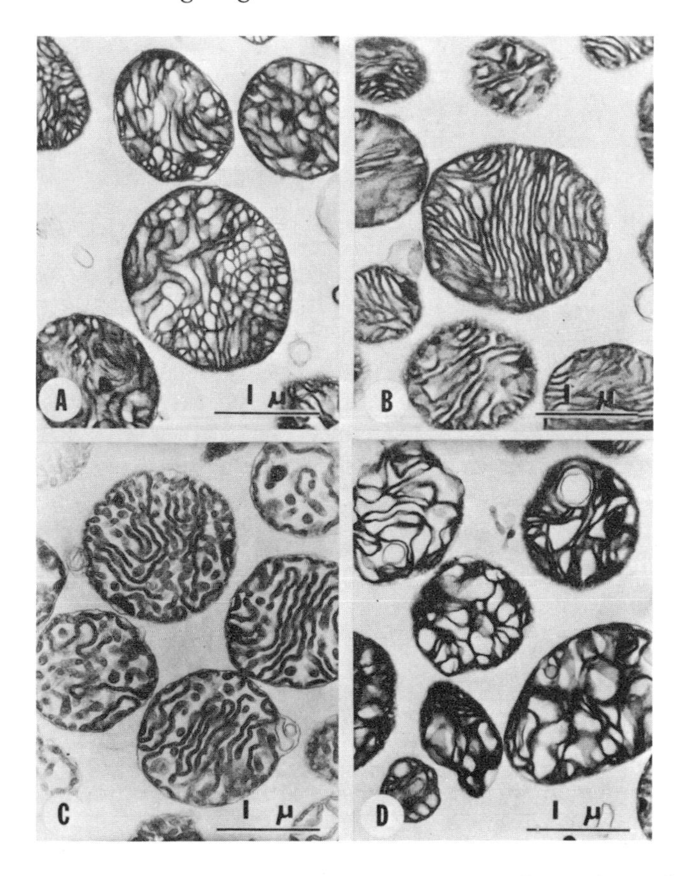


conditions (Fig. 12.2). In the nonenergized configuration, the cristal membranes are expanded. The densely stained junction of two apposed, distended cristal membranes is what is visualized in the electron micrographs. The junction shows up as two tightly compressed membranes with no lumen separating the two (Fig. 12.2A). When substrate is added to the mitochondria, the cristae assume the energized configuration. The junction made by the two membranes is now significantly expanded and a lumen can often be seen separating the two membranes (Fig. 12.2B). On further addition of inorganic phosphate, the cristal membranes assume the energized-twisted configuration which is characterized by helical tubes that arise from the apposed membranes by a process of tubularization (Fig. 12.2C). The distance between the two apposed membranes is least in the nonenergized configuration and greatest in the energized-twisted configuration. Finally the addition of an uncoupler such as metachlorocarbonyl cyanide phenylhydrazone (m-ClCCP) discharges the energized configuration and regenerates the nonenergized configuration of the cristal membrane (Fig. 12.2*D*). We have thus completed one turn of the configurational cycle.

It must be borne in mind that we take the configuration of the membrane to be an expression directly of the preponderant conformation of the component repeating units. We are in fact studying configurational changes in the membrane as an index of conformational changes in the repeating units. If all the repeating units of the cristal membrane are in the same conformation, then configurational change in the membrane provides an accurate index of the conformational states of the repeating units.

## CORRELATION OF CONFIGURATIONAL CHANGES WITH CHANGES IN THE FUNCTIONAL STATE OF THE MITOCHONDRION

From the relations shown in the energy diagram (Fig. 12.3) a large number of predictions can be made as to how the configuration of the cristal membrane will be affected by each of a set of specific reagents if, in fact, the configuration of the cristal membrane is an accurate index of the energized state. For example, substrate or ATP should induce the energized configuration; uncouplers and the reagents required for work performances should discharge the energized configuration; the generation of the energized configuration should be prevented by inhibitors of electron transfer when substrate is used, and by inhibitors of ATP hydrolysis when ATP is used. We have tested very thoroughly this correlation between the configuration of the cristal membrane and the energized state of the mitochondrion. Two independent methods were used to test the correlation—electron microscopy (Green, Asai, Harris, and Penniston, 1968) and the method of 90° light scattering (Harris, Asbell, Jolly, Asai, and Green, 1969). By the first method, we deter-

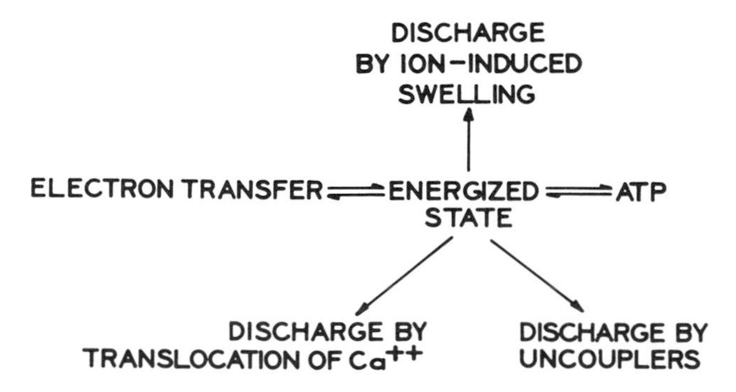


*Fig. 12.3.* The energy diagram.

mine the state of the cristal membranes in fixed and sectioned mitochondria; by the second method, we scan the reflecting surfaces of the mitochondrion and record gross changes in the geometry of the cristal membranes. Fig. 12.4 is a record of a typical correlative study between the configurational state of the cristal membrane determined electron-micro-

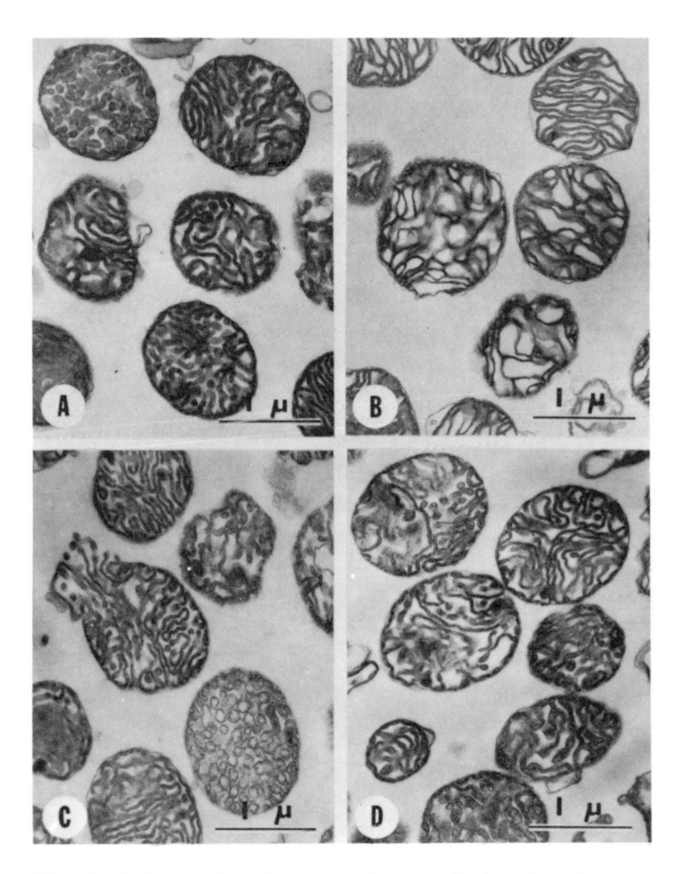


*Fig. 12.4.* Prevention by atractyloside of the ADP-induced discharge of the energized-twisted configuration. Incubation medium, as in Fig. 12.2; *A,* energized-twisted configuration obtaining in the presence of succinate (5 mM), $P_i$, *p*H 7.4 (10 mM), and rotenone (2 μg/mg protein); *B,* mixed configurations of nonenergized and energized induced by the addition of ADP to mitochondria in the energized-twisted configuration; *C,* energized-twisted configuration obtaining in the presence of succinate, $P_i$, rotenone, and atractyloside; and *D,* energized-twisted configuration still obtaining after addition of ADP to atractyloside-inhibited mitochondria in the energized-twisted configuration.

scopically and the functional state of the mitochondrion. We have already described in considerable detail a full set of such correlative electron micrographic studies (Harris, Asbell, Jolly, Asai, and Green, 1969). Sufficient to say that the correlation has been found to be exact in all cases. Fig. 12.5 contains a set of correlative experiments in which changes in the light-scattering function of the mitochondrial suspension are related to the energy state. The light-scattering change from the nonenergized to the energized configuration, induced by addition of substrate, is very small compared to the change from the energized to the energized-twisted configuration, induced by the addition of inorganic phosphate. The

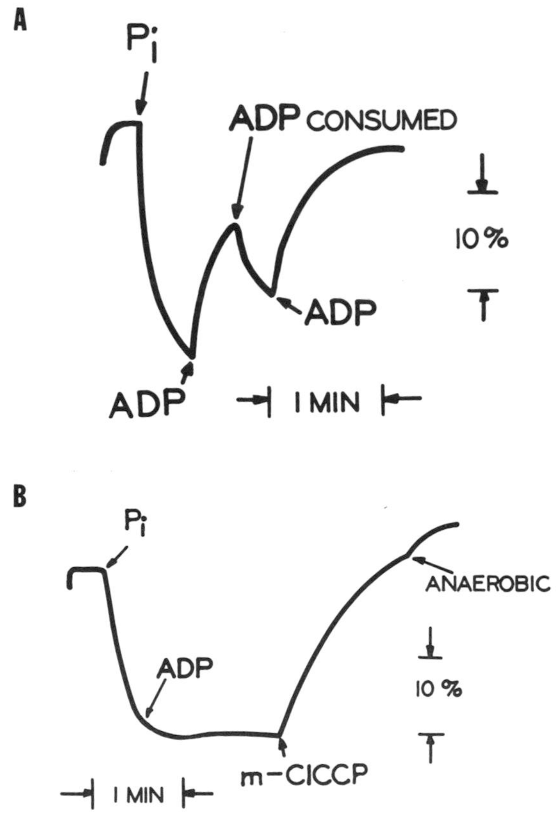


*Fig. 12.5.* Light-scattering changes observed with change in energy state of beef heart mitochondria. Incubation medium, as in Fig. 12.2; *A,* discharge of the energized-twisted configuration by the initiation of oxidative phosphorylation; *B,* failure of ADP addition to discharge the energized-twisted configuration in the presence of atractyloside (0.17 mM).

second configurational change is quantitatively much greater than the first and thus more easily recorded by the light-scattering method. The correlation between the light-scattering changes and the energy state of the mitochondrion is precisely what is predictable from the energy diagram. Thus, the two independent methods of recording configurational changes in mitochondria (the electron microscopic and light-scattering measurements) both lead to the same correlation of configurational change with the functional state of the mitochondrion.

### THE CONFORMATIONAL CYCLE

Electron micrographic studies at high resolution have made it possible to visualize changes in the conformation of the repeating units of the

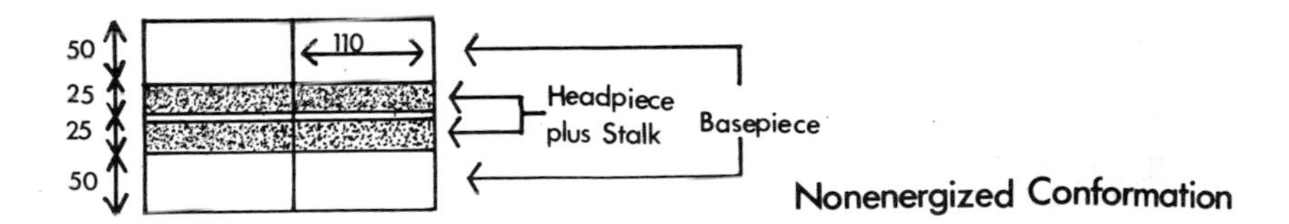


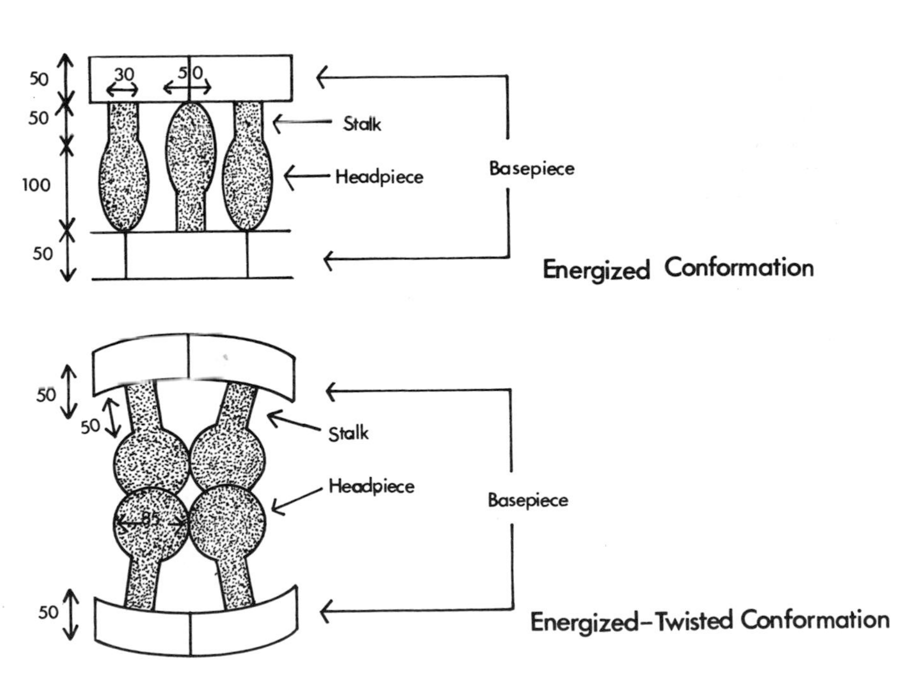


*Fig. 12.6.* The conformation of the repeating units of the cristal membrane during the energy cycle.

cristal membrane during the energy cycle (Fig. 12.6). In the nonenergized conformation of isolated mitochondria the headpiece-stalk sector appears to be collapsed or relaxed (Vail, Korman, and Green, unpublished observations). When the repeating units are energized by substrate or ATP the headpiece-stalk sector extends away from its own basepiece to a basepiece on the apposed membrane. Finally with the introduction of inorganic phosphate into the system, the headpieces separate away from the apposed basepieces and take up a position in the interior of the lumen. The tendency of headpieces to associate is probably related to the strikingly helical character of the energized-twisted conformation (G. Vanderkooi and Green, unpublished observations).

Although the details are not established, the essential ultrastructural events may prove to be the extension of the headpiece-stalk sector in the energized conformation and the relaxation of this sector in the nonenergized conformation. There is some theoretical basis for postulating that in the nonenergized conformation the headpiece-stalk sectors are more nearly neutral whereas in the energized conformation these sectors are charged and tend to repel one another. The energized binding of inorganic phosphate further increases the charge on the headpieces. Despite this increase in electrostatic repulsion, the conformational change has the effect of minimizing the separation of headpieces. The tendency of the basepieces to undergo an increase in curvature is opposed by the electrostatic repulsion as the headpieces are pushed closer together (Korman, Vanderkooi, and Green, unpublished observations).

### SPEED OF CONFIGURATIONAL AND CONFORMATIONAL CHANGE

The thesis of the primacy of conformational change in the conservation of energy in the mitochondrion requires that conformational change be no slower than the rate of the overall work performance or the rate of electron transfer. What is the experimental evidence that bears on this question of the rate of conformational change? As yet no very precise measurements have been made but some approximate answers are possible. The configurational change induced by uncoupler (m-ClCCP) in a submitochondrial particle that had been energized by substrate was complete in about 2 sec at room temperature, i.e., within the response time of the recorder used in the measurement of light-scattering change (Harris, Asbell, and Green, 1969). Bearing in mind that configurational change will necessarily be slower than conformational change, the observed speed of light-scattering change represents a significant percentage of the speed (>20%) with which the electron transfer components turn over under coupled conditions. The rate of change of the *p*H in mitochondria or submitochondrial particles exposed to substrate is the nearest approach to the direct measurement of conformational change. J. Penniston and J. Southard in our laboratory have measured the speed of mitochondria, energized by substrate. They found a remarkably close correspondence between the rate of proton release and the rate of electron flow under the same experimental conditions. By the criterion of the speed of proton ejection we may conclude that conformational change is as rapid as electron transfer under coupling conditions.

### CONFIGURATIONAL CHANGES IN SITU VERSUS IN VITRO

Mitochondria *in situ* are generally in the orthodox mode in the nonenergized conformation (Harris, Williams, Caldwell, Green, and Valdivia, 1969; Williams, Harris, Vail, Caldwell, Green, and Valdivia, 1970), whereas mitochondria *in vitro* (i.e., as isolated by standard procedures) are generally in the aggregated mode (Penniston, Harris, Asai, and Green, 1968). But there are indeed instances of mitochondria *in situ* being in the aggregated mode in the nonenergized conformation. Moreover, isolated mitochondria can be manipulated experimentally to as-

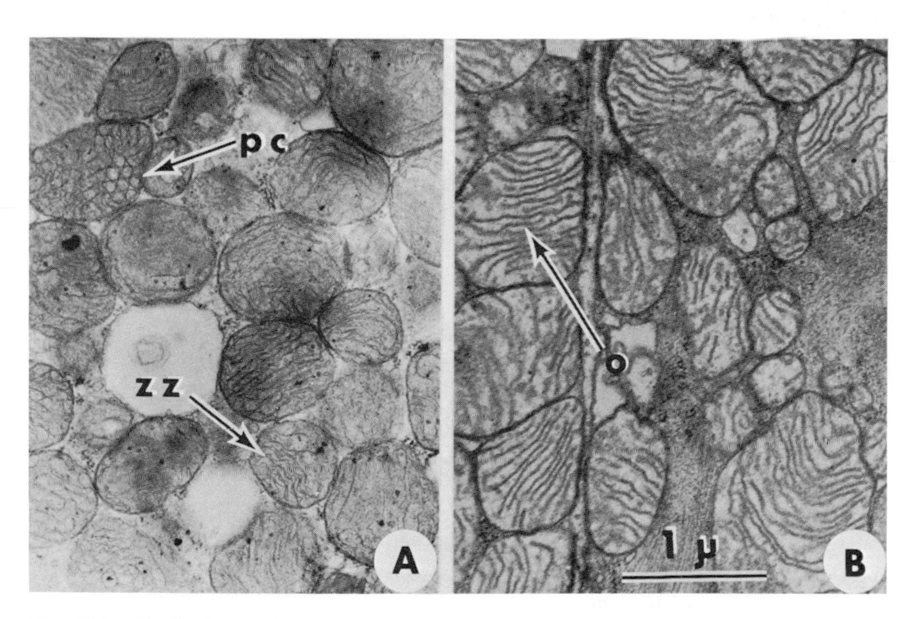


*Fig. 12.7.* The basic incubation medium was a modified Krebs-Ringer-phosphate solution (6 ml) which contained rotenone (17 µg/ml), rutamycin (67 µg/ml), and sodium iodoacetate (1 mM). The samples were incubated for 2 min in a water bath at 20° C: *A,* mitochondria in the energized configuration observed in the presence of 10 mM sodium succinate; *B,* mitochondria in the nonenergized configuration obtained in the presence of 10 mM sodium succinate and 0.4 mM dinitrophenol. *zz,* zigzag; *pc,* paracrystalline array; *o,* orthodox configuration.

sume the orthodox mode (Harris, Harris, and Green, 1968; Allmann and Munroe, unpublished observations; Hackenbrock, 1968; Deamer, Utsumi, and Packer, 1967). The two energized configurational states are the same whether the mitochondria are *in situ* or isolated. In other words, coupling requires the aggregated mode and this requirement is independent of the locale of the mitochondrion whether *in situ* or in isolation (Williams, Vail, Harris, Green, Caldwell and Valdivia, 1970; Harris, Harris, and Green, 1968; Allmann and Munroe, unpublished observations). The nonenergized and energized configurational states of heart mitochondria *in situ* are presented in Fig. 12.7.

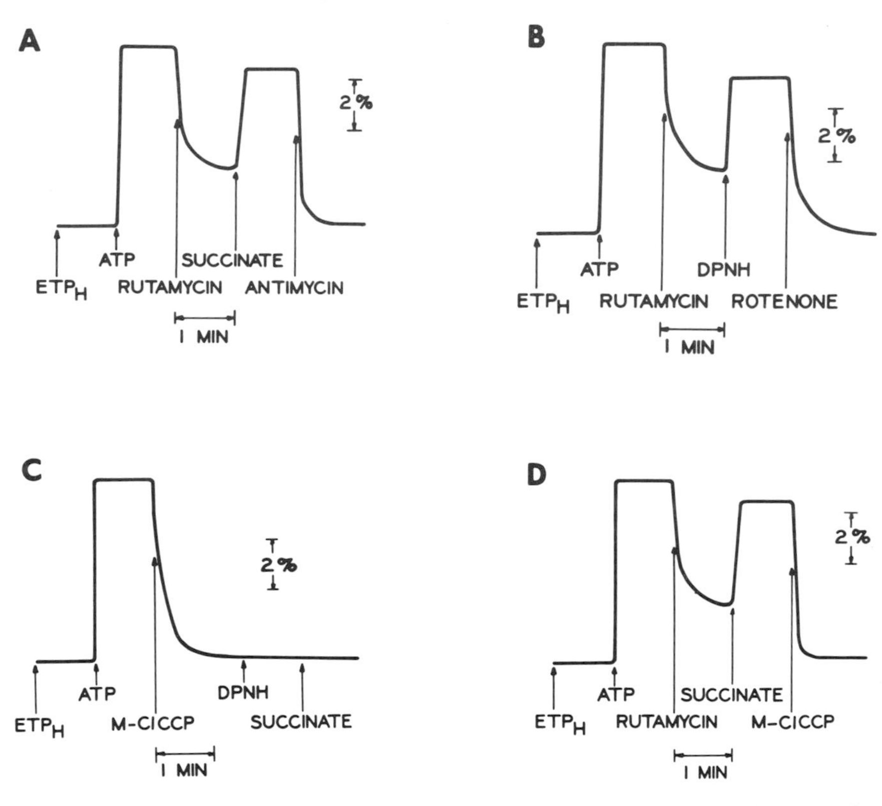


*Fig. 12.8.* Correlation of light-scattering changes with the functional state of $ETP_H$. The experiments were carried out at 30° C in a medium (3 ml) which was 0.25 M in sucrose, 10 mM in Tris-Cl, *p*H 7.4, and 10 mM in potassium phosphate, *p*H 7.4. The medium contained 0.5 mg of $ETP_H$ in experiments *A–D*. The final concentrations of the other reagents are shown in parentheses: ATP (1 mM), m-ClCCP (carbonylcyanide m-chlorophenylhydrazone) ($10^{-6}$), DPNH (0.1 mM), succinate (2.5 mM), rutamycin (2 $\mu$g/mg of protein), antimycin (2 $\mu$g/mg of protein), and rotenone (2 $\mu$g/mg of protein). An upward deflection of the light-scattering trace indicated an increase in the amount of light scattered.

### CONFIGURATIONAL AND CONFORMATIONAL CHANGES IN SUBMITOCHONDRIAL PARTICLES

Submitochondrial particles are either membranous fragments of cristae which are spherical (ETP) or rod-shaped ($ETP_H$). For obvious reasons configurational changes are difficult to recognize electron-microscopically but they are readily recorded by light-scattering measurements of $ETP_H$ (Fig. 12.8). These measurements show the configurational transition from the nonenergized to the energized-twisted configuration induced by ATP hydrolysis or by succinate oxidation. Electron microscopic examination of $ETP_H$ under energizing conditions (exposed to substrate + inorganic phosphate) reveals that the particles contain vesicular invaginations (Fig. 12.9) (Asbell, Vail, Harris, Korman, and Green, 1970). ETP, the spherical particle, stands in relation to $ETP_H$, the rod-shaped particle, as does the orthodox mode to the aggregated mode. ETP does not exhibit configurational change and does not efficiently couple electron transfer

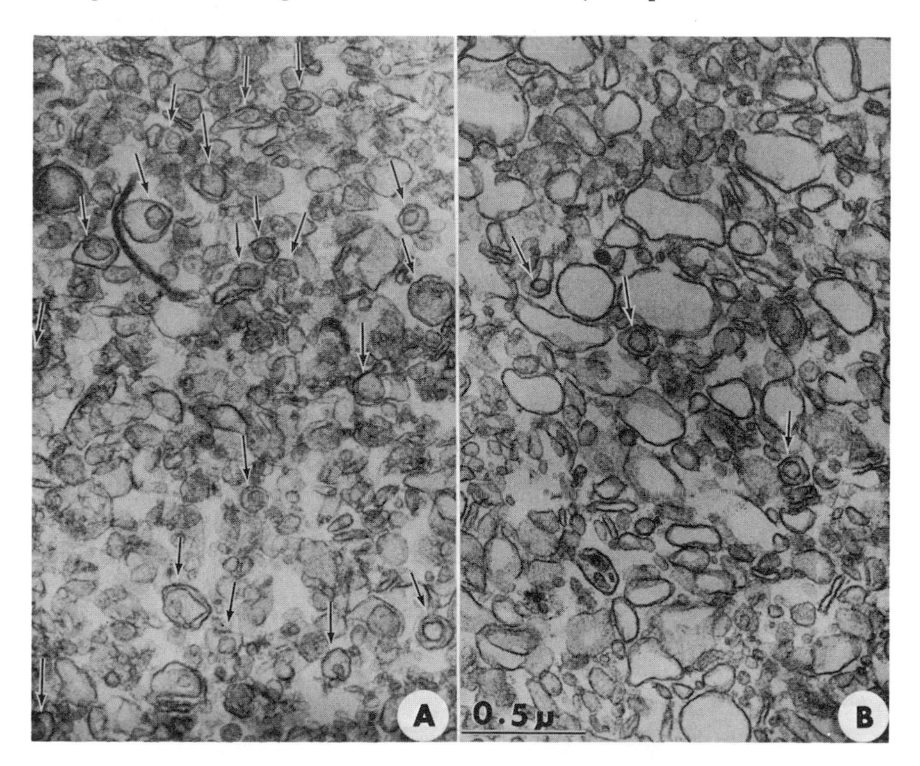


*Fig. 12.9.* Correlation of configurational state with the medium described in Fig. 12.8: *A,* $ETP_H$ energized with ATP plus succinate; *B,* $ETP_H$ discharged after energization by the addition of m-ClCCP. Arrows point to regions of invaginated membrane.

to synthesis of ATP, but does show proton movement during electron transfer.

### TOPOLOGY OF THE MEMBRANES DURING THE ENERGY CYCLE

The inner membrane is a composite of the inner boundary membrane and the attached cristae (Fig. 12.10). The topology of this membrane remains unchanged throughout the energy cycle (Penniston, Harris, Asai, and Green, 1968; Korman, Addink, Wakabayashi, and Green, 1970). Basically the energizing involves the interplay of the space within the cristae (intracristal space) and the space between cristae (matrix space). In the orthodox mode, the volume of the intracristal space is minimal; the volume of the matrix space is maximal. In the aggregated mode, the reverse is true. Tubularization involves an invagination of a tube originating from one of a pair of apposed cristae into the intracristal space (Korman, Addink, Wakabayashi, and Green, 1970). When the tubularization proceeds to completion the tubes can fill the whole of the intramitochondrial space (Penniston, Harris, Asai, and Green, 1968; Korman, Addink, Wakabayashi, and Green, 1970). Since the matrix space is contained within the tubules, tubularization represents, in fact, the expansion of the matrix space and the corresponding decrease of the intracristal space (Green, Asai, Harris, and Penniston, 1968; Korman, Addink, Wakabayashi, and Green, 1970).

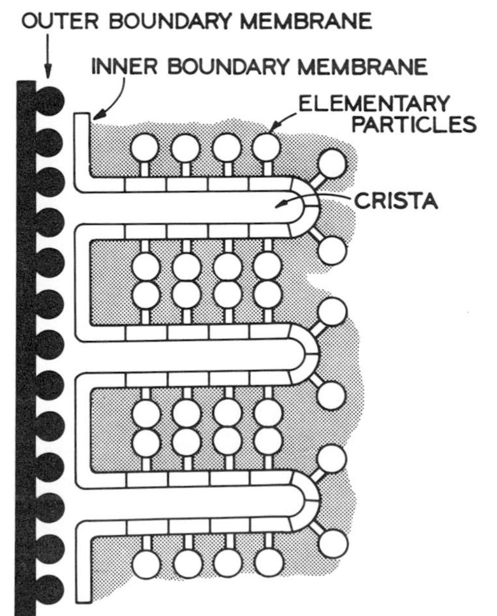


*Fig. 12.10.* The membranes of the mitochondrion.

PSEUDOENERGIZING OF THE CRISTAL MEMBRANES

Appropriate salts can induce the same configurational changes in beef heart mitochondria as do electron transfer and ATP hydrolysis; moreover these changes are unaffected by the presence of uncouplers and inhibitors either of electron transfer or hydrolysis of ATP (Blondin, Vail, and Green, 1969; Asai, Blondin, Vail, and Green, 1969). We can go a step further and postulate that the active salts must induce the same geometric changes in the repeating units of the cristal membrane as do electron transfer and ATP hydrolysis. But again these changes are insensitive to any of the inhibitors of energized conformational changes. It has to be pointed out that the salt-induced configurational or conformational change is a one-shot performance. Once established, the conformation or configuration is extremely stable. As long as the salt is present in the medium the induced conformational change is not reversed with time.

Not all salts will accomplish this induction of configurational change. For example, appropriate sodium salts are active but potassium salts generally are not active. The competent sodium salts are the salts of weak acids (salts of acetate, propionate, phenylacetate, phosphate, etc.); the incompetent sodium salts are the salts of strong acids (salts of chloride, sulfate, etc.). But ion facilitators, such as gramicidin or valinomycin, can convert any incompetent ion pair to a competent pair (Blondin, Vail, and Green, 1969; Asai, Blondin, Vail, and Green, 1969). Thus, given the proper facilitating agents, almost any salt is competent to induce pseudoenergizing of the membrane.

At first acquaintance, the fact of pseudoenergizing of the membrane would appear to undermine the concept of the conformational basis of energy transduction in the mitochondria. If any salt can induce the same conformational changes as are observed in energized processes, does this not suggest that these changes are not the significant events in energized processes in mitochondria? It must be borne in mind that we are equating the salt-induced conformational change with the energy-induced conformational change only in a gross geometric sense. The geometry of the repeating units is the same in the two cases but the tertiary structure is clearly not identical. In a strict sense, it is inappropriate to equate two conformational states on the basis of geometry alone. We should consider the salt-induced conformational state to be intrinsically different energetically from the energy-induced conformation.

All pseudoenergized membranes in the energized-twisted configuration tend to swell. This swelling is an indication that the inducing salts have increased the net charge in the matrix space; the increase in net charge leads inexorably to ion and water movements. Pseudoenergizing can also be achieved in a medium which is salt-free and contains only distilled water. When the salt level in the mitochondrion is drastically reduced, as in a medium containing distilled water, the net charge in the mitochondrial spaces may be modified significantly and this change in charge may lead to energy-independent conformational and configurational transitions.

### ENGAGEMENT AND TUBULARIZATION

Providing cristae are initially in the aggregated mode or can assume the aggregated mode, energizing of the repeating units leads to the engagement of apposed cristae. As a working hypothesis we might consider engagement as an expression of some interaction between the headpiece-stalk sector of a repeating unit and the basepiece of an apposed repeating unit. The conformational change in the transition from the nonenergized to the energized conformation leads to the extension of the headpiece-stalk and the possible "hooking" of this extended sector to the apposed basepiece. We might consider the hooking phenomenon as a stabilization of the conformational change by some sort of interaction, the nature of which is still unspecified. When hooking is excluded for geometric reasons (for example, when the cristae are constrained in the orthodox mode), then engagement is impossible and all the work performances that depend upon the engagement of apposed membranes (ATP synthesis and energized ion movements) may be excluded.

There are two stages in the engagement phenomenon—the generation of the energized configuration and the generation of the energized-

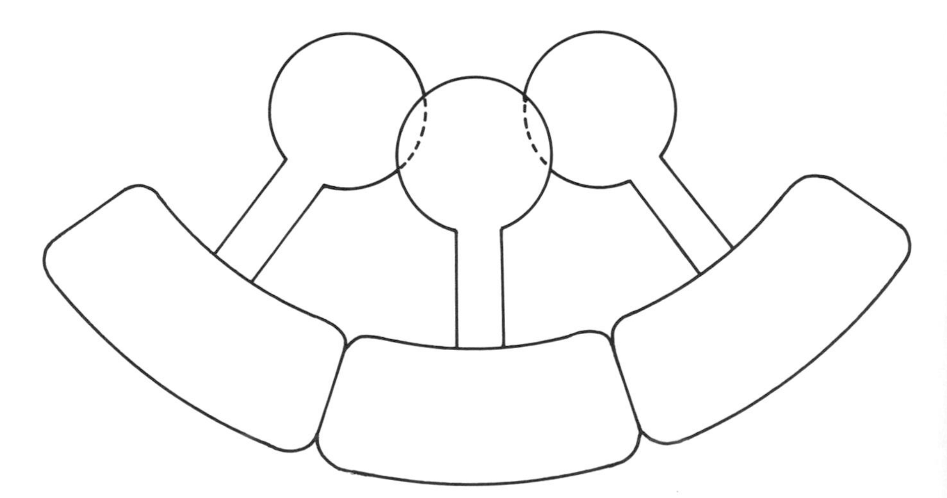

*Fig. 12.11.* Association of headpieces leading to curvature of the basepieces of the repeating units in the energized-twisted conformation.

twisted configuration. The energized configuration is geometrically compatible with the engagement of two extended, flat membranes (see Fig. 12.6). As soon as inorganic phosphate is introduced into the system, a profound geometric rearrangement takes place. The headpiece-stalk sectors appear to move into the lumen between the two engaged membranes (Vail, Korman, and Green, unpublished observations; Vanderkooi and Korman, unpublished observations). Moreover, headpiece-stalk sectors from neighboring repeating units tend to clump together in sets and this tight association of sets of headpiece-stalk sectors compels curvature of the basepieces (Fig. 12.11). The new geometry of the repeating units in the energized-twisted conformation is not compatible with the engagement of two extended, flat membranes. The introduction of phosphate creates an instability which is relieved by the formation of a helical tube, which arises by invagination of repeating units lining the two apposed membranes (Korman, Addink, Wakabayashi, and Green, 1970) (Fig. 12.12). This invagination can proceed until the entire crista becomes tubularized.

### VESICULAR VS. AGGREGATED CRISTA IN ADRENAL MITOCHONDRIA.

Adrenal mitochondria *in situ* can exist in one of two modes—the vesicular and the aggregated (Fig. 12.13) (Allmann and Munroe, unpublished observations; Allmann, Munroe, Hechter, and Matsuba, 1969). The same two options are open to isolated adrenal mitochondria (Fig. 12.14). The vesicular cristae are in fact in the orthodox mode and accordingly uncoupled with respect to work performances such as ATP synthesis or ion movements (Allmann, Wakabayashi, Korman, and Green, 1970; Allmann, Munroe, Hechter, and Matsuba, 1969). Mitochondria with vesicular cristae (induced, for example, by addition of $Ca^{++}$) can carry on electron transfer or hydrolysis of ATP, but cannot couple these functions to any useful work performance. $Ca^{++}$ induces the orthodox mode while $Mg^{++}$ tends to hold the aggregated mode (Allmann, Munroe, Wakabayashi, and Green, 1970). When both are present, the ratio of $Ca^{++}$ to $Mg^{++}$ is a critical factor in determining the proportion of the mitochondrial population that will be in either the orthodox or aggregated mode. The capacity for steroidogenesis which adrenal mitochondria possess is manifest whether the cristae are in the orthodox or aggregated mode, but the activity is greatest in mitochondria which have vesicular cristae (Allmann, Munroe, Wakabayashi, and Green, 1970; Allman, Munroe, Hechter, and Matsuba, 1969).

$Ca^{++}$ is not the only reagent that induces the orthodox mode, and similarly $Mg^{++}$ is not the only reagent that stabilizes the aggregated mode. They happen to be the most important physiological reagents and per-

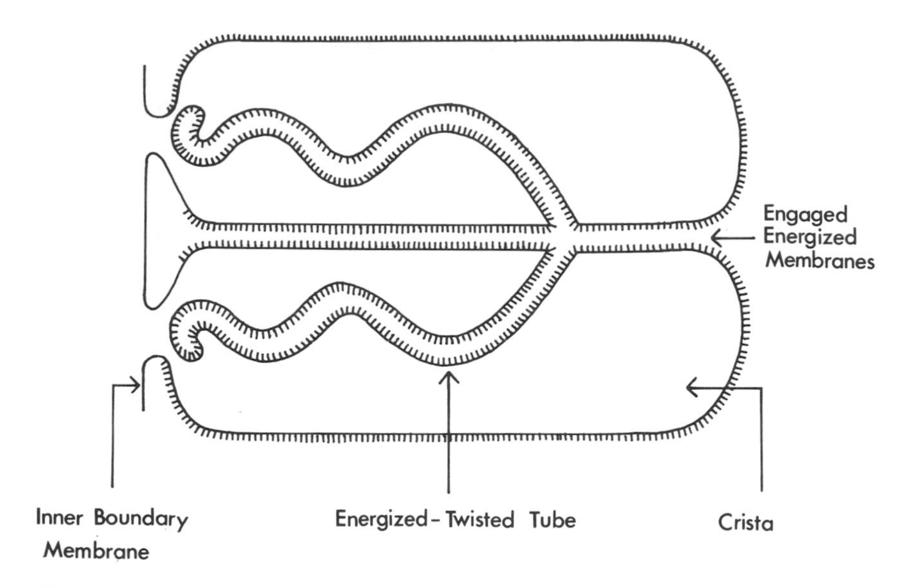


*Fig. 12.12.* Tubularization in the transition from the energized to the energized-twisted conformation of the repeating units.

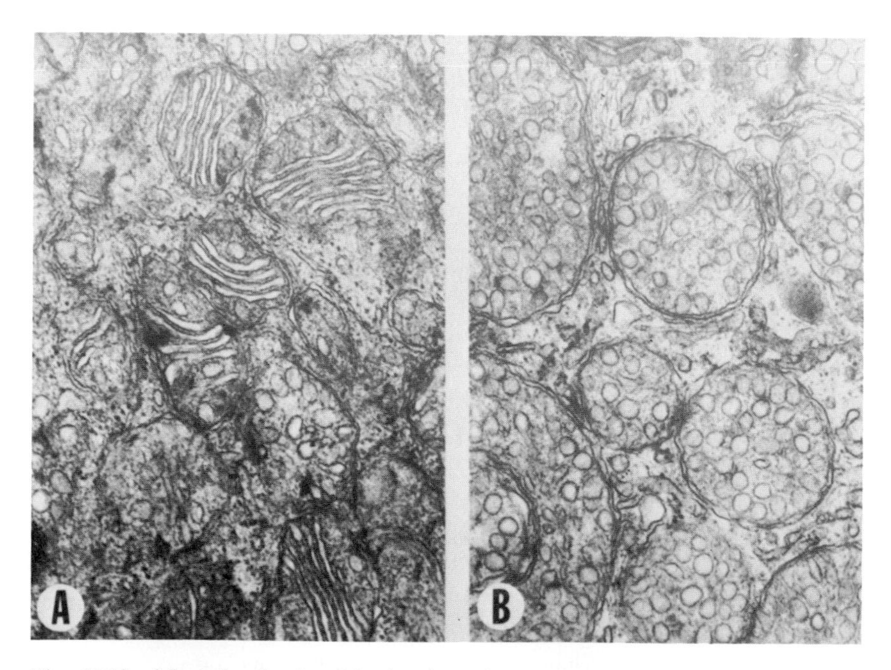


*Fig. 12.13.* Adrenal mitochondria *in situ; A,* aggregated modality; *B,* orthodox modality. × 30,000 Electron micrographs provided by D. W. Allmann, J. Munroe, and T. Wakabayashi.

haps among the most effective. In theory, any reagent that increases the charge density of the proteins abutting into the intracristal space will induce the orthodox or vesicular mode while any reagent that increases the charge density of the proteins abutting into the matrix space will induce the aggregated mode (Vanderkooi and Green, unpublished observations). Increase in charge density leads to ion movements and the ion movements are accompanied by water movements.

The $Ca^{++}$-$Mg^{++}$ ratio is also a determinant factor of the ratio of the orthodox to the aggregated mode in mitochondria from beef heart muscle and beef liver (Allmann and Munroe, unpublished observations). The level of $Ca^{++}$ required to induce the orthodox mode is much higher in beef heart mitochondria than in adrenal mitochondria. Liver mitochondria resemble more closely adrenal mitochondria in this respect.

TRANSITION OF CRISTAE FROM THE AGGREGATED TO THE ORTHODOX MODE

A whole range of phenomena have been recognized as variations on the theme of the orthodox-aggregated dichotomy. For example, the endotoxin of *Bordetella bronchiseptica* functionally uncouples beef heart mitochondria by inducing the orthodox mode of the cristae (Harris,

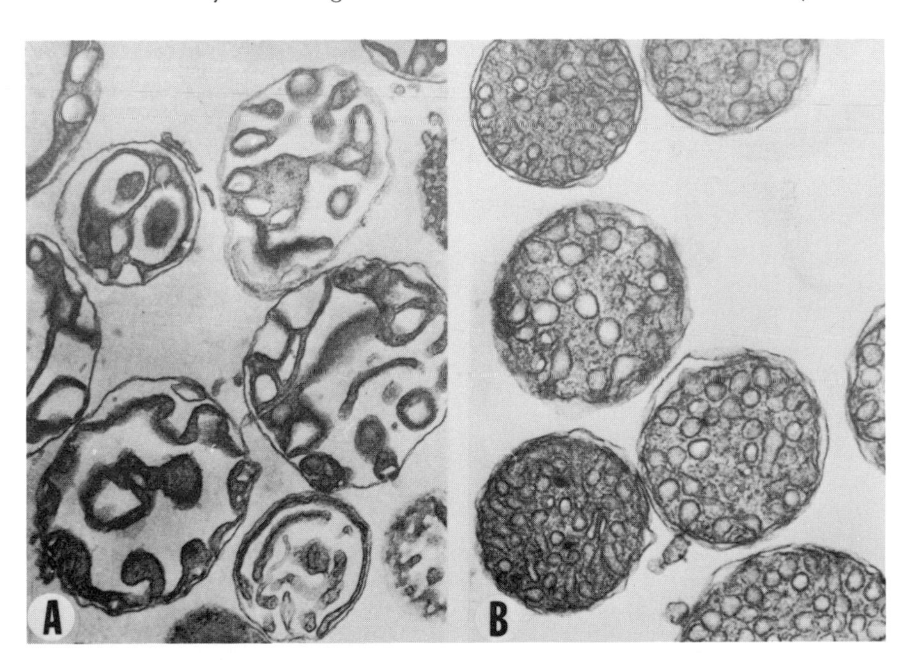


*Fig. 12.14.* Adrenal mitochondria *in vitro; A,* aggregated modality; *B,* orthodox modality. Electron micrographs provided by D. W. Allmann, J. Munroe, and T. Wakabayashi.

Harris, and Green, 1968). The reduced form of the complex anion, silicomolybdate, is rapidly oxidized by liver mitochondria, but the oxidation cannot be coupled to synthesis of ATP. When the oxidation is carried out in presence of $Mn^{++}$, $Mg^{++}$, $Ba^{++}$, and $Sr^{++}$ (cations which stabilize the aggregated state) coupling is restored (Jacobs, 1956). Apparently, reduced silicomolybdate induces the orthodox mode and this effect is reversed by addition of appropriate divalent metal ions. Anes-

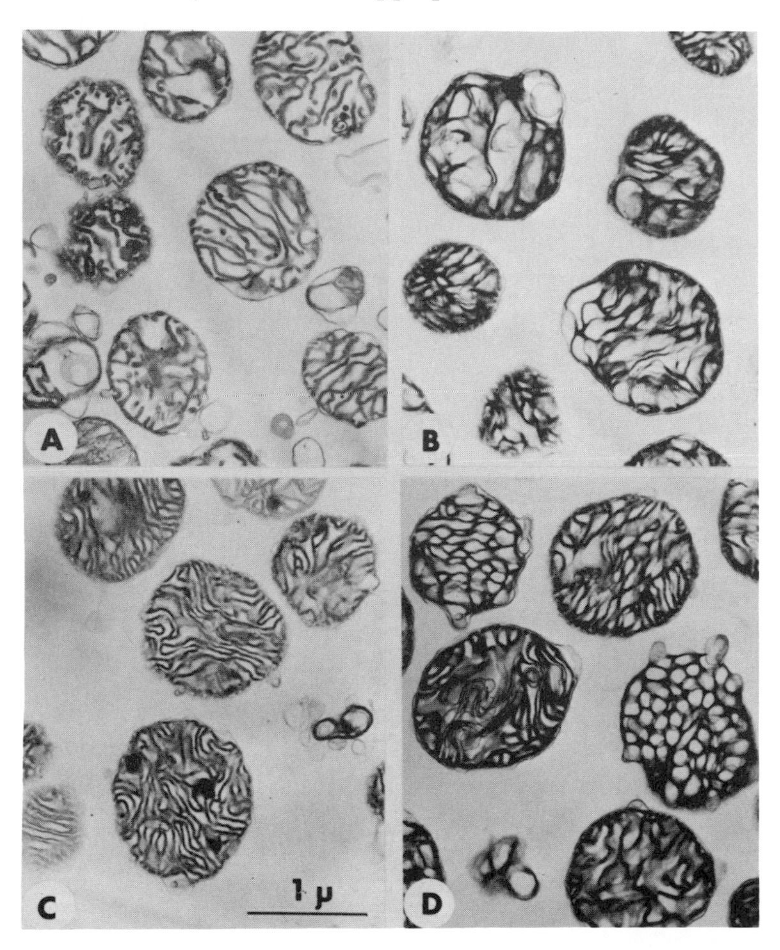


*Fig. 12.15.* Effect of high-medium sucrose concentration upon the configurational states of beef heart mitochondria: *A,* mitochondria incubated in a medium containing 0.5 M sucrose, 5 mM succinate, 10 mM $P_i$, *p*H 7.4, and rotenone at 2 μg/mg of mitochondrial protein; *B,* m-ClCCP ($10^{-6}$ M) was added to mitochondria incubated under conditions described in A; *C,* same conditions as A, except sucrose concentration was 0.88 M; and *D,* m-ClCCP was added to the mitochondria of C.

thetics such as halothane or barbiturates also induce the orthodox mode of beef heart mitochondria and thereby suppress coupling (Taylor, Harris, Williams, and Green, 1969).

The antagonism between $Ca^{++}$ and $Mg^{++}$ in the medium in which beating heart cells are suspended may also be referred to the effect of these reagents alone and in combination in controlling the ratio of mitochondria in the uncoupled orthodox and coupled aggregated configurational states (Williams, unpublished observations).

### OSMOTICALLY-SENSITIVE STEP IN THE CONFIGURATIONAL CYCLE

High concentrations of sucrose or other nonpermeant solutes in the suspending medium can prevent completely the transition of the cristal membrane from the energized to the energized-twisted configuration, although without appreciable effect on the transition from the non-energized to the energized configuration (Harris, Williams, Jolly, Asai, and Green, 1969) (Fig. 12.15). Mitochondria in sucrose media above 0.5 M lose the capability to carry out coupled synthesis of ATP or coupled translocation of ions but show nonetheless a proton jump when energized by substrate. By contrast, levels of sucrose which completely abolish coupled phosphorylation in mitochondria have only a marginal effect on coupled phosphorylation carried out by submitochondrial particles such as $ETP_H$.

When mitochondria are exposed to a medium high in sucrose concentration, the expanded cristae are tightly pressed together, and an intense electrostatic field develops in the matrix space between the apposed cristae. The second configurational transition requires the binding of anions (phosphate, acetate, etc.) to the headpieces. This binding of anions is suppressed in a medium high in sucrose concentration. It is this suppression of anion binding which underlies the inhibition of configurational change and the inhibition of coupled reactions such as active transport and oxidative phosphorylation. Binding of anions to mitochondria requires penetration of the cristal membrane by these ions whereas the

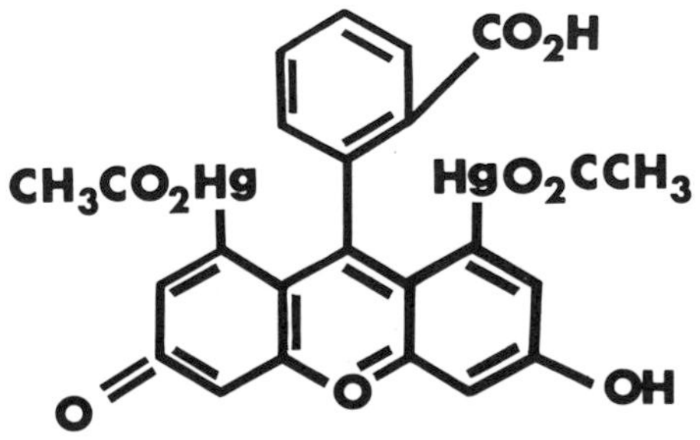

$CO_2H$   $CH_3CO_2Hg$   $HgO_2CCH_3$   O   O   OH

*Fig. 12.16.* The structure of fluorescein mercuric acetate.

penetration problem is eliminated in $ETP_H$—the headpieces abut directly into the external medium. It is this difference in orientation of headpieces with respect to the external medium that accounts for the relative insensitivity of coupled processes in $ETP_H$ to high levels of sucrose. At the present time, two possible interpretations of the sucrose effect must be considered. First, the osmotic compression of cristae creates an electrostatic barrier to anion binding. Second, sucrose in high concentrations has a direct action on the ionophore which transfers phosphate ions through the cristal membrane.

### UNCOUPLING ACTION OF FLUORESCEIN MERCURIC ACETATE (FMA)

FMA, like media high in sucrose concentration, can suppress the transition of the cristal membrane from the energized to energized-twisted configuration, though without apparent effect on the transition from the nonenergized to energized configuration (Lee, Harris, and Green, 1969). The same suppression also applies to the corresponding conformational transition. The concentration of FMA (structure shown in Fig. 12.16) required to uncouple beef heart mitochondria is about $10^{-5}$ M. Sulfhydryl (-SH) reducing agents such as dithiothreitol can completely reverse the inhibitory action of FMA. While FMA can abolish all coupled work functions involving phosphate, it increases the proton jump accompanying electron transfer. FMA acts not by inducing the orthodox mode of the cristal membrane, but rather by decreasing phosphate binding and by potentiating an increase in the permeability of the cristal membrane to ions. In that sense, FMA behaves just like valinomycin and gramicidin, but differs from these two antibiotic facilitating agents in that it suppresses the second of the two configurational transitions in the energy cycle.

### *Types of chemical uncouplers*

The knowledge of the conformational cycle has facilitated the recognition of the multiple ways by which uncoupling can be achieved. m-Chlorocarbonylcyanide phenylhydrazone (m-ClCCP) suppresses the transition from the nonenergized to the energized conformation and also the *p*H jump accompanying electron transfer. This is the most complete type of uncoupling since coupling is abolished at the primary event, namely the first conformational change. Fluorescein mercuric acetate (FMA) suppresses the second conformational change and has no effect on all processes which are dependent exclusively on the first conformational change (e.g., the *p*H jump and the energized binding of $Ca^{++}$) (Lee, Harris, and Green, 1969). Reagents which induce the orthodox mode of the cristal membrane are in effect uncouplers, but the uncoupling achieved by this means may have to do with the failure of engagement rather than interference with the conformational changes. Reagents such as valinomycin or gramicidin can prevent the coupling of electron transfer to synthesis of ATP by limiting the coupling potential to the coupling of electron transfer to the movement of specific ions. These polypeptide reagents are not general uncouplers but specific reagents for suppressing the coupling of electron transfer to synthesis of ATP.

The uncoupling action of both m-ClCCP and of FMA can be reversed by addition of dithiothreitol (Lee, Harris, and Green, 1969). The fact that the action of these two reagents which affect different parts of the

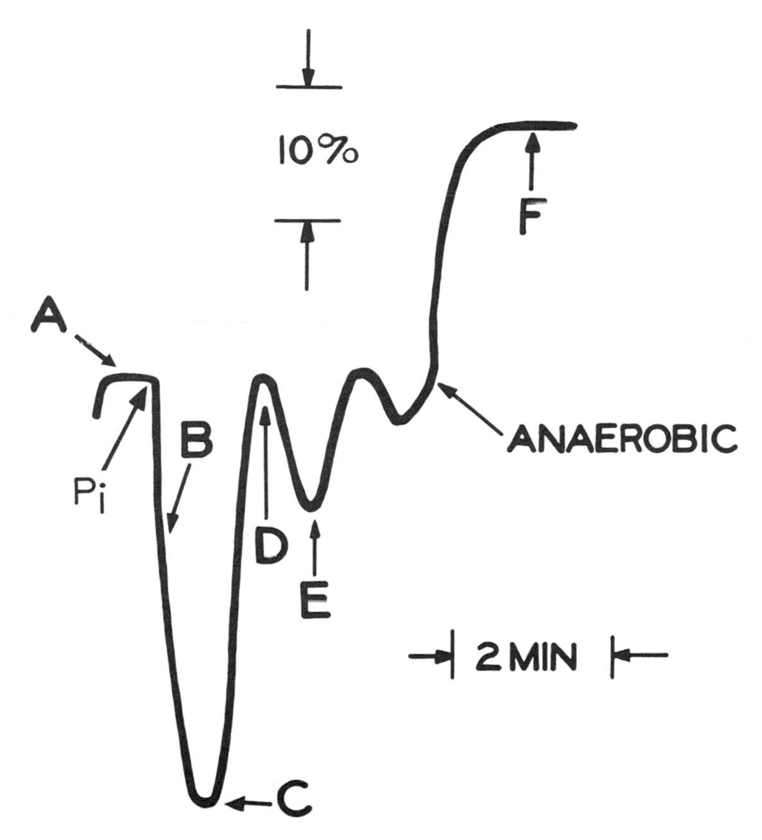


*Fig. 12.17.* Oscillations of the configurational state of beef heart mitochondria as measured by light scattering. The sucrose-Tris incubation medium was fortified with succinate (5 m), rotenone (2 μg/mg mitochondrial protein), valinomycin ($10^{-9}$ M), and $P_i$ (20 mM at *p*H 6.8). The temperature was maintained at 20° C.

coupling mechanism should be reversed by -SH reagents is suggestive that specific -SH groups probably play a key role in the conformational cycle. Uncouplers such as $Cd^{++}$ may in fact be reagents which combine with one or more of the -SH groups which are essential for the coupling cycle.

A large number of uncouplers of oxidative phosphorylation have been described in the biochemical literature. As yet only a few have been studied from the standpoint of where these act in the conformational cycle. In time, a wide variety of reagents should be available to interrupt the coupling mechanism at any one of the various critical transitions.

### RESPIRATORY CONTROL

Respiratory control may be defined as the ratio of the rate of oxygen consumption when ATP is being synthesized to the rate of oxygen consumption when ADP is not available as a phosphoryl acceptor. Defined in this manner, the ratio is in essence determined by the degree of stability of the energized-twisted conformation. If it were very stable, the ratio would be very large. If it had no stability, the ratio would be 1. Under optimal conditions, the respiratory control ratio for mitochondria varies between the limits of 4 and 20. It is possible to show that variation in the respiratory control of undamaged mitochondrial suspension is, in part, determined by the ratio of mitochondria in the two respective modes (orthodox and aggregated).

Submitochondrial particles do not show evidence of respiratory control as defined above although these particles may show theoretical P/O ratios. Obviously, respiratory control is not essential for perfect coupling. We may think of two kinds of stability of the energized-twisted conformation—long-term, which confers respiratory control, and short-term, which makes possible perfect coupling. In submitochondrial particles, such as $ETP_H$, the stability is for a period sufficient to make possible perfect coupling but not for a period of time sufficient for the demonstration of respiratory control.

Why is the energized-twisted conformation in the repeating units of submitochondrial particles less stable than in the repeating units of mitochondria? There is, of course, an important difference in the geometry of the membranes between mitochondria and submitochondrial particles. But this by itself could hardly account for the complete loss of respiratory control. Another important difference lies in the presence of protein within the spaces of the mitochondrion, i.e., in the intracristal and matrix spaces, and the complete absence of this protein in submitochondrial particles. The role of the internal proteins within the spaces of the mitochondria on the stabilization of the energized-twisted configuration has yet to be assessed experimentally.

### SEQUELLAE OF CONFORMATIONAL AND CONFIGURATIONAL CHANGES

While it is certainly true that the molecular unit of the energy cycle is the tripartite repeating unit of the cristal membrane, it is also true that the individual tripartite repeating unit is not the operational unit. The two-dimensional polymeric form of the tripartite repeating unit, namely the cristal membrane, is in fact the operational unit. The conformational cycle is an expression of the molecular changes in the individual tripartite repeating unit, whereas the configurational cycle is an expression of the integration of the conformational changes in all the repeating units of a given membrane. Thus, it is impossible to separate the membrane

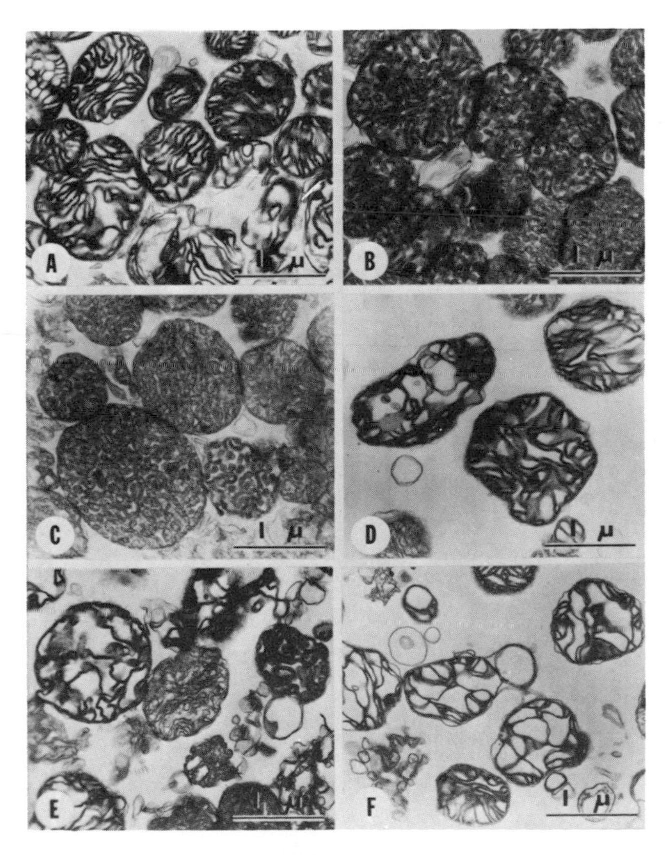


*Fig. 12.18.* Electron microscopic analysis of the oscillatory states of beef heart mitochondria. Samples of the mitochondrial suspension used for the light experiment shown in Fig. 12.17 were fixed for electron microscopy at the points indicated by the letters *A* through *F*.

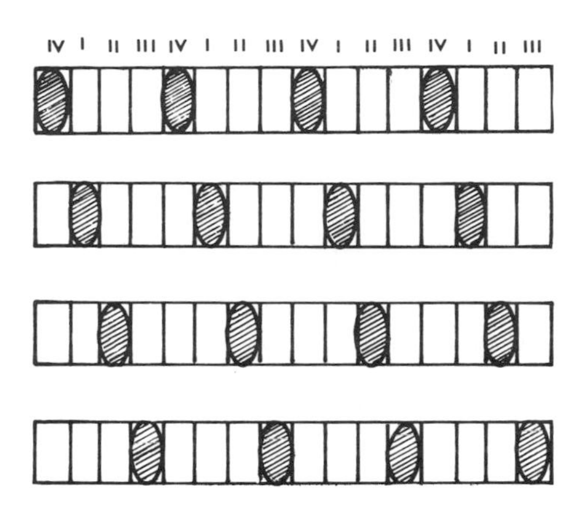


*Fig. 12.19.* Transmission of the energized state in a membrane. The sequential arrangement of the different complexes is purely arbitrary and is invoked merely for illustration of the movement of the energized state through the membrane. The diagram shows how the energized state generated at basepieces containing Complex IV is transmitted to the other complexes by an exchange reaction. The transmission is probably random and not unidirectional as shown in the diagram.

parameters from the molecular parameters in considering the energy cycle of the mitochondrion. When mitochondria undergo energizing by substrate to the energized configuration, there is a series of membrane changes that are characteristic of this transition. The matrix space becomes expanded; the intracristal space becomes compressed; and water and ions flow from the intracristal space to the matrix space. This may be because the macro-ion charge that underlies the Donnan equilibrium becomes greater in the matrix space than in the intracristal space (Vanderkooi and Green, unpublished observations). All of these membrane phenomena are postulated to be sequellae of the transition of the repeating units from the nonenergized to the energized conformation, and appear to be intrinsic to the energy cycle. Since the membrane is the *de facto* operational unit, not the individual repeating unit, it is not possible to segregate the conformational from the configurational changes and to consider the conformational changes in the repeating units as the only relevant primary processes. In soluble enzyme systems, the molecular unit and the operational unit also need not be the same. The operational unit can be the dimer, trimer, or tetramer but only rarely more polymeric than the tetrameric. In membrane systems, the operational unit can contain hundreds or thousands of the molecular units, i.e., the repeating units.

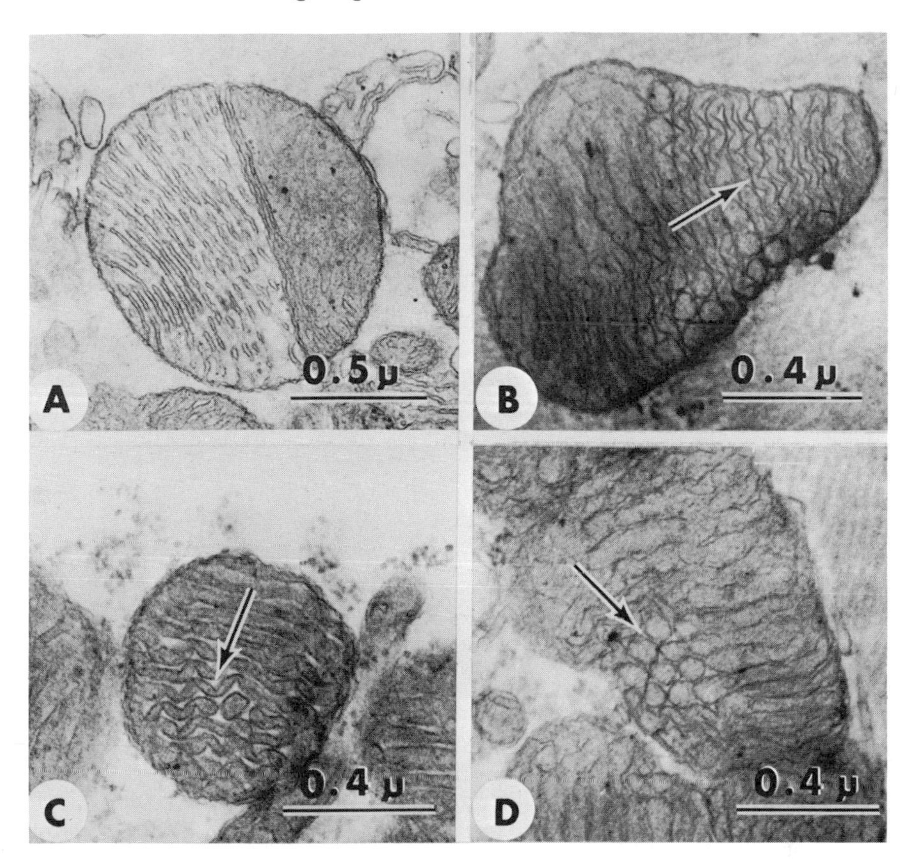


*Fig. 12.20.* Paracrystalline arrays of the cristal membrane observed with beef heart mitochondria *in situ.* Basic incubation medium, as in Fig. 12.7. *A,* a mitochondrion in a transitional state observed in the presence of antimycin (33 μg/ml) and sodium cyanide (7 mM) ; *B,* mitochondrion with energized zigzag configuration of the cristal membrane observed in the presence of 10 mM sodium succinate; *C,* mitochondrion with energized twisted configuration of the cristal membrane obtained in the presence of 10 mM sodium succinate; *D,* mitochondrion with a paracrystalline array of the cristal membrane observed in the presence of 10 mM sodium succinate.

When a repeating unit is a part of a membrane, the concept of asymmetry must at once be introduced. Events on one side of a repeating unit oriented in a membrane are necessarily different from events on the other side, not only because the repeating unit may be asymmetric but also because there need not be the same concentration of ions on the two sides of the membrane. The membrane may act as a barrier to the movement of certain ions. When conformational change of the repeating unit takes place, the ionic charge on the two sides of the membrane may be very different, and this differential can lead to differences in the value

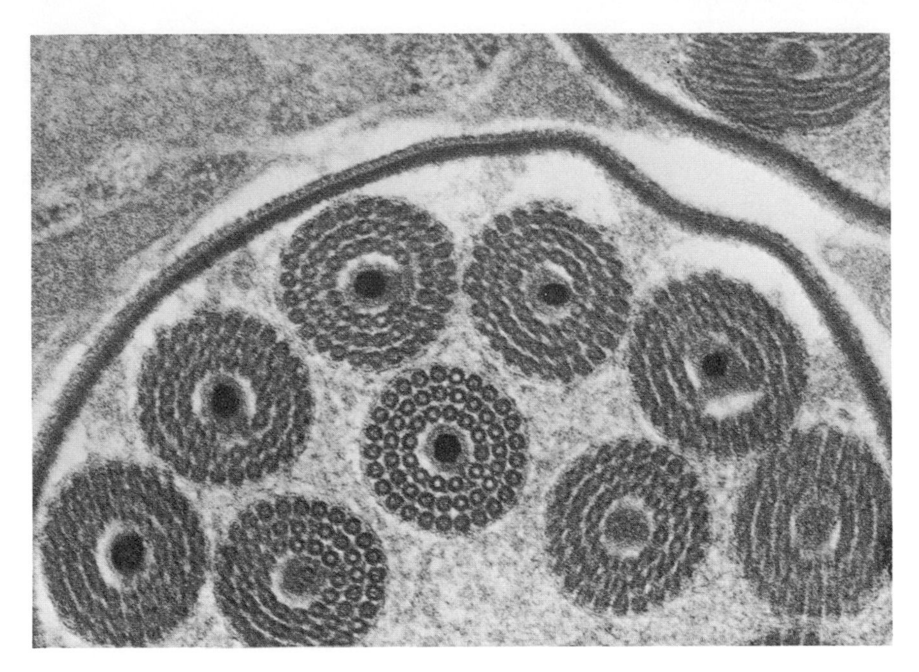

*Fig. 12.21.* Arrangement of mitochondria in the sperm of the mealy bug. Each of the tubular doughnut-like structures surrounding the central dark core is a cross section of a single elongated crista that wraps around the central core. When the mitochondrion becomes energized, the tubular cristae tilt from the vertical to the horizontal and the energized-twisted tubules are then seen lengthwise in side view. Electron micrograph provided by W. J. Robison.

of the Donnan equilibrium (Vanderkooi and Green, unpublished observations; Blondin, unpublished observations). Hence, conformational change can trigger movement of water and ions between the media on the two sides of the membrane. Equally important, the permeability of the membrane may become profoundly altered in the transition from the nonenergized to the energized configuration.

### CONFORMATIONAL VERSUS CONFIGURATIONAL CHANGE

In the experiments designed to show a correlation between the configurational state of the cristal membrane and the energy state of the mitochondrion (Green, Asai, Harris, and Penniston, 1968), precautions were taken to try to insure that the conformation of the repeating units and the configuration of the membrane were identical. One important precaution necessary to achieve this end was to eliminate the possibility of the system doing work. As soon as the energized membrane can do work, i.e., synthesize ATP or move ions, then a steady state situation is set up and the synchrony of conformational and configurational states may no longer be assumed. It is with this steady state situation that we are presently concerned. What are the ground rules for a flip in configuration when the repeating units are undergoing rapid cycles of conformational change and at any instant are in different phases of the cycle? It is obvious that the configuration of the membrane will reflect the predominant conformation of the repeating unit. It is also obvious that the configuration of the membrane can remain unchanged even though the repeating units are pulsating in their conformational cycle. Probably, under physiological conditions, the membrane is either in the nonenergized configuration (when a low level of activity is reached) or in the energized or energized-twisted configuration (when a high level of activity is reached) (Williams, Vail, Harris, Caldwell, Valdivia, and Green, 1970). Thus, the configuration of the membrane changes only occasionally, whereas the conformation of the repeating units changes all the time.

A dramatic example of this dichotomy between configuration and conformation is provided by the oscillation phenomenon (Harris, Asbell, Jolly, Asai, and Green, 1969; Deamer, Utsumi, and Packer, 1967) (Fig. 12.17). The mitochondrion in presence of substrate, inorganic phosphate, potassium salts, and valinomycin goes through a phase of accumulating $K^+$ followed by a phase of discharging the accumulated $K^+$. This oscillation can repeat itself several times but each successive oscillation becomes smaller until finally the phenomenon disappears. The electron microscope (Fig. 12.18) shows that when the mitochondria are actively accumulating $K^+$ the cristae are in the energized-twisted configuration, whereas when they are discharging $K^+$, they are in the nonenergized configuration. Note that at the beginning of the oscillation cycle, all the mitochondria are in the same configuration. As oscillation becomes damped, the mitochondrial population becomes heterogeneous and out of phase.

### TRANSMISSION OF THE ENERGIZED STATE

The source of the electrons that generate the energized state can be restricted to one substrate and to one complex, e.g., Complex IV. Despite this restriction the entire cristal membrane becomes energized (Green, Asai, Harris, and Penniston, 1968). Many variations on the same theme have been tested and shown to lead to the same result. The membrane is either energized or not energized. There is never partial or incomplete energizing. This must mean that the energized state can be transmitted through the membrane by something akin to an exchange reaction (Fig. 12.19), which must be extremely fast because it has never been possible

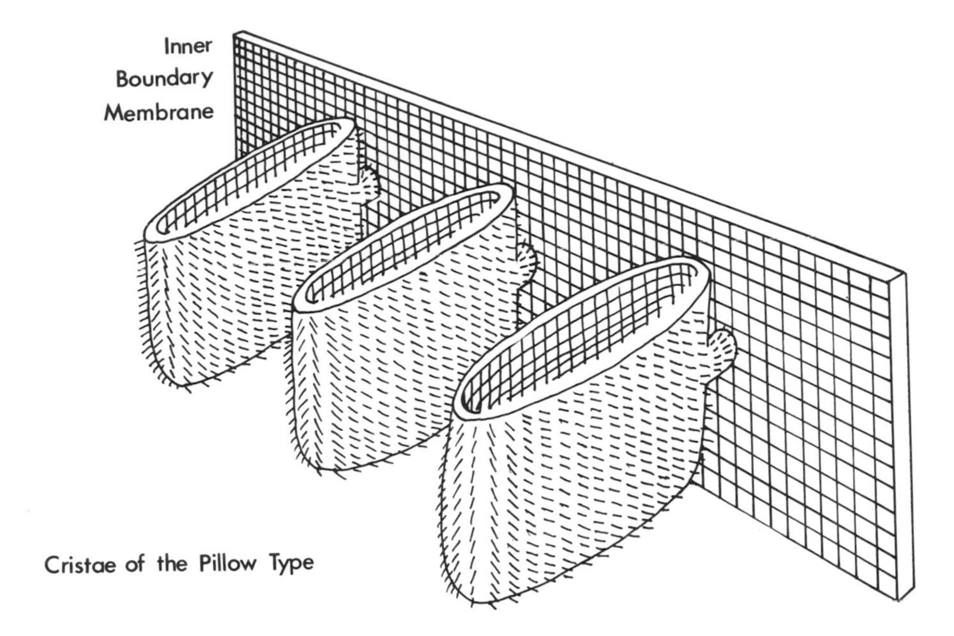


*Fig. 12.22.* The cristal membrane of the mitochondrion, in the flattened pillow-like configuration.

to find the membrane in an intermediate configuration, regardless of the mode of energizing.

It is this phenomenon of the transmission of the energized state which probably underlies the substoichiometry of uncouplers such as m-ClCCP (Margolis, Lenaz, and Baum, 1967). These uncouplers can uncouple the membrane at concentration levels at which there is one molecule of uncoupler per 100 repeating units. Given the transmission of the energized state, each repeating unit that binds the uncoupler can act as an energy sink and contribute to the discharge of a large number of repeating units in the energized conformation.

When mitochondria are damaged by lyophilization, freezing and thawing, or other means, some of the repeating units become incompetent in the sense that these discharge the energized state as if combined with uncoupler. Each such damaged repeating unit can then act as an energy sink and thus prevent the membrane from maintaining an energized configuration (Jolly, Harris, Asai, Lenaz, and Green, 1969). Experiment has shown that lyophilization leads to the detachment of headpiece-stalks and to the liberation of phospholipid. The amount of ultrastructural dislocation is admittedly small but the consequences are severe because of the transmission of the energized state. When mitochondria which

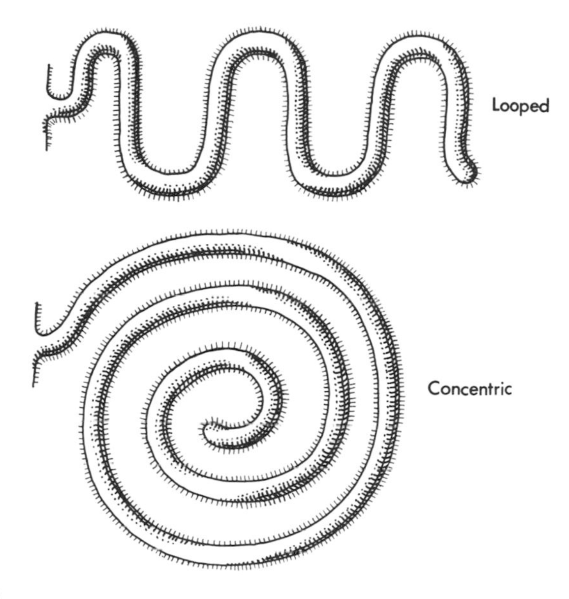


*Fig. 12.23.* The cristal membrane in the tubular configuration.

have lost their capability for coupled synthesis of ATP by virtue of lyophilization have been fragmented by sonic irradiation, the derivative submitochondrial particles show the full capacity for coupling. Probably sonic irradiation leads to the reconstitution of the damaged repeating units.

### PARACRYSTALLINE ARRAYS IN MITOCHONDRIA

In recent years electron micrographic evidence has accumulated (Pappas and Brandt, 1959; Slautterback, 1965; Ross and Robison, 1969) for the presence in mitochondria of unusual paracrystalline arrays (Figs. 12.20–21). The detailed study (Korman, Harris, Williams, Wakabayashi, Green, and Valdivia, 1970) of these arrays has revealed some new ultrastructural facets of the energizing process. The cristal membranes of mitochondria can be either discoid, like flattened pillows (Fig. 12.22), or tubular, like water hose (Fig. 12.23). The engagement of apposed cristae, as we have described it above, applies to the pillow type of crista. This engagement involves the apposition of two flat membranes when the repeating units are in the energized conformation and then tubularization when the repeating units are in the energized-twisted conformation. If the tubularization is very precise and regular, Star of David patterns (Harris, Williams, Caldwell, Green, and Valdivia, 1969; Korman, Harris, Williams,

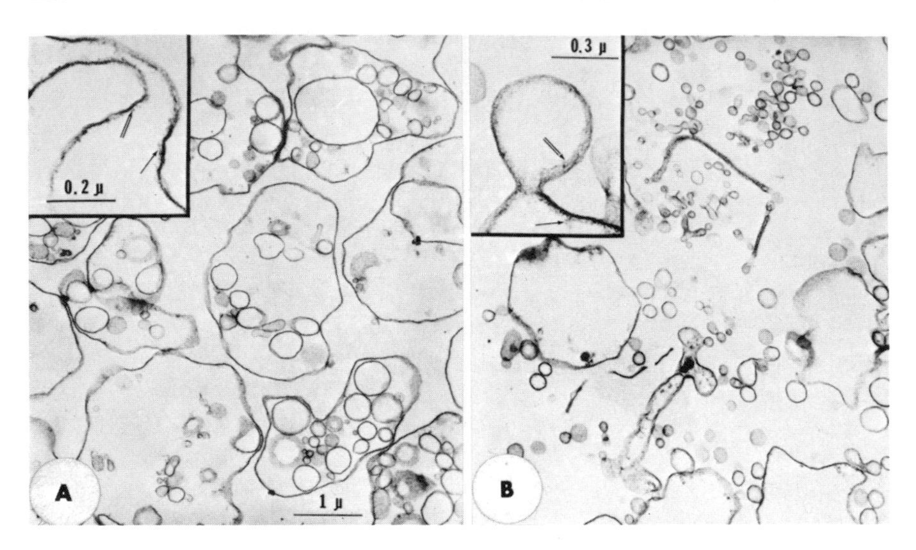


*Fig. 12.24.* The energy cycle of a ghost. *A,* energized configuration of cow erythrocyte ghosts. The vesicles are in the interior of the erythrocyte ghosts. Note that the filamentous material of the boundary membrane faces inward *(single arrow)*, while the filamentous material of the vesicle faces outward *(double arrow)*; *B,* nonenergized configuration of cow erythrocyte ghosts. The vesicles are exterior to the erythrocyte ghosts. Note that the filamentous material of both the boundary membrane *(single arrow)* and of the vesicle *(double arrow)* faces inward. Electron micrographs provided by J. T. Penniston.

Wakabayashi, Green, and Valdivia, 1970) such as are shown in Fig. 12.20 can be seen in the mitochondria of canary heart muscle and of the jumping muscle of the locust (Goglia, 1969). We have previously referred to the mitochondria of the mealy bug sperm which are in essence elongated mitochondria with rows of cristae in the orthodox mode that run parallel with the long axis of the mitochondrion (see Fig. 12.21). The parallel cristae of the mitochondria of the mealy bug sperm wind helically around a central core, presumably of nucleic acid. How the cristae in such mitochondria undergo configurational change remains to be determined.

### THE CONDENSED STATE OF ENERGIZED CRISTAE

When pillow type cristae are randomly arranged rather than in neat parallel arrays (e.g., in isolated liver mitochondria), or when tubular cristae are randomly folded rather than in neat concentric or folded alignment (e.g., in isolated adrenal mitochondria), engagement of apposed cristae or tubular folds, when the repeating units are in the energized conformation, leads to a conglomerate or condensed state in which the outlines of the contributing membranes are blurred (Figs. 12.

13–14) (Allmann, Munroe, Hechter, and Matsuba, 1969). In the isolation of liver and adrenal mitochondria, the regularity of the cristae or the cristal folds is lost or diminished and the engagement leads to the characteristic condensed state. The cristae in the same mitochondria *in situ* are usually arranged in more regular fashion and engagement rarely leads to the condensed state.

### CONFIGURATIONAL CHANGES IN OTHER MEMBRANES

On the basis of some limited studies with other membrane systems (ghost membranes of the red blood corpuscle [Penniston and Green, 1968], membrane of the chromatophores of *Chromatia,* the inner membranes of the chloroplast, and the membrane of the sarcoplasmic reticulum [Asbell, unpublished observations]) we have the impression that the concept of the conformational basis of energy transduction may have validity for all membrane systems which carry out energized work performances. The form of the configurational changes appears to be variable from membrane to membrane. For example, in the ghost membrane of the red blood corpuscle, the configurational cycle can be described as a pinocytotic cycle (Fig. 12.24) (Penniston and Green, 1968). In other membranes, such as the chloroplast, the configurational cycle takes the form of the alternate expansion and contraction of the inner membranes. If we are in fact dealing with a universal mechanism of energy transduction in membrane systems, then the principles first recognized by the study of the mitochondrial cristal membranes should be applicable, *mutatis mutandis,* to all other membrane systems.

## *References*

Allmann, D. W., J. Munroe, D Hechter, and M. Matsuba. 1969. Ultrastructural basis for the regulation of steroidogenesis in bovine adrenal cortex mitochondria. *Fed. Proc. 28:*2274.

Allmann, D. W., J. Munroe, T. Wakabayashi, and D. E. Green. 1970. Studies on the transition of the cristal membrane from the orthodox to the aggregated configuration III. Loss of coupling capability of adrenal cortex mitochondria in the orthodox configuration. *J. Bioenergetics.* (In press.)

Allmann, D. W., T. Wakabayashi, E. F. Korman, and D. E. Green. 1970. Studies on the transition of the cristal membrane from the orthodox to the aggregated configuration I. Topology of bovine adrenal cortex mitochondria in the orthodox configuration. *J. Bioenergetics.* (In press.)

Asai, J., G. A. Blondin, W. J. Vail, and D. E. Green. 1969. The mechanism of mitochondrial swelling IV. Configurational changes during mitochondrial swelling. *Arch. Biochem. Biophys. 132:*524.

Asbell, M. A., W. J. Vail, R. A. Harris, E. F. Korman, and D. E. Green. 1970. The conformational basis of energy conservation in membrane systems IX. Energy dependent configurational change of submitochondrial particles. *J. Bioenergetics.* (In press.)

Blondin, G. A., W. J. Vail, and D. E. Green. 1969. The mechanism of mitochondrial swelling II. Pseudoenergized swelling in the presence of alkali metal salts. *Arch. Biochem. Biophys. 129:*158.

Deamer, D. W., K. Utsumi, and L. Packer. 1967. Oscillatory states of mitochondria III. Ultrastructure of trapped conformational states. *Arch. Biochem. Biophys. 121:*641.

Goglia, G. 1969. *Acta Med. Romana.* (In press.)

Green, D. E., D. W. Allmann, E. Bachmann, H. Baum, K. Kopaczyk, E. F. Korman, S. H. Lipton, D. H. MacLennan, D. G. McConnell, J. F. Perdue, J. S. Rieske, and A. Tzagoloff. 1967. Formation of membranes by repeating units. *Arch. Biochem. Biophys. 119:*312.

Green, D. E., J. Asai, R. A. Harris, and J. T. Penniston. 1968. The conformational basis of energy transductions in membrane systems III. Configurational changes in the mitochondrial inner membrane induced by changes in functional states. *Arch. Biochem. Biophys. 125:*684.

Hackenbrock, C. R. 1968. Ultrastructural bases for metabolically linked mechanical activity in mitochondria I. Reversible ultrastructural changes with change in metabolic steady state in isolated mitochondria. *J. Cell Biol. 37:*345.

Harris, R. A., M. A. Asbell, and D. E. Green. 1969. Correlation of changes in light scattering with changes in energy state of phosphorylating submitochondrial particles. *Arch. Biochem. Biophys. 131:*316.

Harris, R. A., M. A. Asbell, W. W. Jolly, J. Asai, and D. E. Green. 1969. The conformational basis of energy conservation in membrane systems VI. Correlation of changes in light scattering with changes in energy and configurational state of the inner membrane of beef heart mitochondria. *Arch. Biochem. Biophys. 132:*545.

Harris, R. A., D. L. Harris, and D. E. Green. 1968. Effect of *Bordetella* endotoxin upon mitochondrial respiration and energized processes. *Arch. Biochem. Biophys. 128:*219.

Harris, R. A., J. T. Penniston, J. Asai, and D. E. Green. 1968. The conformational basis of energy conservation in membrane systems II. Correlation between conformation change and functional states. *Proc. Nat. Acad. Sci. 59:*830.

Harris, R. A., C. H. Williams, M. Caldwell, D. E. Green, and E. Valdivia. 1969. Energized configurations of heart mitochondria *in situ. Science 165:*700.

Harris, R. A., C. H. Williams, W. W. Jolly, J. Asai, and D. E. Green. 1970. The conformational basis of energy transformations in membrane systems VII. Effect of the external osmotic pressure upon the energized processes and the energized configurational states of heart mitochondria. *Arch. Biochem. Biophys.* (In press.)

Jacobs, E. 1956. Phosphorylation coupled to the oxidation of complex inorganic anions by rat liver mitochondria. *Biochim. Biophys. Acta 22*:583.

Jolly, W. W., R. A. Harris, J. Asai, G. Lenaz, and D. E. Green. 1969. Studies on ultrastructural dislocations in mitochondria II. On the dislocation induced by lyophilization and the mechanism of uncoupling. *Arch. Biochem. Biophys. 130*:191.

Kagawa, Y., and E. Racker. 1966. Partial resolution of the enzymes catalyzing oxidative phosphorylation X. Correlation of morphology and function in submitochondrial particles. *J. Biol. Chem. 241*:2475.

Kopaczyk, K., J. Asai, and D. E. Green. 1968. Reconstitution of the repeating unit of the mitochondrial inner membrane. *Arch. Biochem. Biophys. 126*:358.

Korman, E. F., A. D. F. Addink, T. Wakabayashi, and D. E. Green. 1970. A unified model of mitochondrial morphology. *J. Bioenergetics.* (In press.)

Korman, E. F., R. A. Harris, C. H. Williams, T. Wakabayashi, D. E. Green, and E. Valdivia. 1970. Paracrystalline arrays in mitochondria. *J. Bioenergetics.* (In press.)

Lee, M., R. A. Harris, and D. E. Green. 1969. Action of fluorescein mercuric acetate upon the energized processes of beef heart mitochondria. *Biochem. Biophys. Res. Comm. 36*:937.

Margolis, S. A., G. Lenaz, and H. Baum. 1967. Stoichiometric aspects of the uncoupling of oxidative phosphorylation by carbonylcyanide phenylhydrazones. *Arch. Biochem. Biophys. 118*:224.

Pappas, G. D., and P. W. Brandt. 1959. Mitochondria I. Fine structure of the complex patterns in the mitochondria of *Pelomyxa carolinensis* Wilson (Chaos chaos L.) . *J. Biophys. Biochem. Cytol. 6*:85.

Penniston, J. T., and D. E. Green. 1968. The conformational basis of energy transformations in membrane systems IV. Energized states of pinocytosis in erythrocyte ghosts. *Arch. Biochem. Biophys. 128*:339.

Penniston, J. T., R. A. Harris, J. Asai, and D. E. Green. 1968. The conformational basis of energy transformations in membrane systems I. Conformational changes in mitochondria. *Proc. Nat. Acad. Sci. 59*:624.

Ross, J., and W. J. Robison. 1969. Unusual microtubular patterns and three dimensional movement of mealybug sperm and sperm bundles. *J. Cell Biol. 40*:426.

Slautterback, D. B. 1965. Mitochondria in cardiac muscle cells of the canary and some other birds. *J. Cell Biol. 24*:1.

Taylor, C. A., R. A. Harris, C. H. Williams, and D. E. Green. 1969. Effect of halothane upon the energized state and energized processes of beef heart mitochondria. (Abstr.) *Amer. Soc. Anesthesiologists,* September, San Francisco.

Williams, C. H., W. J. Vail, R. A. Harris, M. Caldwell, D. E. Green, and E. Valdivia. 1970. Conformational basis of energy transduction in membrane systems VIII. Configurational changes of mitochondria *in situ* and *in vitro. J. Bioenergetics.* (In press.)

# Chapter 13

# *Form of the Sarcoplasmic Reticulum and T System of Striated Muscle*

L. D. PEACHEY

The main features of the structure of the sarcoplasmic reticulum (SR) and transverse tubules (or T system) of striated muscle have been known for several years now, and were discussed by D. B. Slautterback in Chapter 5 of *The Physiology and Biochemistry of Muscle as a Food* (1966). Several other recent reviews have dealt with this samc subjcct (Sandow, 1965; Smith, 1966; Peachey, 1968; Hoyle, 1969), and I will not here enter upon an extended discussion of the literature on the field. The reviews mentioned above and additional works cited in the list of references should serve the needs of anyone interested in pursuing this subject in greater depth.

I propose to summarize the structure of the SR and T system in some representative vertebrate skeletal muscles, and to show some aspects of the variation of these structures in other muscles from various animals, frequently using for illustrations those muscles that have been studied more recently.* I shall then discuss some aspects of the development of these membrane systems, and their experimental disruption, since these have not been extensively reviewed before.

One relatively minor point of terminology needs clarification first. Recent evidence, which I will discuss in more detail later, proves that the T system is derived from the surface or plasma membrane of the muscle cell, and remains attached to the cell surface in the adult muscle cell. The SR, on the other hand, has an intracellular origin, being derived from the rough-surfaced endoplasmic reticulum during development, and maintains an essentially intracellular existence in the mature cell. I prefer, therefore, to consider these as two distinct and separate membrane systems, rather than to include the T system as part of the SR, as done by some authors.

* None of the micrographs presented here has been published before.

A few words about methodology are in order here. In his introduction to the published volume from the first of these conferences, Dr. G. F. Stewart (1966) expressed the hope that post mortem changes in muscle and the technology of converting muscle to meat products would receive major attention. Electron microscopists can, I think, claim to be specialists in this area. For most of the preparations illustrated here, a fresh muscle was fixed in an aldehyde followed by further fixation in osmium tetroxide. It was then dehydrated and embedded in an epoxy resin, and sections 0.1 $\mu$ in thickness or thinner were cut. These subsequently were stained with heavy-metal-containing stains such as uranyl acetate and lead citrate to provide contrast in the electron microscope image. The illustrations consist of micrographs of these stained sections magnified from 1,000–20,000 times; most of the time, therefore, the illustrations show only a tiny fraction of a single muscle cell.

Fig. 13.1 shows a transverse section of a portion of a mammalian (cat) extraocular muscle as seen in the light microscope. This is a skeletal muscle, although somewhat smaller in total size and containing smaller fibers than the muscles of the limbs. If the illustration looks different from the usual sort of light micrograph of muscle, it is probably because this muscle was prepared by methods usually used for electron microscopy except that a section from 1–2 $\mu$ thick was cut and photographed in a light microscope, using phase optics. Illustrations such as this are helpful as maps to keep track of one's location in the muscle while using the electron microscope. Fig. 13.1 constitutes a useful starting point in a gradual transition to higher magnifications.

About 50 muscle fibers, of several thousand in the muscle, are seen in Fig. 13.1, along with a bundle of nerve fibers and a blood vessel. Variation in the appearance of different muscle fibers reflects the mixed nature of this muscle. The largest fiber in this field has a diameter of about 30 $\mu$, and the smallest measures only 5 $\mu$ across.

Fig. 13.2 represents an electron micrograph, but the magnification is only slightly increased over Fig. 13.1. The appearance is roughly the same, although close inspection will show differences arising from the differences in resolution and the source of optical contrast in the two kinds of microscope. We now resolve more clearly the small blood capillaries and some of the internal structure of the muscle fibers.

As the magnification is further increased, certain features of the sections are seen more clearly. For example, in Fig. 13.3 the small blood capillaries lying between the muscle fibers are even more obvious. Each is formed by an endothelial cell wrapped around the lumen of the vessel, whose content of serum proteins appears as a greyish mass. Outside each

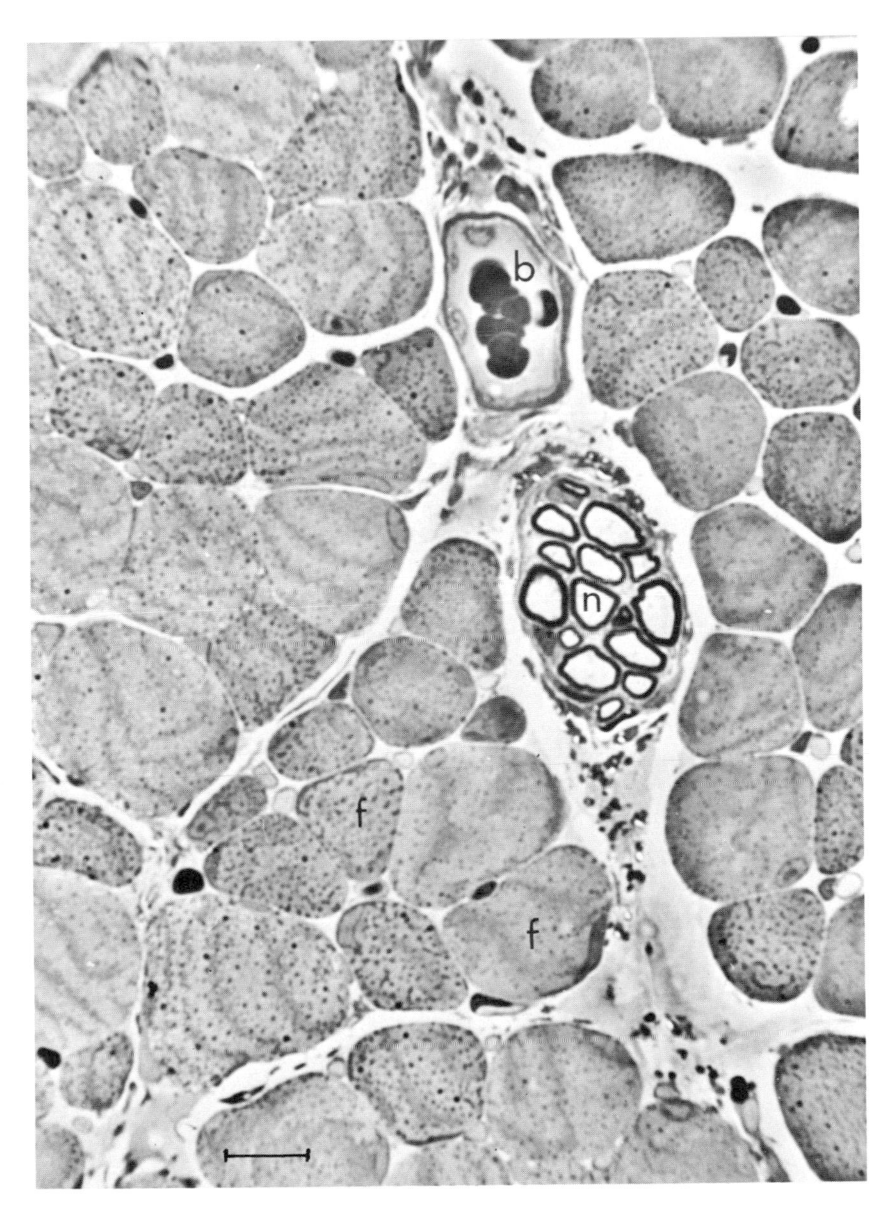


*Fig. 13.1.* Light micrograph of a transverse section of a superior rectus muscle of an adult cat. A variety of muscle fiber types (*f*) is seen, along with a blood vessel (*b*) and a nerve bundle (*n*). Line represents 0.01 mm.

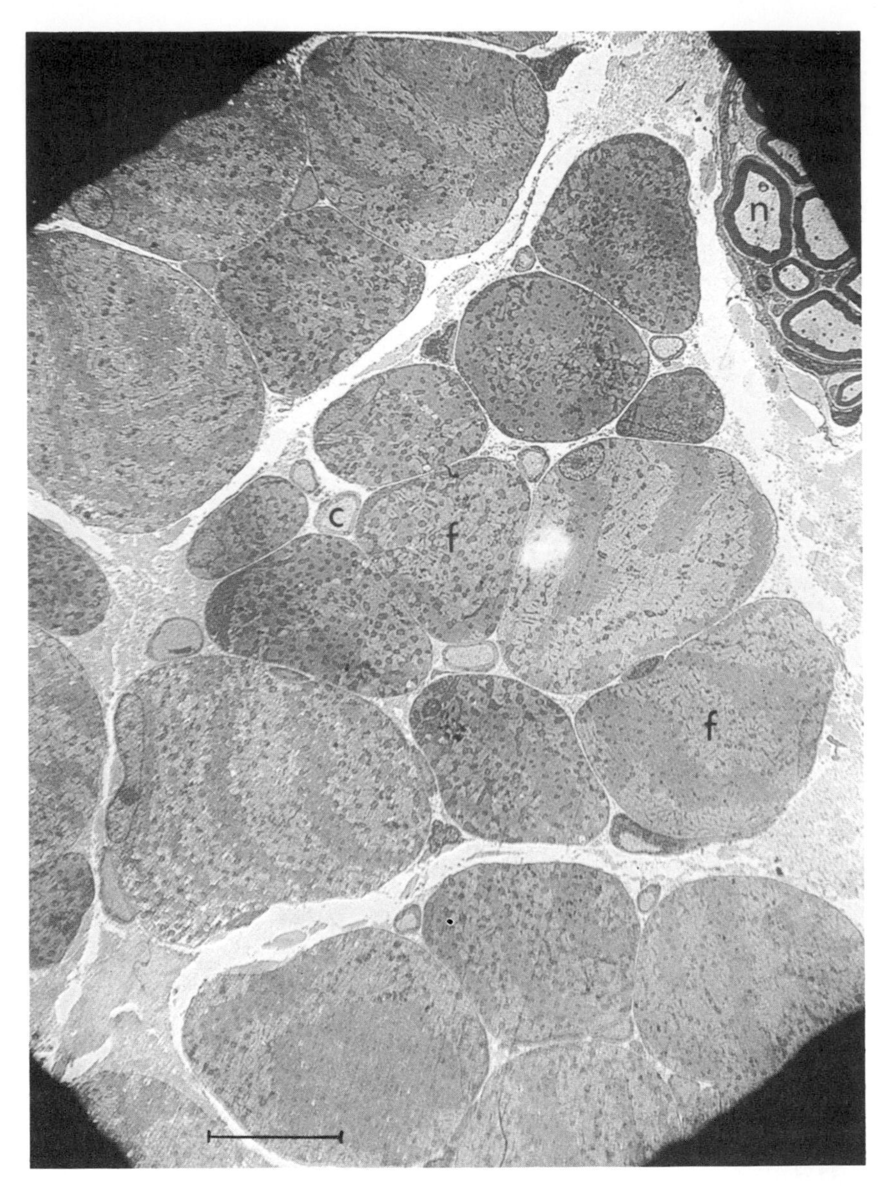


*Fig. 13.2.* Electron micrograph of a transverse section through the same portion of the same preparation as shown in the light micrograph in Fig. 13.1. Muscle fibers (*f*), small capillaries (*c*) and part of the nerve bundle (*n*) are seen. Line represents 0.01 mm.

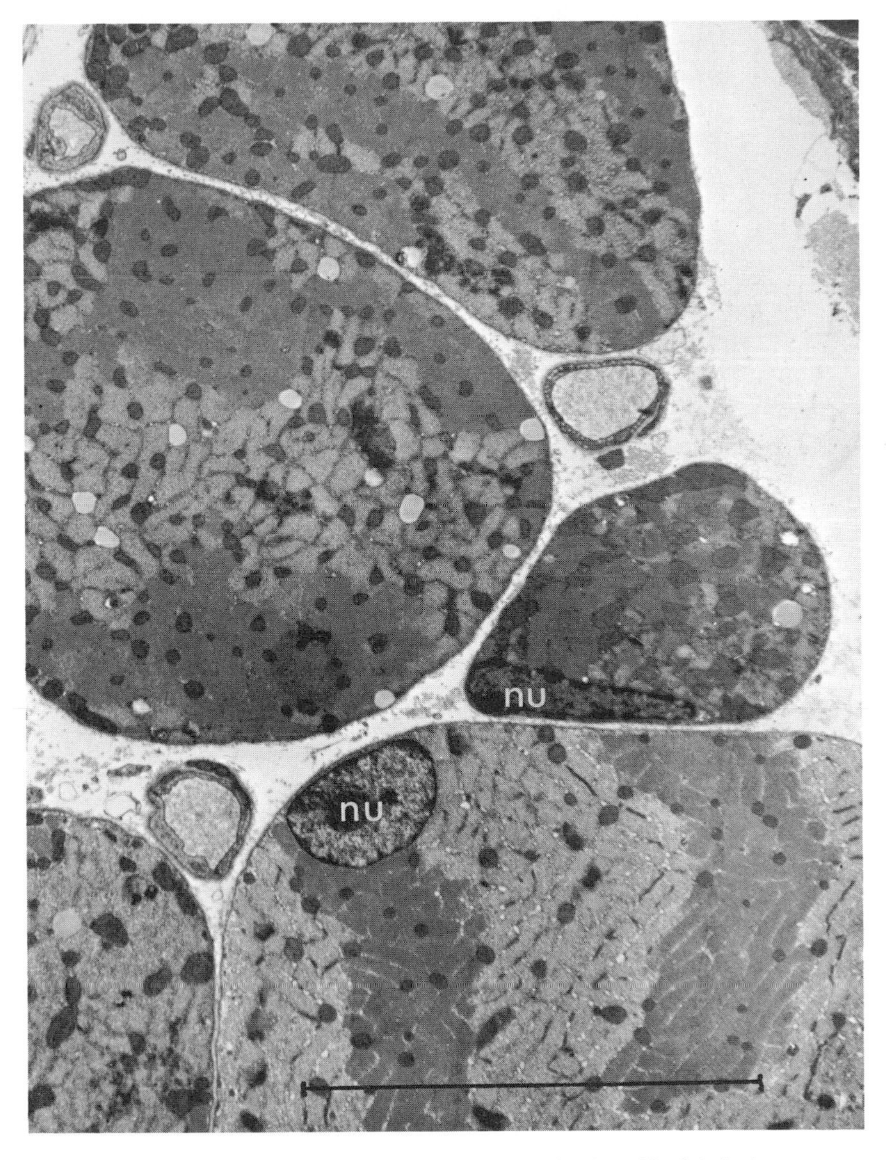


*Fig. 13.3.* Similar to Fig. 13.2, but at higher magnification. Nuclei (*nu*) are seen. Line represents 0.01 mm.

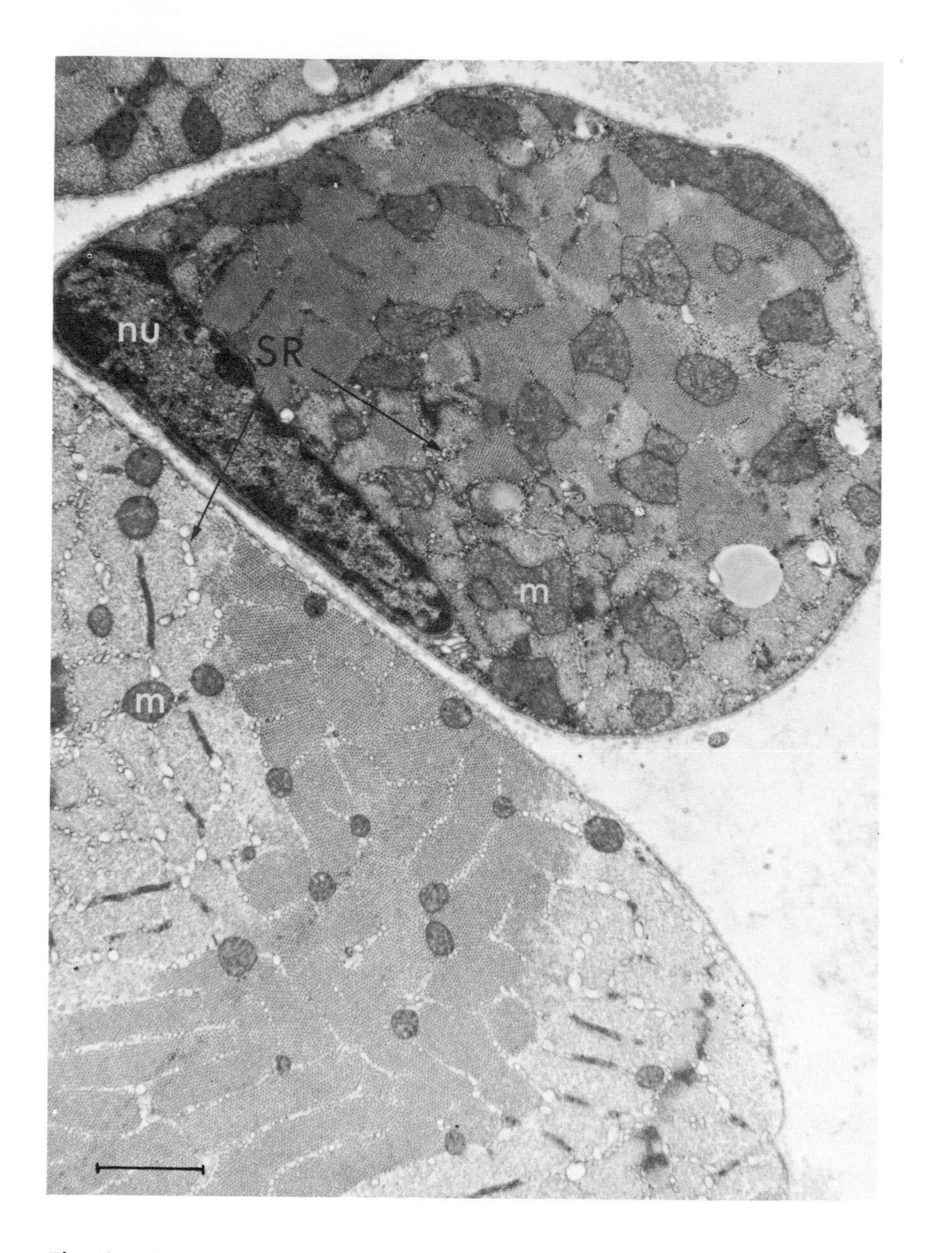


*Fig. 13.4.* Again similar to Fig. 13.2, but at even higher magnification than Fig. 3. At this magnification one can see mitochondria (*m*), and membranes of SR between the myofibrils. A nucleus is just under the surface of one fiber (*nu*). Line represents 1 μ.

*Fig. 13.5.* Electron micrograph of a transverse section of the trunk musculature of amphioxus (*Branchiostoma caribaeum*). Each vertical band is a single muscle fiber (*f*), covered with its plasma membrane (*p*), and showing the punctate appearance of the myofilaments cut transversely. The dark patch in one fiber is a Z line. Line represents 1 $\mu$.

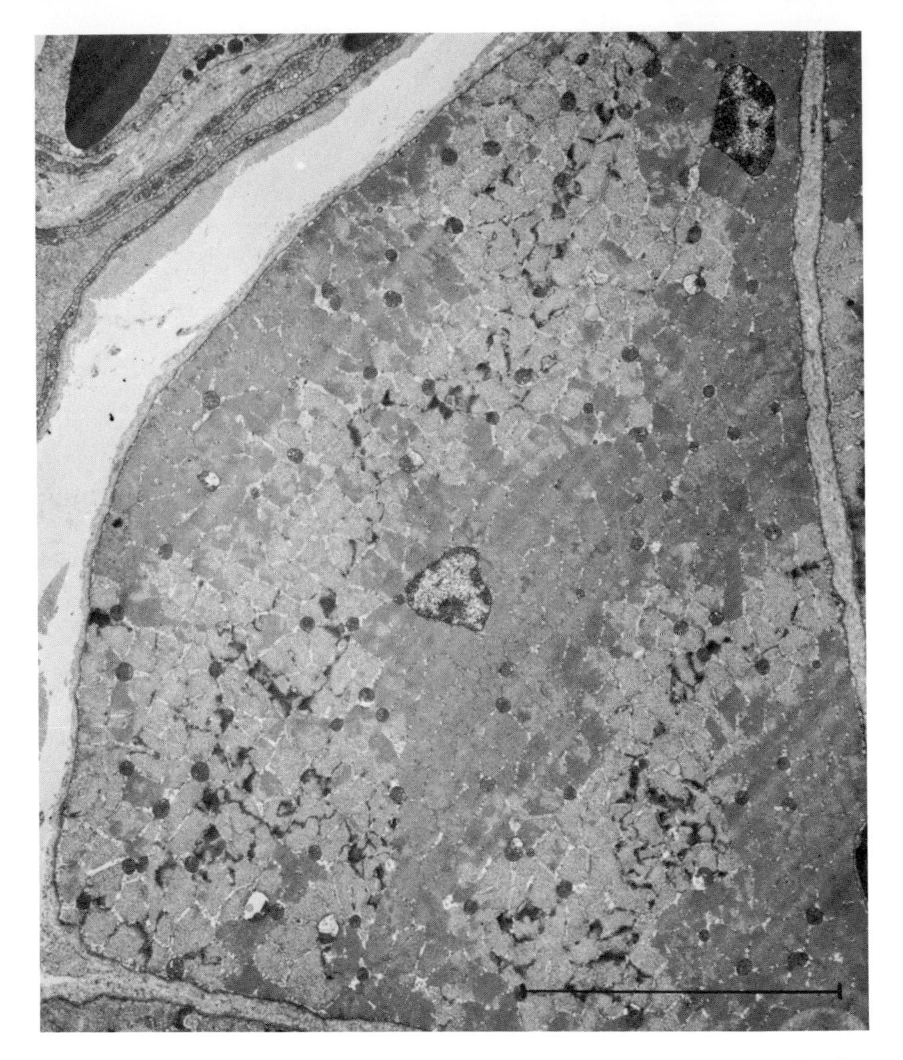

*Fig. 13.6.* Electron micrograph of a transverse section of a sartorius muscle of a frog (*Rana pipiens*). Most of the cross section of one fiber is seen. Line represents 0.01 mm.

capillary is a thin basement membrane, which also is present on the surface of each muscle cell.

At somewhat higher magnification, as in Fig. 13.4, we can see the bundles of collagen fibrils that comprise the connective tissue network in the muscle. We also now begin to resolve the sarcolemma, which I take to include both the basement membrane, mentioned earlier, and the plasma membrane completely covering each muscle cell.

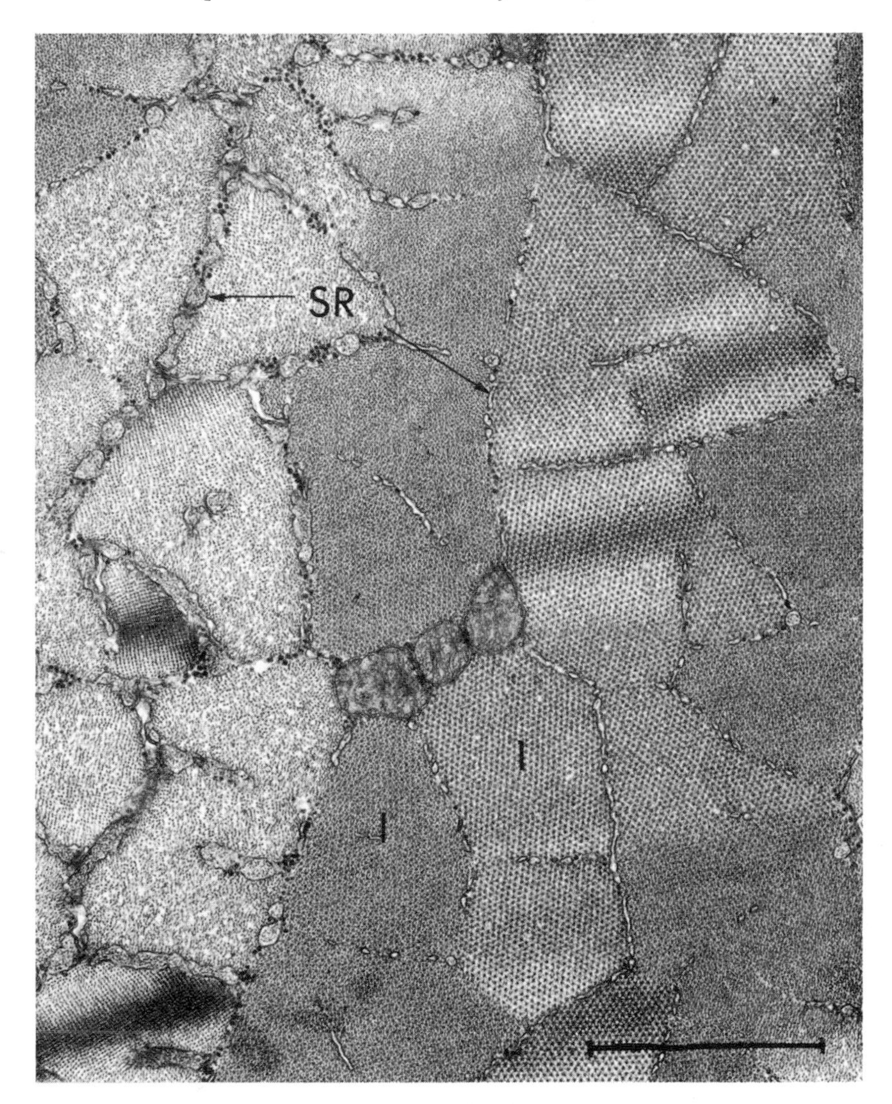


*Fig. 13.7.* Similar to Fig. 13.6 but at higher magnification. Membranes of SR lie in the spaces between the myofibrils (*l*). Line represents 1 $\mu$.

The large dark and light patches seen within the fibers in these micrographs arise from the section passing sometimes through the A (dark) bands and sometimes through the I (light) bands. The dark, round, or oval organelles seen in varying sizes and numbers in different fibers are the mitochondria, which carry out the biochemical reactions of oxidative metabolism. The larger organelle seen in one fiber in Figs. 13.3

and 13.4 is a nucleus, one of many in each muscle fiber. Light, roughly circular areas about the size of mitochondria are regions where lipid droplets were present in the fibers, but these have been extracted during preparation, leaving behind empty spaces. The rest of the area of each fiber profile is occupied by the myofibrils and, between the myofibrils, the membranes of the SR and T system. These form three-dimensional tubular networks in the sarcoplasmic spaces between the fibrils. It is the structure of these networks that will occupy the remainder of this discussion.

While there are certain similarities in the form of the SR and T system in muscles of different species of animals, the differences are considerable and presumably are important in the physiology of the individual muscles. I shall describe only a few examples of muscles from a large number that now have been examined. But, before getting down to specific structural details, I shall consider some general aspects of the distribution and variation of the SR and T system.

One of the parameters one can use to describe the SR and T system in a particular type of striated muscle is the quantity present in a given cell volume. Considerable variation has been found in cells of different types, and in fact this variation has been used to support one hypothesis of the functional significance of these membrane systems (Huxley and Taylor, 1958; Peachey and Porter, 1959; Hodgkin and Horowicz, 1960). Generally, if not universally, one finds that the larger a muscle cell is in its smallest dimension (usually diameter), and the faster its cycle of contraction and relaxation, the more extensively developed are its SR and T system. This has been taken as evidence for a role in the inward spread of contraction activation, thought to involve $Ca^{++}$ release, and a role in the removal of $Ca^{++}$ from the myofibrils to induce relaxation following the peak of contraction. In muscle cells with little SR and T system, this function must be served largely by the surface membrane. It follows from this argument that such a cell, with little or no SR or T system, must be small if it is to contract and relax quickly, because of the time limitation of inward and outward diffusion of $Ca^{++}$. Conversely, if such a cell with little or no SR or T system is large in size, it must be slow in contraction. Various combinations of large and small, and slow and rapid, have been found, and in each case the extent of development of the relevant membrane systems has supported this now widely accepted hypothesis for the function of the SR and T system.

I can illustrate this point with a few micrographs. Fig. 13.5 shows a simple case. The muscle fibers of the primitive chordate, amphioxus (Peachey, 1961; Flood, 1968), are flat plates with a thickness that is very small, less than 1 $\mu$. Thus, each of the vertical bands seen in Fig. 13.5 is

*Fig. 13.8.* Electron micrograph of a longitudinal section of a sartorius muscle fiber of a frog (*Rana pipiens*). In a band through the center of the figure, the SR and T system are cut in "face-view" and show various regions of the SR: fenestrated collar (*F*), longitudinal tubules (*L*), and triads (*T*). Elsewhere, triads are seen between the myofibrils. Dense granules are glycogen. Line represents 1 $\mu$.

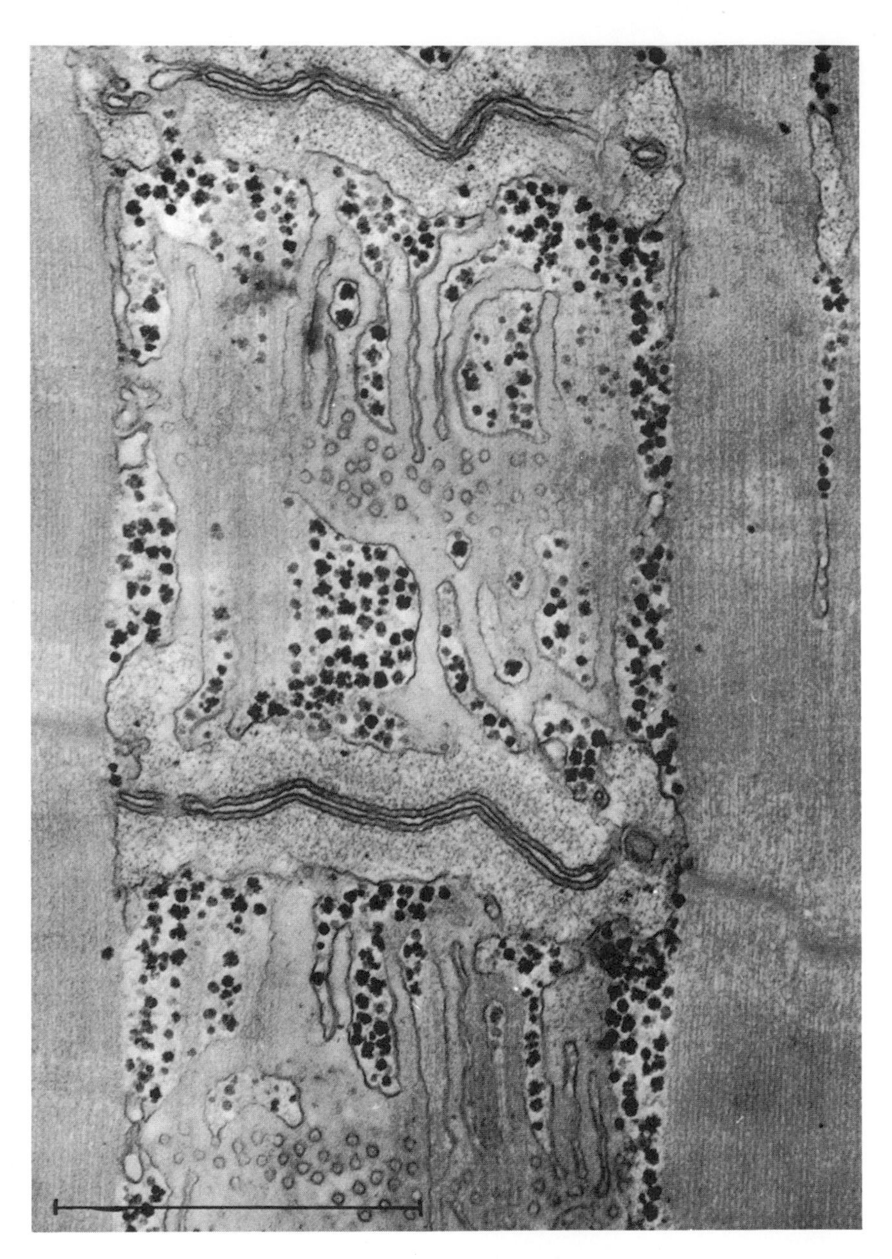

*Fig. 13.9.* Similar to Fig. 13.8, but at higher magnification. Line represents 1 $\mu$.

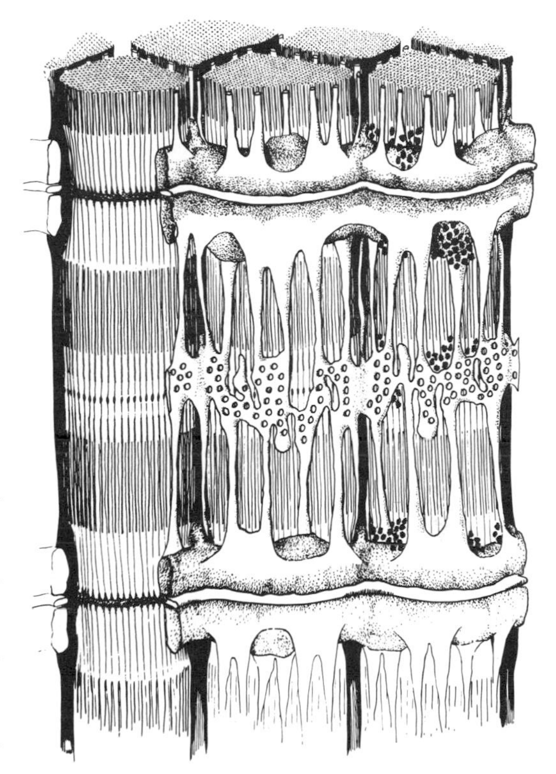

*Fig. 13.10.* Three-dimensional reconstruction of the SR and T system of frog muscle. Vertical columns are myofibrils, formed of bundles of myofilaments. Part of the SR and T system is shown cut away at the left to show triads. Reproduced with permission from the *Journal of Cell Biology* (Peachey, 1965).

part of a separate muscle cell, and the membranes lying between them are the surface membranes of these cells. Surface-to-interior diffusion distances are correspondingly small, and diffusion times for ordinary ions would be measured in fractions of a millisecond. We have no very good measure of the speed of these fibers, but I think they can be judged to be relatively fast since the animal swims quite quickly and effectively. Little SR and no T system is seen in most of these fibers, and according to our hypothesis, none is needed because of the one small dimension of the fibers. Presumably $Ca^{++}$ is released at or near the surface membranes of the lamellar cells and diffuses a very short distance into the contractile fibrils.

On the other hand, we have the much larger skeletal muscle fibers found in higher vertebrates and in many invertebrates. Many of these, such as the frog sartorius fiber shown in Figs. 13.6–9, are also relatively fast in their contraction and relaxation. Fig. 13.6 is at relatively low magnification, and Fig. 13.7 shows only a small portion of one cell, the whole cell being perhaps 100 $\mu$ in diameter. The membranes seen between the fibrils in this case are those of the SR and T system. To be both large and fast, these fibers should require extensive internal membrane systems, and, as seen in Fig. 13.7, electron microscopy bears this out.

Slow striated muscle fibers of the frog, on the other hand, contract much more slowly, although they are also quite large. There is quantitatively much less SR and T system in the slow fibers than in the fast fibers of the same muscles (Peachey and Huxley, 1962; Page, 1965). The very fast muscle fibers of the toad fish swim bladder and of the bat cricothyroid muscle contain very remarkable quantities of SR and T system in keeping with their high speed (Fawcett and Revel, 1961; Revel, 1962).

It seems appropriate at this point to analyze one of these muscle types in more detail to illustrate the form of the SR and T system in a reasonably typical vertebrate muscle fiber of the twitch or fast type. I have chosen for this purpose the sartorius muscle of the frog (Birks, 1965; Page, 1965; Peachey, 1965), shown already in Figs. 13.6–7 in transverse section. In such a section, one can already see evidence that the SR, which lies between the contractile myofibrils, varies in its form at different levels of the sarcomere. In the I-band region (at the left in Fig. 13.7), the SR is in the form of dilated cisternae, filled with a dense, granular material. These were called the "terminal cisternae" by Porter and Palade (1957). Adjacent to the A bands of the myofibrils (to the right in Fig. 13.7), the SR appears, in transverse sections, as narrow flattened cisternae, or as circular profiles which are in fact cross sections of tubules about 300 Å in diameter.

Longitudinal sections, as in Fig. 13.8, shed more light on the form of these membranes at various levels of the sarcomere banding. Across the center of this figure, the SR is cut in "face view" as it would be seen if we could look at the SR lying against one side of a fibril. The SR can be divided into several morphologically distinct regions on the basis of views such as this (see also Fig. 13.9). At the center of the A band, the SR is in the form of a perforated sheet, usually called a fenestrated collar. The holes seen in this region do not penetrate into the interior of the SR, but rather are formed by fusion of the two membranes on either side of the SR. The function, if there is one, of these holes is certainly not known, although they may have the effect of increasing the surface area of the SR in this region.

On either side of the fenestrated collar region of the SR, toward the ends of the A bands, the SR is in the form of longitudinally oriented tubules. These tubules lead, often through a flattened, intermediate cisterna in the I-band region, into the terminal cisternae located near the Z lines. Thus two terminal cisternae are present near each Z line, one from the SR on each side of the Z line. Between these adjacent terminal

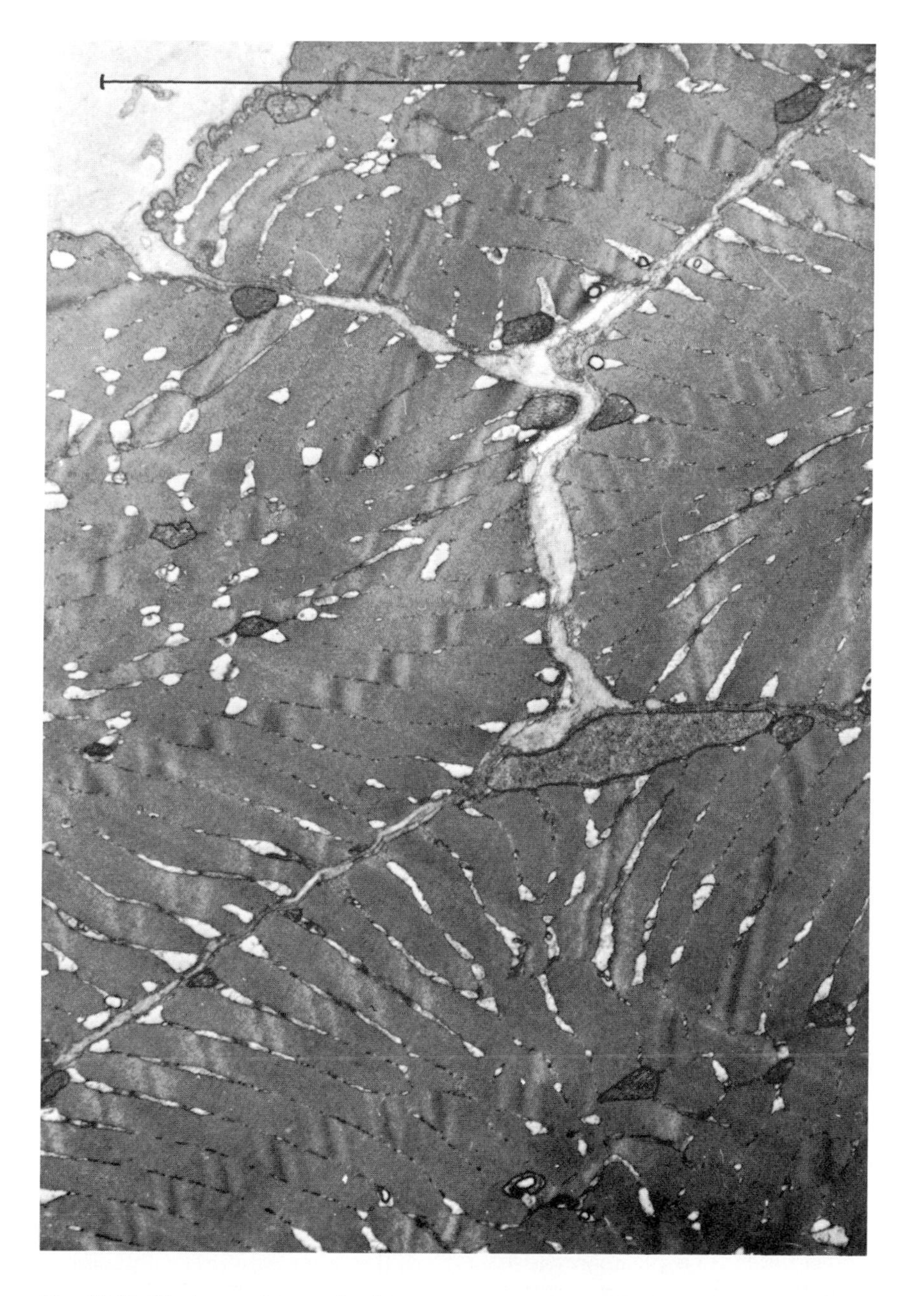

*Fig. 13.11.* Electron micrograph of a transverse section of body white muscle fibers of a fish (*Mollienisia* sp.). Myofibrils under the surface of the fiber are lamellar, while those near the center of the fiber are approximately round in cross section. Membranes of SR separate the myofibrils. Line represents 10 $\mu$.

cisternae is a central tubule, the transverse tubule, and together these three elements form a structure known as a triad. I will have more to say about the central element of the triad, the transverse tubule, later in my presentation. For the present it will suffice to say that the transverse tubules, taken together, comprise the T system.

Perhaps it would be useful at this point to consider the SR and T system surrounding one myofibril for a length of one sarcomere, from Z line to Z line, as a sort of unit cell. The term unit cell means something rather precise to the crystallographer, I realize, whereas here there is some variation in exact structure from one such unit to another. But it still may be a useful concept in understanding the relationship between the SR around one fibril and that of the whole cell. Each sarcomere of each myofibril is surrounded by this collar of membranes, differentiated into terminal and intermediate cisternae, longitudinal tubules, and a fenestrated collar. The next sarcomere along the same fibril has a similar collar, and so on along the whole length of the fibril. Adjacent fibrils also are similarly surrounded along their whole length, so we have in essence a series and parallel arrangement of repeating units of SR. If we know the structure of one of these units, and visualize how they are stacked up and arranged next to each other, then we can in principal visualize the entire SR and T system network in the whole fiber. Fig. 13.10 is a three-dimensional drawing of a few of these units; note that each fibril does not have its own separate collar, but rather shares portions of its collar with adjacent fibrils.

It might be appropriate to consider how many such units are found in a given amount of frog muscle fiber. Taking a typical fiber as 100 $\mu$ in diameter, the average fibril cross-sectional area as 0.5 $\mu^2$, and a sarcomere length of 2 $\mu$, one calculates that there are 75 million of these units in a centimeter length of muscle fiber.

Now, while this roughly cylindrical unit cell idea is useful in gaining an understanding of the structure of the SR and T system in relation to fibrils, and while it is in fact correct for many types of muscles which do have cylindrical fibrils, there are exceptions where this concept breaks down. In many of the muscle fibers of fish, for example, some of the fibrils are not even approximately cylindrical, but are flat or sheetlike and oriented radially in the muscle fiber (Nakajima, 1969), as seen in Fig. 13.11. The SR and T system are thus constrained to a planar arrangement, except near the center of the fiber where the fibrils are smaller and more cylindrical. This is shown at somewhat higher magnification in Fig. 13.12. Many muscles of insects (Smith, 1966) also have fibrils that are grossly noncylindrical, and the SR and T system are found in a roughly planar arrangement.

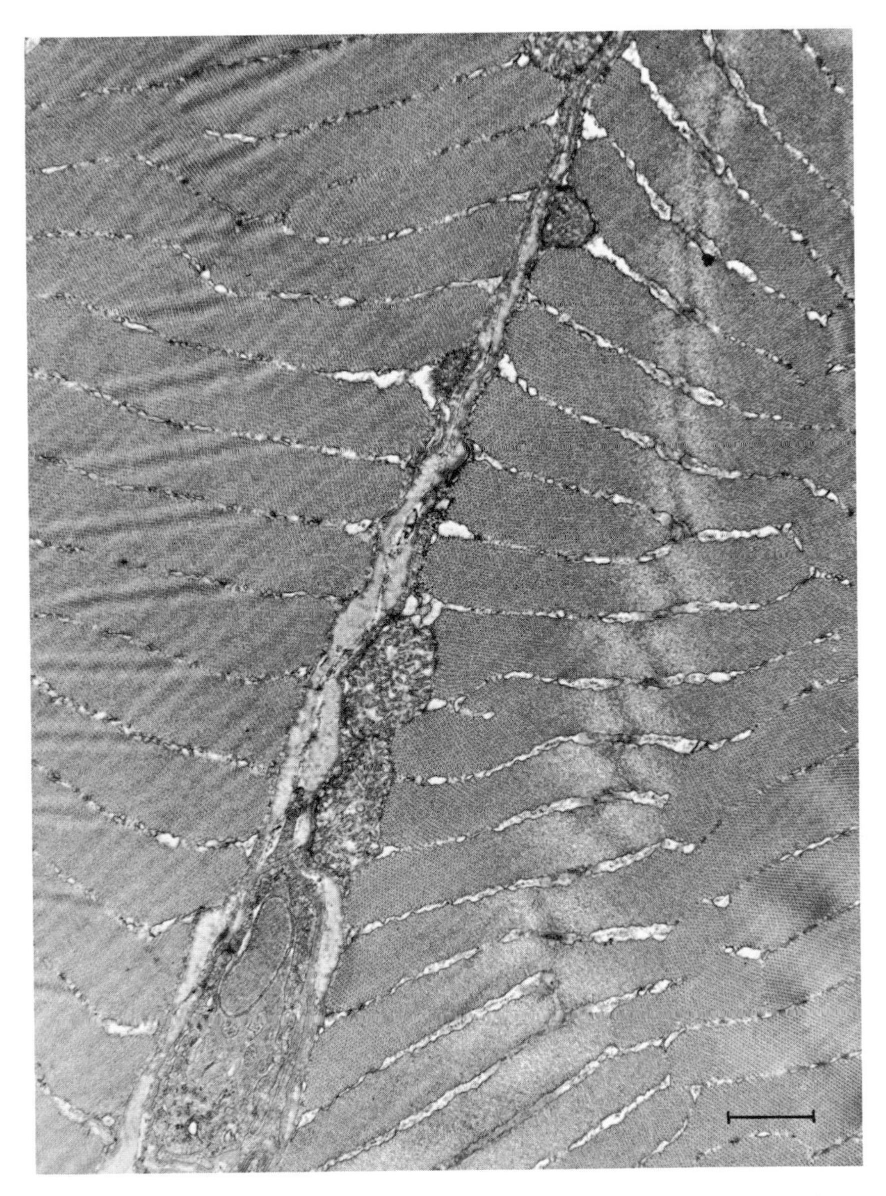

*Fig. 13.12.* Similar to Fig. 13.11, but at higher magnification. Portions of two fibers are shown, separated at the middle of the figure by extracellular space. Line represents 1 $\mu$.

*Fig. 13.13.* Electron micrograph of a longitudinal section of quadriceps femoris muscle of human. The biopsy is from a patient with myositis ossificans, but this region of the muscle has no apparent abnormalities of fine structure. Triads (T) in this and all mammalian skeletal muscles are located midway between the Z lines and the centers of the A bands. Line represents 1 $\mu$.

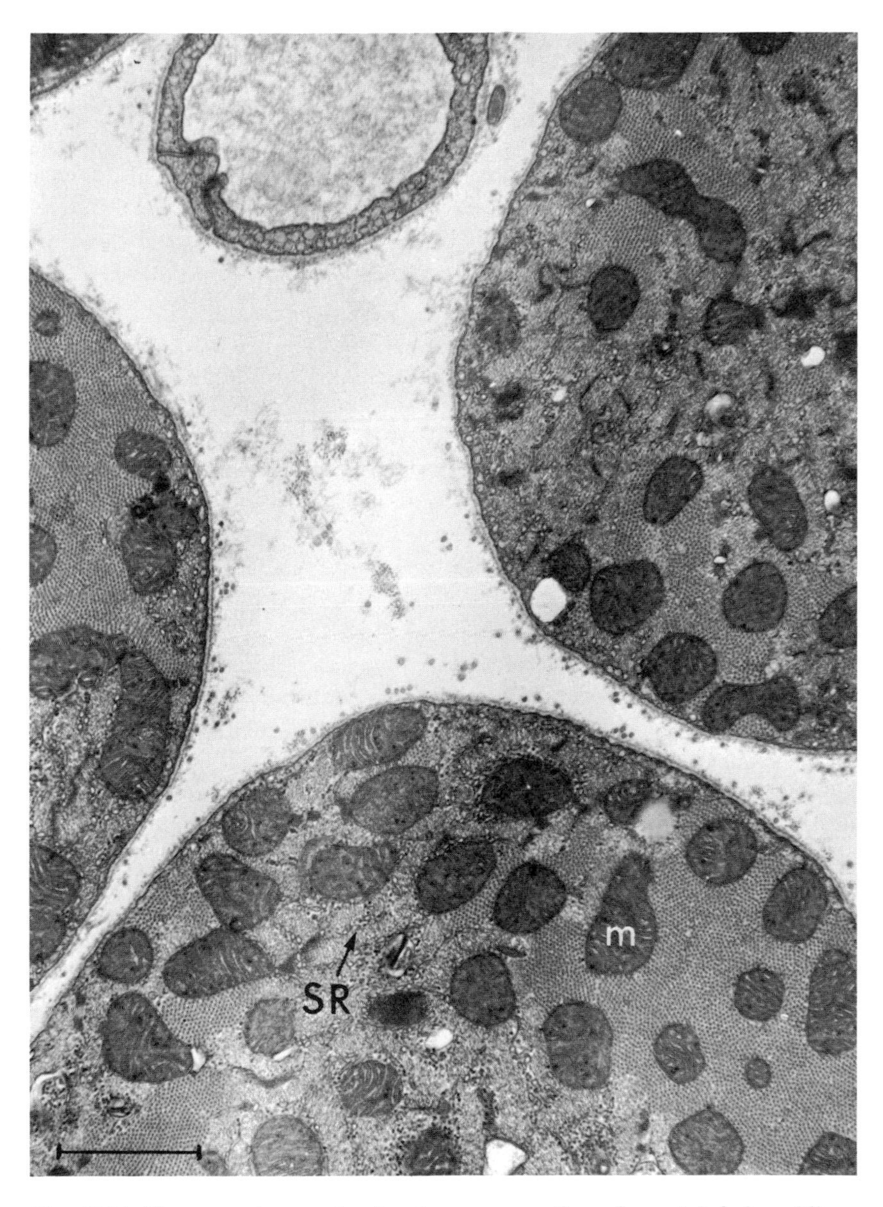


*Fig. 13.14.* Electron micrograph of a transverse section of a cat inferior oblique muscle. This region is from the red, orbital surface of the muscle, and shows small diameter fibers with many mitochondria (*m*) and a moderate amount of SR, appearing as small, circular profiles of tubules, especially in the I bands. Line represents 1 $\mu$.

Another variation in structure of the SR is the appearance of triads at a location other than adjacent to the Z lines of the fibrils. In all mammalian skeletal muscles, in some fish muscles, and in at least the few reptilian muscles that have been studied, each sarcomere contains two triads, and these are found near the ends of the A bands, usually approximately midway between the Z lines and the center of the A band. Fig. 13.13, a longitudinal section of human skeletal muscle, shows this triad position. This arrangement is typical of mammalian skeletal muscle of all types.

In the slow striated fibers of the frog, mentioned earlier as having very little SR or T system, what few triads there are have no particular orientation or distribution with respect to the myofibrillar striations (Page, 1965). Morphologically equivalent fibers have not as yet been found in mammals, however. So far, slowly contracting muscle fibers of mammals have regularly arranged triads near the ends of the A bands, although there may be very few of these in the very slow fibers.

It would seem appropriate at this point to consider the differences in structure between so-called fast and slow muscles of mammals, which might have some bearing on the theme of this conference. It is common to classify "white" muscles as "fast" and "red" muscles as "slow," a classification that has some validity and probably is very useful in meat products research. There are exceptions to this relationship, however, and I object to the use of the simple designation "red" and "white" for expressing physiological classification with respect to speed of contraction (Peachey, 1968). It seems clear that redness of color reflects a high content of myoglobin and cytochromes in the muscle, which in turn measures the capability of the muscle to do oxidative metabolism. It is true that a muscle that does its work using oxidative metabolism to a great extent (and therefore is red) often has a rather tonic function, and this also usually means that the muscle is a relatively slow one. Conversely, phasic muscles usually live on glycolysis (have few mitochondria and therefore are white), and operate in short bursts of rapid contraction. But there is no necessary relation between these three distinct parameters of muscle composition and function, i.e., color (red vs. white), mode of use (tonic vs. phasic), and speed of contraction (slow vs. fast). For the near future at least, it would seem wise not to classify a muscle by a single parameter, e.g., color, and thereby imply a number of other properties of the muscle, e.g., its speed, but rather to use as many different parameters as possible to describe the muscle. It may be that all muscles eventually will fall into a small number of classes, but in view of many recent results, and even the possibility experimentally of changing one or more properties of a given muscle (Buller, Eccles, and Eccles,

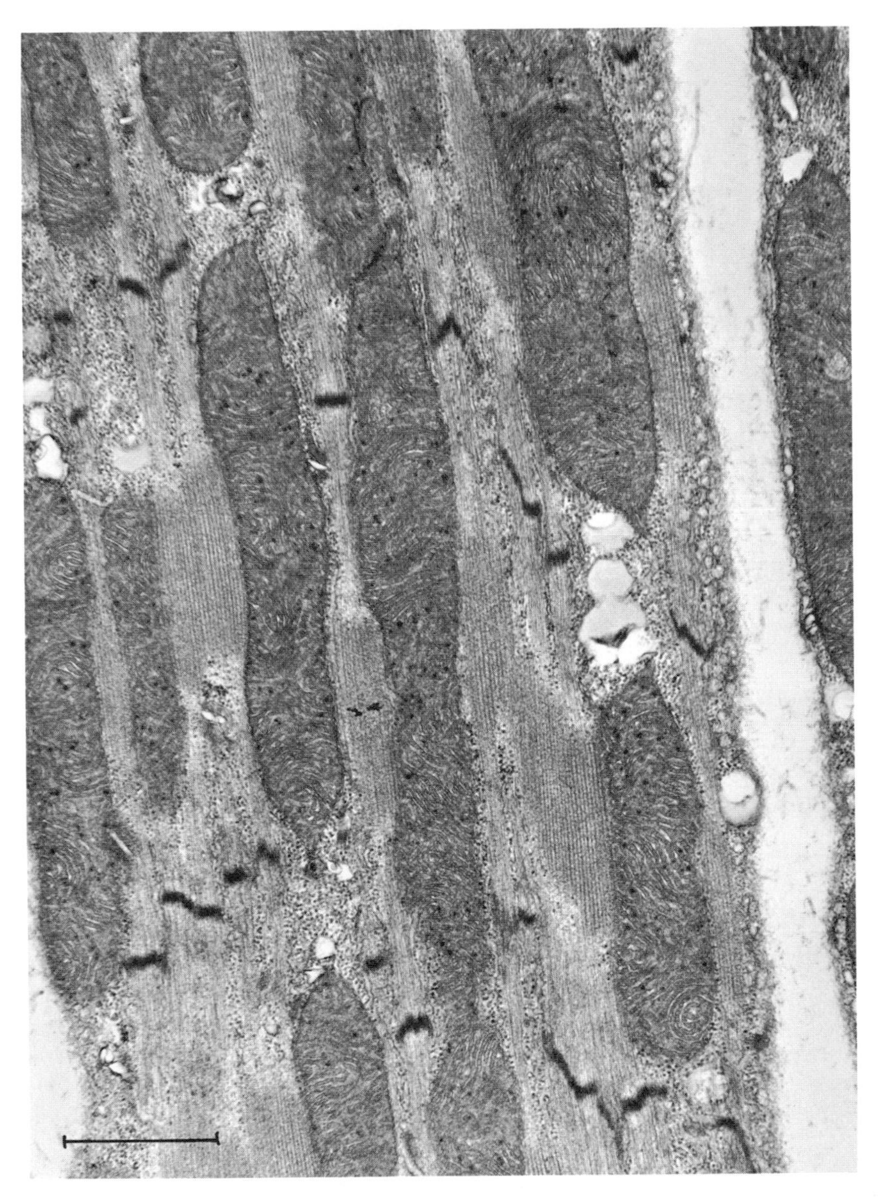

*Fig. 13.15.* Same as Fig. 13.14, except longitudinal section. Line represents 1 μ.

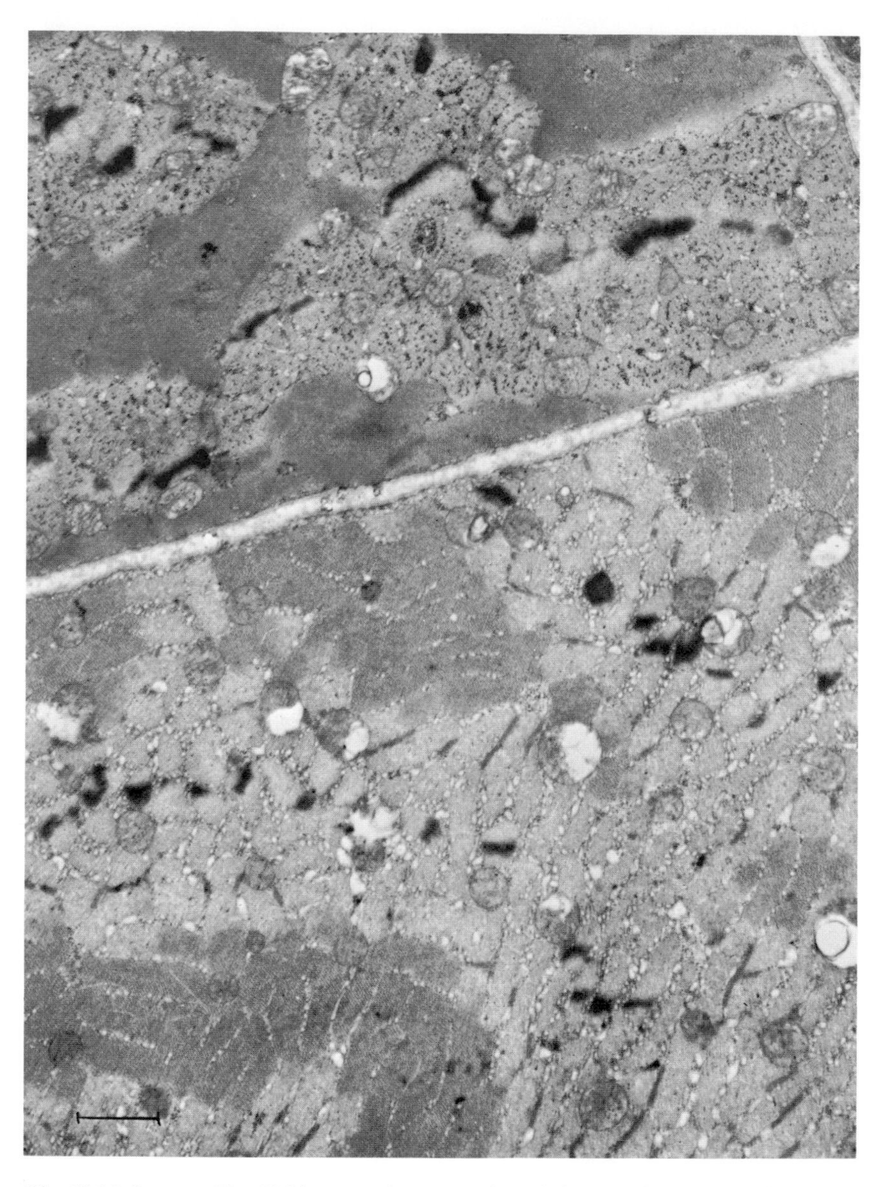

*Fig. 13.16.* Same as Fig. 13.14 except from a region of the muscle closer to the globe. Two fiber types are seen here. The lower fiber has much SR, and small, flat fibrils. The upper fiber has much less SR and larger fibrils with indistinct outlines. Line represents 1 $\mu$.

*Fig. 13.17.* Longitudinal section of the fiber type seen in the lower half of Fig. 13.16. These fibers contain much SR in the I-band regions. Line represents 1 $\mu$.

1960; Dubowitz, 1967; Hník, Jirmanovà, Vyklický, and Zelená, 1967; Close and Hoh, 1968), I doubt that a simple system of classification will arise.

One can illustrate this point with micrographs of mammalian red and white muscle fibers. The fibers shown in Figs. 13.14–15 are from the orbital surface of a cat extraocular muscle. While there is some doubt about the membrane properties of these fibers (Pilar, 1967), it seems likely that they are relatively slow in their contraction time (Hess and Pilar, 1963; Bach-y-Rita and Ito, 1966). This surface of the muscle is red in color, and as seen in the figures these fibers have a high mitochondrial content. We are unable to recognize myoglobin in the electron microscope, but such masses of mitochondria would alone be expected to give the fibers a red color.

In contrast, deeper in the muscle toward the globe where the muscle is white, we have a population of fibers with fewer and smaller mitochondria (Figs. 13.16–17), and physiologically these fibers are among the fastest known. Morphologically they have a highly developed SR and T system. These fibers would certainly be classified as white and fast. In the same region of these muscles, however, we have another population of fibers with equally few mitochondria (Figs. 13.16, 13.18). These fibers differ from the previous ones most strikingly by having a very small complement of SR and T system. On the basis of fine structure, therefore, we expect these to be slowly contracting fibers, although direct physiological information is lacking. If this identification is correct, however, then these fibers are both white in color and slow in contraction, an example of a clear violation of the dogma that white is always fast, or that slow is always red.

So far I have mentioned the T system simply as forming the central elements of the triads. Since recent evidence very strongly suggests the T system as the pathway for the inward spread of depolarization from the cell surface, the detailed morphology of the T system has taken on great physiological importance. It is now firmly established that the T system is directly continuous with the plasma membrane at the outer surface of the fiber. This was not at all clear in the early days of electron microscopy of this system (Porter and Palade, 1957), although physiologists seem not to have doubted that the morphologists eventually would establish this continuity (Hodgkin and Horowicz, 1960). The first demonstration of the continuity in vertebrate skeletal muscle was made by Franzini-Armstrong and Porter (1964) in a fish muscle. Their micrographs clearly showed that the T system is a network of invaginations of the plasma membrane extending into the fiber and forming the central elements of

*Fig. 13.18.* Longitudinal section of the fiber type seen in the upper half of Fig. 13.17. These fibers contain less SR, but show triads (*T*) clearly. Line represents 1 μ.

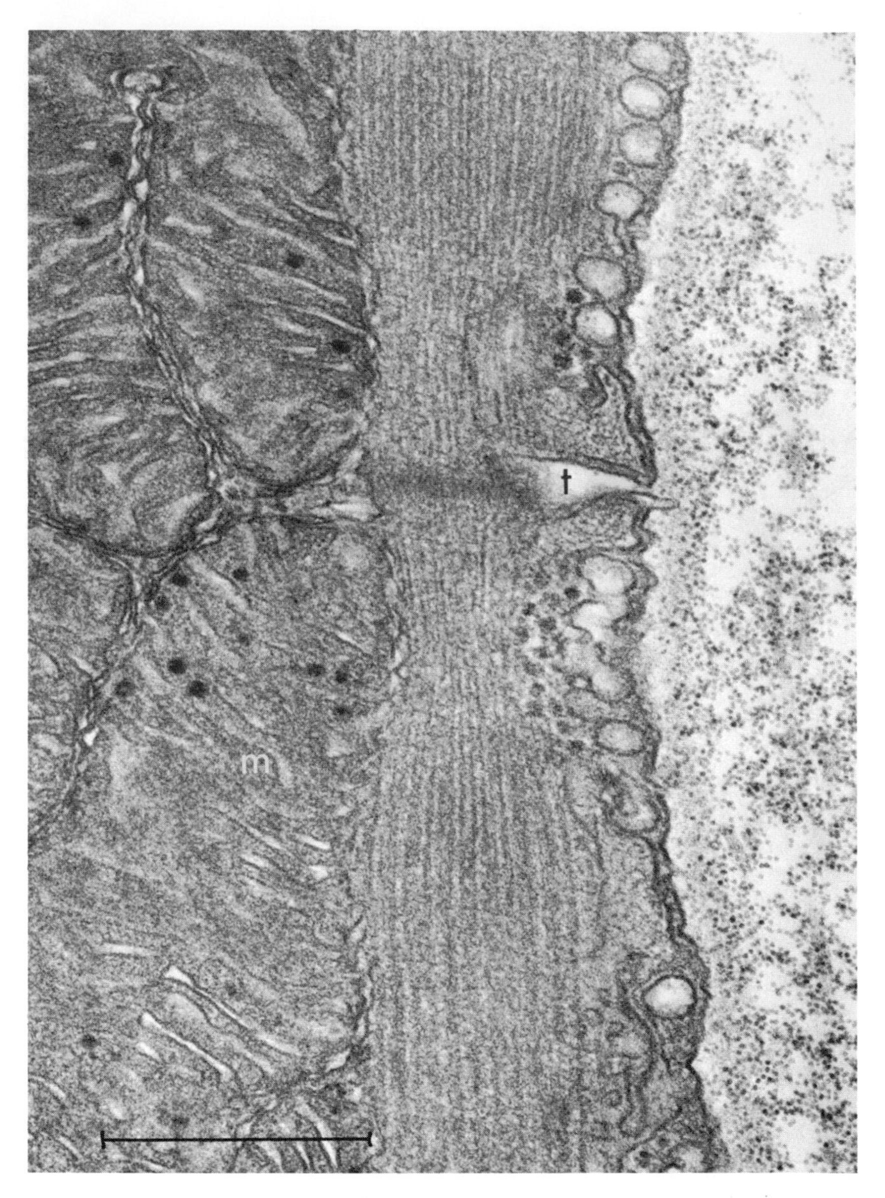


*Fig. 13.19.* Electron micrograph of the surface region of a frog (*Rana temporaria*) muscle fiber, cut longitudinally. A triad just under the sarcolemma has as its central element a transverse tubule (*t*) apparently open to the exterior. Mitochondria (*m*) are seen. Line represents 0.5 $\mu$.

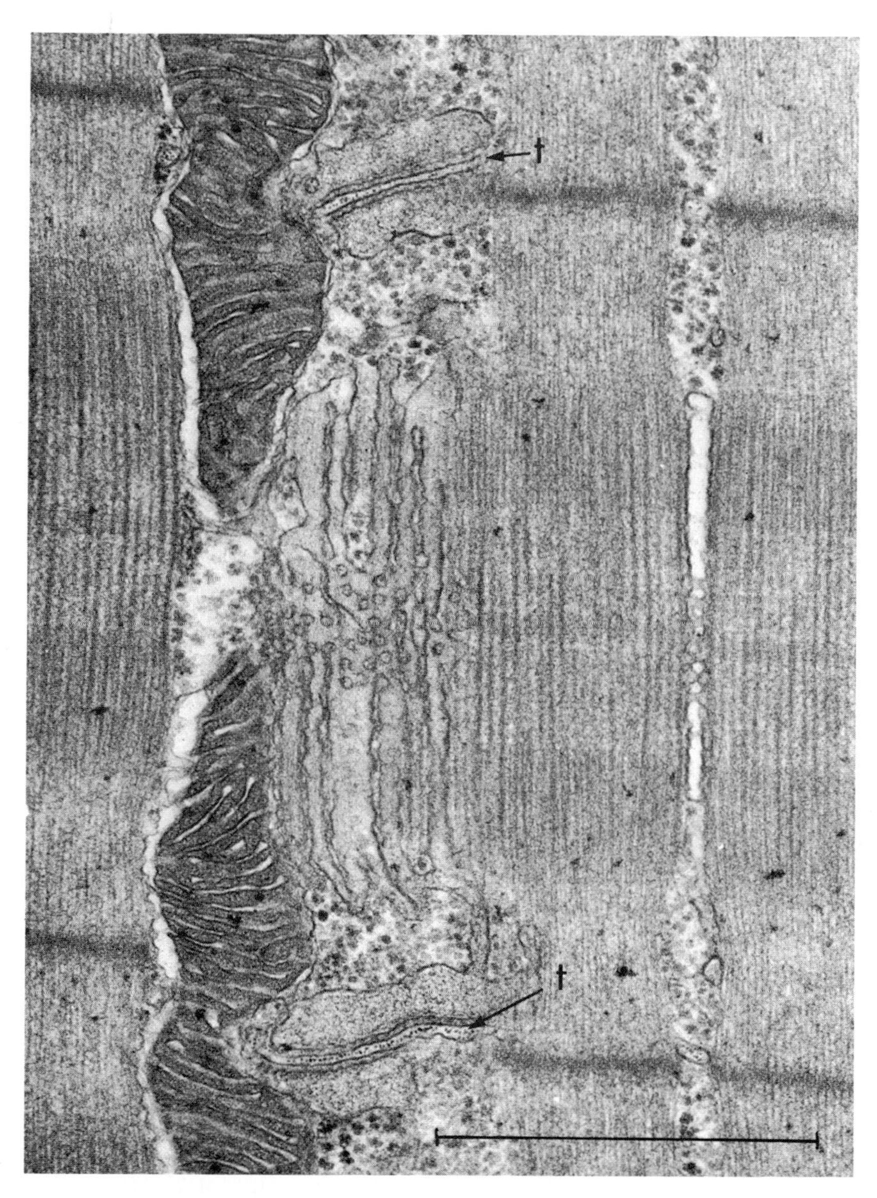


*Fig. 13.20.* Electron micrograph of a longitudinal section of a frog semitendinosus muscle (*Rana temporaria*) soaked in ferritin prior to fixation. Ferritin seen inside the T system (*t*) is evidence that the T system is open to the extracellular space. Line represents 1 $\mu$.

the triads. The equivalent picture in frog or any mammalian muscle has been frustratingly elusive. The most convincing picture I have ever taken is shown in Fig. 13.19, and this is much less satisfying than the elegant pictures obtainable from fish muscle. In 1964, however, both H. E. Huxley and S. Page published reports of experiments in which a frog muscle was soaked in ferritin prior to fixing it for electron microscopy. Ferritin is an iron-containing protein, unable to cross intact membranes, and visible in the electron microscope. Therefore its presence in the T system throughout a muscle fiber can be taken as showing that the lumen of the T system is directly continuous with the external fluid surrounding the fiber, and this is convincing evidence for the continuity of the membranes forming the T system with the fiber surface membrane. Fig. 13.20 shows the result of a similar experiment. Subsequently, the entry of another protein, peroxidase (Eisenberg and Eisenberg, 1968), into the T system, and of a fluorescent dye into the proper region of the sarcomere (Endo, 1966) also supported this idea. The entry of ferritin into frog and toad muscles has been quantitated along the sarcomere (Peachey and Schild, 1968), and shown to be concentrated in the Z-line region.

It is interesting that the T system is not only attached to the fiber surface in the mature muscle fiber; it is derived from the cell surface during cell differentiation. This has been demonstrated recently in cultured chick myoblasts by Ezerman and Ishikawa (1967) (see Fig. 13.21). Muscle cultures were exposed to ferritin prior to fixation, and one sees a system of dilated infoldings of the plasma membrane forming a primitive T system, filled with ferritin (Fig. 13.21A). In Fig. 13.21B, one sees evidence that the early SR network arises in connection with rough-surfaced endoplasmic reticulum in these cells. Later, these two separately derived membrane systems join to form triads characteristic of chicken muscles in the adult state. The important point is the intracellular origin and position of the SR in contrast to the essentially extracellular or surface origin and attachment of the T system.

A very productive approach to the function of the T system in the muscle fiber has been the study of fibers whose T system has been disrupted by a particular method of treating the muscle with solutions made hypertonic by the addition of glycerol (Howell and Jenden, 1967). Such fibers conduct action potentials, but do not contract, and this was presumed to be the result either of a disconnection of the T system from the fiber surface or a more drastic disruption of the T system. The experiments of Eisenberg and Eisenberg (1968) show that peroxidase, an extracellular marker mentioned earlier, can no longer enter the T system

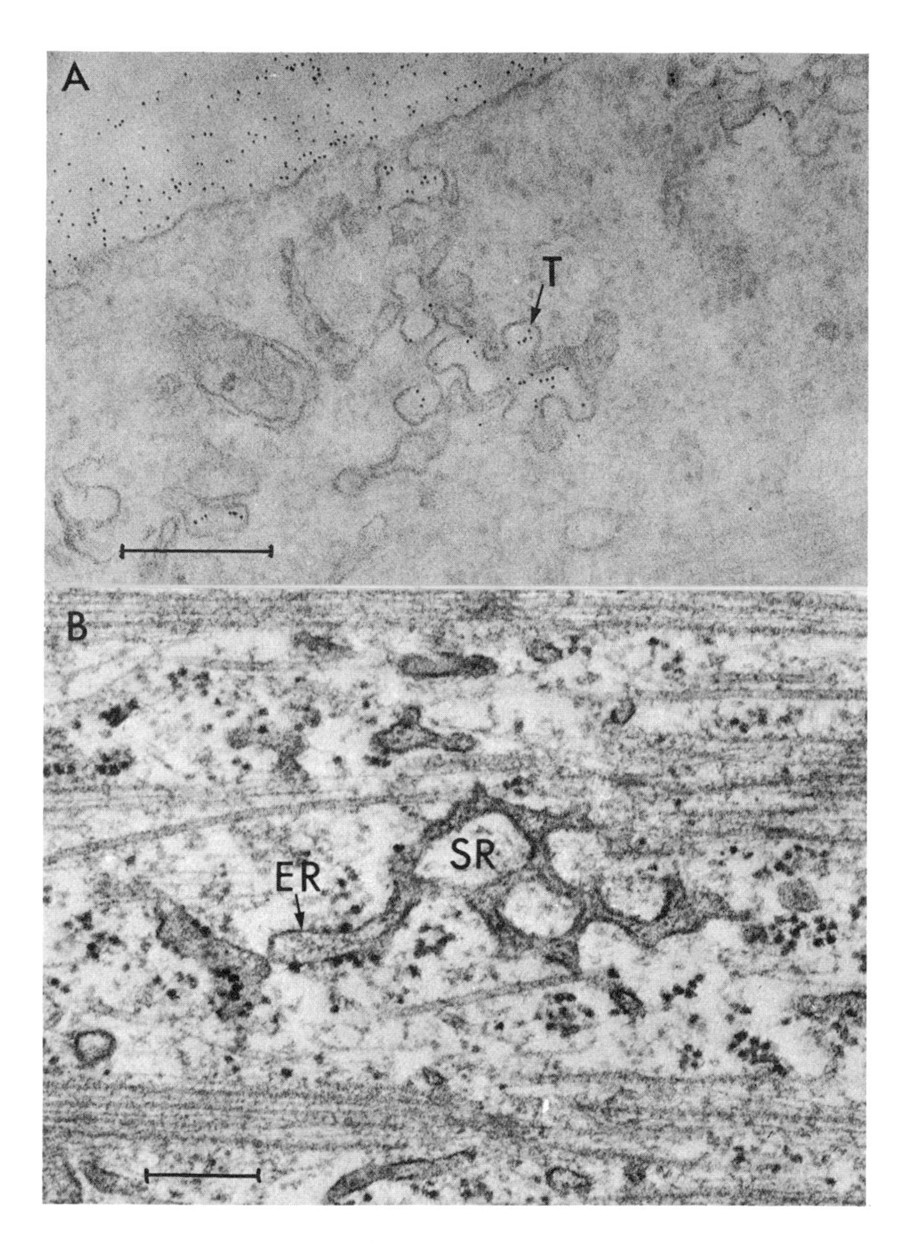


*Fig. 13.21.* Electron micrographs of sections of chick myoblasts soaked in ferritin after 6 days in tissue culture. *A,* developing T system, invaginating from the plasma membrane of the cell and containing ferritin; *B,* early SR network attached to rough-surfaced endoplasmic reticulum (*ER*). Lines represent 1 $\mu$. Micrographs supplied by Dr. H. Ishikawa.

*Fig. 13.22.* Electron micrograph of a longitudinal section of a frog muscle (*Rana temporaria*) treated with glycerol in such a way as to functionally disrupt the T system. Ferritin, seen outside the fiber (on the right) can no longer enter the T system. Line represents 1 $\mu$.

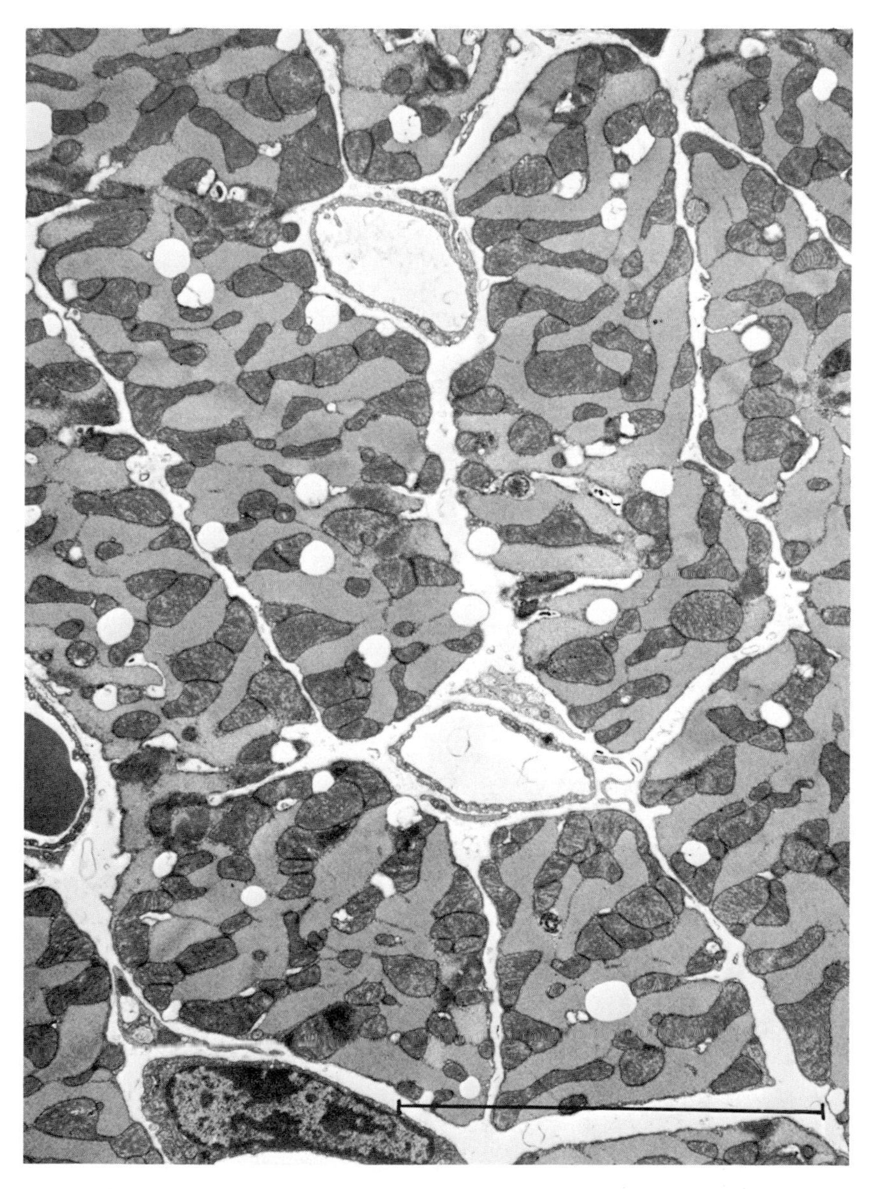

*Fig. 13.23.* Electron micrograph of a transverse section of Syrian golden hamster cardiac muscle (ventricle). The fibers are relatively small, and contain many mitochondria. Line represents 10 $\mu$. Micrograph supplied by Dr. R. Waugh.

after this treatment, and Nakajima, Nakajima, and Peachey (1968) obtained a similar result using ferritin (Fig. 13.22).

These fibers, with disrupted T system, show a lack of a number of electrical properties that have been associated with the T system. For example, the slow change in potential during a long hyperpolarizing pulse seen by Adrian and Freygang (1962) is absent from glycerol-treated fibers (Eisenberg and Gage, 1967), supporting the idea that this slow potential change arises in the T system. Also, the slow potential change seen following a rapid decrease in external $K^+$ concentration (Hodgkin and Horowicz, 1960) is not seen after glycerol treatment (Nakajima, Nakajima, and Peachey; 1968), presumably for the same reason. Perhaps most important, the high capacitance of frog muscle fibers, thought to arise partly in the T system (Falk and Fatt, 1964) is greatly reduced after glycerol treatment (Gage and Eisenberg, 1969). These results, I think, greatly strengthen our feeling, based partly on electron micrographs, that the T system of normal fibers is electrically coupled to the surface membrane of the fiber and plays some part in the electrical events occurring at the time of the action potential.

Mammalian cardiac muscle has a rather simpler arrangement of T system and SR than found in most of the muscles previously illustrated. Figs. 13.23–25 show sections of golden hamster ventricle. Fig. 13.23, at low magnification, demonstrates the small size and high mitochondrial content of these cells. As in some of the earlier figures, circular blank spaces are left by extraction of lipid droplets. The T system in these fibers is larger in caliber than we have seen in skeletal muscle, and carries with it a fairly extensive basement membrane layer. This is shown in longitudinal section in Figs. 13.24–25. The rather meager SR found in these and in most cardiac muscle cells forms dyad or triad contacts along the flat surfaces of these T-system-like surface invaginations (Fig. 13.25). The cisternae of the SR are so narrow that their dense content is constrained into a single layer of granules.

Finally, here are some data on the quantity of SR and T system present in frog skeletal muscle. Table 13.1 shows the surface area and volume of

TABLE 13.1
*Volume and surface areas of SR and T system*

| | *Fractional fiber volume* | *Surface area per cm² outer surface area of fiber 100 μ in diameter* |
|---|---|---|
| SR | 0.003 | 140 cm² |
| T system | 0.13 | 7 cm² |

From Peachey (1965).

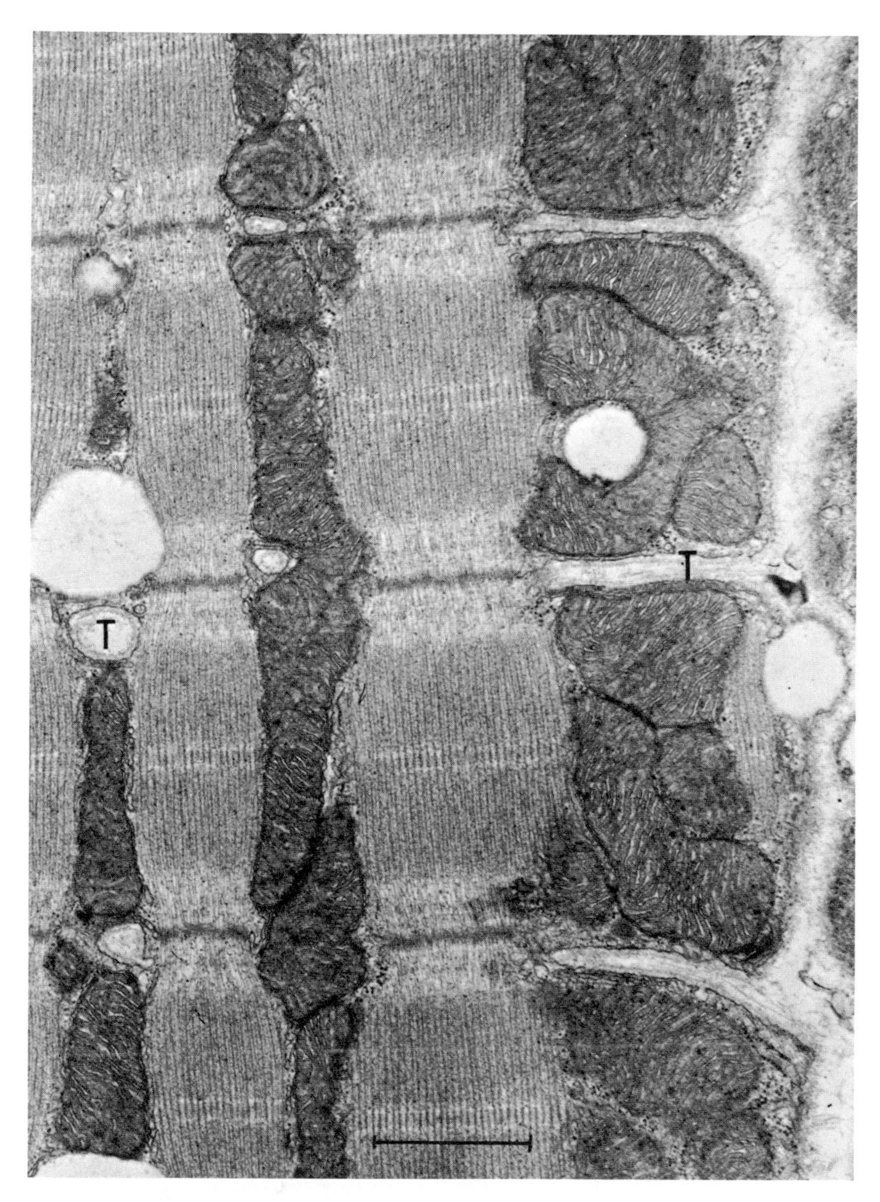


*Fig. 13.24.* Same as Fig. 13.23, but longitudinal. The T system of these fibers is larger in caliber than that of skeletal muscle, and contains basement membrane material. Line represents 1 $\mu$. Micrograph supplied by Dr. R. Waugh.

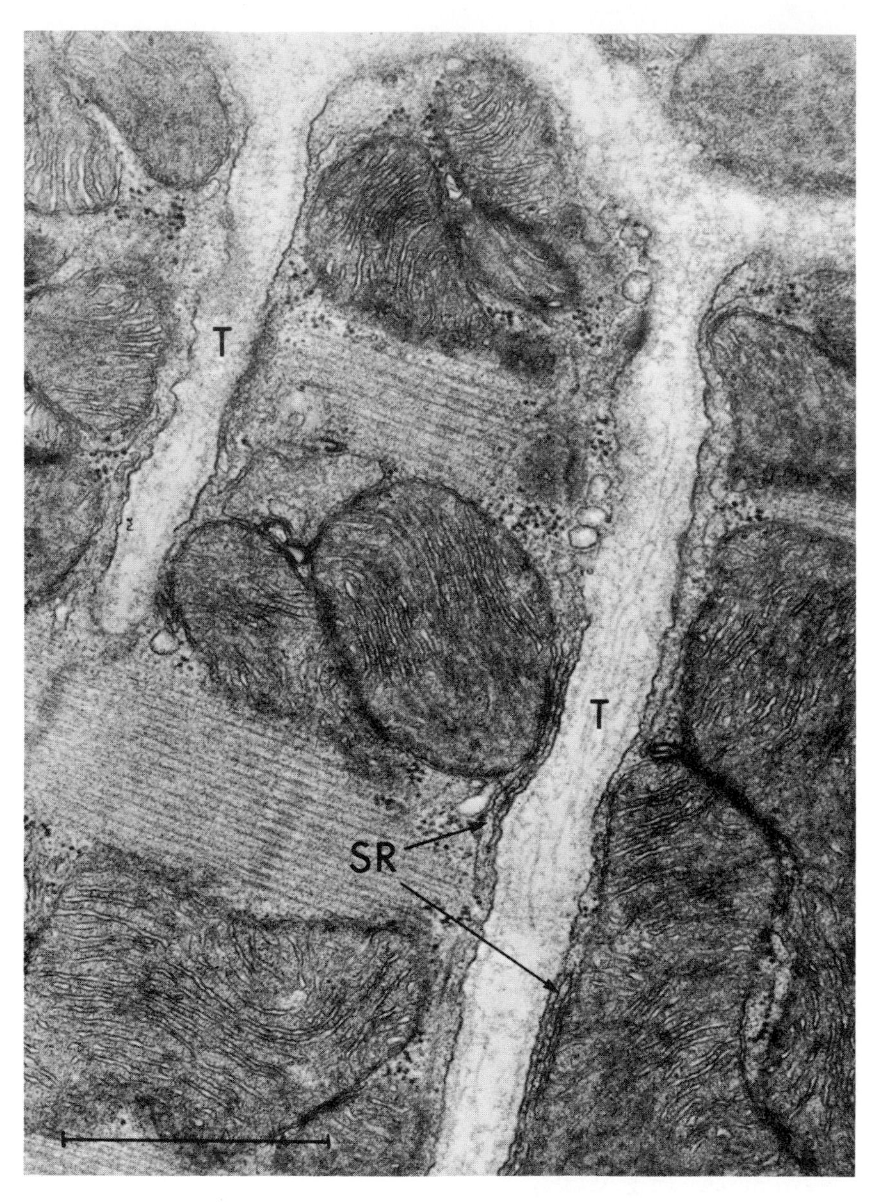


***Fig. 13.25.*** Same as Fig. 13.24 but at higher magnification, showing SR lying against the walls of the T system. Line represents 1 μ. Micrograph supplied by Dr. R. Waugh.

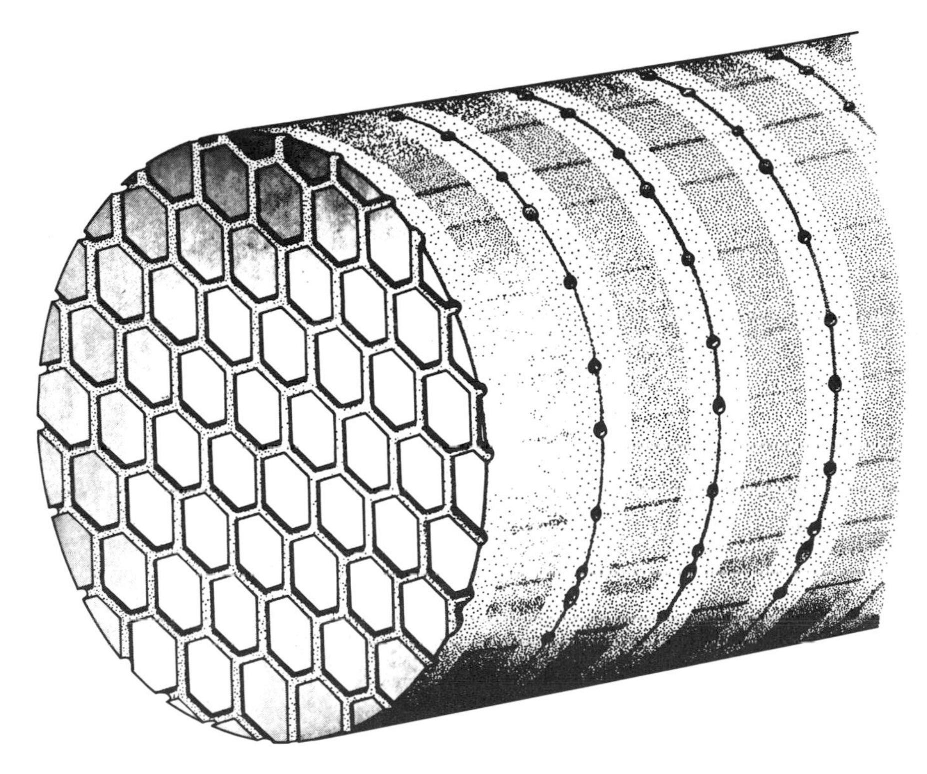

*Fig. 13.26.* Schematic representation showing a transverse cut through the T system in one Z line plane of a muscle fiber such as those of frogs. Many invaginating tubules join into a network across the whole diameter of the fiber. Openings of other T system networks in the fiber are seen along the fiber surface.

both systems in frog muscles as estimated by electron microscopy. While the volume of the T system is very small, its surface area is very considerable in relation to the surface area of the fiber, and this has been correlated in a few different types of fibers with the electrical capacity measured between an intracellular electrode and the exterior of the cell. The SR, on the other hand, represents a considerable portion of the cell volume, about 13% in this case. The surface area of the SR is very considerable, but an electrical capacity commensurate with this surface area has never been detected.

In summary, the SR varies in quantity and form in muscle cells of different types and different species. It is basically an intracellular membrane system, and can be highly developed and arranged in close association with the contractile myofibrils. It is thought to be the storage site for $Ca^{++}$ in a resting fiber, and to be the intracellular organelle responsible for its release during activation of contraction as well as for its uptake during relaxation. The T system, in contrast, is part of the surface or plasma membrane of the muscle cell, and extends into the interior of the fiber as a branching system of tubules, forming close contacts with the SR. The final figure (Fig. 13.26) shows a somewhat stylized visualization of the T system of a frog twitch muscle fiber, and is discussed by R. H. Adrian in Chapter 14.

## *References*


Adrian, R. H., and W. H. Freygang. 1962. The potassium and chloride conductance of frog muscle membrane. *J. Physiol. 163*:61.

Bach-y-Rita, P., and F. Ito. 1966. *In vivo* studies on fast and slow muscle fibers in cat extraocular muscles. *J. Gen. Physiol. 49*:1177.

Birks, R. I. 1965. The sarcoplasmic reticulum of twitch fibres in frog sartorius, p. 199. *In* W. M. Paul, E. E. Daniel, C. M. Kay, and G. Monckton (eds.), *Muscle*. Pergamon, Oxford.

Buller, A. J., J. C. Eccles, and R. M. Eccles. 1960. Interaction between motoneurones and muscles in respect of the characteristic speeds of their responses. *J. Physiol. 150*:417.

Close, R., and J. F. Y. Hoh. 1968. Effects of nerve cross-union on fast-twitch and slow-graded muscle fibres in the toad. *J. Physiol. 198*:103.

Dubowitz, V. 1967. Cross-innervated mammalian skeletal muscle: histochemical, physiological and biochemical observations. *J. Physiol. 193*:481.

Eisenberg, B., and R. S. Eisenberg. 1968. Selective disruption of the sarcotubular system in frog sartorius muscle. A quantitative study with exogenous peroxidase as a marker. *J. Cell Biol. 39*:451.

Eisenberg, R. S., and P. W. Gage. 1967. Frog skeletal muscle fibers: changes in electrical properties after disruption of transverse tubular system. *Science 158*:1700.

Endo, M. 1966. Entry of fluorescent dyes into the sarcotubular system of frog muscle. *J. Physiol. 185*:224.

Ezerman, E. B., and H. Ishikawa. 1967. Differentiation of the sarcoplasmic reticulum and T system in developing chick skeletal muscle *in vitro*. *J. Cell Biol. 35*:405.

Falk, G., and P. Fatt. 1964. Linear electrical properties of striated muscle fibres observed with intracellular electrodes. *Proc. Roy. Soc. B. 160*:69.

Fawcett, D. W., and J. P. Revel. 1961. The sarcoplasmic reticulum of a fast-acting fish muscle. *J. Biophys. Biochem. Cytol. 10,* No. 4 (suppl.) : 89.

Flood, P. R. 1968. Structure of the segmental trunk muscle in amphioxus. *Z. Zellforsch. 84*:389.

Franzini-Armstrong, C., and K. R. Porter. 1964. Sarcolemmal invaginations constituting the T system in fish muscle fibers. *J. Cell Biol. 22*:675.

Gage, P. W., and R. S. Eisenberg. 1969. Capacitance of the surface and transverse tubular membranes of frog sartorius muscle fibers. *J. Gen. Physiol. 53*:265.

Hess, A., and G. Pilar. 1963. Slow fibres in the extraocular muscles of the cat. *J. Physiol. 169*:780.

Hník, P., I. Jirmanovà, L. Vyklický, and J. Zelená. 1967. Fast and slow muscles of the chick after nerve cross-union. *J. Physiol. 193*:309.

Hodgkin, A. L., and P. Horowicz. 1960. The effect of sudden changes in ionic concentrations on the membrane potential of single muscle fibres. *J. Physiol. 153*:370.

Howell, J. N., and D. J. Jenden. 1967. T-tubules of skeletal muscle: morphological alterations which interrupt excitation-contraction coupling. *Fed. Proc. 26*:553.

Hoyle, G. 1969. Comparative aspects of muscle. *Ann. Rev. Physiol. 17*:43.

Huxley, A. F., and R. E. Taylor. 1958. Local activation of striated muscle fibres. *J. Physiol. 141*:426.

Huxley, H. E. 1964. Evidence for continuity between the central elements of triads and extracellular spaces in frog sartorius muscle. *Nature 202*:1067.

Nakajima, S., Y. Nakajima, and L. D. Peachey. 1968. Speed of repolarization and morphology of glycerol-treated muscle fibres. *J. Physiol. 200*:115P.

Nakajima, Y. 1969. Fine structure of red and white muscle fibers and their neuromuscular junctions in the snake fish (*Ophiocephalus argus*). *Tissue and Cell 1*:299.

Page, S. 1964. The organization of the sarcoplasmic reticulum in frog muscle. *J. Physiol. 175*:10P.

———. 1965. A comparison of the fine structure of frog slow and twitch muscle fibers. *J. Cell Biol. 26*:477.

Peachey, L. D. 1961. Structure of the longitudinal body muscles of amphioxus. *J. Biophys. Biochem. Cytol. 10,* No. 4 (suppl.) : 159.

———. 1965. The sarcoplasmic reticulum and transverse tubules of the frog's sartorius. *J. Cell Biol. 25*:209.

———. 1968. Muscle. *Ann. Rev. Physiol. 30*:401.

Peachey, L. D., and A. F. Huxley. 1962. Structural identification of twitch and slow striated muscle fibers of the frog. *J. Cell Biol. 13*:177.

Peachey, L. D., and K. R. Porter. 1959. Intracellular impulse conduction in muscle cells. *Science 129*:721.

Peachey, L. D., and R. F. Schild. 1968. The distribution of the T-system along sarcomeres of frog and toad sartorius muscles. *J. Physiol. 194*:249.

Pilar, G. 1967. Further study of the electrical and mechanical responses of slow fibers in cat extraocular muscles. *J. Gen. Physiol. 50*:2289.

Porter, K. R., and G. E. Palade. 1957. Studies on the endoplasmic reticulum III. Its form and distribution in striated muscle cells. *J. Biophys. Biochem. Cytol. 3*:299.

Revel, J. P. 1962. The sarcoplasmic reticulum of the bat cricothyroid muscle. *J. Cell Biol. 12*:571.

Sandow, A. 1965. Excitation-contraction coupling in skeletal muscle. *Pharmacol. Rev. 17*:265.

Slautterback, D. B. 1966. The ultrastructure of cardiac and skeletal muscle, p. 39. *In* E. J. Briskey, R. G. Cassens, and J. C. Trautman (eds.), *The Physiology and Biochemistry of Muscle as a Food.* Univ. of Wis.

Smith, D. S. 1966. The organization and function of the sarcoplasmic reticulum and T-system of muscle cells. *Progr. Biophys. 16*:107.

Stewart, G. F. 1966. "Postmortem" perspectives, p. 3. *In* E. J. Briskey, R. G. Cassens, and J. C. Trautman (eds.), *The Physiology and Biochemistry of Muscle as a Food.* Univ. of Wis.

# Chapter 14

# *The Electrophysiology of Contraction Activation*

R. H. ADRIAN

The release and subsequent reabsorption of $Ca^{++}$ by the sarcoplasmic reticulum are now so well established that we are apt to overlook some of the gaps in our knowledge of how the regulation of this cycle of events takes place in the intact muscle fiber. Ebashi and Endo (1969) have presented a most clear and complete account of our present knowledge in this field. In the vertebrate twitch fiber a propagating action potential at the fiber surface, lasting a few milliseconds, is the first of the series of events which leads to the generation of tension by the myofibrils throughout the fiber. But almost at once we lose the thread of these events, partly because they are taking place in such very small structures; so small that the tip of the conventional intracellular microelectrode is an order of magnitude larger than the structures from which we should like to be able to record electrical changes. We do not know whether the surface action potential triggers a similar kind of active propagated disturbance in the transverse tubules, or whether the inward transmission of the electrical signal is a passive electrotonic spread of the kind which takes place in a core-conductor cable. If we know very little about the nature of the potential change across the tubular wall, we know even less about the way in which this potential change brings about the release of $Ca^{++}$ from the lateral elements of the triad. Now that the structure of the sarcoplasmic reticulum, and in particular the structure of the transverse tubular system, has been so clearly defined by electron microscopy, there are plausible electrophysiological predictions that can be made and some of these can be tested.

A starting point is to consider that the tubular system is a transversely disposed tubular network of some reasonably regular geometry, and with a lumen containing a fluid of essentially the composition of extracellular fluid. Fig. 13.26 in this volume shows such an idealized representation of the tubular system in the cross section of a muscle fiber. The continuity of the tubular lumen and the extracellular fluid is very strongly suggested by the evidence that ferritin (Huxley, 1964; Page, 1965) and dyes (Endo, 1966) have relatively free access to the tubules. If this is the case there will be a resting potential difference of some 90 mv across the wall of the tubules, the lumen being positive with respect to the sarcoplasm. When ionic currents flow in such a transverse network, produced either by an action potential in the fiber surface, or by an intracellular current electrode, it is possible, assuming linear circuit elements, to predict the distribution of current and potential across the tubular wall by obtaining solutions of the "cable equations" in a circular coordinate system, analogous to the one-dimensional solutions used to derive the passive distribution of current across the surface membrane of nerve or muscle fibers.

Several authors have considered this problem (Falk and Fatt, 1964; Falk, 1964; Adrian, Chandler, and Hodgkin, 1969). In a very general way it can be stated that for a steady current the potential change produced across the wall of the transverse tubular system is greater near the surface of the muscle fiber than at the center of the fiber. The degree of this central attenuation of a steady surface potential depends upon the ratio of the effective radial resistance of the tubular lumen in 1 cm of fiber ($R_L$ in Ω cm) and the membrane resistance of the tubule in unit volume of fiber ($R_w$ in Ω $cm^3$). The potential across the wall of the tube ($u$) at any distance $r$ from the center of the fiber ($r < a$ = fiber radius) as a fraction of the potential at the circumference of a fiber is given by the equation:

$$\frac{u}{u_a} = \frac{I_0(r/\lambda_T)}{I_0(a/\lambda_T)}$$

where $I_0$( ) are hyperbolic Bessel functions of the first kind and order zero. $\lambda_T$ is a space constant analogous to the space constant of a linear cable and is equal to $\sqrt{R_w/R_L}$. Fig. 14.1 shows this relation calculated for two values of $\lambda_T$ and for a surface depolarization of +35 mv from a resting potential of −90 mv.

For brief currents the central attenuation of the surface potential changes will be greater, and will depend on the membrane capacity as well as on the luminal and membrane resistance of the transverse tubules. For a potential change as rapid as an action potential at room temperature, the central potential change is relatively independent of the membrane resistance; the important determinants of the degree of attenuation are the membrane capacity and the luminal resistance.

The measurements of impedance of frog muscle fibers as a function of frequency made by Falk and Fatt (1964) have suggested that the capacities of a square centimeter of tubular wall and a square centimeter of

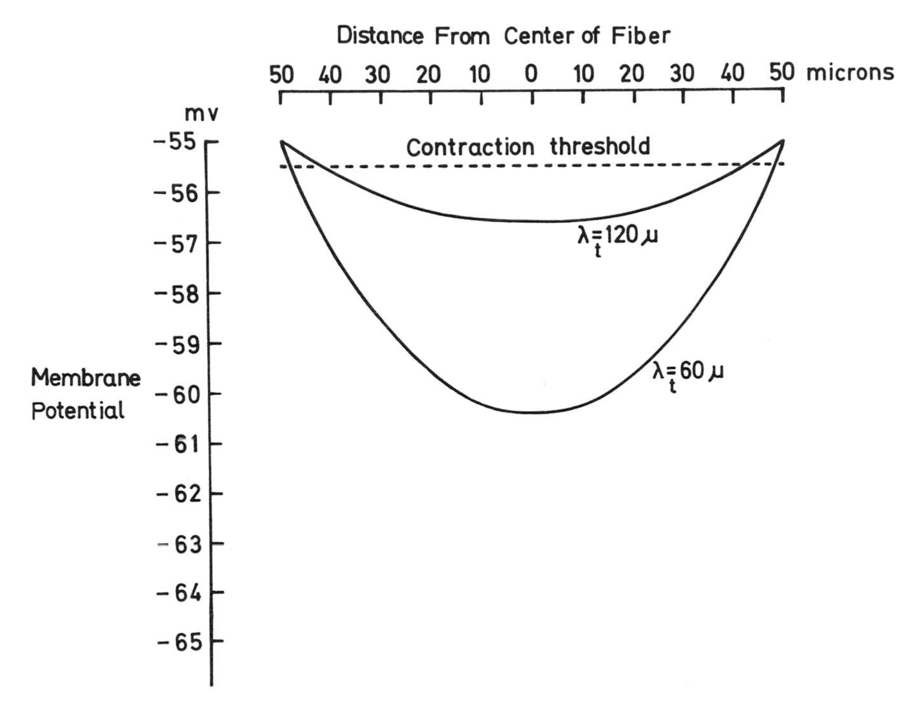


*Fig. 14.1.* The potential profile across the wall of the transverse tubular network along a diameter of a 100-$\mu$ fiber. The surface membrane is depolarized from —90 mv to —55 mv ($u_a = 35$ mv). The potential profile is given for two values of the tubular space constant. Contraction is supposed to occur where ever the potential difference across the wall of the tubules is less than 55.5 mv. If $\lambda_T = 120$ $\mu$ the difference between depolarizations which cause superficial and central contraction should be about 1.5 mv. If $\lambda_T = 60$ $\mu$ this difference should be about 5.5 mv.

surface membrane are about the same and that probably neither capacity is very much greater than the generally accepted value of 1 $\mu F/cm^2$. On these measurements and the estimates of the area of surface and tubular membranes (Peachey, 1965; Page, 1965) the transverse tubular system contributes more than two-thirds of the capacity of unit length of fiber. This distribution of capacity between the surface and the transverse tubular system has been confirmed by Gage and Eisenberg (1969), who measured the capacity of intact frog muscle fibers as 6.1 $\mu F/cm^2$ (*total* capacity of the fiber calculated for 1 $cm^2$ of fiber surface). They also measured the capacity of fibers exposed to a solution made hypertonic with glycerol and returned to an isotonic frog Ringer solution. This treatment apparently disrupts the connection of the transverse tubular system with the extracellular fluid (Howell and Jenden, 1967; Eisenberg and Gage, 1967; Nakajima, Nakajima, and Peachey, 1969), producing fibers which are functionally "detubulated" and which do not contract when the membrane is depolarized. The same treatment reduces the capacity of the fiber to 2.2 $\mu F/cm^2$, giving 3.9 $\mu F/cm^2$ as the capacity of the tubular walls in the volume of muscle fiber associated with each square centimeter of fiber surface.

Less is known about the resistance of the tubular wall than about its capacity. In a muscle, as in nerve, there are several ionic channels which show voltage- and time-dependent changes. Some of the characteristics of these channels are known, others are being analyzed in voltage clamp experiments (Adrian, Chandler, and Hodgkin, 1966). But it is difficult to distribute, with any certainty, these various channels between the fiber surface and the tubular wall. There are many suggestions in the literature about the distribution of these individual ionic channels (Hodgkin and Horowicz, 1959; Adrian and Freygang, 1962; Freygang, Goldstein, Hellam, and Peachy, 1964), but the recent measurements of Eisenberg and Gage (1969) present perhaps the most convincing evidence for the distribution of the resting potassium and chloride conductance. Using detubulated muscles and making use of the fact (Hutter and Warner, 1967) that *p*H affects the chloride, but not the potassium, conductance, they conclude that all the chloride conductance is in the surface, and that about two-thirds of the potassium conductance (55 $\mu mho/cm^2$) is in the tubular wall. If this is the case and the conductivity of the tubular lumen is the same as Ringer fluid, the tubular length constant ($\lambda_T$) would be more than 200 $\mu$ (the volume fraction of the tubules has been taken as 0.003). If $\lambda_T$ is 200 $\mu$ there would be only a small attenuation at the center of a 100 $\mu$ fiber of a steady potential change at the surface:

$$\frac{u_0}{u_a} = \frac{1}{I_0(0.25)} = 0.985 .$$

The natural stimulus for contraction is an action potential, but it is not the necessary stimulus, nor is it a very suitable tool with which to investigate the earliest stages of the contraction-activation cycle. Its explosive and all-or-nothing character needs controlling. The work of Hodgkin and Horowicz (1960) made it clear that tension is generated when the membrane potential difference is reduced for times longer than about 0.1 sec to less than about 50 mv. Many agents are known which affect the precise value of this mechanical threshold potential. The lyotropic anions move the threshold potential towards the resting potential (Kao and Stanfield, 1968); calcium ions move the threshold away from the resting potential (Costantin, 1968). Hodgkin & Horowicz (1960) also showed that there was a steep, but not apparently discontinuous, relation between the tension produced and the membrane potential. Once the threshold potential was reached the tension rose steeply with potential to reach a saturating level over a potential range of 12–15 mv. Gonzalez-Serratos (1965), who used solutions with raised potassium concentrations to depolarize, as did Hodgkin and Horowicz (1960), suggested that the submaximal tensions might be due to activation of superficial myofibrils only, and he showed microscopic evidence to support this idea. This kind of interpretation of the relation between tension and potential demands substantial attenuation of a steady surface potential at the center of the fiber.

By microscopical observation of isolated frog muscle fibers, we have been able to investigate the characteristics of electrical stimuli which cause contraction of superficial and central myofibrils (Adrian, Costantin, and Peachey, 1969). The action potential of the fiber under observation was prevented with tetrodotoxin and control of the membrane potential at a point was achieved by inserting two capillary microelectrodes into the fiber: one electrode to record the potential and the other to pass feedback current. A thin optical section across the diameter of the fiber in the immediate vicinity of the electrodes was observed and photographed. Fig. 14.2 is a diagram of the apparatus. Using long (200 msec) depolarizing pulses it was seen that the threshold depolarization

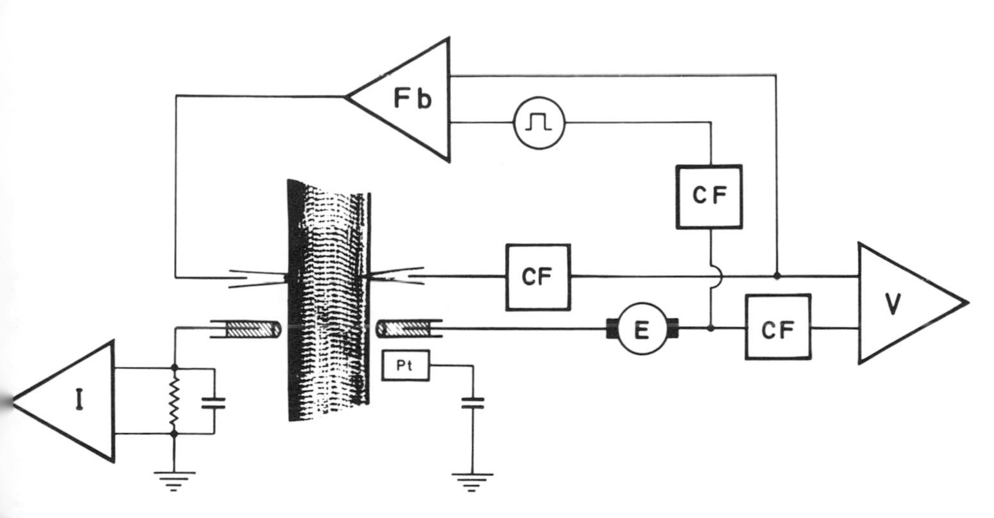


*Fig. 14.2.* Diagram of "point-clamp" apparatus used on isolated muscle fibers under microscopic observation. The "fiber" is a photomicrograph from the cine film record of an experiment. *CF,* cathode follows. *FB,* feedback amplifier: command pulses at its input produce identical alterations of the membrane potential at the right-hand microelectrode. *E,* voltage source for backing off the resting potential. *V* and *I,* the amplifier of an oscilloscope for recording the changes in feedback current and membrane potential.

for the contraction of the myofibrils at the surface of the fiber was several mv less than the depolarization required to make the central myofibrils shorten. Fig. 14.3 shows photographic records from an experiment of this kind. If one makes the assumption that the threshold potential across the tubular wall is the same at the surface and the center, these observations suggested that the potential change at the center of the fiber is about 90% of the steady change at the surface of the fiber. From this figure one can derive a value for the tubular length constant $\lambda_T$. Fig. 14.4 shows the results and calculated values of $\lambda_T$ for 18 fibers. We found that $\lambda_T$ was approximately equal to the fiber radius, a value substantially less than we had expected. If the conductance of the tubular

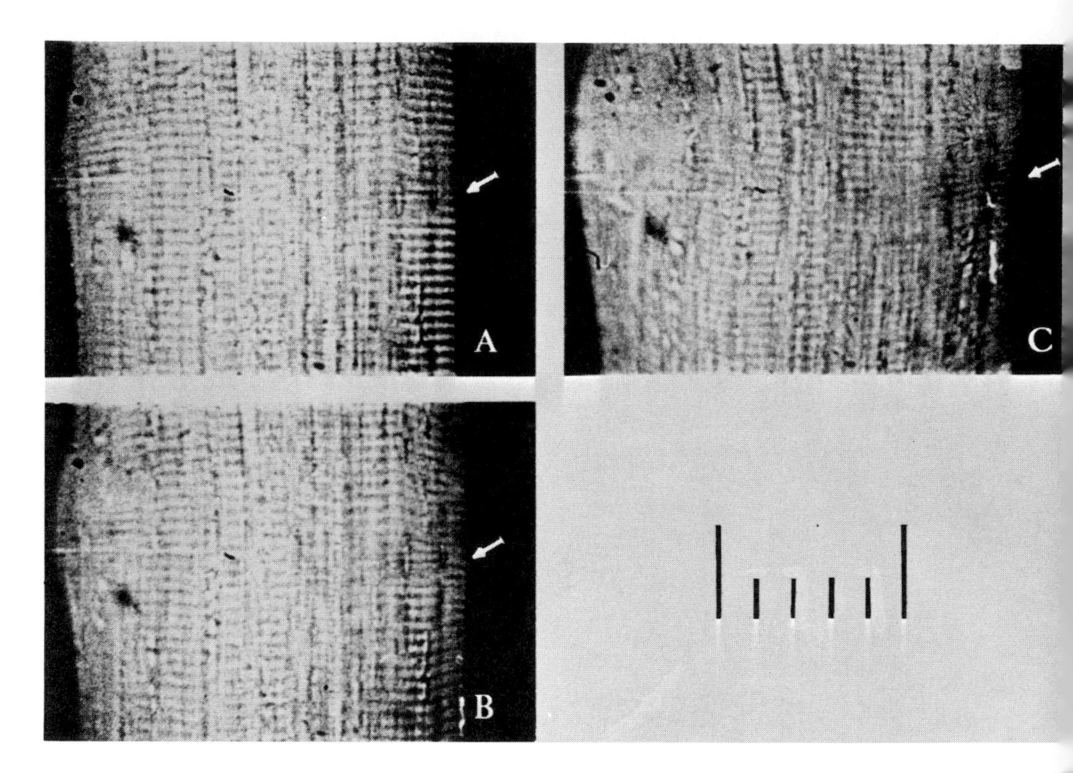


*Fig. 14.3.* Stills from a cinefilm of a relaxed fiber (*A*) and during 200 msec depolarizing pulses of +39 mv (*B*) and +43 mv (*C*) from the resting potential of −90 mv. The voltage-recording electrode is at the arrow, but out of the plane of focus. The current-passing electrode is at the left-hand edge of the fiber: there is a localized contraction at this electrode, an explanation for which is given by Adrian, Costantin, and Peachey (1969). At right hand edge, when the potential is accurately controlled, B shows a just detectable contraction of the most superficial myofibrils; C shows a contraction which spreads inwards from the arrow for between one-third and one-half of the radius. The scale is 50 $\mu$ in 10 $\mu$ divisions.

lumen and the volume fraction of the transverse tubular system are known, $\lambda_T$ can be used to derive the resistance of unit length of fiber. We measured this resistance in the fibers under observation and found that if the fluid in the tubular lumen has the same specific conductance as Ringer all the resting conductance of the fiber must be in the walls of the transverse tubular system. Such a conclusion appears contrary to the

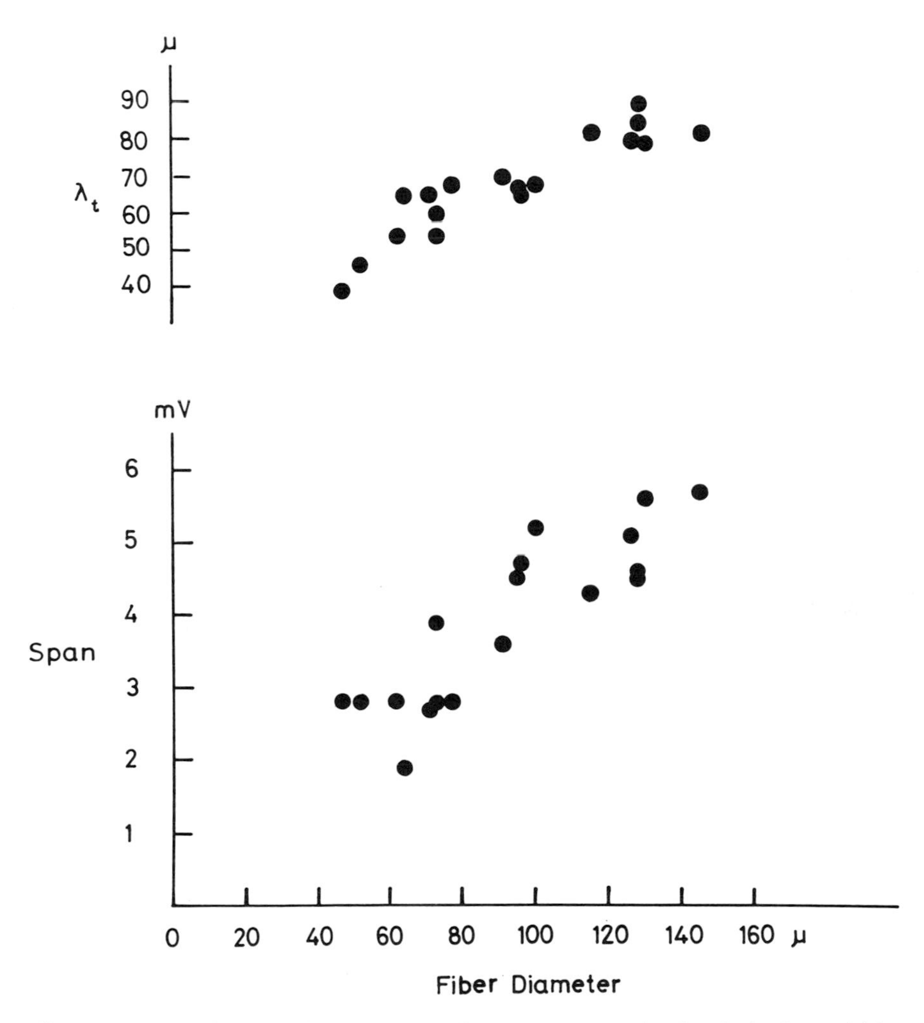


*Fig. 14.4.* Values from 18 fibers of the difference between the depolarizations which caused contraction of superficial myofibrils and shortening of central myofibrils (span). The values are plotted against the fiber diameter. In the upper part of the figure the derived values of the tubular space constant ($\lambda_T$) are also plotted against the diameter of each fiber.

findings of Gage and Eisenberg (1969) mentioned earlier. Greater weight should probably be put on their more direct method of measurement. The discrepancy certainly casts doubt on the simplified cable model of the transverse tubular system. The values of the volume fraction and the luminal conductivity used are also open to question. Perhaps just as disturbing was our inability to alter the space constant of the transverse tubules by conditions which might be expected to alter the resistance of the tubular wall or the tubular lumen. At any rate there are grounds for supposing that the tubule may not be a simple tube filled with whatever solution surrounds the fiber. Uncertainties about the conductivity and the volume of the tubular lumen make the interpretation of the observed $\lambda_T$ difficult. Freygang, Goldstein, Hellam, and Peachey (1964) have already demonstrated conditions where the volume of the transverse tubular system can vary over a wide range, and it is clearly possible that fixed charges could exist in the walls or at the mouth of the tubular system (Fatt, 1964).

A brief depolarization of the surface was found to be a less effective stimulus to contraction of the central myofibrils than a long-lasting depolarization; the shorter the depolarization the greater it must be to produce contraction. This is to be expected because of the effect of the capacity of the tubular wall, and one can ask the question whether the action potential can activate the central myofibrils by purely electrotonic spread along the transverse tubular system. Gonzalez-Serratos (1966) has shown that under normal conditions the central myofibrils in a frog muscle fiber do indeed contract during a single twitch, but under his conditions there may have been a regenerative potential change in the tubular system. By recording action potentials from one muscle fiber and imposing an attenuated version of the recorded action potentials, by a voltage-clamp method, on a second fiber poisoned with tetrodotoxin, we were able to measure the threshold amplitudes of the action potential imposed on the surface membrane of the fiber which caused contraction in the superficial and central myofibrils. We found that the imposed "action potential" had to be the full size (130 mv) to produce contraction in the central myofibrils. An action potential of about two-thirds the normal size was sufficient to activate the superficial myofibrils. If there is no regenerative potential change in the transverse tubular system, the safety factor for activation of the central myofibrils by a surface action potential at 20° C appears to be no greater than 1. This is perhaps an ambiguous result; it might suggest a regenerative tubular potential change under normal conditions, though it does not require one.

We found by using short step depolarizations that there was a strength-duration relation for the activation of the most superficial myofibrils. This finding cannot, of course, be explained in terms of the capacitative attenuation in the tubular network. We were however able to show that the *difference* between the strength-duration relations for activation of superficial myofibrils and central myofibrils was just what would be expected from the capacitative attenuation of the simple tubular network model. Again, however, reduction of the conductivity of the extracellular fluid, which might be supposed to increase the central attenuation of brief pulses by decreasing the luminal conductivity, had no effect on the difference between the strength-duration curves for the superficial and central myofibrils.

In some earlier experiments with A. L. Hodgkin and W. K. Chandler the strength-duration relation was studied in the hopes that it might give some information about the kinetics of the early stages of activation release (Adrian, Chandler, and Hodgkin, 1969). The contractions of a fiber were observed with rather low magnification and we probably did not see the very superficial contractions. Threshold contractions in these experiments probably represent contractions occurring in the greater part of the cross section of the fiber. The muscle fibers were poisoned with tetrodotoxin, and the membrane potential, controlled with a two-electrode clamp, was altered with rectangular pulses of varying amplitude and duration. These experiments were carried out at 2° C. At this temperature a depolarization of 190 mv to an internal potential of +100 mv was required for a contraction when the pulse duration was about 1 msec. We did not investigate shorter durations, but within the range examined there was no suggestion that there was any minimum duration of depolarization. A useful empirical threshold criterion for short pulses proved to be the integral of the potential-time record over the time for which the internal potential (V) was more positive than −30 mv. For potentials greater than about −10 mv, contraction occurred when this integral exceeded about 150 mv.msec. The implication of this finding is that the rate constant for this very early step of the activation process is directly proportional to (V + 30) mv at least for potentials greater than −10 mv. At 2° C the relevant area of the action potential is about 250 mv.msec, so that by this criterion of effectiveness the action potential has a safety factor of nearly 2. We have not examined the effect of temperature on the strength-duration curve, but the rate constants for the processes underlying it would have to be large if the action potential at room temperature were to remain as effective a stimulus as it is in the cold.

We can account for the observed strength-duration curve by a model which involves a store of activator and a site of action of the activator; movement of the activator between the store and the site of action is a first-order process controlled by potential-dependent rate constants ($\alpha$

and $\beta$). If there is a large excess of activator in the store, its concentration there ($\bar{y}$) will be constant and the rate of change of activator concentration (y) at the site of action will be

$$\frac{dy}{dt} = \alpha\bar{y} - \beta y$$

We suppose that contraction occurs when the activator concentration y reaches some critical concentration $y_c$. The time taken to reach $y_c$ is given by

$$t' = -\frac{1}{\beta}\ln\left(1 - \frac{\beta yc}{\alpha\bar{y}}\right).$$

The exact shape of the strength-duration curve depends upon how $\alpha$ and $\beta$ are related to the potential. Quite good agreement between experimental and calculated curves can be achieved with a fairly wide choice of potential functions for $\alpha$ and $\beta$. A quite successful fit can be achieved by supposing that the activator is a univalent anion held in the store by a resting potential across a membrane which bounds the store. This membrane is assumed to have a constant permeability to the activating ion. Reduction of the potential difference between the interior of the store and the sarcoplasm would lead to a movement of the activating ion into the sarcoplasm. At least as good a fit could be achieved by a model where the activating ion is divalent and the voltage-dependent rate constants $\alpha$ and $\beta$ are equal at about −30 mv. To this extent the model is unhelpful, though it is surprising that such a simple system does so well. It also suggests experiments to measure directly the potential dependence of the rate constants, and to test more rigorously whether the first order system is adequate.

Can the postulated movement of activator be related to the movement of $Ca^{++}$ from the sarcoplasmic reticulum? There are immediate difficulties. At least in its simplest form the scheme demands a negative charge (or charges) on the activating molecule, which is moved from a store to the sarcoplasm by a positive change in the sarcoplasmic potential. Another difficulty with the identification are the observations of Jöbsis and O'Connor (1966). They have observed the time course of $Ca^{++}$ concentration within a toad muscle fiber using the colored Ca-murexide complex to detect the presence of $Ca^{++}$. It appears that the concentration of Ca-murexide continues to increase for 50–75 msecs after an action potential, which suggests that the release of $Ca^{++}$ continues over a considerably longer period than the release of "activator" in the model used to describe the strength-duration findings. If the activator release were first order its concentration at the "site of action" should not continue to increase after the end of the depolarizing pulse. It may be that the $Ca^{++}$ release, once initiated, is no longer influenced by the potential difference across the surface membrane and the tubular wall, and that the "movement of activator" postulated to explain the strength-duration relation is in fact the initiation of an all-or-nothing process which results in the release of $Ca^{++}$. These questions will be resolved by experiments in which alterations of sarcoplasmic $Ca^{++}$ concentration can be followed during controlled alteration of the membrane potential.

## *References*

Adrian, R. H., W. K. Chandler, and A. L. Hodgkin. 1966. Voltage clamp experiments in skeletal muscle fibres. *J. Physiol. 186:*51P.

———. 1969. The kinetics of mechanical activation in frog muscle. *J. Physiol. 204:*207.

Adrian, R. H., L. L. Costantin, and L. D. Peachey. 1969. Radial spread of contraction in frog muscle fibres. *J. Physiol. 204:*231.

Adrian, R. H., and W. H. Freygang. 1962. The potassium and chloride conductance of frog muscle membrane. *J. Physiol. 163:*61.

Costantin, L. L. 1968. The effect of calcium on contraction and conductance thresholds in frog skeletal muscle. *J. Physiol. 195:*119.

Ebashi, S., and M. Endo. 1969. Calcium ion and muscle contraction. *Progr. Biophys. Mol. Biol. 18:*123.

Eisenberg, R. S., and P. W. Gage. 1967. Frog skeletal muscle fibers: changes in electrical properties after disruption of transverse tubular system. *Science 158:*1700.

———. 1969. Ionic conductances of the surface and transverse tubular membrane of frog sartorius muscle. *J. Gen. Physiol. 53:*279.

Endo, M. 1966. Entry of fluorescent dyes into the sarcotubular system of the frog muscle. *J. Physiol. 185:*224.

Falk, G. 1968. Predicted delays in the activation of the contractile system. *Biophys. J. 8:*608.

Falk, G., and P. Fatt. 1964. Linear electrical properties of striated muscle fibres observed with intracellular electrodes. *Proc. Roy. Soc. B 160:*69.

Fatt, P. 1964. An analysis of the transverse electrical impedance of striated muscle. *Proc. Roy. Soc. B 159:*606.

Freygang, W. H., D. A. Goldstein, D. C. Hellam, and L. D. Peachey. 1964. The relation between the late after potential and the size of the transverse tubular system of frog muscle. *J. Gen. Physiol. 48:*235.

Gage, P. W., and R. S. Eisenberg. 1969. Capacitance and transverse tubular membranes of frog sartorius muscle. *J. Gen. Physiol. 53:*265.

Gonzalez-Serratos, H. 1965. Differential shortening of myofibrils during contractures of single muscle fibres. *J. Physiol. 179:*12P.

———. 1966. Inward spread of contraction during a twitch. *J. Physiol. 185:*20P.

Hodgkin, A. L., and P. Horowicz. 1959. The influence of potassium and chloride ions on the membrane potential of single muscle fibres. *J. Physiol. 148:*127.

———. 1960. Potassium contractures in single muscle fibres. *J. Physiol. 153:*386.

Howell, J. N., and D. J. Jenden. 1967. T-tubules of skeletal muscle: morphological alterations which interrupt excitation-contraction coupling. *Fed. Proc. 26:*553.

Hutter, O. F., and Anne E. Warner. 1967. The *p*H sensitivity of the chloride conductance of frog skeletal muscle. *J. Physiol. 187:*403.

Huxley, H. E. 1964. Evidence for continuity between the central element of the triads and extracellular space in frog sartorius muscle. *Nature 202:*1067.

Jöbsis, F. F., and M. J. O'Connor. 1966. Calcium release and reabsorption in the sartorius muscle of the toad. *Biochem. Biophys. Res. Comm. 25:*246.

Kao, E., and P. R. Stanfield. 1968. Actions of some anions on electrical properties and mechanical threshold of frog twitch muscle. *J. Physiol. 198:*291.

Nakajima, S., Yasuko Nakajima, and L. D. Peachey. 1969. Speed of repolarization and morphology of glycerol-treated muscle fibres. *J. Physiol. 200:*115P.

Page, S. 1965. A comparison of the fine structure of frog slow and twitch muscle fibers. *J. Cell Biol. 26:*477.

Peachey, L. D. 1965. The sarcoplasmic reticulum and transverse tubules of the frog's sartorius. *J. Cell Biol. 25:*209.

# Chapter 15

# *Summary and Discussion of Part 3*

PANEL MEMBERS: D. B. SLAUTTERBACK, *Chairman*
J. B. PETER
M. GREASER
S. KORNGUTH
L. LORAND

*Slautterback:* Much progress has been made in the study of mitochondria, muscles, and membranes in the last 20 years, and the chapters by Green and Harris, Peachey, and Adrian not only give ample testimony to those advances in the area of membranes in striated muscle cells, but also suggest the pace at which developments continue to emerge. In pointing to past progress, however, I would indeed be derelict if I did not also comment on the large task which lies ahead.

At the risk of repeating a cliché, I would like to put first on my list of concerns the molecular organization of membranes. I do not believe that all of the problems of the biosphere will suddenly fall away on the day that the problem of the molecular organization of membranes is finally solved to everyone's satisfaction (if that day ever comes). Nonetheless, many questions now difficult would clearly become more accessible to investigation if we knew more about how molecules are put together to form membranes with specific properties and functions. Drs. Green and Harris have given us an elegant example of how far a membrane model, ingenious experiment, and bold interpretation can go in the development of new understanding. Stimulating as the many aspects of their concept are, this dynamic view of supramolecular form varying from second to second and not only reflecting molecular events but indeed permitting them and in a sense giving them a *raison d'être,* is what appeals to me most. In this, it has the quality of the sliding filament concept. Furthermore, embodied in the recognition of rapidly changing form may be the reconciliation of the number of proposed membrane models, now approaching infinity. It is clear that the appearance of membranes in electron micrographs represents at best only an instant in the life-history of the structure, and at worst is further confounded by the effects of preservation. Advances in this area will surely be made by well-chosen experiments, but may be accelerated by sophisticated new techniques and instrumentation currently coming into general use. Among these are freeze-etching and deep freezing as a means of preservation, the steady improvement of resolution in electron microscopes and electrical recording, and the possibilities of very small selected-area electron diffraction in the high-voltage electron microscopes. I stress the matter of resolution here because it is precisely in the area of macro- to supramolecular transitions that our knowledge is weakest. Even though our instruments often permit study in this area, our specimens usually will not.

The origin, assembly, and control of membranes is another challenging problem, or complex of problems. It is clear that the components of sarcoplasmic reticulum tubules, sarcolemma, transverse tubules, micropinocytosis vesicles, Golgi complex, mitochondria, and nuclear envelope must be synthesized, assembled in a characteristic pattern, transported to the appropriate locus, and assembled into the functional organelle. The way in which all this is accomplished and control effected is not understood. We do not even know if the stated sequence is correct. The obvious local variations in the structure of the sarcoplasmic reticulum that Dr. Peachey described, and the likelihood of an accompanying variation in function, offer a good illustration of the complexity of the problem. Similarly the relative scarcity of visible T-tubule invaginations in mammalian skeletal muscle remains enigmatic.

Perhaps these anastomoses are transient. Certainly this coupling of the T tubule at the sarcolemma, as well as the coupling (possibly compartmentalized) between T tubules and lateral cisternae of the sarcoplasmic reticulum, must have a considerable effect on electrical events and the interpretation of the current path in stimulated muscle cells. It may not be too much to hope that the path of ion movements relative to coupled membranes will come to be better understood.

Another area of considerable interest is the supraorganellar control of mitochondria. As you have heard, such investigations are under way and are proving productive. A recent publication advances an old argument that the bizarre shapes of some muscle mitochondria are caused by their attempts to meet the demands of enlarging myofibrils for a proximate ATP source. What is the signal to which these mitochondria respond and what "motor" propels them to the appropriate locus?

As Dr. Peachey said, there is a remarkable, though perhaps superficial, similarity in the elaboration of sarcoplasmic reticula, but what correlations can be made with the striking differences? Why are some cardiac reticula in register and others not? Are the myriad small vesicles of the smooth muscle cell representative of a T system? And what explains the bizarre longitudinal T system of some invertebrates?

To return for an instant to the sarcolemma—we want to know more about what happens at myoneural junctions, but we do not really know much about myotendinous junctions—presumably a matter of more than passing interest to the meat scientist.

I have listed a few problems of muscle cell membrane systems which I regard as interesting. Each of you could make a different and longer list. It is encouraging and exciting that our greatest advances are likely to be made in the area considered in Part 3, that is, the electrical, chemical, and morphological manifestations of fluctuating macro- and supramolecular relations.

## DISCUSSION

*Greaser:* Electron microscopic examination of membranes that have been fixed and sectioned has revealed two dense lines separated by a zone of lesser electron density. Using the negative staining technique, however, mitochondrial, sarcoplasmic reticulum, and several other membranes display the basepiece, stalk, and headpiece arrangement. Do you envisage the headpieces and basepieces as forming the two dense lines or do you have another explanation for the positions of the membrane constituents?

*Green:* Some recent studies in our laboratory by W. J. Vail have made it possible to interpret the trilaminar structure of the cristal membrane. The basepieces which form the membrane continuum are exclusively responsible for the trilaminar structure. The two dark lines correspond to the darkly staining and exteriorily exposed polar sectors of the phospholipid which covers the top and bottom of the membrane and includes the "edge" of the protein surface. The electron transparent zone between the two dark lines corresponds to the inaccessible interiors of the proteins which make up the basepieces as well as to the fatty residues of the phospholipid molecules which are aligned in channels between proteins at right angles to the direction of the membrane. In favorable photographs it can be seen that the headpieces and stalks project from the trilaminar membrane. I suspect this interpretation would apply with equal force to all membranes with tripartite repeating units.

*Greaser:* Is there any evidence that the tripartite repeating units are all identical and that each is capable of performing all the functions of the inner membrane?

*Green:* There is no evidence at present to suggest that the headpiece-stalk sectors of the tripartite repeating units of the cristal membrane are not all identical. We have looked for activities besides adenosine triphosphatase activity and found that none of these appears to be localized in the headpiece-stalk sectors. However, the door is still open to another possibility. The repeating units of the inner boundary membrane have been shown in our laboratory to have tripartite character. The headpiece-stalk sectors intrinsic to the repeating units of this membrane could possibly be different from the headpiece-stalk sectors of the cristal membrane repeating units.

*Lorand:* I wonder if Dr. Green, in using the term "conformational change" for the tripartite units, envisages rearrangement of component subunits rather than what would customarily be referred to as a conformational change in a polypeptide chain?

*Green:* Each repeating unit is a mosaic of three fitting parts—basepiece, stalk, and headpiece. Both the basepiece and headpiece are multiprotein units each with a mass of some 300,000 in terms of protein. The repeating unit in the three conformations has a different geometry. By means of high-resolution electron microscopy we can recognize the basis for this change in geometry. The relation of the three fitting parts is different in each of the three conformational states. For example, the headpiece is flattened in the nonenergized conformation, more extended in the energized, and fully extended in its spherical form in the energized-twisted conformation. The stalk is collapsed in the nonenergized conformation and fully extended in the other two conformations. The basepiece is squarish in the nonenergized and energized conformations and curved in the energized-twisted conformation. Ultimately all these changes in geometry are reflections of changes in the conformation of individual proteins. The massive geometric changes actually observed are set in motion by the original small conformational changes in single polypeptide chains. We must not be surprised to find in the mitochondrion a new dimension of conformational change because the mitochondrion is after all concerned with the conservation of energy via conformational change. This conservation requires a special way of modulating conformational change—hence the rather unique way in which the repeating units undergo their cycle of conformational change.

*Kornguth:* In the light-scattering experiments how does one distinguish between configurational shifts occurring within a given mitochondrial unit and the association of several mitochondria with no overall change in shape?

*Green:* You are quite right in pointing out that the light-scattering changes could be expressions either of configurational changes within each electron transfer particle or of aggregation of particles without configurational change being involved. But we have excluded the latter possibility by electron microscopic examination of the particles. There is no evidence of gross aggregation, and there is unambiguous evidence of configurational change, namely the evidence of tubularization. That is to say that we can observe a qualitative change in the configuration of the electron transfer particles that accounts satisfactorily for the direction and magnitude of the light-scattering change.

*Peter:* Does Dr. Green have any illustrations that support the notion that conformational changes in the repeating units bring about the configurational changes in the whole membrane? That is to say, do the repeating units actually change their geometry or their relationship to adjacent repeating units?

*Green:* Fig. 12.2 illustrates the conformation of the repeating units in the three configurational states. The change in conformation in the transition from nonenergized to energized cristae and from energized to energized-twisted cristae is readily demonstrable electron-microscopically in positively stained thin sections of phospholipase C-treated mitochondrial suspensions, fixed with glutaraldehyde before this treatment and stained, dehydrated, embedded, and sectioned after the treatment. In the first configurational transition, the headpiece-stalk sectors become extended from a collapsed state; in the second transition, the headpiece-stalk sectors separate from the apposed basepieces and the headpieces associate with one another in clusters. Also the basepieces undergo a transition from a cuboidal to a curved geometry. This emergent curvature of the basepieces underlies the helical character of the energized-twisted configuration of the membrane.

*Peter:* What proportion of mitochondria in a given sample will change their configuration under any of the various conditions described? Do the electron microscope changes relate in a direct rather than a random fashion to changes in the medium? I am especially interested in the energized to energized-twisted change.

*Green:* When the experimental conditions are properly controlled, all the intact mitochondria in a given population behave in exactly the same fashion; the uniformity of response applies to both configurational transitions. If the membranes of the mitochondria are obviously damaged such mitochondria will not respond in the same way as the intact mitochondria. We have specified in our publications the conditions which have to be controlled to assure this synchronous behavior of mitochondria in a given suspension. Not only must the determinants of the energizing process be freely available but precautions have to be taken to exclude side reactions which dissipate the energized state. The use of suitable inhibitors such as antimycin, rotenone, and rutamycin is highly effective in suppressing side reactions.

*Peter:* Is there a steric factor (e.g., $ETP_H$ being too small) which inhibits the energized-twisted form and hence high respiration control in $ETP_H$?

*Green:* $ETP_H$ can be shown to undergo configurational change during the transition from energized to energized-twisted, not only by light-scattering measurements but also by electron microscopy. There are two kinds of respiratory control—one that is measured by the rate of oxygen uptake ± uncoupler and one that is measured by the rate of oxygen uptake ± ATP. $ETP_H$ has the first but not the second kind of respiratory control. With respect to the critical size below which configurational change may be interdicted, we have no evidence that would support the concept of a critical size for a membrane vesicle.

*Peter:* Has actual expansion of the matrix space been measured with "tubularization" of mitochondria?

*Green:* The tubular membranes in the energized-twisted configuration are some 100 Å wider than the corresponding membranes in the energized configuration.

*Peter:* What is being measured with light scattering in $ETP_H$?

*Green:* Light scattering measures the configurational change in $ETP_H$, which is expressed electron-microscopically as an invagination of the membrane. The invagination leads to a change in shape of the vesicle as well as a change in the light-scattering properties of the vesicle.

*Greaser:* What types of conformational changes do you believe occur in the sarcoplasmic reticulum, and what relationship might these changes have to muscle function?

*Green:* We have not studied, in detail, configurational changes induced by ATP in the sarcoplasmic reticulum. Our impression from limited electron micrographic evidence is that such changes do in fact take place, but these changes are not as obvious or at least not as striking as the configurational changes in mitochondria. It is our working hypothesis that all energized membrane functions must involve the translation of the free energy released during electron transfer or during hydrolysis of ATP into conformational energy.

*Greaser:* We did some studies at the University of Wisconsin on $Ca^{++}$ uptake by isolated fragments of sarcoplasmic reticulum and we also examined them with negative staining (Greaser, Cassens, Hoekstra, and

Briskey, 1969). We found that only about 15% or 20% of the particles actually accumulated demonstrable calcium oxalate crystals.

*Green:* I think the basic difficulty is that in a mixed population it is difficult to show the significant configurational changes because some of the particles do not respond.

*Kornguth:* It has been reported that polycationic proteins, such as histone, cause energy-dependent swelling of isolated mitochondria. Inorganic phosphate is necessary for maximal swelling. Since polycations, such as polylysine, can act as electron transfer units, do you imagine that there exists a very basic protein in the mitochondrial membrane which plays a role in the function of this organelle? This is of particular interest since you have indicated that inorganic phosphate reduces the charge on headpieces, indicating that they may be cationic.

*Green:* The action of histones and polybases in facilitating or inducing energized swelling is very intriguing, but we have ourselves not studied this action in detail. Certainly there are basic proteins in the mitochondrion, particularly cytochrome *c* and some of the proteins of the citric acid cycle. You may well be right in suggesting a role for these proteins in energized swelling.

*Kornguth:* Do polycations confer configurational effects on isolated electron transport particles when $P_i$ and substrate are present?

*Green:* The combined presence of $P_i$ and substrate will induce configurational change in electron transfer particles without any further additions. Polycations could of course modulate or suppress this configurational change. We have no information on this point.

*Greaser:* What do you believe is the role of mitochondrial translocation of ions? Do you think there is any relationship of this ion transport to the process of muscle contraction?

*Green:* I would find it difficult to believe that a capability for translocating ions intrinsic to mitochondria generally is not put to use physiologically. In our laboratory, D. Allmann has been able to show that divalent ions such as $Ca^{++}$ can control the configurational mode of mitochondria and in that way determine the coupling capability of the mitochondria. Greater emphasis in some instances may have to be placed on ion movements rather than on translocation. Nonetheless, I believe the available evidence is consistent with the postulate that during muscular contraction the mitochondria undergo an ebb and flow of function synchronized with the cycle of $Ca^{++}$ release and withdrawal. The mitochondria do not necessarily participate in the $Ca^{++}$ cycle in the sense that they determine it, but they are profoundly affected in function and configuration by the cycle.

*Kornguth:* In view of the postulated conformational and configurational changes in mitochondria following addition of substrate and inorganic phosphate it might be interesting to examine the optical rotatory properties of mitochondria complexed with a dye. The technique of optical rotatory dispersion and circular dichroism has proved useful in detecting small changes in conformations of proteins that have a chromophore attached.

*Green:* Several laboratories are, indeed, exploring the use of conformational probes, i.e., molecules which can bind to the repeating unit and report by means of changes in fluorescence or changes in electron spin resonance the fact of conformational change in the repeating unit. We have seriously considered using these probes and have tentatively concluded that they are not likely to tell us more than we know already, namely that there is conformational and configurational change. The molecular events that underlie conformational change will have to be ferreted out by direct chemical studies. I doubt that conformational probes will aid materially in the next phase of our investigations, the chemical phase.

*Slautterback:* In light of your understanding of mitochondrial construction and the changes with functional variation, can you suggest what the substrate-ATP traffic pattern looks like?

*Green:* You have raised a very intriguing question which has two parameters. What determines the concentration of substrate and adenine nucleotides in the medium surrounding the mitochondrion, and by what route do these critical molecules reach the sites with which they must react? The answer to the first question has to do with events external to the mitochondrion, and an attempt at analysis would take us too far afield. The answer to the second question can at best be only a partial answer. The movement of substrate and adenine nucleotides into the mitochondrion involves more than passive diffusion. There is evidence of directed movements. For example, the transfer of a phosphoryl group from internal ATP to external ADP depends upon a pathway which is sensitive to atractyloside and which has been shown to involve a nucleotide monophosphokinase. The rate of oxidation of citric cycle intermediates is profoundly affected by the osmotic pressure of the medium and by the energy state of the mitochondrion. The nucleotides of the mitochondrion leak out when the cristae are in the orthodox configuration and are retained when the cristae are in the aggregated configuration. These observations are cited to point up the complexity of the problem and to emphasize that diffusion of molecules across passive membrane barriers cannot possibly be the complete answer to the traffic of key molecules in and out of the mitochondrion.

*Slautterback:* In your definition of "respiratory control" as a ratio of oxygen consumption would you explain how ADP availability relates to, or is determined by, configuration?

*Green:* The cristal membrane in the energized-twisted configuration can be discharged to the nonenergized configuration when ADP is present in the external medium. This discharge is accompanied by ATP synthesis. Thus it is the presence or absence of ADP in the external medium which can determine the configuration of the cristal membranes. If ADP is absent, then the energized-twisted configuration will be generated in presence of substrate and inorganic phosphate. If ADP is present, the energized-twisted membrane will be discharged.

*Sybesma:* You referred to uncoupling of mitochondria as a possibility due to $Ca^{++}$-binding during the contraction. Does this mean that under certain physiological circumstances the mitochondria are not able to function?

*Green:* $Ca^{++}$ can induce the orthodox configuration of the mitochondrion and stabilize the mitochondrion in this configuration. Such a mitochondrion is indeed uncoupled since coupling requires engagement of cristae, and engagement is excluded when the cristae are stabilized in the orthodox configuration.

*Lewis:* How was the *p*H determined on the lyophilized mitochondria? Did these contain any moisture?

*Green:* The *p*H was determined with a glass electrode immersed in the medium in which the lyophilized mitochondria were suspended. Lyophilization was carried out to the point at which some 90% of the total water was removed. Beyond this point of dehydration, irreversible changes set in which impair not only the coupling function but even the capacity for electron transfer.

*Jay:* How long do mitochondria persist as morphologic entities in muscle post mortem; is there any evidence of oxidative activity on the part of these structures in dead animals?

*Green:* If excised cardiac muscle tissue is kept at 0° C, active mitochondria can be isolated even after 12–24 hours. Excised tissues vary greatly with respect to the permissible time limit within which the mitochondria can be preserved. The degree of washing of minced tissue and the nature of the suspending medium as well as the temperature are critical factors in such preservation. But as a general rule, the retention of mitochondrial functionality in muscle post mortem is a well-established fact.

*Cassens:* We feel that the maintenance of morphological features in mitochondria varies greatly depending on the characteristics of the muscle during the very early post mortem periods. Greaser, Cassens, Briskey, and Hoekstra (1969) have shown very acceptable-appearing mitochondria isolated from pig muscle that would undergo normal changes in post mortem metabolism. Some muscles, however, are prone to undergo an extremely rapid glycolysis during post mortem change, and there is a consistent difference recognizable in mitochondria from these muscles. These mitochondria have a conformational pattern like those from normal muscle but some are markedly swollen and have a decreased matrix density; also a smaller proportion of them are intact. In both cases the mitochondria are prepared immediately after the animal is exsanguinated. We conclude then that there must be a morphological change in the mitochondria of certain muscles that begins immediately when the animal is killed, or that the mitochondria may become more susceptible to damage during homogenization.

*Slautterback:* How do you account for the regular interval of bending in zigzag cristae and the square ends and triangular cross sections sometimes reported?

*Green:* In the energized-twisted configuration, the headpieces of the repeating units in each of the apposing membranes associate in sets (two or three in a set). This association of the headpieces leads to curvature of the basepieces. The periodicity of the curvature (zigzag or sine-wave character) is rooted in the geometric problem of interleaving periodic sets of associated headpieces from the two apposed membranes. In the zig portion, the headpieces of each membrane point one way; and in the zag portion the headpieces point the opposite way. The triangles are readily accounted for as the extension of the intracristal space while the square ends represent the flat end of the crista. If one thinks of a single crista, the geometric relationships are more readily grasped. Imagine a rectangular crista with zigzag on the top and bottom of two of the four long faces. The zigzags of top and bottom faces are in a mirror-image relation to one another. The origin of the triangles (between the zig and zag) and square ends (the short faces of the crista) is immediately apparent. To complete the picture, another crista is set above and below the first. These cristae then interdigitate the zigzag of their long faces with the zigzag of the first.

*Slautterback:* If zigzag cristae are formed by tubular invagination as you suggest, how do you account for the common observation that the lumen of a zigzag is directly continuous with the space between inner and outer membranes, rather than with the matrix space as in your diagram (Fig. 12.12)?

*Green:* You are quite right in suggesting that the zigzag cristae in canary heart mitochondria are probably not tubular since the characteristic bulbar ends of tubular invaginations are missing. Tubularization and zigzagging are two alternative configurational forms of expression of the transition from the energized to the energized-twisted configuration. The former involves one of the two apposed membranes; the latter is a symmetrical change in both apposed membranes. Moreover, tubularization represents an invagination whereas zigzagging involves helical rippling of sheet membranes. The lumen within the apposed zigzagging membranes is the matrix space and Figure 12.12 is correctly drawn.

*Slautterback:* The mitochondria of cells of the seminiferous epithelium, including the sertoli cells, commonly are preserved in tissue sections in a form very similar to that which you call "energized," that is, with a relatively large intracrista space and matrix faces approaching close apposition. Can you offer an explanation for this observation?

*Green:* Investigators who take precautions to fix the tissues they are examining before anaerobiosis sets in are likely to find mitochondria in the energized configuration more often than those who do not. I have the impression that the states found in the mitochondria of the seminiferous epithelium are normal for energized mitochondria and can be found in any mitochondria in any tissue, providing the proper precautions are taken and the requisite speed of processing is achieved.

*Slautterback:* Mitochondria from hypoxic tissues of the lung show a collapsed matrix space and an enlarged intracrista space. Can you explain this observation in terms of your model?

*Green:* The configurational state of the cristal membrane will be determined by a variety of factors—the osmotic pressure of the membrane, the concentration of $Ca^{++}$, $Mg^{++}$, and monovalent ions, the oxygen tension, and the concentration of substrate and inorganic phosphate. While it may seem strange that in hypoxic tissues, the mitochondria under steady state conditions may be in the aggregated or energized configuration (I deduce this from your description of the state of the cristae), it must be remembered that oxygen is not the only determinant of the configurational state. We have to know how hypoxia affects all the other relevant parameters before concluding that the configurational state is inconsistent with the physiological state.

*Goldspink:* When one examines individual mitochondria in electron micrographs of muscle fibers, their appearance may differ considerably even within the same muscle fiber. Is it probable that mitochondria even within the same fiber or cell may, at any one time, be in different states, i.e., energized or nonenergized?

*Green:* It is indeed not uncommon to find mitochondria within the same muscle fiber or the same cell in different configurational states. The concentration of oxygen, substrate, ADP, and inorganic phosphate—the determinants of the energizing process—may not be at all times equal in different parts of the same muscle fiber or cell. Some sort of gradient of these determinants may well underlie the fact that a fraction of the mitochondria is energized while the rest is not. When the suspending medium is fortified with these determinants, uniformity of configurational state can readily be achieved.

*Slautterback:* In cardiac muscle the T tubule is usually of very large diameter and its membrane is often accompanied by the basement membrane. What is the significance of the difference in bore size, and does the basement membrane, or any part of it, accompany the T-tubule membrane in skeletal muscle?

*Peachey:* The invaginating tubules of cardiac muscle, as you say, are larger in diameter than in skeletal muscle, and carry basement membrane with them. Physiologically the large bore size would aid the inward spread of depolarization from the surface, and with such large and short tubules as are usually seen in cardiac muscle, there would be very little delay or attenuation of the surface action potential as it spread into the tubules, even if the tubules were passive. The basement membrane of skeletal muscles does not enter the T system, except in those cases, as in crab muscle, where large-bore tubules like the cardiac tubules are found. It may be that we should not refer to these large-bore tubules with basement membrane content as "T system," but should distinguish them from the usual T system. I have not made up my mind on this.

*Dhalla:* Some published reports in the literature reveal connections between mitochondria and sarcoplasmic reticulum. Would Dr. Peachey comment on this observation and particularly on the view that mitochondria play an important role in the regulation of $Ca^{++}$ in addition to that of the sarcoplasmic reticulum?

*Peachey:* I am not convinced, by any published reports I know of, that such connections have been established. Unless such connections or some sort of communication exist, I do not see how mitochondria could act along with (or instead of) the sarcoplasmic reticulum in releasing $Ca^{++}$ during mechanical activation of the muscle fiber. Certainly mitochondria could act as a $Ca^{++}$ buffer, which might have some long-term benefit for the cell. I suppose it is possible that they could be induced to release $Ca^{++}$ under the influence of a chemical transmitter—even $Ca^{++}$ itself, released during activation—but I do not think there is any evidence for such activity on the part of mitochondria. One possibility is that mitochondria pick up some of the $Ca^{++}$ released by the sarcoplasmic reticulum and respond to this with some alteration of their oxidative activity, eventually releasing this $Ca^{++}$ for return to the sarcoplasmic reticulum. This is merely speculation at the present.

*Mommaerts:* Could it not be that certain structures can be both a mitochondrion and a sarco-reticular sac? In some muscles there might be reasons for this, perhaps like in the muscles of the croaking fish in which the sarco-reticular sacs are very prominent and mitochondria are not too obvious. The two structures have very many things in common, for example, sensitivity to certain antibiotics and certain exchange reactions or at least the rudiments of this. It could be that the sarco-reticular sac is a rudiment of the mitochondrion which has lost its oxidative reactions.

*Green:* That would be unlikely; to accept it would be to deny chemistry. The structure of the mitochondrion is specific. Its chemistry is entirely different from that of sarcoplasmic reticulum, and cannot in my opinion be changed so easily. That would be like converting hemoglobin into collagen. I suppose it could be done, but it is rather unlikely. This brings up a point—often the proximity of one membrane to another is interpreted by some morphologists as an indication of a chemical kinship. To the biochemist this is heresy, because of the implication that proximity can lead to chemical alchemy—that somehow the endoplasmic reticulum can end up as a mitochondrion or that mitochondrion can end up as the sarcoplasmic reticulum just by togetherness. Profound changes in phospholipids and protein composition and structure would have to be brought about to make possible such a transformation of one membrane into another.

*Mommaerts:* I was thinking in terms of homology, which would certainly allow further chemical differentiation.

*Peachey:* I think that in muscles of the adult frog or adult mammals, mitochondrion and sarcoplasmic reticulum seem to be quite morphologically distinct. They not only look different; one can also separate them. In such isolated preparations, one finds oxidative enzymes associated with the mitochondria and glycolytic enzymes and $Ca^{++}$ uptake ability associated with the sarcoplasmic reticulum. It has been said in morphology that the morphologist names the structure and then he thinks he understands it. It would be difficult to give up the names mitochondrion and sarcoplasmic reticulum, together with the meaning they convey to us. It does not strike me as impossible that there might be, in some special muscles, a membrane system which does incorporate both of these properties. I think one would want to study this morphologically and isolate it biochemically, and if you came up with such an organelle which had oxidative activity, glycolytic activity, and $Ca^{++}$-binding activity, then I would be perfectly willing to say that you have discovered sarcoplasmiceon or something like that.

*Kornguth:* Does ferritin or peroxidase pass from fibril to fibril via the fenestrated collar pores if introduced into the muscle cells?

*Peachey:* Ferritin that gets into damaged muscle fibers during soaking in a solution of ferritin is found throughout the muscle fiber sarcoplasm. I do not think we can tell if it passed through the holes in the fenestrated collar, or went around the sarcoplasmic reticulum by another route, since we see only static pictures in the electron microscope.

*Slautterback:* You pointed out the existence of fenestrations or perforations of the sarcoplasmic reticulum over the central region of the sarcomere, and remarked that this specialization may be associated with a need for greater membrane surface there. Can you explain why increased surface might be useful? Also, in connection with the same region, it seems to me that the increased surface is best seen by contrast with the lateral cisternae, since such sac-like enlargements are seldom if ever seen over the H band. Might it not be equally accurate to say that surface is lost for the sake of increased volume of the reticulum lumen?

*Peachey:* The only reason I thought that increased area might be useful in the fenestrated collar region of the sarcoplasmic reticulum was to provide more membrane through which $Ca^{++}$ could be pumped or to which $Ca^{++}$ could be absorbed. I suppose the relevant parameter should be surface-to-volume ratio.

*Bailey:* If, in fact, the sarcoplasmic reticulum and the T tubule are two separate systems with no physical connection, how is the sarcoplasmic reticulum excited by stimulation through the tubules?

*Peachey:* I meant to say merely that the sarcoplasmic reticulum and the T tubules are two separate membrane systems, implying that they are not directly continuous. The membrane of one does not run directly into the other in the way, for example, that the plasma membrane at the surface of the cell leads directly into the membrane of the T tubule. Please do not infer from this that the sarcoplasmic reticulum and T tubules are not physically connected in any way: clearly they are, or the triad would have no structural stability. The two membranes are attached over a considerable area in the triads (I have estimated this contact area as 80% of the tubular area in frog muscle), and this attachment zone has an interesting morphology, suggestive of some sort of electrical coupling (see Peachey, 1965). Presumably it is this junction that couples the depolarization of the T system, discussed by Dr. Adrian in Chapter 14, to the sarcoplasmic reticulum and leads to $Ca^{++}$ release, but it must be admitted that we know nothing about the nature of this coupling at the present time.

*Chrystall:* Is there evidence for the continuity of the sarcoplasmic reticulum system between sarcomeres of the same fibers? If so, what is its physiological significance?

*Peachey:* Yes, it seems clear from electron microscopy that the sarcoplasmic reticulum of adjacent fibrils in the same sarcomere is linked laterally into a continuous unit across the diameter of the fiber. One also sees connections between terminal cisternae from one sarcomere to the next along the same fibril in frog muscle, and it may well be that the whole sarcoplasmic reticulum is linked into one unit filling the entire fiber. The physiological significance of this is not known but it might aid the equilibration of $Ca^{++}$ ions or other substances held inside the sarcoplasmic reticulum in different regions of the fiber.

*Goll:* Have you ever observed or do you believe that there may be direct connections between the sarcoplasmic reticulum (or the T system) and the myofibril (e.g., perhaps at the Z line)? Is there any evidence that the Z line may be attached to the sarcolemma?

*Peachey:* These are difficult questions to answer, in that they depend on the interpretation of close proximities and suggestive densities in electron micrographs. Speaking conservatively, I would say that no convincing case has been made for either of the attachments you suggest, except in a couple of isolated examples of unusual muscles, but it must be admitted that the sarcoplasmic reticulum and fibrils stay in essentially perfect register in the longitudinal direction, and this implies some sort of interaction, although not necessarily at the Z line. I might also mention that one needs to be careful to separate the terms "Z line" as used by the electron microscopist, and "Z membrane" as used by light microscopists prior to the study of muscle cells in the electron microscope. The light microscopists' "Z membrane" was a thin, dense disc he saw after his particular methods of preparation, and probably included both the Z lines of the myofibrils and some parts of the sarcoplasmic reticulum (the triads in those muscles having triads at the level of the Z line). Electron microscopists, using their own particular methods of specimen preparation, distinguish between sarcoplasmic and fibrillar structures at this level of the striations, using the term "Z line" for the latter.

*Greaser:* You mentioned that the terminal cisternae often contained a dense granular material. In electron microscopic studies of isolated sarcoplasmic reticulum fragments at the University of Wisconsin we (Greaser, Cassens, Briskey, and Hoekstra, 1969) often observed similar material in an appreciable number of vesicles. Are you aware of any studies concerning the composition of this material and whether it has any possible role in $Ca^{++}$ binding or storage?

*Peachey:* We do not know. All we know about the material is that it has the appearance of being finely granular and often fairly dense, but we have introduced density into the preparation with osmium and with the heavy metal stains. In most muscles it is confined to the terminal cisternae and does not enter the longitudinal tubules or the other portions of the reticulum. In some muscles, one sees very clearly a narrow constriction that is just at the boundary between the terminal cisternae and longitudinal tubules. The granular material found on one side of the constriction is not found on the other side. I think one could also comment that the region where this granular material is found is the region which is implicated in audio-radiographic studies as the site of storage and release of $Ca^{++}$ (Winegrad, 1965).

*Kornguth:* If ions are passed into the myofibrillar area of one muscle fiber, can changes be detected rapidly in adjacent fibers by the Lowenstein procedure?

*Peachey:* No such experiment of looking for passage of ions from one skeletal muscle cell to another has been done to my knowledge, but I think we can rule out this possibility by noting that current passed into one fiber from an internal microelectrode does not lead to depolarization and contraction of adjacent fibers.

*Kornguth:* During the studies of the ontogenesis of the sarcoplasmic reticulum, are ribosomes seen over the areas defined as presumptive sarcoplasmic reticulum? Is there any evidence for the secretion of newly synthesized protein into the lumen of sarcoplasmic reticulum in adult muscle?

*Peachey:* Ribosomes are seen attached to vesicular membranes in early cultured muscle cells, but by the time these membranes can be recognized as primitive sarcoplasmic reticulum networks, there are no attached ribosomes. We cannot say whether the ribosomes detach as the membranes transform from vesicles into networks, or if the networks are extruded from the vesicles between the ribosomes. In adult muscle, most of the ribosomes are free, that is, not attached to membranes, and I know of no evidence for the segregation of newly synthesized protein into the lumen of the sarcoplasmic reticulum. I believe that several authors have commented that such segregation is typical of cells synthesizing proteins for "export," that is, for secretion outside the cell, and muscle seems to synthesize its protein only for domestic or intracellular use.

*Kornguth:* Have attempts been made to separate the sarcoplasmic reticulum from the rough endoplasmic reticulum by subcellular fractionation procedures? If so, do they have different densities?

*Peachey:* Attempts have been made to separate subfractions of the sarcoplasmic reticulum according to differences in sedimentability. I do not believe it has been possible to relate the various fractions to different regions of intact sarcoplasmic reticulum.

*Lawrie:* Is there any evidence to suggest that the contraction of a muscle helps serve other needs—such as aiding the transport of nutrients or the removal of waste products?

*Peachey:* I know of no such evidence.

*Lorand:* What proportion of the total surface area of the sarcoplasmic reticulum would you estimate to be terminal cisternae? Do various muscles display different ratios between the surface areas of terminal cisternae and the longitudinal tubules of the sarcoplasmic reticulum?

*Peachey:* I estimate that the terminal cisternae represent about one-third of the surface area of the sarcoplasmic reticulum in frog muscle. I suspect that differences in this figure will be found in different fiber types, but numerical values have been obtained only in the one case.

*Greaser:* As is stated in Chapter 13, the correlation between the amount of sarcoplasmic reticulum and T-system membranes and the speed of contraction appears to be quite high. Are there any muscles that have been found with large amounts of sarcoplasmic reticulum and limited T systems or vice versa? Which of the two systems is most related to contraction speed?

*Peachey:* Generally the two membrane systems vary in a rather parallel way. In other words, I do not know of any muscles with a large amount of T system and little sarcoplasmic reticulum, or vice versa, although I am quite prepared to be corrected on this. According to current ideas, one would expect that a large amount of T system would be needed for a muscle to turn on its contraction rapidly, whereas a large amount of sarcoplasmic reticulum would be related to rapid relaxation. Usually, however, these two speeds also vary in parallel, that is the whole cycle of contraction and relaxation is slow or fast.

*Gunther:* Is there a species difference in the location of the transverse tubular system (TTS)? It occurs at the Z band in frog muscle. I have seen electron micrographs that show the T system at the A-I junction in bovine longissimus muscle.

*Peachey:* The species difference you mention is a real one, and many examples of the two different locations you mentioned are now known. The A-I junction location is typical of mammalian muscle, for example. I do not believe we know anything about the reasons for the difference, although the A-I location could imply a shorter diffusion distance for $Ca^{++}$ and thus a more rapid activation.

*Adrian:* In vertebrates the TTS is found either at the Z line or at the A-I junction. Apart from the examples in the question, the TTS occurs at the A-I junction in lizard muscle. Where this happens there will be 2 triads per sarcomere instead of only 1, and $Ca^{++}$ released from the triad would have to diffuse a shorter distance to the overlap region of the actin and myosin filaments. The TTS is usually found at the A-I junction in rapidly contracting twitch fibers, at the Z line in slowly contracting twitch fibers.

*Lawrie:* It was interesting to hear Dr. Peachey put the proposition that the classification of muscles as "red" or "white" cannot explain all the physiological findings. Similar views arise from meat investigations. Thus, muscles which are similarly "red" or "white" often react quite differently to processing and in preservation.

What are the physiological characteristics of those fibers in cat ocular muscles which have few mitochondria (low oxidative capacity) and little sarcoplasmic reticulum development (little capacity for speed or relaxation); and what properties of the muscles do they represent?

*Peachey:* The particular fibers you mentioned from the cat extraocular muscles have morphological properties that suggest that they are slow in their contraction cycle. To say that they *are* slow would be dangerously close to circular reasoning, since we base our interpretation of "slow morphology" on the idea that long diffusion distances and little sarcoplasmic reticulum must be slow according to our current ideas of the mechanism of excitation-contraction coupling. I do think, however, that any such correlation we can make is useful as a test of these ideas, and currently we are looking at these muscles electron-microscopically and physiologically with just this in mind. Unfortunately, we have no evidence as yet that we can be sure relates to the physiology of this particular group of fibers. I would guess that, when the evidence is available, these fibers will be found to play a role in some of the tonic functions of the eye muscles, such as convergence.

*Greaser:* Although I agree with your objections to directly equating "red" and "white" with "slow" and "fast" respectively, I wonder if there are any direct contradictions between the physiologically measured speed and the muscle color. The example you gave was based on microscopic evidence alone.

*Peachey:* You are quite right in challenging my choice of example for a muscle fiber type violating the red/white, slow/fast dogma. I chose this particular example simply because it represents a new observation, not published before. But, as you say, it is based on microscopic evidence alone. The slow fibers of frogs represent a well-documented violation of dogma, in that they are known to be physiologically slow, and yet they do not appear to be red, nor do they contain many mitochondria (Peachey and Huxley, 1962). Conversely, the pectoralis and supracoracoideus muscles of the rubythroated hummingbird contain only red fibers. These are the main flight muscles, and these birds have very high wing-beat frequencies (my recollection is 80 per sec) so obviously these red muscles are very fast. It would appear that redness here is associated with the ability for sustained flight, since these tiny birds are reported to fly across the Gulf of Mexico, a distance of over 500 miles, and to use stored fat for energy on these flights, requiring a major oxidative effort (George and Berger, 1966). Perhaps these two examples better document my case, which is, simply, that color is not necessarily associated with, nor is it sufficient evidence for, a particular speed of contraction. When you observe the color of a muscle, or the histochemical properties of a muscle fiber, you may know something of its biochemical capabilities, but you do not necessarily know its speed of contraction. To say that a fiber or a muscle is red and therefore it is slow may be right most of the time, but it clearly will be wrong some of the time.

*Peter:* Is it true that most of the exceptions to the red fiber = slow fiber equations deal with those fibers which have rather small diameters?

*Peachey:* We do not have complete information on this. It is true of the extraocular fibers. I think, in the pigeon breast muscle, which is mixed red and white, the fibers are smaller. I do not know, offhand, whether the fibers which are carrying out the fast function but which seem to be red are reduced in size in those animals that have all red muscle in the breast.

*Lawrie:* What other muscles have such fibers? Are there, for example, similar morphological contradictions in the fibrillation of muscles of insect thorax?

*Peachey:* Fibers similar to the frog slow fibers have been found in other amphibians and in reptiles, but not to my knowledge in mammals outside the extraocular group. No insect muscles I know of correspond closely in morphology to these fibers.

*Slautterback:* Is it possible at this time to say anything about the nature of the coupling of the ionic events in the transverse tubule membrane and the contiguous membrane of the lateral cistern of the sarcoplasmic reticulum?

*Adrian:* Very little is known about this crucial step in the contraction-activation process. As in cardiac muscle, a number of ionic conductance mechanisms are identifiable in striated muscle. But there is little agreement, and a good deal of conflicting evidence, about how these channels are distributed between the fiber surface and the wall of the TTS. Any ionic channels in the membrane of the lateral cistern would presumably be in series with and obscured by the ionic channels in the walls of the TTS. A parallel between the tubular-cisternal junction and the neuromuscular junction, has been suggested. There seem to be no grounds for supposing that the tubular-cisternal events involve the quantal release of a transmitter substance.

*Greaser:* Do you feel the close apposition of the T-tubule membranes with the terminal cisternae of the sarcoplasmic reticulum is a necessary prerequisite for the completion of the electrical events during a contraction cycle?

*Adrian:* Yes, such a juxtaposition seems to be required if the $Ca^{++}$ for activation is to come from an internal store. The terminal cisternae of the longitudinal reticulum seem the most probable site of the $Ca^{++}$ store. The volume and surface area of the TTS make it unlikely that $Ca^{++}$ is either stored in or bound to that system.

*Greaser:* There is some evidence that the $Ca^{++}$ involved in activation may be bound to sites on the sarcoplasmic reticulum rather than transported across the membranes. Would this be more compatible than membrane transport in the model you have suggested?

*Adrian:* Rather little is known about the way in which the potential difference across a membrane might affect binding of an ion to a site on the membrane surface. Many mechanisms would be possible and they could doubtless be devised to fit the rather crude strength-duration results.

*Bailey:* What is the mechanism of the transduction of electrical energy to chemical change, associated with transfer of $Ca^{++}$ across the membrane of the sarcoplasmic reticulum?

*Adrian:* It is unknown. One can say, however, that some amplification stage is required; the potential change across the wall of the TTS appears to trigger a release of $Ca^{++}$.

*Greaser:* It has been postulated that in muscle fibers with small diameters the $Ca^{++}$, which presumably triggers contraction, diffuses in from the cell exterior. Have there been any electrophysiological experiments with controlled extracellular $Ca^{++}$ levels to test this hypothesis?

*Peachey:* I do not recall any experiments specifically directed toward establishing a relationship between fiber size and source of activating $Ca^{++}$ in skeletal muscle.

*Kornguth:* You indicated that $Ca^{++}$ moves the threshold away from the resting potential. Is this a consequence of a mitochondrial effect (i.e., can uncouplers or adenosine triphosphatase inhibitors shift the threshold)?

*Adrian:* In all probability the $Ca^{++}$ is acting on the luminal surface of the wall of the TTS; at what is topologically the exterior surface of the fiber. $Ca^{++}$ has similar effects on the threshold for electrical excitability.

*Kornguth:* Have any attempts been made to fractionate the tubular systems of the muscle and determine whether they have the intrinsic capacity to concentrate $Ca^{++}$?

*Adrian:* I do not know of any such attempt.

*Greaser:* There are several reports suggesting that the interior of the T tubules is in communication with the extracellular space. What evidence is there that this lumen fluid has the same ionic composition as that found in other areas of the extracellular space?

*Adrian:* There is no good evidence that the lumen of the TTS contains a fluid with the same ionic composition as the extracellular fluid. But since very large molecules such as ferritin and dyes are not excluded from the lumen, there seems little reason to suppose that the extracellular ions will be excluded, unless there is a high fixed charge density on the walls of the TTS. This cannot be excluded at present, but there is little evidence for a significantly large charge density.

*Kornguth:* I am interested in the postulate that fixed charges exist at the mouth of the tubular system. Do large-molecular-weight polyelectrolytes affect the threshold potential? If so they could be tagged and examined in the electron microscope to see whether they are localized to the surface tubular system.

*Adrian:* The use of high-molecular-weight polyelectrolytes which can be visualized in the electron microscope might give very useful information about the presence of surface charges, if one could exclude the other reasons for the surface adsorption of the polyelectrolyte.

*Lorand:* Would Dr. Adrian speculate on the possible electrophoretic implication of the radial field in his postulate?

*Adrian:* If you are considering that this radial field might move material into the T system, then that is a question which has exercised a number of people. I think possibly the more serious question is not so much the electrophoretic transport of material into and out of the T system, but rather the electrokinetic movement. We think there are relatively large currents flowing in very small tubular structures. We do not know, but we can probably put an upper limit to the surface charge density on the inside of the tubular system. If there were any surface charges attached to the internal surface of the tubule, the passage of a current back and forth along the tubule would probably cause large bulk movements of fluid and volume changes in the T system when a current passes. I can only say that we are aware of this possibility but have not yet been really forced into going into what I think is an exceedingly difficult field to explore.

*Slautterback:* Do I understand correctly that your model assumes homogeneity in the current path from sarcolemma to activator store? What would happen to your model if you assumed inhomogeneity at the contact between the membranes of the triad?

*Adrian:* The model proposed for the strength-duration results cannot be taken too literally. There are many ways of reconstructing strength-duration relationships of the kind that gives very tolerable fits to the experimental data. In the particular model we used, the current which is the movement of the activator might be only part of the total current flowing between the inside and the outside of the fiber by the many possible pathways. To this extent the model does not assume a single pathway from sarcolemma to activator store. Since the concentration of the activator and the volumes of the "site of action" are undefined in absolute terms, it is not possible to say what fraction of the total ionic current is represented by the movement of the "activator."

*Kornguth:* Is the rate of decline in capacitance from 6 to 2 $\mu F/cm^2$ similar to the rate of detubulation?

*Adrian:* Experimentally it is not really possible to determine the rate of detubulation in a muscle fiber.

*Fedde:* Does the individual twitch muscle fiber contract (*in situ*) in an all-or-nothing manner with the stimulus of a single action potential, or can there be grading of contraction depending on the frequency of stimulating action potentials?

*Adrian:* In response to a single action potential a muscle fiber *in situ* gives a single all-or-nothing twitch. This twitch probably involves all the myofibrils in the fiber. As the frequency of the stimulating action potentials rises the twitches become an unfused tetanus and finally a fused tetanus.

*Greaser:* Most studies on the electrophysiology of muscle have been conducted using frog muscle fibers. Could you briefly explain the reasons for the scarcity of this type of study with mammalian muscle?

*Adrian:* The requirement of working at 37° C is a considerable experimental limitation. The events of contraction and its activation are rapid and being able to slow them by cooling to 0° C is a great advantage. Frog muscle works well at 0° C and moreover even at room temperature its metabolic rate is such that reasonably small muscles can be adequately supplied with oxygen by diffusion alone.

*Audience:* What is the effect of glycerol treatment on slow fibers of frog muscle?

*Adrian:* Stephani and Steinbach (1968) have tried the glycerol treatment on the tonus bundle of the frog iliofibularis muscle. This contains both fast and slow muscle fibers. They report that the electrical properties of the slow fibers are unaltered by the glycerol treatment, neither is the contraction in response to depolarization abolished. Their results are consistent with the notion that in slow fibers the very sparse T system does not contribute much to the measured membrane capacity, and does not play much part in excitation-contraction coupling.

*Duggan:* In association with Dr. Anthony Martonosi, we have been investigating some of the permeability properties of rabbit skeletal muscle microsomes. While much work has been done on $Ca^{++}$ uptake by these vesicles—a function which mimics that of relaxation in the intact muscle—little attention has been devoted to the mechanism of $Ca^{++}$ release, the trigger for muscle contraction. This release is a very rapid event occurring in a few milliseconds, whereas the recovery or reaccumulation of $Ca^{++}$ takes a much longer time. Also, in isolated vesicles, the $Ca^{++}$ which is taken up leaks out at a low rate, so that inhibition of uptake is not sufficient to account for the observed rate of activation of muscle.

Using inulin carboxyl-$^{14}C$ as a marker of permeability, we measured the volume of microsomal water into which inulin could not penetrate in sediments at 61,000 *g*. This amounted normally to 3.5–5.0 $\mu$l per mg of protein. On treatment with EDTA at *p*H 7.2, this inulin-inaccessible water volume did not change, but at *p*H 8.0 it diminished to 1.5–2.0 $\mu$l per mg protein in 1 mM EDTA. Simultaneously with the penetration in inulin with increasing EDTA concentrations, the $Ca^{++}$-uptake rate decreased and the $Ca^{++}$-dependent adenosine triphosphatase activity rose by about 200%. There is a diminution in bound $Mg^{++}$ and $Ca^{++}$ in EDTA-treated vesicles; EGTA is about as effective as EDTA in affecting the permeability to inulin, though only the $Ca^{++}$ content falls. While the EDTA treatment is, of course, not a physiological situation, it provides a technique for studying permeability changes in muscle microsomes, which may be the actual event in muscle activation.

## *References*

George, J. C. and A. J. Berger. 1966. *Avian Myology*. Academic Press, New York.

Greaser, M. L., R. G. Cassens, E. J. Briskey, and W. G. Hoekstra. 1969. Post-mortem changes in subcellular fractions from normal and pale, soft, exudative porcine muscle 2. Electron microscopy. *J. Food Sci. 34*:125.

Greaser, M. L., R. G. Cassens, W. G. Hoekstra, and E. J. Briskey. 1969. Purification and ultrastructural properties of the calcium accumulating membranes in isolated sarcoplasmic reticulum preparations from skeletal muscle. *J. Cell. Physiol. 74*:37.

Peachey, L. D. 1965. The sarcoplasmic reticulum and transverse tubules of the frog's sartorius. *J. Cell Biol. 25*:209.

Peachey, L. D. and H. Huxley. 1962. Structural identification of twitch and slow striated muscle fibers of the frog. *J. Cell Biol. 13*:177.

Stephani, E. and A. Steinbach. 1968. Persistence of excitation contraction coupling in "slow" muscle fibres after a treatment that destroys transverse tubules in "twitch" fibres. *Nature 218*:681.

Winegrad, S. 1965. The location of muscle calcium with respect to the myofibrils. *J. Gen. Physiol. 48*:997.

# PART 4

## Myofibrillar Proteins

# Chapter 16

# *Interaction of Major Myofibrillar Proteins*

J. GERGELY

In looking over the proceedings of the first symposium on muscle as a food one cannot help being impressed by the progress made during the last 4 years, not only in our total knowledge of muscles but particularly in the area of interaction of the muscle proteins which forms the molecular basis of muscle contraction and muscle relaxation.

It is now clear, through the work especially of S. Ebashi and his colleagues, that the interaction of myosin and actin, the myofibrillar proteins that have been known for many years to be involved in contraction, is regulated by some other proteins. The picture has become somewhat simpler during the past 4 years, although our knowledge of both the major, as well as the so-called minor, regulatory proteins has expanded. Four years ago it may have seemed that the newer proteins were integrally linked with the interaction of actin and myosin, and particularly the role of $Ca^{++}$ in this interaction looked much more complicated. Whereas Davies' (1963) important paper in *Nature* considered $Ca^{++}$ as being a link between actin and myosin, today it seems safe to say that this component of the myofibrillar contractile apparatus interacts with the regulatory proteins, and that interaction of actin and myosin can be discussed without the complications that might have arisen from the $Ca^{++}$ requirement for the actin-myosin interaction (Ebashi and Endo, 1968).

In order to discuss the interaction of the "major" proteins, actin and myosin, it is first important to discuss their localization within the myofibril, describe their molecular properties, then consider how the properties of the individual proteins can throw light on their interaction *in vitro* and, finally, consider the way in which the *in vitro* interactions might have a bearing on the interactions in the actual living system.

There is today general agreement on the correctness of what has come to be known as the sliding filament theory of muscle; it has both morphological and functional implications. Morphologically it refers to the fact that the contractile apparatus is organized in two sets of filaments, each forming a hexagonal array, running parallel to the axis of the muscle fiber (Fig. 16.1). Thick ones, approximately 120 Å in diameter and having a length of 1.6 $\mu$, occupy the central part of the sarcomere; thin ones, of diameter 80 Å and length 2 $\mu$, are symmetrically disposed about the Z band (Huxley, 1960). Functionally the sliding filament theory states that contraction depends on a relative sliding of these filaments without length changes in the filaments themselves. This view is supported by recent X-ray work to which I shall return—particularly that of Huxley and Brown (1967), in confirmation of earlier electron microscopic and phase microscopic observations (Huxley and Niedergerke, 1954; Huxley and Hanson, 1954; Page and Huxley, 1963). Thus what appears as contraction on the level of the whole muscle or on the level of the sarcomere does not have a counterpart in terms of overall contraction of molecular constituents.

Thick filaments of vertebrate muscle are thought to consist almost entirely of myosin, with the exception of the M line. They have an important feature: the cross bridges are seen along the whole length of the thick filament except the central 0.2 $\mu$ stretch. Although in the early stages of the development of these concepts it appeared that cross bridges were spaced every 70 Å or so (Huxley, 1960) it seems clear now that their separation along the filament axis is 143 Å and there are two diametrically opposed cross bridges at each level (Fig. 16.2). Pairs of cross bridges

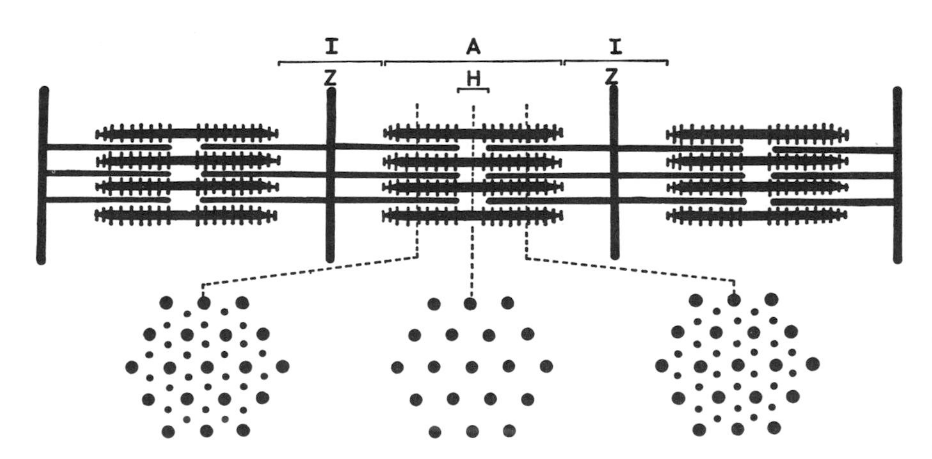


*Fig. 16.1.* Diagram showing arrangement of filaments in skeletal muscle. Thick filaments contain myosin; thin ones contain actin. Note cross bridges on myosin filaments with the exception of the 0.2 $\mu$ central portion. From Huxley (1969).

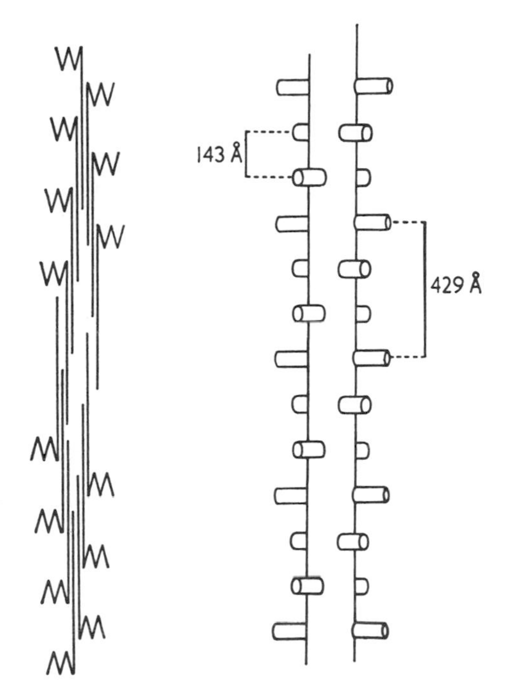


*Fig. 16.2.* Aggregation of myosin and arrangement of heavy meromyosin projections. *Left,* Schematic representation of filament formation. Note reversal of polarity on either side of the center. The central portion contains only light meromyosin rods, but no heavy meromyosin projections. From Huxley (1963). *Right,* Helical arrangement of projections on myosin filaments as deduced from X-ray data. From Huxley and Brown (1967).

at successive levels are rotated by 120° with respect to each other and thus the whole cross-bridge system constitutes a 6/2 helix (Huxley and Brown, 1967). As we shall see shortly, these cross bridges correspond to that portion of the myosin molecule that carries the site or sites responsible both for the interaction with ATP and with actin.

The thin filaments, mainly actin, reveal a globular substructure (Fig. 16.3). Hanson and Lowy (1963) were first to show that the F-actin filaments consist of a double helix of globular units, each globular unit corresponding to one G-actin molecule. The pitch of the actin helix is about 750 Å but the data so far obtained do not permit an unambiguous assignment for this value *in vivo* (Hanson, 1967).

The sliding filament theory and the role of the cross bridges in tension production is further supported by the good agreement between the experimentally determined tension of single muscle fibers as a function of length and the tension which would follow from the sliding theory on the assumption that tension is proportional to the number of links formed between the thick and thin filaments (Gordon, Huxley, and Julian, 1966) (Fig. 16.4).

After much discussion in the literature on the structure of myosin, there begins to emerge a reasonably well-supported picture of the mole-

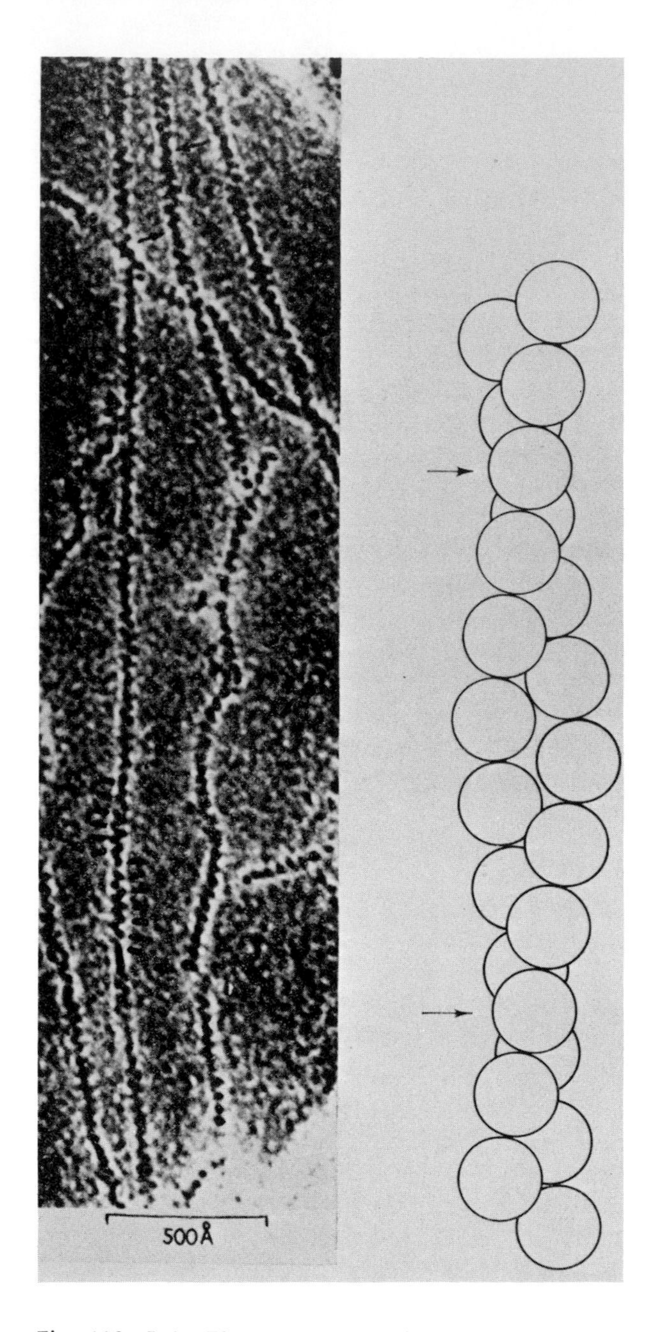


*Fig. 16.3. Left,* Electron micrograph of actin filaments. *Right,* Schematic representation of actin filaments. Arrows indicate crossover points of strands. The subunits, corresponding to G-actin molecules, are spheres of 55 Å. From Hanson and Lowy (1963).

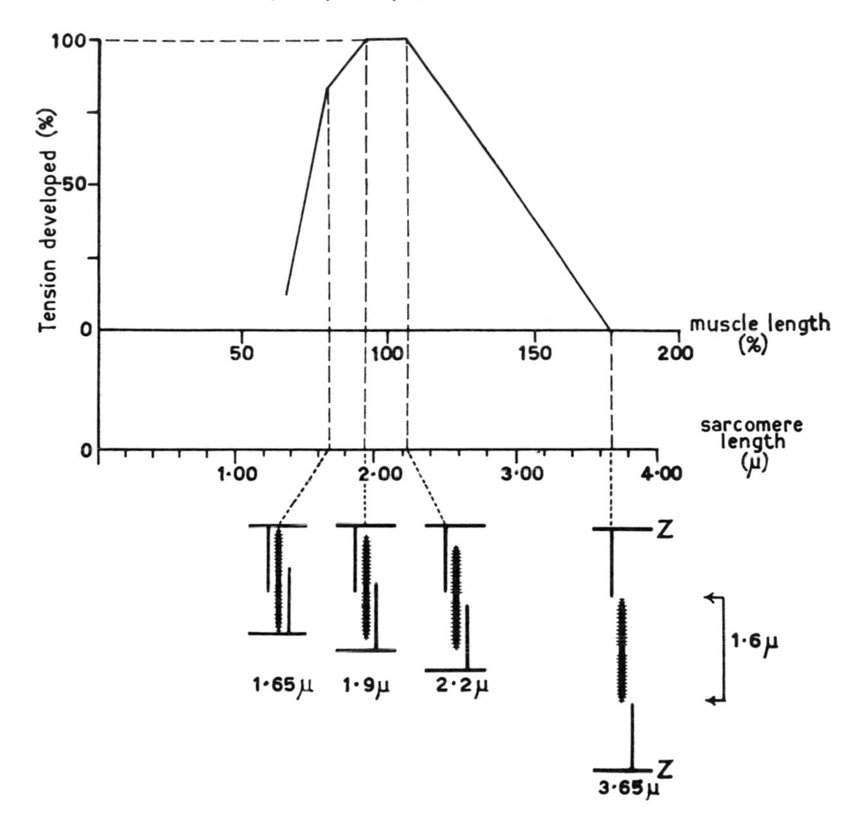


*Fig. 16.4.* Relation between the amount of tension produced actively by frog muscle and the extent of overlap of actin (thin) and myosin (thick) filaments. The upper part of the figure shows the tension produced by a single fiber plotted against muscle length (in per cent of a standard length) and also against sarcomere length, determined by observing the fiber in an interference microscope. The lower part of the figure shows the relative positions of the myosin filaments (1.6 μ long) and the actin filaments (1.0 μ long) at four selected sarcomere lengths. The length-tension relation is taken from Gordon, Huxley, and Julian (1963), the thin filament length from Page and Huxley (1963). From Hanson and Lowy (1965).

cule. One of the lucky breaks in the elucidation of the structure of myosin has come from the unexpected finding that myosin can undergo mild proteolytic digestion without losing its biological activities. I remember my surprise just 20 years ago, in 1949, when I began studying the effect of trypsin on myosin (Gergely, 1950, 1953). I got into this work in a rather roundabout way when Albert Szent-Györgyi suggested that it might be worth looking at the effect of trypsin on heat-denatured myosin since, as we then assumed, the increased digestibility of the denatured molecule might give us some clues to the structure of proteins in general and to the electronic continuum in the intact protein in particular (Evans and Gergely, 1949). In one sense this study came to nothing, but in another it opened up many new possibilities, when it appeared that myosin could be digested with trypsin even in the native state yet, strangely, although there was a profound change in structure, the enzymic activity remained intact. I reported these findings first 20 years ago in Albert Szent-Györgyi's Woods Hole house when the then rather small group of investigators active in this field came together to review the problems at the frontiers of muscle research.

As these problems unfolded, Andrew Szent-Györgyi introduced the term meromyosin and has greatly contributed to the elucidation of the nature of the fragments obtained after proteolytic digestion of myosin. Light meromyosin (LMM) was separated as a highly α-helical fragment from heavy meromyosin (HMM) which carries the site of ATP and actin interaction (Szent-Györgyi, 1953). We were able to show later that not only trypsin but also chymotrypsin can produce the very limited split in the myosin molecule, separating those regions that have different functional roles (Gergely, Gouvea, and Karibian, 1955). Other proteolytic enzymes were also found to produce limited proteolysis of myosin, and during recent years papain (Kominz, Mitchell, Nihei, and Kay, 1965) has become a very useful tool, particularly in the hands of Lowey, Slayter, Weeds, and Baker (1969). This paper provides the best picture of the myosin molecule, showing that it consists of two major polypeptide chains which run through the whole length of the myosin molecule (Fig. 16.5). Proteolytic enzymes are able to split it at two points, thus producing the original α-helical fragment, L-meromyosin, a second highly helical fragment, subfragment-2, and two rather globular particles—first observed by Mueller and Perry (1962)—that carry adenosine triphosphatase activity (see below) and the ability to combine with actin (HMM subfragment-1). These two properties, the ability to interact with ATP and with actin, play a key role in the molecular processes that underlie muscle contraction. The essentially two-subunit nature of myosin is not only revealed by physico-chemical measurements (Lowey, Slayter, Weeds, and Baker, 1969), the study of the binding of nucleotides (Schliselfeld and Bárány, 1968; Young, 1967) and of their analogues, such as pyrophosphate (Nauss, Kitagawa, and Gergely, 1969) to myosin and its fragments, but also, perhaps most beautifully, through the electron microscopic observations of Slayter and Lowey (1967). Electron micrographs of preparations shadowed while the specimen was rotated directly

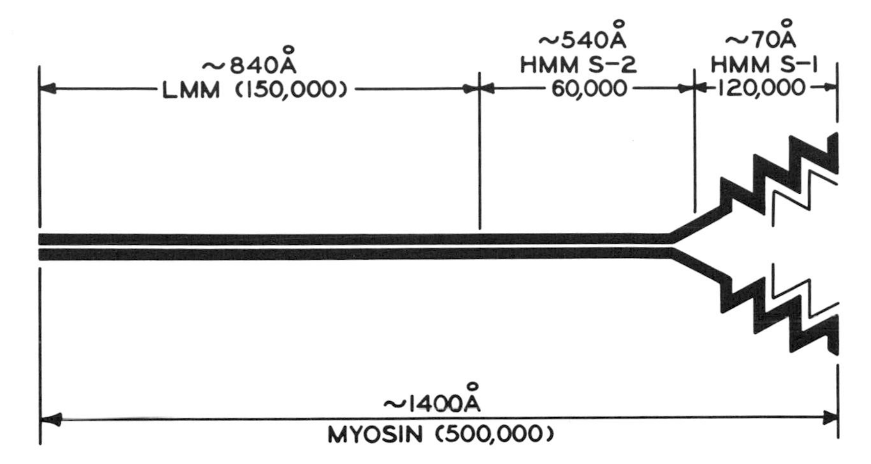


*Fig. 16.5.* Schematic representation of the myosin molecule. The numbers in parentheses indicate molecular weights. Two light chains are indicated, one adjacent to each HMM S-1. From Lowey, Slayter, Weeds, and Baker (1969).

reveal the presence of two so-called heads in the myosin molecule. Each of these heads can be identified with one of the subfragment-1's.

In the light of the structure of the myosin molecule the structure of the thick filaments is readily understood. As Huxley (1963) first pointed out, myosin molecules have a tendency to form end-to-end aggregates involving the LMM rods, which then grow into larger structures, the polarity of the myosin molecules being reversed on either side of the center. The globular ends of the myosin molecules form projections on these aggregates, the appearance of which bears a remarkable resemblance to the thick filaments. Pepe's recent work (1967*a, b*), combining the use of fluorescent antibodies with electron microscopic studies, has greatly elucidated the structure of the myosin filaments. I shall return to it later.

Another aspect of the myosin molecule which may play an important role in the regulation of enzymic activity and its interaction with actin is found in the so-called small subunits or short chains of the myosin molecule. These are small proteins having a molecular weight of about 20,000 (Lowey, Slayter, Weeds, and Baker, 1969; Dreizen, Gershman, Trotta, and Stracher, 1967)—possibly three different molecular species (Weeds, 1969). They are released from the parent myosin on mild heating, on exposure to alkali, or on treatment with acylating reagents such as succinic anhydride. According to published reports these subunits can be reversibly removed and the enzymic activity restored on recombination (Frederiksen and Holtzer, 1968; Stracher, 1969). These observations, if confirmed, will dispel doubts concerning the essential role of the small subunits in the structure and function of myosin (Gaetjens, Bárány, Bailin, Oppenheimer, and Bárány, 1968).

Before discussing in detail the enzymic properties of myosin, it seems convenient to consider briefly some of the properties of actin. The discovery of actin, as is probably well known today, resulted from the work in Albert Szent-Györgyi's laboratory in the early 1940's in Szeged, Hungary. It stemmed from observations that the response to ATP of muscle extracts, then referred to as myosin,* varied depending on the length of the extraction. It soon became apparent that these extracts really contained two components, one of which became identified as myosin and the other as actin (Banga and Szent-Györgyi, 1942). The direct extraction of actin from muscle is rather difficult, but from an acetone-dried muscle residue actin can be extracted with water (Straub, 1942). These extracts contain actin in a globular form with a molecular weight, according to most recent estimates, of 48,000. This molecular weight is based on physical chemical measurements, on the stoichiometry of nucleotide binding to actin (Rees and Young, 1967) first shown by Straub and Feuer (1950), and on the amounts of an unusual amino acid —3-methyl histidine—present in actin (Johnson, Harris, and Perry, 1967).

One of the striking properties of actin is its polymerizability; that is, the ability of the actin monomers to form fibrous aggregates on addition of salts (Fig. 16.6). This polymerized form is referred to as F-actin and the electron microscopic appearance of these *in-vitro*-formed fibrous proteins is identical with that of the thin filaments found in muscle (Hanson and Lowy, 1963; Huxley, 1963). The ability of actin to polymerize is not only of interest as far as the structure of muscle is concerned, but provides an extremely useful tool for the purification of actin. Mommaerts (1951) showed that polymers can be sedimented in the high-speed preparative centrifuge leading to separation from many—but not all, as it turned out—impurities present in actin extracts. In fact, some of these "impurities" are members of the regulatory system to be discussed in the next paper. They are virtually absent if actin is extracted at low temperature (Drabikowski and Gergely, 1962).

Actin contains a tightly bound divalent metal—one mole per mole of monomer—which, depending on the milieu to which actin is exposed in the course of the preparative procedure, is either $Ca^{++}$ or $Mg^{++}$, usually a

* In the earlier terminology myosin stood for what we would now term natural actomyosin. The Szent-Györgyi school introduced the term myosin B for natural actomyosin, myosin A denoting myosin proper.

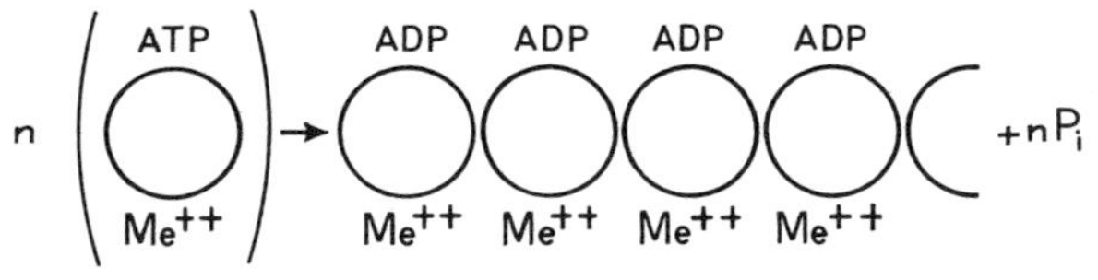

*Fig. 16.6.* Polymerization of G-actin on addition of salts. Note the stoichiometric liberation of inorganic phosphate.

mixture of both (Bárány, Finkelman, and Therattil-Antony, 1962; Drabikowski and Strzelecka-Golaszewska, 1963; Kasai and Oosawa, 1968). Kasai (1969) has recently suggested that the thin filaments *in vivo* contain $Mg^{++}$ as the bound divalent cation. On polymerization of actin monomers, ATP is dephosphorylated to ADP with liberation of phosphate (Straub and Feuer, 1950; Laki, Bowen, and Clark, 1950; Mommaerts, 1952) and under the usual conditions of polymerization this dephosphorylation stops with polymerization so that a stoichiometric amount of phosphate is liberated. The dephosphorylation of actin-bound ATP on polymerization has intrigued many workers since it appeared attractive to tie this dephosphorylation—and a reversible polymerization and depolymerization of actin—to the splitting of ATP associated with muscle contraction (Straub and Feuer, 1950). The evidence to date, however, does not favor the view that the dephosphorylation of ATP observed on polymerization of actin is in any way connected with the chemical events accompanying or underlying muscle contraction or that a reversible change in the state of polymerization of actin takes place (Martonosi, Gouvea, and Gergely, 1960*a;* West, Nagy, and Gergely, 1967). Indeed, *in vitro* polymerization can take place with actin preparations that contain ADP instead of ATP (Hayashi and Rosenbluth, 1960), or even those that lack the nucleotide and the divalent metal (Kasai, Nakano, and Oosawa, 1965; Bárány, Tucci, and Conover, 1966).

At present all available evidence suggests that the divalent metal and the nucleotide attached to G-actin contribute to the stability of the globular form (Nagy and Jencks, 1962), and once polymerization has taken place they do not seem to play a functional role. Both features are useful for detecting possible changes in the properties of the actin molecule. The nucleotide and the metal are exchangeable in the globular form but quite stable, both kinetically and thermodynamically, in the fibrous form (Martonosi, Gouvea, and Gergely, 1960*b*). On interaction with myosin, increased exchangeability of both nucleotide (Szent-Györgyi and Prior, 1966) and metal (Oosawa, Asakura, Asai, Kasai, Kobayashi, Mihashi, Ooi, Taniguchi, and Nakano, 1964) have been reported, but these changes may be rather secondary (Kitagawa, Drabikowski, and Gergely, 1968; Moos and Eisenberg, 1969; Moos, Eisenberg, and Estes, 1967) reflecting changes subsequent to the essential interaction between myosin and actin that leads to contraction. Mechanical effects such as vigorous treatment in a homogenizer (Kakol and Weber, 1965) or exposure to ultrasound (Asakura, Taniguchi, and Oosawa, 1963) produce increased exchangeability which is also observed on raising the temperature (Asai and Tawada, 1966). It is interesting that under conditions that lead to increased exchangeability of the nucleotide, there is an accompanying hydrolysis which suggests that actin is a would-have-been enzyme which, however, is prevented from continuing its activity once polymerization has taken place.

The interaction between actin and myosin is readily observed *in vitro* and, depending on the conditions under which the observation is made, can assume various forms. Thus under conditions of high ionic strength —about 0.5—there is an increase in viscosity; at low ionic strength—resembling that prevailing in living muscle—a fine suspension is formed with actin and myosin combined (Szent-Györgyi, 1945). One of the early and most remarkable observations on the *in vitro* system was made with the use of ATP. ATP is not only enzymically dephosphorylated by myosin or actomyosin, but it also changes the state of actomyosin (Engelhardt and Ljubimowa, 1939; Needham, Chen, Needham, and Laurence, 1941). At high ionic strength dissociation takes place (Gergely, 1956), whereas the phenomena at low ionic strength depend on a number of conditions. It may be immediate superprecipitation—that is, the formation of a flocculant precipitate—or it may be preceded by clearing (Spicer, 1952) corresponding to the dissociation of actomyosin shown by physicochemical measurements (Maruyama and Gergely, 1962) as well as by direct electron microscopic observations (Ikemoto, Kitagawa, and Gergely, 1966) (Fig. 16.7). These phenomena, superprecipitation and clearing, may be regarded as the *in vitro* analogues of contraction and relaxation.

The interaction of actin with myosin has a built-in directionality. The structural polarity of the myosin filament has already been mentioned. Huxley has also shown that the combination of HMM with actin brings to light a polarity residing in the actin filament itself. The combination of heavy meromyosin with actin filaments gives rise to the so-called arrowhead structures; these point in the same direction along the same actin filament. If combination takes place with actin filaments still attached to a Z band, arrowheads will point away from it on both sides, suggesting that the polarity of the actin filaments is reversed at the Z

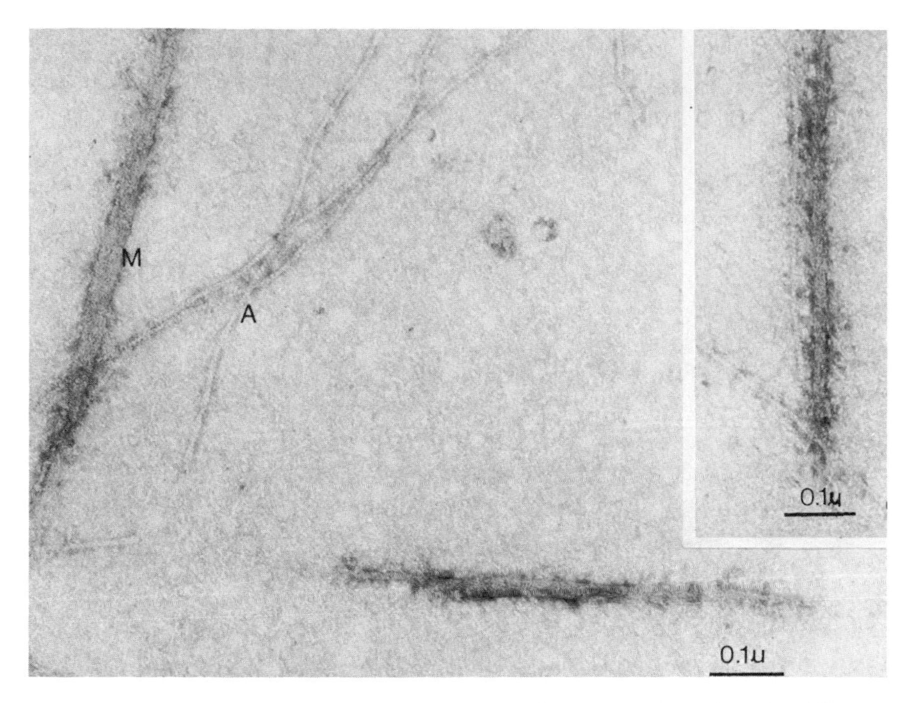


*Fig. 16.7.* Clearing of reconstituted actomyosin. Note separation of actin (*A*) and myosin (*M*) filaments. The insert shows a rarely found association in clear actomyosin of single actin filament with a myosin filament. From Ikemoto, Kitagawa, and Gergely (1966).

band. The polarization of the filaments is the basis of the directionality necessary for contraction.

A remarkable property of myosin is that its low adenosine triphosphatase activity in the presence of $Mg^{++}$ is greatly stimulated by the interaction with actin (Szent-Györgyi, 1945). In modern terminology this is likely to be anallosteric effect. The adenosine triphosphatase activity of myosin itself can be activated by either $Ca^{++}$ or $K^{+}$ (Seidel, 1969*a*). The ionic concentrations required to produce $Ca^{++}$-activated adenosine triphosphatase or $K^{+}$-activated adenosine triphosphatase are, however, much higher than those existing under physiological conditions. Because of the inhibitory effect of $Mg^{++}$ on the $K^{+}$-activated adenosine triphosphatase activity of myosin, the $K^{+}$ activation can only be observed if a chelator such as EDTA is present (Mühlrad, Fábián, and Biró, 1964; Offer, 1964). This seems to be the correct explanation of the long known EDTA activation (Friess, 1954; Bowen and Kerwin, 1954) of myosin adenosine triphosphatase at high $K^{+}$ concentrations. EDTA is only required to remove the inhibition that would be produced by traces of $Mg^{++}$ present.

The fact that activation by actin of the adenosine triphosphatase activity of myosin under physiological ionic conditions is accompanied by superprecipitation suggests that the interaction between myosin and actin is the basis of the increased hydrolysis of ATP observed *in vivo* when muscles contract. While there is good correlation between the amount of creatine phosphate or ATP broken down and the work done by a muscle (Cain, Infante, and Davies, 1962; Mommaerts, Seraydarian, and Maréchal, 1962; Carlson, Hardy, and Wilkie, 1963), no such correlation appears to exist with shortening of the muscle (Cain, Infante, and Davies, 1962). So far no chemical equivalent has been found for the so-called shortening heat (Davies, Kushmerick, and Larson, 1967). Indeed, Hill's earlier formulation of the various energy-terms involved in muscle contraction have been revised by himself (Hill, 1964).

The inhibitory effect of EDTA on the actomyosin system at low ionic strength (Bozler, 1954; Watanabe, 1955) was at first a puzzling phenomenon, but became clarified largely owing to the work of S. Ebashi and A. Weber and their colleagues (Weber, 1959; Weber and Winicur, 1961; Ebashi, Ebashi, and Fujie, 1960; Ebashi and Endo, 1968). At this point it is sufficient to say that inhibition of actomyosin adenosine triphosphatase activity is due to removal of traces of $Ca^{++}$ which are required for the interaction of actin and myosin in the presence of the so-called regulatory protein complex consisting of tropomyosin and troponin (Ebashi and Endo, 1968). Studies with purified actin and myosin or with the desensitized actomyosin of Perry and his colleagues (Johnson, Harris, and Perry, 1967), which they like to refer to as DAM, suggest that $Ca^{++}$ does not play an intrinsic role in the absence of the regulatory protein complex which can be added to purified actomyosin *in vitro* and which is, of course, present *in vivo* (Ebashi and Endo, 1968).

The interaction of actin and myosin involves sulfhydryl (-SH) groups, both in the actin and in the myosin molecule (see, e.g., Gergely, 1966; Bailin and Bárány, 1967), and one may think that those regions of the molecule that contain these groups are the sites of protein-protein or protein-ATP interactions. A word of caution is, however, in order. In the light of recent detailed information on the behaviour of the hemoglobin molecule, it is increasingly clear that groups that seem to be involved in some function need not be near the region where the interaction occurs. In the case of hemoglobin, modification or replacement of groups not in the immediate vicinity of the oxygen-combining center has a profound influence on oxygen binding (Perutz and Lehmann, 1968; Perutz, 1967). Thus the finding that certain -SH groups play an important role in the enzymic activity of myosin or in the interaction between actin and myosin does not justify the conclusion that they are in the immediate vicinity of the regions in which the interactions take place. There are two -SH groups for each active center that is intimately involved in the enzymic activity of myosin. Blocking of the first increases calcium activation but inhibits $K^+$ activation. Blocking of an additional group with thiol reagents inhibits both activities (for a review see, e.g., Gergely, 1966). Seidel (1969b) has recently shown that although two -SH groups are clearly distinct in terms of their reactivity toward thiol reagents they occupy what one might call a symmetrical position with respect to the adenosine triphosphatase center in that selective blocking of either of these group has the same effect on enzymic activity. Trotta, Dreizen, and Stracher (1968) have reported that it is possible to produce by proteolytic enzymes fragments that lack -SH groups, yet are able to carry on enzymic activity. More work is clearly needed on this extremely interesting aspect that suggests that -SH groups are not in the active center proper but rather located in a separate regulatory site.

X-ray work (Huxley and Brown, 1967; Huxley, Brown, and Holmes, 1965; Elliott, Lowy and Millman, 1965) has shown that the cross bridges of the thick filaments undergo a transient movement when muscle contracts. Differences in the orientation of the cross bridges could be seen clearly when electron micrographs of relaxed and rigor insect muscle were compared (Fig. 16.8) (Reedy, Holmes, and Tregear, 1965). It should be remembered that the region of myosin that has full ability to interact with actin is a small portion of the myosin molecule that consists of two globules, each having a diameter of less than 100 Å. From experiments with the isolated globules—subfragment-1—it appears that each of the two enzymically reactive units found in myosin molecules can interact with actin (Mueller and Perry, 1962; Eisenberg, Zobel, and Moos, 1968). It seems, however, that when the intact myosin molecule or that fragment of myosin which still contains both globules, heavy meromyosin, interacts with actin, only one head can be activated in each myosin molecule (Eisenberg, Barouch, and Moos, 1969). Whether this is due to steric hindrance or whether the interaction of one head of the myosin molecule with actin and ATP produces a change in the other, which in turn inhibits interaction, is not known. Aspects of the inhibition of the interaction of actin and myosin in the presence of regulatory proteins and on the effect of nucleotides and ions on this inhibition have been considered by Weber (1969).

The term conformational change has been very much in the forefront of modern protein chemistry. Of course a great deal depends on what one means by protein conformation and changes in it. Klotz (1966), for

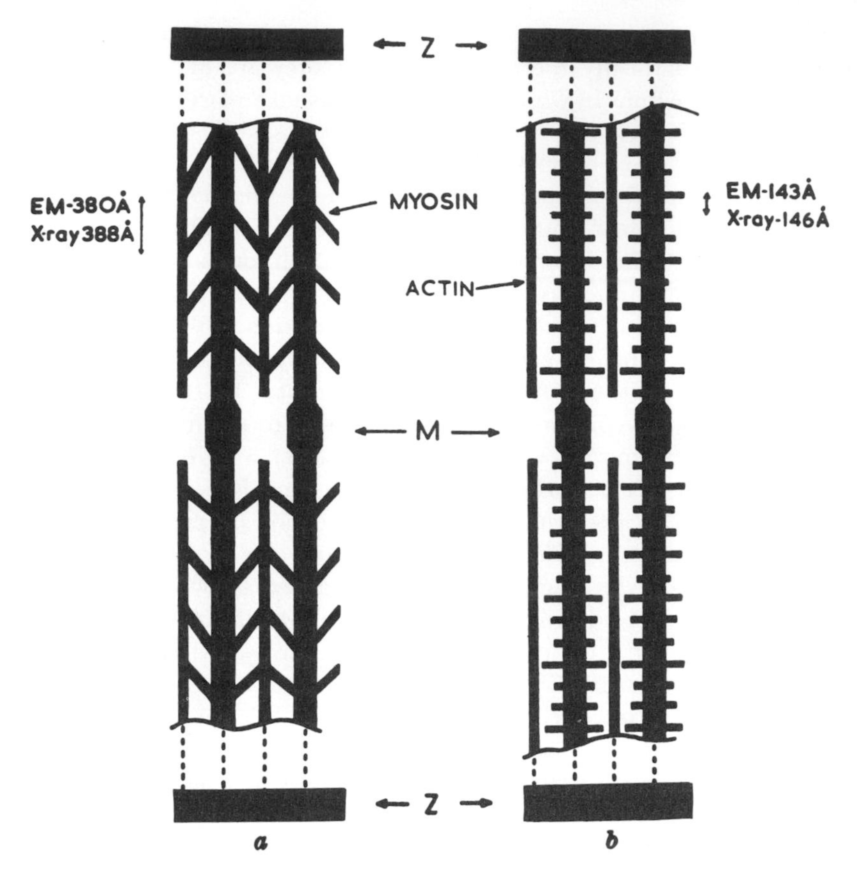


*Fig. 16.8.* Diagrams of actin and myosin filaments showing cross-bridge positions found to be characteristic of (*a*) rigor and (*b*) relaxed states of glycerinated flight muscle. Thick filaments represent myosin from which cross bridges extend toward actin (thin) filaments. From Reedy, Holmes, and Tregear (1965).

instance, has suggested that almost any interaction of a protein with a substrate or another protein has to involve a conformational change; for any chemical reaction involves a change in free energy, and free energy changes imply some rearrangement in the reaction of the atoms to each other. Others, however, would require a major rearrangement of the protein structure to justify the use of the term.

The actual molecular changes taking place on the interaction of myosin with ATP—or some other nucleoside triphosphate—or of myosin with actin have been rather elusive. It seems clear that no major rearrangement of the α-helical structure of myosin is detectable by optical rotatory or circular dichroism measurements, at least not under *in vitro* conditions (Gratzer and Lowey, 1969). Spectral changes have been reported on the interaction of ATP and ADP with myosin and heavy meromyosin (Morita, 1967, 1969). These changes probably represent a small rearrangement in the area in which ATP combines, but whether they are of sufficient magnitude to be related to the motion that has to be generated in order to produce contraction is not clear. A number of other techniques are available to show rearrangement in the molecular structure. The increased reactivity of the so-called $S_2$-SH group towards thiol reagents in the presence of ATP or ADP (Yamaguchi and Sekine, 1966), and the increases in chemical reactivity towards dinitrofluorobenzene (Bárány, Bailin, and Bárány, 1969) found on interaction of myosin with ATP and with actin, indicate conformational alterations in the vicinity of at least a few groups. Fluorescence changes observed on combining 8-anilino-1-naphthol-sulfonate may also be used to probe changes in conformation (Cheung and Morales, 1969). Spin labeling (see, e.g., Hamilton and McConnell, 1968) has recently found its way into biochemical-biophysical research. This technique involves the attachment of a stable free radical compound to a protein molecule. The mobility of the free radical, which is revealed by the pattern of the electron spin resonance spectrum, is determined by the conformation of the protein in which the molecule is attached. Changes in this conformation are shown by alterations of the region in the electron spin resonance spectrum. Since it is possible to produce spin label analogues of various specific reagents, one can explore the neighborhood of specific areas of the molecule such as -SH groups, tryosine residues, etc. So far the chief use has been made of spin labels attached to -SH groups. Morales and his colleagues first used spin labels for the study of myosin and actin. Some changes have been found in actin on polymerization (Stone and Botts, 1969) and in the actin-myosin-regulatory protein system (Tonomura, Watanabe, and Morales, 1969) on interaction with $Ca^{++}$. In our work the spin labels have proved useful in distinguishing various classes of -SH groups in myosin. When ATP or pyrophosphate combines with myosin, changes occur (Fig. 16.9) in the state of the spin label attached to -SH groups that are involved in the activation of the adenosine triphosphatase activity (Seidel, Chopek, and Gergely, 1969; Chopek, Seidel, and Gergely, 1969).

It is not clear whether indications of conformational changes found *in vitro* when myosin interacts with ATP or actin are large enough to explain the considerable change that has to take place in order to produce the sliding motion discussed earlier. That fairly large changes

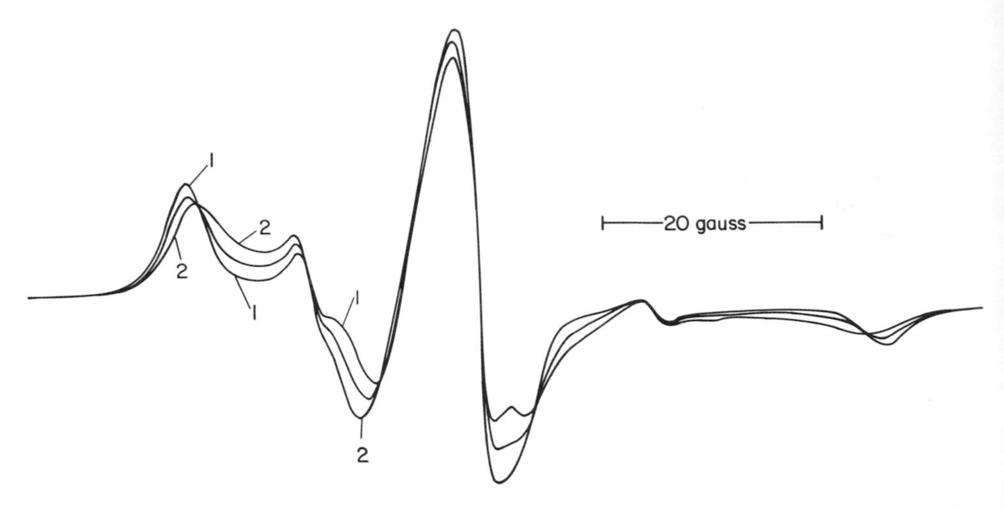


*Fig. 16.9.* Effect of pyrophosphate (PP) on electron spin resonance spectrum of iodoacetamide analog spin label attached to heavy meromyosin. The changes in the spectrum indicate increased mobility of the label owing to conformational changes in the protein. Curves: *1,* no PP; *unmarked,* 0.1 mM PP; *2,* 0.2 mM PP. Heavy meromyosin concentration $1.1 \times 10^{-4}$M. No further change was observed on increasing the pyrophosphate concentration. This suggests combination of two moles of pyrophosphate per mole of heavy meromyosin. From Seidel, Chopek, and Gergely (1970).

occur in the arrangement of the myosin molecules without, however, shifting their position longitudinally is also shown by Pepe (1967*b*). He found that certain portions of the myosin molecule in the middle of the helical shaft which are not accessible to antibodies specific for that portion of the molecule become available as interaction with actin takes place. This flexibility of the myosin molecule (Fig. 16.10) would remove the difficulty inherent in the X-ray data showing an increasing separation of actin and myosin filaments with decreasing sacromere length (Elliott, Lowy, and Worthington, 1963; Huxley, 1968, 1969). The role of electrostatic and van der Waals forces in maintaining the separation of the filaments has recently been stressed by Elliott (1968) and Rome (1968).

Various theoretical treatments have been suggested, starting with A. F. Huxley's important article (1957) and including the recent series of papers by T. L. Hill (1968*a, b, c, d*), in order to account for the utilization of ATP as the cross bridges interact with actin and to explain the mechanical and energetic properties of living muscles in terms of these elementary events. A high energy phosphate-protein intermediate —with either actin or myosin—has been a recurrent theme. Although vigorous attempts to demonstrate such a phosphorylated intermediate

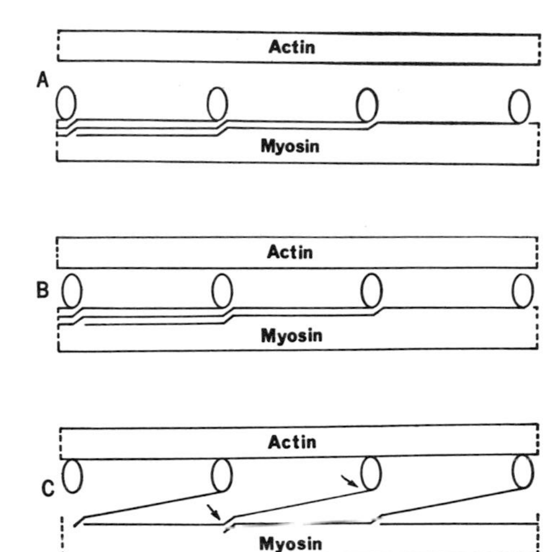


*Fig. 16.10.* Diagram showing relative positions of filaments and cross bridges at two different interfilament spacings. (*A*) 250 Å and (*B*) 200 Å, corresponding in frog sartorius muscle to sarcomere lengths of 2.0 and 3.1 μ. During contraction, or in rigor, the cross bridges could attach to the actin filament by bending at two flexible junctions, as shown by arrows in (*C*). From Huxley (1969).

have failed (Sartorelli, Fromm, Benson, and Boyer, 1966), recent work based on isotope exchange still points to this possibility with regard to myosin (Nakamura and Tonomura, 1968).

From this brief and sketchy survey it will, I hope, be clear that our knowledge of the molecular architecture of the muscle proteins, of their arrangement in filaments, and even of some of the detailed features of the interaction of actin and myosin, has greatly expanded during the last few years. Many questions, however, remain to be answered. What is the precise change that takes place when ATP interacts with actin and myosin? How can we explain the constant force exerted per cross bridge at varying sarcomere length when the interfilament distance varies? Do the two heads of the myosin molecule interact jointly or singly with actin? Is the region in which a change in conformation may take place near or at the site of interaction of actin, or is it near one of the hinge regions within the myosin molecule (Huxley, 1969)? All these questions require answers.

## ACKNOWLEDGMENTS

This work was supported by grants H-5949 from the National Heart Institute (NIH), and FR-05527 from the General Research Support Branch (NIH); the National Science Foundation; the Muscular Dystrophy Associations of America, Inc.; the Life Insurance Medical Research Fund; and the American Heart Association.

## *References*

Asai, H., and K. Tawada. 1966. Enzymic nature of F-actin at high temperature. *J. Mol. Biol. 20*:403.

Asakura, S., M. Taniguchi, and F. Oosawa. 1963. The effect of sonic vibration on exchangeability and reactivity of the bound ADP of F-actin. *Biochim. Biophys. Acta 74*:140.

Bailin, G., and M. Bárány. 1967. Studies on actin-actin and actin-myosin interactions. *Biochim. Biophys. Acta 140*:208.

Banga, I., and A. Szent-Györgyi. 1942. Preparation and properties of myosin A and B. *Stud. Inst. Med. Chem.* (Univ. Szeged) *1*:5.

Bárány, M., G. Bailin, and K. Bárány. 1969. Reaction of myosin with 1-fluoro-2, 4-dinitrobenzene at low ionic strength. *J. Biol. Chem. 244*:648.

Bárány, M., F. Finkelman, and T. Therattil-Antony. 1962. Studies on the bound calcium of actin. *Arch. Biochem. Biophys. 98*:185.

Bárány, M., A. F. Tucci, and T. E. Conover. 1966. The removal of the bound ADP of F-actin. *J. Mol. Biol. 19*:483.

Bowen, W. J., and T. C. Kerwin. 1954. A study of the effects of ethylene diaminetetraacetic acid on myosin adenosinetriphosphatase. *J. Biol. Chem. 211*:237.

Bozler, E. 1954. Relaxation in extracted muscle fibers. *J. Gen. Physiol. 38*:149.

Cain, D. F., A. A. Infante, and R. E. Davies. 1962. Chemistry of muscle contraction. *Nature 196*:214.

Carlson, F. D., D. J. Hardy, and D. R. Wilkie. 1963. Total energy production and phosphocreatine hydrolysis in the isotonic twitch. *J. Gen. Physiol. 46*:851.

Cheung, H. C., and M. F. Morales. 1969. Studies of myosin conformation by fluorescent techniques. *Biochemistry 8*:2177.

Chopek, M., J. C. Seidel, and J. Gergely. 1969. Stoichiometry of the reaction of spin labels with myosin, p. 191. *Abstr., 3rd Int. Biophys. Congr.,* Cambridge, Sept., 1969.

Davies, R. E. 1963. A molecular theory of muscle contraction of calcium-dependent contractions with hydrogen bond formation plus ATP-dependent extension of part of the myosin-actin cross bridges. *Nature 199*:1068.

Davies, R. E., M. J. Kushmerick, and R. E. Larson. 1967. ATP, activation and the heat of shortening of muscle. *Nature 214*:148.

Drabikowski, W., and J. Gergely. 1962. The effect of the temperature of extraction on the tropomyosin content in actin. *J. Biol. Chem. 237*:3412.

Drabikowski, W., and H. Strezelecka-Golaszewska. 1963. The exchange of actin bound calcium with various bivalent cations. *Biochim. Biophys. Acta 71*:486.

Dreizen, P., L. C. Gershman, P. P. Trotta, and A. Stracher. 1967. Myosin subunits and their interactions. *J. Gen. Physiol. 50*:85.

Ebashi, S., F. Ebashi, and Y. Fujie. 1960. The effect of EDTA and its analogues on glycerinated muscle fibers and myosin adenosine triphosphatase. *J. Biochem. 47*:54.

Ebashi, S., and M. Endo. 1968. Calcium ion and muscle contraction. *Progr. Biophys. Mol. Biol. 18*:123.

Eisenberg, E., W. W. Barouch, and C. Moos. 1969. Binding of MgATP and actin to myosin, heavy meromyosin (HMM) and subfragment-1. *Fed. Proc. 28*:536.

Eisenberg, E., C. R. Zobel, and C. Moos. 1968. Subfragment-1 of myosin: ATPase activation by actin. *Biochemistry 7*:3186.

Elliott, G. F. 1968. Force-balance and stability in hexagonally-packed polyelectrolyte systems. *J. Theoret. Biol. 21*:71.

Elliott, G. F., J. Lowy, and B. M. Millman. 1965. X-ray diffraction from living striated muscle during contraction. *Nature 206*:1357.

Elliott, G. F., J. Lowy, and C. R. Worthington. 1963. An X-ray and light-diffraction study of the filament lattice of striated muscle in the living state and in rigor. *J. Mol. Biol. 6*:295.

Engelhardt, W. A., and M. N. Ljubimowa. 1939. Myosin and adenosine triphosphatase. *Nature 144*:668.

Evans, M. G., and J. Gergely. 1949. A discussion of the possibility of bands of energy levels in proteins. Electronic interactions in non-bonded systems. *Biochim. Biophys. Acta 3*:188.

Frederiksen, D. W., and A. Holtzer. 1968. The substructure of the myosin molecule: production and properties of the alkali subunits. *Biochemistry 7*:3935.

Friess, E. T. 1954. The effect of a chelating agent on myosin ATPase. *Arch. Biochem. Biophys. 51*:17.

Gaetjens, E., K. Bárány, G. Bailin, H. Oppenheimer, and M. Bárány. 1968. Studies on the low molecular weight protein components in rabbit skeletal myosin. *Arch. Biochem. Biophys. 123*:82.

Gergely, J. 1950. On the relationship between myosin and ATPase. *Fed. Proc. 9*:176.

———. 1953. Studies on myosin-adenosinetriphosphatase. *J. Biol. Chem. 200*:543.

———. 1956. The interaction between actomyosin and adenosine triphosphate. Light scattering studies. *J. Biol. Chem. 220*:917.

———. 1966. Contractile proteins. *Ann. Rev. Biochem. 35*:691.

Gergely, J., M. A. Gouvea, and D. Karibian. 1955. Fragmentation of myosin by chymotrypsin. *J. Biol. Chem. 212*:165.

Gordon, A. M., A. F. Huxley, and F. J. Julian. 1963. Apparatus for mechanical investigations on isolated muscle fibres. *J. Physiol. 167*:42P.

———. 1966. The variation in isometric tension with sarcomere length in vertebrate muscle fibres. *J. Physiol. 184*:170.

Gratzer, W. B., and S. Lowey. 1969. Effect of substrate on the conformation of myosin. *J. Biol. Chem. 244*:22.

Hamilton, C. L., and H. M. McConnell. 1968. Spin labels, p. 115. *In* A. Rich and N. Davidson (eds.), *Structural Chemistry and Molecular Biology*. W. H. Freeman, San Francisco.

Hanson, J. 1967. Axial periods of actin filaments: electron microscopic studies. *Nature 213*:353.

Hanson, J., and J. Lowy. 1963. The structure of F-actin and of actin filaments isolated from muscle. *J. Mol. Biol. 6*:46.

———. 1965. Molecular basis of contractility in muscle. *Brit. Med. Bull. 21*:264.

Hayashi, T., and R. Rosenbluth. 1960. Studies on actin I. The properties of G-ADP actin. *Biol. Bull. 119*:294.

Hill, A. V. 1964. The effect of load on the heat of shortening of muscle. *Proc. Roy. Soc. B 159*:297.

Hill, T. L. 1968*a*. Phase transition in the sliding filament model of muscular contraction. *Proc. Nat. Acad. Sci. 59*:1194.

———. 1968*b*. On the sliding filament model of muscular contraction II. *Proc. Nat. Acad. Sci. 61*:98.

———. 1968*c*. On the sliding filament model of muscular contraction III. Kinetics of cross-bridge fluctuations in configuration. *Proc. Nat. Acad. Sci. 61*:514.

———. 1968*d*. On the sliding filament model of muscular contraction IV. Calculation of force-velocity curves. *Proc. Nat. Acad. Sci. 61*:889.

Huxley, A. F. 1957. Muscle structure and theories of contraction. *Progr. Biophys. Biophys. Chem. 7*:255.

Huxley, A. F., and R. Niedergerke. 1954. Structural changes in muscle during contraction. *Nature 173*:971.

Huxley, H. E. 1960. Muscle cells; p. 365. *In* J. Brachet and A. E. Mirsky (eds.), *The Cell,* Vol. 4. Academic Press, New York.

———. 1963. Electron microscope studies on the structure of natural and synthetic protein filaments from striated muscle. *J. Mol. Biol. 7*:281.

———. 1968. Structural difference between resting and rigor muscle; evidence from intensity changes in the low-angle equatorial X-ray diagram. *J. Mol. Biol. 37*:507.

———. 1969. The mechanism of muscular contraction. *Science 164*:1356.

Huxley, H. E., and W. Brown. 1967. The low-angle X-ray diagram of vertebrate striated muscle and its behavior during contraction and rigor. *J. Mol. Biol. 30*:383.

Huxley, H. E., W. Brown, and K. C. Holmes. 1965. Constancy of axial spacings in frog sartorius muscle. *Nature 206*:1358.

Huxley, H. E., and J. Hanson. 1954. Changes in the cross-striations of muscle during contraction and stretch and their structural interpretation. *Nature 173*:973.

Ikemoto, N., S. Kitagawa, and J. Gergely. 1966. Electron microscopic investigation of the interaction of actin and myosin. *Biochem. Z. 345*:410.

Johnson, P., C. I. Harris, and S. V. Perry. 1967. 3-Methylhistidine in actin and other muscle proteins. *Biochem. J. 105*:361.

Kakol, I., and H. H. Weber. 1965. Reaktionen und relative Affinitäten zwischen verschiedenen Nucleotiden und F-actin unter Ultraschall. *Z. Naturforsch. 20b*:977.

Kasai, M. 1969. The divalent cation bound to actin and thin filament. *Biochim. Biophys. Acta 172*:171.

Kasai, M., E. Nakano, and F. Oosawa. 1965. Polymerization of actin free from nucleotides and divalent cations. *Biochim. Biophys. Acta 94*:494.

Kasai, M., and F. Oosawa. 1968. The exchangeability of actin-bound calcium with various divalent cations. *Biochim. Biophys. Acta 154*:520.

Kitagawa, S., W. Drabikowski, and J. Gergely. 1968. The exchange and release of the bound nucleotide of F-actin. *Arch. Biochem. Biophys. 125*:706.

Klotz, I. M. 1966. Protein conformation, autoplastic and alloplastic effects. *Arch. Biochem. Biophys. 116*:92.

Kominz, D. R., E. R. Mitchell, T. Nihei, and C. M. Kay. 1965. The papain digestion of skeletal myosin A. *Biochemistry 4*:2373.

Laki, K., W. J. Bowen, and A. Clark. 1950. The polymerization of proteins. Adenosine triphosphate and the polymerization of actin. *J. Gen. Physiol. 33*:437.

Lowey, S., H. S. Slayter, A. G. Weeds, and H. Baker. 1969. Substructure of the myosin molecule I. Subfragments of myosin by enzymic degradation. *J. Mol. Biol. 42*:1.

Martonosi, A., M. A. Gouvea, and J. Gergely. 1960*a*. Studies on actin III. Transformation of actin and muscular contraction. *J. Biol. Chem. 235*:1707.

———. 1960*b*. Studies on actin I. The interaction of $C^{14}$-labeled adenine nucleotides with actin. *J. Biol. Chem. 235*:1700.

Maruyama, K., and J. Gergely. 1962. Interaction of actomyosin with adenosine triphosphate at low ionic strength I. Dissociation of actomyosin during the clear phase. *J. Biol. Chem. 237*:1095.

Mommaerts, W. F. H. M. 1951. Reversible polymerization and ultracentrifugal purification of actin. *J. Biol. Chem. 188*:559.

———. 1952. The molecular transformation of actin III. The participation of nucleotides. *J. Biol. Chem. 198*:469.

Mommaerts, W. F. H. M., K. Seraydarian, and G. Maréchal. 1962. Work and chemical change in isotonic muscle contractions. *Biochim. Biophys. Acta 56*:1.

Moos, C., and E. Eisenberg. 1969. Effect of myosin on bound nucleotide exchange in F-actin; enhancement of bound ADP release in the absence of ATP. *Biophys. J. 9*:A237.

Moos, C., E. Eisenberg, and J. E. Estes. 1967. Bound nucleotide exchange in actin and actomyosin. *Biochim. Biophys. Acta 147*:536.

Morita, F. 1967. Interaction of heavy meromyosin with substrate I. Difference in ultraviolet absorption spectrum between heavy meromyosin and its Michaelis-Menten constant. *J. Biol. Chem. 242*:4501.

———. 1969. Interaction of heavy meromyosin with substrate II. Rate of the formation of ATP induced ultraviolet difference spectrum of heavy meromyosin measured by the stop-flow method. *Biochim. Biophys. Acta 172*:319.

Mueller, H., and S. V. Perry. 1962. The degradation of heavy meromyosin by trypsin. *Biochem. J. 85*:431.

Mühlrad, A., F. Fábián, and N. A. Biró. 1964. On the activation of myosin ATPase by EDTA. *Biochim. Biophys. Acta 89*:186.

Nagy, B., and W. P. Jencks. 1962. Optical rotary dispersion of G-actin. *Biochemistry 1*:987.

Nakamura, H., and Y. Tonomura. 1968. The pre-steady state of the myosin-adenosine triphosphate system V. Evidence for a phosphate exchange reaction between adenosine triphosphate and the reactive myosin phosphate complex. *J. Biochem. 63*:279.

Nauss, K. M., S. Kitagawa, and J. Gergely. 1969. Pyrophosphate binding to and ATPase activity of myosin and its proteolytic fragments—implications for the substructure of myosin. *J. Biol. Chem. 244*:755.

Needham, J., S. L. Chen, D. M. Needham, and A. S. G. Laurence. 1941. Myosin birefringence and adenylpyrophosphate. *Nature 147*:766.

Offer, G. W. 1964. The antagonistic action of magnesium ions and ethylenediamine tetraacetate on myosin A ATPase (potassium activated). *Biochim. Biophys. Acta 89*:566.

Oosawa, F., S. Asakura, H. Asai, M. Kasai, S. Kobayashi, K. Mihashi, F. Ooi, M. Taniguchi, and E. Nakano. 1964. Structure and function of actin polymers, p. 158. *In* J. Gergely (ed.), *Biochemistry of Muscle Contraction.* Little Brown, Boston.

Page, S., and H. E. Huxley. 1963. Filament length in striated muscle. *J. Cell Biol. 19*:369.

Pepe, F. A. 1967*a*. The myosin filament I. Structural organization from antibody staining observed in electron microscopy. *J. Mol. Biol. 27*:203.

———. 1967*b*. The myosin filament II. Interaction between myosin and actin filaments observed using antibody staining in fluorescent and electron microscopy. *J. Mol. Biol. 27*:227.

Perutz, M. F. 1967. X-ray analysis, structure and function of enzymes. *Europe. J. Biochem. 8*:455.

Perutz, M. F., and H. Lehmann. 1968. Molecular pathology of human haemoglobin. *Nature 219*:902.

Reedy, M. K., K. C. Holmes, and R. T. Tregear. 1965. Induced changes in orientation of the cross bridges of glycerinated insect flight muscle. *Nature 207*:1276.

Rees, M. K., and M. Young. 1967. Studies on the isolation and molecular properties of homogeneous globular actin. Evidence for a single polypeptide chain structure. *J. Biol. Chem. 242*:4449.

Rome, E. 1968. X-ray diffraction studies of the filament lattice of striated muscle in various bathing media. *J. Mol. Biol. 37*:331.

Sartorelli, L., H. J. Fromm, R. W. Benson, and P. D. Boyer. 1966. Direct and $^{18}O$-exchange measurements relevant to possible activated or phosphorylated states of myosin. *Biochemistry 5*:2877.

Schliselfeld, L. H., and M. Bárány. 1968. The binding of ATP to myosin. *Biochemistry 7*:3206.

Seidel, J. C. 1969*a*. Effect of salts of monovalent ions on the ATPase activities of myosin. *J. Biol. Chem. 244*:1142.

———. 1969*b*. Similar effects on enzymic activity due to chemical modification of either of two sulfhydryl groups of myosin. *Biochim. Biophys. Acta 180*:216.

Seidel, J. C., M. Chopek, and J. Gergely. 1970. Spin labelling of myosin: number of reacting groups and effect of ATP. *Biochemistry*. (In press.)

Slayter, H. S., and S. Lowey. 1967. Substructure of the myosin molecule as visualized by electron microscopy. *Proc. Nat. Acad. Sci. 58:*1611.

Spicer, S. S. 1952. The clearing response of actomyosin to adenosine triphosphate. *J. Biol. Chem. 199:*289.

Stone, D. B., and J. Botts. 1969. Actin-actin and actin-myosin interaction studied by the spin-labeling technique. *Biophys. J. 9:*A93.

Stracher, A. 1969. Evidence for the involvement of light chains in the biological functioning of myosin. *Biochem. Biophys. Res. Comm. 35:*519.

Straub, F. B. 1942. Actin. *Stud. Inst. Med. Chem.*, (Univ. Szeged) *2:*3.

Straub, F. B., and G. Feuer. 1950. Adenosinetriphosphate—the functional group of actin. *Biochim. Biophys. Acta 4:*455.

Szent-Györgyi, A. 1945. Studies on muscle. *Acta Physiol. Scand. 9* (suppl. XXV).

Szent-Györgyi, A. G. 1953. Meromyosins, the subunits of myosin. *Arch. Biochem. Biophys. 42:*305.

Szent-Györgyi, A. G., and G. Prior. 1966. Exchange of adenosine diphosphate bound to actin in superprecipitated actomyosin and contracted myofibrils. *J. Mol. Biol. 15:*515.

Tonomura, Y., S. Watanabe, and M. Morales. 1969. Conformational changes in the molecular control of muscle contraction. *Biochemistry 8:*2171.

Trotta, P. P., P. Dreizen, and A. Stracher. 1968. Studies on subfragment-1, a biologically active fragment of myosin. *Proc. Nat. Acad. Sci. 61:*659.

Watanabe, S. 1955. Relaxing effects of EDTA on glycerol-treated muscle fibers. *Arch. Biochem. Biophys. 54:*559.

Weber, A. 1959. On the role of calcium in the activity of adenosine-5′-triphosphate hydrolysis by actomyosin. *J. Biol. Chem. 234:*2764.

———. 1969. Interaction between Ca, nucleotide triphosphates and troponin. *Biophys. J. 9:*A15.

Weber, A., and S. Winicur. 1961. The role of calcium in the superprecipitation of actomyosin. *J. Biol. Chem. 236:*3198.

Weeds, A. G. 1969. The stoichiometry of the light chains of myosin. *3rd Int. Biophys. Congr.,* Cambridge, IID3.

West, J. J., R. Nagy, and J. Gergely. 1967. Free ADP as an intermediary in the phosphorylation by creatine phosphate of ADP bound to actin. *J. Biol. Chem. 242:*1140.

Yamaguchi, M., and T. Sekine. 1966. Sulfhydryl groups involved in the active sites of myosin A adenosine triphosphatase I. Specific blocking of the SH group responsible for the inhibitory phase in "biphasic response" of the catalytic activity. *J. Biochem. 59:*24.

Young, M. 1967. On the interaction of adenosine diphosphate with myosin and its enzymically active fragments. Evidence for three identical catalytic sites per myosin molecule. *J. Biol. Chem. 242:*2790.

*Chapter 17*

# Regulatory Proteins of Muscle

K. MARUYAMA AND S. EBASHI

It has been well established that the interaction of myosin and F-actin in the presence of MgATP plays an essential role in the molecular mechanism underlying muscular contraction. Recently, there have been discovered in our laboratories several new structural proteins which exert regulatory functions on the ATP-myosin-actin system, directly or indirectly. Although their functions are so diverse, we call them tentatively "regulatory proteins."

From the physiological point of view, troponin (Ebashi and Kodama, 1966; Ebashi, Kodama, and Ebashi, 1968) is of the utmost importance. It is the $Ca^{++}$-receptive protein and is responsible for the $Ca^{++}$ sensitivity of the actomyosin system in collaboration with tropomyosin (Bailey, 1948). The triggering action of $Ca^{++}$ on muscular contraction is ascribed to the troponin-tropomyosin complex (Ebashi and Kodama, 1966), which is distributed along the entire length of the thin filament (Endo, Nonomura, Masaki, Ohtsuki, and Ebashi, 1966). Troponin has been shown to be located along the entire length of the thin filament with 400-Å periodicity (Ohtsuki, Masaki, Nonomura, and Ebashi, 1967). In view of the fact that troponin molecules bind to a specified region of a period of tropomyosin paracrystals (Nonomura, Drabikowski, and Ebashi, 1968) (Fig. 17.1) as well as tropomyosin crystals (Higashi and Ooi, 1968), this 400-Å periodicity is ascribed to tropomyosin molecules, which have a length of over 400 Å and are bound to F-actin filaments in a side-by-side way (Maruyama, 1964). A model regarding the fine structure of the thin filament is presented in Fig. 17.2 (Ebashi, Endo, and Ohtsuki, 1969).

A remarkable fact is that in the absence of $Ca^{++}$, troponin exerts a depressing effect on the interaction of myosin and actin in the presence of MgATP. The triggering action of $Ca^{++}$ therefore is considered as removing this depression. A question then arises as to how troponin controls the interaction of myosin and F-actin. It was assumed that troponin would modify this interaction by altering the structure of

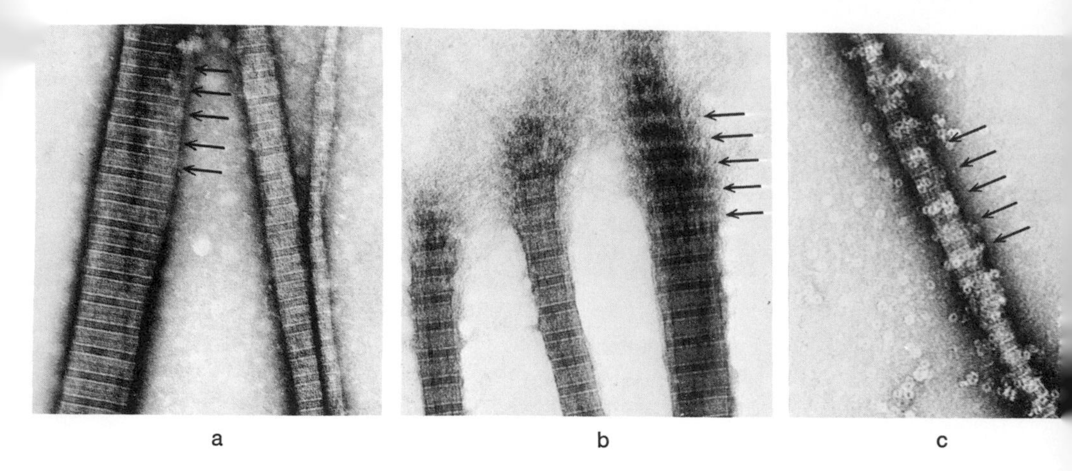


*Fig. 17.1.* Localization of troponin in tropomyosin paracrystals.
(*a*) Tropomyosin paracrystals (Cohen and Longley, 1966). Arrows indicate the darkish line in the cen of the light band.
(*b*) Paracrystals of troponin-tropomyosin complex (native tropomyosin). Notice that the darkish li in the center of the light band, shown in (a), is converted into a broader light line, as indicated by rows.
(*c*) Paracrystals of ferritin-labeled troponin-tropomyosin complex. Ferritin-labeled troponins are found the middle of the light band as shown by arrows, indicating that the broad light line in (b) is due to binding of troponin to this location. From Nonomura, Drabikowski, and Ebashi (1968).

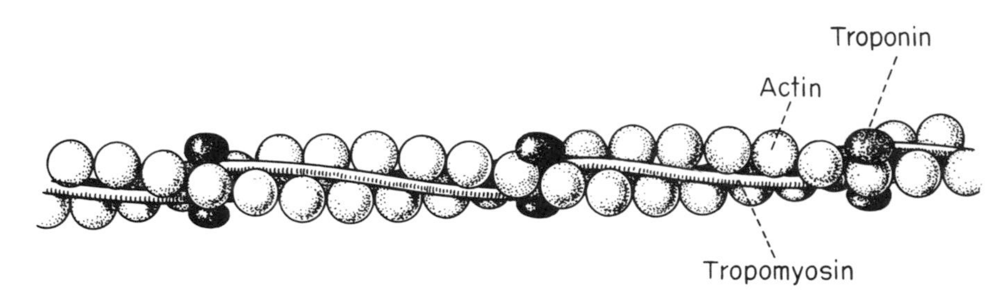


*Fig. 17.2.* A model for the fine structure of the thin filament. From Ebashi, Endo, and Ohtsuki (1969).

F-actin, and that tropomyosin would be the mediator of this modifying action (Ebashi and Endo, 1968; Ebashi, Kodama, and Ebashi, 1968). For this mechanism, troponin must undergo a $Ca^{++}$-dependent conformational change. Indeed, it has been shown that $Ca^{+}$ ions result in the dissociation of dimers of troponin formed in the absence of $Ca^{++}$ (Wakabayashi and Ebashi, 1968). Furthermore, using spin-label technique, some conformational change of F-actin has been detected on changing $Ca^{++}$ concentrations, if both troponin and tropomyosin coexist with F-actin (Tonomura, Watanabe, and Morales, 1969).

The work on troponin has been widely carried out in various laboratories (Hartshorne and Mueller, 1968, 1969; Hartshorne, Theiner, and Mueller, 1969; Fuchs and Briggs, 1968; Drabikowski, Barylko, Dabrowska, and Nowak, 1968; Arai and Watanabe, 1968*a, b*). It is worthy of note that Hartshorne and Mueller (1968, 1969) separated troponin into two components, troponin A and B, under rather drastic conditions, *p*H 1. According to them, troponin A is the $Ca^{++}$-receptive protein and B is an inhibitor of the $Mg^{++}$-activated adenosine triphosphatase activity of actomyosin (cf. Schaub, Hartshorne, and Perry, 1967; Hartshorne, Perry and Schaub, 1967).

Originally α-actinin (Ebashi and Ebashi, 1965) was regarded as a contraction-promoting factor of actomyosin. It was also found that α-actinin results in gelation of F-actin particles (Maruyama and Ebashi, 1965), and this action appears to be responsible for enhancing the speed and extent of superprecipitation of actomyosin. The actions of α-actinin have further been studied in Mommaerts's laboratory (Seraydarian, Briskey, and Mommaerts, 1967, 1968; Briskey, Seraydarian, and Mommaerts, 1967*a, b;* Goll, Mommaerts, Reedy, and Seraydarian, 1968). Now it is established that α-actinin consists of two components, 6 S and 10 S (Nonomura, 1967) and the 6 S component is the active principle (Masaki and Takaiti, 1969). The 6 S component is located in the Z band (Masaki, Endo, and Ebashi, 1967; cf. Goll, Mommaerts, Reedy, and Seraydarian, 1969). It was found that the 6 S component of α-actinin results in the lateral association of F-actin particles, as shown in Fig. 17.3 (Kawamura, Masaki, Nonomura, and Maruyama, 1970). The formation of such an aggregate confirms the previous view (Maruyama and Ebashi, 1965; Briskey, Seraydarian, and Mommaerts, 1967*b*) that α-actinin produces a cross linking of F-actin particles. Very probably, α-actinin plays a role as a cementing substance in holding thin filaments at the Z band. The function of 10 S α-actinin has not yet been found.

It was first considered that the particle length of F-actin could be regulated to be 1 μ under the influence of β-actinin (Maruyama, 1965*a, b*). However, it has been observed that β-actinin only reduces the length of F-actin filaments without changing its heterogeneity (Maruyama and Kawamura, 1968). It must be emphasized that β-actinin strongly inhibits the intermolecular interaction of preformed F-actin particles in solution (Maruyama, 1965*a*).

F-actin forms a network in solution, resulting in a weak rigidity. In a recent study (Maruyama, 1970), it has been observed that F-actin, 1 mg/ml, shows a value of 8 dyne/$cm^2$ for rigidity. This rigidity was calculated to correspond to one cross-linking unit per 3,000,000 g. This value means that there is one cross linking per about 60 actin monomers,

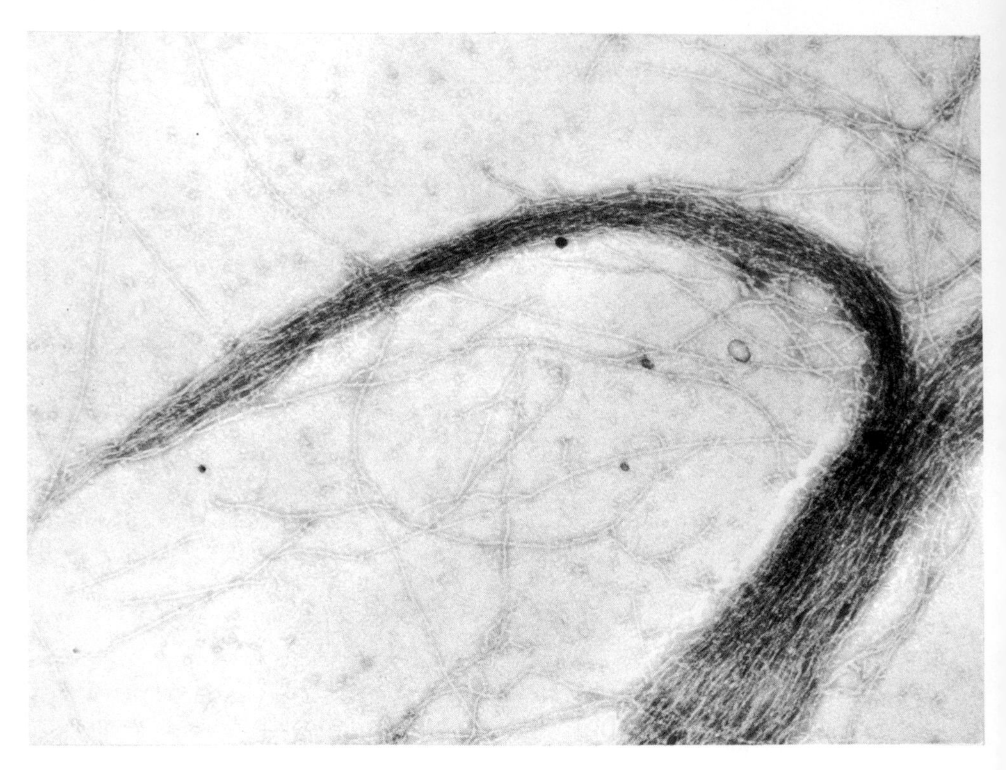

*Fig. 17.3.* Aggregate of F-actin formed by 6 S α-actinin. From Kawamura, Masaki, Nonomura, and Maruyama (1970).

suggesting one link per 4 pitches of actin helices. As shown in Fig. 17.4, only 5% of β-actinin by weight decreased the rigidity markedly.

It might be of some interest to note the possible relationship between actin and actinins. The amino acid compositions of actin, 10 S α-actinin, and β-actinin are similar to each other (Table 17.3), while that of 6 S α-actinin is considerably different from that of actin, especially in the values for Thr, Ileu, Met, and Tyr. Examinations of fingerprints of these proteins have revealed that 10 S α-actinin and actin are almost identical; β-actinin, to some extent, and 6 S α-actinin, to a larger extent, are different from the above two proteins (Horiuchi, Hama, Yamasaki, and Maruyama, 1970). Although it is premature to draw a conclusion for the present, it is likely that actin and actinins have been evolved from a common ancestor protein.

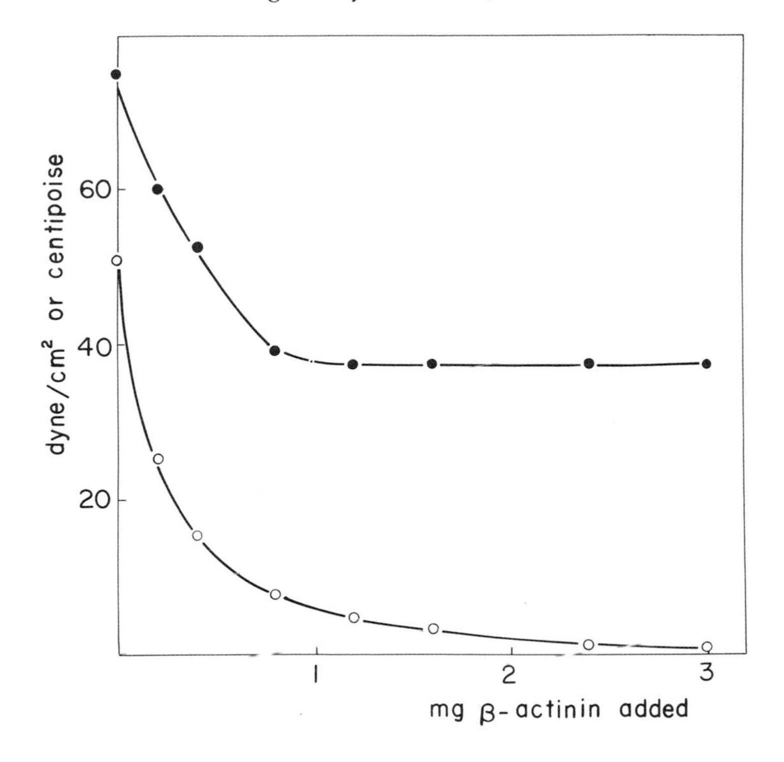


*Fig. 17.4.* Effect of β-actinin on the dynamic rigidity and viscosity of F-actin (Maruyama, 1970). F-actin 6 mg/ml, 0.1 M KCl, 0.02 M tris buffer, *p*H 7.2; 25° C. Total volume 2.5 ml (15 mg F-actin). Dynamic rigidity and viscosity were measured in an apparatus of Fukada and Date at an oscillation of 10 cycles per sec and amplitude of 35 μ (velocity gradient of 2.2 $sec^{-1}$). *Open circle,* rigidity ($dyne/cm^2$); *solid circle,* viscosity (centipoise).

TABLE 17.1
*Protein composition of myofibrils*

| | *Localization* | *% of the total myofibrillar protein* |
|---|---|---|
| Myosin | Thick filament | 55 |
| Actin | Thin filament | 20 |
| Tropomyosin | Thin filament | 5 |
| Troponin | Thin filament | 3 |
| 6 S α-actinin | Z band | 2 |
| 10 S α-actinin | ? | 5 |
| β-actinin | Thin filament | <1 |
| M protein | M line | <1 |

TABLE 17.2
*Some physicochemical properties of regulatory proteins of muscle*

| | Sed. const. $S_{20,w}$ | Intrinsic viscosity | Molecular weight | Helical content | $E^{1\%}_{278}$ | $\frac{E_{278}}{E_{255}}$ | Effect of salt |
|---|---|---|---|---|---|---|---|
| Troponin | 3.5$^{S}$ | 0.05$^{dl/g}$ | 5×10$^{4}$ | —% | 4.0 | 2.0 | no change |
| 6 S α-actinin | 6.8 | 0.14 | 16 | 41 | 10.0 | 1.9 | lightly aggr. |
| 10 S α-actinin | 10 | 0.14 | — | 13 | 9.4 | 2.1 | heavily aggr. |
| β-actinin | 4.5 | 0.05 | 6 | 13 | 9.3 | 2.2 | lightly aggr. |
| M protein | 8 | 0.06 | — | — | 10.4 | 2.1 | slowly aggr. |

TABLE 17.3
*Amino acid composition of muscle structural proteins*

| | Actin | 10 S α-actinin | 6 S α-actinin | β-actinin | Myosin | Troponin | M protein | Tropomyosin |
|---|---|---|---|---|---|---|---|---|
| Asp | 82 | 82 | 90 | 84 | 85 | 83 | 86 | 89 |
| Thr | 58 | 56 | 38 | 48 | 41 | 22 | 37 | 26 |
| Ser | 50 | 49 | 39 | 47 | 41 | 31 | 42 | 40 |
| Glu | 96 | 98 | 108 | 98 | 155 | 159 | 85 | 212 |
| Pro | 44 | 48 | 44 | 46 | 22 | 26 | 39 | 0 |
| Gly | 69 | 67 | 64 | 57 | 39 | 43 | 52 | 11 |
| Ala | 72 | 69 | 78 | 62 | 78 | 74 | 65 | 108 |
| Cys | 8 | — | — | 5 | 8.6 | — | — | 6.5 |
| Val | 48 | 53 | 51 | 47 | 42 | 37 | 57 | 27 |
| Met | 35 | 34 | 20 | 28 | 22 | 27 | 16 | 16 |
| Ileu | 65 | 61 | 34 | 54 | 42 | 33 | 38 | 29 |
| Leu | 61 | 64 | 63 | 62 | 79 | 65 | 73 | 95 |
| Tyr | 36 | 36 | 15 | 31 | 18 | 12 | 30 | 15 |
| Phe | 28 | 29 | 28 | 27 | 27 | 23 | 32 | 4 |
| Lys | 46 | 45 | 42 | 52 | 85 | 100 | 44 | 113 |
| His | 19 | 18 | 16 | 16 | 16 | 17 | 22 | 5 |
| Arg | 40 | 43 | 43 | 38 | 41 | 66 | 53 | 41 |
| Try | 10 | — | — | — | — | — | — | 0 |
| ($NH_3$) | (66) | — | — | (75) | (86) | (60) | (67) | (64) |

Recently a protein was isolated from the Hasselbach-Schneider extract of washed muscle mince and identified with the substance constituting the M line of myofibrils (Masaki, Takaiti, and Ebashi, 1968). The M protein, thus designated, is slowly aggregated by the addition of 0.1 M

KCl, and the aggregate markedly accelerated the lateral association of myosin aggregate in 0.1 M KCl.

The content of these regulatory proteins in myofibrils (rabbit skeletal muscle) is summarized in Table 17.1, together with that of the major structural proteins myosin and actin.

Some physico-chemical properties of the regulatory proteins are listed in Table 17.2. Some of them are revised values of the published data, as purification procedures have been improved. The amino acid compositions are compared with each other in Table 17.3.

## *References*

Arai, K., and S. Watanabe. 1968*a*. Troponin, tropomyosin, and the relaxing protein; myofibrillar proteins of rabbit skeletal muscle. *J. Biochem. 64*:69.

———. 1968*b*. A study of troponin, a myofibrillar protein from rabbit skeletal muscle. *J. Biol. Chem. 243*:5670.

Bailey, K. 1948. Tropomyosin: a new asymmetric protein component of the muscle fibril. *Biochem. J. 43*:271.

Briskey, E. J., K. Seraydarian, and W. F. H. M. Mommaerts. 1967*a*. The modification of actomyosin by α-actinin II. The effect of α-actinin upon contractility. *Biochim. Biophys. Acta 133*:412.

———. 1967*b*. The modification of actomyosin by α-actinin III. The interaction between α-actinin and actin. *Biochim. Biophys. Acta 133*:424.

Cohen, C., and W. Longley. 1966. Tropomyosin paracrystals formed by divalent cations. *Science 152*:794.

Drabikowski, W., B. Barylko, R. Dabrowska, and E. Nowak. 1968. Binding of calcium to troponin. *Bull. Polish Acad. Sci. 16*:397.

Ebashi, S., and F. Ebashi. 1965. α-actinin, a new structural protein from striated muscle I. Preparation and action on actomyosin-ATP interaction. *J. Biochem. 58*:7.

Ebashi, S., and M. Endo. 1968. Calcium ion and muscle contraction. *Progr. Biophys. Mol. Biol. 18*:123.

Ebashi, S., M. Endo, and I. Ohtsuki. 1969. Control of muscle contraction. *Quart. Rev. Biophys.* (In press.)

Ebashi, S., and A. Kodama. 1966. Interaction of troponin with F-actin in the presence of tropomyosin. *J. Biochem. 59*:425.

Ebashi, S., A. Kodama, and F. Ebashi. 1968. Troponin I. Preparation and physiological function. *J. Biochem. 64*:465.

Endo, M., Y. Nonomura, T. Masaki, I. Ohtsuki, and S. Ebashi. 1966. Localization of native tropomyosin in relation to striation patterns. *J. Biochem. 60*:605.

Fuchs, F., and F. N. Briggs. 1968. The site of calcium binding in relation to the activation of myofibrillar contraction. *J. Gen. Physiol. 51*:655.

Goll, D. E., W. F. H. M. Mommaerts, M. K. Reedy, and K. Seraydarian. 1969. Studies on α-actinin-like proteins liberated during trypsin digestion of α-actinin and of myofibrils. *Biochim. Biophys. Acta 175*:174.

Hartshorne, D. J., and H. Mueller. 1968. Fractionation of troponin into two distinct proteins. *Biochem. Biophys. Res. Comm. 31*:647.

———. 1969. The preparation of tropomyosin and troponin from natural actomyosin. *Biochim. Biophys. Acta 175*:301.

Hartshorne, D. J., S. V. Perry, and M. C. Schaub. 1967. A protein factor inhibiting the magnesium-activated adenosine triphosphatase of desensitized actomyosin. *Biochem. J. 104*:907.

Hartshorne, D. J., M. Theiner, and H. Mueller. 1969. Studies on troponin. *Biochim. Biophys. Acta 175*:301.

Higashi, S., and T. Ooi. 1968. Crystals of tropomyosin and native tropomyosin. *J. Mol. Biol. 34*:699.

Horiuchi, T., H. Hama, M. Yamasaki, and K. Maruyama. 1969. Peptide maps of actin and actinins. *J. Biochem.* (In press.)

Kawamura, Y., T. Masaki, M. Nonomura, and K. Maruyama. 1970. An electro microscopic study of the action of the 6 S component of α-actinin on F-actin. *J. Biochem.* (In press.)

Maruyama, K. 1964. Interaction of tropomyosin with actin. A flow birefringence study. *Arch. Biochem. Biophys. 105*:142.

———. 1965*a*. A new protein-factor hindering network formation of F-actin in solution. *Biochim. Biophys. Acta 94*:208.

———. 1965*b*. Some physico-chemical properties of β-actinin, "actin-factor," isolated from striated muscle. *Biochim. Biophys. Acta 102*:542.

———. 1970. A study of β-actinin. *J. Biochem.* (In press.)

Maruyama, K., and S. Ebashi. 1965. α-actinin, a new structural protein from striated muscle II. Action on actin. *J. Biochem. 58*:13.

Maruyama, K., and M. Kawamura. 1968. Effect of H-meromyosin and β-actinin on the particle length of F-actin. *J. Biochem. 64*:731.

Masaki, T., M. Endo, and S. Ebashi. 1967. Localization of 6 S component of α-actinin at Z-band. *J. Biochem. 62*:630.

Masaki, T., and O. Takaiti. 1969. Some properties of chicken α-actinin. *J. Biochem. 66*:637.

Masaki, T., O. Takaiti, and S. Ebashi. 1968. "M-substance," a new protein constituting the M-line of myofibrils. *J. Biochem. 64*:909.

Nonomura, Y. 1967. A study of the physico-chemical properties of α-actinin. *J. Biochem. 61*:796.

Nonomura, Y., W. Drabikowski, and S. Ebashi. 1968. The localization of troponin in tropomyosin paracrystals. *J. Biochem. 64*:419.

Ohtsuki, I., T. Masaki, Y. Nonomura, and S. Ebashi. 1967. Periodic distribution of troponin along the thin filament. *J. Biochem. 61*:817.

Schaub, M. C., D. J. Hartshorne, and S. V. Perry. 1967. The adenosine triphosphatase activity of desensitized actomyosin. *Biochem. J. 104*:263.

Seraydarian, K., E. J. Briskey, and W. F. H. M. Mommaerts. 1967. The modification of actomyosin by α-actinin I. A survey of experimental conditions. *Biochim. Biophys. Acta 133*:399.

———. 1968. The modification of actomyosin α-actinin IV. The role of sulfhydryl groups. *Biochim. Biophys. Acta 162*:424.

Tonomura, Y., S. Watanabe, and M. Morales. 1969. Conformational changes in the molecular control of muscular contraction. *Biochemistry 8*:2171.

Wakabayashi, T., and S. Ebashi. 1968. Reversible changes in the physical state of troponin induced by calcium ion. *J. Biochem. 64*:731.

# Chapter 18

# *The Dependence of Relaxation on the Saturation of Myosin with Adenosine Triphosphate*

ANNEMARIE WEBER

Although in 1943 Szent-Györgyi assembled an *in vitro* system capable of contraction (cf. Szent-Györgyi, 1947), it took another 20 years of study in many laboratories before a system capable of relaxation could be reconstituted *in vitro.* Pure actin and myosin in the presence of Mg and ATP do not relax but remain contracted and hydrolyze ATP until all of the ATP is exhausted. Without ATP actomyosin is not relaxed but is in a state of rigor characterized by a stable complex between actin and myosin.

In the resting state, actin and myosin are dissociated and there is little interaction between them. To maintain dissociation requires a more complex protein system. Ebashi and his collaborators (Ebashi and Ebashi, 1964; Ebashi and Kodama, 1965) discovered that under physiological conditions of *p*H and ionic strength the interaction between actin and myosin is prevented by two proteins, troponin and tropomyosin. A combination of these proteins is necessary for relaxation; troponin alone directs the alternation between contraction and relaxation. If, in the presence of ATP, troponin does not contain bound $Ca^{++}$, actin and myosin are dissociated; if troponin is saturated with $Ca^{++}$, actomyosin contracts (Fuchs and Briggs, 1968; Ebashi, Ebashi, and Kodama, 1967; Ebashi, Kodama, and Ebashi, 1968).

The mechanism by which troponin and tropomyosin regulate the state of actomyosin is not understood. In the following I shall enlarge on some findings that reflect on certain aspects of the mechanism.

Ebashi and Ebashi (1964) as well as Kominz and Maruyama (1967) demonstrated that troponin and tropomyosin bind to actin and not to myosin. Consequently one may consider the possibility that in the absence of $Ca^{++}$ the two proteins prevent the interaction of myosin with actin by masking the active site of actin. This simple concept, however, is not compatible with some well-established facts. First, as mentioned before, ATP is necessary for actomyosin dissociation. This "plasticizing" effect of ATP has been known for a long time (Weber and Portzehl, 1952). In the absence of ATP, myosin combines with actin also when troponin is quite free of bound $Ca^{++}$. Second, troponin and tropmyosin do not even suppress ATP-supported processes, e.g., syneresis and actin-activated myosin–adenosine-triphosphatase activity, if the concentration of MgATP is low (Fig. 18.1) (Weber and Herz, 1963; Levy and Ryan, 1965; Endo, 1964; Weber, Herz, and Reiss, 1969). At concentrations of MgATP below 2 $\mu$M syneresis and adenosine triphosphatase activity are identical with and without bound $Ca^{++}$. If in the absence of $Ca^{++}$ the ATP concentration is gradually raised to 10 $\mu$M the interaction of myosin with actin diminishes gradually. Only at concentrations above 50 $\mu$M and of other nucleoside triphosphates at higher values (Weber, 1969) is syneresis completely inhibited and the adenosine triphosphatase activity reduced to its minimum rate.

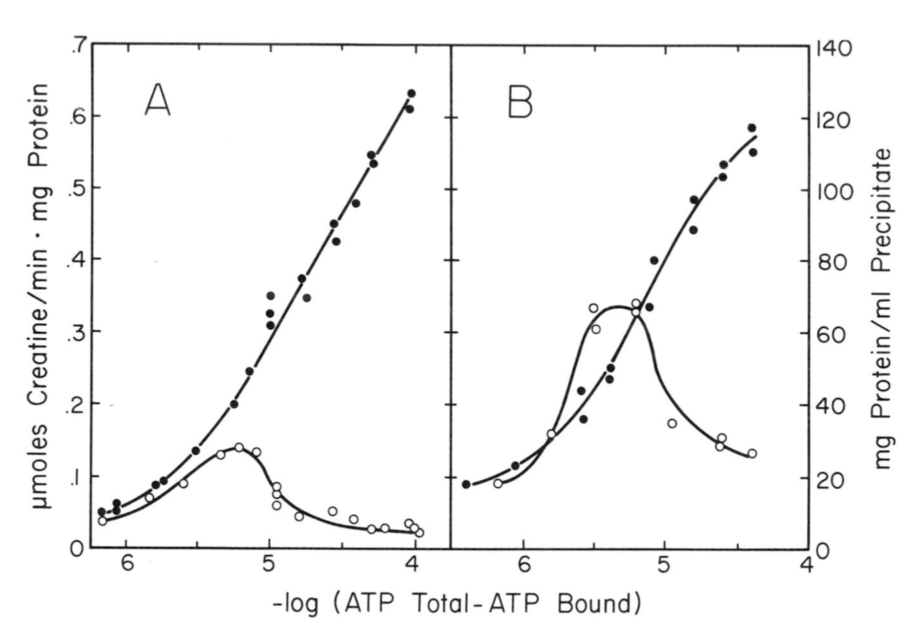


*Fig. 18.1.* Syneresis and adenosine triphosphatase activity as a function of the ATP concentration in the absence (*open circles*) and presence (*solid circles*) of $Ca^{++}$. *A,* adenosine triphosphatase activity; in the presence of 8 mM creatine phosphate and 2 mg/ml creatinephosphokinase, *p*H 7.0 and 0.08 ionic strength; *open circles,* 4 mM **EGTA,** ***closed circles,*** 40 $\mu$M $Ca^{++}$. *B,* syneresis in the presence of 10 mM creatine phosphate and 0.6 mg/ml kinase, *p*H 7.0 and 0.11 ionic strength; *open circles,* 2 mM EGTA; *closed circles,* 40 $\mu$M $Ca^{++}$. 25° C. From Weber (1969).

Possibly troponin and tropomyosin are incapable of causing relaxation unless a nucleoside triphosphate is bound to either protein. Troponin or tropomyosin may contain the binding site for the "inhibitory" ATP that has been postulated by several authors (Weber, Herz, and Reiss, 1964; Levy and Ryan, 1965). Under the assumption that the binding of ATP converts these proteins into inhibitors of actin-myosin interaction, $Ca^{++}$ must reverse this ATP-induced reaction. With this assumption one may expect that $Ca^{++}$ interferes with the binding of ATP on the one hand and that ATP interferes with the binding of $Ca^{++}$ on the other.

The following findings seem to meet these expectations. Fig. 18.2 shows that ATP inhibits the activation of adenosine triphosphatase activity by

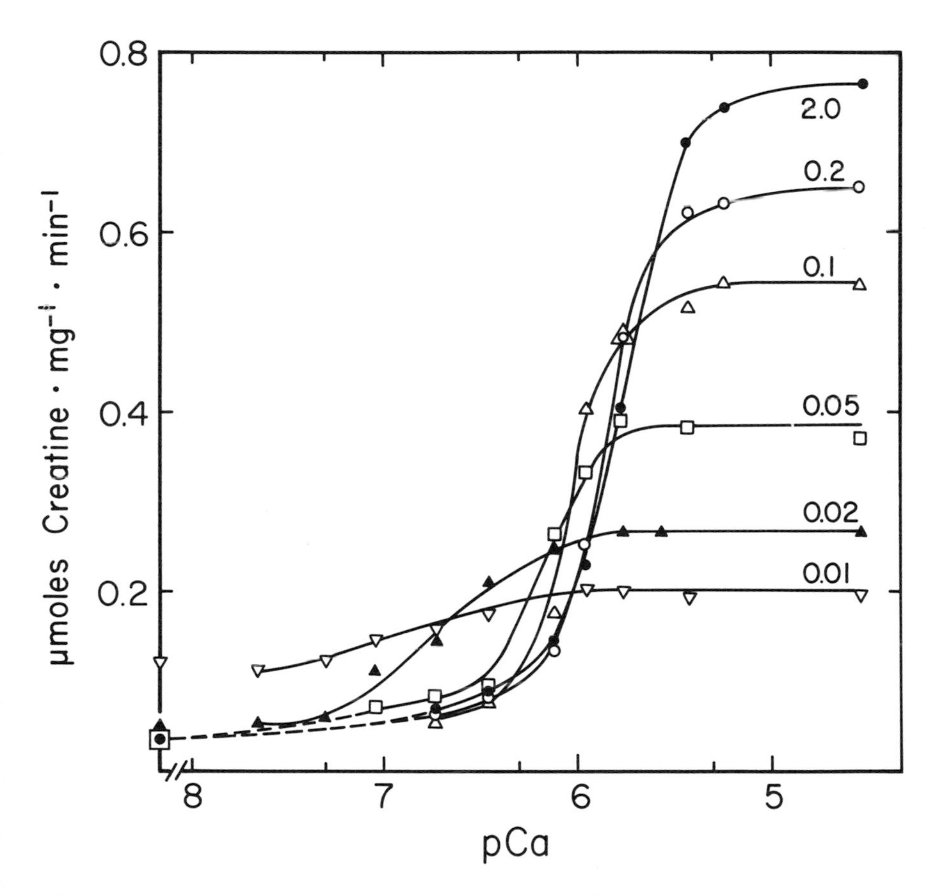


*Fig. 18.2.* Adenosine triphosphatase activity of myofibrils as a function of pCa in the presence of increasing concentrations of MgATP. Numbers on curves refer to mM ATP concentrations. 4 mM creatine phosphate, and 1.5 mg/ml kinase; *p*H 7.0; 5 mM $Mg^{++}$; 0.09 ionic strength. EGTA + CaEGTA 4 mM; 26.4° C.

$Ca^{++}$. When the ATP concentration was raised from 10 to 100 μM a tenfold increase in the concentration of ionized $Ca^{++}$ was necessary to obtain half maximal adenosine triphosphatase activity. Similarly when relaxation had been caused by various concentrations of ITP (inosine triphosphate), contraction required that increasing concentrations of ITP be compensated for by increasing concentrations of $Ca^{++}$ (Weber, 1969). In addition to the concentration of the nucleoside triphosphate, the activating $Ca^{++}$ concentration depended on the nature of the base (Weber, 1969). Less $Ca^{++}$ was needed in the presence of ITP than in the presence of an equal concentration of ATP.

K. Maruyama and I investigated whether $Ca^{++}$ interferes with the binding of ATP to myofibrils. In the limited range of ATP concentrations in which binding can be measured with some accuracy, the amount of bound ATP was found to be greater in the absence of $Ca^{++}$ than in its presence (Fig. 18.3). These findings are quite reproducible. From a double reciprocal plot of bound ATP versus free ATP it appeared that $Ca^{++}$ reduced the maximal amount of bound ATP by about 0.4 μmoles/g

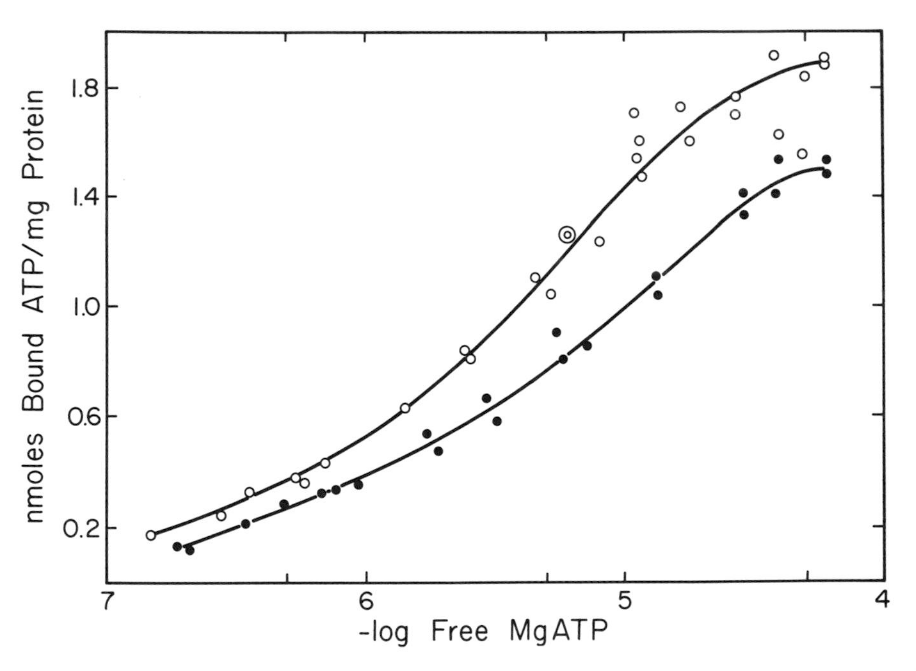


*Fig. 18.3.* Binding of ATP to myofibrils in the presence and absence of $Ca^{++}$ as a function of increasing ATP concentrations. 8 mM creatine phosphate and 2 mg/ml kinase; *p*H 7.0; 1 mM $Mg^{++}$; 0.08 ionic strength. *Open circles,* 4 mM EGTA; *closed circles,* 40 μM $Ca^{++}$. Binding was measured with $^{14}C$-ATP.

myofibrillar protein. Since 1 g myofibrillar protein contains 4 $\mu$moles of actin (cf. Weber, Herz, and Reiss, 1969) 0.4 $\mu$moles is close to the concentration of troponin as suggested by Ebashi, Kodama, and Ebashi (1968), i.e., 1 troponin and 1 tropomyosin for 7 actin, or 0.6 troponin for 4 actin. Thus it is tempting to assume that in the absence of $Ca^{++}$, troponin contained bound ATP that was displaced as a result of $Ca^{+}$ binding to troponin. Nevertheless, 0.4 $\mu$moles/g out of about 2.0 $\mu$moles/g is too small a fraction to be certain that the maximal amount of bound ATP is really reduced by the presence of $Ca^{++}$. Because of the limited range of ATP concentrations investigated, it cannot be excluded that the $Ca^{++}$-induced reduction of bound ATP over a range of nonsaturating ATP concentrations may be caused by a reduction in the binding constant without a change in the maximal amount of bound ATP. This possibility is strongly supported by an especially elaborate binding experiment undertaken to obtain a Scatchard plot (Scatchard, Coleman, and Shen, 1957) with a reasonably defined slope (Fig. 18.4).

This uncertainty indicated a need to measure directly the binding of ATP to the tropomyosin-troponin system. In collaboration with Robert Bremel the binding of ATP to actin containing troponin and tropomy-

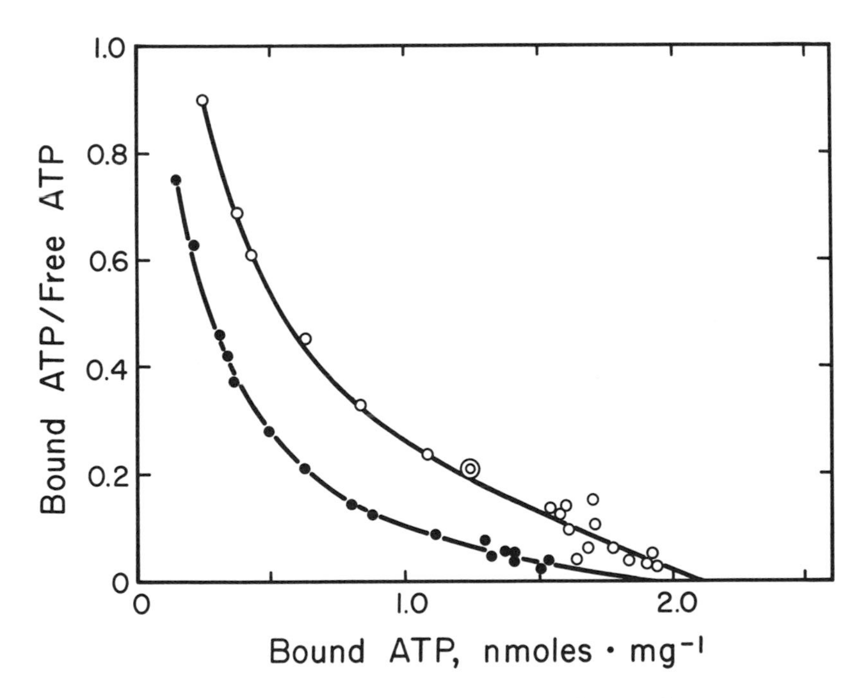


*Fig. 18.4.* Scatchard plot of the data of Fig. 18.3.

osin was measured. The actin was nearly saturated with tropomyosin and troponin because in the absence of $Ca^{++}$ the activation of the myosin adenosine triphosphatase by actin was inhibited by 85% and 92% respectively, compared with 95% in the intact isolated myofibril. With ATP concentrations between 20 and 100 $\mu$M maximally 0.2 nmoles ATP/mg protein were bound in one experiment and 0.2 nmoles in the other (Table 18.1). One mg protein contained 13 to 15 nmoles of actin (i.e., tightly bound ADP) and about 2 nmoles of troponin and tropomyosin each, if it is assumed (Ebashi, Kodama, and Ebashi, 1968) that 1 troponin and tropomyosin each for 7 actin are sufficient for complete relaxation. Since according to all investigators (Fuchs and Briggs, 1968; Ebashi, Kodama, and Ebashi, 1968) only troponin, of the three proteins present, binds $Ca^{++}$ specifically in appreciable amounts, an estimate of the troponin content can also be gained from the amount of bound $Ca^{++}$. With 4.6 and 6.8 nmoles/mg protein the previous estimate is confirmed with 2.3 to 3.4 nmoles troponin/mg protein, if one assumes 2 $Ca^{++}$ per troponin (Ebashi, Kodama, and Ebashi, 1968; see also below). Therefore it appears that the amount of bound ATP did not exceed 1 ATP per 10 troponin or tropomyosin or 1 ATP per 75 actin. This bound ATP therefore cannot be responsible for relaxation. Furthermore this bound ATP was not in ready equilibrium with the ATP of the medium because it was not removed when the ATP was washed out. The relaxing effect of ATP is quite reversible and disappears on lowering the ATP concentra-

TABLE 18.1

| Exper. no. | nmoles ATP bound / mg protein | Free ATP ($\mu$M) | nmoles actin / mg protein | nmoles $Ca^{++}$ bound / mg protein | % ATPase inhibition at 0.001 $\mu$M $Ca^{2+}$ |
|---|---|---|---|---|---|
| 1 | 0.16 | 25 | 15 | 4.6 | 85 |
| | 0.16 | 38 | | | 90 |
| | 0.22 | 44 | | | 85 |
| 2 | 0.019 | 43 | 13 | 6.8 | 92 |
| | 0.019 | 68 | | | |
| | 0.00 | 100 | | | |
| Control | — | — | 20 | 0.3 | 0 |

The assay for ATP binding contained 2–3 mg/ml actin; 3 mM $Mg^{2+}$; 3 mM creatine phosphate, 1.5 mg/ml kinase; 1 mM EGTA; 7 mM imidazole, *p*H 7.0 and $^{14}C$-ATP as indicated. The $Ca^{++}$-binding assay contained 1.8 mg/ml actin; 6 mM imidazole, *p*H 7.0; 2.5 mM $Mg^{++}$ and 16 and 19 $\mu$M $Ca^{++}$ respectively. For the control experiment actin was freed of troponin and tropomyosin (Martonosi, 1962). To measure binding the actin was spun down for 3 hours at 100,000 g at room temperature. Adenosine triphosphatase activity after the addition of myosin was measured after an equal exposure to room temperature.

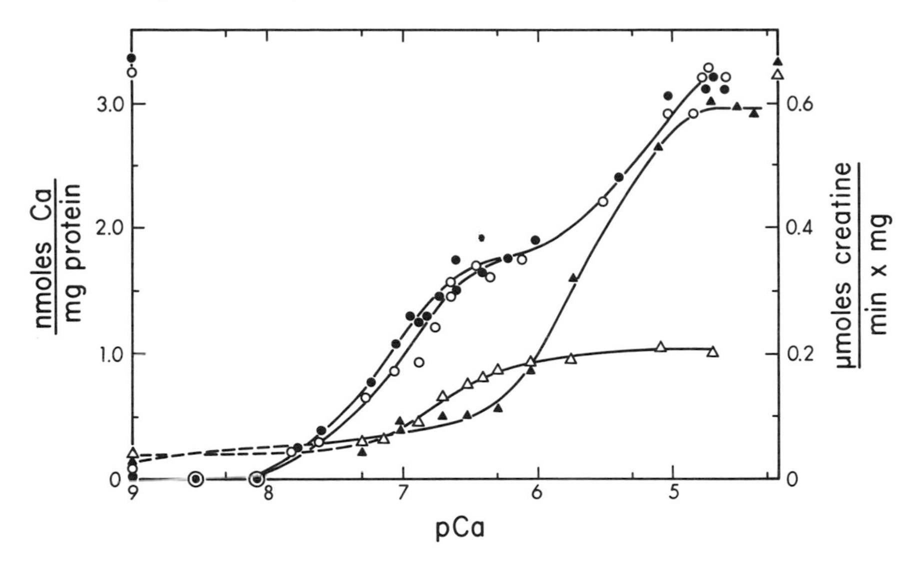


*Fig. 18.5.* $Ca^{++}$ binding to myofibrils and adenosine triphosphatase activity as a function of pCa in the presence of 2 mM *(closed symbols)* and at a low MgATP concentration *(open symbols)*. *Circles,* binding; *triangles,* adenosine triphosphatase. *In binding experiments* low ATP = 0; 4 mM creatine phosphate and 0.6 mg/ml kinase when 2 mM ATP was present. Where kinase and ATP were omitted the contaminating $Ca^{++}$ was replaced by an equal amount of $CaCl_2$. $Ca^{++}$ binding was calculated from the binding of $^{45}Ca$. *In adenosine triphosphatase assays* low ATP not clearly defined because of ADP contamination in creatine phosphate. It was about 10 µM. In both measurements 5 mM $Mg^{++}$; *p*H 7.0; 0.085 ionic strength. *p*Ca calculated from CaEGTA/EGTA ratio with an apparent binding constant at *p*H 7.0 of $5.3 \times 10^6$. In binding experiments EGTA 600–0 µM; in adenosine triphosphatase assays EGTA + CaEGTA = 4 mM.

tion. Since it is known that a small fraction of the ADP bound to F-actin is exchangeable (Martonosi, Gouvea, and Gergely, 1960) it is likely that the bound ADP had exchanged with the tightly bound ADP of F-actin.

The finding that neither actin, tropomyosin, or troponin bind ATP in equilibrium with the ATP of the medium is in good agreement with the finding of Ebashi, Kodama, and Ebashi (1968) that the binding of $Ca^{++}$ to isolated troponin is not influenced by ATP over a wide range of concentrations. It follows therefore that the antagonism between $Ca^{++}$ and ATP must be explained in another manner not involving the binding of ATP to the troponin-tropomyosin system.

No definite explanation can be given for the finding that more ATP appears to be bound in the absence of $Ca^{++}$ than in its presence before more data have been assembled, allowing a more extensive analysis. It is not surprising, however, to find such differences in the ATP binding if one considers that in the presence of $Ca^{++}$ ATP may be thought to bind to actomyosin because actin and myosin interact, whereas ATP would be expected to bind to free myosin in the relaxed system where myosin and actin are dissociated.

Since ATP does not bind to troponin, the inhibition of the $Ca^{++}$ activation of adenosine triphosphatase activity by ATP is probably not caused by an interference with the binding of $Ca^{++}$. In fact, the binding of $Ca^{++}$ to myofibrils is the same in the presence and absence of ATP (Fig. 18.5). In addition, Fig. 18.5 suggests that there is more than one category of binding sites for $Ca^{++}$. This becomes quite evident from Fig. 18.6, a Scatchard plot of the same data that indicates two types of sites. Fig. 18.5 also describes the adenosine triphosphatase activity of the same myofibrils at high and at low ATP concentrations. A comparison between the $Ca^{++}$ binding and the adenosine triphosphatase activity explains the basis for the decrease in the response to $Ca^{++}$ at high ATP. Whereas at low concentrations of ATP, adenosine triphosphatase activity increases with the binding of the first $Ca^{++}$, at high concentrations of ATP the second $Ca^{++}$ must also be bound before activity increases significantly.

Presumably both $Ca^{++}$ are bound to troponin because their binding was eliminated by short trypsin treatment (Fig. 18.7). This treatment removed only troponin but not tropomyosin, and did not alter the adenosine triphosphatase activity in the presence of $Ca^{++}$, thus indicating that it did not affect myosin. Such an interpretation is not in agreement with the data on $Ca^{++}$ binding by isolated troponin presented by Fuchs and Briggs (1968) as well as by Arai and Watanabe (1968), who found only one binding site for troponin. Ebashi, Kodama, and Ebashi (1968), however, obtained the best fit of their data by assuming two binding sites. Their constant for the high-affinity site and the only constant by Fuchs and Briggs are comparable within a factor of 2, with the constant for the high-affinity site calculated from Fig. 18.6 ($1.5 \times 10^7$ M$^{-1}$) if one takes into account the different methods of calculation.* The same is true for Ebashi's low-affinity constant and that estimated from Fig. 18.6 ($2$ to $4 \times 10^5$ M$^{-1}$). Our constant is directly comparable to the only constant obtained by Arai and Watanabe (1968).

In conclusion, one may state that the role of ATP in relaxation is not accounted for by its binding to troponin, tropomyosin, or actin. That

* Our data are calculated with the dissociation constant

$$\frac{Ca \times (EGTA^{4-} + EGTA^{3-} + EGTA^{2-})}{CaEGTA} = K = 2 \times 10^{-7}$$

for $p$H 7.0, calculated from the binding constants for $Ca^{++}$ and $H^{+}$ and quoted in Chaberek and Martell (1959), whereas the method used by the other two groups arrives at a dissociation constant five to seven times higher (Briggs and Fleischman, 1965).

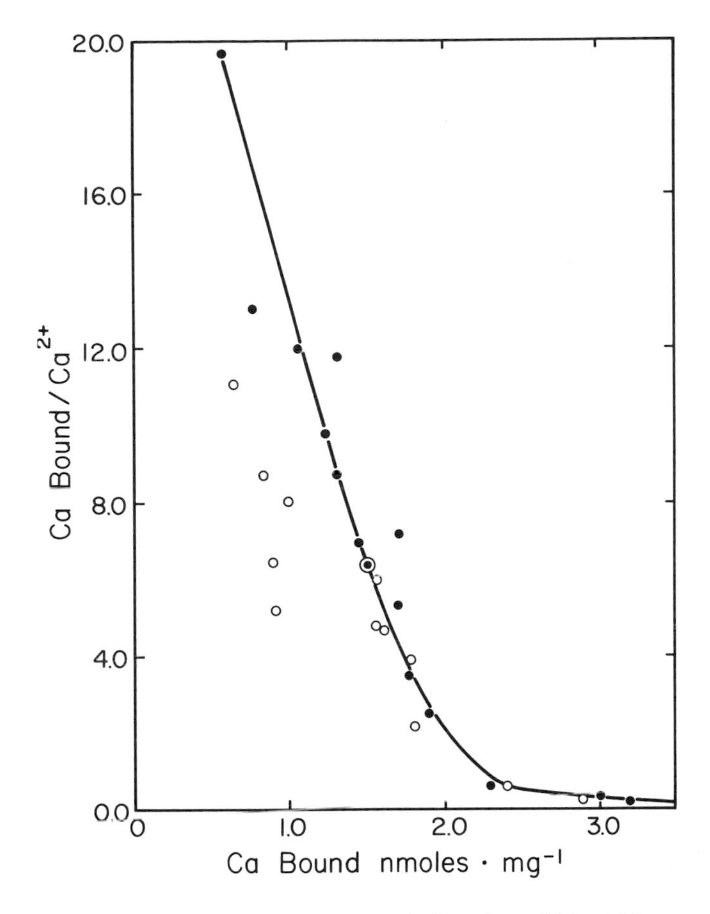


*Fig. 18.6.* Scatchard plot of the binding data of Fig. 18.5.

leaves myosin as the only binding site for ATP on the contractile proteins. It appears then that relaxation requires a certain ATP content of myosin. At this point it cannot be decided whether relaxation depends on the binding of ATP to a special site apart from the active site for hydrolysis or whether it depends on the cooperative effect of ATP bound to several active sites, e.g., both active sites of each myosin molecule (Slayter and Lowey, 1967). A comparison of the relaxing effect of four different nucleoside triphosphates showed a parallelism between the concentration required for maximal rate of hydrolysis and that for maximal relaxation (Weber, 1969). This finding may be interpreted in favor of a cooperative effect of nucleoside-triphosphate-saturated hydrolytic sites. However, it is not impossible that a second site has a configura-

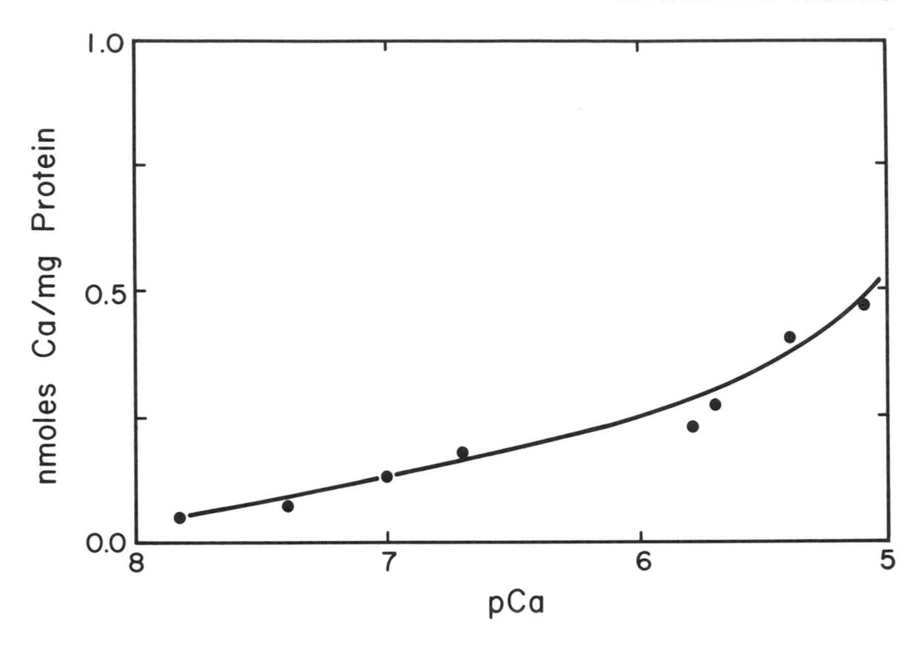


*Fig. 18.7.* Binding of $Ca^{++}$ to trypsin-treated myofibrils as a function of pCa. Conditions as in Fig. 18.5.

tion so similar to the hydrolytic site that it has very similar binding constants for the four different triphosphates.

Relaxation may be discussed in terms of an antagonism between myosin and troponin. The degree of antagonism is dependent on the binding of $Ca^{++}$ to troponin on the one hand and the binding of ATP to myosin on the other. When the ATP concentration is so low that the binding of ATP to myosin is less than 20% of maximal there is no antagonism quite independent of the $Ca^{++}$ saturation of troponin.

Over the range of increasing ATP saturation of myosin (2–50 $\mu$M), where the interaction with actin decreases to a minimum if troponin is free of $Ca^{++}$, the binding of only 1 $Ca^{++}$ to troponin is effective in overcoming the inhibition of interaction. Under conditions where myosin is fully saturated with ATP, however, both classes of $Ca^{++}$ binding sites of troponin must be saturated with $Ca^{++}$ to overcome the antagonism and to allow any significant activation of myosin adenosine triphosphatase by actin to occur.

To understand the mechanism for the antagonism of the two proteins, we must resolve the problem of how one troponin can make unavailable to myosin several active sites on actin, most of which are separated from troponin by a considerable distance.

ACKNOWLEDGMENT

The research reported in this paper was supported by grant GM–14034 from the National Institute of General Medical Sciences (NIH).

## *References*

Arai, K., and S. Watanabe. 1968. A study of troponin, a myofibrillar protein from rabbit skeletal muscle. *J. Biol. Chem. 243:*5670.

Briggs, F. N., and M. Fleischman. 1965. Calcium binding by particle-free supernatants of homogenates of skeletal muscle. *J. Gen. Physiol. 49:*131.

Chaberek, S., and A. E. Martell. 1959. *Organic Sequestering Agents.* John Wiley, New York.

Ebashi, S., and F. Ebashi. 1964. A new protein component participating in the superprecipitation of myosin B. *J. Biochem. 55:*604.

Ebashi, S., F. Ebashi, and A. Kodama. 1967. Troponin as the $Ca^{++}$ receptive protein in the contractile system. *J. Biochem. 62:*137.

Ebashi, S., and A. Kodama. 1965. A new factor promoting aggregation of tropomyosin. *J. Biochem. 58:*107.

Ebashi, S., A. Kodama, and F. Ebashi. 1968. Troponin I. Preparation and physiological function. *J. Biochem. 64:*465.

Endo, M. 1964. The superprecipitation of actomyosin and its ATP-ase activity in low concentration of ATP. *J. Biochem. 55:*614.

Fuchs, F., and F. N. Briggs. 1968. The site of calcium binding in relation to the activation of myofibrillar contraction. *J. Gen. Physiol. 51:*655.

Kominz, D. R., and K. Maruyama. 1967. Does native tropomyosin bind to myosin? *J. Biochem. 61:*269.

Levy, H. M., and A. M. Ryan. 1965. Evidence that calcium activates the contraction of actomyosin by overcoming substrate inhibition. *Nature 205:*703.

Martonosi, A. 1962. Studies on actin VII. Ultracentrifugal analysis of partially polymerized actin solutions. *J. Biol. Chem. 237:*2795.

Martonosi, A., M. A. Gouvea, and J. Gergely. 1960. Studies on actin I. The interaction of $C^{14}$-labeled adenine nucleotides with actin. *J. Biol. Chem. 235:*1700.

Scatchard, G., J. S. Coleman, and A. L. Shen. 1957. The attractions of proteins for small molecules and ions. *J. Amer. Chem. Soc. 79:*12.

Slayter, H. S., and S. Lowey. 1967. Substructure of the myosin molecule as visualized by electron microscopy. *Proc. Nat. Acad. Sci. 58:*1611.

Szent-Györgyi, A. 1947. *Chemistry of Muscular Contraction.* Academic Press, New York.

Weber, A. 1969. Parallel response of myofibrillar contraction and relaxation to four different nucleoside triphosphates. *J. Gen. Physiol. 53:*781.

Weber, A., and R. Herz. 1963. The binding of calcium to actomyosin systems in relation to their biological activity. *J. Biol. Chem. 238*:599.

Weber, A., R. Herz, and I. Reiss. 1964. The regulation of myofibrillar activity by calcium. *Proc. Roy. Soc. B 160*:489.

———. 1969. The role of magnesium in the relaxation of myofibrils. *Biochemistry 8*:2266.

Weber, H. H., and H. Portzehl. 1952. Kontraction, ATP-Cyclus und fibrilläre Proteine des Muskels. *Ergeb. Physiol. Biol. Chem. Exp. Pharmakol. 47*:369.

# Chapter 19

# *Morphological and Biophysical Properties of the Myofibril*

T. FUKUZAWA AND E. J. BRISKEY

Scientists the world over have used the myofibril for direct and indirect studies of the contractile mechanism in skeletal muscle. While the fine structure of the myofibril is continuing to emerge with almost unquestionable clarity, its chemical morphology is now only starting to achieve some measure of delineation. Although the predominant constituents of the A and I bands have been recognized for decades, we are in need of further clarification of the molecular architecture of these regions, and particularly of some features of the I and Z bands. Maruyama and Ebashi (1969) have described the recent discovery of several regulatory or accessory proteins. Knowledge of these proteins has greatly enriched our understanding of the intact myofibril. The constituents of the I bands are obviously involved in the muscle's physiological response to $Ca^{++}$ (Maruyama and Ebashi, 1970), a response which must be influenced or regulated in some way by the chemical and structural features of the Z band; both are of more than minor importance to the post mortem changes in muscle and its use as a food.

Most workers who have attempted to disassemble the myofibril have used a high ionic strength extraction, followed by either acetone treatment or enzyme digestion.

In this report we wish to describe various morphological and biophysical aspects of the myofibril: (1) as influenced by natural, mild post mortem changes and (2) as influenced by mild extraction at low temperatures. We wish to introduce this subject by giving a detailed description of recent observations on the fragmentation of the myofibril.

### SINGLE SARCOMERE AND FORMATION OF MYOFIBRILLAR FRAGMENTS

The periodic pattern of striated myofibrils has been shown by Huxley and Hanson (1957) to be attributable to the regular distribution of actin and myosin. Szent-Györgyi's (1951) glycerol-treated muscle and Perry and Corsi's (1958) isolated myofibrils serve as muscle models, and although they are intermediate between actomyosin (actin and myosin) threads on the one hand and living muscle on the other, they both contract upon the addition of ATP in a way which is similar to the contraction of living muscle, during its association with the ATP-splitting process.

The functional unit of the myofibril has been designated as a "sarcomere." Therefore, isolated sarcomeres and small fragments of myofibrils (a few sarcomeres) can serve as still two more models from which much fundamental information can be gleaned. The properties of single sarcomeres and myofibrillar fragments have been investigated in detail by Takahashi, Mori, Nakamura, and Tonomura (1965) and Fukazawa, Hashimoto, and Tonomura (1963). Single sarcomeres can be induced to contract upon the addition of MgATP; this contraction can occur even in sarcomeres which have lost their Z lines. To isolate a single sarcomere, however, the degradation of at least part of the Z band is unavoidable. Detailed studies have been conducted on the ultrastructure of the Z line (Knappeis and Carlsen, 1962; Franzini-Armstrong and Porter, 1964). As knowledge on the chemical morphology of these Z bands is being enriched, we are establishing a firmer foundation for understanding the degradation of the Z-I junction and Z-band material. Many workers (Takahashi, Fukazawa, and Yasui, 1967; Stromer and Goll, 1967; Stromer, Goll, and Roth, 1967; Cook and Wright, 1966) have reported that the fragmentation of post mortem myofibrils appears to occur principally at the level of the Z line, suggesting that this point has been weakened so that the myofibrils can break or rupture during the homogenization procedure. Such a degradation of the Z line can never be achieved during the preparation of myofibrils from tissue taken immediately after death of the animal.

Stromer, Goll, and Roth (1967) have shown that the treatment of myofibrils with 0.05% trypsin results in a rapid loss of Z lines. According to Corsi and Perry (1958), the Z line of rabbit myofibrils disappears after 20 days of extraction at low ionic strength and neutral *p*H. Very recently, Stromer, Hartshorne, Mueller, and Rice (1969) reported an electron microscopic study of Z- and M-line reconstitution from various protein fractions. They concluded that the extraction of glycerinated bundles of rabbit psoas muscle with low ionic strength solvent for 7 days resulted first in the removal of M lines and secondly, in the removal of Z lines. Goll, Mommaerts, Reedy, and Seraydarian (1969) found that $\alpha$-actinin-like proteins could be liberated during the removal of the Z band by trypsin digestion. They concluded that $\alpha$-actinin was a constituent of the

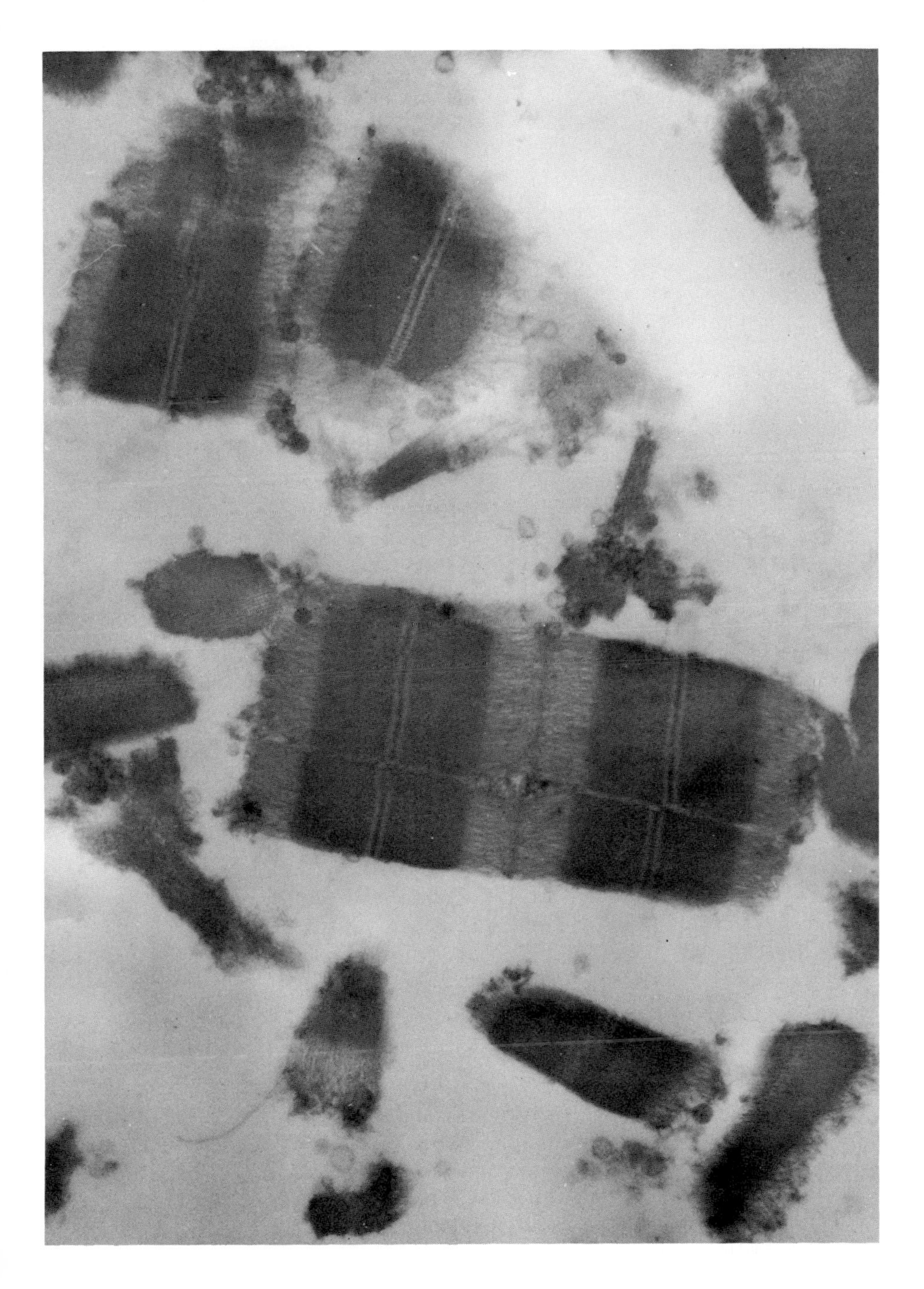

*Fig. 19.1.* Low-power magnification of sarcomeres and myofibrillar fragments of chicken pectoral muscle prepared by method 3 (see text, p. 402) and stored for 24 hours at 5° C. × 21,000.

Z band and of adjacent regions in the I filaments of the myofibril. At the present time we cannot describe the reason that fragmentation becomes feasible after some chemical changes have undoubtedly occurred in the Z lines with time post mortem, but studies of the morphology of myofibrils prepared from chicken pectoral muscle during or after storage post mortem have revealed at least the following two types of changes in the Z lines of sarcomeres and myofibrillar fragments: (1) degradation and/or disappearance of Z lines; and (2) breakdown of the junction of Z line and I filaments. Fig. 19.1 shows the morphology of sarcomeres and myofibrillar fragments which were prepared from pectoral muscles retained on the skeleton for 24 hours at 5° C, after the chickens had been scalded and plucked. Fig. 19.2 also shows breakage at the junctions of the Z lines and I filaments. Most of the myofibrils break into fragments of 1–4 sarcomeres, depending to some extent upon the degree of washing and blendorizing that has been employed in the preparation procedures. Although the zigzag configuration of the Z line is evident in intact myofibrils isolated from muscle tissue immediately after death, it has never been observed in myofibrils of muscle which had been stored for 24 hours post mortem. Several papers (Huxley, 1963; Masaki, Endo, and Ebashi, 1967; Briskey, Seraydarian, and Mommaerts, 1967; Goll, Mommaerts, Reedy, and Seraydarian, 1969) have implicated the localization of tropomyosin and α-actinin in the Z line of myofibrils. One point that seems certain is that post mortem changes must occur in the Z line and at the Z-I junction before fragmentation can occur.

Fukazawa, Briskey, Takahashi, and Yasui (1969) have completed an extensive study on the properties of protein obtained through extracting intact myofibrils, sarcomeres, and myofibrillar fragments. Suffice at this point to say that the gentle extraction of intact myofibrils does not disintegrate or dissolve the Z line and, when tested on reconstituted actomyosin, the protein obtained through washing does not convey any α-actinin activity, but rather merely sensitizes the reconstituted actomyosin to the removal of $Ca^{++}$. Conversely, when sarcomeres with partially disintegrated Z lines are extracted, a marked α-actinin activity can be detected in the protein obtained through the extraction procedure.

Fukazawa, Nakai, Ohki, and Yasui (1970) also employed the room-temperature extraction procedure of Ebashi and Ebashi (1964) on chicken pectoral muscle immediately after death and on muscle after it had been held at 5° C for 48 hours post mortem. When the α-actinin fractions of the protein isolates were added to trypsin-treated myosin B, only the 48-hour sample enhanced the onset and extent of superprecipitation (Fig. 19.3).

Since there are reports that α-actinin exists in the Z line (Masaki, Endo, and Ebashi, 1967; Briskey, Seraydarian, and Mommaerts, 1967; Goll,

*Fig. 19.2.* Electron micrograph of sarcomeres obtained from chicken pectoral muscle prepared by method 4 (see text, p. 402), and stored at 5° C for 24 hours. × 42,500.

Mommaerts, Reedy, and Seraydarian, 1969) and that the Z line disintegrates as the myofibrils are fragmented after the rigor-mortis-associated changes have occurred, it can be clearly stated that there occur in the Z line post mortem changes that render that 6 S component of α-actinin more easily solubilized.

When myofibrils, which have been prepared from chicken pectoral muscle 48 hours post mortem, are treated with a Guba-Straub solution, the Z line disappears (Fig. 19.4). This indicates that the Z lines which had been held intact at 0 hour post mortem, even after myosin A extraction (Fig. 19.12), now are easily solubilized at 48 hours post

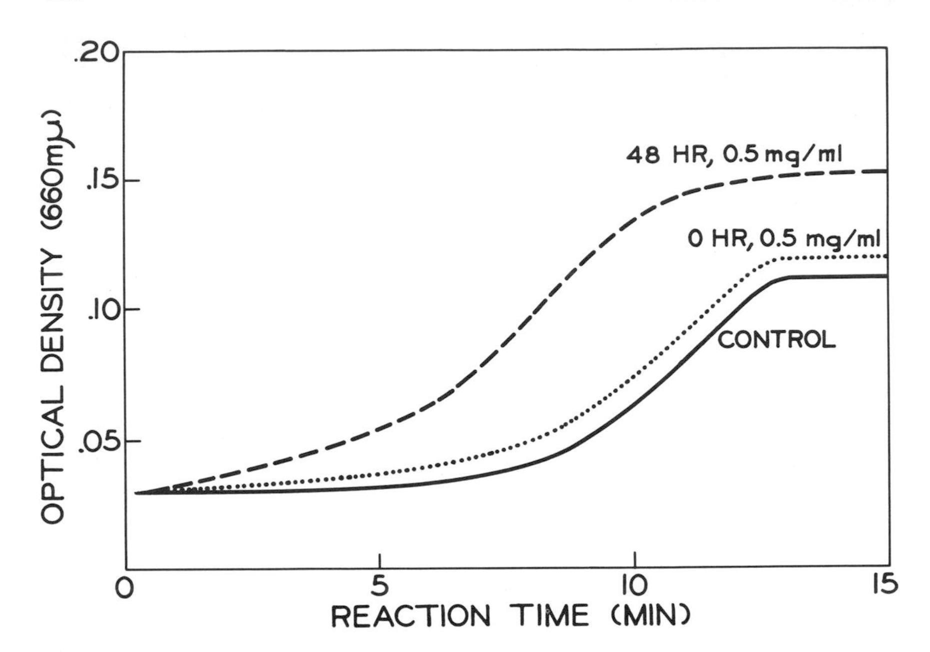


*Fig. 19.3.* Influence of α-actinin fractions, isolated from chicken pectoral muscle immediately after death and after storage for 48 hours post mortem, upon the ATP-induced superprecipitation of trypsin-treated myosin B. Final concentration of reaction mixture: myosin B, 0.37 mg/ml, tris-maleate buffer (*p*H 6.8), 20 mM, KCl, 0.03 M; $MgCl_2$, 1 mM; ATP, 1 mM; EGTA, 0.025 mM.

mortem, with the same Guba-Straub solution. It is well known that α-actinin induces a marked gelation of F-actin preparations. When the Z-line protein that is extracted from the 48-hour post mortem myofibril with Guba-Straub solution is added to F-actin preparations, the gelation is markedly enhanced.

### REVERSIBLE CONTRACTION AND FORMATION OF IRREGULAR INTERFILAMENTAL GRANULARITY

It seems certain at this point that the resolution of rigor mortis is associated with the post mortem tenderization of muscle tissue. Although the chemical changes associated with rigor resolution are not completely understood, there is considerable evidence (Goll, 1968) to support the concept that rigor resolution does in fact occur, and that during the development of this phenomenon the sarcomeres return to a more highly relaxed banding pattern. Stromer, Goll, and Roth (1967) have reported that when ATP-contracted myofibrils are subjected to short-term treatments with trypsin, they undergo a lengthening or relaxation of the supercontracted state.

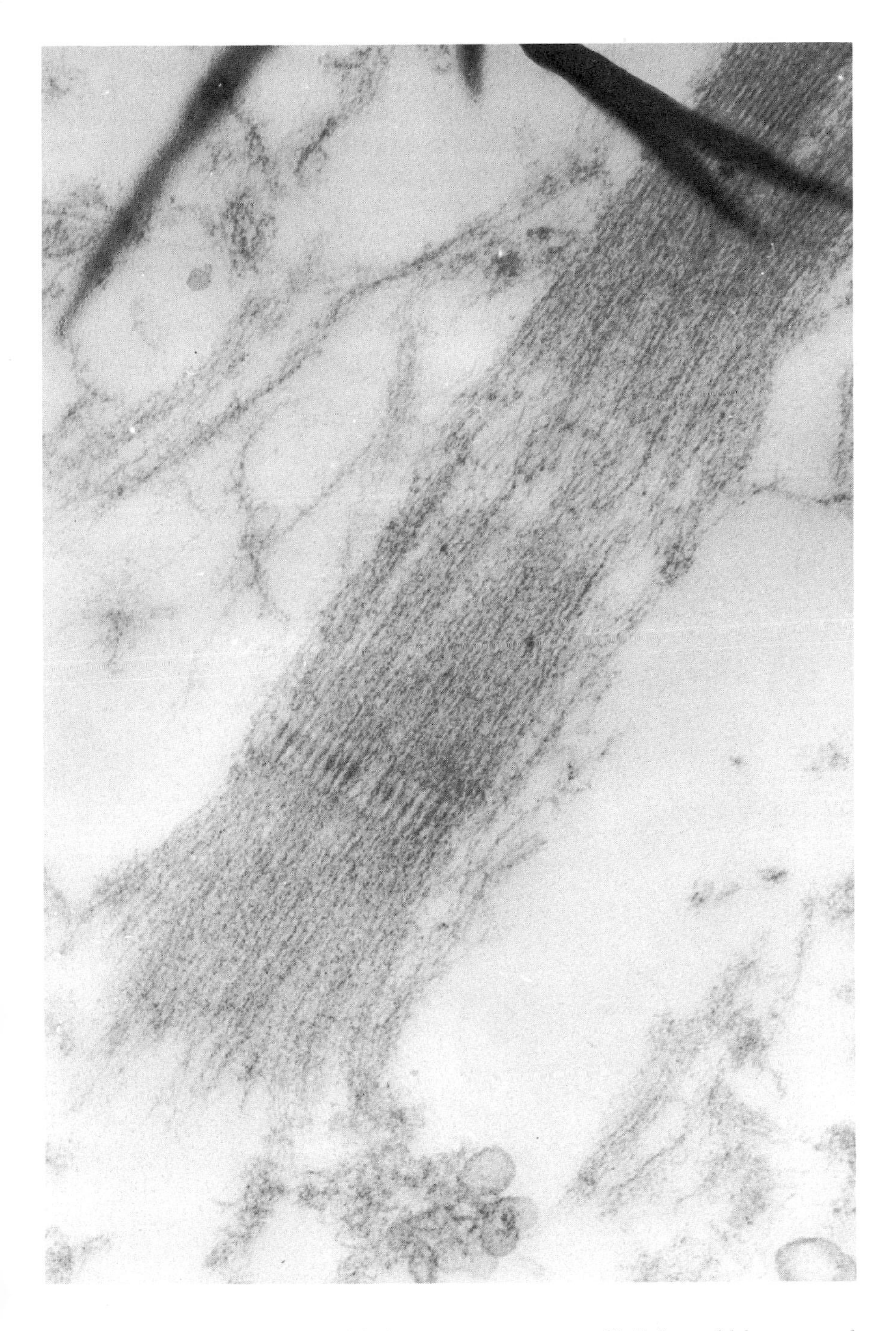

*Fig. 19.4.* Electron micrograph of 48-hour post mortem myofibril from chicken pectoral muscle after it has been washed with a Guba-Straub solution. × 60,000.

We wish to extend this discussion with a description of the irregular, interfilamental granularity (Fukazawa and Yasui, 1967) which develops in chicken myofibrils during the resolution of rigor, as long as the muscles remain on the skeleton with a reasonable amount of tension. First of all, it is essential to describe the experimental treatments which were employed to obtain varying degrees of lengthening or rigor resolution. The following four treatments were utilized in this study: (1) immediately after death the carcasses were plucked without scalding and stored at 5° C; (2) immediately after death the carcasses were plucked without scalding, and the pectoral muscles excised as soon as possible and stored at 5° C; (3) immediately after death the carcasses were scalded, plucked, and stored at 5° C; (4) immediately after death the carcasses were scalded and plucked, and the pectoral muscles excised as soon as possible and stored at 5° C. The sarcomere lengths taken from intact and

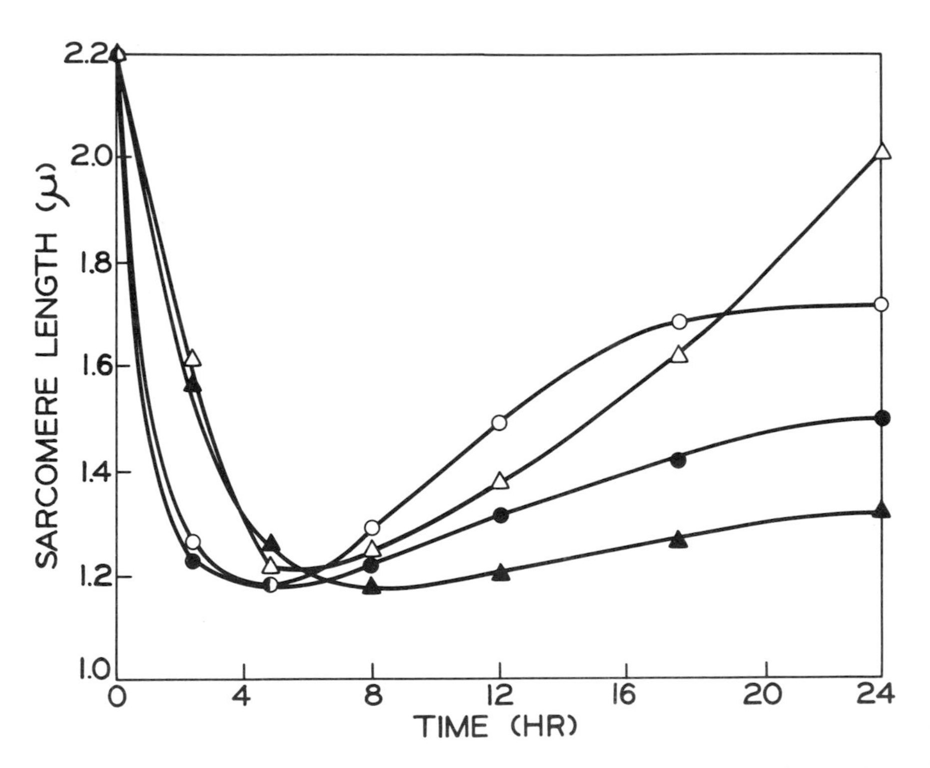


*Fig. 19.5.* Sarcomere length of myofibrils from chicken pectoral muscle after various periods of storage post mortem at 5° C. Each point represents the mean value in over 30 different myofibrillar fragments. *Open circles,* method 1 (see text, p. 402) ; *solid circles,* method 2; *open triangles,* method 3; *solid triangles,* method 4.

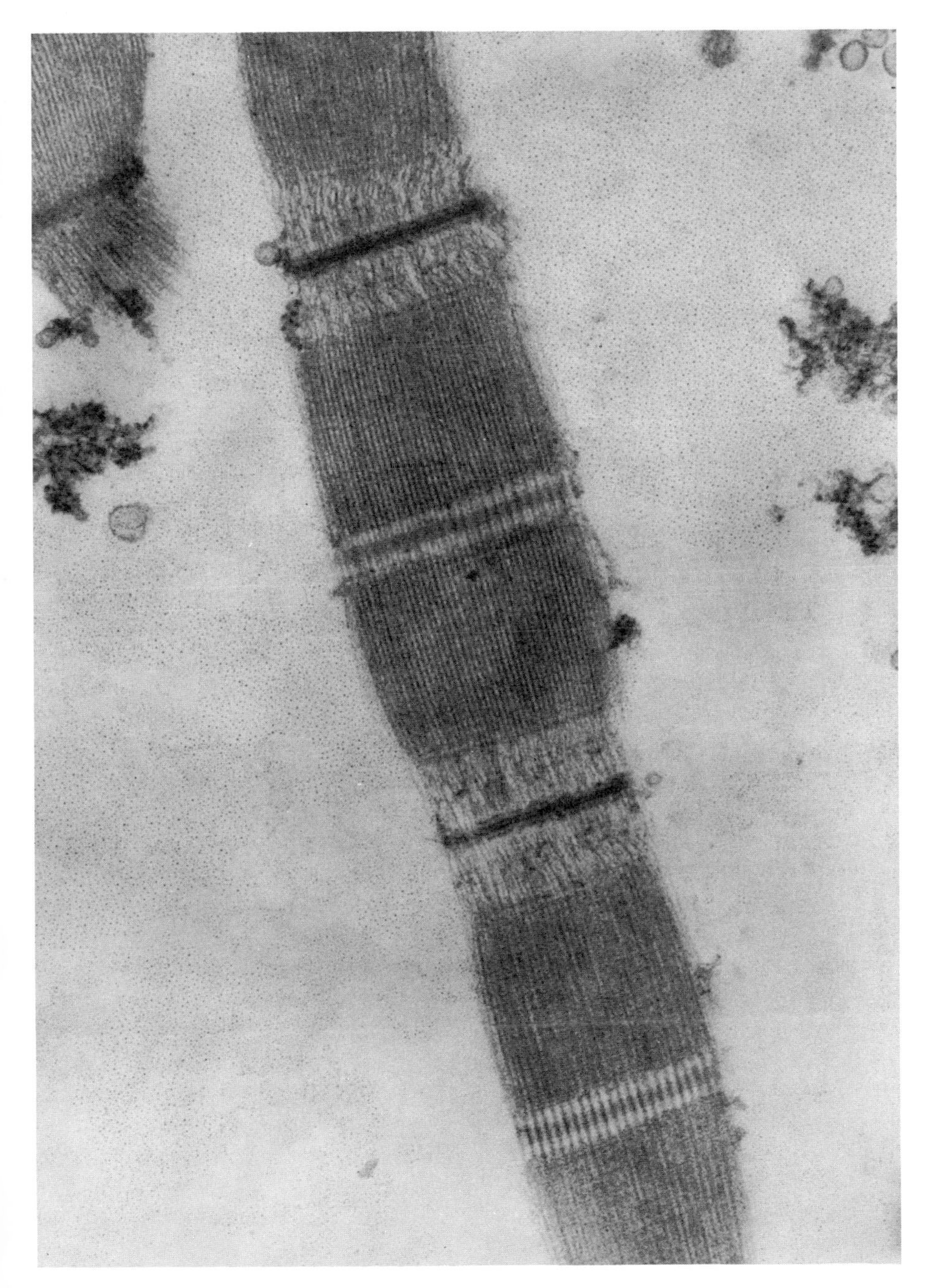

*Fig. 19.6.* Electron micrograph of typical myofibrils of chicken pectoral muscle prepared by method 1 (see text, p. 402) . × 42,500.

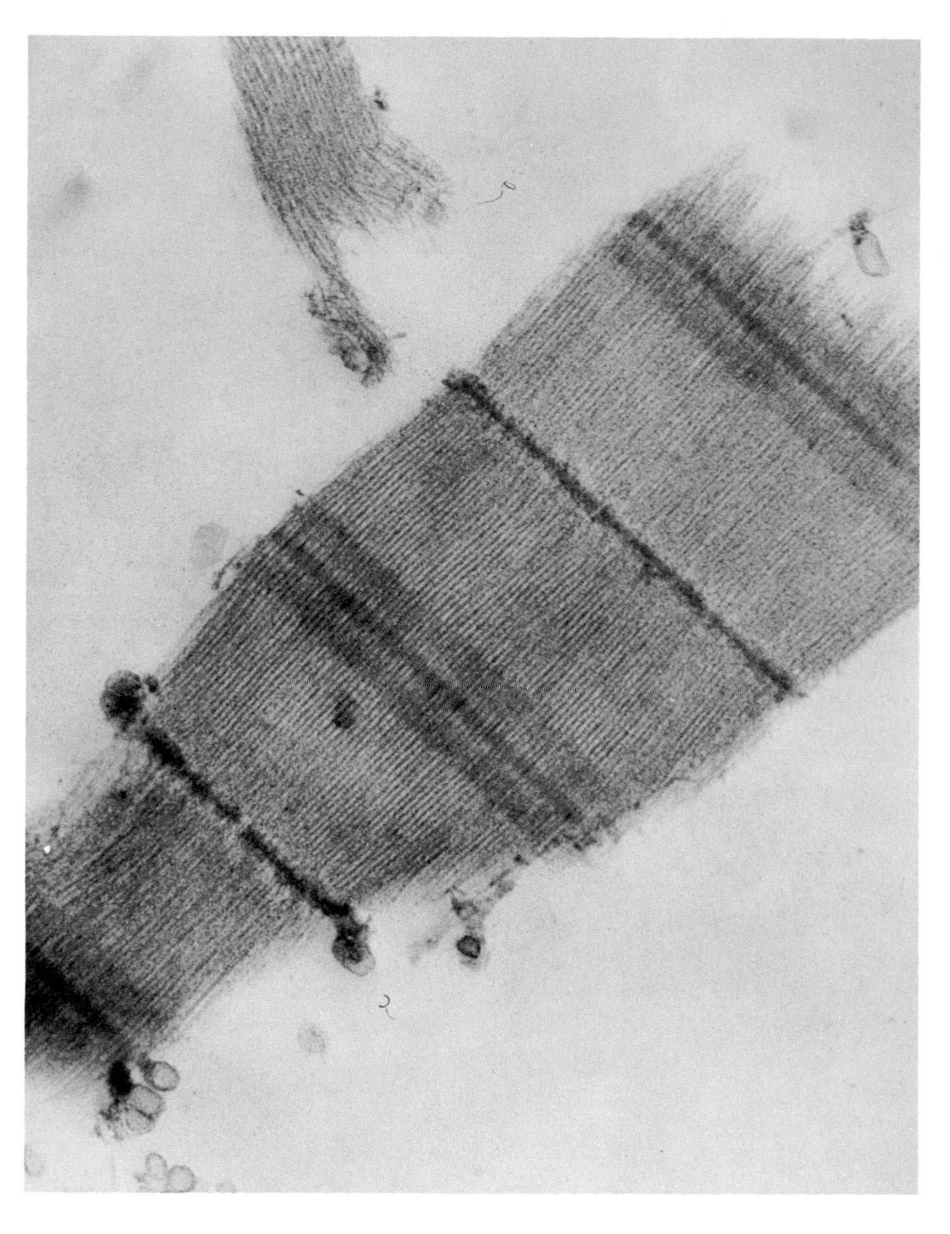

*Fig. 19.7.* Electron micrograph of typical myofibrils of chicken pectoral muscle prepared by method 2 (see text, p. 402) . × 42,500.

excised muscles at various times post mortem are shown in Fig. 19.5. It can be clearly seen that all muscles undergo a very severe shortening post mortem, after which varying amounts of lengthening occur. We wish to term this phenomenon "reversible contraction." Those muscles which were excised immediately after death showed very little reversal of the contraction that had occurred. Conversely, when the chickens had been scalded and the muscles retained on the skeleton, the amount of reversible contraction which developed resulted in almost complete restoration of the resting sarcomere length. Electron micrographs of typical banding patterns in the muscles from these four treatments 24 hours after death can be seen in Figs. 19.6–9. The well-established fact that the amount of overlap of the I filaments in the A bands governs the length of the sarcomeres can be clearly seen in these pictures. This investigation clearly establishes that the banding pattern is highly influenced by at least two factors: tension and temperature. Temperature appears beneficial if there is, in fact, some tension on the muscle, but appears highly detrimental if the muscles are subsequently excised and remain free of tension. The sarcomeres of muscles which had been retained on the carcasses of scalded birds were the only ones which nearly regained their resting length; furthermore, these were the only sarcomeres which showed the irregular interfilamental granularity (Fig. 19.8). Additionally, because of this fact, the irregular granularity appears to be related to the ultimate length of the sarcomere; Fig. 19.10 shows the relationship between the ultimate sarcomere length and the width of the area showing the interfilamental granularity. These results, when summarized, can be interpreted to mean that sarcomeres which contract during the development of rigor mortis proceed toward the relaxed state during the resolution of rigor; the ultimate sarcomere length, therefore, that is achieved is highly related to the width of the area showing this kind of granularity. Fig. 19.11 shows the morphological changes which can be induced in the sarcomeres by addition of ATP. Before the addition of ATP, the sarcomeres are near rest length and have considerable irregular interfilamental granularity; however, after ATP is added, the I filaments pull into the A bands and the width of the I bands shortens—through this shortening the irregular granularity disappears on both sides of the H zone.

### DISASSEMBLING THE MYOFIBRIL WITH MILD EXTRACTANTS AT LOW TEMPERATURE

Next, we wish to briefly describe some investigations which were conducted with the purpose of gently disassembling the myofibril. It was thought that this kind of approach could bring us closer to an under-

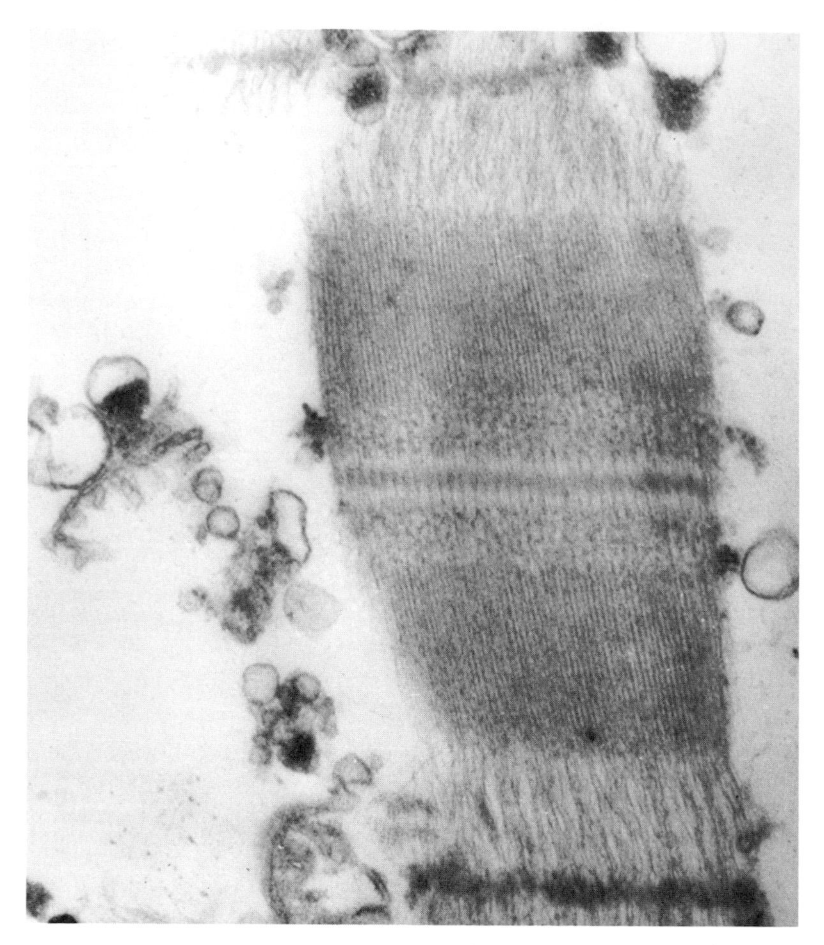

*Fig. 19.8.* Electron micrograph of typical myofibrils of chicken pectoral muscle prepared by method 3 (see text, p. 402) . X 42,500.

standing of the native state of the proteins in the myofibril. The results of these experiments are reported elsewhere (Fukazawa, Briskey, and Mommaerts, 1970*a, b;* Fukazawa and Briskey, 1970; Briskey and Fukazawa, 1970) our remarks here will be confined to generalizations on the fabrication of the myofibril.

We have previously indicated that the retention of myosin influences the extraction of protein from the I filament. Further, the use of acetone and/or high-temperature extractions represent rigorous handling of protein systems and may materially alter them from their native forms. Our approach therefore, was to prepare (Fig. 19.12) I-Z-I brushes (Fukazawa, Briskey, and Mommaerts, 1969*b*) which were completely free of myosin

*Fig. 19.9.* Electron micrograph of typical myofibrils from chicken pectoral muscle prepared by method 4 (see text, p. 402) . × 42,500.

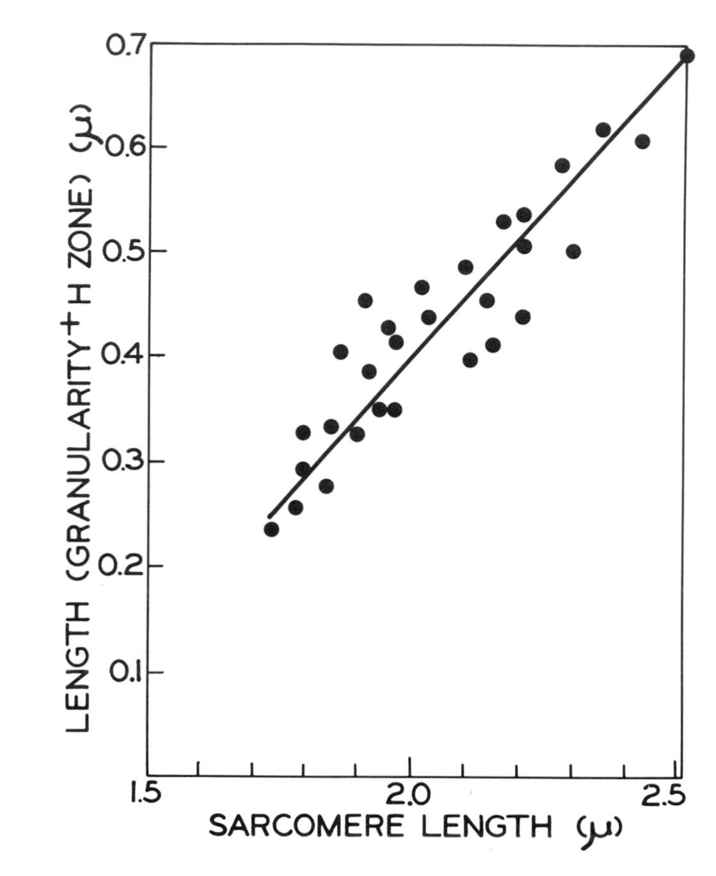


*Fig. 19.10.* Relationship between sarcomere length and the width of the H-zone region which appears with irregular interfilamental granularity.

A, and to exhaustively disassemble these I-Z-I brushes with mild solvents at 0° C. Table 19.1 gives an example of protein yields with each extraction step until only very thin filaments were remaining. After two extractions with mild solvents, one could note a definite loosening of the I-filaments. Fig. 19.13 shows the appearance of the I-Z-I brush after five extractions. Note that the I bands appear very loose but the zigzag configuration of the Z line has not disappeared. After eight extractions the Z line has completely disappeared and only thin filaments, presumably actin, can be seen. The extensive discussion of the properties of all of these extracts will appear elsewhere. Nevertheless, we wish to discuss some points of general interest. While extracts with α-actinin activity are obtainable, the initial extracts have been studied most extensively (Fuka-

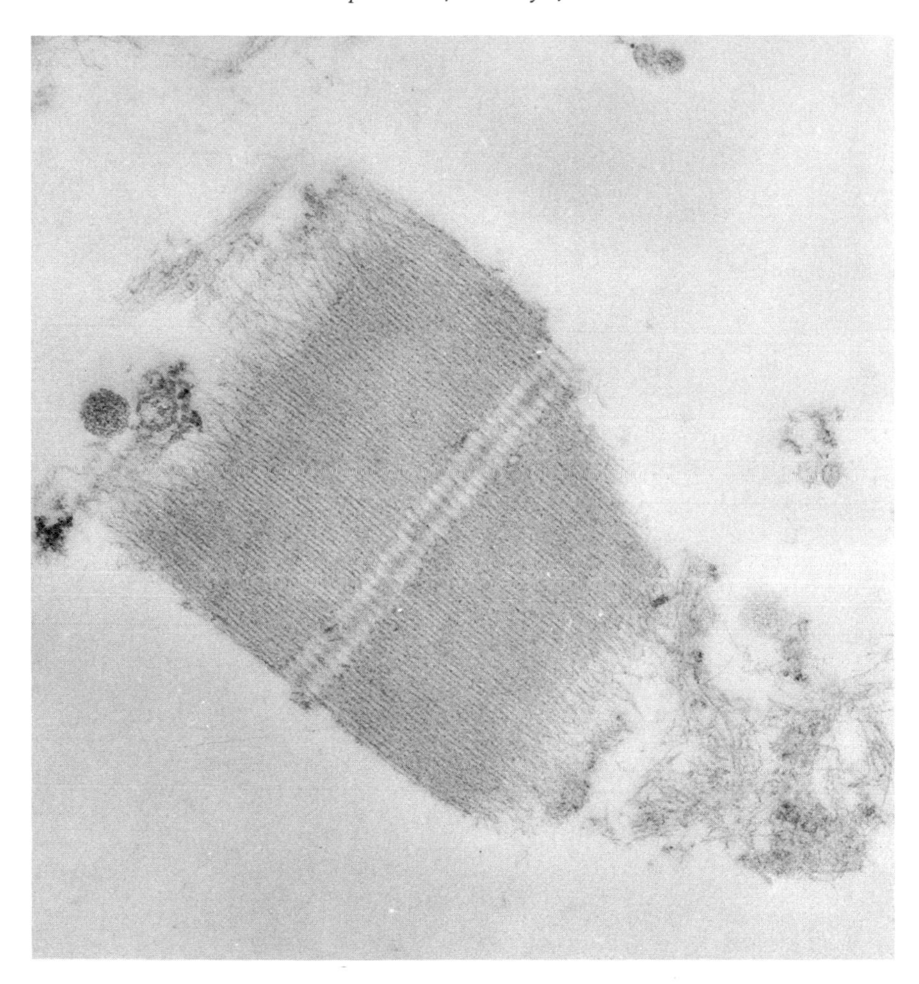

*Fig. 19.11.* Electron micrograph of a single sarcomere after the addition of ATP. × 40,000.

zawa, Briskey, and Mommaerts, 1970*b*). Through the procedures we have used, it has not been possible to depolymerize actin and remove it without simultaneously extracting Z-line material. Stromer, Hartshorne, Mueller, and Rice (1969) have described the extraction of thin-glycerinated bundles of rabbit psoas muscle with a low ionic strength solvent, wherein the M lines and Z lines were removed exclusive of the rest of the myofibril. In their experiments, they used 2 mM tris (*p*H 7.6) and 1 mM dithiothreitol as the extracting solvent. Even though they used glycerinated psoas and we used I-Z-I brushes, it is nevertheless interesting that

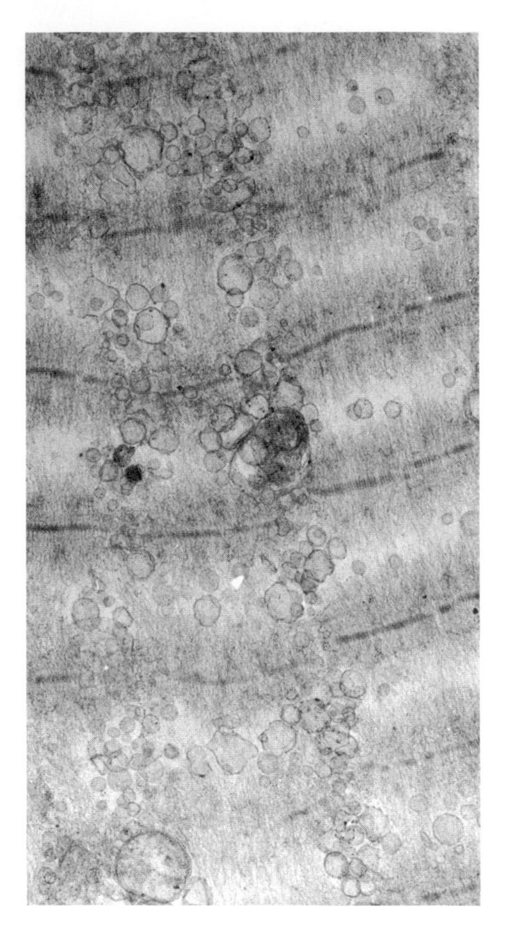

*Fig. 19.12.* Electron micrograph of I-Z-I brushes prepared from rabbit myofibrils. × 15,600.

whereas they selectively removed Z and M lines, we could not selectively extract the Z lines. As indicated earlier, when we used intact myofibrils (prepared at death) we could only extract small amounts of protein—predominantly of the $Ca^{++}$-sensitizing type. While some extracted protein could be precipitated with 3.3 M KCl, as α-actinin-like protein, it was not active and probably represented protein other than the 6 S variety. Therefore, the only difference between the protein coming out in the extracts of Stromer, Hartshorne, Mueller, and Rice (1969) and ours was in the Z-line material. After 24 hours of post mortem changes the Z-line material, with α-actinin activity, was extractable. Perhaps the glycerination process, by itself, had an effect on the Z-line material or Z-I junction in order to render it easily extractable—a strenuous process which we did not undertake in our experiments.

TABLE 19.1

*The relationship between the extraction times and the amounts of extracted proteins from I-Z-I brushes with diluted ATP (0.2 mM), ascorbate (0.2 mM), pH 7.6, solution at 0° C (per 180 g of muscle)*

| Extraction number | Extracted protein | % (in total) |
|---|---|---|
| 1 | 308.0 | 12.1 |
| 2 | 589.7 | 23.2 |
| 3 | 451.8 | 17.8 |
| 4 | 272.6 | 10.7 |
| 5 | 206.4 | 8.1 |
| 6 | 267.2 | 10.5 |
| 7 | 134.2 | 5.3 |
| 8 | 194.6 | 7.7 |
| 9 | 66.2 | 2.6 |
| 10 | 31.5 | 1.2 |
| 11 | 17.0 | 0.7 |
| Total | 2539.2 | 99.9 |

Now we will return to a few general statements on the protein which does, in fact, come out in the first two extractions of our I-Z-I brushes. This protein shows a high degree of purity through ultracentrifugation and is outstanding in its $Ca^{++}$-sensitizing ability. Through the employment of several techniques, we concluded that this preparation represented a tighter and more effective complex of tropomyosin-troponin than occurs in the native tropomyosin (NT). We have termed this preparation the troponin-tropomyosin factor (TF) (Fukazawa, Briskey, and Mommaerts, 1970*a*).

In the preparation of NT, Ebashi and Ebashi (1964) extracted myosin A, and then made a $H_2O$ extraction for 4 hours at room temperature—the contents of the extract were thought to consist of NT, α-actinin, and denatured actin. The major difference between their procedure and ours has been the use of 0° C in all of our work. Drabikowski and Gergely (1962) demonstrated that when actin was extracted from acetone powder at room temperature it contained considerable amounts of tropomyosin, while if extracted at 0° C, it was relatively free of tropomyosin. Ebashi and Ebashi (1964) showed the extraction of NT, which contained tropomyosin and troponin, by use of room temperature procedures, a step which is still recommended by Masaki, Wakabayashi, and Takaichi (1969). Our preparation has many similar properties to NT and even better $Ca^{++}$-sensitizing activity, but is completely extractable by short-term low-temperature (0° C) extraction steps.

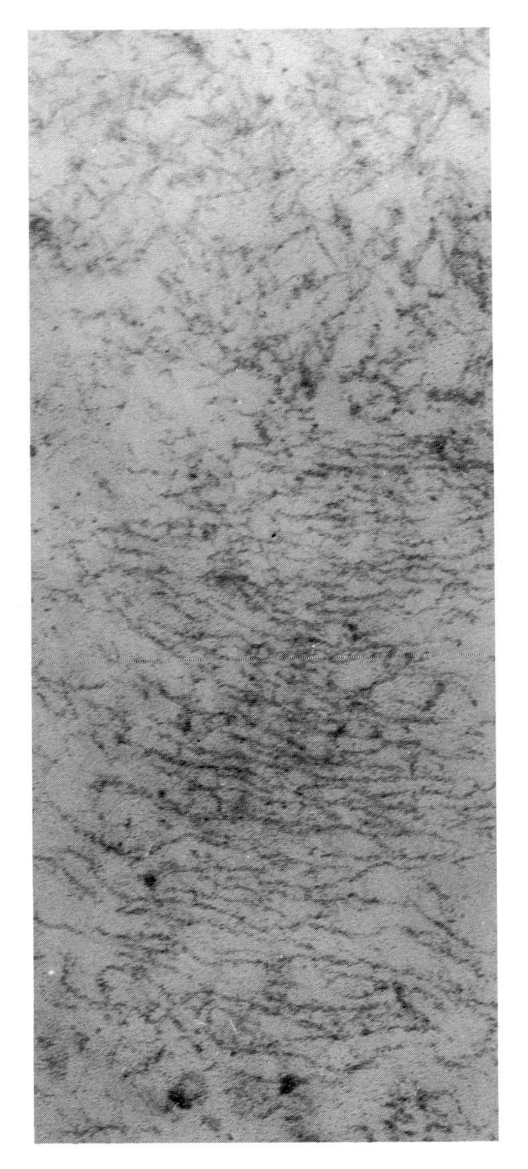

*Fig. 19.13.* Electron micrograph of I-Z-I brushes after they had been extracted 8 times with an ATP-ascorbate solution (*p*H 7.6) at 0° C. × 78,000.

PROPERTIES OF TF PREPARATION

An extensive study of the properties of TF has been conducted; its results will appear elsewhere. For purposes of this chapter, however, suffice to say that through the use of the $(NH_4)_2SO_4$ fractionation procedures of Watanabe and Staprans (1966), the isoelectric precipitation procedures of Ebashi, Kodama, and Ebashi (1968), and through other procedures, we feel that TF probably contains tropomyosin and troponin, but in a tighter network than in NT.

Upon treatment with ammonium sulfate, TF shows fractions with different sedimenting characteristics, but when they are recombined they sediment as a homogenous entity.

The TF preparation crystallizes under conditions where tropomyosin (TM) also crystallizes (Bailey, 1948). The crystals are usually in the form of coffin-shapes, which are among the forms recognized for TM (Casper, Cohen, and Longley, 1969). However, occasionally one sees a new shape which has never been previously shown in any tropomyosin preparations.

TABLE 19.2

*Fractionation of the extractable component preparation by ammonium sulfate salting-out*

| Fraction | 1 | 11 |
|---|---|---|
| $(NH_4)_2SO_4$ (g/dl) | 22.5–30.0 | 30.0–33.8 |
| Yield (%)* | 14.0 | 23.9 |
| (Watanabe and Staprans, 1966) | 0.52 | 26.0 |

* Yield is expressed in percentage of the total protein in the crude extract. By adding 22.5 g $(NH_4)_2SO_4$/100 ml to the extract, about 47% of the total protein was precipitated.

Electron-micrographically, the crystals consist of planes built up of the square or tetragonal network lattices described for TM by Huxley (1963), Hodge (1959), Higashi and Ooi (1969), and Casper, Cohen, and Longley (1969). They differ, however, by the presence of dots placed midway on certain, but not all, of the ribs of the lattice. These dots have not been shown, to date, in preparations of NT, and it is conceivable that they represent the binding sites in this protein complex. Whether this represents a more natural form of this structural and regulatory entity will be considered elsewhere.

## References

Bailey, K. 1948. Tropomyosin: a new asymmetric protein component of the muscle fibril. *Biochem. J.* *43*:271.

Briskey, E. J., and T. Fukazawa. 1970. Low temperature extraction of α-actinin-like protein from the I-Z-I brush of striated muscle. *Biochim. Biophys. Acta.* (In press.)

Briskey, E. J., K. Seraydarian, and W. F. H. M. Mommaerts. 1967. The modification of actomyosin by α-actinin III. The interaction between α-actinin and actin. *Biochim. Biophys. Acta 133*:424.

Casper, D. L. D., C. Cohen, and W. Longley. 1969. Tropomyosin: crystal structure, polymorphism and molecular interaction. *J. Mol. Biol. 41*:87.

Cook, C. F., and R. G. Wright. 1966. Alterations in contracture band patterns of unfrozen and prerigor frozen bovine muscle due to variations in post-mortem incubation temperature. *J. Food Sci. 31*:801.

Corsi, A., and S. V. Perry. 1958. Some observations on the localization of myosin, actin and tropomyosin in the rabbit myofibril. *Biochem. J. 68*:12.

Drabikowski, W., and J. Gergely. 1962. The effect of the temperature of extraction on the tropomyosin content in actin. *J. Biol. Chem. 247*:3412.

Ebashi, S., and F. Ebashi. 1964. A new protein component participating in the superprecipitation of myosin B. *J. Biochem. 55*:604.

Ebashi, S., A. Kodama, and F. Ebashi. 1968. Troponin I. Preparation and physiological function. *J. Biochem. 64*:465.

Franzini-Armstrong, C., and K. R. Porter. 1964. The Z-disc of skeletal muscle fibrils. *Z. Zellforsch. 66*:661.

Fukuzawa, T., and E. J. Briskey. 1970. Exhaustive extraction of the troponin-tropomyosin factor from the I-Z-I brush of striated muscle. *Biochim. Biophys. Acta.* (In press.)

Fukazawa, T., E. J. Briskey, and W. H. F. M. Mommaerts. 1970*a*. A new form of native tropomyosin. *J. Biochem. 67*:147.

———. 1970*b*. Low temperature extraction of a troponin-tropomyosin factor from the I-Z-I brush of striated muscle. *Arch. Biochem. Biophys.* (In press.)

Fukazawa, T., E. J. Briskey, F. Takahashi, and T. Yasui. 1969. Effect of treatment and post mortem aging on the Z-line of myofibrils from chicken pectoral muscle. *J. Food Sci. 34*:606.

Fukazawa, T., Y. Hashimoto, and Y. Tonomura. 1963. Isolation of single sarcomere and its contraction on addition of ATP. *Biochim. Biophys. Acta 75*:234.

Fukazawa, T., H. Nakai, S. Ohki, and T. Yasui. 1970. Some properties of myofibrillar proteins obtained from low ionic strength extracts of washed myofibrils from pre- and post-rigor chicken pectoral muscle. *J. Food Sci.* (In press.)

Fukazawa, T., and T. Yasui. 1967. The change in zigzag configuration of the Z-line of myofibrils. *Biochim. Biophys. Acta 140*:534.

Goll, D. E. 1968. The resolution of rigor mortis, p. 16. *Proc. 21st Reciprocal Meat Conf.*, Chicago, Ill.

Goll, D. E., W. F. H. M. Mommaerts, M. K. Reedy, and K. Seraydarian. 1969. Studies on α-actinin-like proteins liberated during trypsin digestion of α-actinin and of myofibrils. *Biochim. Biophys. Acta 175*:174.

Higashi, S., and T. Ooi. 1968. Crystals of tropomyosin and native tropomyosin. *J. Mol. Biol. 34*:699.

Hodge, A. J. 1959. Fibrous proteins of muscle. *Reb. Mod. Phys. 31*:409.

Huxley, H. E. 1963. Electron microscope studies on the structure of natural and synthetic protein filaments from striated muscle. *J. Mol. Biol. 7*:281.

Huxley, H. E., and J. Hanson. 1957. Quantitative studies on the structure of cross-striated myofibrils. *Biochim. Biophys. Acta 23*:229.

Knappeis, G. G., and F. Carlsen. 1962. Ultrastructure of the Z disc in skeletal muscle. *J. Cell Biol. 13*:323.

Maruyama, K., and S. Ebashi. 1970. Regulatory proteins. *This volume,* p. 373.

Masaki, T., M. Endo, and S. Ebashi. 1967. Localization of 6 S component of α-actinin at Z-band. *J. Biochem. 62*:630.

Masaki, T., T. Wakabayashi, and T. Takaichi. 1969. Molecular biology of muscular contraction III. (In Japanese.) *Kagaku 39*:268.

Nonomura, Y., W. Drabikowski, and S. Ebashi. 1968. The localization of tropomyosin paracrystals. *J. Biochem. 64*:419.

Perry, S. V., and A. Corsi. 1958. Extraction of proteins other than myosin from the isolated rabbit myofibril. *Biochem. J. 68*:5.

Stromer, M. H., and D. E. Goll. 1967. Molecular properties of post-mortem muscle III. Electron microscopy of myofibrils. *J. Food Sci. 32*:386.

Stromer, M. H., D. E. Goll, and L. E. Roth. 1967. Morphology of rigor-shortened bovine muscle and the effect of trypsin on pre- and post-rigor myofibrils. *J. Cell Biol. 34*:431.

Stromer, M. H., D. J. Hartshorne, H. Mueller, and R. V. Rice. 1969. The effect of various protein fractions on Z- and M-line reconstitution. *J. Cell Biol. 40*:167.

Szent-Györgyi, A. 1951. *Chemistry of Muscular Contraction* (2d ed.). Academic Press, New York.

Takahashi, K., T. Fukazawa, and T. Yasui. 1967. Formation of myofibrillar fragments and reversible contraction of sarcomeres in chicken pectoral muscle. *J. Food Sci. 32*:409.

Takahashi, K., T. Mori, H. Nakamura, and Y. Tonomura. 1965. ATP induced contraction of sarcomeres. *J. Biochem. 57*:637.

Watanabe, S., and I. Staprans. 1966. Purification of the relaxing protein of rabbit skeletal muscle. *Proc. Nat. Acad. Sci. 56*:572.

# Chapter 20

# Summary and Discussion of Part 4

PANEL MEMBERS: W. F. H. M. MOMMAERTS, *Chairman*
M. REEDY
R. E. DAVIES
L. LORAND
R. N. SAYRE
M. STROMER

*Mommaerts:* The first chapter in Part 4 is a broad review by J. Gergely of the myofibrillar proteins and their interactions; besides the information it contains by itself, this chapter also sets the stage for the more specific papers that follow. The major development in this field in the last few years has been the discovery, mostly by Ebashi and his associates (Ebashi and Endo, 1968), of accessory proteins, as opposed to myosin and actin, which might still be called the major proteins on account of their bulk and the directness of their chemomechanical involvement. It might be said that myosin and actin are the direct agents of the transducer process which is contraction, that the troponin-tropomyosin complex prevents this interaction until released by $Ca^{++}$, and that there are a few more minor proteins whose functions are not yet satisfactorily elucidated.

As to the structural organization of the major proteins, and the nature of their participation in the contractile event, the sliding filament model originated by Hanson and H. Huxley (1955), and by A. F. Huxley alone (1957), and added to significantly by H. Huxley (1963, 1969), by Hanson and Lowy (1963), and by Reedy, Holmes, and Tregear (1965), has now found wellnigh general acceptance. It is rare among biological theories in that it has come out so strongly after only a decade or two of testing, though objectivity requires one to state that challenging views are still held by a few investigators of prestige.

Dr. Gergely's review deals with the various facts extensively, and in a manner which would be hard to condense further. But, if I may enter one personal note, while it is true that purified actin in the past has often contained members of the accessory protein family, as contaminants, it would be incorrect to leave the impression that my original actin was so contaminated, inasmuch as it was extracted at low temperature, which was later found to be essential. Concerning the chemistry of actin, it has become known through the work of Oosawa, Asakura, Higashi, Kasai, Kobayashi, Nakano, Ohnishi, and Taniguchi (1965) that, notwithstanding the great interest in the association of actin with ATP and the breakdown of this nucleotide in polymerization, the ATP participation is not an essential part of the polymerization reaction. This by itself does not exclude that such a nucleotide splitting might well occur *in vivo,* but this has been eliminated by Gergely and his associates. But even this is not necessarily the last word on the issue: there is reason to believe that the ATP-splitting reaction *in vivo,* presumably occurring once in each cycle of a fully interacting cross bridge, is a cooperative property between myosin and actin, for only so can the $Mg^{++}$-activation *in vivo* or in myofibrils be understood. While in view of demonstrations by Bárány (1967) and by Buller, Mommaerts, and Seraydarian (1969) it is the myosin moiety which sets the pace, it is still not known just what the role of actin in this process is. The term "allostery" comes to mind, but would have to be more specifically developed to be elucidating. At the same time, Gergely points to additional values of an awareness of allosteric effects, for example, the role of sulfhydryl (-SH) groups in many of the interactions need not be one of direct participation. Closely related is the question of conformational changes; among revealing lines of approach besides direct optical ones, I single out Pepe's (1966) work on the changing reactivity of myosin toward its antibodies.

Drs. Maruyama and Ebashi deal with the regulatory proteins in general in Chapter 17. Among these, the discovery of troponin is surely the most significant event; curiously, it does not act alone but only in combination with tropomyosin, so that this protein, known for so long mainly as something with an interesting crystallography, now also has a functional role. This protein pair confers $Ca^{++}$ sensitivity upon actomyosin. Rephrasing the authors' views and our own, we would now state the case in these terms: in resting muscle, actin and myosin do not interact, hence do not contract nor exert their $Mg^{++}$-activated adenosine triphosphatase activity, because the troponin-tropomyosin complex prevents this. When $Ca^{++}$ is added (and this occurs physiologically as a consequence of the excitatory event), this suppression is relieved, and actin and myosin interact completely. Possible mechanisms of these effects are being explored, and again the thought presents itself that one protein exerts upon the other some conformation- or reactivity-modifying influence. Contrary to the case of troponin and its combination with tropomyosin, no clear function has emerged for α-actinin; the original supposition that it is essential for contraction has not held up. There is agreement between the Ebashi group, Dr. Briskey's, and ours, on different experimental foundations, that α-actinin may be a structural component of the Z band or of nearby regions of the I filaments. It can, however, exert a distinct activating effect upon actomyosin–adenosine triphosphatase, and this would not have an immediately visible meaning if it were exclusively located at the maximal distance of where the actomyosin–adenosine triphosphatase is.

A. Weber's chapter deals with one important problem touched upon above: the manner in which the $Ca^{++}$ participates in the interactions in the indicated manner. Certainly an important feature is that troponin has an affinity (with tropomyosin) to actin, and likewise to $Ca^{++}$. Somehow among these interactions, the mechanism of the regulatory process is concealed, and efforts are under way to penetrate into the mystery. Dr. Weber's paper contributes a number of important facts among the ones that will have to be known in this connection: the detailed presentations must be consulted in full. This author likewise proposes that relaxation, in her words "may be discussed in terms of an antagonism between myosin and troponin. The degree of antagonism is dependent on the binding of $Ca^{++}$ to troponin on the one hand and the binding of ATP to myosin on the other." The latter is a significant addition to the picture we had already recognized. Her closing sentence is also thought-provoking: "the problem must be solved of how one troponin can make unavailable to myosin several active sites on actin, most of which are separated from troponin by a considerable distance." We shall return to this problem further below.

In Chapter 19, Drs. Fukuzawa and Briskey discuss the structure of the myofibril, first giving an extensive review, including a discussion of post mortem changes, then going on to deal in particular with the I filaments, investigated by detaching constituents piecemeal with gentle successive extractions. These investigations lend detailed support to the view that this filament is not merely an actin-polymer, but contains the accessory proteins to form an organized microcosm. One novel finding is that the troponin-tropomyosin complex was obtained in a form more native perhaps than "native tropomyosin." Likewise constituting a troponin-tropomyosin complex, it differs in that the two components are less easily separable, so that the complex might occur in a higher form of integrity. This complex, too, crystallizes much like tropomyosin. In tropomyosin, which is unique among fibrous proteins in that it crystallizes in a nonfibrous manner, the crystal lattice can be investigated electron-microscopically, and is found to consist of a square wire-mesh weave with several possible variations. The new form of the complex has this same basic structure, but in addition carries globular appositions at the centers of the long edges of the kite-shaped modification of the square lattice. In the analysis of Casper, Cohen, and Longley (1969) the arms of the lattice consist of two helical strands of tropomyosin each, in an antiparallel arrangement. The long arms are those where successive helices are joined end to end. Hence, it seems to be at the long ribs that the troponin units are primarily apposited.

This would seem to be, then, the explanation of the periodic arrangement of troponin in the I filaments. It brings us back to the question raised by Dr. Weber: how does troponin determine the reactivity, vis-à-vis myosin, of actin monomers or doublets that may be up to 200 Å away from it? The beaded actin strand is not likely to provide that continuity, but perhaps tropomyosin could, in which case we would be dealing with an interesting transmission of an allosteric influence over a distance. Or is there another possibility? A. F. Huxley's (1957) theory contains a calculation allowing one to determine the distance between successive I-filament sites with which the cross bridge can interact, and this calculation, if revised in the light of some new data, gives a result close to the value of the troponin periodicity (Mommaerts, 1969). Could this mean that the cross bridges can attach only at the troponin binding sites? That would eliminate Dr. Weber's dilemma, though it would contradict a large number of data, among them the knowledge that troponin and tropomyosin are not essential for contraction.

Clearly, while our structural knowledge of the filaments has increased considerably, a new set of problems is coming up in connection with their component interaction, and the role this plays in the regulatory aspects of the transduction process.

## DISCUSSION

*Guenther:* What types of bonds are involved in the spontaneous aggregation of myosin? In what ratio does actin combine with myosin and on what unit basis is this ratio based?

*Gergely:* The fact that both ionic strength and *p*H play an important role in the aggregation of myosin suggests that electrostatic forces are involved (Kaminer and Bell, 1966; Josephs and Harrington, 1966). The combining weight ratio of myosin to actin *in vitro* has been found to be 4:1 (Spicer and Gergely, 1951); for a summary of the evidence see Perry, 1967. This ratio would roughly correspond to the combination of one myosin molecule with two monomeric units of actin (Eisenberg and Moos, 1967). More recent work (Eisenberg, Barouch, and Moos, 1969) suggests that the adenosine triphosphatase activity of only one head of the myosin molecule can be activated by actin.

*King:* What molecular or geographical relation is there between the myosin site for adenosine triphosphatase activity and that for binding to actin? Are -SH or amino groups involved in the myosin site for binding to actin?

*Gergely:* Since actin stimulates the adenosine triphosphatase activity of myosin, it follows that ATP and actin can bind simultaneously. Both sites are located in the subfragment-1 region; the precise relationship of the two sites has not been determined. It is not unlikely that they are rather near each other.

-SH groups are indeed involved in the binding of myosin to actin. Stracher (1964) has reported that some -SH groups can be protected by ATP, others by actin, against disulfide interchange. This suggests that there are two distinct kinds of -SH groups.

*Sayre:* From the chapters in Part 4, it appears that the authors favor the complete two-chain structure of myosin. The structure consisting of three globular heads is also presented in the literature, however. Would you please comment on the arguments for these two structural proposals?

*Gergely:* Space would not permit a complete review of this complicated question. I would refer you to some recent papers (Nauss, Kitagawa, and Gergely, 1969; Slayter and Lowey, 1967; Lowey, Slayter, Weeds, and Baker, 1969), and the review by Young (1969), not wholly oriented to the two-chain theory.

*Sanger:* You mentioned that the head part of the myosin molecule is composed of two parts or chains. What is the relationship of these two head chains to the cross bridges observed in electron micrographs of thick filaments? Do we see two chains as one cross bridge in the electron micrographs? Can each of these head chains bind to actin?

*Gergely:* Although most people believe that each cross bridge corresponds to one myosin molecule and contains both chains, there is no rigorous proof for this view (Pepe, 1966, 1967; Reedy, Fischman, and Bahr, 1969). Perhaps Dr. Reedy should comment on this aspect. It seems that all the subfragment-1 derived from heavy meromyosin, hence material from both chains, can combine with actin (Mueller and Perry, 1962; Young, Himmelfarb, and Harrington, 1965).

*Sayre:* If tilting of the globular head of myosin in relation to a rigid attachment site on actin is the basis for movement as proposed by Huxley (1969), would the binding limitation of only one globular region per active site conflict with this proposal?

*Gergely:* I have no information on that question.

*Lorand:* Could Dr. Gergely draw the immunoglobin type of representation of the myosin molecule, that is the two heavy and the two light chains in parallel, and then summarize the existing knowledge in respect to N and C termini?

*Gergely:* Fig. 16.5 is such a representation. In view of the presence of the smaller subunits, most likely three per myosin molecule, the end group situation is somewhat unclear. N-acetyl-serine has been described at the N terminus of myosin (Offer, 1965). Subsequently Offer and Starr (1968) reported that this residue is located in the long chain in the heavy meromyosin portion. The N-terminal histidine reported by Kielley, Kimura, and Cooke (1964) may be located in one of the light chains. According to Offer the C terminus of the long chains would be in the light meromyosin moiety. The C-terminal isoleucine reported by Locker (1954) and by Sarno, Tarendash, and Stracher (1965) would have to be assigned to the light chain.

*Lorand:* Did I understand Dr. Gergely correctly in that one can block all -SH groups in myosin, then digest the modified myosin with hydrolytic enzymes and isolate the subfragment-1 ($S_1$) as a functioning adenosine triphosphatase?

*Gergely:* This has been reported by Trotta, Dreizen, and Stracher (1968), as I understand it, with respect to the functionally important -SH groups. Our own studies with Dr. Seidel and Mr. Chopek with the use of spin labels seem to indicate that as long as subfragment-1 has enzymic activity, the label attached to the $S_1$-SH group remains attached to it.

*Lorand:* I recall the fluorescence studies of Sekine in Japan in regard to possible conformational changes in myosin. Does Dr. Gergely know if that work has progressed to permit conclusions really indicating such conformational changes?

*Gergely:* The work of Yamaguchi and Sekine (1966) showed that one of the -SH groups ($S_2$) involved in adenosine triphosphatase activity became more reactive towards thiol reagents if ATP is present and thus suggests a conformational change. Changes in fluorescence observed on adding ATP to myosin labeled on the same -SH group with a fluorescent thiol reagent, reported by Sekine, Kanaoka, Kameyama, Machida, and Takada (1967), are in line with the earlier work referred to above. I am not aware of a definitive paper dealing with the fluorescence work.

*Passbach:* What is the influence of *p*H on the binding of ATP by myosin?

*Weber:* That has not been investigated.

*Perry:* Does your scheme for the mechanism of the regulatory proteins depend upon the site in the myosin molecule for actin interaction and ATP splitting being identical? Would it be valid if separate sites existed for each of these activities?

*Weber:* It would be valid if the two sites were different. It was conceived under the notion that they were different.

*Chrystall:* Actin discovery has been attributed to Szent-Györgyi, but it seems very similar to "myosin ferment" of Halliburton (1887) as pointed out recently by Finck (1968).

*Gergely:* Actin was actually first described by Straub (1942). The suggestion that what had been at the time described as myosin actually consisted of two components emerged from the work of Banga and Szent-Györgyi (1942) (see also Szent-Györgyi, 1942). Although it appears that Halliburton described a substance that had the properties of actin, it may not be correct to attribute to him the discovery of actin in the modern sense. It would seem that Halliburton regarded his "myosin ferment" as an enzyme acting on myosin, rather than another protein reacting in a stoichiometric fashion. This may represent a situation similar to, but perhaps even in a more exaggerated way, that discussed by Kuhn (1965) with regard to the "discovery" of oxygen.

*Sayre:* How does actin, polymerized in the absence of nucleotide or Ca-Mg, behave in association with myosin, tropomyosin, or troponin? Are nucleotides or Ca-Mg easily bound to F-actin polymerized free of these substances?

*Gergely:* According to Bárány, Tucci, and Conover (1966), the reaction of nucleotide-deficient actin with myosin—superprecipitation and stimulation of adenosine triphosphatase activity—is unchanged. As far as I know the interaction of troponin and tropomyosin with nucleotide-free actin has not been studied.

*Sayre:* Spin-labeling techniques have shown that a conformational change in actin results from binding of $Ca^{++}$ to troponin. Is there evidence of a different conformational change in actin as a result of either one or two moles of $Ca^{++}$ being bound to troponin? Would this conformational change necessarily indicate a change in availability of the binding site of actin to myosin?

*Weber:* That has not been measured. I would assume that the conformational change decreases the reactivity of the actin site for myosin.

*Sayre:* Low temperature promotes depolymerization of F-actin in solution. Is there evidence for this action in intact muscle?

*Gergely:* I am not aware of any fact pointing to *in vivo* depolymerization of actin. Even *in vitro* (e.g., Kasai, 1969) lowering of the temperature affects chiefly the rate of polymerization and the so-called critical actin concentration. At higher actin concentrations the final degree of polymerization would not change greatly.

*Davies:* Dr. Gergely said there was a dogma that the myosin and actin interaction was similar to the situation occurring in the cyclic contraction, but Dr. Weber suggested that perhaps these were different. There was also another possibility of a discrepancy in that Dr. Gergely said that involvement of $Ca^{++}$ in the binding of the thick and thin filament could now be excluded, while Dr. Weber's chapter suggested that the $Ca^{++}$ was overcoming a repulsion (presumably an electrostatic repulsion) of the thick and thin filaments. In fact, can we exclude the direct involvement of $Ca^{++}$ in the binding of the thick and thin filament, *in situ,* in the sarcomeres?

*Weber:* The $Ca^{++}$ is bound to troponin, as shown by Ebashi, Kodama, and Ebashi (1968) and by Fuchs and Briggs (1968). If during contraction the thick and thin filaments were linked to each other by $Ca^{++}$, i.e., if $Ca^{++}$ were linked to both troponin and myosin, then contraction would be dependent on troponin rather than on actin alone. However, contraction does not require the presence of troponin but only that of actin.

*Mommaerts:* You mean electrostatic in respect to what?

*Weber:* During relaxation, repulsion by negative charges between myosin and the thin filament and during contraction, neutralization of the negative charges by $Ca^{++}$. Hartshorne (1970) has data that are not compatible with such a concept.

*Gergely:* At this point perhaps it is worth pointing out that Elliot (1968) and Rome (1968) have been stressing the role of electrostatic forces as well as Van der Waals interaction in keeping filaments at their appropriate distance. I think that the gross shift in the electrostatic balance is unlikely since, as Huxley (1968) has recently shown, the interfilamental separation does not depend on whether the muscle is relaxed or in rigor. So this additional interaction which occurs between the cross bridges and the thin filament does not change the overall separation and therefore there is probably no gross change in the charge. However, this may still leave some small charge effect from the cross bridge itself.

*Davies:* That does not follow at all, because the interpretation of recent X-ray pictures (Huxley, 1969) show that the whole head-end of the myosin, instead of vibrating around near the thick filament, moves right over and so is picked right up on the X-ray diagram as part of the actin lattice. There has to be some link, either covalent or one involving electrostatic charges; otherwise one is not left with much except magnetism.

*Mommaerts:* Yes, but you do not need $Ca^{++}$ ions for it.

*Davies:* That remains to be seen.

*Adrian:* Dr. Weber, in connection with the electrostatic idea, is the effect of ionic strength being investigated? Obviously one could not change it in a large degree but I wondered whether the rates were, in fact, dependent on ionic strength, as I think they would be if the repulsion or distance apart of the two interacting sites was dependent only on electrostatic repulsion?

*Weber:* Contraction can take place in a range of ionic strength between 0.04 to 0.1, i.e., a fairly wide range extending to a rather high ionic strength. The argument has been made though– based on the paper by Rome (1968) —that the ions are not between the filaments.

*Perry:* Does the repulsive fit hypothesis implicitly demand that there is only one site for actin-combining and adenosine triphosphatase activity? Is it correct that they are identical sites?

*Weber:* No. The hypothesis demands inaccessibility of that site on actin that interacts with myosin during the hydrolysis of ATP.

*Perry:* If you have two sites on the myosin, and I believe there are, I think you can avoid the steric objections.

*Weber:* No, the existence of two sites does not change anything. I would like to point out that complete dissociation of individual actin-myosin links occurs also in the complete absence of tropomyosin and troponin. That is a prerequisite for shortening. Each cycle of actin-myosin interaction must be followed by complete dissociation so that myosin can attach again to a new site on the thin filament. In this manner the thin filament is moved inwards into the lattice. Therefore the dissociation of an individual actin-myosin link does not require the regulatory proteins. Troponin and tropomyosin exert their effect by preventing the reassociation of myosin and actin. The complete dissociation of the whole filaments from each other results.

*Gergely:* I would like Dr. Weber to clarify the points she made in connection with her last two illustrations (Figs. 18.6–7). She showed that there are presumably two distinct $Ca^{++}$ binding sites on troponin, and somehow related this to the fact that at low ATP concentration when there was partial ATP saturation a different $Ca^{++}$ concentration was required from that at high ATP concentration. Does this mean that when both ATP heads are saturated, then you need both $Ca^{++}$ to overcome it, and if one is saturated then one $Ca^{++}$ is needed?

*Weber:* When Dr. Maruyama and I measured the binding of ATP to myofibrils we found that maximally 2 nmoles ATP were bound per mg protein, i.e., per 1 nmole of myosin. One may thus conclude that each head binds ATP. And on that basis one could say that when each head carries ATP both classes of $Ca^{++}$ binding sites must be saturated for contraction. There is no contraction when only the high affinity $Ca^{++}$ site is saturated. If myosin is saturated with ATP only to 30–40%, contraction starts when only the high affinity site is occupied by $Ca^{++}$.

*Gergely:* You do not need $Ca^{++}$ at all at very low ATP concentrations?

*Weber:* That is the problem: There are three conditions for contraction: no $Ca^{++}$, 1 bound $Ca^{++}$, 2 bound $Ca^{++}$ and only 2 heads.

*Sayre:* Only nucleotide triphosphate in the presence of $Ca^{++}$ causes contraction; however, it has been reported that ADP or even pyrophosphate prevent interaction between actin and myosin. If ADP could be retained in the muscle as ATP was depleted, would it be possible to retain some degree of extensibility in the myofibrils?

*Weber:* One must distinguish between the effect of ADP in 0.6 M KCl and at low salt. I do not think that it is known whether ADP can cause complete dissociation of actin and myosin in low KCl.

*Reedy:* It has been shown that the constant volume behavior of living resting muscle (Elliott, Lowy, and Worthington, 1963) can be imitated *in vitro* by glycerinated muscle only at ionic strengths corresponding to 0.015–0.020 M KCl (Rome, 1968). It seems to me that this might prompt the muscle biochemists to study their favorite *in vitro* systems or phenomena at this low ionic strength, rather than exclusively working at higher salt concentrations. Has this tendency become at all prominent in very recent work? Where would you expect it to clarify any problems of the relationship between *in vitro* and living systems?

*Gergely:* I agree with you that the study of the actomyosin system—whether in suspension, thread, myofibril or glycerinated fiber—at low ionic strength can be more informative than studies at high ionic strength. In fact, considerable work has been done under these conditions, particularly with regard to the relaxing system, discussed in Part 5.

*Galloway:* By what mechanism is the ATP, which is apparently bound to myosin, coupled to the mitochondria and other presumably compartmentalized energy-producing pathways? Is it simply linked via diffusion, and the ATP bound randomly?

*Weber:* I do not think that there is any evidence for a mechanism other than diffusion.

*King:* In Fig. 18.7, myosin aggregates were separate from actin filaments. Is there a difference between the composition of this extract and the classical system in which superprecipitation is demonstrated?

*Gergely:* In the presence of tropomyosin-troponin, superprecipitation depends on the presence of $Ca^{++}$. Chelation of $Ca^{++}$ by EGTA or ATP, in excess over $Mg^{++}$, leads to clearing.

*Jay:* What is the chemical composition of the Z-line material?

*Gergely:* A few years ago it appeared that tropomyosin was the chief constituent of the Z band (Knappeis and Carlsen, 1962; Reedy, 1964; Franzini-Armstrong and Porter, 1964), although fluorescent antitropomyosin failed to react with it (Pepe, 1966). Recently, evidence has been put forward in favor of the presence of $\alpha$-actinin in the Z band (Masaki, Endo and Ebashi, 1967; Stromer, Hartshorne, Mueller and Rice, 1969).

*Sayre:* Since the Z line disintegrates only after several hours of aging and the myofibrillar breaks in muscle homogenized after little or no aging are at the I-Z junction, doesn't the I-Z junction appear to be the weak part of the myofibril rather than the electron-dense area of the Z line?

*Fukazawa:* The I-Z junction is probably the weakest point in the sarcomere.

*Sayre:* When the I-Z-I brushes are extracted, do the I filaments from either side of the Z line tend to separate as the Z line is progressively removed?

*Fukazawa:* This aspect of our work has been discussed by Fukazawa and Briskey (1970). The I filaments loosen during the removal of troponin-tropomyosin factor and associated proteins.

*Guenther:* How much overlap is there between thick and thin filaments in resting bovine muscle?

*Gergely:* I have no information on the subject. I expect it would depend on the length at which muscle was allowed to relax and it would vary between 0 and about 1 $\mu$.

*Chrystall:* When muscle contracts to sarcomere lengths less than 1.5 $\mu$, the myosin filaments must penetrate Z bands. Is there any electron microscopic evidence for this?

*Gergely:* Usually extreme shortening leads to the development of so-called contraction bands or disorganization of the ordered array of filaments. In some giant barnacle muscles penetration of the Z band by thick filaments has been described (Hoyle, McAlear, and Silverston, 1965).

*Reedy:* What can your data tell us about the following model for relaxation in the absence of $Ca^{++}$? Suppose that in 2 mM MgATP each cross bridge can bind at least 2 molecules of MgATP, but that this maintains the cross bridge in a ("fat") form which sterically cannot get past the tropomyosin-troponin guard to reach the actin molecule. Then suppose that at very low MgATP concentration, all but one bound nucleoside is stripped from every cross bridge, leading to a ("thin") myosin form which can slip past the tropomyosin-troponin guard and combine with actin to make actomyosin adenosine triphosphatase, which now proceeds through a conventional hydrolytic and mechanical contractile event.

*Weber:* In that case stretching would move myosin away from the guard and thus allow reattachment at the unguarded actin monomers (1 troponin may be every 400 Å). Therefore slow stretching during rest should encounter about as much resistance as slow stretching during contraction. It does not, I think. Furthermore, spin-labeling experiments (Tonomura, Watanabe, and Morales, 1969) have demonstrated that conformational changes occur in the actin molecules resulting from $Ca^{++}$ binding to troponin.

*Reedy:* At the lower ATP levels you have explored, you indicate that myosin is so far from saturated with ATP that perhaps half of the cross bridges in your myofibrils have no ATP bond at all. I would expect such bridges to attach firmly to actin in rigor relationship; yet you observe definite contraction while these ATP levels are maintained? I do not understand. Either the ATP-free bridges are not actin-bound, or else contraction occurs regardless of their attachment, which should tear them loose or crumple some filaments. Can you explain this situation? Is it possible that at such very low ATP concentrations every actin-combining site on every myosin has at least one "plasticizing" ATP bond, even though only a fraction of the bridges have ATP bound in such a manner that they are capable of actomyosin hydrolytic activity? Another approach to this problem is as follows. In rigor, all possible cross bridges are attached to actin. There must be some very low range of MgATP concentrations, where only a small fraction of the bridges in any one sarcomere are detached and active at any moment. The bridges remaining attached at any moment should stabilize the sarcomere against any shortening. Have your experiments involved this situation? If not, then what maximum MgATP concentrations would you consider appropriate to keep, say, approximately half of the cross bridges attached to actin at all times? I am curious about the structural and mechanical consequences of maintaining such MgATP levels for long periods; have you information on this?

*Weber:* Shortening should be possible when the number of ATP-powered force-generating links is sufficient to break the rigor attachments that persist because of a lack of ATP saturation. Shortening of myofibrils begins when about 0.3 to 0.4 ATP are bound per myosin. The remaining binding sites for ATP (1.6 to 2 per myosin) need not be in the rigor configuration, however. They could be in a high-energy intermediate state. A comparison between shortening at low and at high ATP may be quite complex because a transition from loaded (internal resistance) to unloaded shortening may occur.

It is, however, completely unknown how much force may be needed to break rigor links when they are pushed in the direction of shortening. The force required to break links when the actin filaments are pushed toward the center of the sarcomere may be much smaller than that required to break links when the fiber is stretched, i.e., when the filaments are pulled away from the center.

*Mommaerts:* When one produces iodoacetate rigor, isometrically, and then frees the muscle, it shortens. Is this not the same problem?

*Reedy:* If one develops rigor mortis in insect muscle just by withdrawing ATP, then a 1.5% shortening can still be realized after rigor.

*Mommaerts:* Yes, but it never shortens more than 1.5%.

*Reedy:* It will shorten 10% if it has sufficient ATP and $Ca^{++}$ and no load, before it crunches into the ends of the A filaments.

If fewer than half of the cross bridges are bearing ATP at the time contraction is occurring, are they hanging on to actin and being torn loose from their attachment? I fail to understand this if the ATP level available is insufficient to plasticize all the heads.

*Weber:* I do not think that there are enough data to answer the question. It is possible and cannot be ruled out that shortening occurs only when all actin-myosin links have undergone a reaction with ATP and are energized, e.g., according to Tonomura, Kitagawa, and Yoshimura (1962), phosphorylated. That would make shortening at low ATP comparable to a contraction with maximal isometric force.

*Reedy:* Do you think that the idea of rigor has the same meaning to all of us? I ask this because I sometimes feel that investigators of post mortem processes and of drugs which lead to the rigor state intend the term "rigor" to cover a more complicated and diversified set of phenomena than the final setting of myosin-actin cross links when the last ATP has been hydrolyzed.

*Gergely:* As far as rigor is concerned, I prefer to think of it as a freezing of the myosin-actin cross links. I am not sure whether it is necessary to hydrolyze all the ATP before rigor sets in. Perhaps there is a critical low level of ATP at which $Ca^{++}$ is released from the sarcoplasmic reticulum and no relaxation can take place.

*Weber:* One must be careful with the word "rigor." On the one hand the term iodoacetate rigor describes a very slow normal muscle contraction. It may be slow because $Ca^{++}$ is released at a slow rate. On the other hand the word "rigor" is used to describe the association of myosin and actin in the absence of ATP.

*Davies:* There are real problems about this iodoacetate rigor because the 1-fluoro-2,4-dinitrobenzene (FDNB) rigor and the thaw rigor are really two phases. One first has a situation where $Ca^{++}$ becomes liberated and this leads to a contraction that uses ATP. There is ATP present if the muscle moves more than the distance appropriate for one cycle of the bridge, which in the insect muscle is about 1% or 2% of the sarcomere. If it moves more than one cycle, and the movement associated with the bridge is not more than 100 Å, then you must have ATP present. This is what happens in the beginning of the iodoacetate rigor. If you hold it firmly, it will not move. If you hold it firmly until there is no ATP, it will not move more than this 100 Å. If you take a muscle and pull it so there is no overlap, or treat it with FDNB, pull it so there is no overlap of the thick and thin filaments, wait till there is no ATP left in the muscle, and then let go, you will have a perfectly soft and floppy muscle that never ever goes stiff because the first actin that hits the first myosin forms a rigor link that does not break. The rest of the filaments just form a floppy thing like permanent waves and the muscle never gets rigid. So it is theoretically in rigor mortis, but it is perfectly soft.

You stated that the interaction of pure actin and pure myosin can occur without ATP or $Ca^{++}$; however, this interaction is not necessarily the same as occurs *in vivo* but is what happens in muscle in rigor mortis. Cannot troponin be considered as the agent conferring specificity on $Ca^{++}$ for this interaction in the presence of ATP and $Mg^{++}$, just as validly as it is taken as an inhibitor of interaction in the absence of $Ca^{++}$? What is meant by the statement that "$Ca^{++}$ ions do not play an intrinsic role in the presence of the regulatory protein complex"? As stated earlier, $Ca^{++}$ is required for the interaction of actin and myosin in the presence of this regulatory protein complex to produce an adenosine triphosphatase. In its absence, specificity is lost, and $Mg^{++}$ alone will do. When you use the term "the active center proper," is this for the actin-myosin interaction or for the adenosine triphosphatase?

*Gergely:* Since activation of myosin adenosine triphosphatase by actin can take place—in the absence of the tropomyosin-troponin system—without $Ca^{++}$ ions, it would seem that an interaction between myosin and actin has taken place. With the use of Occam's popular razor one could say that this interaction is the same as that occurring in the presence of tropomyosin and troponin and $Ca^{++}$. I would certainly agree that more work should be done on this, particularly to clarify the relation of the actin-myosin links in rigor to those formed during physiological contraction. I use the term "active center" to indicate those regions of the myosin molecule that contain both the ATP-binding and the actin-binding site.

*Hamm:* In studies on post mortem changes of -SH groups of myofibrillar proteins it would be necessary to prevent the onset of rigor mortis. If the onset of rigor is not prevented, it cannot be excluded that during the determination of -SH groups of the myofibrillar proteins *in situ* these groups are changing and, therefore, the proteins are not analyzed in the prerigor state. Is there any way to prevent rigor mortis by application of a reagent which does not react with -SH groups?

*Gergely:* I suppose exposure of the muscle to a chelator such as EGTA —perhaps after mild glycerol treatment to produce a permeable membrane—could prevent development of rigor by chelation of $Ca^{++}$. I am not aware that an experiment of this type has been done.

*Wierbicki:* Muscle as a food is basically protein and water in the ratio close to 1 part protein: 4 parts $H_2O$. Can you comment on the mechanism of interaction between ions and proteins on one side and $H_2O$ on the other during the rigor, prerigor and relaxation of the muscle fibers? Quality of the muscle as a food (quality of meat) depends greatly on the way water is associated with protein.

*Weber:* Our knowledge is as yet insufficient for a detailed analysis of the interaction of the contractile proteins with ions and water.

*Sayre:* Would you expect less tension development during the onset of rigor mortis if conditions could be maintained which prevent $Ca^{++}$ release from the sarcoplasmic reticulum until after the bulk of ATP has been removed?

*Weber:* Yes.

*Melton:* I am working on a muscle biopsy procedure and trying to find solutions of ions, perhaps $Mg^{++}$ and ATP, that will keep the excised sample in a relaxed state throughout the histological processing. I would like to know if you know of concentrations of these two, (ATP and $Mg^{++}$), that will maintain the prerigor state. I am thinking of solutions either to inject locally or to soak the sample in after excision.

*Weber:* Does histological processing not usually involve fixation, i.e., protein denaturation? The structure of the relaxed state should be preserved if the biopsy is mechanically restrained from shortening until fixation.

To maintain a biopsy of more than 100 $\mu$ relaxed is a difficult problem because relaxation depends on an adequate ATP concentration. ATP diffusing inwards, however, is constantly broken down, so that the ATP concentration near the center of the tissue never attains a high enough level for relaxation. Optimum conditions would be: low temperature, 5 mM MgATP, 10 mM creatine phosphate, 5 mg creatine phosphokinase per g tissue, 5 mM EGTA, 1 mM $MgCl_2$.

*Guenther:* Please comment on the "resolution" of rigor in terms of dissociation of actomyosin. Does actomyosin actually dissociate post mortem or is this protein simply cleaved into smaller units of actomyosin?

*Briskey:* This question has not been unequivocally answered in the literature. In fact, it has only been since the first symposium in 1965 that workers in this field have been willing to accept the fact that there may be a "resolution" of rigor (Marsh, 1966). The pictures of interfilamental granularity show such a nice relationship to sarcomere length that one might be tempted to speculate the withdrawal of the actin filament from this region, leaving the myosin heads disrupted. There are, of course, other interpretations that could be given to these findings—none, however, with absolute certainty. Further, Dr. Fukazawa has shown that the I-Z junctions are particularly vulnerable to breakage and it would seem likely that some of the so-called resolution may actually be due to cleavage at this point.

*Jungk:* Please elaborate on your findings of the resolution pattern which occurs in muscles which go into rigor in the stretched condition, specifically regarding sarcomere length.

*Fukazawa:* This question is at present under investigation in our laboratory.

*Sayre:* How do you account for the fact that the gross length of muscle attached to the carcass remains constant if the sarcomeres shorten and subsequently lengthen during aging? If some sarcomeres shorten then others must be stretched to greater than rest length.

*Fukazawa:* Carcasses do not remain constant in form, and, in fact, may shorten or be altered to quite an extent during the development of rigor mortis.

*Sayre:* What prevents the short sarcomere lengths observed in prerigor muscle from developing during the process of homogenization? Three factors would prevent EDTA from blocking contraction: (1) Excision of the muscle and any mincing prior to homogenization would stimulate the muscle to contraction; (2) EDTA will not pass the sarcolemma of an intact fiber and would not efficiently chelate the $Ca^{++}$; (3) even if $Ca^{++}$ was removed, the ATP level in the homogenate may well be diluted to the level where, according to Dr. Weber, $Ca^{++}$ is not required for adenosine triphosphatase activity and contraction.

*Fukazawa:* In all experiments, at least 500 sarcomeres are counted in over 20 fields. We believe that the potential problem that you have mentioned is overcome when the averages are considered.

*Reedy:* Since a muscle retained on the skeleton cannot shorten without stretching antagonistic muscles, can we assume that the reversible contraction in chicken pectoral muscle is matched by a reversible extension in antagonistic muscles?

*Fukazawa:* We assume that it occurs to some extent.

*Reedy:* At what points in "reversible contraction" is the chicken breast muscle actually in rigor, in terms of exhausted ATP and inextensibility?

*Fukazawa:* This is an extremely valuable point, and must be given high priority in future work.

*Reedy:* How did you sample the sarcomere lengths of reversible contraction? Is it likely that contracture takes place during isolation and mincing of muscle for fibril samples, and that it simply runs out of ATP at lesser degrees of shortening as the muscle gets older?

*Fukazawa:* All sampling and isolations were carried out under low KCl concentration, buffered to a neutral *p*H and in the presence of from 1 to 5 mM EDTA. Under these conditions the sarcomere lengths do not change.

*Reedy:* What nucleoside or polyionic substance comes along in post mortem muscle to dissociate cross bridges and permit a shortened rigor muscle to relax and stretch?

*Fukazawa:* Adenosine diphosphate and adenosine monophosphate are two possibilities.

*Reedy:* Glycerol-extracted muscle does not shorten during rigor induction if it is restrained. Does form restraint prevent the severe post mortem shortening you observed in chicken breast muscles?

*Fukazawa:* Yes, but reference is also made to the work of Herring, Cassens, and Briskey (1967), wherein with bovine muscle, varying levels of restraint gave appropriately altered sarcomeres.

*Reedy:* Can filtered juice from muscles at various post mortem stages cause glycerol-extracted muscle (fibrils or bundles) to display any of the phases of the "reversible contraction" you describe in chicken breast muscle?

*Fukazawa:* This is a good point for future study.

*Sayre:* Stromer, Hartshorne, Mueller, and Rice (1969) have shown that extraction of the electron-dense Z-line material does not result in an obvious break between the thin filaments from two sarcomeres. Which, if any, known protein could account for this linkage?

*Maruyama:* As is well known, there is a lattice structure in the Z line which connects the I filaments. What protein constitutes this has not yet been elucidated. α-Actinin appears to be a cementing substance in the Z membrane, which would certainly be responsible for the density there.

*Stromer:* Have you done experiments to determine the ultrastructural localization of either the 6 S or the 10 S components of α-actinin to see whether or not one of these might be localized in the Z-line backbone itself, or if it is located in the less structured amorphous material of the Z line?

*Maruyama:* We think that the 6 S α-actinin is located in the less organized amorphous material of the Z membrane. As for the 10 S α-actinin, we do not know its localization, because it does not produce antibody, just like actin.

*Sayre:* Goll, Mommaerts, Reedy, and Seraydarian (1969) noted a stiff or rigid section of the thin filaments 700 Å long next to the Z line; it was removed by short trypsin digestion. It has also been noted that myofibrils break in this same region after aging. Since α-actinin has been reported to be relatively resistant to tryptic digestion, please comment on the attachment region of thin filaments to the Z line.

*Maruyama:* All I can say is that trypsin treatment of myofibrils results in the release of intact α-actinin from the Z membrane, as Goll, Mommaerts, Reedy, and Seraydarian (1969) have reported.

*Sayre:* Is there any interaction of $Ca^{++}$ or $Mg^{++}$ with α-actinin, since both of these substances have been reported to promote actin or thin filament release from intact muscle?

*Maruyama:* We have not observed any interaction between α-actinin and $Mg^{++}$; $Ca^{++}$ does not affect the action of α-actinin at all.

*Reedy:* Does α-actinin promote lateral aggregation of F actin regardless of whether the actin is pure or is saturated with native tropomyosin?

*Maruyama:* Yes, it does.

*King:* I have a question regarding the length of the α-actinin–F-actin aggregates and their dependence on *p*H. If one obtains a long aggregate at low *p*H, can it be shortened by raising the *p*H? Is this process reversible? Is there a clue to the functional groups in this aggregation?

*Maruyama:* A longer aggregate was formed at *p*H 6 than at *p*H 8.5. We have not tested the reversibility. We do not know anything about the functional groups in the aggregate formation. It would be interesting to test the effect of urea, and higher KCl concentration. Also the effect of -SH groups should be examined. Seraydarian, Briskey, and Mommaerts (1968) have noted some involvement of -SH groups in α-actinin–actomyosin interaction.

*Reedy:* Has there been any indication yet from fluorescent antibody work as to where β-actinin is concentrated in the myofibril?

*Maruyama:* Preliminary experiments with crude anti–β-actinin suggest that it is located in the A-band region of the myofibrils (Masaki, unpublished observations). Since β-actinin does not interact with myosin, I believe that it is located in the I filaments—not along the entire length, but in the part near its free end. Recently, it has been shown that a β-actinin-like protein or "end-blocking factor" exists at the free end of the I filament (Kasai and Hama, 1969).

*Reedy:* If β-actinin is not a length determinant but inhibits interaction between actin filaments, can you propose a role for it in the intact myofibril?

*Maruyama:* β-actinin may be still a length determinant of F-actin—it would stop further elongation at the end of F-actin.

F-actin, pure or contaminated by tropomyosin, etc., has a strong interparticulate interaction in solution. However, the isolated I filaments (natural F-actin) do not form any network in solution. When β-actinin is removed from natural F-actin, it becomes just like a usual F-actin of the Straub type. Therefore, I consider that β-actinin somehow acts as preventing the interfilamental interaction of the I filaments, resulting in the maintenance of the organized hexagonal position in myofibrils.

*Sayre:* If β-actinin prevents interaction between thin filaments in solution, could it function to maintain the hexagonal array of the thick and thin filaments in muscle?

*Maruyama:* Your comment is exactly what I would like to emphasize. During the sliding process, the I filaments would be bent so as to attach to other I filaments; β-actinin would inhibit the interaction between the I filaments. I think that tropomyosin plays a role in making actin filaments rigid too.

*Lorand:* Is the inhibition of actin aggregation (since there is no evidence of covalent bonding between the monomers, I would rather not use the term polymerization to this process) by β-actinin analogous to the effect which can be seen when partially digested fibrin is present in a clotting system of normal fibrin? By inference, could one regard β-actinin as a somewhat degraded or modified actin which, having lost one of its "aggregating centers" (of which there must have been at least two), now acts as a chain terminator?

*Maruyama:* This is a very interesting comment. The peptide map analyses showed a certain extent of similarity between actin and β-actinin. At present, I think that the two proteins have a common structure, but with different functional sites. However, as you have suggested, β-actinin could be a modified form of actin. So far, I have observed that the treatment by trypsin or denaturation by heat did not result in the change of actin to $\beta$-actinin. Further examinations using proteases such as chymotrypsin, etc., may be of interest.

*Davies:* Could $\beta$-actinin be released at a given time during F-actin formation to inhibit further growth?

*Maruyama:* I think that $\beta$-actinin can be released from F-actin at any time. When $\beta$-actinin was added to polymerizing actin, the rate of polymerization was greatly retarded, but eventually a full polymerization could be achieved. Some of the $\beta$-actinin bound to F-actin can be released into solution, when such complex was solubilized in 0.1 M KCl and then ultracentrifuged to precipitate the complex.

*Davies:* Your preparation is stated to have "even better $Ca^{++}$ sensitizing activity than native tropomyosin." Does this mean that it binds more $Ca^{++}$ per molecule or that less of it will confer sensitivity to $Ca^{++}$?

*Fukazawa:* We know that there is a difference between native tropomyosin and our preparation in their relative amounts of troponin and tropomyosin. Further, we have considerable evidence to support the argument that our preparation is a tighter complex and that the -SH groups are in better condition than in native tropomyosin. Additionally, the $Ca^{++}$ binding of our preparation, *per se,* is very high. Beyond these points we cannot give a specific answer to this question.

*Reedy:* Do you plan to make optical diffraction patterns from electron micrographs of your TF [troponin-tropomyosin factor] crystals in order to compare them with those patterns from pure tropomyosin published by Caspar, Cohen, and Longley (1969)?

*Fukazawa:* Yes, we plan to make these comparisons.

*Reedy:* Have you tried making the needle-shaped paracrystals from your TF, after Cohen and Longley (1966), in order to see whether they resemble the tropomyosin tactoids treated with troponin which were described by Nonomura, Drabikowski, and Ebashi (1968)?

*Fukazawa:* This question is still under investigation.

*Karmas:* You stated that when troponin is changed by $Ca^{++}$, muscle contraction occurs. What is the change that takes place in troponin? How does tropomyosin affect muscle contraction?

*Weber:* It is not known what change $Ca^{++}$ produces in troponin. Wakabayashi and Ebashi (1968) report some changes in the pattern of ultracentrifugation and acrylamide electrophoresis. The interpretation of these data, however, may be better postponed until further, more detailed study. Tropomyosin by itself does not appear to affect any process correlated to contraction. For relaxation both tropomyosin and troponin together are required. Without tropomyosin, troponin is not bound to the actin filaments (Endo, Nonomura, Masaki, Ohtsuki, and Ebashi, 1966).

*King:* Regarding the hypothesis by which tropomyosin and troponin repulse ATP from binding with myosin, what is the effect of substituting $Mg^{++}$ (or other divalent cation) for $Ca^{+}$ on TP?

*Weber:* I am afraid there is a misunderstanding. Tropomyosin and troponin do not inhibit the binding of ATP to myosin. They inhibit the interaction of ATP-containing myosin with actin. This interaction is reactivated only by $Ca^{++}$ and $Sr^{++}$.

*Kastenschmidt:* Are you postulating an allosteric change in the troponin-tropomyosin due to $Ca^{++}$? You get two binding sites with a Scatchard analysis of $Ca^{++}$ binding with a high affinity for $Ca^{++}$. What is the significance of the other sites with lower $Ca^{++}$-binding affinity on the troponin molecule?

*Weber:* In the broader sense one may consider the effect of $Ca^{++}$ allosteric. A change in the troponin molecule caused by $Ca^{++}$ is conducted to actin, rendering its active site available to interaction with ATP-saturated myosin molecules. When spin-labeled actin is combined with tropomyosin and troponin, $Ca^{++}$ binding to troponin caused changes in the environment of the spin label (Tonomura, Watanabe, and Morales, 1969).

I am not aware that troponin has more than two classes of binding sites. Interaction with ATP-saturated myosin is possible only when both classes are saturated with $Ca^{++}$.

*Stromer:* Does the tropomyosin-troponin complex inhibit contraction by actually preventing the actin-myosin interaction, or does it inhibit contractile response even though actin and myosin may actually be interacted? For example, is it possible that the kind of interaction described by Eisenberg and Moos (1967), where a viscosity increase occurs in the absence of any turbidity rise, can also occur for the actin-myosin-tropomyosin-troponin system?

*Weber:* I would like to point out that a turbidity rise and a contractile response cannot be equated. Turbidity may not increase at all under conditions where actomyosin contracts maximally, according to measurements of syneresis and adenosine triphosphatase activity (Briskey, Seraydarian, and Mommaerts, 1967). Relaxation has been characterized by its lack of actin-myosin interaction (for summary see Weber, 1966). Since relaxation is caused by troponin together with tropomyosin, these proteins must prevent actin-myosin interaction.

*Stromer:* Drabikowski and Nonomura (1968) have suggested that troponin can bind to F-actin in the absence of tropomyosin and that this troponin binding causes formation of an insoluble complex. Have you observed troponin binding to F-actin in the absence of tropomyosin, and if so, is it possible that troponin does in fact bind directly to F-actin but that this binding is constrained to certain specific sites by the simultaneous presence of tropomyosin?

*Weber:* The effect observed by Drabikowski and Nonomura (1968) depends on very high troponin concentrations. It is doubtful if this effect is related to the activity of troponin, since actin tends to coprecipitate with proteins. We have not been able to confirm the effect with purified troponin. Endo, Nonomura, Masaki, Ohtsuki, and Ebashi (1966) have shown that tropomyosin is necessary for the binding of troponin to actin.

*Davies:* You state that "one troponin makes unavailable to myosin several actins, most of which are separated from troponins by a considerable distance." How long is troponin *in situ* on the thin filament? Are all actins in contact with part of a troponin?

*Weber:* Although the physical constants for troponin as given by Ebashi and Endo (1968) are not compatible with globular shape, they also do not allow for an asymmetry comparable to that of tropomyosin which does contact all of the inactivated actins. In view of the complex nature of troponin, which is a complex of two proteins (Hartshorne and Mueller, 1968, 1969), a detailed characterization is still lacking. However, it seems reasonably certain that only tropomyosin and not troponin communicates with all of the actins.

*Stromer:* Have you any evidence which might indicate how or where troponin is bound to the F-actin–tropomyosin complex? Is it at the end of the tropomyosin molecule but away from the F-actin strand, or is it between the tropomyosin molecule and therefore also bound to F-actin?

*Maruyama:* This is a question which we have tried to answer. However, for unknown reasons, we have not yet succeeded in showing the localization of troponin in the F-actin–troponin–tropomyosin complex by a use of ferritin-labeled antitroponin (Ohtsuki, unpublished observations).

*Stromer:* Would you comment on the specific role of $Mg^{++}$ in the native tropomyosin complex to achieve the $Mg^{++}$-concentration-dependent effect on $Ca^{++}$ sensitization of actomyosin?

*Maruyama:* We think that MgATP is essential to the onset of contraction of actomyosin, and more MgATP is needed to produce relaxation, whereas $Ca^{++}$ induces contraction in the presence of troponin-tropomyosin complex. In this connection, Kaminer (1968) has found the binding of $Mg^{++}$ ions to the native tropomyosin complex.

*Stromer:* Since the length of the tropomyosin molecule is 400 Å and the periodicity of troponin is 400 Å by the antibody technique, it might be interesting to speculate on how they might be arranged. Is there any information to show that the troponin molecule is, in fact, distal to the tropomyosin molecule on the thin filament?

*Mommaerts:* We could draw some kind of an answer from the work of Fukazawa, Briskey, and Mommaerts (1970). Those of you who know the paper by Casper, Cohen, and Longley (1969) will know that they have identified the long ribs of the 200-Å tetragon lattice as the place where the tropomyosin molecules join end to end. It is just in the long ribs that the additional globules are found which are characteristic for the TF crystals as opposed to the tropomyosin crystals. At any rate, this suggests that these knobs place themselves at the ends where the tropomyosin chain joins.

*Gergely:* Does this not simply follow from the fact that the antibodies are spaced 400 Å apart and that corresponds to the tropomyosin length?

*Mommaerts:* We are learning now that in order to convey $Ca^{++}$ sensitivity upon the contractile system we need three different things together. Who knows if it is going to be four or five? Under certain conditions, as pointed out in the last paper by Drs. Fukazawa and Briskey, it behaves as one protein and it precipitates as one protein. But with only a small change it precipitates as two or three. A very similar circumstance is evolving with work on the nerve growth factor of Levi-Montalcini and Angeletti (1968).

*Lorand:* Could β-actinin be a form of actin which interferes with aggregation just as fibrin degradation products interfere with the aggregation of fibrin?

*Maruyama:* Do you mean that denatured actin would interfere with the polymerization of native actin?

*Lorand:* Or it could be partially digested actin.

*Maruyama:* Partially digested actin or denatured actin, due to aging, results in the overall shortening of the resultant F-actin. When such material is subjected to sonication there is no more change, however, so apparently the effect of such denatured or partially digested actin is different from that of β-actinin.

*Perry:* In collaboration with Dr. M. C. Schaub in our laboratory it has been shown recently that troponin preparations previously purified by SE Sephadex chromatography can be resolved into two components by further chromatography on SE Sephadex in 6 M urea (Fig. 20.1).

These components which we designate as the $Ca^{++}$-sensitizing factor and the inhibitory factor will not independently replace troponin in relaxing protein-system activity. The inhibitory factor is eluted from SE Sephadex at higher ionic strength, inhibits the $Mg^{++}$-activated adenosine triphosphatase of desensitized actomyosin in the absence of EGTA, and is

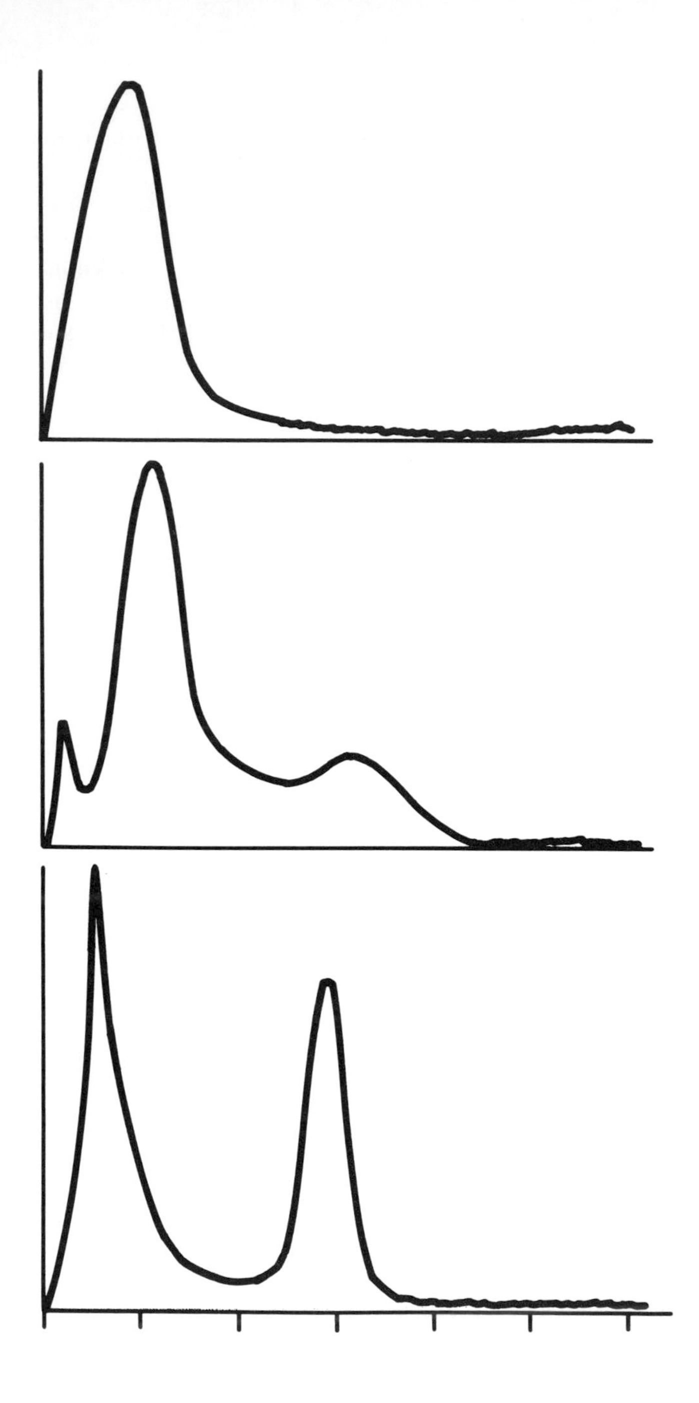

*Fig. 20.1.* Densitometric traces of electrophorograms of column-purified troponin complex. 200–250 μg protein applied to 8% acrylamide gels in tris-glycine buffer, *p*H 8.5, containing 15 mM 2-mercaptoethanol. (*a*) gel containing 10% glycerol; (*b*) gel containing 2.5 M urea; (*c*) gel containing 6 M urea.

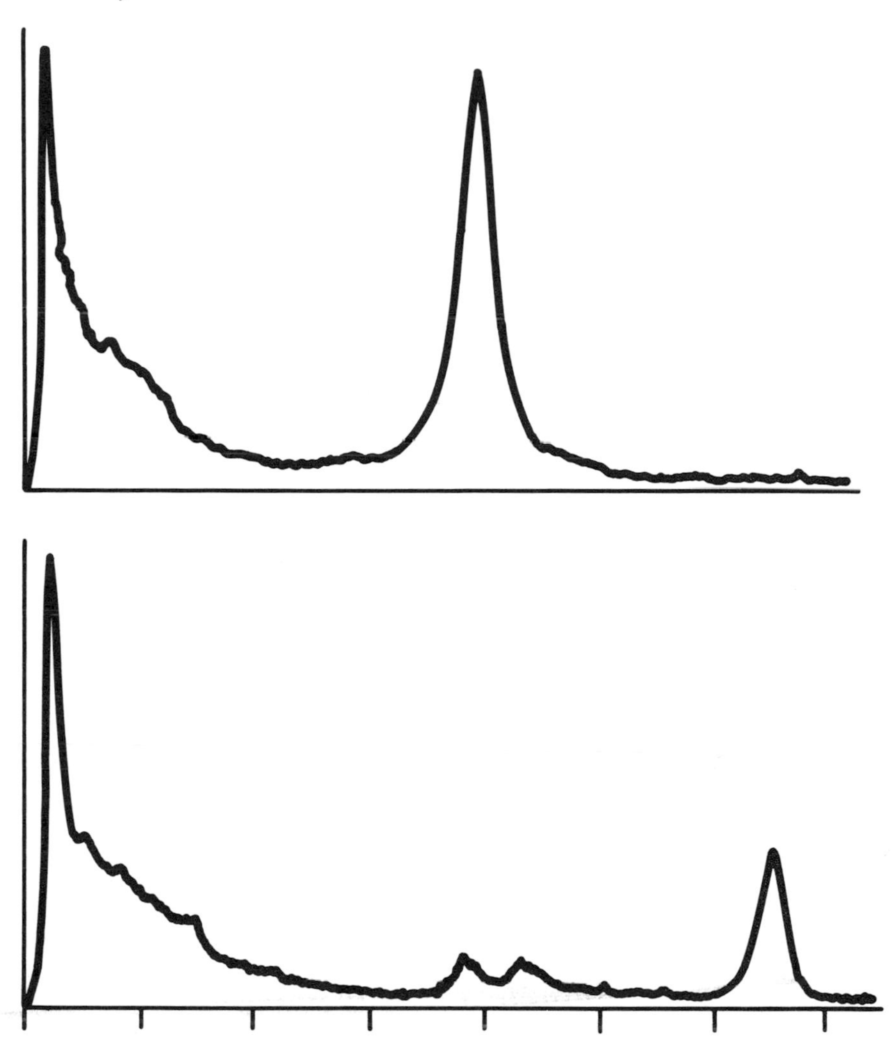

*Fig. 20.2.* Densitometric traces of electrophorograms of troponin complex in the presence of EGTA. Polyacrylamide gels as in Fig. 20.1c. (*a*) about 200 $\mu$g of troponin complex without EGTA; (*b*) about 250 $\mu$g of troponin complex containing 2 mM EGTA.

probably identical with the inhibitory factor earlier isolated from myofibrils under conditions of extraction at high ionic strength (Hartshorne, Perry, and Schaub, 1967). It corresponds to the troponin B of Hartshorne, Theiner, and Mueller (1969). On addition of the $Ca^{++}$-sensitizing factor the inhibitory factor no longer inhibits on its own, but requires the presence of EGTA. Thus the two components recombine to restore the original troponin activity when tested in the presence of tropomyosin.

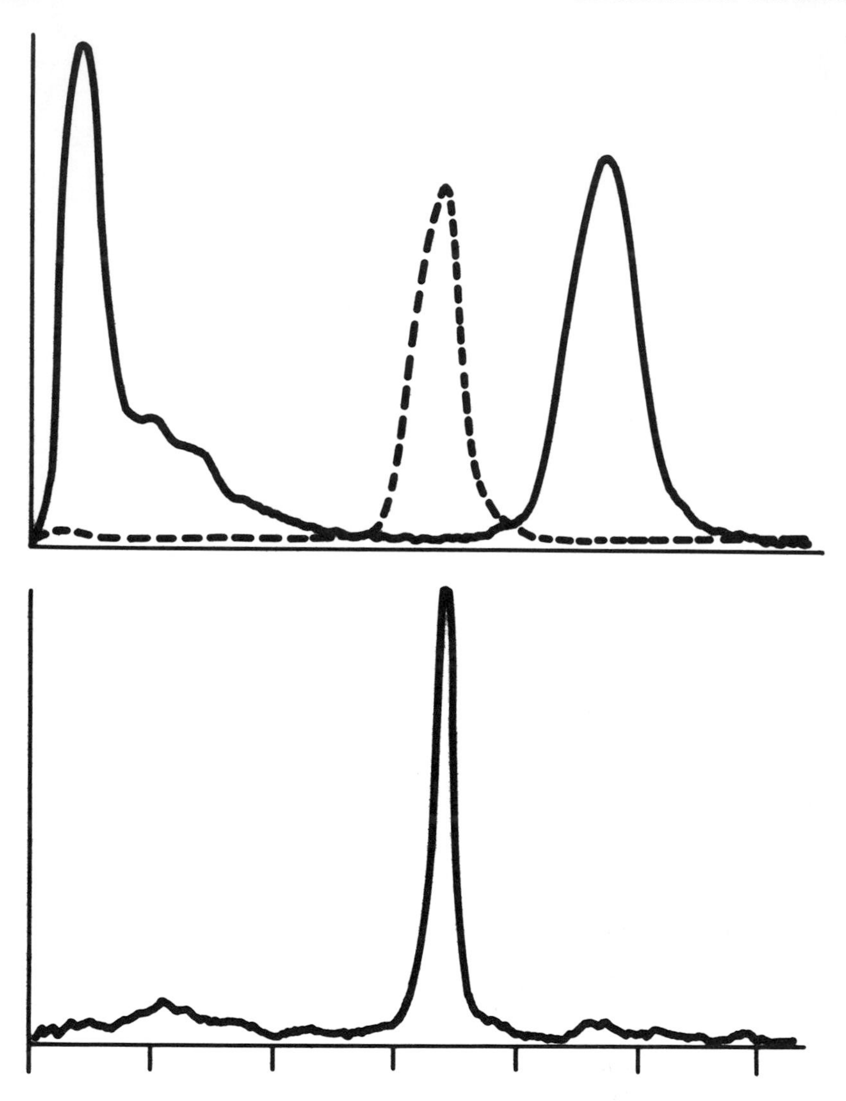

*Fig. 20.3.* Densitometric traces of electrophorograms of fraction eluted at low ionic strength from SE Sephadex. 200–300 μg of protein on polyacrylamide gels as in Fig. 20.1c. (*a*) troponin complex (*solid line*) and tropomyosin (*dashed line*); (*b*) first peak eluted at low ionic strength from SE Sephadex loaded with troponin complex.

When run on acrylamide gel electrophoresis the inhibitory factor moves slowly and the $Ca^{++}$-sensitizing factor much faster. Neither band is identical with tropomyosin which under the same conditions moved with a band of intermediate mobility (Fig. 20.2).

In the presence of 2 mM EGTA the $Ca^{++}$-sensitizing factor, corresponding to the troponin A of Hartshorne, Theiner, and Mueller (1969), moves with increased mobility at *p*H 8.6 (Fig. 20.3).

The mobilities of the inhibitory factor and tropomyosin are not modified by EGTA, indicating that the $Ca^{++}$-sensitizing factor probably contains tightly bound $Ca^{++}$, which is removed by the chelating agent with a consequent increase in charge. It is considered that all troponin preparations consist of a complex of the inhibitory and $Ca^{++}$-sensitizing factors which, together with tropomyosin or possibly some as yet unidentified component present in these preparations (Schaub and Perry, 1969), constitute the system that regulates the $Mg^{++}$ activated adenosine triphosphatase of myofibrils or natural actomyosin, the relaxing protein system.

*Gergely:* Dr. Greaser, in our laboratory, has found that the so-called purified troponin preparations have even more components than two.

*Weber:* Most people get a half-dozen bands when they take the usual preparation.

*Gergely:* Dr. Perry's preparation seemed very pure indeed.

*Greaser:* We have tried different purification procedures that are given in the literature; some of these claim to produce a preparation with gel electrophoresis homogeneity. We use Coomassie blue stain, which is more sensitive than Amido black. We also use 8 M urea, which is a little bit more disruptive than 6 M, but I do not think significantly so. We usually get 3 to 4 major components. In our hands, the separation with the acid treatment is not a complete one but yields preparations that show all the bands of the original material but just differing ratio of them.

*Stromer:* Does the M-line substance in your experiments promote the aggregation of myosin molecules to form synthetic thick filaments and also promote the ordering or alignment of these thick filaments into relative positions resembling that seen in the A band? What techniques were used to observe these effects and to determine that the aggregating effect was exerted only on LMM and not HMM?

*Maruyama:* The M protein promoted the lateral aggregation of myosin filaments in a random way as determined by electron microscopy. The extent of aggregation was checked by optical density measurements.

*Reedy:* Does the Ebashi group have any comment or experience with regard to the protein fibrillen described by Guba and Garamvolgyi (1967)?

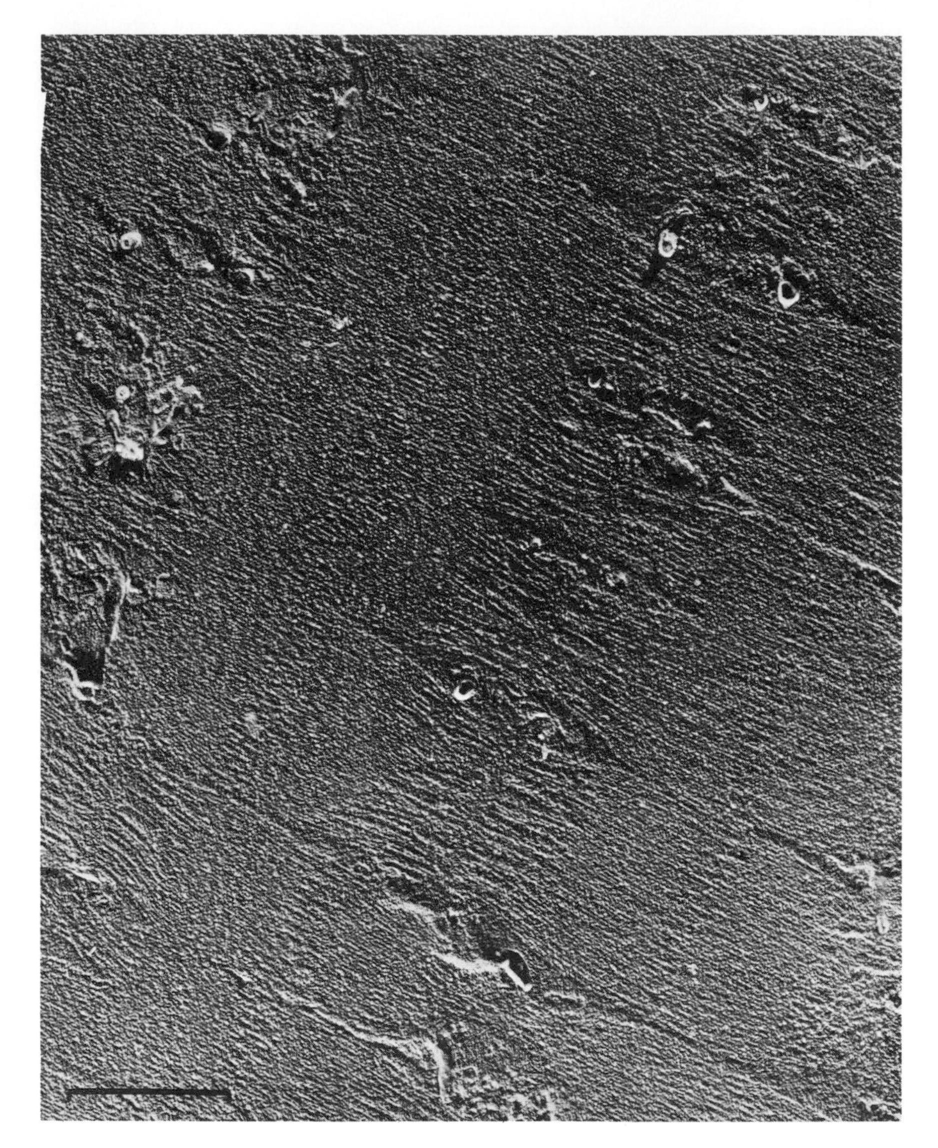

*Fig. 20.4.* Mouse gastrocnemius fixed with glutaraldehyde-veronal buffer and treated with 20% aqueous glycerol before freezing. Line = 1 $\mu$. From Reed, Houston, Todd, Kennelly, Frazier, Flawith, and Boyes (1969).

*Maruyama:* We have been interested in Guba's fibrillen. When we followed his procedure, we experienced difficulty in extracting G-actin with a dilute ATP solution from myosin-free myofibrils. $\alpha$-Actinin was obtained instead of actin.

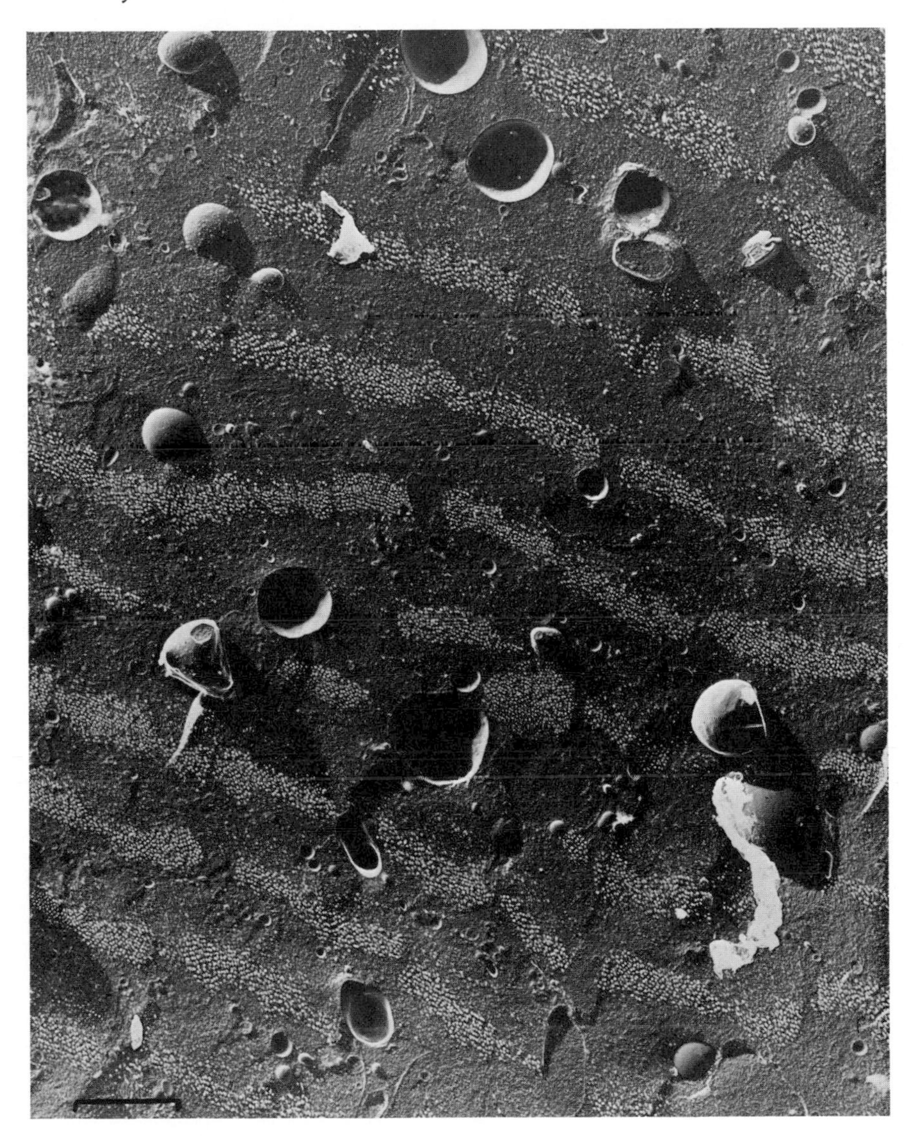

*Fig. 20.5.* Fresh mouse gastrocnemius, glycerinated and freeze-etched. Line = 1$\mu$. From Reed, Houston, Todd, Kennelly, Frazier, Flawith, and Boyes (1969).

*Reedy:* Your protein bookkeeping leaves almost 10% of the myofibril unaccounted for. How do you expect that this will eventually be accounted for?

*Maruyama:* It may well be fibrillen and other minor structural proteins such as fibrous protein constituting the backbone structure of the Z

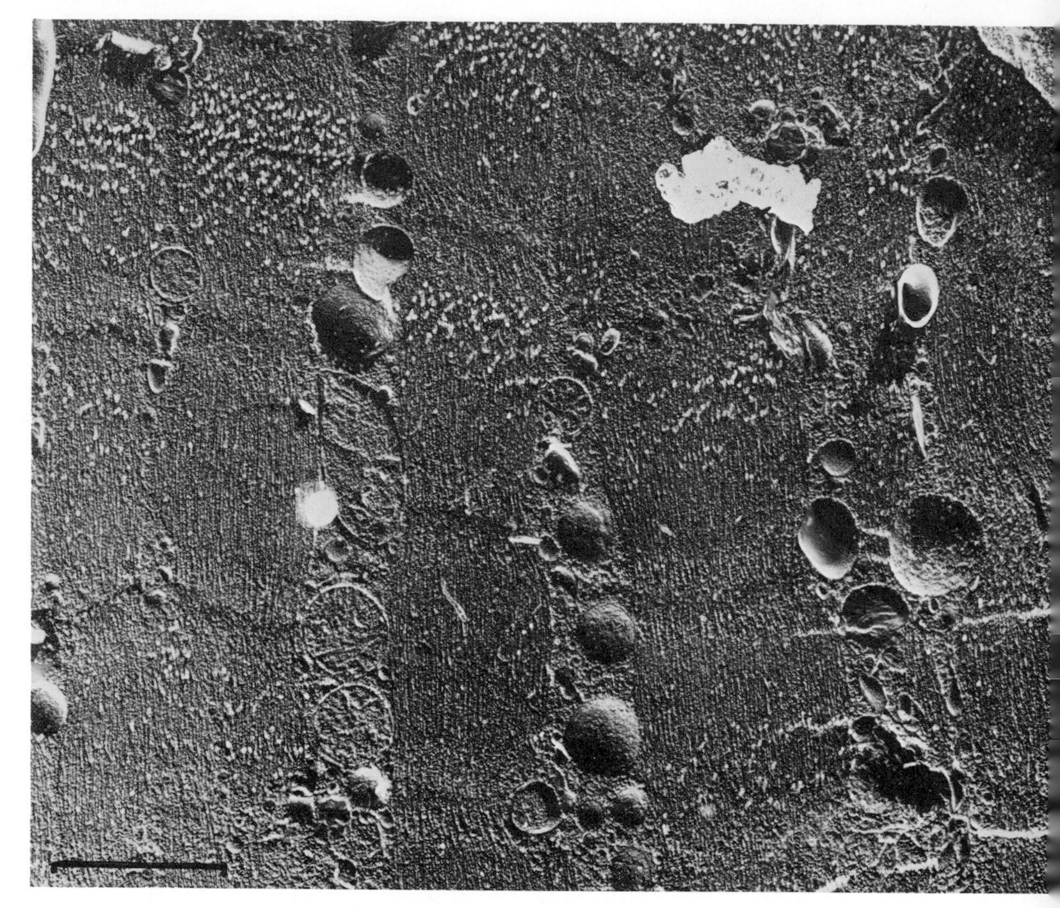

*Fig. 20.6* Fresh pig muscle (peroneus tertius) treated with 2% sodium chloride before freeze-etch Line = 1 μ. From Reed, Houston, Todd, Kennelly, Frazier, Flawith, and Boyes (1969).

membrane. I might add a comment that there will be changes in the percentage of the listed proteins as more examinations are made.

*Todd:* The following electron micrographs show muscle prepared by Reed, Houston, Todd, Kennelly, Frazier, Flawith, and Boyes (1969) using the freeze-etch method of specimen preparation.

In this method the specimen is frozen at rates greater than 1000° C/sec to −150° C, embedding the tissue in a glass of its own fluids. This has the advantage that no specimen alteration is caused by chemical action or dehydration as with normal preparative methods. The still-frozen sample is fractured and a slight sublimation of the ice is carried out at −100° C, exposing the cellular structures. These are then replicated for electron microscopic examination.

Figure 20.4 shows the now familiar parallel-filament array; the A, I, Z, and M bands are easily distinguished together with parts of the sarcoplasmic reticulum. To obtain this result the muscle was prefixed in glutaraldehyde. This may seem strange with a method designed to obviate chemical fixation until Fig. 20.5 is examined. This is of fresh mouse muscle, treated with 20% glycerol prior to freeze-etching. It is difficult to correlate the structures seen in this specimen, namely the random distribution of mitochondria and lack of filamentous structure, with the normal fixed appearance. This "random" structure is seen in fresh glycerinated or biopsy muscle samples. One explanation may be that muscle is a more dynamic system than was previously thought, and is highly sensitive to changes in ionic strength and chemical treatment. The random array seen in Fig. 20.5, which is obtained regardless of specimen orientation, can be converted into a parallel system by treatments such as fixation (Fig. 20.4), development of rigor, or the action of ionic solutions; for example, the effect of 2% sodium chloride on fresh muscle is shown in Fig. 20.6. The lining up of the myofilaments and mitochondria can be seen clearly.

This method should be of value in following the changes in the fine structure of muscle during the development and resolution of rigor, and also provide an insight into the ways that different processing regimes affect the meat, as no dehydration is involved in the preparation of specimens.

*Davies:* If those are biopsy specimens from mammalian muscles, then even your rate of freezing at 1000° C/sec is by no means fast enough to prevent them from being highly contracted at the time you are looking at them, because just cooling them from 37° C down to 0° C rapidly will cause a contraction. You have to excise a biopsy specimen in some manner. It seems you should take something like a frog sartorius, cool it down to 0° C, and then freeze it. It freezes without contracting, and then you might compare this by freezing it at 0° C, either in tetanus or relaxed, and find out whether you can see any of the filaments. I think you must be aware of this problem before really making a statement to the effect that there are no filaments present.

*Todd:* Our biopsy procedure was designed to obtain material in as near the natural state as possible.

*Davies:* I would suggest you do a dissection under anesthetic and squirt on cold Freon at high speed. That will freeze it and then you can cut it off from the frozen muscle *in situ.*

*Todd:* I think that the method itself is reasonably acceptable, because you are not getting any alteration in appearance, as with a glutaraldehyde-fixed specimen. So there is no structural alteration caused by this method.

*Davies:* Well, the glutaraldehyde ones are already cross-linked by the properties of the two aldehyde groups. It is important to use an animal that has a relaxed muscle at 0° C, and that is not a pig or a man.

*Todd:* We also used a mouse.

*Reedy:* Dr. Davies is trying to emphasize that the muscle was free to undergo contracture, and if you can answer that you will, to some extent, answer what he is saying. Even if you held the muscle, it is possible that zones of it underwent contracture. The zones of abysmally undefined structure could, in fact, be zones of contracture which always look dreadful however visualized.

*Todd:* These are rather upsetting pictures to get, but we find them very consistently. It is also worth noting that, both in our own work and that of Rayns, Simpson, and Bertand (1968) on striated muscle, the sarcolemma is not thrown into folds as would be expected if violent freeze contraction had taken place.

*Slautterback:* I agree with Dr. Davies that the probability of contracture is great. Furthermore, the formation of a eutectic substance between the glycerol and water has no doubt contributed to the obscurity of the myofilaments. Close inspection of the micrographs reveals the presence of filaments, however. It must be remembered that fine details are easily lost in this technique, both in the process of replica formation and in the shadowing. Of more interest here than the technical problems of this new and interesting technique is the prominence of globular units in the I band. I presume these are the same as are often seen in chemically fixed muscle where they are less conspicuous.

## *References*


Banga, I., and A. Szent-Györgyi. 1942. Preparation and properties of myosin A and B. *Stud. Inst. of Med. Chem.* (Univ. Szeged) *1*:5.

Bárány, M. 1967. Adenosine triphosphatase activity of myosin correlated with speed of muscle shortening. *J. Gen. Physiol. 50*:197.

Bárány, M., A. F. Tucci, and T. E. Conover. 1966. The removal of the bound ADP of F-actin. *J. Mol. Biol. 19*:483.

Briskey, E. J., K. Seraydarian and W. F. H. M. Mommaerts. 1967. The modification of actomyosin by α-actinin II. The effect of α-actinin upon contractility. *Biochim. Biophys. Acta 133*:412.

Buller, A. J., W. F. H. M. Mommaerts, and K. Seraydarian. (1969). Enzymic properties of myosin in fast and slow twitch muscles of the cat following cross-innervation. *J. Physiol. 205*:581.

Casper, D. L. D., C. Cohen, and W. Longley. 1969. Tropomyosin: crystal structure, polymorphism and molecular interactions. *J. Mol. Biol. 41*:87.

Cohen, C., and W. Longley. 1966. Tropomyosin paracrystals formed by divalent cations. *Science 152*:794.

Drabikowski, W. and Y. Nonomura. 1968. The interaction of troponin with F-actin and its abolition by tropomyosin. *Biochim. Biophys. Acta 160*:129.

Ebashi, S. and M. Endo. 1968. Calcium ion and muscle contraction. *Progr. Biophys. Mol. Biol. 18*:123.

Ebashi, S., A. Kodama, and F. Ebashi. 1968. Troponin I. Preparation and physiological function. *J. Biochem. 64*:465.

Eisenberg, E., W. W. Barouch, and C. Moos. 1969. Binding of MgATP and actin to myosin, heavy meromyosin (HMM) and subfragment-1. *Fed. Proc. 28*:536.

Eisenberg, E. and C. Moos. 1967. The interaction of actin with myosin and heavy meromyosin in solution at low ionic strength. *J. Biol. Chem. 242*:2945.

Elliott, G. F. 1968. Force-balance and stability in hexagonally-packed polyelectrolyte systems. *J. Theoret. Biol. 21*:71.

Elliott, G. F., J. Lowy, and C. R. Worthington. 1963. An X-ray and light-diffraction study of the filament lattice of striated muscle in the living state and in rigor. *J. Mol. Biol. 6*:295.

Endo, M., Y. Nonomura, T. Masaki, I. Ohtsuki, and S. Ebashi. 1966. Localization of native tropomyosin in relation to striation patterns. *J. Biochem. 60*:605.

Finck, H. 1968. On the discovery of actin. *Science 160*:332.

Franzini-Armstrong, C. and K. R. Porter. 1964. The Z-disc of skeletal muscle fibrils. *Z. Zellforsch. 61*:661.

Fuchs, F. and F. N. Briggs. 1968. The site of calcium binding in relation to the activation of myofibrillar contraction. *J. Gen. Physiol. 51*:655.

Fukazawa, T. and E. J. Briskey. 1970. Exhaustive extraction of the troponin-tropomyosin factor from the I-Z-I brush of striated muscle. *Biochim. Biophys. Acta.* (In press.)

Fukazawa, T., E. J. Briskey, and W. F. H. M. Mommaerts. 1969. A new form of native tropomyosin. *J. Biochem.* (In press.)

Goll, D. E., W. F. H. M. Mommaerts, M. K. Reedy, and K. Seraydarian. 1969. Studies on $\alpha$-actinin-like proteins liberated during trypsin digestion of $\alpha$-actinin and of myofibrils. *Biochim. Biophys. Acta 175*:174.

Guba, F. and N. Garamvolgyi. 1967. A myofibrillar protein probably localized in the Z-lines. *Acta Biochim. et Biophys. Acad. Sci. Hung. 2*:417.

Halliburton, W. D. 1887. On muscle plasma. *J. Physiol. 8*:133.

Hanson, J., and H. E. Huxley. 1955. The structural basis of contraction in striated muscle. *Symp. Soc. Exp. Biol. 9*:228.

Hanson, J. and J. Lowy. 1963. The structure of F-actin and of actin filaments isolated from muscle. *J. Mol. Biol. 6*:46.

Hartshorne, D. J. 1970. Interactions of desensitized actomyosin with tropomyosin, troponin A, troponin B, and polyanions. *J. Gen. Physiol.* (In press.)

Hartshorne, D. J. and H. Mueller. 1968. Fractionation of troponin into two distinct proteins. *Biochem. Biophys. Res. Comm. 31*:647.

———. 1969. The preparation of tropomyosin and troponin from natural actomyosin. *Biochim. Biophys. Acta 175*:301.

Hartshorne, D. J., S. V. Perry, and M. C. Schaub. 1967. A protein factor inhibiting the magnesium-activated adenosine triphosphatase of desensitized actomyosin. *Biochem. J. 104*:907.

Hartshorne, D. J., M. Theiner, and H. Mueller. 1969. Studies on troponin. *Biochim. Biophys. Acta 175*:320.

Herring, H. K., R. G. Cassens, and E. J. Briskey. 1967. Tenderness and associated characteristics of stretched and contracted bovine muscles. *J. Food Sci. 32*:317.

Hoyle, G., J. H. McAlear, and A. Silverston. 1965. Mechanism of supercontraction in a striated muscle. *J. Cell Biol. 26*:621.

Huxley, A. F. 1957. Muscle structure and theories of contraction. *Progr. Biophys. Biophys. Chem. 7*:257.

Huxley, H. E. 1963. Electron microscope studies on the structure of natural and synthetic protein filaments from striated muscle. *J. Mol. Biol. 7*:281.

———. 1968. Structural differences between resting and rigor muscle; evidence from intensity changes in the low-angle equatorial X-ray diagram. *J. Mol. Biol. 37*:507.

———. 1969. The mechanism of muscular contraction. *Science 164*:1356.

Josephs, R. and W. F. Harrington. 1966. Studies on the formation and physical chemical properties of synthetic myosin filaments. *Biochemistry 5*:3474.

Kaminer, B. 1968. Magnesium and the regulatory protein of "calcium-sensitizing factor" in muscle. *Biochem. Biophys. Res. Comm. 33*:542.

Kaminer, B. and A. L. Bell. 1966. Myosin filamentogenesis: effects of *p*H and ionic concentration. *J. Mol. Biol. 20*:391.

Kasai, M. 1969. Thermodynamical aspects of G-F transformations of actin. *Biochim. Biophys. Acta 180*:399.

Kasai, M. and H. Hama. 1969. Natural F-actin III. Natural F-actin as inactive polymer. *Biochim. Biophys. Acta 180*:550.

Kielly, W. W., M. Kimura, and J. P. Cooke. 1964. The active site of myosin. *Abstr. 6th Int. Cong. Biochem. 8*:634.

Knappeis, G. G. and F. Carlsen. 1962. The ultrastructure of the Z disc in skeletal muscle. *J. Cell Biol. 13*:323.

Kuhn, T. S. 1965. *The Structure of Scientific Revolutions,* pp. 53–56. Phoenix Books, Univ. Chi.

Levi-Montalcini, R. and P. U. Angeletti. 1968. Nerve growth factor. *Physiol. Rev. 48*:534.

Locker, R. H. 1954. C-terminal groups in myosin, tropomyosin and actin. *Biochim. Biophys. Acta 14*:533.

Lowey, S., H. S. Slayter, A. G. Weeds, and H. Baker. 1969. Substructure of the myosin molecule I. Subfragments of myosin by enzymic degradation. *J. Mol. Biol. 42*:1.

Marsh, B. B. 1966. Discussion, p. 271. *In* E. J. Briskey, R. G. Cassens, and J. C. Trautman (eds.), *Physiology and Biochemistry of Muscle as a Food*. Univ. of Wis.

Masaki, T., M. Endo, and S. Ebashi. 1967. Localization of 6 S component of α-actinin at Z-band. *J. Biochem. 62*:630.

Mommaerts, W. F. H. M. 1969. Energetics of muscular contraction. *Physiol. Rev. 49*:427.

Mueller, H. and S. V. Perry. 1962. The degradation of heavy meromyosin by trypsin. *Biochem. J. 85*:431.

Nauss, K. M., S. Kitagawa, and J. Gergely. 1969. Pyrophosphate binding to and ATPase activity of myosin and its proteolytic fragments—implications for the substructure of myosin. *J. Biol. Chem. 244*:755.

Nonomura, Y., W. Drabikowski, and S. Ebashi. 1968. The localization of tropomyosin paracrystals. *J. Biochem. 64*:419.

Offer, G. W. 1965. I. Studies on N-acetyl peptides from a pronase digest of myosin. *Biochim. Biophys. Acta 111*:191.

Offer, G. W. and R. L. Starr. 1968. A parallel two chain structure for the myosin rod. (Abstr.) *Interaction between Subunits of Biological Macromolecules* (Symp. Int. Union Pure Appl. Biophys., Comm. Mol. Biophys., Cambridge, England).

Oosawa, F., S. Asakura, S. Higashi, M. Kasai, S. Kobayashi, E. Nakano, T. Ohnishi, and M. Taniguchi. 1965. Morphogenesis and notility of active polymers, p. 77. *In* S. Ebashi, F. Oosawa, T. Sekine, and Y. Tonomura (eds.), *Molecular Biology of Muscular Contraction*. Igaku Shoin, Tokyo.

Pepe, F. A. 1966. Some aspects of the structural organization of the myofibril as revealed by antibody-staining methods. *J. Cell Biol. 28*:505.

———. 1967. The myosin filament I. Structural organization from antibody staining observed in electron microscopy. *J. Mol. Biol. 27*:203.

Perry, S. V. 1967. The structure and interactions of myosin. *Progr. Biophys. Mol. Biol. 17*:325.

Rayns, D. G., F. O. Simpson, and W. S. Bertand. 1968. Surface features of striated muscle. *J. Cell Sci. 3*:467.

Reed, R., T. W. Houston, P. M. Todd, J. S. Kennelly, J. A. Frazier, S. M. Flawith, and T. D. Boyes. 1969. The structure of muscle tissue as seen in freeze-etched preparations. *Proc. Leeds Phil. Lit. Soc. 10*:17.

Reedy, M. K. 1964. The structure of actin filaments and the origin of the axial periodicity in the I-substance of vertebrate striated muscle. *Proc. Roy. Soc. B. 160*:458.

Reedy, M. K., D. A. Fischman, and G. F. Bahr. 1969. The correspondence between myosin content and cross bridge population in sarcomeres of insect flight muscle. *Biophys. J. 9A*:A95.

Reedy, M. K., K. C. Holmes, and R. T. Tregear. 1965. Induced changes in orientation of the cross bridges of glycerinated insect flight muscle. *Nature 207*:1276.

Rome, E. 1968. X-ray diffraction studies of the filament lattice of striated muscle in various bathing media. *J. Mol. Biol. 37*:331.

Sarno, J., A. Tarendash, and A. Stracher. 1965. Carboxyl terminal residues of myosin and heavy meromyosin. *Arch. Biochem. Biophys. 112*:378.

Schaub, M. C. and S. V. Perry. 1969. The relaxing protein system of striated muscle: resolution of the troponin complex into inhibitory and calcium-ion-sensitizing factors and their relationship to tropomyosin. *Biochem. J. 115*:993.

Sekine, T., Y. Kanaoka, T. Kameyama, M. Machida, and A. Takada. 1967. Studies on local conformational change at the active site of myosin A ATPase: use of fluorescent reagent as a chemical reporter. *Abstr. 7th Int. Congr. Biochem.* F-97:773.

Seraydarian, K., E. J. Briskey, and W. F. H. M. Mommaerts. 1968. The modification of actomyosin by α-actinin IV. The role of sulfhydryl groups. *Biochim. Biophys. Acta 162*:424.

Slayter, H. S. and S. Lowey. 1967. Substructure of the myosin molecule as visualized by electron microscopy. *Proc. Nat. Acad. Sci. 58*:1611.

Spicer, S. S. and J. Gergely. 1951. Studies on the combination of myosin with actin. *J. Biol. Chem. 188*:179.

Stracher, A. 1964. Disulfide-sulfhydryl interchange studies on myosin A. *J. Biol. Chem. 239*:1118.

Straub, F. B. 1942. Actin. *Stud. Inst. Med. Chem.* (Univ. Szeged) *2*:3.

Stromer, M. H., D. J. Hartshorne, H. Mueller, and R. V. Rice. 1969. The effect of various protein fractions on Z- and M-line reconstitution. *J. Cell Biol. 40*:167.

Szent-Györgyi, A. G. 1942. Muscle studies. *Stud. Inst. Med. Chem.* (Univ. Szeged) *1*:4.

Tonomura, Y., S. Kitagawa, and J. Yoshimura. 1962. The initial phase of myosin ATPase and the possible phosphorylation of myosin A. *J. Biol. Chem. 237*:3660.

Tonomura, Y., S. Watanabe, and M. Morales. 1969. Conformational changes in the molecular control of muscle contraction. *Biochemistry 8*:2171.

Trotta, D. D., P. Dreizen, and A. Stracher. 1968. Studies on subfragment-1, a biologically active fragment of myosin. *Proc. Nat. Acad. Sci. 61*:659.

Wakabayashi, T. and S. Ebashi. 1968. Reversible change in physical state of troponin induced by calcium ions. *J. Biochem. 64*:731.

Weber, A. 1966. Energized calcium transport and relaxing factors. *Current Topics in Bioenergetics 1*:203.

Yamaguchi, M. and T. Sekine. 1966. Sulfhydryl groups involved in the active site of myosin A adenosine triphosphatase I. Specific blocking of the SH group responsible for the inhibitory phase in "biphasic response" of the catalytic activity. *J. Biochem. 59*:24.

Young, D. M., S. Himmelfarb, and W. F. Harrington. 1965. On the structural assembly of the polypeptide chains of heavy meromyosin. *J. Biol. Chem. 240*:2428.

Young, M. 1969. The molecular basis of muscle contraction. *Ann. Rev. Biochem. 38*:913.

# PART 5

## Stromal and Sarcoplasmic Proteins

# Chapter 21

# *Collagen*

A. VEIS

The 4 years since the previous conference have seen remarkable progress in the chemistry and biology of collagen. Elastin chemistry, which had reached an exciting peak at that time, has not developed too much further. In the collagen field, most progress has been made in three areas: the sequences of the component peptide chains; the chemistry of the aldehyde cross linkages of intramolecular character; and the mode of assembly of the fibrillar systems. These topics will be considered below. First, however, it seems appropriate to review briefly some of the basic features of the collagen system.

## THE STRUCTURE AND COMPOSITION OF COLLAGEN MOLECULES

Collagen is the major protein component of the white connective tissue fibers. In addition to the primary connective tissue elements of the skin, bone, teeth, cornea, and sclera, finer networks of collagen fibers pervade almost every tissue and organ, including the muscle tissue. Overall, collagen may represent up to 30% of the total protein in some mammals. The collagens vary slightly with respect to composition from species to species but, considering the wide phylogenetic range of organisms containing collagen, the variations are remarkably small and quite conservative. It is also probable that collagens in different tissues of the same animal may show some small variations in composition. Representative analyses for human collagens from skin, bone, and dentin are given in Table 21.1 Noteworthy features of these data are the facts that glycine comprises nearly 33% of the amino acid residues; that the imino acids proline and hydroxyproline represent about 25 residue %; and that collagen contains two amino acids of unique character, hydroxyproline and hydroxylysine. Glycine not only comprises one-third of the residues, it appears as the first member of a triplet -gly-X-Y- which places glycine at every third residue position throughout most of each polypeptide chain. Proline may

TABLE 21.1
*Composition of human collagens*

| Amino acid | Skin* | Bone* | Dentin* |
|---|---|---|---|
| Hydroxyproline | 90.9 | 100 | 99 |
| Aspartic acid | 47.2 | 47.0 | 46 |
| Threonine | 18.3 | 18.4 | 17 |
| Serine | 36.9 | 35.9 | 33 |
| Glutamic acid | 77.7 | 72.2 | 74 |
| Proline | 125 | 123 | 116 |
| Glycine | 324 | 319 | 329 |
| Alanine | 114 | 113 | 112 |
| Valine | 24.5 | 23.6 | 25 |
| Cystine | trace | — | — |
| Methionine | 7.0 | 5.3 | 5.3 |
| Isoleucine | 10.4 | 13.3 | 9.3 |
| Leucine | 24.8 | 25.5 | 24 |
| Tyrosine | 3.5 | 4.5 | 6.4 |
| Phenylalanine | 12.6 | 13.9 | 16 |
| Hydroxylysine | 5.9 | 3.5 | 9.6 |
| Lysine | 26.6 | 28.0 | 22 |
| Histidine | 5.4 | 5.8 | 4.7 |
| Arginine | 49.0 | 47.1 | 52 |

* Reported as residues of amino acid per 1000 total residues. Data from Eastoe (1967).

be found at position X or Y but hydroxyproline is found only at position Y. Not indicated in the table is the fact that collagen is a glycoprotein, containing small quantities of galactose and glucose (Butler and Cunnigham, 1966).

X-ray diffraction studies show that the peptide chains in collagen share the backbone conformations common to poly-L-proline II and polyglycine II (Rich and Crick, 1955; Cowan, McGavin, and North, 1955). The molecular unit is comprised of three chains, each helical, and wound about each other in a gentle helix of opposite screw sense (Ramachandran and Kartha, 1955). Two systems of hydrogen bonding have been proposed, the so-called one- (Rich and Crick, 1961) and two-bonded (Lakshmanan, Ramakrishnan, Sasisekharan, and Thathachari, 1962) structures, but some doubt remains as to which structure is correct. The most recent evidence (Andreeva, Esipova, Millionova, Rogulenkova, and Shibnev, 1967) indicates that the Rich-Crick (1961) one-bonded structure II is the correct one, at least for the proline-and hydroyproline-rich chain sequences (Engel, 1967).

The precise molecular weight of an intact collagen molecule is not known with certainty. It is, however, close to 300,000. The molecule can be represented as a long rod with an axial ratio close to 200:1 (Boedtker and Doty, 1956). The molecular length has been estimated as 2990 Å (Hodge and Petruska, 1963), the bulk of which is in the triple helical conformation. A small portion at one end has been treated as being nonhelical and of atypical composition (e.g., glycine not in every third residue position in each chain).

Upon denaturation, soluble collagen falls apart into its three constituent polypeptide chains or firmly linked aggregates of these. Piez, Eigner, and Lewis (1963) fractionated these denatured polypeptides by carboxymethyl cellulose chromatography and coined the terminology presently in use. In this system, a single intact polypeptide chain is called an $\alpha$ component, a dimer of two $\alpha$'s joined by a covalent cross linkage is called a $\beta$ component. The trimer, corresponding in molecular weight to the intact collagen monomer unit, and containing at least two covalent linkages joining the 3 $\alpha$ chains into a single unit, is called the $\gamma$ component. In most of the collagens studied since the nomenclature was established two of the three $\alpha$ chains appear to be similar while the third has a distinctly different composition. The similar chains are designated as $\alpha 1$, the third chain as $\alpha 2$. The formula for a collagen molecule is thus $(\alpha 1)_2 \alpha_2$, and the total $\alpha 1$ content in the tissue is present in an amount twice that of $\alpha 2$. Depending upon the specific tissue, the solvent used for extraction, and the age of the host organism, the relative amounts of the $\alpha$, $\beta$, and $\gamma$ components may vary. The polymer components ($\beta$, $\gamma$) are present in greatest amounts in collagen of greatest age and in lesser amounts in the most newly synthesized collagen.

The amino acid compositions of native intact bovine corium collagen and its $\alpha 1$ and $\alpha 2$ components are given in Table 21.2 in terms of residues of amino acid per 1000 total residues. The $\alpha 2$ component is somewhat more basic than $\alpha 1$ and contains less methionine. Sequence studies on the collagens are extremely difficult because of the large amounts of glycine, proline, and hydroxyproline and because of the repetitive triplet (-gly-X-Y-) nature of the residue distribution. Piez and his colleagues have exploited the specific cleavage of the $\alpha$ chains of rat skin collagen by cyanogen bromide (CB) at methionine residues, to simplify the problem (Bornstein and Piez, 1965, 1966; Bornstein, Kang, and Piez, 1966*a;* Butler, Piez, and Bornstein, 1967). The $\alpha 1$ chains contain only 8 methionine residues and hence upon cleavage yield only 9 distinct peptides. Similarly, $\alpha 2$ chains yield only 6 peptides. These peptides can be separated and the smaller ones have already been sequenced.

The cyanogen bromide peptides from the $\alpha 1$ chain of rat skin collagen have been separated and analyzed by Butler, Piez, and Bornstein (1967) and the $\alpha 2$ cyanogen bromide peptides by Fietzek and Piez (1969). The

TABLE 21.2
*The amino acid composition of bovine corium collagen and its α-component subunits*

| Amino acid | Intact collagen* | α1* | α2* |
|---|---|---|---|
| Hydroxyproline | 92 | 102 | 88 |
| Aspartic acid | 46 | 45 | 49 |
| Threonine | 16.8 | 17.3 | 18 |
| Serine | 32 | 35 | 35 |
| Glutamic acid | 74 | 80 | 74 |
| Proline | 127 | 140 | 117 |
| Glycine | 338 | 308 | 328 |
| Alanine | 105 | 114 | 98 |
| Valine | 21.6 | 16.4 | 31 |
| Methionine | 4.9 | 6.5 | 3.7 |
| Isoleucine | 11.9 | 9.4 | 16.3 |
| Leucine | 25.4 | 21.4 | 32 |
| Tyrosine | 3.8 | 2.8 | 2.3 |
| Phenylalanine | 12.3 | 12.6 | 13.7 |
| Hydroxylysine | 6.5 | 5.7 | 8.8 |
| Lysine | 27 | 30 | 23 |
| Histidine | 4.6 | 2.6 | 6.2 |
| Arginine | 50 | 52 | 55 |
| (Ammonia) | (46) | (53) | (64) |

* Reported as residues of amino acid per 1000 total residues.
Data from Clark and Veis (unpublished).

compositions of these peptides are given in Tables 21.3 and 21.4. The order of the α1 peptides, from the amino terminus, is α1-CB 0, 1, 2, 4, 5, 8, 3, 7, 6. The order of the α2 peptides has not been determined. Aside from its intrinsic interest, this work is of major importance since the unique character of each of the peptides appears to preclude a subunit assembly system for the individual chains (Butler, Piez, and Bornstein, 1967).

A second feature of major importance in this work came from the examination of the $\beta_{12}$ component.* A new cyanogen bromide peptide appeared containing two homoserine lactone groups. At the same time α1-CB1 and α2-CB1 disappeared. Since homoserine is the product of methionine reaction it was evident that the new peptide represented a joint pair and probably contained an intramolecular cross-linkage point. The peptides were identified as corresponding in composition to the sum

* The notation for polymers of α chains is to use the symbol β, γ or δ to represent the degree of polymerization, i.e., n = 2, 3, or 4, while the subscripts indicate the kinds of α chains in the polymer. For example, $\beta_{12} = (\alpha 1) - (\alpha 2)$.

TABLE 21.3

*Amino acid composition of the cyanogen-bromide (CNBr) peptides from α1 chains of rat skin collagen**

| Amino acid | α1-CB1 | α1-CB2 | α1-CB3 | α1-CB4 | α1-CB5 | α1-CB6 | α1-CB7 | α1-CB8 | Total CNBr peptides | α1 found |
|---|---|---|---|---|---|---|---|---|---|---|
| 3-Hydroxyproline | 0 | 0 | 0 | 0 | 0 | 1(1.2) | 0 | 0 | 1 | 0.8 |
| 4-Hydroxyproline | 0 | 5 | 14 | 6(6.0) | 3(2.7) | 16 | 27 | 27 | 98 | 104 |
| Aspartic acid | 1 | 0 | 7(7.0) | 3(3.0) | 3(3.0) | 10 | 12 | 10 | 46 | 47 |
| Threonine | 0 | 0 | 2(2.0) | 1(1.0) | 1(1.0) | 4(4.2) | 7(6.7) | 5(4.7) | 20 | 20.6 |
| Serine | 2 | 2 | 3(3.1) | 0 | 2(2.0) | 12 | 9(8.7) | 10 | 40 | 40 |
| Glutamic acid | 1 | 4 | 16 | 3(3.0) | 3(3.0) | 13 | 16 | 19 | 75 | 76 |
| Proline | 2 | 7 | 14 | 6(5.8) | 2(1.9) | 31 | 39 | 31 | 132 | 126 |
| Glycine | 3 | 12 | 51 | 15 | 12 | 66 | 92 | 87 | 338 | 356 |
| Alanine | 1 | 2 | 20 | 3(3.0) | 3(3.1) | 19 | 29 | 36 | 113 | 118 |
| Valine | 2 | 0 | 4(4.2) | 0 | 0 | 2(1.5) | 8(7.8) | 5(4.9) | 21 | 21.4 |
| Isoleucine | 0 | 0 | 0 | 0 | 0 | 3(2.9) | 3(3.0) | 1(1.1) | 7 | 8.0 |
| Leucine | 0 | 1 | 3(3.0) | 2(2.0) | 1(1.0) | 4(4.1) | 4(4.2) | 4(4.0) | 19 | 20.5 |
| Tyrosine | 1 | 0 | 0 | 0 | 0 | 1(0.6) | 0 | 0 | 2 | 1.9 |
| Phenylalanine | 0 | 1 | 3(2.9) | 0 | 1(0.9) | 1(1.1) | 3(3.1) | 3(3.0) | 12 | 12.2 |
| Hydroxylysine | 0 | 0 | 0(0.2) | 0 | 1(1.1) | 2(1.8) | 1(0.7) | 1(1.0) | 5 | 4.3 |
| Lysine | 1 | 0 | 5(4.7) | 2(2.0) | 2(1.8) | 3(3.3) | 9(9.0) | 9(8.9) | 31 | 30 |
| Histidine | 0 | 0 | 0 | 0 | 1(1.0) | 1(0.9) | 0 | 0 | 2 | 2.2 |
| Arginine | 0 | 1 | 6(6.1) | 4(4.1) | 1(1.0) | 11 | 13 | 14 | 50 | 52 |
| Homoserine | 1 | 1 | 1(0.9) | 1(1.0) | 1(0.9) | 0 | 1(0.9) | 1(1.0) | 7 | 6.6 |
| Total | 15 | 36 | 149 | 46 | 37 | 200 | 273 | 263 | 1019 | 1048 |

* Reported as residues per peptide.
Data of Butler, Piez, and Bornstein (1967).

TABLE 21.4

*Amino acid composition of CNBr peptides of the α2 chain of rat skin collagen**

| Amino acid | α2-CB0 | α2-CB1 | α1-CB2 | α1-CB3 | α1-CB4 | α1-CB5 | Total CNBr peptides | α2 |
|---|---|---|---|---|---|---|---|---|
| 4-Hydroxyproline | 0 | 0 | 3 | 26 | 30 | 27 | 86 | 89 |
| Aspartic acid | 0 | 1 | 3 | 14 | 12 | 13 | 43 | 44 |
| Threonine | 0 | 0 | 1 | 5(4.8) | 8(7.5) | 8(8.1) | 22 | 22 |
| Serine | 0 | 2 | 1 | 14 | 13 | 14 | 44 | 44 |
| Glutamic acid | 0 | 1 | 1 | 24 | 25 | 20 | 71 | 71 |
| Proline | 0 | 2 | 3 | 33 | 31 | 33 | 102 | 109 |
| Glycine | 1(1.0) | 3 | 10 | 105 | 107 | 106 | 332 | 346 |
| Alanine | 0 | 1 | 2 | 36 | 37 | 30 | 106 | 106 |
| Valine | 0 | 1 | 1 | 8(7.6) | 13 | 12 | 35 | 35 |
| Isoleucine | 0 | 0 | 0 | 6(5.9) | 4(4.1) | 7(7.1) | 17 | 19 |
| Leucine | 1(1.0) | 0 | 1 | 7(7.3) | 13 | 13 | 35 | 36 |
| Tyrosine | 0 | 1 | 0 | 0 | 0 | 2(1.5) | 3 | 3(3.2) |
| Phenylalanine | 0 | 0 | 0 | 4(3.7) | 4(4.1) | 3(2.7) | 11 | 10(10.0) |
| Hydroxylysine | 0 | 0 | 0 | 2.4 | 3.7 | 2.8 | 8.9 | 9.5 |
| Lysine | 0 | 1 | 0 | 9.0 | 6.8 | 5.9 | 23 | 26 |
| Histidine | 0 | 0 | 0 | 1(1.0) | 2(2.3) | 6(5.6) | 9(8.9) | 10(9.5) |
| Arginine | 0 | 0 | 3 | 15 | 17 | 17 | 52 | 54 |
| Homoserine | 1(0.9) | 1 | 1 | 1(0.8) | 1(0.9) | 0 | 5 | 5(4.9) |
| | 3 | 14 | 30 | 310 | 328 | 320 | 1005 | 1039 |

* Reported as residues per peptide.
Data of Fietzek and Piez (1969).

of the $\alpha$1-CB1 and $\alpha$2-CB1 peptides (Bornstein and Piez, 1966) with the difference being that the lysine at residue 5 in each peptide was absent and a new compound, an $\alpha$, $\beta$-unsaturated aldehyde appeared. The mechanism proposed is that the lysyl residues are enzymically oxidatively deaminated to aldehydes which then condense via an aldol-type reaction to produce the unsaturated aldehyde. The precursor aldehyde, lysyl-derived $\delta$-semialdehyde of $\alpha$-aminoadipic acid in peptide linkage, has been isolated and identified (Bornstein, Kang, and Piez, 1966*b*). Thus it was argued that the chemistry of intramolecular cross linking in collagen was similar to the elastin cross-linking chemistry as described by Partridge (Partridge, Elsden, and Thomas, 1963; Partridge, Elsden, Thomas, Dorfman, Teiser, and Ho, 1964, 1965).

Recently, the question of the identity of the two $\alpha$1 chains has been raised. Stark and Kühn (1968) have suggested that the carboxy terminal region of one of the "$\alpha$1" chains may be different from the other. The reason for the difference and its significance is not clear at this time.

## THE ASSEMBLY OF COLLAGEN FIBERS AND INTERMOLECULAR CROSS LINKING

Native collagen fibers, formed from the aggregation of monomer units with their long axes parallel to each other, exhibit an axial periodicity of ~ 700 Å in the electron microscope. The reconciliation of this periodicity with the molecular length of ~ 3000 Å has been a major challenge. Hodge and Schmitt (1960) resolved the problem in part by suggesting that each molecule was shifted by ~ one-quarter of its length with respect to its nearest lateral neighbors. Hodge and Petruska (1963) refined this model further, concluding that, since each molecule is 4.4 times the

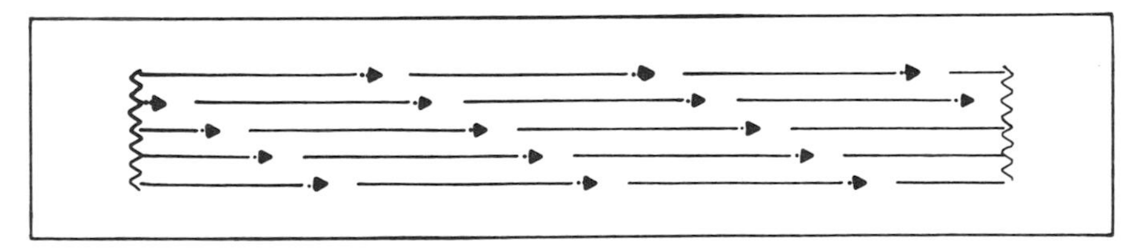

*Fig. 21.1.* A two-dimensional representation of the Hodge-Petruska quarter-stagger hole-overlap structure for the construction of native collagen fibers. The arrow heads represent the amino terminus of each molecule, the dashed portion indicates that this region is not in the triple helical organized structure characteristic of the entire molecule. The "overlap" region is thus between the amino terminus of one molecule and an equivalent length of the B end of the neighboring molecule. From Veis, Anesey, and Mussell (1967).

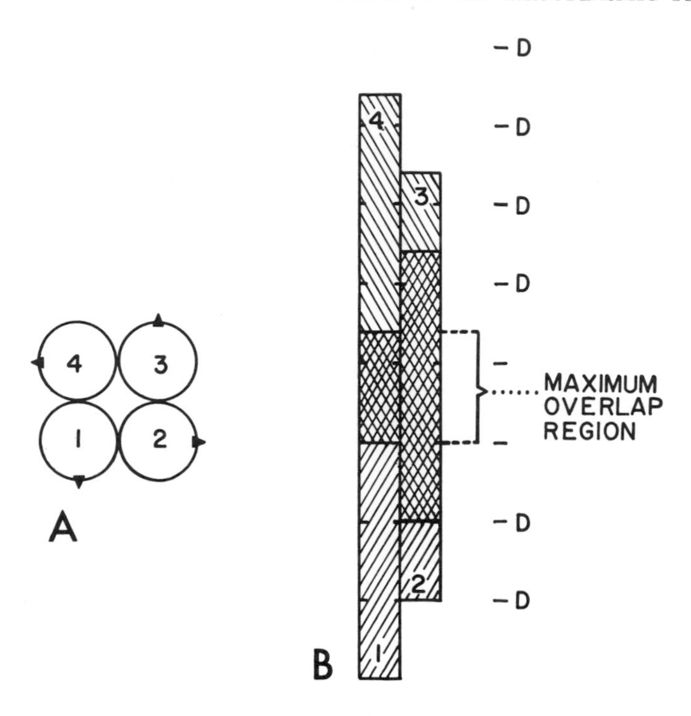


*Fig. 21.2.* The unit fibril assembly according to the model of Veis, Anesey, and Mussell (1967). (*a*) An end view of the tetrad unit indicating the proposed cubic packing of the monomers; (*b*) A side view projection along the fiber axis of a tetrad showing the 1 D unit stagger of successive molecules from the B end. A–B overlaps in this model are not exclusively restricted to the region close to the B end. From Veis, Bhatnagar, Shuttleworth, and Mussell (1970).

length of the repeat period (700 Å = 1 D) there must be a gap of 0.6 D between molecules along each molecular axis. This arrangement, with a regular array of holes or structure defects, is illustrated in Fig. 21.1. It is evident from the figure that there is an overlap of 0.4 D of the amino terminal of one monomer with the carboxy terminal of the next. Several investigators have pointed out the inadequacy of this model in three-dimensional representations (Grant, Horne, and Cox, 1965; Smith, 1965; Veis, Anesey, and Mussell, 1967). Veis, Anesey, and Mussell (1967), Veis, Bhatnagar, Shuttleworth, and Mussell (1969), and Smith (1968) have proposed alternative "unit assembly," "limiting microfibril" models in which the fundamental grouping is a 4- or 5-mer. Beginning with the first monomer, the four units in the tetrad model of Veis, Anesey, and Mussell

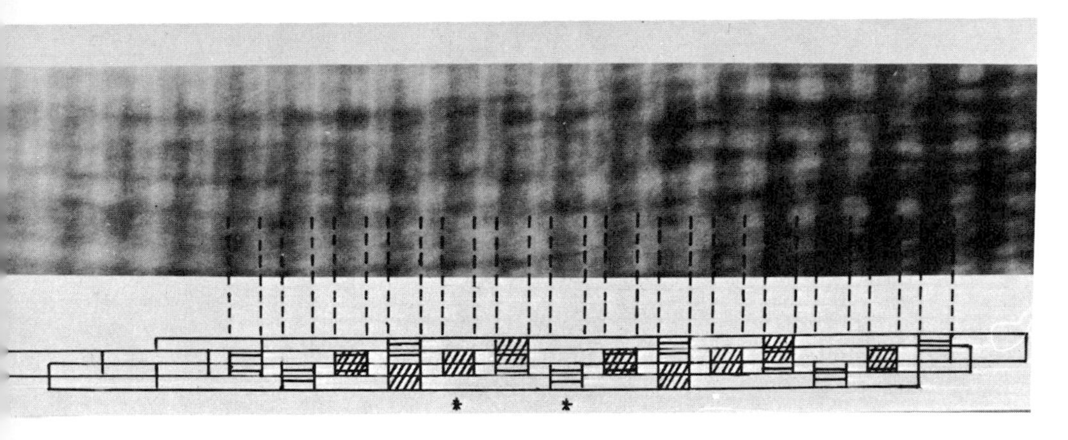

g. 21.3. The relationship between the band pattern of native collagen and the hole regions in the uni sembly system. From Veis, Spector, and Carmichael (1970).

(1967) are staggered and packed as depicted in the side view and cross-section representations shown in Fig. 21.2. The limiting fibril is formed by the overlap between sequential tetrad units. The native fiber is formed from randomly staggered side-by-side alignments of the limiting fibrils. In this model every fibril has a series of regularly disposed surface defects or gaps 0.6 D in length. Fig. 21.3 shows the relationship between the gaps in the limiting fibrils and the variations in fiber density seen in the electron microscope. From this point of view it is obvious that collagen fibers are porous and that their interiors are relatively easily accessible to solvent and small molecules. This has important consequences for the reactivity and mineralization of the collagen fiber system. The point of interest in the present discussion, however, is the relationship between the limiting fibril structure and the intermolecular cross linkage of the collagen.

Veis and Anesey (1965) demonstrated the presence of $\gamma$ components and higher polymers in mature bovine corium collagen. The $\gamma$ components were particularly interesting since at least two species $\gamma_{111}$ and $\gamma_{222}$ were present in addition to the expected intramolecular polymer $\gamma_{112}$. In fact, these were present in greater amount than the $\gamma_{112}$ component, indicating that intermolecular cross linking was the predominant reaction in the insolubilization of collagen. Studies on the renaturation properties of $\gamma_{111}$ and $\gamma_{222}$ suggested (Veis and Anesey, 1965; Veis, Anesey, and Mussell, 1967) that $\gamma_{111}$ was composed from side-by-side arrangements of $\alpha 1$ chains with their ends in register whereas in $\gamma_{222}$ the $\alpha 2$ chains were staggered as in the native molecule. The question thus arose as to how the covalent intermolecular bonds might be placed to produce two such sets of cross-linked polymers.

The approach taken was to attempt to disperse the insoluble collagen by nonhydrolytic means into the limiting fibrils of the model outlined above. The procedure developed by Steven (1967) was modified to delete the enzyme (crude amylase) and to depend upon thermal tempering and lyotropic swelling to loosen the structure. Subsequent acid swelling and homogenization of the tissue in the cold brought a substantial fraction of the collagen (~ 20%) into solution as high-molecular-weight aggregates (Veis, Bhatnagar, Shuttleworth, and Mussell, 1969). Electron microscopy on the dispersed collagen (Fig. 21.4) demonstrated that the solubilized collagen was present as long thin filaments with diameters of the order of 50 Å, a value entirely consistent with the limiting microfibril model.

Carboxymethylcellulose chromatography and disc electrophoretic analyses of the soluble collagen after denaturation showed that at least 70% of the collagen was present as polymers with molecular weights higher than that of the $\gamma$ component (Veis, Bhatnagar, Shuttleworth, and Mussell, 1969). Moreover, there was a smaller amount of $\gamma_{111}$ in the low-molecular-weight portion than anticipated from studies on the gelatin released by a 60° C extraction-denaturation procedure (Veis and Anesey, 1965). These data, indicating a high degree of intermolecular cross linking obviously restricted to thin filaments in which the axial stagger structure is maintained, were taken to imply that at least one major system of intermolecular cross linking lies within the limiting fibrils. However, since lyotropic relaxation of the fiber network, followed by acid swelling, does not lead to the total solubilization of the collagen, it is possible that a second system of polymerization links the limiting fibrils together. This second system is probably emphasized in the mineralized tissues (Veis and Schlueter, 1964; Veis, Spector, and Carmichael, 1969).

The placement of intermolecular cross linkages has not been determined. However, it is obvious that in contrast to the in-register placement of bonds intramolecularly, the staggered microfibrillar arrangement requires that the intermolecular bonds involve different parts of the $\alpha$ chains on different molecules.

### THE INTERMOLECULAR CROSS LINKAGES IN COLLAGEN

The insolubility and tensile strength of collagen fibers have been attributed to the presence of a system of intermolecular cross linkages (Levene and Gross, 1959). Collagen from lathyric animals appears to be normal with regard to the organization of the fiber system and the chemical

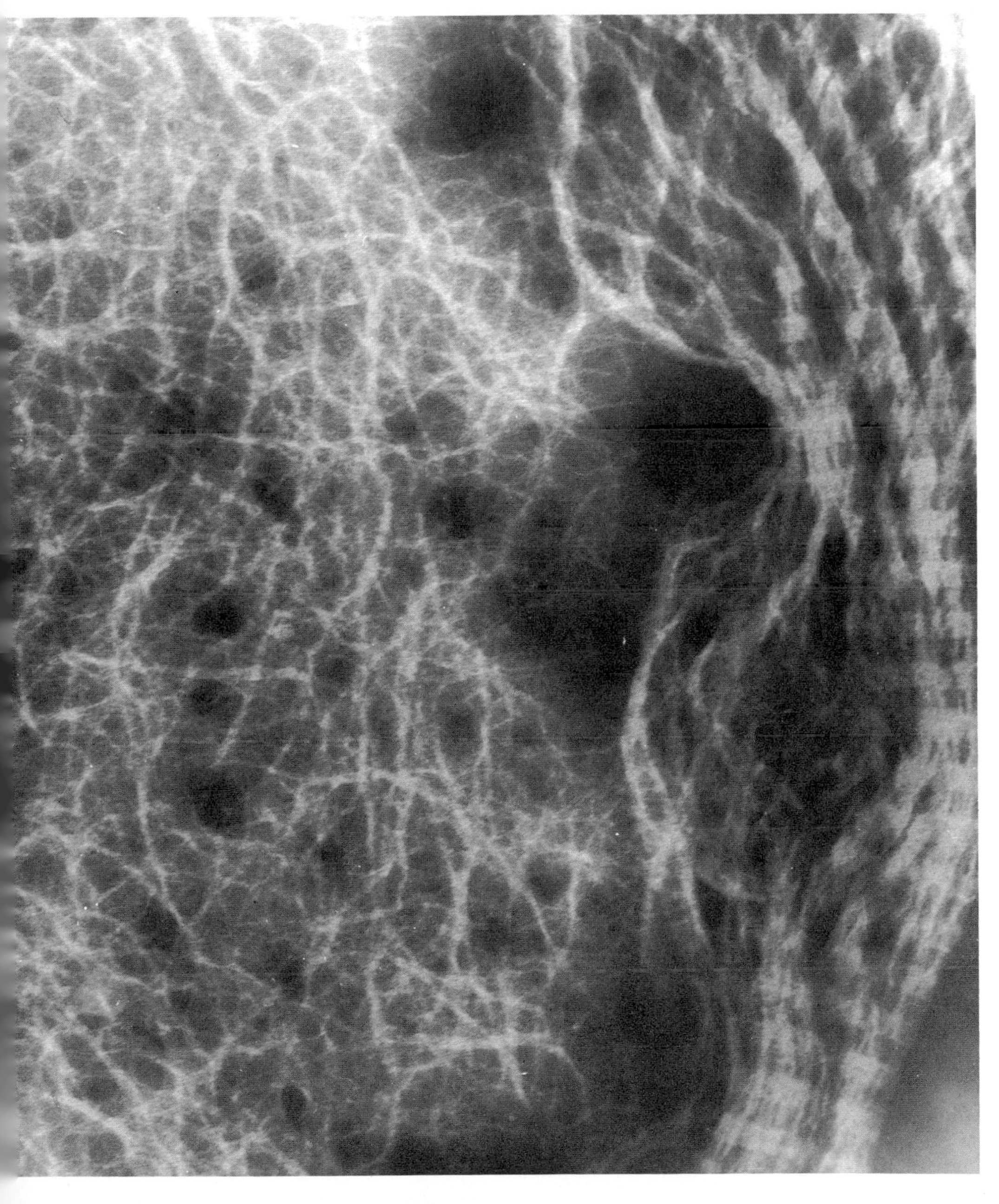

g. *21.4.* The native collagen dispersed by lyotropic swelling showing the fine fibrils and the way in hich they unravel as the native fiber is dispersed. From Veis, Bhatnagar, Shuttleworth, and Mussell 970).

composition of the constituent molecules. In spite of this, lathyric collagen is deficient in tensile strength and does not develop the insolubility characteristic of mature normal collagens. Both of these defects have been attributed to a marked drop in the degree of intermolecular cross linking. It has been clearly shown that the conversion of lysine to its peptide-bound aldehyde form is inhibited in lathyrism (Bornstein and Piez, 1966). Thus, the intramolecular cross links described by Bornstein and Piez (1966) cannot form. Since the development of intermolecular cross linking is also inhibited by the lathyrogens it has been argued that the intermolecular polymerization must also involve lysine-derived aldehydes.

Tanzer, Monroe, and Gross (1966) found that the aldehyde reagent, thiosemicarbazide ($H_2N{-}\overset{\overset{S}{\|}}{C}{-}NH{-}NH_2$), reacted with soluble collagen, presumably to form a thiosemicarbazone ($R{-}C{=}N{-}NH{-}\overset{\overset{S}{\|}}{C}{-}NH_2$). Two moles of thiosemicarbazide were bound per mole of collagen. The incorporation of the thiosemicarbazide led to the inhibition of the insolubilization that occurs following the heat gelation of collagen at neutral *p*H. The thiosemicarbazide binding, however, was independent of the presence of intramolecular $\beta$ components. It was concluded, therefore, that there was no direct relationship between inter- and intramolecular cross linking although aldehydes were involved in both cases. It was suggested that a lysine-derived aldehyde condensed with another $\epsilon$-amino group, to form a peptide-bound Schiff base,

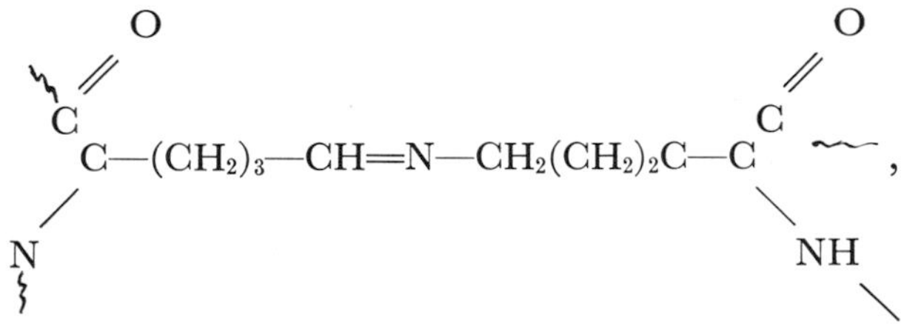

$$\begin{array}{ccc} \sim\overset{O}{\overset{/\!/}{C}} & & \overset{O}{\overset{/\!/}{C}}\sim \\ C{-}(CH_2)_3{-}CH{=}N{-}CH_2(CH_2)_2C{-}C & & \\ / & & \backslash \\ \underset{\wr}{N} & & NH \\ & & \backslash \end{array},$$

which acted as an intermolecular cross link.

Tanzer (1968) proceeded to reduce collagen fibrils, reconstituted from acid-soluble collagen, with sodium borohydride. In this reaction the Schiff base should be saturated to form a stable group. The reaction produced an insoluble collagen. The cross linking was very dependent upon the state of aggregation of the collagen; only those aggregates in native form, with the appropriate quarter-stagger registration between neighboring molecules, were capable of forming the insolubilizing intermolecular cross linkages. Tanzer argued from these data that a specific acceptor site must exist for Schiff base formation. Tanzer could discern no relationship between the inter- and intramolecular cross-linking processes. Reduction did not stabilize intramolecular cross linkages.

Bailey (1968) determined the stability of the intermolecular cross linkages by measuring the breaking load of intact collagen fibers. The more cross-linked a collagen, the higher is the maximum breaking load. He showed that acids, alkalis, and β-aminothiols all reduced the breaking load of collagen but that the effect of these reagents was inhibited if the collagen was reduced with borohydride prior to treatment. Bailey concluded that at least one type of labile intermolecular cross link, probably a Schiff base, was present as an intermediate in the ultimate formation of more stable cross linkages. Tensile strength measurements (Bailey and Lister, 1968) on native, thermally denatured, penicillamine treated and borohydride-reduced collagen fibers showed the reagent-labile bonds to be thermally labile as well. However, measurements on collagen fibers obtained from animals at different stages of maturation showed that only a proportion of the covalent intermolecular cross linkages were present in labile form at any particular time. There does seem to be a progressive decrease in the relative content of thermally labile bonds as the tissue ages. Bailey, Fowler, and Peach (1969) have determined, from studies on the reduction of $^{14}C$ lysine-labeled collagen with tritiated borohydride, that the reducible labile bond does involve a moiety derived from lysine. These studies also indicate the presence of hydroxylysine or leucine and one other compound in the cross link so that hydroxylysine as well as lysine might be involved. The work of Deshmukh and Nimni (1969) on the depolymerization of collagen with aminothiols such as penicillamine is consistent with these observations.

The current state of our knowledge concerning the intermolecular cross linking of collagen may be summarized as follows. Intermolecular cross linkages of several varieties are present in all tissues. Some of these may be Schiff bases involving hydroxylysine or lysine as either the aldehyde donor or the amine donor. The Schiff base form is labile to heat and to a variety of reagents with active amino groups, and to β-aminothiols in particular. The labile bonds are intermediates in the cross-linking scheme and are converted to a stable form. There are always some stable bonds in even the youngest collagens, hence there is no way as yet of ruling out the presence of nonlysine aldehyde-derived covalent intermolecular cross linkages in collagen. From considerations of both spatial arrangement and the chemistries involved, there is probably no direct relationship between the inter- and intramolecular cross-linking processes. The intermolecular bonds play a role in determining the physical properties of the collagen. No role has been discerned for the intramolecular cross linkages.

## *References*


Andreeva, N. S., N. G. Esipova, M. I. Millionova, V. N. Rogulenkova, and V. A. Shibnev. 1967. Polypeptides with regular sequences of amino acids as models of the collagen structure, p. 469. *In* G. N. Ramachandran (ed.), *Conformation of Biopolymers,* Vol. 2. Academic Press, London.

Bailey, A. J. 1968. Intermediate labile intermolecular cross-links in collagen fibers. *Biochim. Biophys. Acta 160:*447.

Bailey, A. J., L. J. Fowler, and C. M. Peach. 1969. Identification of two interchain cross-links of bone and dentine collagen. *Biochem. Biophys. Res. Comm. 35:*663.

Bailey, A. J., and D. Lister. 1968. Thermally labile cross-links in native collagen. *Nature 220:*280.

Boedtker, H., and P. Doty. 1956. The native and denatured states of soluble collagen. *J. Amer. Chem. Soc. 78:*4267.

Bornstein, P., A. H. Kang, and K. A. Piez. 1966*a*. The limited cleavage of native collagen with chymotrypsin, trypsin and cyanogen bromide. *Biochemistry 5:*3803.

———. 1966*b*. The nature and location of intramolecular crosslinks in collagen. *Proc. Nat. Acad. Sci. 55:*417.

Bornstein, P., and K. A. Piez. 1965. Collagen: structural studies based on the cleavage of methionyl bonds. *Science 148:*1353.

———. 1966. The nature of the intramolecular cross-links in collagen. The separation and characterization of peptides from the cross-link region of rat skin collagen. *Biochemistry 5:*3460.

Butler, W. T., and L. W. Cunningham. 1966. Evidence for a linkage of a disaccharide to hydroxylysine in tropocollagen. *J. Biol. Chem. 241:*3882.

Butler, W. T., K. A. Piez, and P. Bornstein. 1967. Isolation and characterization of the cyanogen bromide peptides from the α1 chain of rat skin collagen. *Biochemistry 6:*3771.

Cowan, P. M., S. McGavin, and A. C. T. North. 1955. The polypeptide chain configuration of collagen. *Nature 176:*1062.

Deshmukh, K., and M. E. Nimni. 1969. A defect in the intramolecular and intermolecular cross-linking of collagen caused by penicillamine. *J. Biol. Chem. 244:*1787.

Eastoe, J. E. 1967. Composition of collagen and allied proteins, p. 1. *In* G. N. Ramachandran (ed.), *Treatise on Collagen,* Vol. 1. Academic Press, New York.

Engel, J. 1967. Conformational transitions of poly-l-proline and poly-(l-prolyl-glycyl-l-proline), p. 483. *In* G. N. Ramachandran (ed.), *Conformation of Biopolymers,* Vol. 2. Academic Press, New York.

Fietzek, P. P., and K. A. Piez. 1969. Isolation and characterization of the cyanogen bromide peptides from the α2 chains of rat skin collagen. *Biochemistry 8:*2129.

Grant, R. A., R. W. Horne, and R. W. Cox. 1965. New model for the tropocollagen macromolecule and its mode of aggregation. *Nature 207:*822.

Hodge, A. J., and J. A. Petruska. 1963. Recent studies with the electron microscope on ordered aggregates of the tropocollagen molecule, p. 289. *In* G. N. Ramachandran (ed.), *Aspects of Protein Structure.* Academic Press, New York.

Hodge, A. J., and F. O. Schmitt. 1960. The charge profile of the tropocollagen molecule and the packing arrangement in native-type collagen fibrils. *Proc. Nat. Acad. Sci. 46:*186.

Lakshmanan, B. P., C. Ramakrishnan, V. Sasisekharan, and Y. T. Thathachari. 1962. X-ray diffraction pattern of collagen and the Fourier transform of the collagen structure, p. 117. *In* N. Ramanathan (ed.), *Collagen.* Interscience, New York.

Levene, C. I., and J. Gross. 1959. Alterations in state of molecular aggregation of collagen induced in chick embryos by β-aminopropionitrile (Lathyrus factor). *J. Exp. Med. 110:*771.

Partridge, S. M., D. F. Elsden, and J. Thomas. 1963. Constitution of the cross-linkages in elastin. *Nature 197:*1297.

Partridge, S. M., D. F. Elsden, J. Thomas, A. Dorfman, A. Teiser, and Pei-Lee Ho. 1964. Biosynthesis of the desmosine and isodesmosine cross-bridges in elastin. *Biochem. J. 93:*30c.

———. 1965. Incorporation of labeled lysine into the desmosine cross-bridges of elastin. *Nature 209:*399.

Piez, K. A., E. A. Eigner, and M. S. Lewis. 1963. The chromatographic separation and amino acid composition of the subunits of several collagens. *Biochemistry 2:*58.

Ramachandran, G. N., and G. Kartha. 1955. Structure of collagen. *Nature 176:*593.

Rich, A., and F. H. C. Crick. 1955. The structure of collagen. *Nature 176:*915.

———. 1961. The molecular structure of collagen. *J. Mol. Biol. 3:*483.

Smith, J. W. 1965. Packing arrangement of tropocollagen molecules. *Nature 205:*356.

———. 1968. Molecular pattern in native collagen. *Nature 219:*157.

Stark, M. and K. Kühn. 1968. The properties of molecular fragments obtained on treating calfskin collagen with collagenase from *Clostridium histolylicum. Europe. J. Biochem. 6:*534.

Steven, F. S. 1967. The effect of chelating agents on collagen interfibrillar matrix interactions in connective tissue. *Biochim. Biophys. Acta 140:*522.

Tanzer, M. L. 1968. Intermolecular crosslinks in reconstituted collagen fibrils. Evidence for the nature of the covalent bonds. *J. Biol. Chem. 243:*4045.

Tanzer, M. L., D. Monroe, and J. Gross. 1966. Inhibition of collagen cross-linking by thiosemicarbazide. *Biochemistry 5*:1919.

Veis, A., and J. Anesey. 1965. Modes of intermolecular cross-linking in mature insoluble collagen. *J. Biol. Chem. 240*:3899.

Veis, A., J. Anesey, and S. J. Mussell. 1967. A limiting microfibril model for the three-dimensional arrangement within collagen fibers. *Nature 215*:931.

Veis, A., R. S. Bhatnagar, C. A. Shuttleworth, and S. J. Mussell. 1969. The solubilization of mature, polymeric collagen fibrils by lyotropic relaxation. *Biochim. Biophys. Acta. 200*:97.

Veis, A., and R. J. Schlueter. 1964. The macromolecular organization of dentin matrix collagen I. Characterization of dentin collagen. *Biochemistry 3*:1650.

Veis, A., A. R. Spector, and D. J. Carmichael. 1969. The organization and polymerization of bone and dentin collagen. *Clinical Orthopaedics.* (In press.)

*Chapter 22*

# *Characterization and Study of Sarcoplasmic Proteins*

R. K. SCOPES

Sarcoplasm is the intracellular fluid of muscle. It is not readily obtained free of contamination by extracellular fluids, because even after bleeding and perfusion a certain amount of blood plasma remains in the muscle. This can be largely removed by ultracentrifugation of chunks of muscle (Amberson, Bauer, Philpott, and Roisen, 1964), leaving behind material from which less contaminated sarcoplasm can be extracted. The lowering of muscle *p*H due to glycolytic metabolism during ultracentrifugation is, however, a great disadvantage of this method.

Most work on sarcoplasmic proteins commences with an extract of bled muscle; the extracellular contamination amounts to about 5–10% of the total protein obtained, though rather more may be present in extracts of "slow" muscles with a rich blood supply. Early work was carried out with water extracts, which were then dialysed to very low ionic strength, precipitating the so-called globulins, the albumins remaining in solution. The globulins (globulin X, Weber and Meyer, 1933) were mostly not resoluble in physiological conditions because the *p*H of the dialysate was low (due to dissolved $CO_2$), causing considerable denaturation. Globulin X was probably denatured glyceraldehyde phosphate dehydrogenase (GAPDH) with smaller amounts of other enzymes. The amount of "globulin" is critically dependent on *p*H as well as on ionic strength, because dialysis to very low ionic strength, keeping the *p*H at 7.0 or above, does not cause any significant precipitation.

The albumins remaining in solution were originally divided into "myogen," the major part of which migrated only slowly on free-boundary electrophoresis, and "myoalbumin" (Bate-Smith, 1937) a faster-migrating protein. Jacob (1947) showed that myogen would separate into several fractions on electrophoresis, and Baranowski (1939) was able to crystallize a protein from it which he called "myogen A." It is now clear that myoalbumin was identical with extracellular serum albumin, and that myogen A (when prepared from rabbit muscle) consisted of cocrystallized $\alpha$-glycerophosphate dehydrogenase ($\alpha$GPDH) and aldolase (ALD) (Baranowski and Niederland, 1948). The latter was the main component, but became largely inactivated during the preparation. Baranowski (1940) noted that different crystalline forms were obtained from different species, and it has since been shown that pig myogen A is a different enzyme, lactate dehydrogenase (LDH) (Scopes, unpublished observations).

The term "myogen" has also been used to describe the whole protein extracted at low ionic strength from muscle; see e.g., the excellent review by Czok and Bücher (1960) on the fractionation of myogen by crystallization. In addition, myogens I and II from fish sarcoplasms (Henrotte, 1952, 1960) have been crystallized; it has since been shown (Pechère and Focant, 1965; Gosselin-Rey, Hamoir, and Scopes, 1968) that neither is

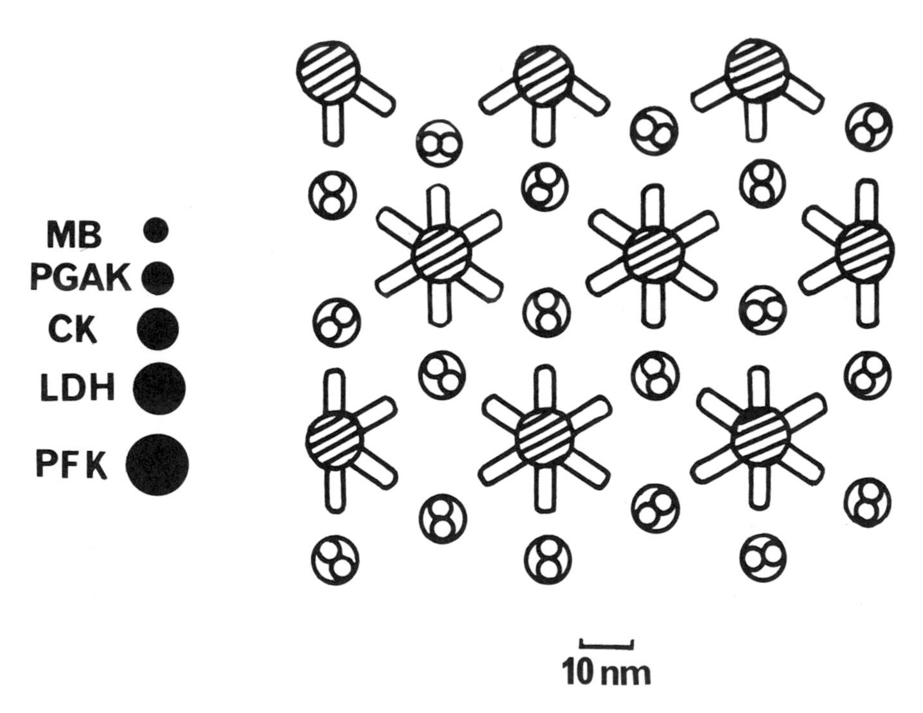


*Fig. 22.1.* Cross-sectional diagram through the A band of muscle, showing sizes of filaments and interacting myosin "feet" approximately to scale. Also shown to scale size, assuming perfectly spherical conformations, are some proteins of the sarcoplasm. *MB,* myoglobin; *PGAK,* phosphoglycerate kinase; *CK,* creatine kinase; *LDH,* lactate dehydrogenase; *PFK,* phosphofructokinase.

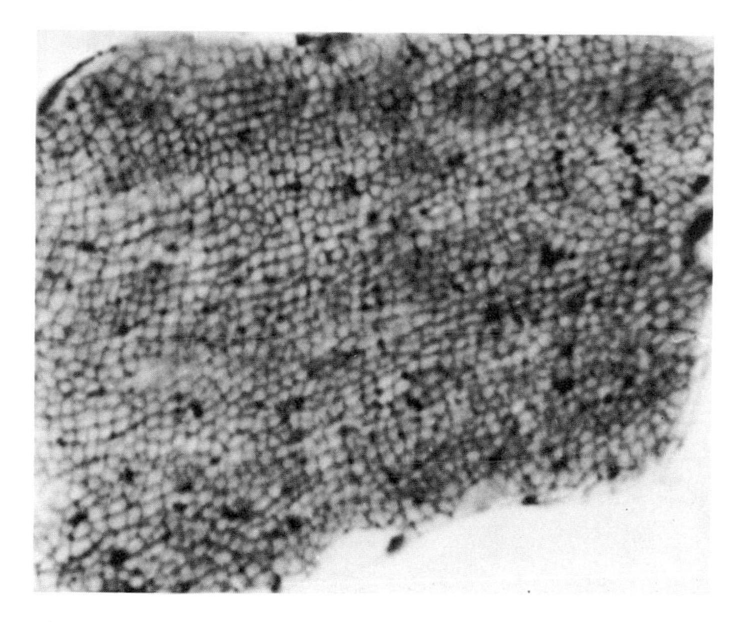

*Fig. 22.2.* Light micrograph of cross section of muscle, stained specifically for sarcoplasmic reticulum. Enclosed spaces represent myofibrils, and dark lines the sarcoplasm. × 2,000.

related to the myogen A of rabbit. It would be helpful if the use of the word myogen were discontinued.

Extracts of muscle made at low ionic strength, with the *p*H at or above 6.8, and preferably made with an isoosmotic extractant, contain up to one-third of the muscle protein (this is later discussed). Most of this protein originates from the extrafibrillar spaces within the muscle fibers, where it exists *in vivo* as a very concentrated solution. It seems improbable that much, if any, of the water-soluble protein exists within the myofibrils, because most of the sarcoplasmic protein molecules are of such a size that, even if assumed to be perfectly spherical, they could only just be fitted between the myosin filaments and heads and the actin-tropomyosin thin filaments (Fig. 22.1). If this were the case, they would severely restrict the efficiency of the contractile process because of the necessity of their being pushed about by the various "feet" of the contractile proteins.

Assuming that the sarcoplasmic proteins are excluded sterically from the myofibrils, the myofibrillar surfaces must then act as an osmotic membrane (even if there is no other membrane completely sheathing the myofibrils), and the water concentration should be similar both within and without. Myofibrils *in situ* consist of about 16% protein and 84%

water, a considerable proportion of which is bound water. Thus 16% is the lower limit for the concentration of the sarcoplasmic proteins and any other dissolved macromolecules *in situ*. Micrographs of cross sections of muscle fibers can give an approximate estimate of the extrafibrillar space, although distortion during fixing and selection of meaningful cross sections introduces great errors. Fig. 22.2 indicates that about 75% of the volume of muscle fibers is taken up by the myofibrils, from which it can be calculated that the probable *in situ* concentration of sarcoplasmic protein is 25–30%. *In vitro,* even at this concentration, sarcoplasmic proteins are quite fluid and lack viscosity. Thus we can think of the two classical divisions of muscle proteins, i.e., those soluble at low ionic strength and those soluble only in more concentrated salt solutions, as a division not only of types but also of spatial origin in the muscle cell, the myofibrillar elements being bathed in, but not permeated by, a sea of sarcoplasmic protein.

### QUANTITATIVE DISTRIBUTION OF SARCOPLASMIC PROTEINS

The quantitative distribution of the muscle proteins is shown in Table 22.1. These figures are approximate only, and refer to a muscle of the "fast" type, such as the longissimus dorsi of rabbit or pig. Other types of muscle would of necessity have different distributions of protein because of their different mechanical and metabolic capacities.

It will be noted that out of a total of 190 mg/g protein, 55 mg/g has been classed as "sarcoplasmic." An extract of prerigor muscle made, for instance, with a dilute tris buffer to ensure that the *p*H of the homogenate does not fall much below 7.0, and centrifuged at 15,000–20,000 *g*, can be defined as an extract of sarcoplasmic proteins. But it would also contain,

TABLE 22.1

*Quantities of various proteins in "fast" mammalian muscle (mg/g)*

Total 190

| *Connective and particulate* | *20* | *Myofibrillar* | *115* | *"Sarcoplasmic"* | *55* |
|---|---|---|---|---|---|
| Collagen | 9.5 | Myosin | 65 | Glycolytic enzymes | 40 |
| Elastin | 0.5 | Actin | 25 | Creatine kinase | 5 |
| Mitochondrial | ~5 | Tropomyosin | ~15 | Myoglobin* | 0.2–2 |
| Other particulate | ~5 | Troponin | ~4 | Extracellular | 3–6 |
| | | α-actinin | ~3 | Others | 8–12 |
| | | Others | ~8 | | |

* Myoglobin values are greater in larger species and older individuals.

in addition to the extracellular proteins mentioned before, some small particles, mainly fragments of the sarcoplasmic reticulum. These can be removed by further centrifugation at high speed, or by coagulation at lower *p*H—a procedure which, although it removes some proteins of interest by isoelectric precipitation, is a useful way of clearing the sarcoplasmic extract of particulate material. The amount of particulate material present initially in the extract will depend on several factors, such as the method of homogenization, number of volumes of extractant, etc.; the value given in Table 22.1 is an estimate of the total amount present in muscle.

Most of the truly soluble protein in the extract consists of the glycolytic enzymes; this, and not an interest in muscle *per se,* is the reason that much of the work on the purification of these ubiquitous enzymes has used muscle as a starting material. The glycolytic and associated enzymes are not evenly distributed in quantity in the sarcoplasm (Table 22.2). Indeed it has been estimated that one enzyme alone, GAPDH, accounts

TABLE 22.2

*Relative proportions of glycolytic and associated enzymes*

| Enzyme | mg/g | Optimal activity, units/g at 30° C |
|---|---|---|
| Phosphorylase *b* (PHb) | 2.5 | 70 |
| Phosphoglucomutase (PGM) | 1.5 | 500 |
| Phosphoglucose isomerase (PGI) | 1.0 | 500 |
| Phosphofructokinase (PFK) | 1.0 | 150 |
| Aldolase (ALD) | 6 | 70 |
| Triose phosphate isomerase (TPI) | 2.0 | 1500 |
| α-glycerophosphate dehydrogenase (αGPDH) | 0.5 | 150 |
| Glyceraldehyde phosphate dehydrogenase (GAPDH) | 12 | 1000 |
| Phosphoglycerate kinase (PGAK) | 1.2 | 3000 |
| Phosphoglycerate mutase (PGAM) | 1.0 | 500 |
| Enolase (EN) | 5 | 800 |
| Pyruvate kinase (PK) | 3 | 1500 |
| Lactate dehydrogenase (LDH) | 4 | 2000 |
| Creatine kinase (CK) | 5 | 1000 |
| Myokinase (AMPK) | 0.5 | 500 |
| AMP deaminase (AMPDA) | 0.2 | 200 |
| Myoglobin (MB) | ~0.5 | – |
| "F protein" (F) | 1.5 | ? |

for over 20% of the sarcoplasmic protein in rabbit muscle (Czok and Bücher, 1960). This and the next four or five most abundant enzymes together make up about half the sarcoplasmic protein. There must nevertheless be many hundreds of other enzymes present, mostly in minute quantity, to carry out the normal functions of cellular metabolism in addition to glycolysis. More than a dozen separate enzymes (several of which can exist in two forms) are involved in the interconversions of glycogen and glucose-1-phosphate.

### LARGE-SCALE FRACTIONATION OF SARCOPLASMIC PROTEINS

The 16 enzymes listed in Table 22.2 carry out a variety of types of reaction, and represent in turn a variety of types of protein, though all are essentially globular in form. This variety makes the separation and purification of individual enzymes a relatively easy matter. Even using only one basic separation method, ammonium sulfate fractionation, Czok and Bücher (1960) were able to crystallize separately nine of these enzymes. All the other major enzymes can be purified and crystallized with the aid of other fractionation methods, such as precipitation with organic solvents, heat denaturation, ion-exchange chromatography, and gel filtration. In Table 22.3, approximate fractionation properties of the individual proteins are indicated; these do not necessarily apply to species other than rabbit. It can be seen that no two proteins behave in exactly the same way. Table 22.4 lists briefly schemes for preparing the various enzymes, all of which have been used successfully by the author. All of the proteins mentioned have been obtained in crystalline form, though this does not necessarily imply complete purity, for in some cases cocrystallization of different proteins may occur, as with ALD and αGPDH, or with triose phosphate isomerase (TPI) and phosphoglycerate kinase (PGAK). Even the enzyme itself often consists of two or more protein forms which cocrystallize.

### SMALL-SCALE SEPARATION OF SARCOPLASMIC PROTEINS

The complexity of the mixture of proteins in a sarcoplasmic extract can be readily demonstrated by gel electrophoresis (Giles, 1962; Hartshorne and Perry, 1962; Scopes, 1964, 1968). Although similar results can be obtained using polyacrylamide gel, starch gel has proved superior because of the effect of electroosmosis, which is much greater in starch. Normally this effect is disadvantageous, but with mammalian sarcoplasmic proteins many of the major components are only slightly charged at *p*H values convenient for electrophoresis, and move only slightly in polyacrylamide, with much spreading and overlapping. In starch, the electroosmotic buffer-streaming tries to move these proteins towards the

TABLE 22.3

*Fractionation behavior of rabbit sarcoplasmic proteins*

| Enzyme | Precipitation range | | Stability* at *p*H 5.5, 55 C | Behavior on DEAE cellulose, *p*H 8.0, I = 0.01† | Behavior on CM cellulose, *p*H 6.5, I = 0.01† | Molecular weight |
|---|---|---|---|---|---|---|
| | in $(NH_4)_2SO_4$, *p*H 5.5, at +10 C; % saturation | in acetone, *p*H 6.5, −5 C; % v/v | | | | |
| PHb | 30–40 | 18–30 | U | A | ? | 180,000 |
| PGM | 50–65 | 45–60 | S | A | N | 62,000 |
| PGI | 55–65 | 35–45 | U | N | WA | 130,000 |
| PFK | 35–50 | 15–25 | U | SA | ? | 360,000 |
| ALD | 45–55 | 30–40 | S | A | A | 160,000 |
| TPI | 60–80 | 45–60 | S | A | N | 45,000 |
| αGPDH | 50–60 | Inactivated | S | SA | A | 60,000 |
| GAPDH | 65–80 | 30–45 | U | N | A | 145,000 |
| PGAK | 60–75 | 45–60 | S | N | A | 36,000 |
| PGAM | 50–70 | 35–45 | S | N | WA | 65,000 |
| EN | 60–75 | 35–45 | U | N | A | 82,000 |
| PK | 55–65 | 25–40 | S | N | A | 240,000 |
| LDH | 50–60 | 25–35 | S | N | A | 145,000 |
| CK | 60–80 | 35–45 | U | WA | N | 82,000 |
| AMPK | 60–75 | 45–60 | U | N | WA | 21,000 |
| AMPDA | 30–50 | 10–20 | U | SA | A | 300,000 |
| MB | 70–90 | 45–60 | U | A | N | 18,000 |
| F | 55–65 | 45–60 | U | N | A | 35,000 |

* S = stable, U = unstable.

† A = adsorbed, W = weakly, S = strongly, N = not adsorbed.

TABLE 22.4

*Purification procedures for enzymes from rabbit muscle sarcoplasm**

| | |
|---|---|
| PHb | AS 0–40%; *p*H 8.8, 37° C; X low ionic strength + MgAMP (Fischer and Krebs, 1958) |
| PGM | Acetone 45–60%; AS 0–65%; *p*H 5.5, 55° C; X in AS |
| PGI | AS 55–65%; acetone 35–45%; CM cellulose; X in AS |
| PFK | Warm isopropanol; cold isopropanol 9–18%; DEAE cellulose; X in AS + ATP (Ling, Marcus, and Lardy, 1965; Parmeggiani, Luft, Love, and Krebs, 1966) |
| ALD | AS 45–55%; X in AS; re-X, see Beisenherz, Bücher, and Garbade, 1955 |
| TPI | Acetone 45–60%; AS 65–80%; *p*H 5.5, 55° C; DEAE cellulose; X in AS |
| αGPDH | AS 50–60%; X in AS; separate ALD X; re-X, see Beisenherz, Bücher, and Garbade, 1955 |
| GAPDH | Remove *p*H 5.4 AS to 75%; *p*H to 8.0, AS to 80%; X in AS |
| PGAK | AS 60–75%; *p*H 5.4, 55° C; CM cellulose; Sephadex G-100; X in AS (Scopes, 1969) |
| PGAM | Acetone 35–45%; *p*H 6.2, 65° C; X in AS |
| EN | Acetone 35–45%; AS 60–75%; X in AS |
| PK | AS 55–65%; Ethanol 12–30%; *p*H 7.0, 60° C; X in AS (Tietz and Ochoa, 1958) |
| LDH | Acetone 0–35%; AS 50–60%; *p*H 5.5, 50° C; X in AS |
| CK | Acetone 35–45%; DEAE cellulose; X in AS or ethanol |
| AMPK | Acetone 45–60%; Acid *p*H 2.5; AS 60–80%; Sephadex G-75; X in AS |
| AMPDA | Adsorb with phospho-cellulose, desorb. X in KCl (Smiley, Berry, and Suelter, 1967) |
| MB | Acetone 45–60%; AS 70–90%; X in AS |
| F | Acetone 45–60%; CM cellulose; X in AS (Scopes, 1966) |

* AS = ammonium sulfate (% saturation range); X = crystals

cathode, but the resistance of the gel opposes this process, thereby enhancing separation. In addition, ALD and GAPDH are partially absorbed in the starch gel, and migrate very little compared with LDH and phosphoglucose isomerase (PGI), which have similar sizes and charges. The results are not ideal, and there is much spreading and streaking of the cathode-migrating bands, compared with the clean, sharp bands on the anode side. Even this can be advantageous, for the shape of a band is often characteristic for a given enzyme, as well as its migration position.

The system used for starch gel electrophoresis incorporates a vertical ice-cooled apparatus operated at low ionic strength (about 0.01); the cooling allows a potential gradient of 20 v/cm in a gel of 6 mm thickness. Other details of the system have been described elsewhere (Scopes, 1968). The separation of samples of whole sarcoplasmic extract is shown in Fig. 22.3, where "fast" and "slow" ("white" and "red") muscles of rabbit are compared. Of interest is the much greater concentration of the major glycolytic enzymes on the cathode side in the fast muscle, and the strong

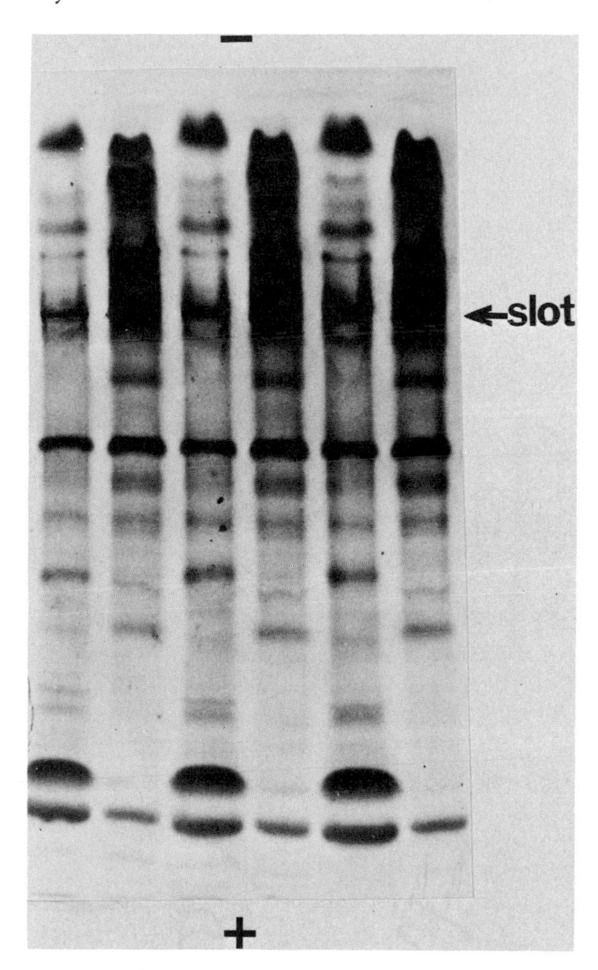


*Fig. 22.3.* Starch gel electrophoretic separation of rabbit muscle sacroplasmic proteins (*p*H 8.2). From left, slots 1, 3, 5, semitendinosus muscle; slots 2, 4, 6, psoas muscle.

band of myoglobin (MB) in the slow muscle. Most differences are quantitative only, but a strongly staining cathodic band in the slow muscle has no counterpart in the fast muscle, and does not correspond to any of the 18 proteins listed in Tables 22.2–4. Almost all of the other major bands that are observed on starch gel electrophoresis correspond to these 18 proteins, together with serum albumin, which is the fastest major anodic band.

## COMPARATIVE BIOCHEMISTRY OF SARCOPLASMIC PROTEINS

So far, we have been talking about sarcoplasmic proteins in general, and those of the rabbit in particular, but when one starts to examine the sarcoplasmic proteins of other species, it is apparent that the story needs

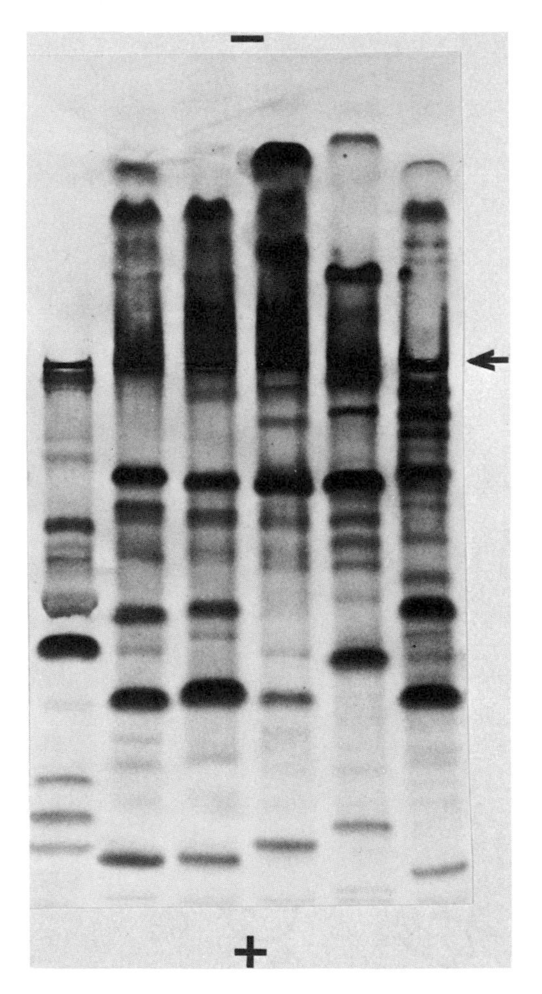


*Fig. 22.4.* Starch gel electrophoretic separation of sacroplasmic proteins of various species. From left to right, crayfish (*Jassus lalandii*), sheep, ox, pig, rat, kangaroo.

modification for each individual species. The wide variation which occurs in the starch gel electrophoretic behavior is demonstrated in Figs. 22.4 and 22.5. Even closely related species such as sheep and ox differ consistently enough and sufficiently to enable an unknown sample to be identified by starch gel electrophoresis. Gel electrophoresis separates both by charge and by size of the proteins. For a given enzyme, significant species variation in molecular size is rare, and comparisons even between rabbit muscle and yeast enzymes often indicate differences of molecular weight not more than 20%. As an example, some recent studies of αGPDH can be quoted. The enzyme prepared from either bee or rat muscle had a molecular weight of only 58,000 compared with the value of 78,000 reported for rabbit (Fondy, Levin, Sollohub, and Ross, 1968).

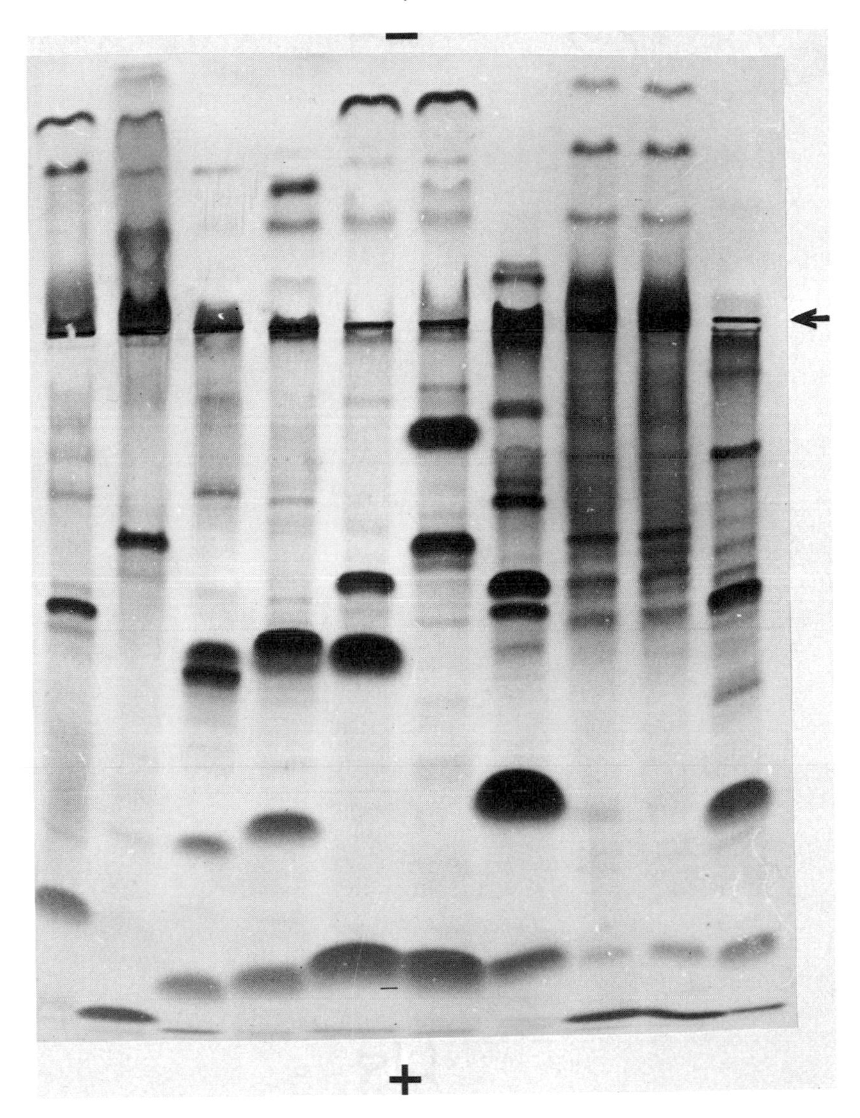


*Fig. 22.5.* Starch gel electrophoretic separation of sarcoplasmic proteins from several fish species. From left to right, dogfish (*Scyllium canicula*), tope (*Galeorhinus galeus*), carp (*Cyprinus carpio*), tench (*Tinca tinca*), roach (*Rutilus rutilus*), chub (*Leuciscus cephalus*), pike (*Esox lucius*), trout (*Salmo fario*), trout, codling (*Gadus callarias*).

This difference was large enough to cast doubt on the value for the rabbit enzyme, and an intensive investigation has in fact revised the figure downwards to little more than 60,000 (Fondy, Ross, and Sollohub, 1969).

Thus it can be stated with reasonable certainty that where the mobility of an enzyme of one species differs from that of another, the difference is mainly due to charge, not to size. This also applies to secondary bands of main components and to polymorphic manifestations, which will be discussed later.

The bewildering variety of patterns obtained by starch gel electrophoresis of sarcoplasmic proteins from different species (e.g., Figs. 22.4 and 22.5) can be sorted out quite easily by using specific staining methods for the various enzymes. It is possible to determine the positions of a dozen different enzymes in a dozen different species—in about that number of days. The methods used (Scopes, 1968) are based on normal assay procedures, involving coupling of the enzyme's activity to other enzymes, and ending in a dehydrogenase which either oxidizes or reduces nicotinamide nucleotide. The dye nitro blue tetrazolium is soluble and yellow in the oxidised form, but insoluble and purple when reduced. NAD (P) H will transfer hydrogen to nitro blue tetrazolium in the presence of a carrier (phenazine methosulphate), and so the site of enzyme activity is revealed. Conservation of expensive reagents and coupling enzymes is achieved by using filter paper as a carrier of the reaction mixture; the reduced dye is then conveniently precipitated within the filter paper pores and the paper can be washed and stored for future reference. In cases where the enzyme being sought manifests itself by oxidation of added NADH, the filter paper is removed after reaction and treated with nitro blue tetrazolium to detect unchanged NADH; the enzyme position then appears as a clear area on a purple background. In other cases, where NAD (P) H is formed by the reactions, nitro blue tetrazolium is included in the mixture and the purple bands can be observed as they develop. Some results of these methods are shown in Fig. 22.6, for an extreme range of species. Fig. 22.7 shows the positions of all the various enzymes detected in pig and rabbit muscles.

When it comes to large-scale fractionation of the sarcoplasmic proteins from species other than rabbit, the applicability of various methods cannot be generalized (apart from gel filtration separations, which depend on molecular size only). In some cases, the rabbit method will work equally well with many other species, e.g., for GAPDH. In particular, the method of Tietz and Ochoa (1958) for pyruvate kinase (PK) actually produces large crystals of the enzyme from pig muscle more readily than from rabbit. In other cases, a method may produce crystals of a different enzyme altogether, as previously mentioned in connection with "Myogen A." As a final example, a preparation procedure for PGAK recently

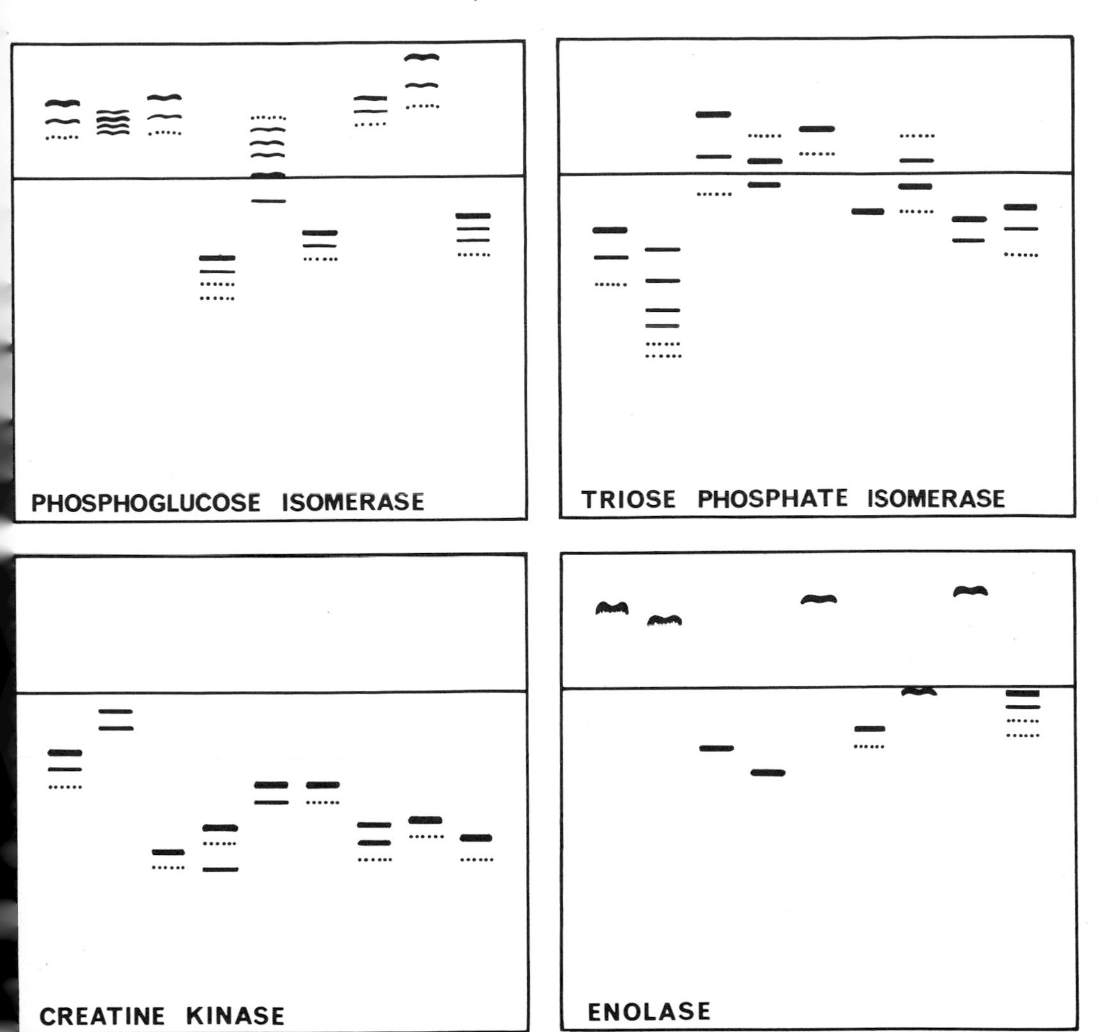


*Fig. 22.6.* The starch gel electrophoretic mobilities of some enzymes from a variety of species. The species compared are, from left to right, pig, kangaroo, carp, conger eel (*Conger conger*), tope, tortoise (*Testudo graeca*), frog (*Rana temporaria*), adder (*Vipera berus*), crayfish.

developed (Scopes, 1969), originally for pig muscle, gave equally good preparations with rabbit muscle, although the penultimate fraction was less pure. But a preparation from ox was less successful (Fig. 22.8). It is clear that in a different species the behavior both of the enzyme being

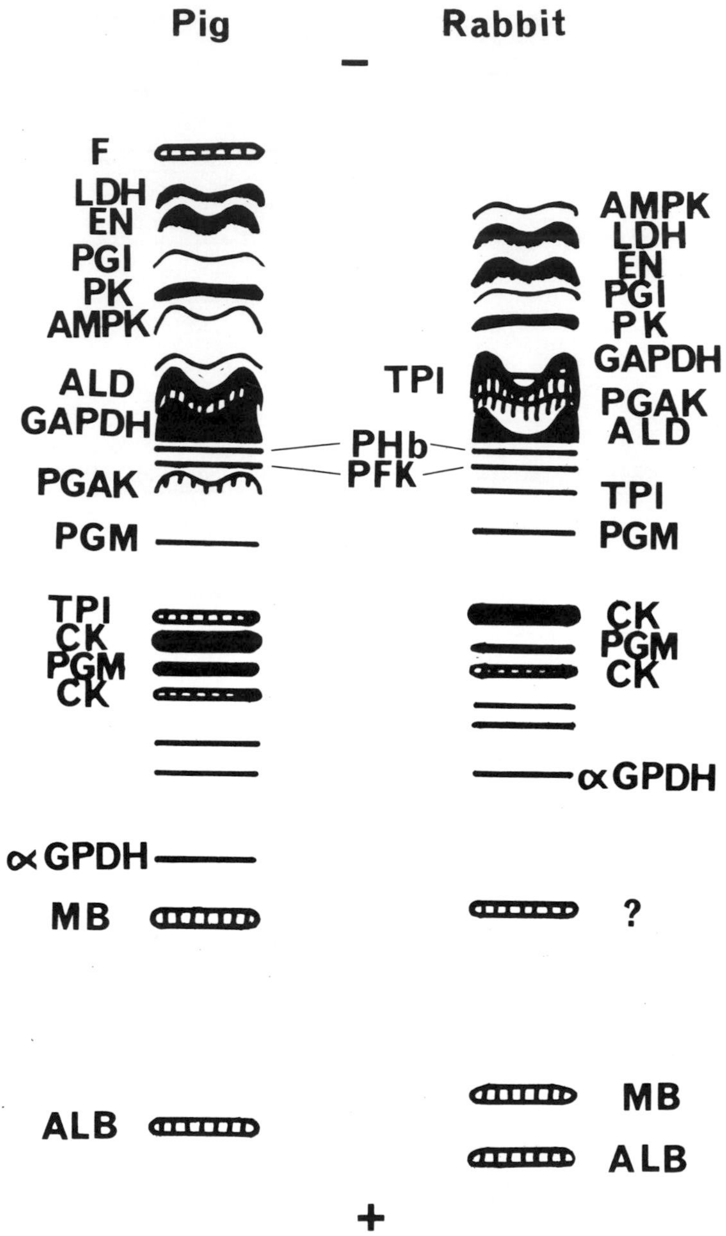

*Fig. 22.7.* Comparative mobilities of the main proteins in the sacroplasmic extracts of pig and rabbit (fast) muscles. For abbreviations, see Table 22.2.

prepared and the proteins being separated from it may be so altered as to necessitate a quite different fractionation scheme.

### BIOCHEMICAL GENETICS AND SARCOPLASMIC PROTEINS

The comparisons between species shown in Figs. 22.4–8 are of particular interest in evolutionary studies and, although only of a generalized nature, can be considered as a useful preliminary to more detailed investigations of the primary structures of the enzymes. The methods used enable a great number of comparisons to be made in a short time, without any purification of the enzymes being necessary.

Mutations which become established in a new species, or which occur as variants within an existing species, frequently result in a protein differing only slightly in charge from the original. Amino acid substitutions which do not lead to any change in charge, size, or shape cannot, of course, be detected by gel electrophoresis; identity of protein structure can then only be proved by exhaustive sequence studies of the purified enzyme. However, this does not necessarily mean that one could not detect many of the possible amino acid substitutions which lead to no net change in charge in themselves, e.g., Ala → Leu, Glu → Asp, Ser → Gly. Such substitutions may, by a variety of means such as the alterations in neighboring hydrogen and hydrophobic bonds, change the electron environment around a nearby histidine or other residue which is only partially ionized at the *p*H of electrophoresis. This would alter its acid dissociation constant, and so cause a slight change in the net charge at a given *p*H. Detectable variation in mobility between closely related species may be equivalent to only 0.1 charge in the whole protein, which is commonly the case, for instance, with mammalian creatine kinase (CK). In Fig. 22.9, the mobilities of CK from several mammalian species are shown, together with the mobilities of rabbit CK, partially and totally reacted with iodoacetamide. It has been shown (Watts, 1964) that the two active sulfydryls in CK are fully ionized at *p*H 8, because reaction of them with iodoacetamide causes a decrease in electrophoretic mobility, whereas iodoacetate has no effect on the mobility. Using the starch gel electrophoretic system described previously, it has been found that fully iodoacetamide-reacted CK has a mobility 10 mm slower than unreacted enzyme and 5 mm slower than enzyme reacted with only one iodoacetamide molecule. From this one concludes that a 5-mm increase in (anodic) mobility of a molecule the size of CK represents one more negative charge per molecule. Thus the variations in mobility between the CKs of, for example, ox and sheep, or rabbit and rat, which are no more than 1 mm, can represent at most only 0.2 charge difference per molecule. All of these variations could be the result of amino acid

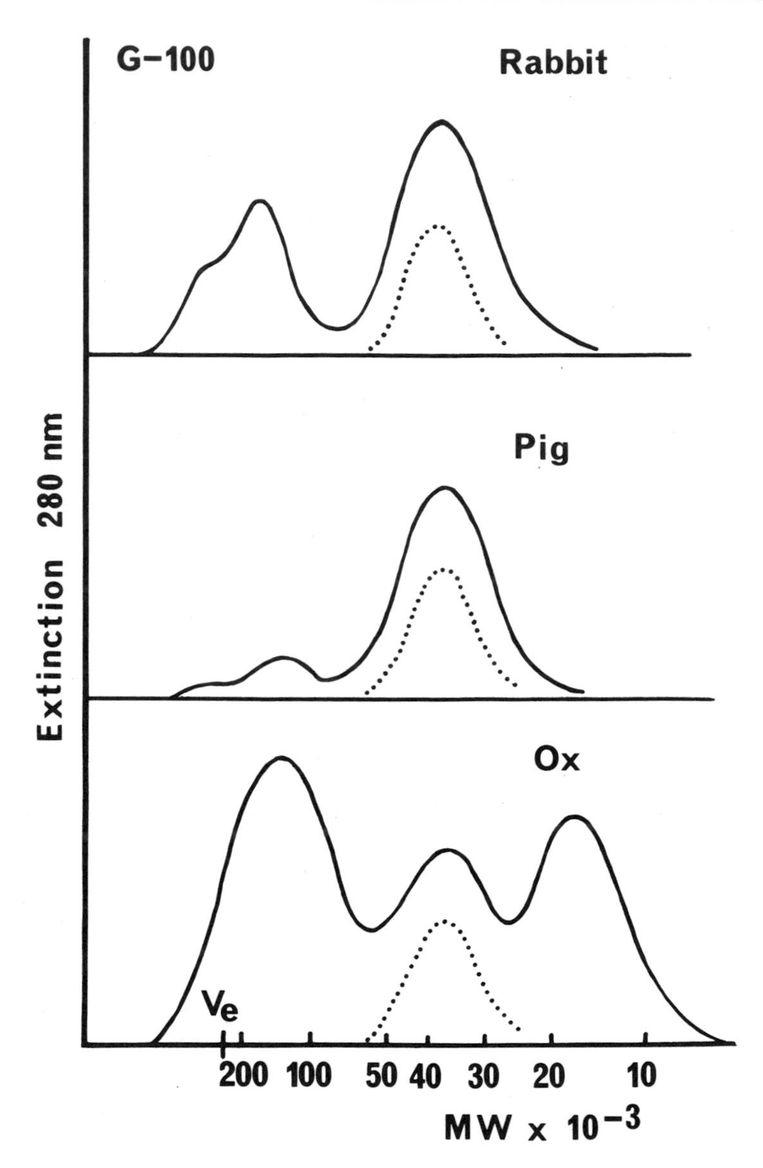


*Fig. 22.8.* Preparation of phosphoglycerate kinase (Scopes, 1969). Behavior of penultimate fraction on Sephadex G–100 column: *solid line,* protein; *dotted line,* enzyme activity.

substitutions involving histidine residues directly, but it seems more probable that some neutral substitutions have occurred, affecting the ionization of nearby histidines, as described above.

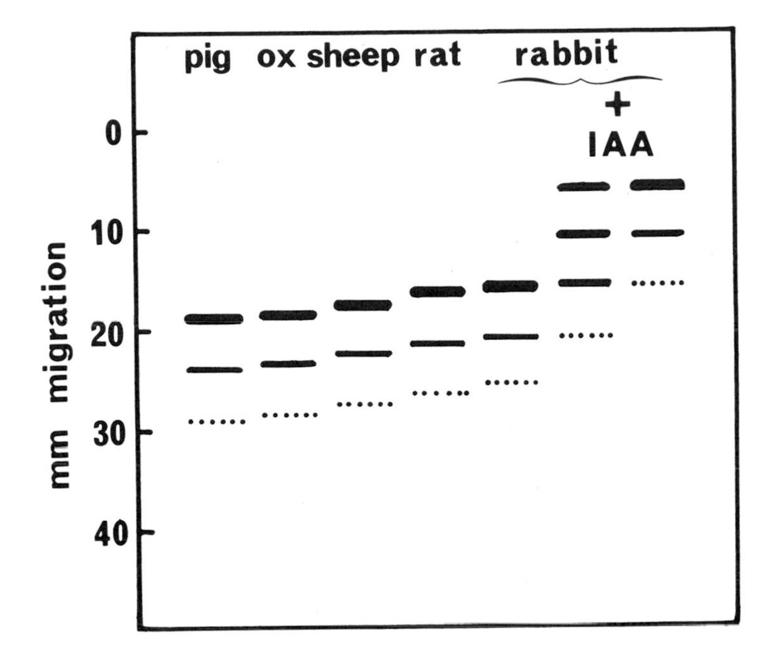


*Fig. 22.9.* Starch gel electrophoretic mobilities of creatine kinase from higher mammals. From left to right, pig, ox, sheep, rat, rabbit, purified rabbit enzyme partially reacted with iodacetamide; purified rabbit enzyme totally reacted with iodacetamide.

Many purified enzymes do not give a single band on starch gel electrophoresis, but rather one main band with minor components, as demonstrated in Figs. 22.6, 22.9. A particularly common occurrence is for the minor components to be more anodic than the major, equally spaced, and decreasing in amount away from the major band. These are generally not artifacts of the purification procedure, for specific staining for the enzymes after electrophoresis of freshly prepared whole extracts gives the same pattern. An extreme example of this is of rabbit TPI (Scopes, 1968), in which no fewer than eight bands are detectable, equally spaced, and of decreasing intensity towards the anode. The spacing between the bands is greater with smaller proteins, and from the results with CK in Fig. 22.9, this spacing is typically equivalent to just one negative charge per molecule. These components can be isolated individually and shown to run as single bands in their respective positions. The question arises as to what these minor components represent. Are they genetically determined as different protein molecules, do they perhaps represent a small proportion of mistakes in the reading of the genetic code, or do they originate from the major component by changes occurring after the protein has been synthesized? The first of these possibilities seems unlikely, since it would necessitate not only separate genes for the production of several molecular forms, but also that these genes be suppressed by varying amounts. In addition, one would expect a greater variety of patterns, rather than the regular spacing usually observed. The second possibility, that of occasional misreading of the genetic code, fits in well with the observed patterns, but one would have to postulate that a certain mistake, resulting in a net increase of one charge in the incorrect molecule (e.g., Val → Glu), was more common than others. Some other facts are difficult to reconcile with this simple explanation.

The third possibility, in which the variants are formed after completion of polypeptide synthesis, is more readily acceptable. Causes could be: (a) proteolysis, a shortening of the polypeptide chain with the loss of charged groups; (b) alteration of individual residues by, for instance, loss of amide ammonia from asparagine or glutamine, or acetylation of exposed lysine amino groups; (c) conformational variants in the tertiary structure, leaving the primary structure unchanged. Cause (a), proteolysis, is improbable, since there would be about as much chance of a net positive gain as there is of the observed almost invariable negative charge gain of the minor components compared with the major one. Cause (b) seems to fit the observations well. However, some careful analysis of amino acids (including ammonia) by Åkeson and Theorell (1960) did not show up any differences in composition between three separated components of horse myoglobin, although tryptic peptide maps differed slightly (acetylation of lysine residues, perhaps under the influence of an acetyl transferase in the sarcoplasm, would not result in any differences in amino acid composition). Cause (c), conformational variation, does not seem to be very likely in the absence of any evidence to favor it, since enzyme activities are greatly dependent on conformation, and the minor components always seem to stain enzymically with an intensity proportional to their protein staining, suggesting identical specific activities despite any other difference. It has been demonstrated, however, with chicken mitochondrial malate dehydrogenase, which exhibits five bands on starch gel electrophoresis, that the differences are in fact conformational (Kitto, Wassarman, and Kaplan, 1966). It was found that the major (i.e., most cathodic) band was thermodynamically the most stable, and contained the most helical structure. The most anodic band could, by reversible acid dissociation, be converted to more cathodic bands, or, by treatment with iodine, which had the effect of opening up the structure, be converted to even more anodic bands. There seems to be no reason why the same would not apply in many cases to the sarcoplasmic enzymes: the minor components would be "conformers," in which an error in tertiary folding has led to a structure with less helix and more negative charge, due to the release of otherwise buried bound protons. But not all cases can be explained in this way, and some examples may be quite complex (see Kaplan, Teeple, Shore, and Beutler, 1968; Dawson and Mitchell, 1969).

Polymorphisms, in which protein variants are clearly linked to genetic mutations, and which can ultimately be demonstrated to be due to amino acid substitutions, are not common in mammalian sarcoplasmic proteins. This makes the electrophoretic fingerprint of the sarcoplasm more useful in evolutionary studies than that of serum proteins, where polymorphism is common. Nevertheless, protein variants have been found in sarcoplasmic extracts, particularly in fish, where polymorphism seems to be commoner. Among the fish enzymes showing the phenomenon are LDH and aspartate transaminase (Odense, Allen, and Leung, 1966), CK (Scopes and Gosselin-Rey, 1968), and enolase (EN) and phosphoglucose isomerase (PGI) (Scopes, unpublished observations). In mammals, more work has been carried out on erythrocyte extracts, and often a polymorphism discovered there will also apply to the corresponding muscle protein, as for example with human myokinase (AMPK) (Fildes and Harris, 1966). PGI in mouse (Carter and Parr, 1967) and phosphoglucomutase (PGM) in human (Spencer, Hopkinson, and Harris, 1964) red cells also show simple polymorphism.

## SARCOPLASMIC PROTEINS AND POST MORTEM METABOLISM

Lactate is produced from glycogen in anaerobic post mortem muscle through the action of the sarcoplasmic enzymes. The rate at which lactate accumulates is, for a normal muscle at 37° C, about 0.3 $\mu$moles/g per min, and for the fastest PSE muscle, about 5 $\mu$moles/g per min. Nevertheless, the potential rate of lactate production (the potential maximum activities of those enzymes in the smallest quantity as determined by activity, i.e., phosphorylase (PH), phosphofructokinase (PFK), and ALD), is about 100 $\mu$moles/g per min. Thus it is clear that the absolute amount of the glycolytic enzymes is less important in determining the rate than the factors controlling the rate-limiting enzymes. These factors are principally the available levels of adenosine mono- and diphosphates, which in turn are determined by the rates of the adenosine triphosphatases acting in the muscle. The metabolic aspects of post mortem muscle are described elsewhere, but the significance of another sarcoplasmic protein, AMP deaminase (AMPDA), should not be overlooked. This enzyme, by deaminating AMP faster the higher the AMP concentration, helps to control the AMP level, so its activity is important in relation to the control of PH and PFK. AMPDA may also have an important role in determining the ultimate *p*H of muscle. The enzyme is readily absorbed on particulate material, and is usually found in association with myofibrils and reticulum as well as in sarcoplasmic extracts; but it seems that it must exist *in vivo* mainly as a soluble protein.

Truly soluble adenosine triphosphatases probably do not exist in the sarcoplasm, but there are several possible combinations of enzymes present which, by acting in a cyclic fashion, like PFK and fructose diphosphatase, or by bypassing a normal ATP synthesis step, like diphosphoglycerate mutase (PGAM) and 2,3-diphosphoglycerate phosphatase (Rapoport, 1968), result in an apparent adenosine triphosphatase activity in the sarcoplasm under post mortem conditions. These could be the main adenosine triphosphatase systems operating post mortem, but with several other particulate adenosine triphosphatases present it is not yet possible to say whether the soluble systems are of any importance.

Tenderization during aging of meat has been ascribed to the action of cathepsins, soluble proteinases in the sarcoplasm of meat, though originating from the lysosomes, and present in only very small amounts compared with the glycolytic enzymes. Another proteolytic enzyme has recently been isolated as a phosphorylase-*b*-kinase-activating factor, though it is not specific for this reaction, and will digest a number of different proteins, at neutral *p*H (Huston and Krebs, 1968).

Mention must also be made of MB, the quantity and color of which are of great importance to the appearance of meat. This pigment varies in quantity, being lowest in fast muscles, young individuals, and small species. But the greatest quantities of MB are found in sea mammals, in which it functions as an oxygen store during deep dives. In the sperm whale, the MB content is as high as 5% by weight, about 70% of the total sarcoplasmic proteins. The other extremes, more familiar at mealtimes, are the pectoral muscles of young broiler fowl, and most species of fish, which have virtually no MB.

It is clear that many of the proteins of the sarcoplasm are important in affecting the quality and appearance of meat. Not only the major protein components, but also trace quantities of certain enzymes can have beneficial or detrimental effects on various meat products.

## *References*

Åkeson, Å., and H. Theorell. 1960. On the microheterogeneity of horse myoglobin. *Arch. Biochem. Biophys. 91*:319.

Amberson, W. R., A. C. Bauer, D. E. Philpott, and F. Roisen. 1964. Proteins and enzyme activities of press juices, obtained by ultracentrifugation of white, red and heart muscles of rabbit. *J. Cell. Comp. Anat. 63*:7.

Baranowski, T. 1939. Crystallizable proteins of extract of rabbit muscle. (In French.) *Compt. Rend. Soc. Biol. 130*:1182.

———. 1940. Crystalline myogen. (In Russian.) *Biokhimiya 5*:174.

Baranowski, T., and T. R. Niederland. 1948. Aldolase activity of myogen A. *J. Biol. Chem. 180*:543.

Bate-Smith, E. C. 1937. Native and denatured Proteins. *Proc. Roy. Soc. B 124*:136.

Beisenherz, G., Th. Bücher, and K.H. Garbade, 1955. α-glycerophosphate dehydrogenase from rabbit muscle, p. 391. *In* S. P. Colowick and N. O. Kaplan (eds.), *Methods in Enzymology*, Vol. 1. Academic Press, New York.

Carter, N. D., and C. W. Parr. 1967. Isoenzymes of phosphoglucose isomerase in mice. *Nature 216*:511.

Czok, R., and Th. Bücher. 1960. Crystallized enzymes from the myogen of rabbit skeletal muscle. *Advance. Protein Chem. 15*:315.

Dawson, D. M., and A. Mitchell. 1969. The isoenzymes of phosphoglucomutase. *Biochemistry 8*:609.

Fildes, R. A., and H. Harris. 1966. Genetically determined variation of adenylate kinase in man. *Nature 209*:261.

Fischer, E. H., and E. G. Krebs. 1958. The isolation and crystallization of rabbit skeletal muscle phosphorylase *b*. *J. Biol. Chem. 231*:65.

Fondy, T. P., L. Levin, S. J. Sollohub, and C. R. Ross. 1968. Structural studies on nicotinamide adenine dinucleotide-linked L-glycerol 3-phosphate dehydrogenase crystallized from rat skeletal muscle. *J. Biol. Chem. 243*:3148.

Fondy, T. P., C. R. Ross, and S. J. Sollohub. 1969. Structural studies on rabbit muscle glycerol 3-phosphate dehydrogenase and a comparison of chemical and physical determinations of its molecular weight. *J. Biol. Chem. 244*:1631.

Giles, B. G. 1962. Species differences observed in the sarcoplasmic proteins of mammalian muscle. *J. Sci. Food Agric. 13*:264.

Gosselin-Rey, C., G. Hamoir, and R. K. Scopes. 1968. Localization of creatine kinase in the starch-gel and moving-boundary electrophoretic patterns in fish muscle. *J. Fish. Res. Bd. Can. 25*:2711.

Hartshorne, D. J., and S. V. Perry. 1962. A chromatographic and electrophoretic study of sarcoplasm from adult- and foetal-rabbit muscles. *Biochem. J. 85*:171.

Henrotte, J. G. 1952. A crystalline constituent from myogen of carp muscles. *Nature 169*:968.

———. 1960. Contribution à l'étude des myogènes de carpe et de plie. *Biochim. Biophys. Acta 39*:103.

Huston, R. B., and E. G. Krebs. 1968. Activation of skeletal muscle phosphorylase kinase by $Ca^{++}$ II. Identification of the kinase activating factor as a proteolytic enzyme. *Biochemistry 7*:2116.

Jacob, J. J. C. 1947. The electrophoretic analysis of protein extracts from striated rabbit muscle. *Biochem. J. 41*:83.

Joyce, B. K., and S. Grisolia. 1959. The purification and properties of muscle diphosphoglycerate mutase. *J. Biol. Chem. 234*:1330.

Kaplan, J. C., L. Teeple, N. Shore, and E. Beutler. 1968. Electrophoretic abnormality in triosephosphate isomerase deficiency. *Biochem. Biophys. Res. Comm. 31*:768.

Kitto, G. B., P. M. Wassarman, and N. O. Kaplan. 1966. Enzymically active conformers of mitochondrial malate dehydrogenase. *Proc. Nat. Acad. Sci. U.S. 56*:578.

Ling, K.-H., F. Marcus, and H. A. Lardy. 1965. Purification and some properties of rabbit skeletal muscle phosphofructokinase. *J. Biol. Chem. 240*:1893.

Odense, P. H., T. M. Allen, and T. C. Leung. 1966. Multiple forms of lactate dehydrogenase and aspartate aminotransaminase in herring. *Can. J. Biochem. 44*:1319.

Parmeggiani, A., J. H. Luft, D. S. Love, and E. G. Krebs. 1966. Crystallization and properties of rabbit skeletal muscle phosphofructokinase. *J. Biol. Chem. 241*:4625.

Pechère, J.-F., and B. Focant. 1965. Carp myogens of white and red muscles. *Biochem. J. 96*:113.

Rapoport, S. 1968. The regulation of glycolysis in mammalian erythrocytes, p. 69. *In* P. N. Campbell and G. D. Greville (eds.), *Essays in Biochemistry,* Vol. 4. Academic Press, London.

Scopes, R. K. 1964. The influence of post-mortem conditions on the solubilities of muscle proteins. *Biochem. J. 91*:201.

———. 1966. Isolation and properties of a basic protein from skeletal-muscle sarcoplasm. *Biochem. J. 98*:193.

———. 1968. Methods for starch-gel electrophoresis of sarcoplasmic proteins. *Biochem. J. 107*:139.

———. 1969. Crystalline 3-phosphoglycerate kinase from skeletal muscle. *Biochem. J. 113*:551.

Scopes, R. K., and C. Gosselin-Rey. 1968. Polymorphism in carp muscle creatine kinase. *J. Fish. Res. Bd. Can. 25*:2715.

Smiley, K. L., A. J. Berry, and C. H. Suelter. 1967. An improved purification, crystallization, and some properties of rabbit muscle 5′-adenylic acid deaminase. *J. Biol. Chem. 242*:2502.

Spencer, N., D. A. Hopkinson, and H. Harris. 1964. Phosphoglucomutase polymorphism in man. *Nature 204*:742.

Tietz, A., and S. Ochoa. 1958. "Fluorokinase" and pyruvic kinase. *Arch. Biochem. Biophys. 78*:477.

Watts, D. C. 1964. Molecular weight determination of creatine kinase from the reaction with iodoacetamide. *Abstr. 1st Meet. Fed. Europe. Biochem. Soc.* (London) A 13.

Weber, H. H., and K. Meyer. 1933. Das kolloidale Verhalten der Muskeleiweisskörper 5. Das Mengenverhältnis der Muskeleiweisskörper in seiner Bedeutung für die Struktur des quergestreiften Kaninchenmuskels. *Biochem Z. 266*:137.

# Chapter 23

# *Summary and Discussion of Part 5*

PANEL MEMBERS: J. R. WHITAKER, *Chairman*
M. W. MONTGOMERY
P. HOPPER
W. LANDMANN
A. MULLINS
J. C. TRAUTMAN

*Whitaker:* Two decades ago, collagen was the protein chemist's nightmare. As it was insoluble, its state of purity was impossible to assess. A few venturesome workers worked half-heartedly on gelatin with the hope it would aid in understanding collagen, and X-ray crystallographers did some work. Then it was learned that tropocollagen is soluble in dilute acids such as acetic acid, and intensive studies were made on this soluble form of collagen, largely obtained from fish. Within the last 8 years, progress in understanding the structure of collagen has been made at an ever-accelerating rate, largely due to workers such as A. Veis, K. A. Piez, and P. M. Gallop. The three-polypeptide structure of the molecule is now clearly established, the two types of polypeptide chains have been successfully separated, the amino acid sequence of these chains is nearly completed, largely because of the successful application of the cyanogen bromide cleavage method, and a great deal is known about the intra- and intermolecular covalent bonds in collagen (those not involving peptide linkages).

The collagen molecule has a molecular weight of 300,000 and is $15 \times 2990$ Å in size. How can such a long molecule be assembled into the ordered structure unique to each of the types of collagen? This organization is not a random one. For example, the physical organization of the collagen molecules is quite different between the soft- and hard-tissue collagens. In soft-tissue collagen, the molecules are aggregated in long parallel bundles. In cornea, the collagen fibrils are stacked parallel in layers with the axes of the fibrils of each layer arranged mutually perpendicular. In hard-tissue collagens, such as bone and dentin collagens, the molecules and fibrils are woven into a highly intertwined network.

After the primary sequence is synthesized in most proteins the only additional covalent linkage added later is the disulfide bond. Not so with the collagens. There is an oxidation of some of the residues of lysine and proline to form hydroxylysine and hydroxyproline. There is the formation of the intra- and intermolecular covalent bonds in collagen. These latter bonds appear to be formed slowly and are found in increasing amounts the older the collagen. There are some 25 lysine residues per 1,000 residues in the collagen polypeptides, yet only one of these lysines is involved in intramolecular cross linkage per polypeptide chain. This lysine must be in a region of the molecule not involved in triple helix formation so that it is readily accessible to oxidative deamination and subsequent Schiff base formation. This region of the polypeptide chain is also quite susceptible to attack by proteolytic enzymes.

There are probably at least four types of intermolecular covalent bonds in the collagen superstructure. These are glutamyl-$\gamma$-carboxyllysyl-$\Sigma$-amino amide bonds, ester bonds involving the carbohydrate moieties, Schiff bases (aldehyde mediated cross linkages), and phosphate di- or triester bonds.

One may well ask about the enzymes involved in these reactions, about the specificity of the reactions, and about the controlling mechanisms of the reactions.

The sarcoplasm of the muscle cell is a rich source of enzymes. Undoubtedly there are hundreds of enzymes present to carry out the necessary metabolic functions of the cell; many of these are present in very small amounts. Other enzymes, particularly some of the glycolytic enzymes, are present in high amounts and are noted for their ease of purification and crystallization. A great deal can be learned about the comparative nature of these enzymes and genetic interrelationships of species by use of gel electrophoretic techniques coupled with specific chromophoric reactions for detecting the enzyme activities in the gel slab. The technique vividly demonstrates the marked differences in the charged properties of a given enzyme among different species. It reinforces what we have all heard in our beginning biochemistry course: that proteins are species-specific. It should also warn us that we cannot identify a given protein from different species merely on the basis of its electrophoretic behavior.

The recognition of multiple molecular forms of an enzyme during separation by gel electrophoresis is both a blessing and a curse. It is interesting and informative as to the nature and number of multiple molecular forms between different organs and between different species.

It may shed light on the molecular structure of the molecule as well as its function and control mechanisms. Yet one must always be concerned about the effect of his preparation and separation techniques in producing artifacts. It was only recently recognized that the same polypeptide chain may exist in different conformers. Since each of these conformers uniquely exposes charged groupings to the surface of the molecule these conformers can be separated by those techniques based largely on charge differences. To have conformers of this type the polypeptide chain must be capable of existing in two or more thermodynamically stable states; that is, the energy barrier separating the conversion of one to the other must be sufficiently high so that the two are not in instantaneous equilibrium with each other. A good check for such conformeric forms is to demonstrate conditions necessary for interconversion of the conformers.

One must ask about the arrangement of the sarcoplasmic proteins in the muscle cell. Are they moving around freely within the cell, excluded from the myofibrillar structure largely because of size? Or is there an internal organization within the cell? Wouldn't it be appropriate to apply *in situ* microtechniques for enzyme activity to properly prepared segments of the cell to determine whether indeed there is an internal organization?. If one is truly to understand the functioning of these enzymes in the muscle cell he must understand their intracellular organization.

## DISCUSSION

*Landmann:* Is it the intra- or intermolecular cross links, or both, which are responsible for the "hardening" of collagen fibers with animal age? Which type do you feel would be of primary significance in increasing toughness of meat?

*Veis:* The intermolecular cross linkages are those responsible for the alterations in the physical properties of the tissues. The role of the intramolecular cross linkages is not clear, in my mind at least, at this time.

*Trautman:* In acid or base swelling of collagen (natured) what are the roles of: hydrogen, intermolecular, and intramolecular bonding?

*Veis:* Intermolecular bonding plays the most important role in regulating swelling of native collagen.

*Landmann:* Are galactose and glucose involved in the cross-link structures, or do they otherwise occur in the primary amino acid chain structure of collagen?

*Veis:* Glucose and galactose are present in collagen as the mono- or disaccharide units gal or gal-glu and are attached to the collagen at a hydroxylysine residue. One such locus of attachment is in the peptide α1 CB5 (William Butler, unpublished observations).

*Hopper:* What effect do forms of energy such as ultrasonics and radiation have on the intermolecular cross linkages in collagen?

*Veis:* Ultrasonic radiation of collagen leads to main-chain degradation and fracture of soluble molecules into shorter rods (one-half or one-quarter length). Irradiation with short-wave-length ultraviolet light leads to either polymerization or degradation depending upon the intensity and duration of the radiation and the presence or absence of oxygen in the system.

*Wipf:* There are several reports concerning the close association of collagen with the mucopolysaccharides. Do these sulfated mucopolysaccharides have any influence upon the type and/or degree of cross linkages in collagen? Are they linked directly to the collagen molecule?

*Veis:* There is no direct evidence that the nature of the glycosaminoglycans in the ground substance are involved in the cross linking of collagen, nor does it appear that they are covalently attached to the collagen. However, as Toole and Lowther (1968) have shown, they do play some role in organizing the collagen fiber systems and the different acidic polysaccharides behave in different ways.

*Mullins:* Do you agree with the report of Mohr and Bendall (1969) that intramuscular collagens are more highly cross-linked than tendonous collagens? If so, can you offer any physiological basis for this?

*Veis:* The data of Mohr and Bendall are very convincing. We know that very closely related collagens are cross-linked to different extents. The physiological basis for these differences, other than the obvious implication that the different types of collagen do have different functions and environments, is not clear to me.

*Mullins:* Would you elaborate briefly on how dermatan sulfate is incorporated into the structure of highly cross-linked collagen molecules since it has been associated specifically to highly cross-linked collagen?

*Veis:* Dermatan sulfate is not incorporated into the structure of highly cross-linked collagen so far as I am aware.

*Mullins:* Can you visualize any relationship between stressor agents (i.e., excitement, fear, and/or starvation) in the animal and the degree of molecular cross linkages in collagen molecules?

*Veis:* In pathological states collagens might well vary in degree of cross linking from the normal situation. I do not know if one would expect emotional stress, particularly for short periods, to have such effects.

*Mullins:* At the present time then you would say that there is no physiological functional requirement?

*Veis:* No, I would say just the opposite. The tissue that has to be highly stabilized with a strong degree of maintenance of form is going to be highly polymerized and highly cross-linked. Those which do not need it apparently are not as extensively cross-linked. So there is a direct physiological connection between cross linking and function.

*Trautman:* Is there sufficient information on aldehyde cross linkages to warrant the development of methods to measure them quantitatively in a muscle system?

*Veis:* The aldehyde content of collagen is low and varies depending upon the tissue. However, it would probably be very useful to determine the aldehyde content and relate it to the physical properties of the tissue.

*Hopper:* You have indicated that the relative amounts of the $\alpha$, $\beta$, and $\gamma$ components of collagen from a specific tissue will vary with the age of the host organism. Would you expect these amounts to vary between animals of the same age, but from different breeds within the same species? For example, is bovine corium collagen from a Hereford animal the same as that from an Angus of the same chronological age?

*Veis:* I would expect some variations in cross linkages of a particular type of collagen from animal to animal and from species to species. This is probably the reason, in my own experience, that there is a variation in yields of soluble collagen in animals presumably of the same age but not litter mates.

I am affiliated with a Veterans Administration laboratory which is studying the problem of aging. We have examined human skin collagen from the back of the neck which is exposed and from the stomach which is not exposed. We have used the range from fetal tissue to men 80 years old. Within a given individual the collagen from one part is identical, as far as we can tell, with that from another. From early stages of growth on, there is no apparent change in degree of polymerization. The increase in polymerization, usually called age-related, begins at a very early stage, right at the end of growth, and for 60 years thereafter there is very little change. There are larger changes in the glycosaminoglycan content in the tissues.

*Montgomery:* Would the higher proline and hydroxyproline content of $\alpha 1$ cause this component to be more rigid than $\alpha 2$?

*Veis:* It is difficult to say that the rigidity of a single chain is affected since the individual chains exist only in the denatured state. However, renatured collagen from pure α1 chains has a higher melting temperature than a renatured α2 collagen.

*Montgomery:* Please comment on the processes involved in the "aging" of collagen, since the β and γ components increase after collagen is synthesized.

*Veis:* The production of the β and γ components may not be parallel events. Recent evidence suggests that although the cross linkages may all involve aldehyde-mediated reactions, the intra (β) - and inter (γ) -molecular cross linkages may take place in different regions of the molecule. All one can say at the moment is that intermolecular cross linking does increase with the age of the host. The rate of cross linking, however, is tissue-dependent.

*Trautman:* Collagens from various anatomical locations of the same animal have been observed to have differing responses to *p*H swelling, ionic strength, and extrusion characteristics. Could this be due to a different ratio of γ112 to γ111 or γ222 species?

*Veis:* These different tissues are cross-linked to differing extents or there might be different placement of the cross linkages.

*Hopper:* Stark and Kühn (1968) describe the difference in the behavior of the α1 and α3 chains observed in the short fragments obtained when calf skin collagen was treated with collagenase. They offer unknown substituents on the side chains of the amino acid residues as a possible explanation. Would you care to comment on this theory?

*Veis:* I believe the data offered by Stark and Kühn (1968) are real and this suggests that some differences do exist at the carboxy terminal region of the α chains. I am not in a position to hazard a guess as to the nature of the chemical composition differences.

*Montgomery:* In Table 21.4 the amino acid composition of CB3, CB4, and CB5 are almost identical. Could this indicate that the sequence in these peptides could be similar and that the sequence may be repeated in each peptide?

*Veis:* The sequences of each of the cyanogen bromide peptides is distinctly different. The apparent similarity arises because glycine always appears in every third residue position.

*Gordon:* I would like to know, aside from the influence of hormones (which may be tied up with physiological age rather than chronological age), how both the ground substance and collagen behave in the intervertebral disc in the common condition of low back? I think the ground substance is very important and should not be left out of the discussion.

*Veis:* The changes that I am aware of are that the fibers get thicker and less highly hydrated. Also there is a change in the distribution of acid mucopolysaccharides. I certainly agree that the ground substance components must play some important role in the tissue, but this is not known at present.

*Gordon:* Does mechanical pressure have some influence on these changes?

*Veis:* I do know that one can change the direction which fibroblasts take (whether they produce bone or cartilage or soft-tissue collagen) by manipulating the oxygen tension on a system and also pressure. I do not know how that expresses itself in terms of the acid mucopolysaccharide content in tissue.

*Davidson:* Would you comment on the conformational and compositional aspects of wound collagen?

*Veis:* As far as I know the collagen produced in healing wounds is similar chemically and structurally (at the molecular level) to normal collagen. Within the scar tissue, however, the fibrils are organized differently than they are in the normal tissue surrounding the wound. This may be due to the nature of the ground substance components, which may influence the way in which the fibrils aggregate. The physical properties of the scar tissue are different from those of the normal surrounding bulk tissue.

*Karmas:* What role does collagen play in water binding of muscle?

*Veis:* Collagen is a typical protein in that it usually contains about 30% firmly bound water. There have been several studies on the nature of the water-collagen relationship, and the probable situation is that the water is bound directly to the carbonyl groups along the peptide backbone since these are particularly accessible in collagen. It is unlikely that the strength of the water binding is very much greater in collagen than in other proteins.

*Landmann:* Are the characteristics of collagen found in the endomysium and perimysium of muscle tissue known? If so, does this type of collagen differ significantly from corium collagen?

*Veis:* I cannot answer this question; however, some ideas may be obtained from the papers of Bailey (1968) and Bailey, Fowler, and Peach (1969).

*Schram:* In negatively stained collagen fibrils a very sharp boundary is seen between the dense and the open segments of the fibril. How does this compare with what one sees if the packing density is plotted against the length of your proposed model fibril?

*Veis:* The model is designed to match the observed packing densities seen in the electron microscope and it does.

*Kastenschmidt:* Would you comment on the specificity of the protocollagen hydroxylase? Does it only bind to a proline adjacent to another proline?

*Veis:* It appears, from recent work by Rhoads and Udenfriend (1969), that protocollagen hydroxylase is very specific for the second proline in the sequence X-pro-pro. X may be glycine, but even in the peptide bradykinin protocollagen hydroxylase will place a hydroxyl on the sequence arg-pro-pro.

*Hopper:* In a recent paper (Warren, Smith, Tellman, and Veis, 1969) you and your associates described a rapid quantitative method for determining the degree of denaturation of insoluble fibrous collagen by bathing the fibers in $D_2O$ and studying the infrared spectrum. Has this method been explored with forms of collagen other than the rat tail tendon and the sheep submucosa?

*Veis:* We are studying collagen-fold formation by the H-D exchange technique, but we have not extended the studies on solid fiber systems beyond those already published.

*Usborne:* Are there proteins, other than enzymes, in the sarcoplasm?

*Scopes:* Excluding the extracellular contaminants in sarcoplasmic extracts, it can be said that at least 80% of the sarcoplasmic protein is composed of enzymes; any other proteins found may well have some specific function. For instance, myoglobin is not strictly an enzyme, but has a specific role in the muscle cell. At least three different proteins have been isolated from sarcoplasmic extracts and crystallized, but have no known function. These are: a nucleotide-containing protein from rabbit (Czok and Bücher, 1960), a basic protein from pig (Scopes, 1966), and the "parvalbumins." The "parvalbumins" are proteins of molecular weights between 9,000 and 13,000 which make up a significant proportion of the sarcoplasmic proteins of fish muscle, and are also present in other aquatic species (Jebson and Hamoir, 1958; Hamoir and Konoshu, 1965; Bhushana Rao, Focant, Gerday, and Hamoir, 1969). One hopes that in time a specific function for each of these proteins will be discovered.

*Usborne:* Are there free amino acids in the sarcoplasm?

*Scopes:* A recent paper gives an analysis of the nonprotein nitrogen in bovine muscle (Parrish, Goll, Newcomb, de Lumen, Chaudhry, and Kline, 1969).

*Landmann:* You state that sarcoplasmic proteins should be extracted from prerigor muscle. What differences would you expect between extracts of pre- and postrigor muscle?

*Scopes:* An extract of postrigor muscle, made at the *p*H of the muscle, would not contain those sarcoplasmic proteins which are isoelectrically precipitated at *p*H 5.5, nor those which are denatured in the post mortem conditions—for instance in PSE muscle of abnormally low *p*H, creatine kinase is denatured (Scopes and Lawrie, 1963). Extracting postrigor muscle at *p*H 7 will resolubilize some of the isoelectrically precipitated proteins, but not the denatured ones. However, the major components observed on starch gel electrophoresis are not normally affected: I actually remove the material precipitating in the cold at *p*H 5.5 when preparing the extracts for electrophoresis, as this fraction tends to clog the pores of the starch gel. I would say that there is very little difference between pre- and postrigor muscle as far as the sarcoplasmic proteins are concerned. Additional information can be found in the paper by Aberle and Merkel (1966).

*Briskey:* We have recently reported that there was no major preferential loss of starch gel bands between the pre- and postrigor periods—even when the muscles developed PSE characteristics. Only when we artificially produced PSE muscle by holding at 37° C for 4 hours were we able to show a significant loss in creatine kinase (Borchert, Powrie, and Briskey, 1969).

*Hamm:* Did you prepare your extracts for electrophoresis from minced tissue? What ions did your buffer solution contain? I think it is very important to define what you mean by "sarcoplasm." Mincing the tissue causes a release of enzymes which are localized in subcellular particles or more or less attached to membranes. Phosphate ions detach enzymes from membranes more easily than, for instance, tris buffers. On the other hand, mincing the tissue may result in an adsorption of proteins by the particles, which were originally not bound. According to my experience, a press juice of nonground muscle tissue prepared at relatively low pressure reflects much better the composition of the sarcoplasmic matrix than the supernatant of an homogenate of minced tissue.

*Scopes:* Dr. Hamm's point is of course important if one is interested in quantitative estimates of the various enzymes in different fractions from muscle. My own interests have been in large-scale fractionation of the proteins, and I have adopted a homogenization extraction technique similar to that described by Czok and Bücher (1960), which optimally solubilizes most of the glycolytic enzymes. They used isoosmotic sucrose, together with some EDTA and triethanolamine, which extracted nearly 100% of each enzyme. I have substituted glycerol for sucrose, and use Tris base. Perhaps the more important fact is that only two volumes of extractant are used; extraction is not complete, and it occurs in the presence of ions at approximately one-third the concentration that existed in the intact muscle.

*Gordon:* One obtains a value of 4 to 5 g % when sarcoplasmic protein content is assessed in muscle of laboratory animals by low ionic strength extraction. If I take Dr. Scopes' results and add myoglobin and myoalbumin, I still cannot realize 5,000 mg. Is there an explanation?

*Scopes:* The figures that were on that table were in mg/g. I took the total content as 6 g/100 g. This is from the prerigor state of muscle such as rabbit psoas or pig longissimus. I think the enzymes that are characterized amount to about 70% of the total. The remaining 30% must be spread out over hundreds of minor enzymes and perhaps nonactive proteins too.

*Montgomery:* Could you detect the difference between a protein and a large polypeptide with your starch gel procedure? I am thinking about polypeptides that may be split off from proteins.

*Scopes:* I think a polypeptide would have such a miscellaneous configuration that it would not run as a band at all. In any case, it would tend to be soluble in the staining conditions.

*Montgomery:* What are your staining conditions?

*Scopes:* It is a classical procedure of staining by denaturing the protein with an acetic acid alcohol mixture and staining with Amido black.

*Veis:* We do all sorts of electrophoresis with low-molecular-weight denatured peptide fragments and they make beautiful bands on starch or polyacrylamide gel.

*Landmann:* In a number of cases, e.g., phosphoglucose isomerase, the number of conformers varies from species to species. For instance, the enzyme from the tope shows six bands, while that from the tortoise shows three. The large differences in mobilities—in one species all bands move toward the anode, while in another, all move toward the cathode—would indicate severe differences in amino acid substitutions, rather than minor differences. In both instances it would seem that extensive charge changes have occurred. Would you comment further on these points?

*Scopes:* The species which are represented in Fig. 22.6 cover an extreme range of creatures, some of whose evolutionary development diverged many hundreds of millions of years ago. In the case of the enzyme phosphoglucose isomerase there may be as many as 600 amino acid residues to each subunit; a large number of these must be mutatable without upsetting the enzyme's activity. The extremes of mobility represented in Fig. 22.6 differ by about 20 charges per molecule, i.e., 10 charges per subunit, and when one considers that two charge mutations, e.g., Glu → Lys, are not only possible, but have been observed as quite common, then the range of mobilities found in these divergent species is not so surprising.

*Landmann:* Does the *pH* of the buffer in which the sample is run affect the number of minor bands or "conformers" for a given component?

*Scopes:* This is a difficult question to answer satisfactorily as I have done very little work with electrophoresis at *pH* values below 7. There are not only denaturation complications at lower *pH* values, but also the general spread of protein bands is much inferior. I can say, however, that rabbit creatine kinase gives the same three-band pattern with the same spacing at *pH* 9.0, 8.0, and 7.2, even though in the latter case the main component runs in the other direction (toward the cathode).

*Jay:* Since some of the sarcoplasmic components are low-molecular-weight compounds, they may be presumed to be rather susceptible to utilization by microorganisms. Were your preparations handled aseptically? Do you know which protein sarcoplasmic components are the most susceptible to attack by microorganisms? Upon long-term aseptic holding of sarcoplasmic preparations, is there a reduction in the number of demonstrable components?

*Scopes:* Sarcoplasmic extracts are indeed excellent growth media for microorganisms. My treatments were not aseptic, but generally, after the last centrifuging the extracts were dialyzed at 1° C for not more than 24 hours before electrophoresis. Any bacterial attack on the proteins in this time must have been negligible. I believe that long-term aseptic storage of meat results in the preferential loss of those proteins which are more readily denatured at *pH* 5.5, e.g., creatine kinase, enolase, etc. (see Table 22.3). Sarcoplasmic extracts do not keep for a great length of time if unfrozen, particularly if the *pH* is outside the range 6.5 to 7.5, and some components are lost. I have not studied this in detail.

*Usborne:* Are any of the sarcoplasmic proteins active during the aging of meat?

*Scopes:* With the exception of a few minor components which are denatured by the low *pH*, all the sarcoplasmic proteins are active for a long period postrigor. A minced preparation, given suitable substrates and nucleotides, will glycolyze rapidly if its *pH* is raised to 7.0.

*Trautman:* It has been suggested that cathepsins do not hydrolyze fibrillar proteins but do hydrolyze sarcoplasmic proteins. Could protein artifacts be produced by cathepsins during the extraction of sarcoplasmic proteins?

*Scopes:* The rate of action of cathepsins, particularly at *pH* 7 and in the presence of EDTA, is too low for significant hydrolysis of sarcoplasmic proteins to occur during preparation of the extracts. No difference in the starch gel electrophoretic patterns is obtained when extracts are made rapidly prerigor, or purposely left to "digest" for some hours.

*Usborne:* What proteolytic enzymes, if any, are present in the sarcoplasm?

*Scopes:* Cathepsins have been isolated from muscle; although probably all originating from lysosomes, they are quickly released into the sarcoplasm after death and so for the purposes of meat studies can be considered to be sarcoplasmic proteins. Their activity even at the ultimate *p*H of meat is very weak, especially on nondenatured protein. Another proteolytic enzyme recently isolated from muscle is mentioned in Chapter 22.

*Trautman:* If sarcoplasmic proteins do exist in the muscle at 25–30% protein concentrations, how is the entire spectrum of enzymic activity maintained without autocatalytic losses? Similar solutions of the pancreatic enzymes and pepsin at physiological temperatures are not stable.

*Scopes:* No doubt a 25% solution of cathepsin D would soon digest itself. However, a 25% solution of sarcoplasmic protein would contain only a very low concentration of proteolytic enzyme. In living tissue, moreover, the cathepsins are mainly compartmentalized in the lysosomes and so are not in contact with the sarcoplasmic protein solution.

*Trautman:* Is there a difference in proteolytic enzyme content or kinetics between fast and slow muscle?

*Scopes:* I have no information on proteolytic enzyme contents of muscles.

*Montgomery:* If a particular enzyme activity could be correlated with the tenderness of meat, could your electrophoretic technique be used to choose animals for breeding purposes?

*Scopes:* The important word in this question is "if"! The only difficulty in using this method on live animals is that it necessitates a muscle biopsy, but only a gram or so is needed.

*Hopper:* Techniques have been developed to tenderize meat by the ante mortem injection of proteases. Is it known whether these enzymes become part of the sarcoplasm or whether they remain in the extracellular fluid?

*Scopes:* I do not know the answer to this question.

*Fox:* Fisher (1960) reported a sarcoplasmic protein fraction in chicken muscle that correlated closely to muscle tenderness. Has this fraction been identified? If so, what is the nature of the protein and what is its relationship to tenderness?

*Scopes:* I have not worked with chicken muscle; it is possible that some protein from the myofibrillar structure could become solubilized, making the muscle more tender. The DEAE cellulose column chromatograms Fisher (1960) presented seem to me too variable to draw any conclusions from; in any case the peak mentioned was certainly a complex mixture of sarcoplasmic proteins.

*Landmann:* Your argument that the sarcoplasmic proteins occur only in the extramyofibrillar space also supports the findings of Bendall and Wismer-Pedersen (1962) that denatured sarcoplasmic proteins coat the myofibrils in PSE pork. If it can be assumed that the sarcoplasmic proteins reach the same state of denaturation in both normal and PSE pork during the course of rigor, it would seem that myofibrils in normal pork should also be coated with these proteins. Since this apparently does not occur, do you have any opinion as to the reason for this difference?

*Scopes:* I do not think you can assume that the sarcoplasmic proteins reach the same state of denaturation in both normal and PSE pork during the course of rigor. The cases of PSE pork which we were observing in Europe some years ago had ultimate *p*H values rather lower than normally occur in Poland China pigs; this caused much more denaturation of the most susceptible sarcoplasmic proteins.

*Aberle:* What evidence do you have that AMP deaminase controls the rate of post mortem ATP degradation?

*Scopes:* This is a slight misunderstanding. The rate of AMP deamination controls not the rate of ATP degradation, but the rate of ATP loss, i.e., the net difference between degradative and synthetic processes. It is fairly clear that deamination and ATP loss go in parallel, since IMP production occurs at the same rate that ATP (+ADP) is lost from the muscle. It is not so obvious that the enzyme AMP deaminase actually controls the rate of ATP loss, and I do not intend to elaborate this theory here. However, I will say, as a piece of evidence, that in a simplified glycolyzing system in which AMP deaminase was lacking, the ATP level remained at its initial value even at *p*H values down to 5.6.

*Greaser:* You mentioned that several soluble enzymes might be able to function as an ATP breakdown system in muscle. Have you made any calculations based on the appropriate enzyme concentrations, substrate concentrations, and $K_m$'s to assess the maximum contribution that these systems might make to the degradation of ATP?

*Scopes:* I have calculated that fructose-1,6-diphosphatase, if acting without inhibition, could account for all the adenosine triphosphatase activity in normal muscle. Thus its contribution may be significant, and this is a possibility which I am at present investigating.

*Davies:* Is there room for myokinase and/or creatine kinase between the filaments? If not, then ADP would have to dissociate at each cycle rather than be regenerated *in situ*. This last process, if it occurs, might explain why muscles move more quickly and use ATP faster than the isolated actomyosin systems.

*Scopes:* There is certainly room for myokinase between the filaments; myokinase is one of the smallest enzymes. There is also room for creatine kinase, but if this molecule is allowed in, all other molecules of similar size or smaller can hardly be excluded. The spaces between the filaments would then become uncomfortably crowded with protein molecules. This is purely speculative, however. It is possible that some information could be obtained by using isolated myofibrils as though they were Sephadex gel particles, so as to find the size of molecules excluded.

*Mullins:* Recently, in our laboratory, we have been successful in producing PSE musculature by injecting aldosterone 30 min prior to slaughter and preventing PSE conditions by dosing animals with an aldosterone blocking agent, aldactazide. We have theorized that this functions through the inactivation of pyruvic acid kinase, thereby preventing the reversal of glycolysis. Would you elaborate on this hypothesis?

*Scopes:* The rate of post mortem glycolysis is determined by the rate of ATP breakdown; PSE occurs because of some physiological effect on muscle membrane systems activating the adenosine triphosphatases, hence the induction or prevention of PSE by various injections. Pyruvate kinase under physiological conditions cannot work in the reverse direction to any significant amount. If its action were blocked, this would apply also to its forward reaction.

*Hopper:* What effect, if any, does the diet of an animal have on the composition of the sarcoplasm?

*Scopes:* Since the proteins' structures are genetically determined, I would imagine that nutrition could only affect quantity.

*Hopper:* What work has been done to study the variation in composition of the sarcoplasmic proteins with age, in a given species?

*Scopes:* This has been partly discussed by Dr. Perry, in whose laboratory most of such work has been carried out. There are large quantitative variations around the time of birth, partly explained in terms of red and white fiber contents of the muscle. There are also some qualitative changes, such as in lactate dehydrogenase, the muscle-type subunit of which only develops when the animal becomes active.

*Woodson:* Is the electrophoretic technique for meat species identification as sensitive as the serological method? Would enzyme content be superior to protein bands in species identification? Do you have any suggestions for species identification of cooked meat?

*Scopes:* To each of Dr. Woodson's questions I could reply no. To the first I would add that the electrophoretic technique requires very little preparation, and can provide an unambiguous answer, whereas serological techniques would give interspecies cross reactions, and may require the preparation of individual proteins from the meat sample.

To the second question I can only say that enzyme content varies more with age and muscle type than it does between species. Quantitative methods for species' identification do not seem to be very useful.

It is possible that solubilizing denatured cooked muscle proteins in guanidine hydrochloride + mercaptoethanol, and subsequent urea gel electrophoresis, may give some indication of the species of the meat.

*Montgomery:* Another possibility for the formation of variants or isozymes of sarcoplasmic enzymes may be the formation of polymers from different monomers. Do you believe this possibility to be feasible in the present case?

*Scopes:* In the majority of cases, if not all, the multiple banding of the various enzymes on gel electrophoresis is due to charge differences. This can be proved by gel filtration methods. In addition, one can argue that polymers would tend to be quantitatively less than the monomer, and so coincide with the weaker bands. However, the weaker bands are usually more mobile in the gel, not less as they would be if polymeric.

## *References*


Aberle, E. D. and R. A. Merkel. 1966. Solubility and electrophoretic behavior of some proteins of post-mortem aged bovine muscle. *J. Food Sci. 31*:151.

Bailey, A. J. 1968. Intermediate labile cross-links in collagen fibers. *Biochim. Biophys Acta 160*:447.

Bailey, A. J., L. J. Fowler and C. M. Peach. 1969. Identification of two interchain cross-links of bone and dentine collagen. *Biochem. Biophys. Res. Comm. 35*:663.

Bendall, J. R., and J. Wismer-Pedersen. 1962. Some properties of the fibrillar proteins of normal and watery pork muscle. *J. Food Sci. 27*:144.

Bhushana Rao, K. S. P., B. Focant, C. Gerday, and G. Hamoir. 1969. Low molecular weight albumins of cod white muscle. *Comp. Biochem. Physiol. 30*:33.

Borchert, L. L., W. D. Powrie, and E. J. Briskey. 1969. A study of the sarcoplasmic proteins of porcine muscle by starch gel electrophoresis. *J. Food Sci. 34*:148.

Czok, R. and T. Bücher. 1960. Crystallized enzymes from the myogen of rabbit skeletal muscle. *Advance. Protein Chem. 15*:315.

Fisher, R. L. 1960. Change in the chemical and physical properties of protein during aging of meat, p. 71. *In Meat Tenderness Symposium* (Campbell Soup Co., Camden, New Jersey).

Hamoir, G. and S. Konoshu. 1965. Carp myogens of white and red muscles. *Biochem. J. 96*:85.

Jebsen, J. W. and G. Hamoir. 1958. Electrophoretic and ultracentrifugal study of plaice myogen. *Acta Chem. Scand. 12*:1851.

Mohr, V. and J. R. Bendall. 1969. Constitution and physical chemical properties of intramuscular connective tissue. *Nature 223*:404.

Parrish, F. C., Jr., D. E. Goll, W. J. Newcomb, B. O. de Lumen, H. M. Chaudhry, and E. A. Kline. 1969. Molecular properties of post-mortem muscle 7. Changes in nonprotein nitrogen and free amino acids of bovine muscle. *J. Food Sci. 34*:196.

Rhoads, R. E. and S. Udenfriend. 1969. Substrate specificity of collagen proline hydroxylase: hydroxylation of a specific proline residue in bradykinin. *Arch. Biochem. Biophys. 133*:108.

Scopes, R. K. 1966. Isolation and properties of a basic protein from skeletal muscle sarcoplasm. *Biochem. J. 98*:193.

Scopes, R. K. and R. A. Lawrie. 1963. Post mortem lability of skeletal muscle proteins. *Nature 197*:1202.

Stark, M. and K. Kühn. 1968. The properties of molecular fragments obtained on treating calfskin with collagenase from *Clostridium histolyticum. Europe. J. Biochem. 6*:534.

Toole, B. P. and D. A. Lowther. 1968. The effect of chondroitin sulfate-protein on the formation of collagen fibrils *in vitro. Biochem. J. 109*:857.

Warren, R. J., W. E. Smith, W. J. Tellman, and A. Veis. 1969. Internal reflectance spectroscopy and the determination of the degree of denaturation of insoluble collagen. *J. Am. Leather Chem. 64*:4.

# PART 6

## Muscle Adaptation

# Chapter 24

# Changes in the Composition of Muscle with Development and Its Adaptation to Undernutrition

E. M. WIDDOWSON

Skeletal muscle changes more in chemical composition during the latter part of fetal and postnatal life than any other soft tissue except perhaps the skin, and the muscles atrophy more than any other tissue except the fat during prolonged undernutrition. It has been known for many years that during development the percentage of water in muscle falls (Needham, 1931). There is at the same time a fall in the concentrations of the extracellular ions sodium and chloride (Kerpel-Fronius, 1937; Yannet and Darrow, 1938) and a rise in the intracellular constituents, protein, potassium, and phosphorus. On the assumption that most if not all of the chloride is in the extracellular phase, this has been interpreted to mean that as the cells develop they come to fill spaces previously occupied by extracellular fluid. Table 24.1 illustrates this in

TABLE 24.1
*Effect of development on the water and electrolytes in skeletal muscle of pigs*

| | Fetus | | Piglet | | |
|---|---|---|---|---|---|
| | 46 days | 90 days | Newborn | 3 weeks | Adult |
| Water (g/100 g) | 92 | 87 | 82 | 78 | 74 |
| Na (meq/kg) | 106 | 77 | 54 | 34 | 24 |
| Cl (meq/kg) | 73 | 46 | 37 | 32 | 21 |
| K (meq/kg) | 47 | 60 | 73 | 102 | 93 |
| "Na space" (g/100 g) | 87 | 59 | 38 | 20 | 17 |
| "Cl space" (g/100 g) | 72 | 45 | 35 | 26 | 17 |
| Intracellular space (g/100 g) (total water-Cl space) | 20 | 42 | 47 | 52 | 57 |

the pig (Dickerson and Widdowson, 1960). In the 46-day fetus the muscle contains 92% of water. Nearly 80% of it is in the chloride space, and is therefore probably outside the cells. By 90 days' gestation the extracellular water accounts for only about half the total, and by term for about 40% of it. By 3 weeks after birth it has fallen to 33% and in adult muscle only 23% of the total water is in the extracellular phase. The sodium space is larger than the chloride space in fetal muscle, so there may be some sodium in fetal muscle cells.

Table 24.2 shows how severe undernutrition of pigs from 10 days of age until they were a year old modified the normal changes in the extracellular-intracellular structure of muscle (Widdowson, Dickerson, and McCance, 1960). These undernourished pigs had a mean weight of 5.8 kg compared with 150 kg for a well-nourished littermate. Their muscles were very small, not only absolutely but also as a percentage of the body weight. Table 24.2 shows that they contained more water and more extracellular water than the muscles of a well-nourished pig 4 weeks old, and, in fact, the proportions were similar to those in a newborn pig. It is clear that the effects of undernutrition on the muscle were in these respects the converse of those taking place during normal development.

The fundamental change that takes place in muscle, as in other tissues, during development is growth first in the number and then in the size of the cells. The muscle of a pig fetus of 46 days' gestation has no true muscle fibers, but myotubes, which have been formed by the fusion of the original myoblasts, and which are long, narrow cells with many nuclei. The nuclei are situated centrally in the myotubes and they are large in proportion to the diameter of the cell. The spaces between the myotubes are filled with the ground substance of the extracellular phase. There are many fibroblasts in it and many fine reticulin fibers (Dickerson and Widdowson, 1960).

TABLE 24.2

*Effect of severe undernutrition on the water and electrolytes in skeletal muscle of pigs*

| | Undernourished (1 year) | Well-nourished (4 weeks) | Well-nourished (1 year) |
|---|---|---|---|
| Weight of pig (kg) | 5.8 | 5.8 | 150 |
| Water (g/100 g) | 83 | 78 | 74 |
| Na (meq/kg) | 58 | 30 | 24 |
| Cl (meq/kg) | 37 | 27 | 21 |
| "Cl space" (g/100 g) | 35 | 22 | 17 |
| Intracellular space (g/100 g) (total water-Cl space) | 48 | 56 | 57 |

Between 46 days and term at 120 days the myotubes are converted into muscle fibers; there is a big increase in their number, and they are now arranged in bundles. There has been little growth yet in the diameter of the fibers, and the nuclei are still relatively large, though they have now migrated to the periphery of the fibers and they lie immediately under the sarcolemma. In adult pig muscle the diameter of the fibers is very much larger, there is little space for extracellular material between them, and the nuclei look comparatively small.

Table 24.3 shows how these morphological changes are reflected in the concentrations of extracellular and intracellular proteins. The percentage of the extracellular proteins, collagen and elastin, increased to a maximum at or shortly after birth and then decreased again. Values for 6-week-old pigs were already lower than those for the 3-week-old ones shown here. These extracellular proteins form the sheaths round the muscle fibers, the bundles of fibers and the whole muscles, and, broadly speaking, the percentage of extracellular protein increased when the fibers were increasing in number, and decreased when they were increasing in size.

The proteins of the sarcoplasm and myofibrils can be separated by extraction of the muscle first with a dilute salt solution and then with 0.1N NaOH (Robinson, 1952). When this was done to muscle from pigs of different ages the results in the last two lines of Table 24.3 were obtained. Total intracellular protein is the sum of the two fractions. At all ages after 46 days' gestation there was more fibrillar than sarcoplasmic protein, and the percentage of both of these fractions of intracellular protein more than doubled during the 3 weeks after birth. This rapid rate of development was brought about by the functional activity of the muscle, for pigs run about soon after they are born, whereas the human infant, whose fibrillar proteins increase much more slowly after birth,

TABLE 24.3

*Effect of development on the extracellular and intracellular proteins of pig muscle**

| | Fetus | | Piglet | | |
|---|---|---|---|---|---|
| | 46 days | 90 days | Newborn | 3 weeks | Adult |
| Total N | 0.80 | 1.15 | 1.36 | 2.85 | 3.32 |
| Extracellular protein N | 0.04 | 0.10 | 0.23 | 0.32 | 0.11 |
| Intracellular protein N | 0.65 | 0.85 | 0.87 | 2.18 | 2.78 |
| Sarcoplasmic protein N | 0.40 | 0.24 | 0.22 | 0.54 | 0.80 |
| Fibrillar protein N | 0.25 | 0.61 | 0.65 | 1.64 | 1.98 |

* g/100 g fresh fat-free muscle.

does not walk for many months. The high value for sarcoplasmic protein in the 46-day fetus may be related to the large number of fibroblasts rather than to muscle cells in pig muscle at this age, for the chemical separation does not differentiate between them.

Table 24.4 shows the intracellular potassium, magnesium, and phosphorus per g of intracellular protein nitrogen. Amounts of all three mineral elements rise during fetal life in proportion to protein and fall again after birth. There are several possible explanations: (1) Fibroblasts may contain more of these elements in relation to protein than do true muscle cells. (2) There is very little information about the composition of the nuclei in muscle, but in calf thymus cells the concentration of potassium has been shown to be 4–5 times higher in the nucleus than in the cytoplasm (Itoh and Schwartz, 1956). Consequently the large nucleus in the small fetal cell may make so great a contribution to its potassium that it raises it above the concentration in the adult cell. (3) Potassium and phosphorus move into the cells with glycogen (Fenn, 1939), and the high concentration of potassium and phosphorus in the muscle cell of the newborn piglet is partly due to this, for the muscles of the pig at birth contain more than 7% of glycogen (McCance and Widdowson, 1959).

The idea was first put forward at the end of the last century that the full number of muscle fibers found in the adult was already present in the newborn animal and baby, and that subsequent growth was entirely due to hypertrophy of existing fibers, both in width and length (MacCallum, 1898). It was also believed for a long time that once the myoblasts had fused to form myotubes there was no further mitosis, so that the number of nuclei in the myotube determined the number of nuclei in the fiber right on into adult life. Most of the histological studies that have been made during the past 70 years on a variety of species of mammal, for example, man, sheep, pig, and rabbit, have supported the first proposition, that the full number of muscle fibers is reached during fetal life or

TABLE 24.4

*Effect of development on the relation between protein and potassium, magnesium, and phosphorus within the muscle cell**

| | Fetus | | Piglet | | |
|---|---|---|---|---|---|
| | 46 days | 90 days | Newborn | 3 weeks | Adult |
| Intracellular K(meq) | 5.1 | 6.7 | 7.0 | 4.5 | 3.5 |
| Mg(meq) | 1.35 | 1.46 | 1.55 | 0.72 | 0.88 |
| P(mmole) | 4.6 | 5.5 | 5.7 | 3.4 | 2.6 |

* Values expressed per g of intracellular protein N.

soon afterwards (McMeekan, 1940; Meara, 1947; Joubert, 1956; Montgomery, 1962; Staun, 1963), though this may not be true for birds (Montgomery, Dickerson, and McCance, 1964). During the past 10 years, however, those concerned with the structure of muscle have been puzzled as to how the comparatively few nuclei that exist in mammalian fetal muscle are sufficient for the bulk of muscle in the adult, and this has now been looked into again using both histological and biochemical approaches. Montgomery (1962) made the point that there is an increase in the sarcolemmal nuclei throughout muscle growth in man, and it is now clear from the results of the Montreal investigators (Enesco and Leblond, 1962; Enesco and Puddy, 1964), and of Winick and Noble (1965) that the number of nuclei in the muscle of the rat increases after birth and that contrary to earlier beliefs the nuclei proliferate by mitotic division (MacConnachie, Enesco, and Leblond, 1964). In whole muscles of rats there is a two- to fourfold increase in the number of nuclei between 16 and 86 days of age. This value was arrived at by determining the DNA and assuming that the DNA content of a nucleus is constant at 6.2 $\mu\mu$g. There is no ploidy in striated muscle nuclei, so that the amount of DNA gives a direct measure of the number of nuclei. Other workers, who have followed the development of rat muscle from earlier in life, have shown that there is a 20-fold increase in DNA between birth and maturity (Cheek, 1968).

Cheek (1968) has used DNA as an index of cellular development in an extensive study of muscle growth in man. He obviously could not obtain weights of individual muscles of living children, so he determined the concentration of DNA in biopsy samples of gluteal muscles, assumed that this was representative of all the muscles in the body, and calculated from this value the amount of DNA in the whole muscle mass. He estimated the muscle mass from the creatinine excretion, using a figure of 1 g of creatinine excreted in 24 hours as equivalent to 20 kg of muscle. His results on 33 boys and 19 girls suggested that there was an increase in the number of muscle nuclei right through human childhood, and that in the boy there is a "spurt" in the rate of nuclear division at the time of puberty, so that between 10 and 16 years the number of nuclei doubles. There is a 14-fold increase in the number of muscle nuclei during growth after birth. In the girl there is also an increase, but a smaller one. Cheek described this as an increase in the number of muscle cells. To most people a muscle cell is a muscle fiber with many nuclei, but Cheek's concept of a muscle cell is one nucleus together with the cytoplasm over which the nucleus has jurisdiction. On this basis a muscle fiber contains many muscle cells, and Cheek's statement that the number of muscle cells increases throughout postnatal growth does not necessarily imply that there is any increase in the number of muscle fibers. Growth in size of a muscle is still believed to be brought about by an increase in size of each muscle fiber, both in width and length. Enesco and Puddy (1964) were able to calculate the weight of one rat muscle fiber, and they showed that this increased 10–23 times while the number of nuclei increased only 2–4 times. Thus the amount of cytoplasm associated with each nucleus also rose. Cheek used the ratio protein:DNA as a measure of the relationship between the cytoplasm and the nucleus, or in his terminology as a measure of the size of the muscle cell. Table 24.5 shows how the value compares in the newborn rat, pig, and human baby, and in adults of these species. The similar value for adults of the two species of very different body size suggests that about 400 mg of protein per mg of DNA, or 2.5 $\mu$g protein per nucleus, may represent the upper limit for striated muscle. The muscle of the human baby has a more mature protein:DNA ratio than the muscle of the rat at birth; the ratio for the newborn pig is intermediate. This is in line with the stage of development reached by the muscles of these three species at birth, as judged by their relative proportions of protein and water (Widdowson and Dickerson, 1964).

As far as I know the use of DNA for the measurement of the number of nuclei and, with protein, for the relation of cytoplasm to nucleus has not

TABLE 24.5

*Amount of protein (mg) associated with each mg of DNA in skeletal muscle*

| | Newborn | Adult |
|---|---|---|
| Rat | 15 | 400 |
| Pig | 53 | 380 |
| Man | 175 | 300 |

TABLE 24.6

*Effect of two weeks' rehabilitation on the amount of DNA and intracellular protein in the quadriceps muscle of severely undernourished pigs*

| | Undernourished | Rehabilitated 2 weeks |
|---|---|---|
| Body weight(kg) | 5.8 | 9.2 |
| Weight of one muscle(g) | 15.9 | 23.9 |
| Intracellular protein N(g/100 g) | 1.43 | 2.01 |
| DNA(mg/100 g) | 60 | 59 |
| DNA(mg in whole muscle) | 9.5 | 14.2 |
| $\frac{\text{Intracellular protein(mg)}}{\text{DNA(mg)}}$ | 150 | 215 |

yet been applied systematically to the muscle of the growing pig or indeed to the muscle of any other meat animal.

Table 24.6 shows results that were obtained some years ago by Browning (1962) and published by Dickerson and McCance (1964) for the muscle of our severely undernourished pigs, and for the muscle of similar animals after 2 weeks' rehabilitation, during which time their body weight had increased by 50%. The weight of the muscle had also increased by 50%. The amount of DNA in the whole muscle and the intracellular protein:DNA ratio also increased, indicating that the growth of the muscle had involved an increase both in number of nuclei and in the ratio of cytoplasm to nucleus. The work of Winick and Noble (1966) suggests that if undernutrition is imposed when the cells are still in the stage of hyperplasia, and normal cell division is interrupted, the tissue is less likely to attain its full genetic size on rehabilitation than if undernutrition is imposed later, when the full number of cells has been reached and the tissue is growing only by an increase in cell size. In muscle the fibers had probably already reached their full number when the pigs were first undernourished, but the nuclei had not. The results of long term rehabilitation of the undernourished pigs indicate that they did not quite reach the size of littermates that had never been undernourished (Lister and McCance, 1967). They had less muscle and less bone, but the relation of the one to the other was the same as in well-nourished control animals of the same age. There was also a suggestion that the concentration of nitrogen in the muscles was lower even after 2–3 years' rehabilitation than in control animals, which again suggests that recovery of the muscle may not have been complete. This is obviously a lead that should be followed.

A pig sometimes becomes undernourished before birth owing to its occupation of an undesirable site in the uterus. It has a smaller placenta and poorer blood supply than the rest of the litter. The runt pig that ensues is a nuisance to the farmer and stockbreeder because it is very likely to die, but it provides an excellent model for the dysmature small-for-dates human baby which is of great interest to pediatricians at the present time. We are using it for this purpose, and among other things we are studying the structure of its muscle, both histologically and chemically. Table 24.7 shows the mean weight of the rectus femoris muscles of 4 newborn runts compared with those from larger piglets from each of the 4 litters, and the composition of the muscles. It also shows corresponding values for the same muscle from 4 normally grown fetuses of 92 days' gestation, chosen to correspond in body weight to the full-term runts. The muscle of the runts was nearer to that of their larger littermates with respect to its water-nitrogen relationships than to the

TABLE 24.7

*DNA and intracellular protein in the rectus femoris muscle of fetal and newborn pigs*

| | Fetus of 92 days' gestation | Runt piglet at term | Large littermate of runt |
|---|---|---|---|
| Body weight(g) | 562 | 577 | 1614 |
| Weight of one muscle(g) | 1.85 | 1.85 | 6.6 |
| Composition of muscle per 100g | | | |
| Water(g) | 88.0 | 82.5 | 80.0 |
| Total N(g) | 1.11 | 1.71 | 1.89 |
| Intracellular protein N(g) | 0.89 | 1.34 | 1.52 |
| DNA(mg) | 324 | 288 | 223 |
| DNA in whole muscle(mg) | 5.9 | 5.3 | 14.8 |
| Intracellular protein(mg) / DNA(mg) | 17.2 | 29.1 | 42.5 |

muscle of fetuses of the same body weight, but the full-term runts had about the same amount of DNA in the whole muscles as the 92-day fetuses, and only about one-third as much as we found in the muscles of the well-developed full-term animals. This suggests that nuclear division had been slowed down or halted during the period of undernutrition *in utero*. The muscle fibers had gone on growing in size, however, and the intracellular protein:DNA ratio was intermediate between the value for the fetal weight control and the well-grown littermate. This effect of undernutrition of the pig before birth is the same as that of undernutrition of the rat after birth. What consequence this has for subsequent muscular development is a problem we are now investigating.

## *References*

Browning, A. 1962. Oxygen consumption and body composition after prolonged undernutrition in pigs and cockerels. M.Sc. thesis, Cambridge University.

Cheek, D. B. 1968. *Human Growth*. Lea & Febiger, Philadelphia.

Dickerson, J. W. T., and R. A. McCance. 1964. The early effects of rehabilitation on the chemical structure of the organs and whole bodies of undernourished pigs and cockerels. *Clin. Sci. 27*:123.

Dickerson, J. W. T., and E. M. Widdowson. 1960. Chemical changes in skeletal muscle during development. *Biochem. J. 74*:247.

Enesco, M., and C. P. Leblond. 1962. Increase in cell number as a factor in the growth of the organs and tissues of the young male rat. *J. Embryol. Exp. Morph. 10*:530.

Enesco, M., and D. Puddy. 1964. Increase in the number of nuclei and weight in skeletal muscle of rats of various ages. *Amer. J. Anat. 114*:235.

Fenn, W. O. 1939. The deposition of potassium and phosphate with glycogen in rat livers. *J. Biol. Chem. 128*:297.

Itoh, S., and I. L. Schwartz. 1956. Sodium and potassium content of isolated nuclei. *Nature 178*:494.

Joubert, D. M. 1956. An analysis of prenatal growth and development in the sheep. *J. Agr. Sci. 47*:382.

Kerpel-Fronius, E. 1937. Über die Besonderheiten der Salz- und Wasserveiteilung im Säuglingskörper. *Z. Kinderheilk. 58*:726.

Lister D., and R. A. McCance. 1967. Severe undernutrition in growing and adult animals 17. The ultimate results of rehabilitation: pigs. *Br. J. Nutr. 21*:787.

MacCallum, J. B. 1898. Histogenesis of the striated muscle fiber and the growth of the human sartorius muscle. *Johns Hopkins Hospital Bull. 9*:208.

McCance, R. A., and E. M. Widdowson. 1959. The effect of lowering the ambient temperature on the metabolism of the new-born pig. *J. Physiol. 147*:124.

MacConnachie, H. F., M. Enesco, and C. Leblond. 1964. The mode of increase in the number of skeletal muscle nuclei in the postnatal rat. *Amer. J. Anat. 114*:245.

McMeekan, C. P. 1940. Growth and development in the pig, with special reference to carcass quality characters. Part I. *J. Agr. Sci. 30*:276.

Meara, P. J. 1947. Postnatal growth and development of muscle as exemplified in the gastrocnemius and psoas muscle of the rabbit. *Onderstepoort J. Vet. Sci. 21*:329.

Montgomery, R. D. 1962. Growth of human striated muscle. *Nature 195*:194.

Montgomery, R. D., J. W. T. Dickerson, and R. A. McCance. 1964. Severe undernutrition in growing and adult animals 13. The morphology and chemistry of development and undernutrition in the sartorius muscle of the fowl. *Br. J. Nutr. 18*:587.

Needham, J. 1931. *Chemical Embryology,* Vol. 3. Cambridge Univ., London.

Robinson, D. S. 1952. Changes in the composition of chick muscle during development. *Biochem. J. 52*:621.

Staun, H. 1963. Various factors affecting the number and size of muscle fibers in the pig. *Acta Agr. Scand. 13*:293.

Widdowson, E. M. 1968. Effects of praematurity and dysmaturity in animals, p. 127. *In* J. H. P. Jonxis, H. K. A. Visser and J. A. Troelstra (eds.), *Nutricia Symposium, Aspects of Praematurity and Dysmaturity.* H. E. Stenfert Kroese, Leiden.

Widdowson, E. M. and J. W. T. Dickerson. 1964. Chemical composition of the body, p. 1. *In* C. L. Comar and F. Bronner (eds.), *Mineral Metabolism,* Vol. 2A. Academic Press, New York.

Widdowson, E. M., J. W. T. Dickerson, and R. A. McCance. 1960. Severe undernutrition in growing and adult animals 4. The impact of severe undernutrition on the chemical composition of the soft tissues of the pig. *Br. J. Nutr. 14*:457.

Winick, M., and A. Noble. 1965. Quantitative changes in DNA, RNA and protein during prenatal and postnatal growth in the rat. *Dev. Biol. 12*:451.

———. 1966. Cellular response in rats during malnutrition at various ages. *J. Nutr. 89*:300.

Yannet, H., and D. C. Darrow. 1938. The effect of growth on the distribution of water and electrolytes in brain, liver and muscle. *J. Biol. Chem. 123*:295.

# Chapter 25

# *Morphological Adaptation due to Growth and Activity*

G. GOLDSPINK

One of the fascinating aspects of striated muscle is its ability to adapt to the type of work that it is habitually required to perform. The scope of this paper is a description of the morphological changes associated with adaptation, although it should be borne in mind that, as far as the animal is concerned, it is the change in the output of the muscle that is the important manifestation of adaptation.

One often talks glibly about the increase in strength or endurance due to particular activity or type of athletic training. But for the physiologist working in the laboratory at the level of the individual muscle, the parameters that can be measured are:

(1) Contractile strength of the muscle, which according to the conditions may be regarded as either the maximum isometric force or the maximum weight that can be lifted through a given distance.

(2) Endurance, which may be regarded as the ability to delay fatigue —in other words, the ability of the muscle to maintain isometric tension or to continue performing external work.

(3) The rate of contraction, which is either the rate at which the muscle develops isometric tension or its maximum velocity of shortening.

(4) The efficiency of the muscle action, which determines the cost in terms of ATP utilization by the muscle in maintaining tension or in performing a given amount of work—in other words, what it costs the animal to carry out a particular type of muscle action.

From merely casual observations of the effect of athletic training it is apparent that the first two parameters can be changed by subjecting the muscle to the appropriate type of exercise. The latter two characteristics change during growth but apparently do not change as a response to any kind of exercise. The change in the speed of contraction of a muscle during growth is really another example of ontogeny recapitulating phylogeny.

After having outlined the physiological parameters which are considered to be important in a study of this kind, it is now possible to discuss the structural aspects of the different types of adaptation.

## CHOICE OF MUSCLE FOR THE STUDY OF ADAPTATION

As with all biological investigations it was necessary first to choose a simple and convenient system for study. We chose to study the growth and adaptive responses of the striated musculature of the laboratory mouse for the following reasons. Several muscles in the mouse have a relatively simple structure with fibers running the full length of the muscle belly, and the small size of mouse muscles made it possible to fix entire muscles *in situ* with the limbs of the animal in a specified position. It was possible, therefore, to take a single transverse section through a predetermined region of the muscle, to count the total number of fibers in the muscle, and to obtain reproducible measurements of fiber size. The muscles of the mouse tend not to be mixed as far as fast and slow fibers are concerned, so that it was possible to study the growth and adaptive responses of a pure fast muscle or a pure slow muscle. With muscles of large animals such as those of meat-producing animals or with human muscles it is very difficult to devise a representative sampling technique, as many of the fibers terminate in the middle of the muscle belly, and the evidence suggests that the different physiological fiber types are intermingled (Dubowitz, 1965; R. G. Cassens and E. J. Briskey, pers. comm., 1968). Mice are in general ideal animals for growth studies as they can be bred in larger numbers at little expense and they have a relatively short life span. Pure inbred strains of mice were used, so that the genetic differences between individuals were kept to a minimum.

## INCREASE IN MUSCLE LENGTH

The limbs of most animals approximately double in length during the course of postnatal growth. Associated with the increase in muscle fiber length there is an increase in the number of sarcomeres in series along the length of the fiber and also an increase in the length of the individual sarcomeres (Goldspink, 1968). It has been suggested that the new sarcomeres are formed at the ends of the myofibrils and that A and I filaments may be arranged into the hexagonal array by the interaction of the cross bridges on the A filaments with the I filaments (Fischman, 1967; Goldspink, 1968). In the biceps brachii of the mouse the sarcomere length, with the limb in the fully extended position, was found to increase from 2.3 $\mu$ to 2.8 $\mu$. The increase was due to a decrease in the percentage overlap of the A and I filaments and not due to any change in the filament lengths (see Figs. 25.1 and 25.2). It is believed to be caused by traction imposed on the muscle due to the rapid longitudinal growth of the bones to which the muscle is attached.

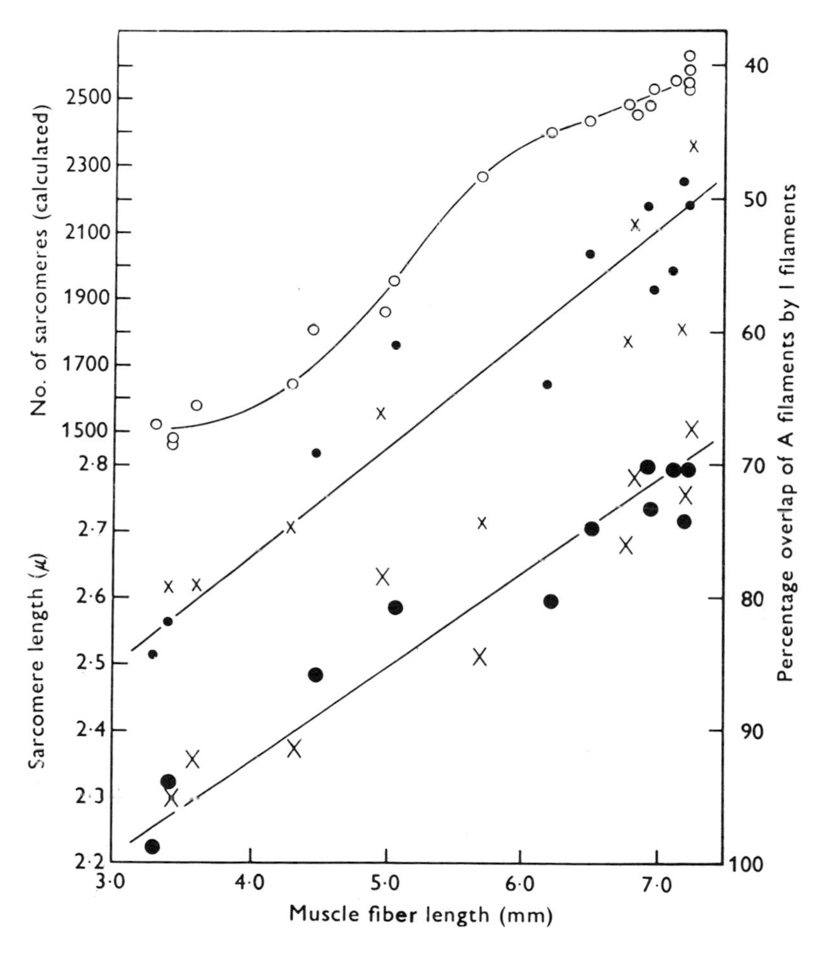


*Fig. 25.1.* Sarcomere length measurements plotted against muscle fiber length: *large solid circle,* muscles fixed immediately after death; *X,* muscles fixed post rigor. The percentage overlap of the total length of the A filaments by the I filaments on each side is also given: *small solid circle,* muscle fixed immediately after death; *x,* muscle fixed post rigor. The plot at the top of the figure gives values for the total number of sarcomeres plotted against the fiber length. The values were obtained by dividing the fiber length by the sarcomere length for each muscle (*open circle*). From Goldspink (1968).

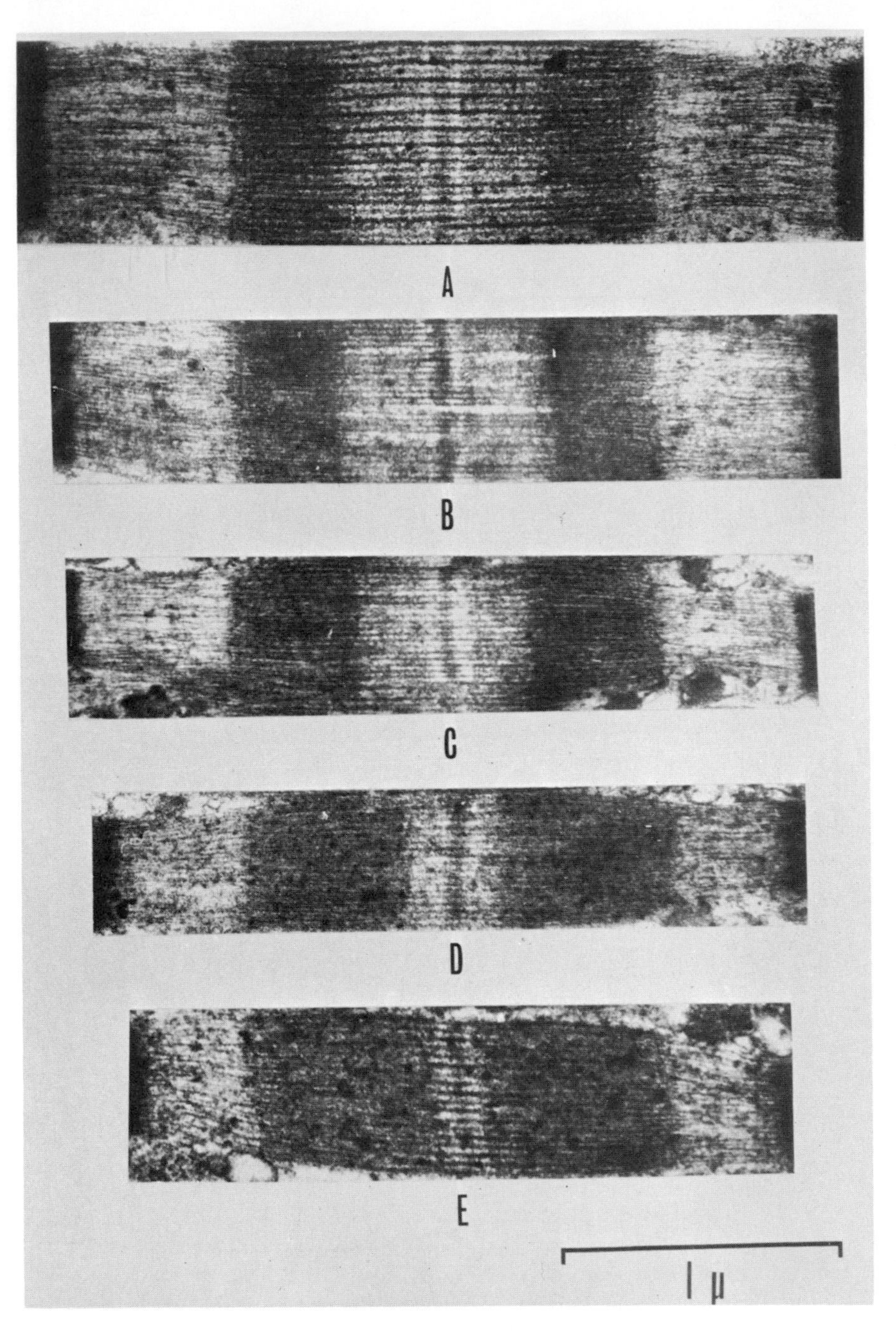


*Fig. 25.2.* Examples of electron micrographs of sarcomeres from muscles of different ages. The body weight of the mice from which these muscles were taken were as follows: *A,* 49 g; *B,* 35 g; *C,* 24 g; *D,* 9 g; and *E,* 2.5 g. From this series it will be seen that the increase in sarcomere length during growth is due to a decrease in the overlap of the A filaments by the I filaments and not to a change in the length of either type of filament. From Goldspink (1968).

The change in the sarcomere length during growth is of physiological importance. The maximum overlap of the A and I filaments in the very young muscles, with the limb fully extended, means that the muscles cannot shorten and develop any considerable force. However, the muscles of very young animals in general are required to develop and maintain isometric tension, whether it be in clinging to the mother or in the act of suckling. They are therefore ideally adapted to this function not only because the sarcomeres have a maximum functional overlap of the A and I filaments but also because the muscles of the young animals are all slow (Buller, Eccles, and Eccles, 1960*a*; Close, 1964).

This latter point is of particular significance as Goldspink, Larson, and Davies (1970) have recently shown that slow muscles use much less ATP in developing and maintaining isometric tension than fast muscles, whereas fast muscles use much less ATP in doing isotonic work. This is believed to be due to the time that the cross bridges remain attached to the actin filaments. If ATP is broken down only when the cross bridges engage and disengage (Davies, 1963), the longer the cross bridges are engaged the more efficient the muscle will be in maintaining isometric tension. However, a long cross-bridge cycle time will be a disadvantage when the muscle is required to shorten and do isotonic work because the cross bridges that are shortening will be working against those that are holding.

When the animal grows a little older, the sarcomeres lengthen and the speed of contraction of many of the muscles increases, apparently owing to the influence of their nerve supply (Buller, 1960*b*; Guth, 1969). The muscles thus become adapted for forceful isotonic contractions and the more efficient performance of external work (M. Z. Awan and Goldspink, unpublished observations, 1970). This type of adaptation has taken place during evolution, and the system has become programmed for the changes to occur during the normal growth process.

### INCREASE IN MUSCLE GIRTH

The muscle fibers constitute between 75% and 92% of the total muscle volume and the extracellular fluid makes up most of the rest of the volume. During normal growth, however, the extracellular compartment is believed to increase in proportion to the fiber mass (Goldspink, 1966), although it may be decreased during rapid hypertrophy of the fibers or increased due to atrophy of fibers resulting from starvation. However, in the main it is the total fiber cross-sectional area that is important in determining muscle girth.

In theory, a change in the total fiber cross-sectional area may be brought about by a change in the total number of fibers or by a change in the size of the existing fibers. It is almost certain that the number of fibers does not normally change once the differentiation of the tissue is complete (MacCallum, 1898; Goldspink, 1962*a*; Rowe and Goldspink, 1969). The total number of fibers that a particular muscle receives is genetically determined (Luff and Goldspink, 1967), whereas the size of the fibers is determined by the physiological conditions to which the muscle is subjected.

The way individual muscle fibers grow during postnatal growth is of particular interest. Once differentiation is complete all the fibers in the muscle (excluding those of the muscle spindles) are of about the same size: 20 $\mu$ in diameter. In some muscles of the mouse (Goldspink and Rowe, 1968; Rowe and Goldspink, 1969), e.g., the soleus (Fig. 25.3) and extensor digitorum longus, the fibers stay at this level of development throughout life; this we call the basic level or the small phase. In other muscles, e.g., the biceps brachii (Fig. 25.3) and anterior tibialis, some of the small phase fibers undergo further development during growth to a diameter of about 40 $\mu$. These, the large phase fibers, can be seen from distribution histograms for the fiber sizes (Fig. 25.4). From the histograms for the biceps brachii the appearance of a second peak becomes evident when the animal is about 3 weeks of age. By the time the animal reaches maturity the second-size population becomes predominant, especially in the biceps brachii of the male.

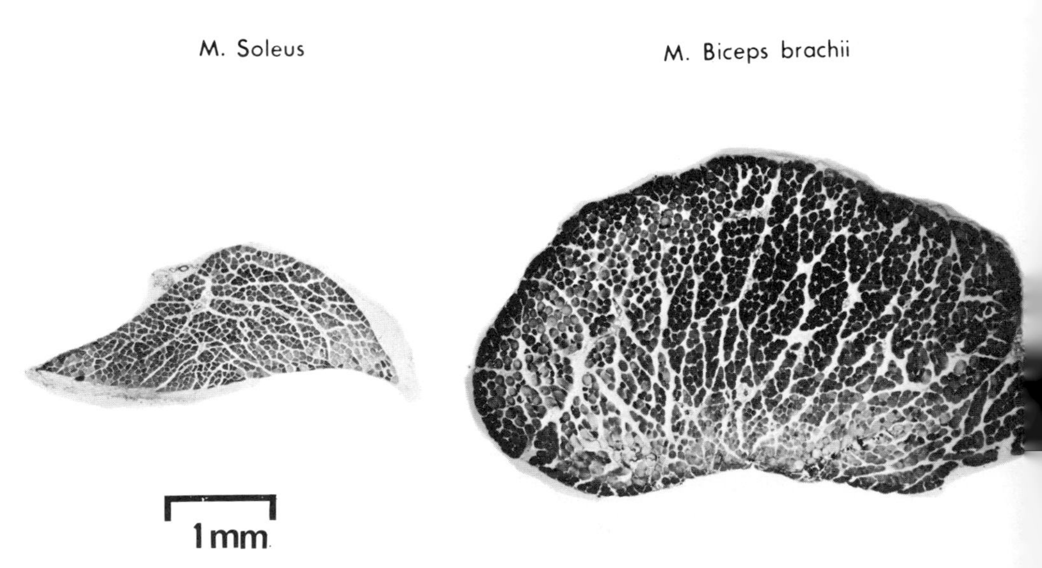


*Fig. 25.3.* Transverse sections of the mouse soleus and biceps brachii muscles. From Goldspink Rowe (1968).

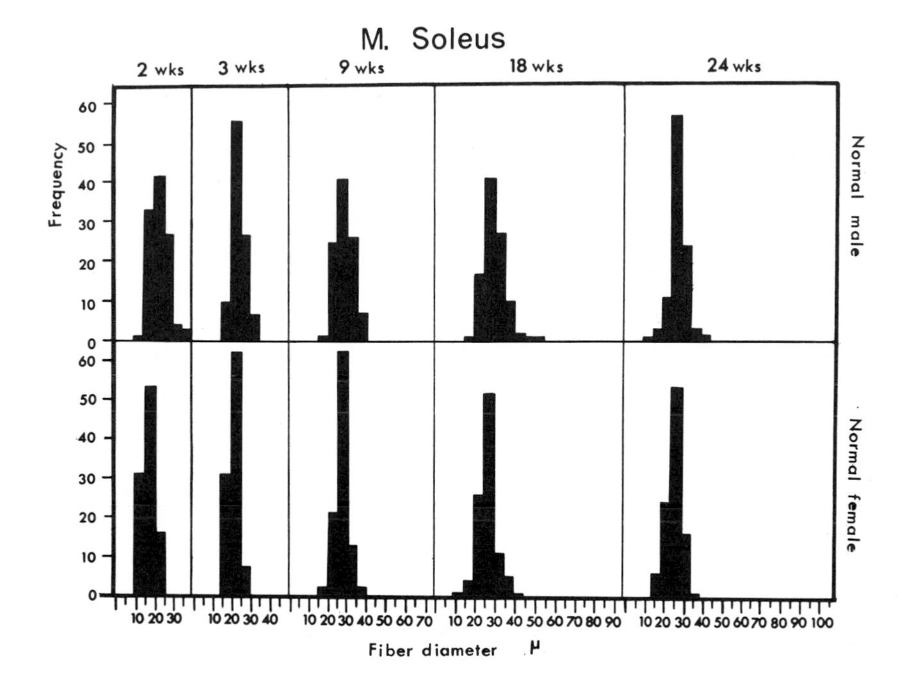


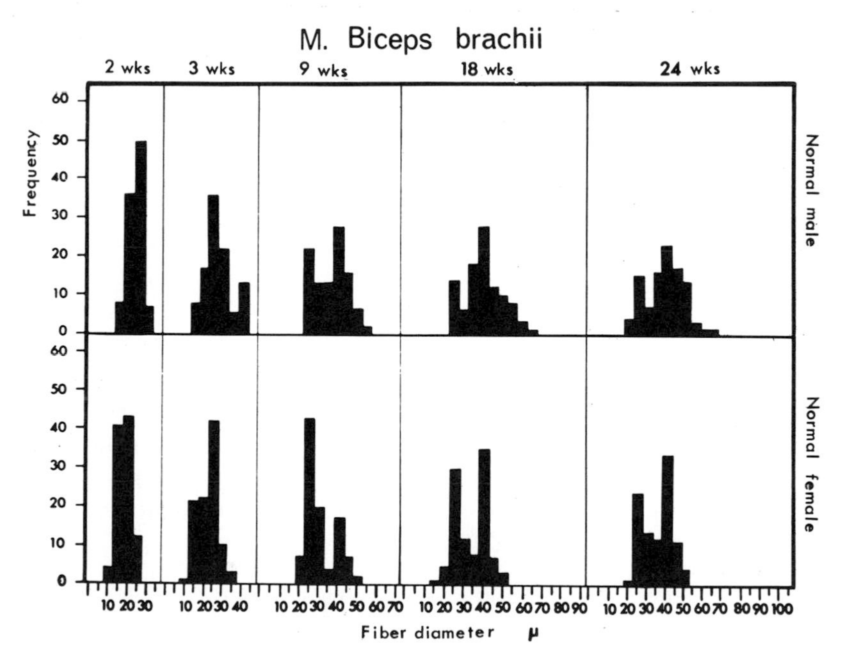


*Fig. 25.4.* Frequency histograms showing the distribution of fiber sizes in the soleus and biceps brachii muscles from mice of both sexes at different ages. Each histogram is for a 100-fiber sample taken at random. From Rowe and Goldspink (1969).

These two phases of fiber development (Fig. 25.5) are of particular importance as far as adaptation is concerned. This becomes apparent when the quantitative structural makeup of fibers in these two phases is examined. The small phase fibers possess a few (approximately 500) small myofibrils with a mean cross-sectional area of 0.3 $\mu^2$. The large fibers on the other hand possess many more myofibrils (approximately 1,250) of 0.8 $\mu^2$ cross-sectional area. The large fibers are about four times the cross-sectional area. Thus when a small fiber is converted into a large fiber there is a disproportionately large increase in the myofibrillar content of this fiber. Indeed, the increase in the girth of the fiber can be explained almost entirely by the increase in the number and size of its myofibrils.

The stimulus which induces myofibril synthesis seems to be the intensity of the work load on the individual fibers. In muscles such as the biceps brachii and anterior tibialis it appears that even during normal growth some of the fibers are functionally overloaded and respond to this situation by synthesizing more myofibrils. It has been shown that the minimum time necessary for a small phase fiber to grow into a large phase fiber is of the order of 1.5 to 2 days (Goldspink, 1965). This is a surprisingly short time considering that during this period the myofibril content of the fiber is increased 6.5 times. It should be realized, however, that at any time only a small percentage of fibers are undergoing this change.

During growth the body weight of the animal increases and this imposes a more intense work load on the muscles. If the intensity of work load becomes very considerable, as it apparently does in some muscles e.g., the biceps brachii and the anterior tibialis, then the muscle adapts by the appropriate percentage of fibers undergoing the jump in size from the small phase into the large phase. In this way the myofibrillar content of the muscle as a whole is increased and it therefore possesses the necessary contractile strength to cope with the increased work load.

The change from the small phase to the large phase has been shown to be quite reversible. Decreasing the food supply causes the large phase fibers to revert back to the small phase, and when the animal is replaced on a high level of nutrition the small phase fibers once again develop into the large phase (Goldspink, 1965).

## ADAPTATION TO REPEATED EXERCISE

Hypertrophy of striated muscle fibers as a result of athletic training has been an accepted fact for many years. Morpurgo (1897), Siebert (1928), and Thörner (1934) attributed hypertrophy of the fibers to an increase in the sarcoplasm and not to an increase in fibrillar material. More

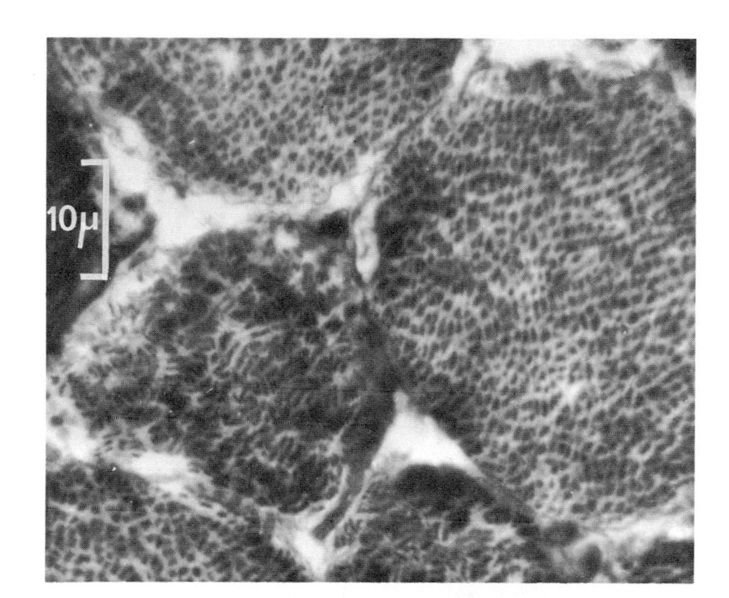


*Fig. 25.5.* Large and small phase fibers viewed with the phase contrast microscope (*top*) and the electron microscope (*bottom*). Note the tighter packing of myofibrils in the large fiber and the greater concentration of mitochondria in the small fiber.

recently, Helander (1961) investigated the effect of exercise and restricted activity on the various protein fractions of the musculature of the rabbit and guinea pig. The conclusions drawn from this work were that exercise increased the myofibrillar protein whereas restricted activity reduced this fraction and increased the proportions of sarcoplasmic proteins. There are many indications that the type of exercise is of great importance in inducing hypertrophy. Steinhaus (1933) cites Siebert (1928) as stating that exercise of endurance (running at low speed over a long distance) leaves the musculature relatively unchanged, whereas exercise involving an intensive effect (running at high speed over a relatively short distance) induces hypertrophy. Certainly it is recognized that athletes who engage in weightlifting develop the largest and most powerful muscles by lifting very heavy weights for only a few times each week. It is, therefore, important when studying adaptation to repeated exercise, to bear in mind the type of activity engaged in: whether in fact the musculature is being trained for endurance or for increased contractile strength.

The method of exercising laboratory animals for studies of this nature is therefore of considerable importance. In the past our interest has been mainly directed towards the adaptation of muscles to intensive exercise. For this reason, a method of exercise based on weightlifting was devised for use with mice and rats. This consisted of a pulley over which a cord was placed with weights at one end and a food cube or food basket at the other end. The animal, in order to obtain its food, had to pull down the cord and at the same time lift the weight. A recording system was attached to the pulley so that the amount of work performed by the animal could be measured each day. This type of intensive exercise is particularly successful not only because it induces quite extensive hypertrophy but also because it does not stress the animal unduly. This latter point is believed to be of particular importance, that is to say, the animals were persuaded to exercise for reward rather than by the administration of punishment (e.g., electric shocks, etc.)

The effect of intensive exercise has been studied at both the gross and the cellular level (Goldspink, 1964). The results showed that the change in muscle weight was not a good indication of the changes that had occurred in fiber size. This finding has been corroborated recently by Gordon, Kowalski, and Fritts (1967). I believe that this is because the rapid increase in the main fiber size results in the partial exclusion of extrafiber compartments. This seems to be the case particularly in muscles which have normally a large extrafiber fluid volume, such as the biceps brachii (Goldspink, 1966). Thus the hypertrophy of the fibers first brings about a consolidation of the tissue, and further hypertrophy of the fibers results in the hypertrophy of the muscle as a whole.

The mechanism of hypertrophy resulting from intensive exercise is interesting in that it seems to be exactly the same as that which produces the increase in fiber girth during growth. Fig. 25.6 shows frequency histograms of fiber size in the biceps brachii muscle of mice that have been exercised by weightlifting for a period of 3 weeks. Also given are histograms for the same muscle of control mice that received no intensive exercise. It will be seen that in this particular experiment, which was incidentally carried out on half-grown mice, the ratio of small to large fibers was approximately 30:70 in the exercised muscle and 60:40 in the control muscles. As the large fibers are about four times the cross-sectional area of the small fibers, this represents a change in the total fiber cross-sectional area of about 40%. Also, as the large fibers possess about 6.5 times more myofibrillar material, it represents an increase in the total myofibril cross-sectional area of about 50%.

Comparable increases in the mean fiber size have been induced in muscles consisting entirely of small red fibers, e.g., the soleus muscle. In these experiments (Rowe and Goldspink, 1968), the work load on this muscle was increased by the surgical incapacitation of the synergetic gastrocnemius. The hypertrophy of this muscle was again due to the conversion of small fibers into large fibers, and not due to an increase in the size of all the fibers.

From the evidence presented above, it is apparent that the morphological adaptation due to intensive exercise results from the acceleration of

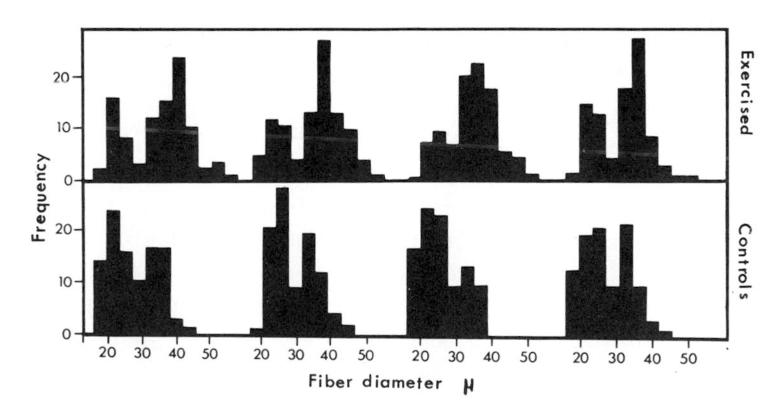


*Fig. 25.6.* Distribution histograms of fiber diameters in the biceps brachii muscles from exercised mice (*top*) and from control mice (*bottom*). Each histogram is for a 100-fiber sample taken at random.

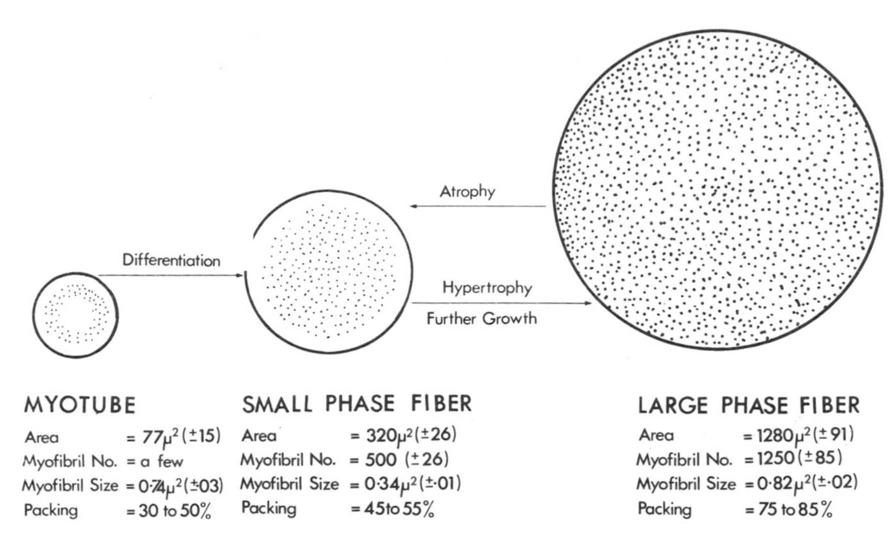


*Fig. 25.7.* The different phases of muscle fiber development during postnatal growth in the mouse.

the rate of change from the small phase to the large phase and by the alteration of the final ratio of small to large fibers. The importance of this change lies in the disproportionately large myofibrillar content of the large phase fibers. However in certain circumstances the change from small phase to large phase may not be entirely beneficial, as there are certain other consequences of hypertrophy. The small fibers are red in color due to the relatively high concentration of myoglobin and cytochromes. They also have a high concentration of mitochondria. When a small fiber undergoes hypertrophy or further growth, the mitochondria do not normally increase in amount, and so become diluted by the additional myofibrillar material (Goldspink, 1962*b*, 1968*b*). This means that, although the fibers have increased in contractile strength, they may lack the ability to contract repeatedly and perform external work for long periods of time. Recently it has been shown that adaptation to exercise of a repetitive kind induces increases in mitochondria (Holloszy, 1967) and in mitochondrial enzymes, but not in the myofibrillar proteins (Gordon, Kowalski, and Fritts, 1967).

An interesting question is presented if all these findings are correct: How is it that intensive exercise switches on the myofibril-synthesizing system, whereas exercise of the repetitive kind switches on the synthesizing systems for mitochondria and the enzymes of the energy-producing pathways? Indeed, what is the chemical link between the mechanical event and the initiation of the different synthetic systems? It is this link which enables the muscle to produce the right sorts of proteins so that it can adapt to the particular conditions to which it is subjected.

### THE MECHANISMS OF MYOFIBRIL PROLIFERATION

As stated above, the number of myofibrils in an individual muscle fiber can increase in number from about 500 to 1,250 in a very short period of time (Fig. 25.7). Studying this system is difficult because the change is all or nothing and at any one time is taking place in only a small percentage of fibers. Recently I (Goldspink, 1969) examined fibers of different sizes in an attempt to elucidate how the number of myofibrils increases during growth and as a response to exercise. In large phase and in fibers of intermediate size many myofibrils observed were apparently in the process of splitting longitudinally. Size measurements showed that the splitting myofibrils were on the average about twice as large as the nonsplitting myofibrils. The measurements also indicated that the myofibrils split more or less down the middle once they attain a certain critical size. In the fork of each split (Fig. 25.8) there were always elements of sarcoplasmic reticulum and sometimes a piece of mitochondrion, so it was very unlikely that the splitting was an artifact due to the mechanical handling of the tissue. A probable explanation for the splitting is that the peripheral I filaments are pulled at an angle slightly oblique to the myofibril axis because of the difference in the spacing of the A filaments and I filaments at the level of the Z disc. When contractile tension is developed by the sarcomeres, a strain is set up in the center of the Z discs, and when the myofibril reaches a certain size the oblique pull of the peripheral I filaments is strong enough to cause the Z discs to rip. Hence two daughter myofibrils are produced from the one original large myofibril. The mechanism of myofibril proliferation is being investigated further.

### SUMMARY

In this paper two types of morphological adaptations have been discussed. The first concerns changes that have taken place during the evolution of the musculature and which are now programmed for during the growth process. The second concerns the ability of the individual muscle fibers to respond to a functional overload by undergoing further development or hypertrophy. The muscle fibers of the mouse act in an all-or-nothing manner in that they jump in diameter from 20 $\mu$ to 40 $\mu$. Associated with this jump in size there is a rapid and disproportionately large increase in the myofibril content of the fiber. It is this which enables the muscle to cope with the increased intensity of work to which it is subjected. The further growth or hypertrophy of muscle fibers has

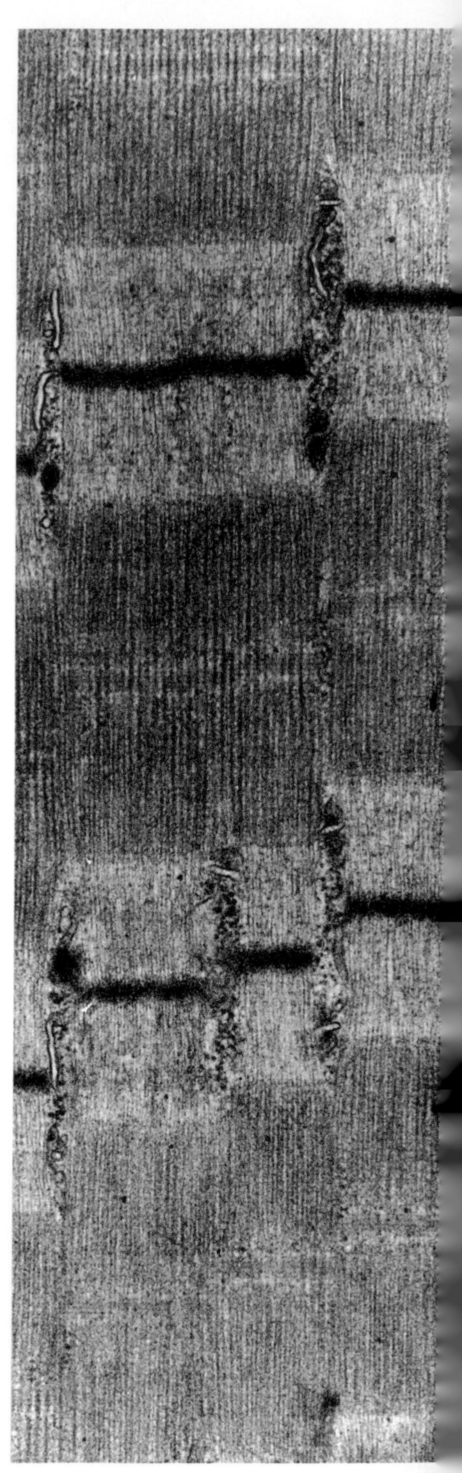

*Fig. 25.8.* Examples of myofibrils which are apparently splitting longitudinally. The one side of the sarcomere can be seen to be intact whereas the Z discs on the other side have split. Elements of the sarcoplasmic reticulum can be seen in the fork of each split. × 32,000.

other consequences which may, in certain circumstances, be undesirable, the main one being the dilution of the mitochondria and the energy supply systems by the additional myofibrillar material. The mitochondria and the energy supply systems in general do not apparently increase as a result of intensive exercise. The mechanism of myofibril proliferation which occurs during growth and as a response to intensive exercise has also been discussed.

## ACKNOWLEDGMENTS

I wish to acknowledge the financial support which I received for this research from the Agricultural Institute, Ireland (An Foras Taluntais), the Muscular Dystrophy Group of Great Britain, and the Agricultural Research Council. I would also like to express my gratitude to J. N. R. Grainger, P. G. 'Espinasse, R. E. Davies, and J. G. Phillips for their encouragement of this work.

## *References*

Buller, A. J., J. C. Eccles, and R. M. Eccles. 1960*a*. Differentiation of fast and slow muscles in the cat hind limb. *J. Physiol. 150:*399.

———. 1960*b*. Interactions between motoneurones and muscles in respect of the characteristic speeds of their responses. *J. Physiol. 150:*417.

Close, R. 1964. Dynamic properties of fast and slow skeletal muscles of the rat during development. *J. Physiol. 173:*74.

Davies, R. E. 1963. A molecular theory of muscle contraction: calcium-dependent contractions with hydrogen bond formation plus ATP-dependent extension of part of the myosin-actin cross-bridges. *Nature 199:*1068.

Dubowitz, V. 1965. Enzyme histochemistry of skeletal muscle I. Developing animal muscle. *J. Neurol. Neurosurg. Psychiat. 28:*516.

Fischman, D. A. 1967. An electron microscope study of myofibril formation in embryonic chick skeletal muscle. *J. Cell Biol. 32:*557.

Goldspink, G. 1962*a*. Studies on postembryonic growth and development of skeletal muscle. *Proc. Roy. Irish Acad. 62B:*135.

———. 1962*b*. Biochemical and physiological changes associated with the postnatal development of the biceps brachii. *Comp. Biochem. Physiol. 7:*157.

———. 1964. The combined effects of exercise and reduced food intake. *J. Cell. Comp. Physiol. 63:*209.

———. 1965. Cytological basis of decrease in muscle strength during starvation. *Amer. J. Physiol. 209:*100.

———. 1966. An attempt at estimating extrafiber fluid in small skeletal muscles by a simple physical method. *Can. J. Physiol. 44:*765.

———. 1968. Sarcomere length during post-natal growth of mammalian muscle fibres. *J. Cell Sci. 3:*539.

———. 1969*a*. The proliferation of myofibrils during post-natal muscle growth. *J. Cell Sci.* (In press.)

———. 1969*b*. Succinic dehydrogenase content of individual muscle fibers. *Life Sci. 8:*791.

Goldspink, G., R. E. Larson, and R. E. Davies. 1970. The immediate energy supply and the cost of maintenance of isometric tension for different muscles in the hamster. *Z. Vergl. Physiol. 66:*389.

Goldspink, G., and R. W. D. Rowe. 1968. Muscle fiber growth in normal and dystrophic mice, p. 116. *In The Fourth Symposium of Current Research in Muscular Dystrophy.* Pitman Medical Press, New York.

Gordon, E. E., K. Kowalski, and M. Fritts. 1967. Adaptations of muscle to various exercises. *J. Amer. Med. Ass. 199:*103.

Guth, L. 1969. Trophic influences of nerve on muscle. *Physiol. Rev. 48:*645.

Helander, E. 1961. Influence of exercise and restricted activity on the protein composition of skeletal muscle. *Biochem. J. 78:*428.

———. 1966. General considerations of muscle development, p. 19. *In* E. J. Briskey, R. G. Cassens, and J. C. Trautman (eds.), *The Physiology and Biochemistry of Muscle as a Food.* Univ. of Wis.

Hoffmann, A. 1947. Weitere Untersuchungen über den Einfluss des Trainings auf die Skelettmuskulatur. *Anat. Anz. 96:*191.

Holloszy, J. O. 1967. Biochemical adaptations in muscle. *J. Biol. Chem. 242:*2278.

Luff, A. R., and G. Goldspink. 1967. Large and small muscles. *Life Sci. 6:*1821.

MacCallum, J. B. 1898. Histogenesis of the striated muscle fiber and the growth of the human sartorius muscle. *Johns Hopkins Hospital Bull. 6:*208.

Morpurgo, B. 1897. Ueber Activitats-Hypertrophie der willkurlicken Muskeln. *Virchow's Arch. 150:*522.

Rowe, R. W. D. 1967. The post-natal growth and development of striated muscle fibers in normal and dystrophic mice. Ph.D. thesis, University of Hull, England.

Rowe, R. W. D., and G. Goldspink. 1968. Surgically induced hypertrophy in skeletal muscles of the laboratory mouse. *Anat. Rec. 161:*69.

———. 1969. Muscle fibre growth in five different muscles in both sexes of mice I. Normal mice. *J. Anat. 104:*519.

Siebert, W. W. 1929. Untersuchungen über Hypertrophie des Skelettmuskels. *Z. Klin. Med. 109:*350.

Steinhaus, A. H. 1933. Chronic effects of exercise. *Physiol. Rev. 15:*103.

Thörner, S. H. 1934. Trainingsversuche an Hunden 3. Histologishe Beobachtungen an Herz- und Skelettmuskel. *Arbeitsphysiologie 8:*359.

# Chapter 26

# *Biochemical Adaptation during Development and Growth in Skeletal Muscle*

S. V. PERRY

In a tissue such as skeletal muscle, which is adapting to a rapidly changing activity pattern during fetal and early postnatal life, it is difficult to distinguish sharply developmental changes from those associated with the rapid growth also occurring during this period. The changes in activity pattern of muscle associated with birth and neonatal development are accompanied by modifications in the biochemical composition of the cell, and at the same time the total amount of muscle mass increases parallel to the growth of the animal. For the purpose of the present discussion attention will be concentrated on the biochemical changes that occur in muscle from the stage at which the tissue can be recognized as composed principally of early polynucleate cells containing some myofibrils. Such skeletal muscle is present in all animals some time before birth and undergoes rapid changes in composition particularly just before birth and during early postnatal life. The discussion will be further confined to the changes associated with the muscle cell itself, but it should not be forgotten that growth in muscle is also accompanied by marked changes in the associated adipose and connective tissues, factors which may be of considerable economic importance.

It is convenient in discussing the changes occurring in muscle to regard the muscle cell as composed of three main biochemical systems, the coordinated activity of which represents the functional activity of the muscle as a whole:

(1) The contractile system localized in the myofibril and consisting principally of myosin, actin, and tropomyosin, and the less well-characterized minor protein components, several of which have a role in regulating the enzymic activity of the actomyosin system. The myofibrillar fraction represents about 55–65% of the total nitrogen in adult skeletal muscle.

(2) The sarcoplasmic reticulum acting as a link between the electrical changes at the membrane and the enzymic activity and hence contractile response of the myofibril. This system, by virtue of its ability to concentrate $Ca^{++}$ within the vesicles, controls the myofibrillar adenosine triphosphatase activity. In adult skeletal rabbit muscle this fraction represents about 2–4% of the total nitrogen content.

(3) The anaerobic and aerobic energy-yielding systems associated with the sarcoplasm and mitochondria respectively.

These systems in total account for up to about 25–35% of the total nitrogen in skeletal muscle and are associated with the sarcoplasmic proteins, except in certain highly aerobic muscles which are very rich in mitochondria.

During growth and development all three systems change both qualitatively and quantitatively, but perhaps the most striking gross feature of the early stages of growth in muscle is the rapid increase in total nitrogen/g wet weight. In the case of rabbit skeletal muscle the nitrogen content increases by a factor of about three over the period from 1 week before birth until 3 weeks after birth. During the same period the total weight of the longissimus dorsi increases by a factor of 25–30 (Holland, 1968). Subsequently little further change in nitrogen content occurs. This period of rapid rise in the dry matter content is reflected in a marked change in the relative proportions of the total nitrogen associated with the myofibril, sarcoplasm, mitochondria, and microsome (sarco-

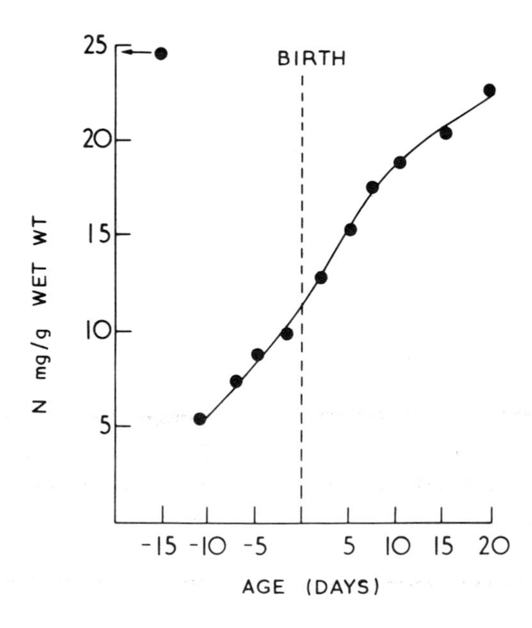


*Fig. 26.1.* Total nitrogen content of the myofibrillar fraction of rabbit longissimus dorsi during growth and development. Muscle homogenized in 0.25 M sucrose and the crude myofibrillar fraction separated by centrifugation for 15 min at 800 g and washed twice.

plasmic reticulum) fractions. It is a time of rapid myofibril formation (Fig. 26.1), and indeed much of the increase in nitrogen is due to the increase in the number of myofibrils per cell, for over this period the percentage of the total nitrogen in the other fractions falls steadily (Fig. 26.2).

There are suggestions that the myofibrils, in addition to increasing in absolute amount, are changing qualitatively during the early phase of growth. In the first place, myofibrils from early neonatal rabbit longissimus dorsi appear to be more fragile than their adult counterparts. Whereas the homogenization procedure used for preparation (Perry and Corsi, 1958) produces satisfactory myofibrils from adult muscle in the neonate, the tissue does not readily break up longitudinally into single myofibrils but tends to break transversely into short bundles. Secondly, although direct analytical data for the protein composition are not available, enzymic studies reveal differences in the nature of the myosin present and the minor components responsible for regulation of the adenosine triphosphatase. Differences in the latter system are evident from the low sensitivity of the $Mg^{++}$-activated adenosine triphosphatase of

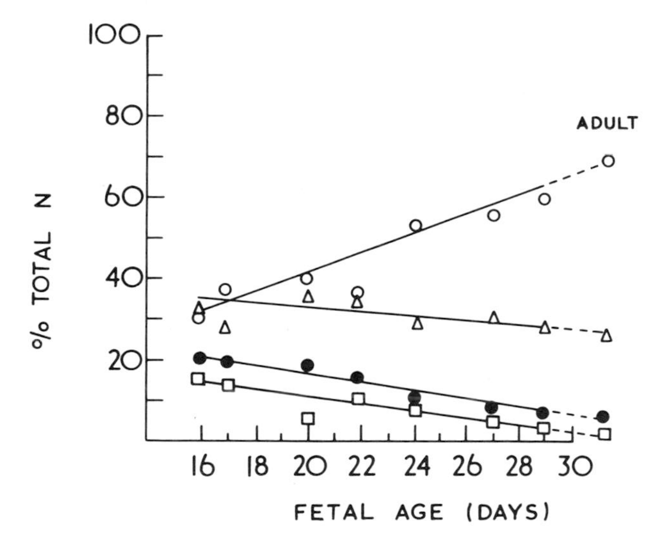


*Fig. 26.2.* Distribution of total nitrogen between the intracellular fractions of rabbit skeletal muscle. Combined back and leg muscles homogenized in 0.25 M sucrose. *Open circle,* myofibrillar-nuclear fraction, 10 min at 1,000 g; *open square,* mitochondrial fraction, 10 min at 12,000 g; *solid circle,* sarcoplasmic reticulum fraction, 1 hr. at 100,000 g; *open triangle,* sarcoplasmic fraction, not sedimented. From Perry and Hartshorne (1962).

myofibrils to 1, 2-bis-(2-dicarboxymethylaminoethoxy-ethane (EGTA) immediately after birth (Fig. 26.3). The effect of EGTA increases with age until the enzymic properties of the myofibril are very similar to those of the adult, in which the chelating agent inhibits the adenosine triphosphatase by 80% under the standard assay conditions.

Myosin is the only major protein component of the myofibril that has been studied in any detail and for which clear evidence has been obtained of structural changes during growth and development of skeletal muscle. The rather limited studies on actin and tropomyosin so far carried out, on the other hand, suggest that the molecular structures of these proteins do not change during growth and development. Myosin isolated from skeletal muscle of fetuses from a number of species is clearly different from that obtained from the corresponding adult tissues, suggesting the existence of a discrete fetal form of the enzyme. The general picture of muscle growth over the late fetal and early postnatal stages, when the myofibrils are rapidly increasing in number, is of a gradual replacement of the fetal type of myosin by a type characteristic of the adult tissue. The rate and extent of this replacement at a given age is

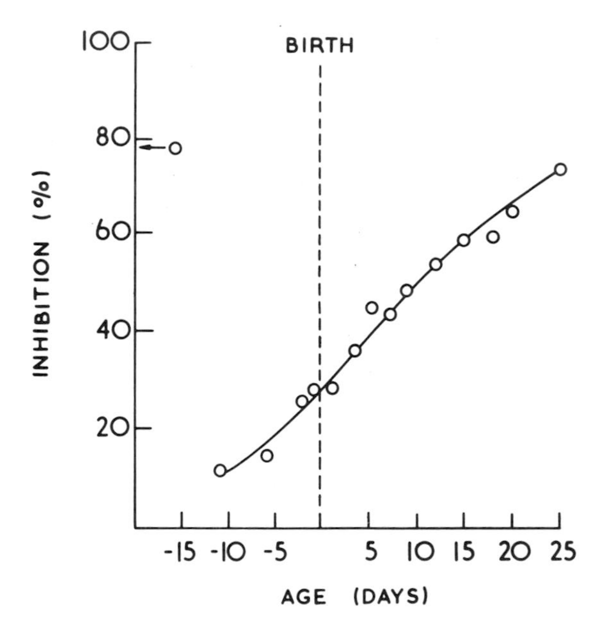


*Fig. 26.3.* Effect of EGTA on the $Mg^{++}$-activated adenosine triphosphatase of myofibrils isolated from rabbit longissimus dorsi. Conditions of assay: 25 mM Tris-HCl, *p*H 7.6, 2.5 mM tris-ATP, 2.5 mM $MgCl_2$, ± 1.0 mM EGTA. Incubated 5 min at 25° C (Holland, 1968).

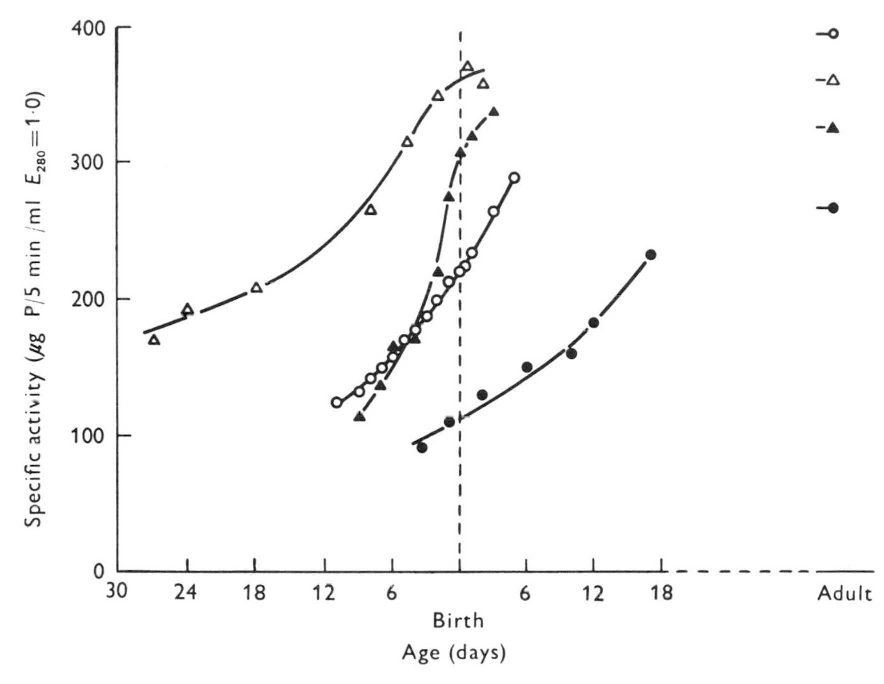


*Fig. 26.4.* Specific adenosine triphosphatase activity of purified myosin isolated from mixed skeletal muscle at different stages of development. Conditions of assay: 5 mM ATP, 5 mM $CaCl_2$, 50 mM Tris-HCl, *p*H 8.2, 0.2 M KCl. Adult values (averages of at least three preparations) indicated by the appropriate symbols (top right). *Open circle,* rabbit; *open triangle,* guinea pig; *solid triangle,* fowl; *solid circle,* rat. From Trayer & Perry (1966).

a feature of the muscle type and the stage of development at birth of the species in which it is studied (Fig. 26.4). In the rabbit, in which most of the studies have been carried out, no differences in molecular weight and other hydrodynamic parameters can be demonstrated (Trayer and Perry, 1967). On the other hand, clear differences can be observed in the specific $Ca^{++}$-activated adenosine triphosphatase, amino acid composition (particularly in the 3-methyl histidine content; Trayer, Harris, and Perry, 1968), immunochemical response (Trayer, 1967), quantum yield of the tryptophan fluorescence (Trayer, Perry, and Teale, 1967), and in the nature of the light subunits obtained on dissociation of myosin (Perrie, Perry, and Stone, 1969).

The significance of the increase in 3-methyl histidine content of myosin from skeletal muscle as development proceeds is not clear (Table 26.1). The amount of 3-methyl histidine in myosin is not simply related to the age of the animal from which the myosin is derived, for lower values are

TABLE 26.1
*3-Methyl histidine content of myosin and actin*

| Protein | Animal | Muscle | Age | No. of Preparations | 3-Methyl histidine (residue/$10^5$ g) |
|---|---|---|---|---|---|
| Myosin | Rabbit | Mixed skeletal | Adult | 5 | 0.327 ± 0.033 |
| | " | " " | 28-day fetus | 5 | Not detectable |
| | " * | Heart | Adult | 2 | 0.27 |
| | Pigeon | Breast | " | 2 | 0.18 |
| Actin | Rabbit | Mixed skeletal | " | 5 | 2.10 ± 0.04 |

*Heavy component isolated by succinylation (Oppenheimer, Bárány, Hamoir, and Fenton, 1966).

found in the red skeletal and cardiac muscles of the adult than are found in the white skeletal muscle of the same animal. The extent of methylation of the histidine of myosin is in some way related to the functional activity of the muscle from which it is extracted as well as to its stage of development. Although the generalization can be made that within a given species, the lower the specific $Ca^{++}$-activated adenosine triphosphatase activity, the less the 3-methyl histidine content, there is no direct evidence that methylation of certain histidine residues of myosin is essential for biological activity. Indeed the evidence obtained by photo-oxidation suggests that it is not (Johnson, Lobley, and Perry, 1969). Further evidence of the different effect of growth and development on myosin compared to actin is evident in the fact that the one histidine per molecule of actin is methylated in the protein isolated from both fetal and adult skeletal rabbit muscle (Johnson, Harris, and Perry, 1967).

From analogy with other systems in which substituted amino acids are found in proteins, it would be expected that the histidines of myosin are methylated after polypeptide-chain synthesis. Preliminary *in vitro* experiments with $^{14}C$-labeled S-adenosyl methionine (Hardy, Perry, and Stone, 1970) support this hypothesis. *In vitro* methylation studies indicate that other methylated amino acids are present in adult rabbit skeletal muscle myosin (Hardy and Perry, 1969). One of the residues has been identified as $\epsilon$-N-methyl lysine. From the studies so far carried out it seems that the lysine residues, unlike the histidines, are methylated also in the myosin isolated from fetal skeletal muscle. Certainly fetal rabbit myosin contains $\epsilon$-N-monomethyl lysine and a methyl amino acid (Hardy and Perry, 1969), now (1970) identified as $\epsilon$-N-trimethyl lysine, in amounts similar to that present in adult myosin.

Another striking difference between myosins from adult and fetal skeletal muscle is in the naure of the low-molecular-weight components present in the molecule. The number of different low-molecular-weight components estimated to be present per molecule of adult myosin has varied among investigators (Dreizen, Gershmann, Trotta, and Stracher, 1967; Locker and Hagyard, 1968; Frederiksen and Holtzer, 1968; Perrie, Perry, and Stone, 1969).

In this laboratory four electrophoretically different low-molecular-weight components have been observed consistently in myosin extracted under specified conditions (Perrie, Perry, and Stone, 1969). Myosin from skeletal muscle of 1-day-old rabbit also contains four components of similar electrophoretic ability to those seen in the myosin isolated from the corresponding adult tissue, although two of them appear to be present in smaller amounts (Fig. 26.5).

In view of this and the other evidence for differences in the myosins from adult and fetal white skeletal muscle, it is suggested that myosin consists of two isozyme forms. One of the isozymes is considered to be the main component in myosin isolated from fetal and cardiac muscles, and

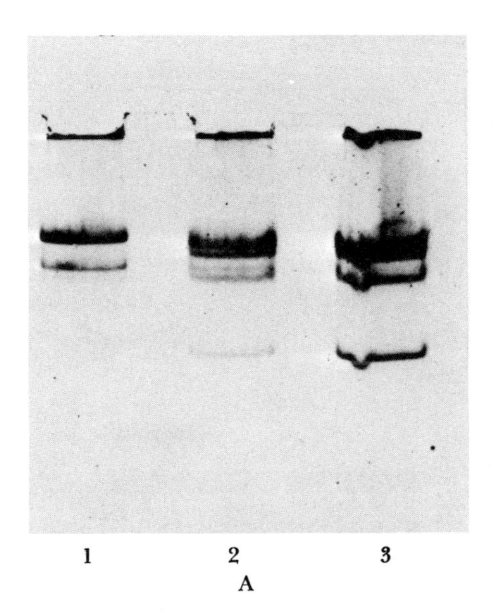


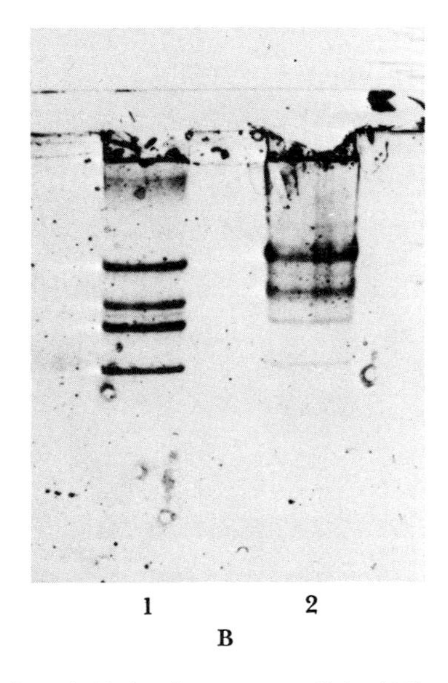


*Fig. 26.5.* Acrylamide electrophoresis of rabbit myosins. *A*, 10% gel cyanogum 41 in 40% glycerol, 20 mM Tris-glycine, *p*H 8.6 in (*1*) cardiac muscle; (*2*) red skeletal muscle; (3) white skeletal muscle. *B*, 10% gel, cyanogum 41 in 8 M urea, 20 mM Tris-glycine, *p*H 8.6 in (*1*) adult white skeletal muscle; (*2*) mixed skeletal muscle of 1-day-old rabbit.

the other the principle isozyme present in adult white muscle. Both isozyme forms are considered to consist of two heavy chains and two light chains (Fig. 26.6). Electrophoretic evidence suggests that the light chains are nonidentical in a given isozyme form and different in the two isozymes. Although it is possible that the two heavy chains are identical in a given isozyme, there may be differences between the heavy chains of the two forms, fetal and adult. Certainly the content of 3-methyl histidine differs in myosin prepared from different muscle types. The fact that the 3-methyl histidine is present in the heavy chain fraction of myosin (Trayer, Harris, and Perry, 1968) and that adult myosin contains 1.6 residues of 3-methyl histidine could be interpreted to suggest that each heavy chain of the adult isozyme contains one 3-methyl histidine and the fetal heavy chains contain none. This would imply that the usual rabbit adult myosin preparation contains 80% of the methylated (adult) and 20% of the unmethylated (fetal) isozyme.

This suggested isozyme distribution is very similar to that already recognized for lactic dehydrogenase and creatine kinase in muscle, although in the case of myosin the difference in specific enzymic activity between the isozyme forms is much more striking than in the case of the soluble enzymes. This special situation is probably a consequence of the need of the myofibril to increase its adenosine triphosphatase capacity per filament containing a constant number of myosin molecules, for it is well known that the speed of contraction increases with development. It does however raise some interesting problems in the synthesis both of myosin and of the myofibril which, as growth progresses (particularly in the early stage of life), is increasing in both abundance per cell and in quality and enzymic capacity per A filament.

Over the stage of rapid growth that occurs immediately post partum the total amount of sarcoplasm does not increase very markedly, the percentage increase in the first 3 weeks of life of the rabbit amounting to

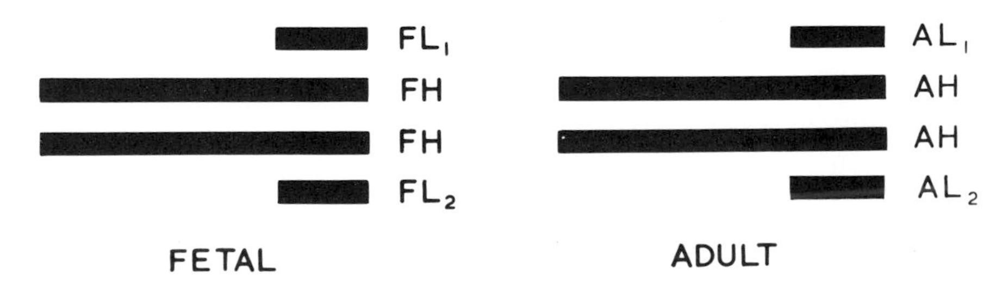


*Fig. 26.6.* Diagrammatic representation of suggested structures of myosin isozymes. *FH* and *FL,* heavy and light chains of the fetal form of myosin; *AH* and *AL,* corresponding chains for the adult form of myosin.

about one-quarter that of the myofibrillar fraction. Nevertheless marked changes in enzyme composition occur. The activities, related to unit weight of N, of enzymes of special significance for muscle activity such as creatine kinase, 5′-AMP deaminase, and adenylate kinase all change in a parallel manner that is characteristic of the muscle activity pattern and the stage of development of the animal at birth (Kendrick-Jones and Perry, 1967). For example the activities of creatine kinase and 5′-AMP deaminase both rise sharply at the stage at which the muscle is rapidly increasing in activity (Figs. 26.7 and 26.8). The activities of both enzymes in rabbit diaphragm muscle rise sharply before birth to reach an adult value immediately after birth, whereas in the leg muscle of the same animal the enzymic activities do not begin to rise sharply until some 8–10 days after birth, when the animal begins to move around actively. The importance of activity in stimulating change in enzymic levels is further suggested by the fact that, if young animals are encouraged to use their limbs earlier than is normal, the creatine kinase responds by increasing activity at an earlier age than in normal unexercised controls (Kendrick-Jones and Perry, 1967).

From the results presented it is apparent that the contractile capacity of the muscle cell is changing during the rapid phase of growth, and

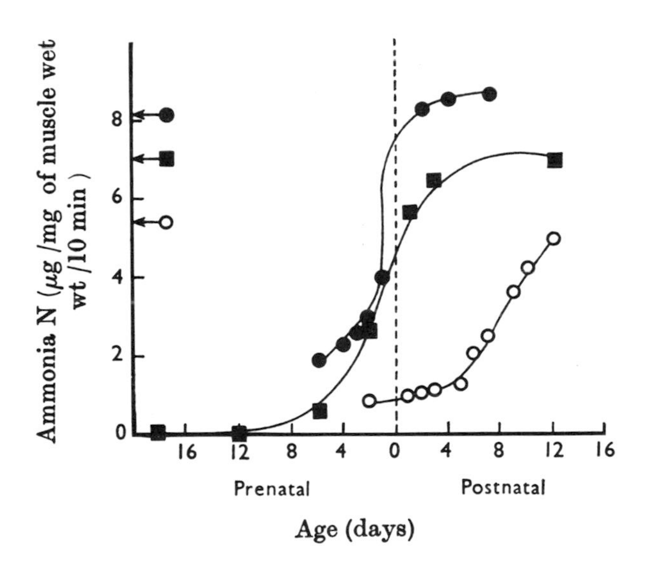


*Fig. 26.7.* 5′-AMP deaminase activity of mixed leg muscle during development. Adult values indicated by arrows on ordinate. *Solid square,* guinea pig; *open circle,* rat; *solid circle,* fowl. From Kendrick-Jones and Perry (1967).

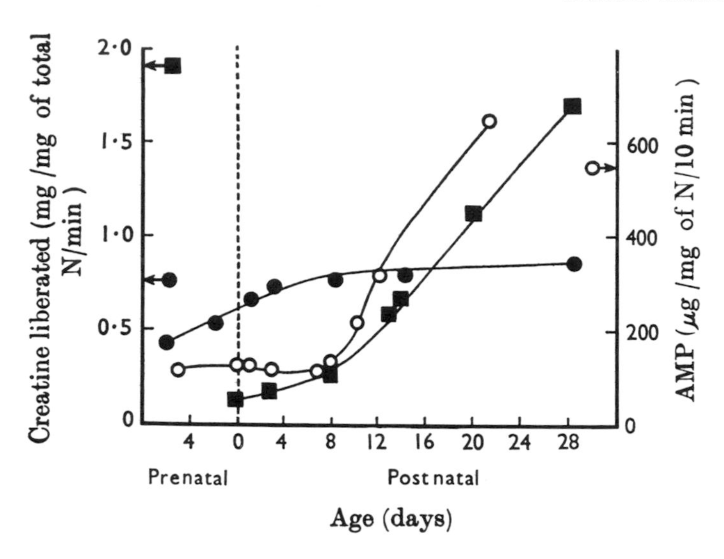


*Fig. 26.8.* Creatine kinase activity of rabbit muscle during development. Adult values indicated by arrows on ordinate. *Solid square,* leg creatine kinase; *solid circle,* cardiac creatine kinase; *open circle,* leg 5′-AMP deaminase. From Kendrick-Jones and Perry (1967).

simultaneously the enzyme systems supplying ATP for myofibrillar use are modified to cope with the increased contractile capacity of the tissue. It might further be expected that the system which regulates the myofibrillar adenosine triphosphatase is modified in step with the functional changes that are particularly marked at the early neonatal stage when growth is rapid.

The regulatory system is composed of the sarcoplasmic reticulum, which operates through the relaxing protein system, consisting of certain of the minor protein components built into the structure of the myofibril, by regulating the $Ca^{++}$ concentration of the sarcoplasm. From the results described earlier (see Fig. 26.3) it is apparent that the relaxing protein system of the myofibril develops somewhat more slowly than the contractile apparatus. At birth the $Mg^{++}$-activated adenosine triphosphatase of the rabbit myofibril is much less sensitive to EGTA and hence presumably to the $Ca^{++}$-sequestering activity of the sarcoplasmic reticulum. As yet there is no evidence as to how the relaxing protein system changes during growth. The system is complex, consisting of at least three components (Hartshorne, Theiner, and Mueller, 1969; Schaub and Perry, 1969), and its net effectiveness depends on the relative amounts of each present. Thus the lowered effectiveness at early ages could be due to the lowered level of all or simply of individual components of the system.

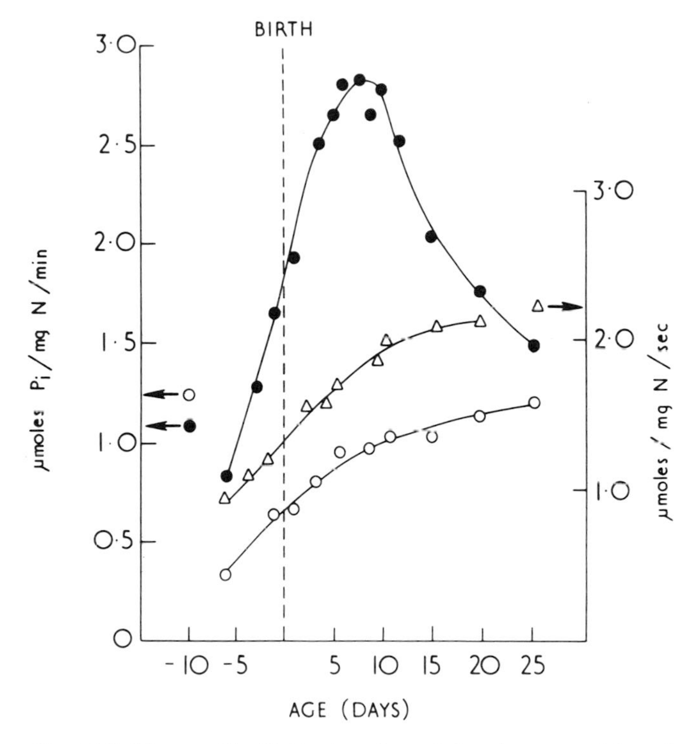


*Fig. 26.9.* The specific $Mg^{++}$ adenosine triphosphatase activity of the mitochondrial and sarcoplasmic reticulum fractions from developing rabbit longissimus dorsi muscle. Fractions were prepared using sucrose zone gradient centrifugation. Adult values indicated by arrows on ordinate. Conditions of adenosine triphosphatase assay: 25 mM Tris-HCl, *p*H 7.6, 2.5 mM $MgCl_2$, 2.5 mM Tris-ATP. Incubated at 25° C for 5 min. *Solid circle,* sarcoplasmic reticulum $Mg^{++}$ adenosine triphosphatase; *open circle,* mitochondrial $Mg^{++}$ adenosine triphosphatase; *open triangle,* mitochondrial cytochrome-c oxidase activity. From Holland and Perry (1969).

The effectiveness of the relaxing protein system in turn depends on the efficiency of the system regulating the $Ca^{++}$ level associated with the sarcoplasmic reticulum. Rapid changes are occurring in the sarcoplasmic reticulum, for when enzymic studies are carried out on the fraction carefully isolated by a gradient centrifugation method, a sharp peak in the specific $Mg^{++}$-activated adenosine triphosphatase, similar to the "basal" adenosine triphosphatase (Hasselbach, 1964), can be observed (Holland and Perry, 1968, 1969). In the case of rabbit longissimus dorsi muscle this peak occurs 8–10 days after birth (Fig. 26.9). Such a sharp peak in the adenosine triphosphatase activity of the sarcoplasmic reticular fraction is characteristic of muscle and is not observed in the microsome fraction isolated from tissues such as liver (Fig. 26.10). The time at which the peak occurs is specific for the particular type of muscle and the species from which it is derived. It occurs at precisely the same time as the sharp increase in the activity of a number of enzymes specific to muscle function referred to earlier (e.g., creatine kinase) is observed. Thus in diaphragm muscle of the rabbit, and in the leg muscles of the

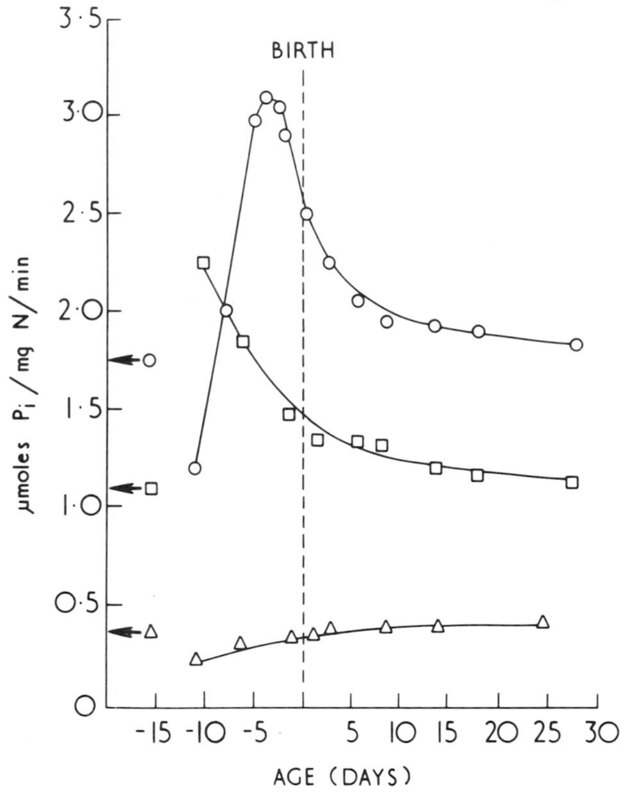


*Fig. 26.10.* $Mg^{++}$-activated ATPase of the "crude granular" fraction sedimented from rabbit muscles and liver. Fraction sedimented after 1 hour at 100,000 g from supernatant prepared by centrifuging whole homogenate for 15 min at 800 g. From muscle this crude granular fraction consists principally of the sarcoplasmic reticulum. Adult values indicated by arrows on ordinates. Conditions of adenosine triphosphatase assay: 25 mM Tris-HCl, *p*H 7.6, 2.5 mM $MgCl_2$, 2.5 mM Tris-ATP. Incubated at 25° C for 5 min. *Open circle,* diaphragm; *open square,* heart; *open triangle,* liver. From Holland and Perry (1969).

BLE 26.2

*nosine triphosphatase activity and $^{45}Ca^{++}$ uptake of the sarcoplasmic reticulum from developing rabbit longissimus dorsi cle*

| e of rabbits ys after birth) | Adenosine triphosphatase activities ($\mu$moles $P_i$/mg N per min) Basal | | Extra | | Basal + Extra | Rate of $Ca^{++}$ transport ($\mu$moles/mg N per min) | | Efficiency (moles $Ca^{++}$/moles ATP) |
|---|---|---|---|---|---|---|---|---|
| 1 day | 1.5 ± 0.1 | (6) | 0.6 ± 0.05 | (6) | 2.1 | 0.18 ± 0.01 | (6) | 0.3 ± 0.02 |
| 8 day | 2.4 + 0.2 | (6) | 1.2 ± 0.09 | (6) | 3.6 | 0.54 ± 0.01 | (5) | 0.45 ± 0.03 |
| Adult | 0.9 ± 0.01 | (9) | 3.0 ± 0.2 | (9) | 3.9 | 4.8 + 0.4 | (7) | 1.6 ± 0.2 |

esults expressed as the means of the initial rates ± SE. Number of experiments in parentheses.
om Holland and Perry (1969).

fowl and the guinea pig, the peak occurs just before birth. In leg muscles of animals in a less developed state, such as the rabbit and the rat, the peak occurs some days post partum.

Whereas the "basal" adenosine triphosphatase increases in this characteristic way, the "extra" adenosine triphosphatase (Hasselbach, 1964)—i.e., that enzymic activity stimulated by the addition of low concentrations of $Ca^{++}$ and presumed to be associated with the transport of this cation—rises in a steady manner. The ability of the sarcoplasmic reticulum to transport $Ca^{++}$, directly determined by radiochemical studies, likewise increased in a steady progressive way up to the maximum value obtained in adult muscle (Table 26.2). On the other hand the total adenosine triphosphatase activity as measured by the sum of "basal" and "extra" values had almost reached the maximum value at 8–10 days. This total activity did not fall with increasing age but slowly rose by another

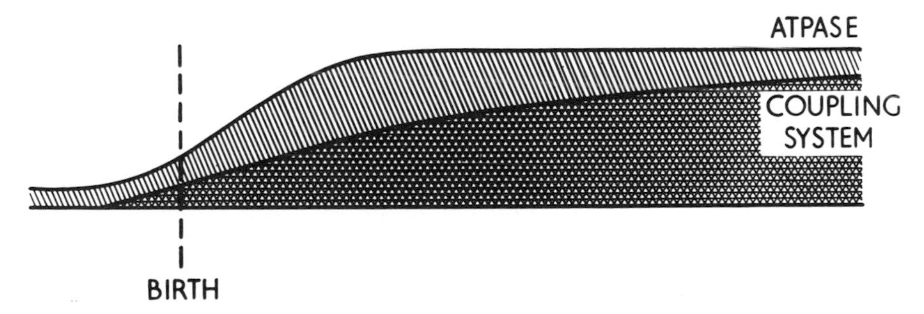


*Fig. 26.11.* Scheme to illustrate the development of sarcoplasmic $Mg^{++}$-activated adenosine triphosphatase and the system coupling it to $Ca^{++}$ transport in the sarcoplasmic reticulum. At a given age the height of the single-hatched area represents the activity of the "basal" adenosine triphosphatase and that of the double-hatched area the "extra" adenosine triphosphatase or capacity to transport $Ca^{++}$.

20–30% to reach the adult value. The results can be interpreted by assuming that the $Ca^{++}$ transport system of the reticulum consists of two components, namely the adenosine triphosphatase and a system that couples this enzymic activity to $Ca^{++}$ transport, which develop at different rates, as illustrated in Fig. 26.11. Thus the adenosine triphosphatase component of the transport system develops first and has almost reached its maximum capacity 8–10 days after birth in the case of rabbit longissimus dorsi. The coupling system, however, is poorly developed at birth and continues to increase in amount steadily after birth, but at a slower rate than the adenosine triphosphatase system. As the "basal" adenosine triphosphatase can be presumed to measure uncoupled enzymic activity and the "extra" adenosine triphosphatase the enzymic activity closely coupled to the transport system, the former rises to a peak and thence falls as more of the enzyme becomes coupled with transport; it is activated only when $Ca^{++}$ is being concentrated in the vesicles of the sarcoplasmic reticulum. At the same time the "extra" adenosine triphosphatase increases and the capacity and efficiency of $Ca^{++}$ transport rises. The fact that the peak of the "basal" adenosine triphosphatase occurs at just the time when enzymes of special significance for muscle also undergo a marked increase in activity suggests that similar mechanisms may be involved in controlling the development of all these systems. Further indications that this might be so were obtained when the effect of enforced exercise on the "basal" adenosine triphosphatase of the sarcoplasmic reticulum was studied in a similar experiment to that described for creatine phosphokinase (Kendrick-Jones and Perry, 1967). The results of this experiment can be seen in Fig. 26.12, which indicates that when the young rabbit is encouraged to use its muscles earlier than is normal, the "basal" adenosine triphosphatase activity rises earlier and the peak occurs at a significantly earlier age than in the case of the control animals.

It is clear from these studies that during the early stages of growth profound changes are taking place in the contractile, regulatory, and energy-yielding systems of muscle. Possibly changes in these systems occur throughout life, but in the main they are largely confined to the early stages of growth when development of the muscle is still not complete. They accompany an increase in absolute mass of the tissue, an aspect that becomes the most dominant in the later stages of growth.

The factors regulating the changes described are clearly complex. Factors endogenous to the organism such as those due to innervation and hormones are clearly important, but there are suggestions, particularly from the experiments designed to increase the use of muscles of the

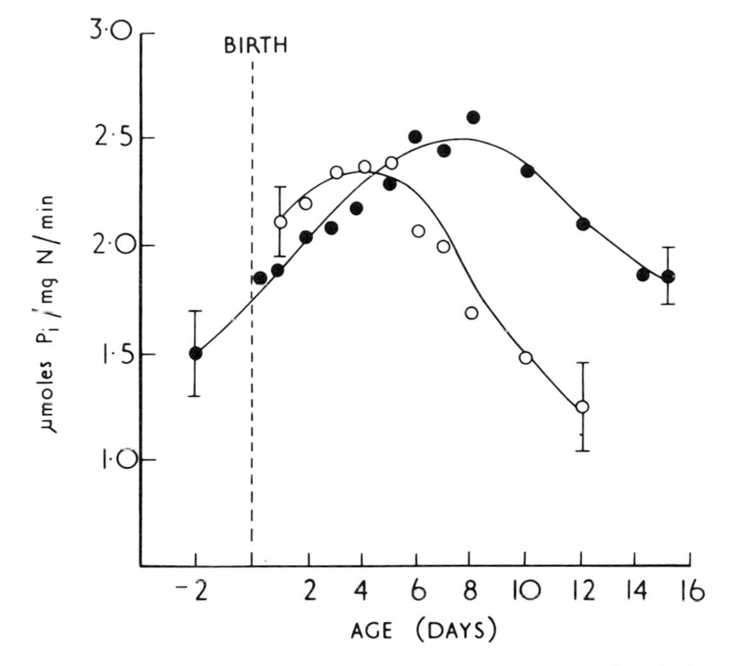


*Fig. 26.12.* The specific $Mg^{++}$ adenosine triphosphatase activity of the crude granular fraction (see Fig. 26.10) isolated from the longissimus dorsi of control and exercised young rabbits. Conditions of assay: 25 mM Tris-HCl, *p*H 7.6, 2.5 mM $MgCl_2$, 2.5 mM Tris-ATP. Incubated at 25° C for 5 min. *Solid circle,* control animals; *open circle,* exercised animals. From Holland and Perry, (1969) .

young animal at an earlier stage than is normal, that the level of contractile activity of the muscle is an important factor. The similarities of the development patterns of components of the different systems studied in each particular muscle type are striking, when allowance is made for the different extent of development of different species at birth. Thus during the rapid growth phase the effectiveness of the myosin system and of the creatine kinase system in handling substrate changes in a somewhat parallel manner. Similar isozyme changes appear to take place, although in the case of myosin the large change in the enzymic activity of the enzyme molecule is a particular feature. Thus it can be concluded that, at least during the rapid phase of growth, many of the biochemical changes which occur arise as adaptation to the changing activity pattern of tissue. Indeed the biochemical makeup of muscle, whether the tissue is growing or not, represents a continuing adaptation to its activity pattern and a reflection of the functional demands made upon it.

ACKNOWLEDGMENTS

The work described has in part been supported by grants from the Medical Research Council and the Muscular Dystrophy Associations of America, Inc.

## *References*

Dreizen, P., L. C. Gershman, P. P. Trotta, and A. Stracher. 1967. Myosin subunits and their interactions, p. 85. *The Contractile Process.* Little, Brown, Boston.

Frederiksen, D. W., and A. Holtzer. 1968. The substructure of the myosin molecule: production and properties of the alkali subunits. *Biochemistry 7:*3935.

Hardy, M. F., and S. V. Perry. 1969. *In vitro* methylation of muscle proteins. *Nature 223:*300.

Hardy, M. F., S. V. Perry, and D. Stone. 1970. ε-N-monomethyl-lysine and trimethyl-lysine in myosin. *Proc. Biochem. Soc.* (London, Dec. 17).

Hartshorne, D. J., M. Theiner, and H. Mueller. 1969. Studies on troponin. *Biochim. Biophys. Acta 175:*320.

Hasselbach, W. 1964. Relaxing factor and the relaxation of muscle. *Progr. Biophys. 14:*167.

Holland, D. L. 1968. Some aspects of skeletal muscle development. Ph.D. dissertation, University of Birmingham.

Holland, D. L., and S. V. Perry. 1968. Changes in the adenosine triphosphatase activity of the sarcoplasmic reticulum during development. *Biochem. J. 108:*13P.

———. 1969. The ATPase and calcium transporting activities of the sarcoplasmic reticulum of developing muscle. *Biochem. J. 114:*161.

Johnson, P., C. I. Harris, and S. V. Perry. 1967. 3 Methylhistidine in actin and other muscle proteins. *Biochem. J. 105:*361.

Johnson, P., G. E. Lobley, and S. V. Perry. 1969. Distribution and biological role of 3-methylhistidine in actin and myosin. *Biochem. J. 115:*993.

Kendrick-Jones, J., and S. V. Perry. 1967. The enzymes of adenine nucleotide metabolism in developing skeletal muscle. *Biochem. J. 103:*207.

Locker, R. H., and C. J. Hagyard. 1968. The myosin of rabbit red muscles. *Arch. Biochem. Biophys. 127:*370.

Oppenheimer, H., K. Bárány, G. Hamoir, and J. Fenton. 1966. Polydispersity of succinated myosin. *Arch. Biochem. Biophys. 115:*233.

Perrie, W. T., S. V. Perry, and D. Stone. 1969. Electrophoretic study of the small subunits of the myosin molecule. *Biochem. J. 113:*28P.

Perry, S. V., and A. Corsi. 1958. Extraction of proteins other than myosin from the isolated rabbit myofibril. *Biochem. J. 68:*5.

Perry, S. V., and D. J. Hartshorne. 1963. The proteins of developing muscle, p. 491. *In* E. Gutman and P. Hník (eds.), *Effect of Use and Disuse on Neuromuscular Functions.* Czechoslovak Acad. Sci. Prague.

Schaub, M. C., and S. V. Perry. 1969. The relaxing protein system of striated muscle: resolution of the troponin complex into inhibitory and calcium-ion–sensitizing factors and their relationship to tropomyosin. *Biochem. J. 115*:993.

Trayer, I. P. 1967. Studies on the myosin of developing muscle. Ph.D. dissertation, University of Birmingham.

Trayer, I. P., C. I. Harris, and S. V. Perry. 1968. 3-Methyl histidine and adult and foetal forms of skeletal muscle myosin. *Nature 217*:452.

Trayer, I. P., and S. V. Perry. 1966. The myosin of developing skeletal muscle. *Biochem. Z. 345*:87.

Trayer, I. P., S. V. Perry, and F. W. J. Teale. 1967. Structural differences between myosins of adult and fetal rabbit skeletal muscle, p. 943. *Abstr. 7th Int. Congr. Biochem.* (Tokyo).

# Chapter 27

# Muscle Fatigue

J. A. FAULKNER

Fatigue occurs rapidly in high intensity exercise. In weightlifting or tasks which involve small localized muscle groups, the duration of exercise prior to fatigue is inversely related to the percentage of maximum strength required (Rohmert, 1959). In running, cycling, or swimming, which involve contractions of large muscle masses, endurance time is inversely related to the percentage of maximum oxygen uptake required (Balke, 1963).

Fatigue is defined operationally in a given experimental context. In isolated or *in situ* muscle preparations, fatigue is the failure of the muscle to maintain twitch or tetanic tension when stimulation intensity and frequency are unchanged. In intact man or animals, voluntary fatigue is the failure to maintain tension or speed of contraction. Voluntary fatigue of a maximum static contraction is equivalent to fatigue of maximum tetanus since supramaximum shocks do not increase the strength or change the fatigue curve of the voluntary contraction (Merton, 1956*a*).

Fatigue might result from failure of the excitation mechanism, of the excitation-contraction coupling, or of the contraction process (Eberstein and Sandow, 1961). Merton (1956*a*) has demonstrated in intact human muscle and Kugelberg and Edström (1968) in *in situ* rat muscle that the decrease in tension results from the failure of the fatigued muscle to respond to a normal action potential. Evidence has not been presented which demonstrates impairment of either the excitation mechanism or excitation-contraction coupling, with the exception of high frequencies of 100 twitches/sec where the refractory periods of nerve and muscle become limiting (Olson and Swett, 1966). Fatigue in the physiological range of stimulation (red fibers 20/sec and white fibers 50–60/sec; Folkow and Halicka, 1968) appears to be a muscular phenomenon that results from a failure of metabolic regenerative processes (Edström and Kugelberg, 1968; Kugelberg and Edström, 1968; Merton, 1956*a, b*).

## METABOLISM IN MUSCULAR EXERCISE

Muscular contraction is an anaerobic process. The immediate energy for muscle contraction is ATP. However, as ATP is hydrolyzed to ADP and $P_i$, ADP is rapidly rephosphorylated to ATP through coupling to the breakdown of creatine phosphate to creatine and phosphate (for review see Needham, 1960). The creatine phosphate concentration of approximately 7 mM/100 g dry weight of muscle provides an energy store equivalent to about 1.5 liters of oxygen in a 70-kg man. The creatine phosphate store may be replenished aerobically by oxidative phosphorylation or anaerobically by glycolysis. The maximum oxygen uptake capacities of adult men vary from 25 ml/kg min to 80 ml/kg min (Astrand and Saltin, 1961*b;* Balke, 1963). The energy supplied by glycolysis may be estimated from the increase in blood lactate concentration. The energy equivalent for 1 g of lactate is 44 ml of oxygen (Margaria, 1967). Since the maximum increase is 1.4 g of lactate per liter of blood, or 1 g/kg of body weight, glycolysis provides an amount of energy equivalent to about 3 liters of oxygen.

The metabolic response to exercise is characterized by an initial transition period and then a pseudo-steady state (Fig. 27.1). The duration of the transition period depends on the magnitude of the change in the metabolism of the skeletal muscle cells. A true steady state is not attained because changes occurring in substrate concentrations can only be restored to resting levels during recovery. At certain metabolic loads, the steady state is further compromised by gradual increases in oxygen uptake, heart rate, cardiac output, and minute ventilation (Saltin and Stenberg, 1964).

In submaximum exercise, up to 50% of maximum oxygen uptake, the oxygen uptake equals or closely approximates the energy requirements of the contracting muscles, and there is little or no increase in blood lactate concentration (Strydom, Wyndham, Radhenand, and Williams, 1962). In maximum work, the energy production from aerobic sources is insufficient to meet the energy requirements of the contracting muscles. Under these circumstances, the creatine phosphate concentration decreases (Hultman, Bergstrom, and Anderson, 1967) and the blood lactate concentration increases (Astrand and Saltin, 1961*a*). When the energy requirement of a task is greater than the maximum oxygen uptake, voluntary fatigue usually occurs in less than 15 min. The raised metabolism during the recovery period (oxygen debt) is roughly proportional to the oxygen uptake at the end of work (Welch and Stainsby, 1967).

The respiratory and circulatory responses to increasing work loads on a bicycle ergometer are presented in Fig. 27.2. Each exercise period was of 10-min duration, and exercise periods were followed by 10-min rest

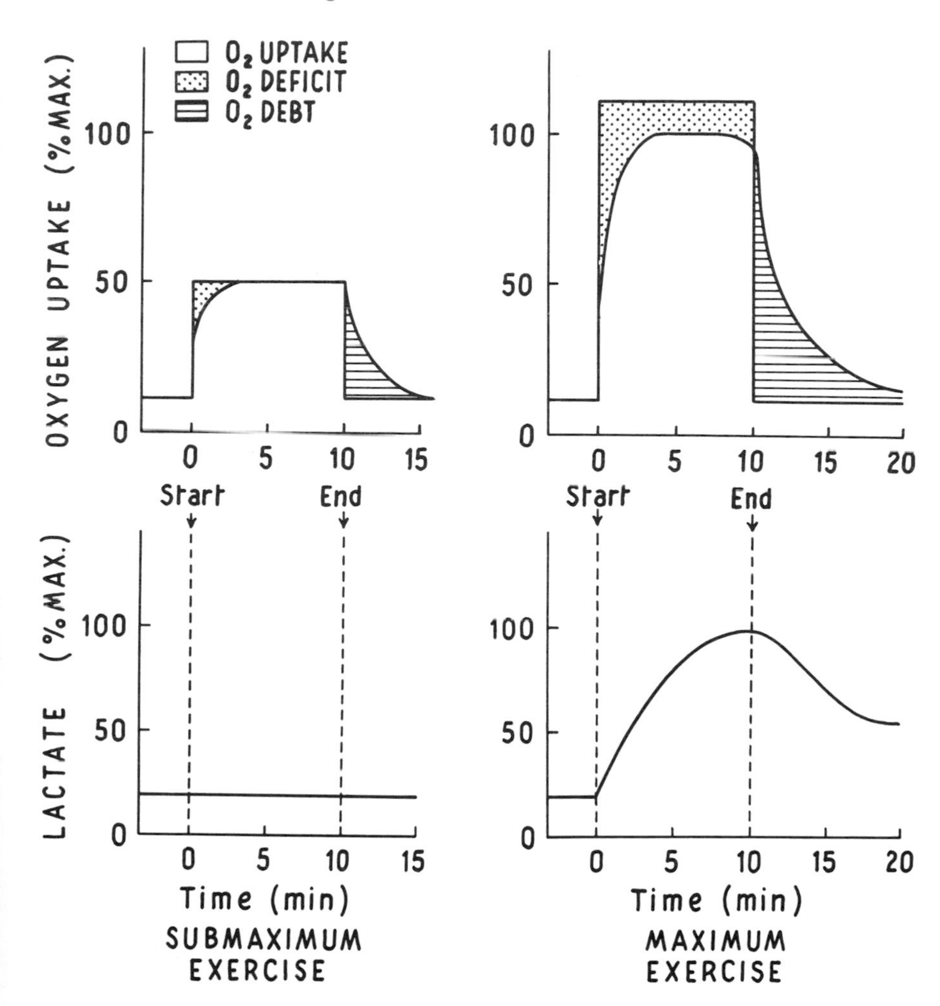


*Fig. 27.1.* The response of oxygen uptake and blood lactate to steady state work loads which required 50% and 150% of the subject's maximum oxygen uptake.

periods. Exercise began at a work load of 600 kg m/min, and the work load was increased 300 kg m/min in each subsequent exercise period until the point of voluntary fatigue was approached. The work increment was then reduced to 150 kg m/min. The data plotted in Fig. 27.2 are the means for 8 subjects who reached voluntary fatigue between 5 and 10 min at a work load of 1650 kg m/min.

The relationship of the variables to work load varies greatly. Stroke volume plateaus at a very low work load, when the cardiac output is only 60% of its maximum value. Further increases in cardiac output are

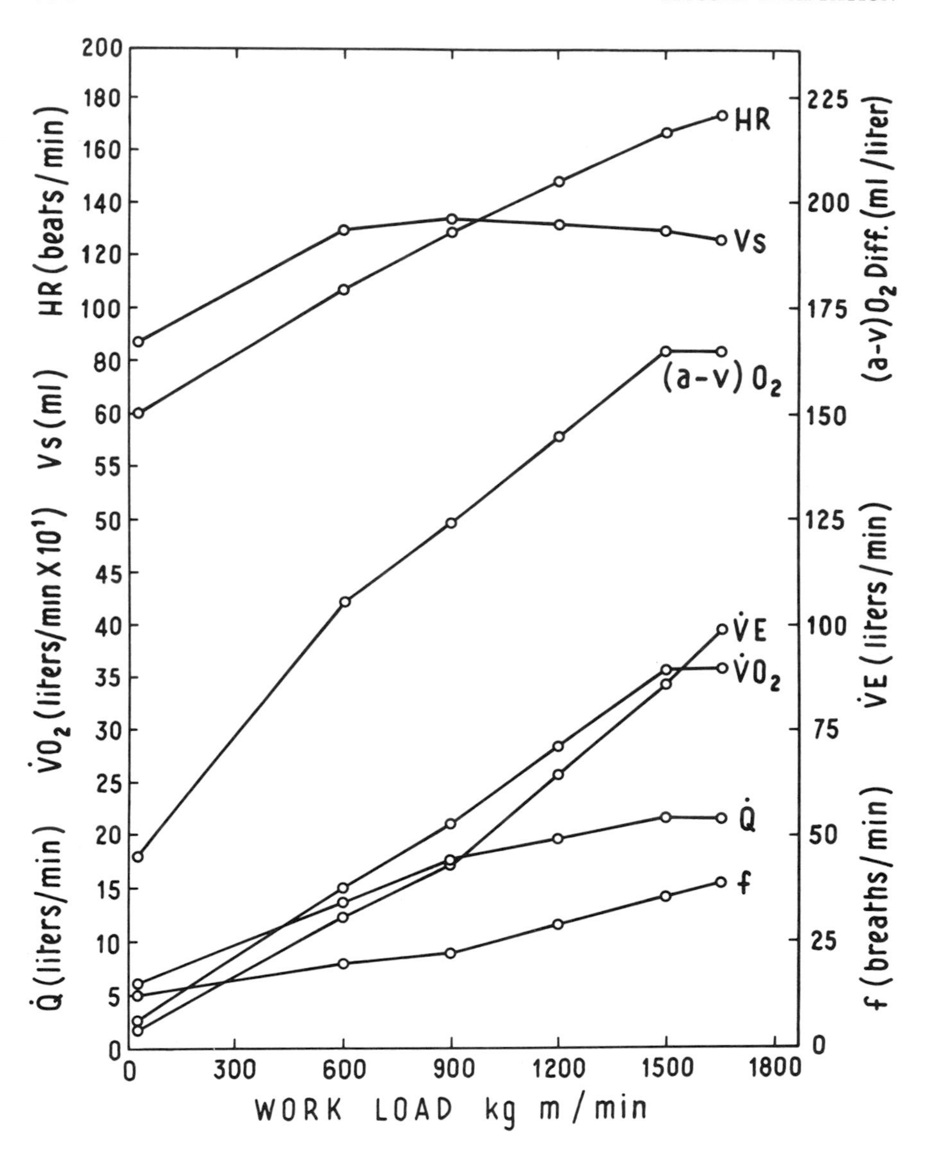


*Fig. 27.2.* Respiratory and cardiovascular response to increasing work loads on a bicycle ergometer. Each work load up to 1500 kg m/min was performed for 10 min. Work duration at 1650 kg m/min to voluntary fatigue ranged from 5 min to 10 min.

directly related to increases in heart rate. Respiratory rate, cardiac output, oxygen uptake, arteriovenous oxygen difference, and heart rate increase linearly with work load up to the maximum cardiac output. When plotted against increasing work loads, oxygen uptake, cardiac output, and arteriovenous oxygen difference appear to reach asymptotes simultaneously at a time when heart rate continues to increase and stroke volume decreases. These data concur with those of Ekblom, Astrand, Saltin, Stenberg, and Wallstrom (1968).

The maximum voluntary ventilation these subjects could sustain for a minute was 150 liters/min (STPD). Since the pulmonary ventilation in maximum exercise averaged 100 liters/min, ventilation does not appear to be a limiting factor in gas exchange. The arteriovenous oxygen difference of 164 ml/liter is higher than the maximum values of 150–159 ml/liter usually cited for trained subjects (Ekelund and Holmgren, 1967; Ekblom, Astrand, Saltin, Stenberg, and Wallstrom, 1968). If the arteriovenous oxygen difference is maximum and the heart rate and stroke volume are interacting in such a way that no further increase can occur in cardiac output, then the oxygen uptake is limited by oxygen transport. Mitchell, Sproule, and Chapman (1958) made a similar interpretation of their data, obtained on untrained men who ran to the point of voluntary fatigue on a treadmill.

The most dramatic difference between the cardiovascular response of trained subjects compared to that of untrained subjects is the relative contributions of stroke volume and heart rate for a given cardiac output. For a given cardiac output, the trained subjects have a higher stroke volume and a lower heart rate than untrained subjects (Hanson and Tabakin, 1965). Since maximum heart rate is the same for both groups, the cardiac output is limited by the stroke volume. Consequently, for untrained subjects maximum arteriovenous oxygen difference and maximum cardiac output are reached, and oxygen transport is rate-limiting, at a lower work load than for trained subjects.

## RATE-LIMITING PROCESSES IN MUSCLE CONTRACTION

Of the five metabolic states of the mitochondrion described by Chance and Williams (1956), states 2, 3, and 5 appear to be possible rate-limiting processes in muscular fatigue (Table 27.1). In humans, muscle glycogen stores are exhausted (state 2) after 60 min of exhaustive work that requires greater than 70% of the subject's maximum oxygen uptake (Hermansen, Hultman, and Saltin, 1967). Muscle glycogen stores are nearly depleted in endurance work of 5 hours' duration (Bergstrom, Hermansen, Hultman, and Saltin, 1967) even though free fatty acids are a major substrate in work of this intensity (Paul, Issekutz, and Miller, 1966). The significant relationship between resting muscle glycogen concentration and the endurance time to voluntary fatigue (Bergstrom, Hermansen, Hultman, and Saltin, 1967), and the lack of glycogen in muscle after exhaustive exercise indicates that substrate availability may

TABLE 27.1
*The metabolic states of mitochondria*

| | Characteristics | | | | | Steady-state percentage reduction of compon[illegible] | | | | |
|---|---|---|---|---|---|---|---|---|---|---|
| State | [$O_2$] | ADP level | Substrate level | Respiration rate | Rate-limiting substance | *a* | *c* | *b* | Flavo-protein* | NAI[illegible] |
| 1 | >0 | low | low | slow | ADP | 0 | 7 | 17 | 21 | ~9 |
| 2 | >0 | high | ~0 | slow | substrate | 0 | 0 | 0 | 0 | |
| 3 | >0 | high | high | fast | respiratory chain | <4 | 6 | 16 | 20 | 5 |
| 4 | >0 | low | high | slow | ADP | 0 | 14 | 35 | 40 | >[illegible] |
| 5 | 0 | high | high | 0 | oxygen | 100 | 100 | 100 | 150 | 1[illegible] |

* These values are based upon the amount of flavoprotein that is reduced upon addition of antimycin A to the mitocho[illegible] in state 2, that is, two-thirds the state 5 value.
From Chance and Williams (1956).

be the major determinant of fatigue when high intensity work is performed for an hour or longer.

Direct evidence is not available on rate-limiting processes in the respiratory chain (state 3). However, Holloszy (1967) noted an increase in electron transport capacity of rats after 12 weeks of treadmill running. The increase in electron transport capacity suggests an overload training stimulus (for review of training stimuli see Faulkner, 1968). Theoretically, if oxygen and substrates are available to the mitochondria in sufficient quantities and yet ATP regeneration is rate-limiting, then the enzyme systems of Krebs cycle or of the respiratory chain would appear to limit the phosphorylation of ATP. Contraction of skeletal muscle would be rate-limited by these enzyme systems at near maximum rates of stimulation after the depletion of creatine phosphate and ATP stores.

Oxygen availability (state 5) is traditionally considered the factor that limits the ability of muscle to contract maximally (Bendall, 1966). If in exercise the rate of oxidative phosphorylation is limited by the availability of oxygen, the percentage of reduced NAD would be increased greatly (Table 27.1). In *in situ* dog gastrocnemius-plantaris muscle contracting at a twitch rate of 5/sec, which is equivalent to its maximum oxygen uptake, the fluorescence decreases relative to that observed in resting muscle (Jöbsis and Stainsby, 1968). The decrease in fluorescence indicates an increase in the amount of oxidized NAD. Stainsby (1965) and Folkow and Halicka (1968) have also reported substantial concentrations of oxygen in the venous blood from *in situ* muscles at maximum rates of stimulation. These data collected from *in situ* muscle preparations are not consistent with the concept of oxygen as the limiting reactant during maximum work.

When skeletal muscle is contracting at a near maximum rate, two conditions may exist in which the rate-limiting processes that result in fatigue are quite different. In isolated or *in situ* muscle preparations or in small muscle groups *in vivo,* autoregulation may fully dilate the skeletal muscle capillary bed (Folkow and Halicka, 1968). Under these circumstances, maximum skeletal muscle blood flow is not limited by cardiac output, oxygen supply is not rate-limiting. In running, however, or in cycling, swimming, and other physical activities that require maximum contractions of large masses of muscle, the cardiac output is insufficient to provide maximum flow in all of the skeletal muscle capillary beds and in the coronary capillary bed simultaneously, and oxygen transport is limiting (Fig. 27.2).

The role of oxygen as the rate-limiting reactant in running is supported by the increase in endurance time to voluntary fatigue observed when subjects breathe higher concentrations of oxygen than are present in room air (Bannister and Cunningham, 1954). Inhalation of the higher concentrations of oxygen during maximum exercise reduces the pulmonary ventilation and the blood lactate. Systematic differences were not observed between inhalation of gas containing 66% and 100% oxygen. The mechanism by which voluntary fatigue is delayed by inhalation of concentrations of oxygen higher than 20.9% is not known. In maximum exercise, however, the partial pressure of arterial oxygen decreases from a normal value of 100 mm Hg to 79 mm Hg (Holmgren and Linderholm, 1958). The inhalation of higher concentrations of oxygen would increase the partial pressure of oxygen in the arterial blood and result in an increased oxygen content through increases in the amount of oxygen combined with hemoglobin and in the amount dissolved in the plasma.

The increase in oxygen content of the arterial blood might increase the ability of skeletal muscle to maintain maximum tension, either directly, by increasing the molecular oxygen available for oxidative phosphorylation at the mitochondrial sites, or indirectly by increasing cardiac output, or most likely by both mechanisms. An increase in cardiac output is a distinct possibility. The myocardium relies almost exclusively on aerobic metabolism. For a 63-kg man at rest with a 300 g heart muscle, Bard (1961) has estimated a myocardial oxygen uptake of 9.7 ml/100 g min, a coronary blood flow of 84 ml/100 g min, and an arteriovenous oxygen difference of 114 ml/liter. Since the critical venous oxygen tension* is approximately 20 mm Hg for the heart (Balke, 1965) compared to 10 mm Hg for contracting skeletal muscle (Stainsby, 1965), the arteriove-

* A critical oxygen tension is the oxygen tension at which oxygen uptake for a tissue is decreased by any further decrease in oxygen tension.

nous oxygen difference is maximum under resting conditions and oxygen uptake can only be increased by an increase in coronary flow. However, in maximum work, coronary flow is eventually impeded by an increasing heart rate and an increasing intramuscular pressure. That the work of the heart is limited by oxygen supply is suggested by the plateau in cardiac output due to an increasing heart rate and a decreasing stroke volume (Fig. 27.2).

### RED AND WHITE SKELETAL MUSCLE CELLS

In mammals, skeletal muscles consist of fibers with different structural and functional characteristics (Henneman and Olson, 1965; Edström and Kugelberg, 1968; Kugelberg and Edström, 1968; Romanul, 1965) by which the fibers may be classified grossly as red, intermediate, or white. The trends for the two extremes are summarized in Table 27.2. Such a distinction between red and white cells is an oversimplification since structural and functional characteristics may vary independently through growth, cross innervation (Buller, Eccles, and Eccles, 1960; Prewitt and Salafsky, 1967), training (Holloszy, 1967; Pattengale and Holloszy, 1967), or hypoxia (Mager, Blatt, Natale, and Blatteis, 1968). A description of red and white fiber characteristics does provide a flexible framework from which to investigate differences in structure and function among different species (George and Naik, 1957), in different muscles in the same species (Henneman and Olson, 1965; Folkow and Halicka, 1968), and in the same muscle adapted to different environmental stresses (Pattengale and Holloszy, 1967; Holloszy, 1967; Mager, Blatt, Natale, and Blatteis, 1968).

Henneman and Olson (1965) observed that the threshold for stimulation of a motoneurone was inversely related to the size of the motoneurone. Therefore, only motor units innervated by small neurones are recruited when the stimulation intensity is low, and motor units innervated by large neurones require higher intensities of stimulation for recruitment. Anatomically, the small neurones tend to innervate motor units composed of small skeletal muscle cells with few cells per motor unit and the large neurones tend to innervate motor units composed of large skeletal muscle cells with many cells per motor unit (Henneman and Olson, 1965). The small motor units are recruited often, have a low natural actomyosin adenosine triphosphatase activity, have a high capacity for oxidative metabolism, and are resistant to fatigue. The large motor units are recruited infrequently, have a high natural actomyosin adenosine triphosphatase activity, have a high capacity for glycolytic metabolism, and fatigue rapidly. The interrelationships among motoneurone size, motoneurone excitability, motor unit recruitment, and the structural

TABLE 27.2

*A summary of the characteristics that tend to be associated with each other in motor units composed of red and white skeletal muscle cells*

| Characteristic | Muscle: Red | Muscle: White |
|---|---|---|
| Motoneurone diameter | Small | Large (Henneman and Olson, 1965) |
| Cells per motor unit | Few | Many (Henneman and Olson, 1965) |
| Recruitment | Often | Seldom (Henneman and Olson, 1965) |
| Contraction time | Slow | Fast (Olson and Swett, 1966) |
| $Ca^{++}$ uptake per unit wt grana (15 min) | Slow | Fast (Gergely et al., 1965) |
| Tension per unit wt protein | Low | High (Henneman and Olson, 1965) |
| Myoglobin concentration | High | Low (Lawrie, 1953) |
| Blood flow | High | Low (Folkow and Halicka, 1968) |
| Mitochondrial density | High | Low (Gergely et al., 1965) |
| Resting metabolism | High | Low (Folkow and Halicka, 1968) |
| Oxidative phosphorylation | High | Low (Romanul, 1965) |
| Glycolytic energy | Low | High (Stubbs and Blanchaer, 1965) |
| Natural actomyosin adenosine triphosphatase activity | Low | High (Gergely et al., 1965) |
| Fatigability | Slow | Fast (Eberstein and Sandow, 1961; Edström and Kugelberg, 1968) |

and functional characteristics of skeletal muscle cells may be used to explain the recruitment pattern in exercise of different intensities. In submaximum work of light intensity, recruitment is primarily of motor units composed of red-type fibers and metabolism is mainly or exclusively aerobic. As the intensity of the work increases, the larger neurones of the motor units composed of the white-type fibers are recruited intermittently. The intermittent recruitment results from fluctuating patterns of excitation and inhibition of cells (Lloyd, 1946). At such a moderate work load, however, stimulation is sufficiently infrequent that the white fibers can deplete creatine phosphate stores during contraction and restore these stores during the relaxation period. As work intensity approaches a maximum, the motor units composed of white-type fibers are recruited so frequently that recovery is not possible and the fibers fatigue.

In the cat, Folkow and Halicka (1968) have investigated the response to graded electrical stimulation of the soleus, primarily composed of red-type fibers, and the gastrocnemius, composed of about 25% red, 25% intermediate, and 50% white fibers (Henneman and Olson, 1965). When contracting aerobically at 1–2 twitches/sec, the soleus had an average oxygen cost per twitch of 20 $\mu$l/100 g compared to 29 $\mu$l/100 g for the gastrocnemius. The potential difference in blood flow between the soleus

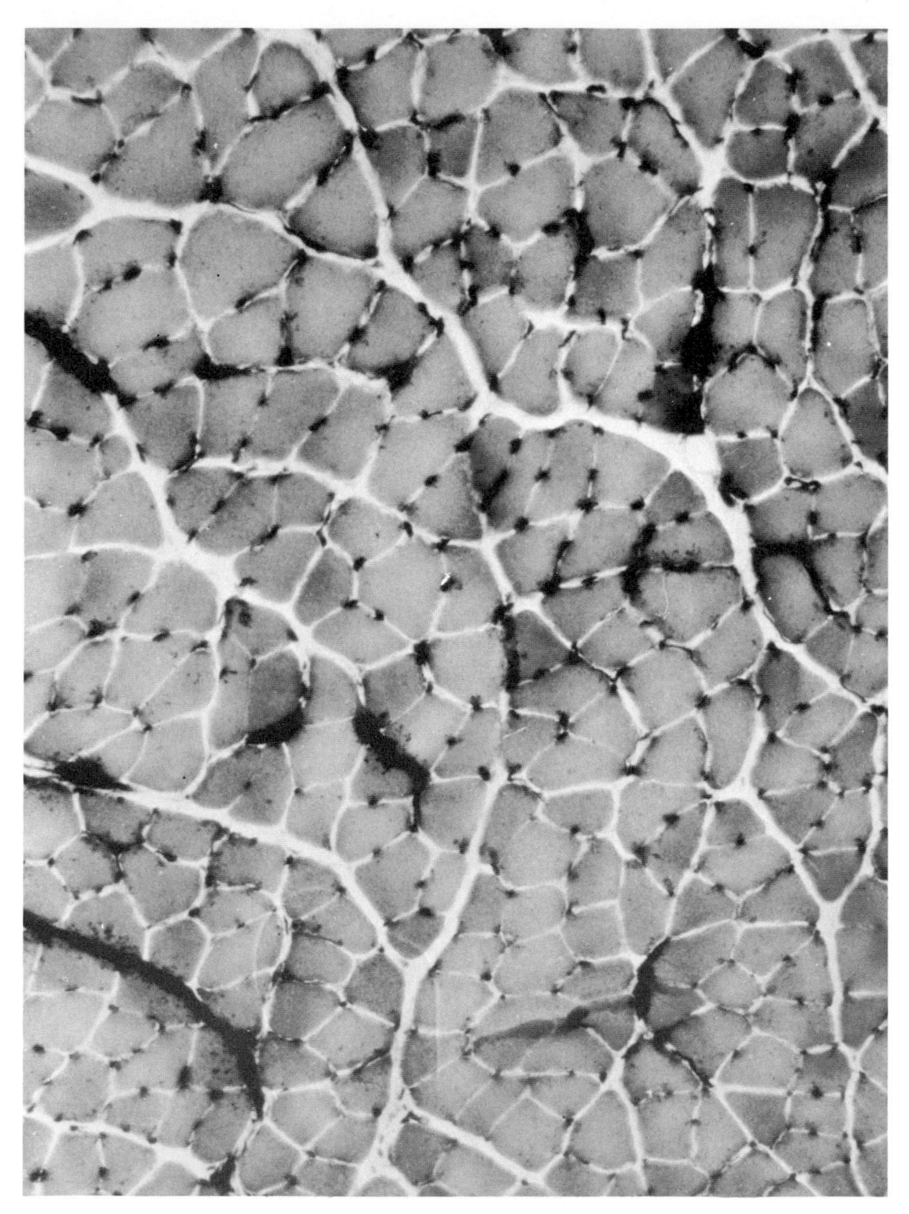

*Fig. 27.3.* Cross section of guinea pig soleus (*left*) and gastrocnemius (*right*) muscles incubated for succinate dehydrogenase and adenosine triphosphatase activity. The

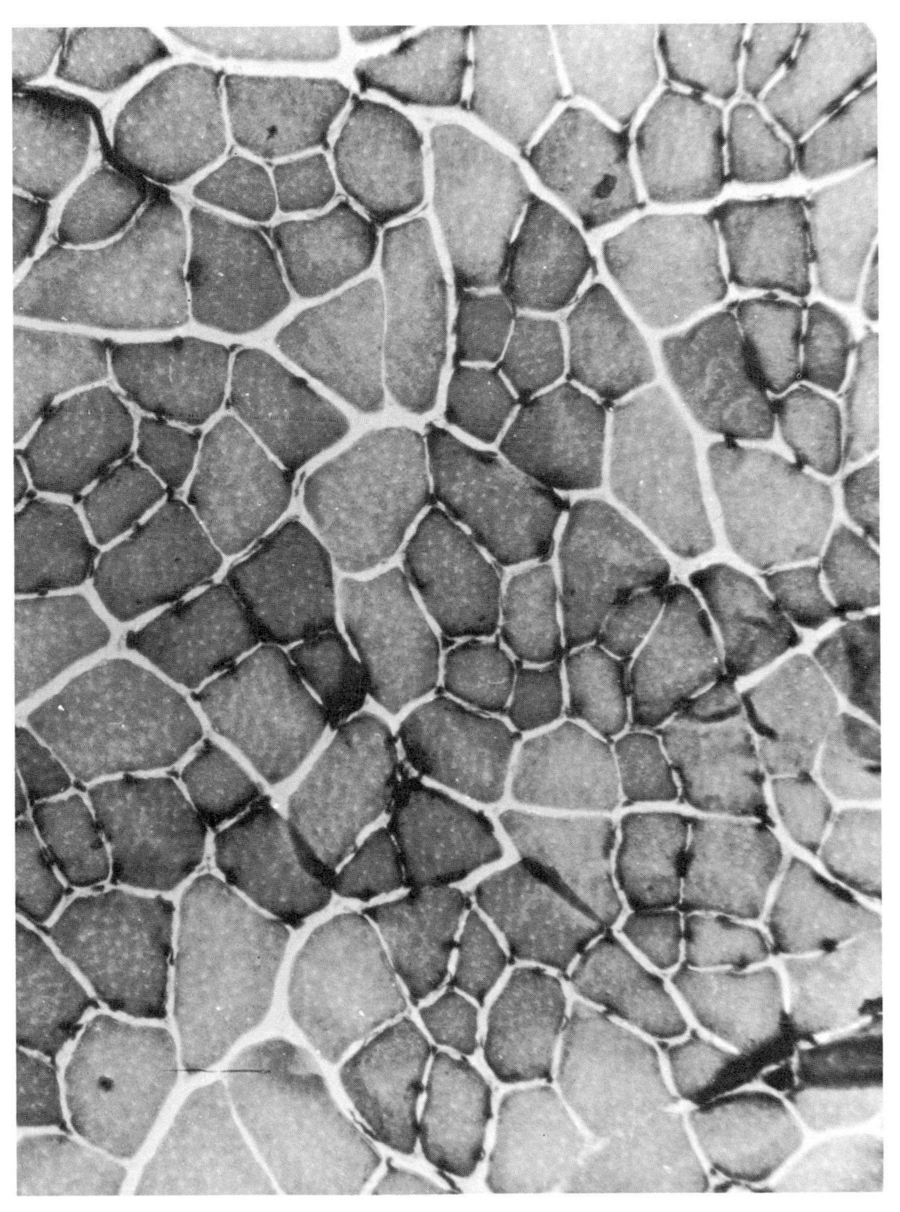

small dark spots around the skeletal muscle cells are the capillaries. (The histochemical sections were prepared by David Brook.)

fibers and the gastrocnemius fibers is evident in histochemical sections in which the capillaries have been stained for succinate dehydrogenase and adenosine triphosphatase activity (Fig. 27.3). The difference in capillary density is reflected in the difference in the relationship between twitch rate and blood flow, both during exercise and during brief interruptions of exercise (Fig. 27.4). Blood flow in the gastrocnemius muscle increases rapidly and reaches a maximum value of 45 ml/100 g min at 4 twitches/sec with or without interruption of exercise. The blood flow of the soleus increases at a slower rate, reaching a peak flow of 60 ml/100 g min at 12 twitches/sec during exercise and 112 ml/100 g min at 20 twitches/sec during brief interruptions of exercise (Folkow and Halicka, 1968).

Impairment of blood flow at high twitch rates occurs in both the soleus and the gastrocnemius muscles (Fig. 27.4). The performance of the

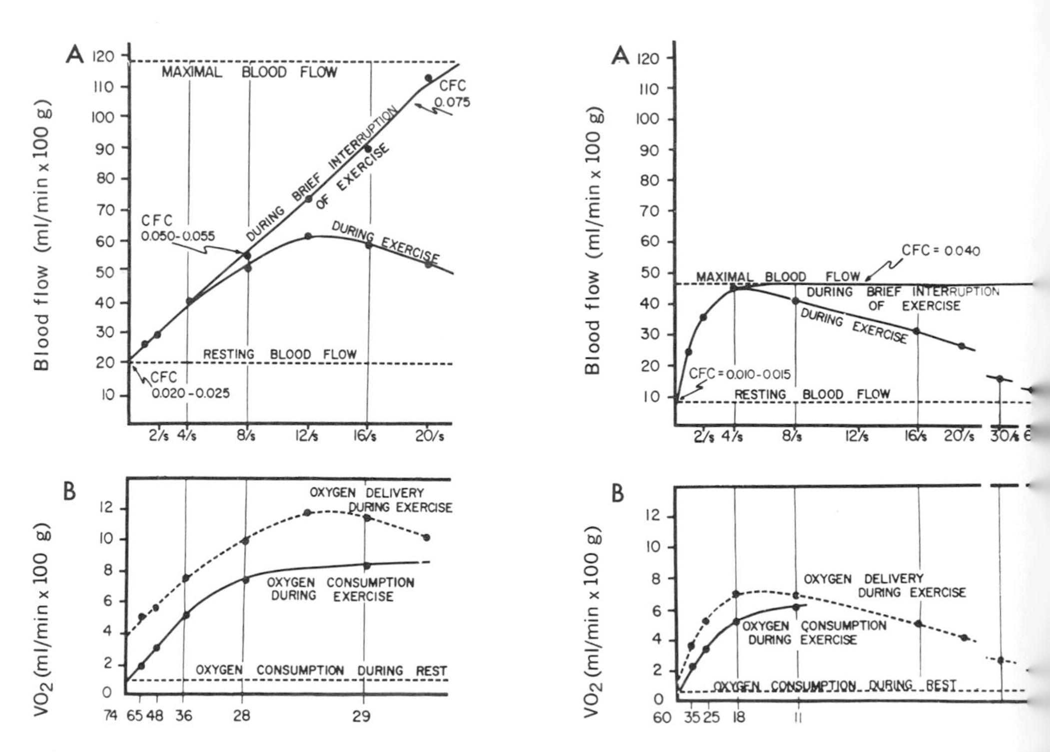


*Fig. 27.4.* Blood flow, oxygen uptake, and oxygen delivered in the soleus and gastrocnemius musc during rest and graded exercise. *Left,* soleus; *right,* gastrocnemius *A,* stimulation frequency; *B,* venc oxygen saturation (%). From Folkow and Halicka (1968).

soleus would, however, be impaired more by reduced circulation than that of the gastrocnemius because it is more dependent on high flow rates for oxygen delivery. The impaired flow results from the twisting, kinking, or local compression of large vessels as they pass through the muscle and the fasciculi (Gray, Carlsson, and Staub, 1967). The shearing forces are severe enough to occlude arteries and veins at a particular site independent of the blood pressure. At different sites, shearing occurs at different muscle tensions. Since the shearing occurs independently of the blood pressure, the pressor effect by which blood pressure is increased by high tension static work (Lind and McNicol, 1967), might not maintain blood flow during contraction. The increased blood pressure would increase blood flow in nonoccluded vessels during contraction and blood flow and speed of recovery between contractions.

In *in situ* muscle preparations, the rephosphorylation of ATP is not limited by insufficient oxygen in either the soleus muscle or, most likely, in the red fibers of the gastrocnemius, owing to their low natural actomyosin adenosine triphosphatase activity, and the high capacity for oxygen delivery of these types of skeletal muscle cells. The role of oxygen in the fatigue of white fibers is difficult to assess. White fibers have neither the capillarization to deliver oxygen nor sufficient mitochondrial enzyme systems for oxidative phosphorylation. The capacities for oxygen delivery and for oxidative phosphorylation play such minor roles in the generation of energy in white fibers that it is difficult to assign much significance to their relative status. Since red and white fibers differ quantitatively in usage, activity of metabolic pathways, and fatigability, the processes limiting the ability of the fibers to contract may be different.

When activated by a supramaximum stimulus at the same twitch rate, the fatigue curves, as estimated by the decline in maximum tension, are very different for red compared to white fibers (Fig. 27.5). Eberstein and Sandow (1961) stimulated small bundles (10 fibers or less) of tonic (red-type) fibers and single phasic (white-type) fibers at 1 twitch/sec. The red-type fibers lost 20% of their maximum tension in 10 min and 40% in 60 min. The white-type fibers displayed a 60% decrease in maximum twitch tension in 4 min and almost zero tension after 20 min. Since oxygen was bubbled through the bath during the experiments, the difference in the fatigue curves is likely to be due to the white fibers hydrolyzing ATP at a much faster rate than the red fibers. The decrease of the white fiber tension to zero would reflect the ability of the white fibers to hydrolyze ATP at a much faster rate than the white fibers are able to regenerate ATP. The difference in the ability of red fibers to hydrolyze and phosphorylate ATP is not so striking and the decline in tension is much less rapid.

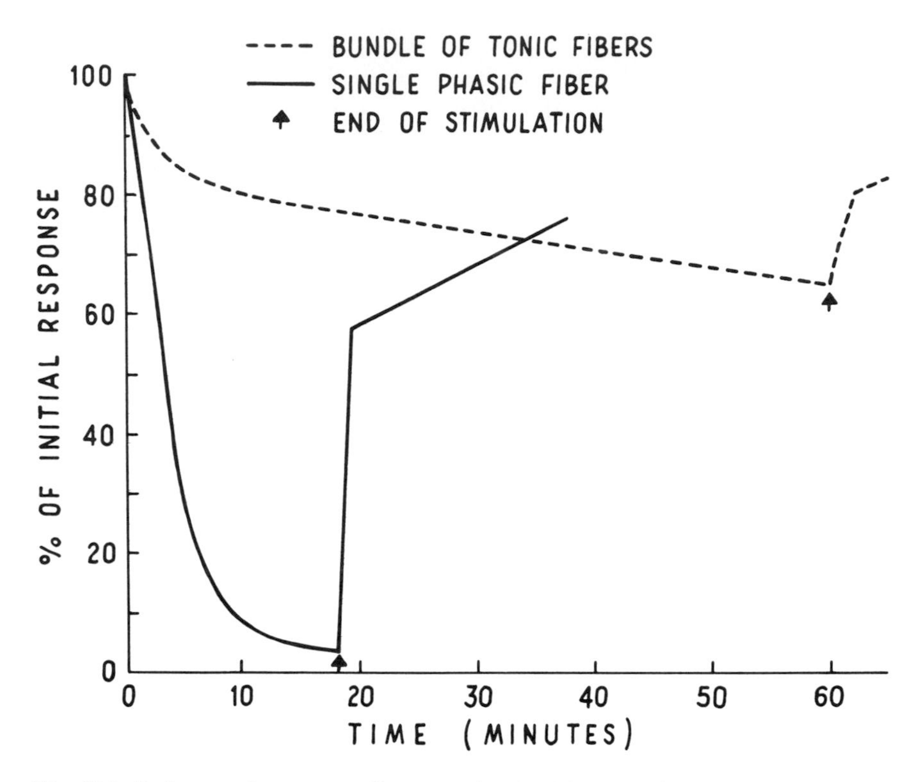


*Fig. 27.5.* Fatigue and recovery effects associated with 1 twitch/sec stimulation of a single phasic fiber and of a bundle of tonic fibers. Modified from Eberstein and Sandow (1961).

For world records in running, the running time in seconds is a logarithmic function of the distance in meters (Lloyd, 1967). Various portions of the relation are surprisingly linear and Lloyd has suggested that the linear portions may represent periods of exhaustive work in which the rate-limiting process is the same. Due to the rapid, powerful stride required in sprinting, the first two portions of the distance-velocity relation are primarily dependent on the contractions of white fibers. Maximum sprinting velocity can be maintained for about 20 sec and fatigue at sprint velocity appears to result from the exhaustion of muscle concentrations of creatine phosphate and ATP. Depletion of muscle creatine phosphate has been observed in short exhaustive runs on a treadmill (Margaria, Oliva, di Prampero, and Cerretelli, 1969).

The second linear portion of the distance-velocity relation is for events of from 20 sec to 8 min duration. Maximum performance in this time period seems to depend on the initial energy stores in the muscle and on the rate at which ATP is resynthesized by both the aerobic and anaerobic pathways. Maximum blood lactates of from 135–175 ml/100 ml are obtained usually several minutes after the work has ceased (Astrand and Saltin, 1961*a*). The magnitude of the blood lactates supports the key role of glycolysis in work of this intensity. If work duration is longer than 60 sec, however, a significant portion (Cunningham and Faulkner, 1969) or 100% of the maximum oxygen uptake will be involved (Astrand and Saltin, 1961*a;* Mitchell, Sproule, and Chapman, 1958). As with work of less than 20 sec duration, indirect evidence suggests that the creatine phosphate store of the contracting muscle is exhausted (Hultman, Bergstrom, and Anderson, 1967).

Increasing the duration of exhaustive work from 8 to 60 min results in a gradually decreasing blood lactate concentration (Astrand, Hallback, Hedman, and Saltin, 1963) and muscle glycogen concentration (Hermansen, Hultman, and Saltin, 1967). *In vivo,* even work intensities that can be continued for hours require the recruitment of some white fiber motor units to supplement the work output of the motor units composed of red fibers. After 30 sec of stimulation, most of the phosphorylase b in white muscle has been converted to phosphorylase a, whereas the enzyme concentrations in red muscle do not change from the resting values (Stubbs and Blanchaer, 1965). The transformation of the enzyme to its active form is appropriate for the white muscle since white muscle relies on glycolysis for energy production. Because the major fuels for red muscle are free fatty acids and the metabolic intermediates, pyruvate and lactate, the lack of phosphorylase activity does not impair the energy production of red muscle cells. The dependence of the white fibers on glycogen for substrate probably explains the eventual depletion of the muscle glycogen stores (Bergstrom, Hermansen, Hultman, and Saltin, 1967; Hermansen, Hultman, and Saltin, 1967). In birds, stimulation of the pectoralis major muscle at 5 twitches/sec results in a complete loss of glycogen from the white fibers and a slight increase in the glycogen content of the red fibers (George and Nene, 1965). With the load constant, the rate at which glycogen is degraded to pyruvate and lactate is dependent on the frequency of contraction (Karpatkin, Helmreich, and Cori, 1964).

## ENDURANCE TRAINING AND MUSCLE FATIGUE

Since endurance training changes the onset and rate of development of fatigue, rate-limiting processes might be the first to adapt to a training regime. Fifteen weeks of treadmill running increased the myoglobin concentration of rat quadriceps muscle by 78% (Pattengale and Holloszy, 1967). Holloszy (1967) has also observed a doubling in the capacity of rat skeletal muscle to oxidize pyruvate due to an increased electron transport capacity in exercised animals relative to control animals. In man, training programs have resulted in increases of 16% in maximum oxygen uptake, 13% in cardiac output, and 7% in arteriovenous oxygen differences (Ekblom, Astrand, Saltin, Stenberg, and Wallstrom, 1968). The increase in cardiac output was due to an increase in stroke volume with no change in maximum heart rate. The increased stroke volume suggests structural and functional changes in the myocardium.

Previously, we compared the maximum oxygen uptakes of trained college swimmers obtained during a tethered swimming test (4.14 liters/min) with that obtained during a treadmill running test (4.18 liters/min) and found no significant difference (Magel and Faulkner, 1967). When the subjects were recreational swimmers who had not trained in competitive swimming, the maximum oxygen uptake running was significantly higher ($P < 0.05$) than it was swimming (Fig. 27.6). The mean maximum oxygen uptake was 4.17 liters/min running and 3.42 liters/min swimming. The untrained swimmers had the same potential for oxygen transport swimming as they had running. Apparently, in the untrained swimmers the muscles of the arms, shoulders, back, and chest did not have either sufficient capillarization to transport the oxygen, or the capacity for oxidative phosphorylation to utilize the oxygen, or they lacked both these capacities. We hypothesize that training in swimming results first in adaptive changes in skeletal muscle cells, enabling the cells to regenerate ATP aerobically at a faster rate than was possible previously. A secondary adaptation might occur in the myocardium in response to increased oxygen requirements of the contracting skeletal muscle cells.

## SUMMARY

Although the rate-limiting factors in muscle fatigue have not been clearly described by systematic investigation, differences in the onset and course of fatigue in different experiments provide some insight in the mechanisms involved. The differences in the structure and function of red and white muscle fibers and particularly in their fatigability also suggest different rate-limiting processes. When the muscle capillary bed is fully dilated and blood flow is not impaired by either cardiac output or intramuscular tension, motor units composed of red fibers are limited only by their ability to hydrolyze ATP and by the eventual depletion of available free fatty acid stores. With occlusion or when cardiac output limits blood flow, oxygen may become the limiting reactant. When contracting at maximum speed and load, motor units composed of white fibers appear to be limited by muscle stores of ATP and creatine phosphate. If the work intensity is reduced so that fatigue occurs in 20 sec to 8

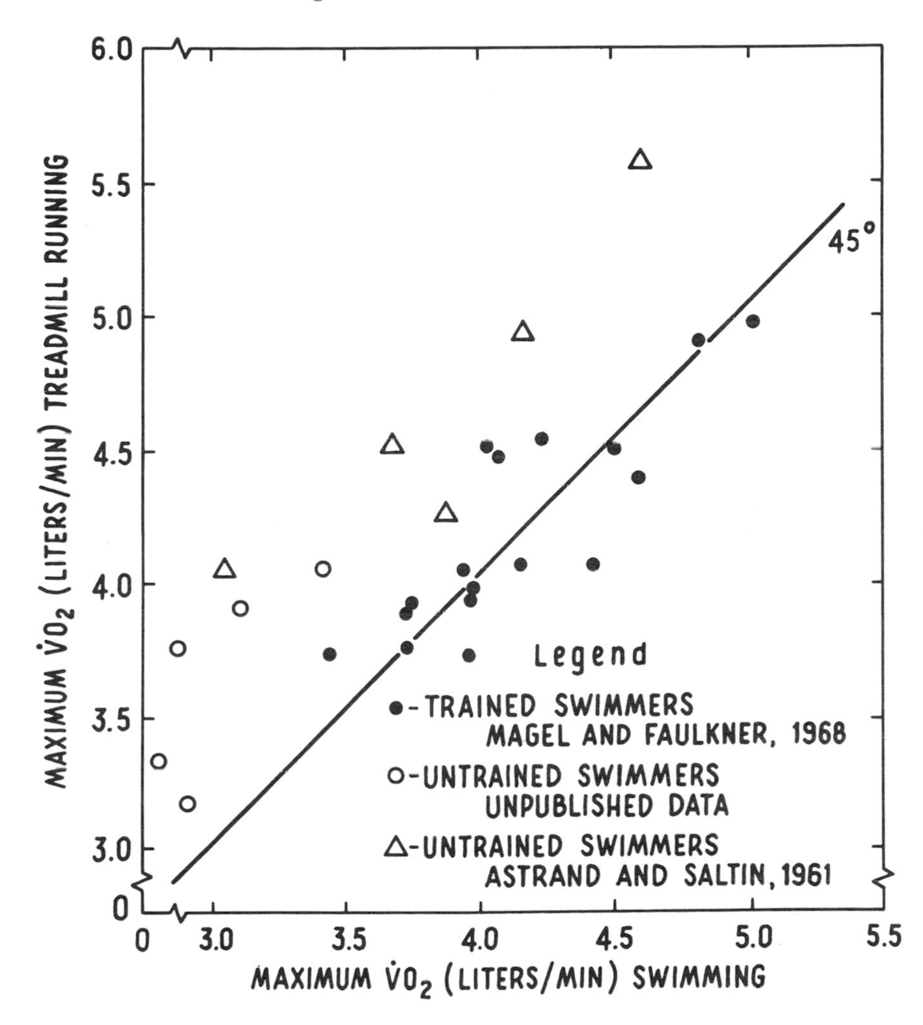


*Fig. 27.6.* The maximum oxygen uptake of trained and untrained swimmers obtained during tethered swimming and treadmill running. Modified from Magel and Faulkner (1967).

min, muscle energy stores and the rate at which ATP can be phosphorylated by both aerobic and anaerobic pathways are major determinants of the fatigue process. In prolonged work in which exhaustion occurs after 60 min or longer, muscle glycogen stores are depleted and the contractile process appears to be substrate-limited. At each work intensity, maximum performance appears to be limited by the processes that result in fatigue of the white skeletal muscle cells rather than any impairment in the ability of the red fibers.

## ACKNOWLEDGMENTS

This paper is based on data collected with grant support from the Michigan Heart Association and the Western Electric Company. I thank Jack K. Barclay, James H. Sherman, and Hugh G. Welch for critical evaluation of the manuscript. Data were collected in collaboration with John R. Magel, Leo C. Maxwell, and David A. Brook.

## *References*

Astrand, P.-O., I. Hallback, R. Hedman, and B. Saltin. 1963. Blood lactates after prolonged severe exercise. *J. Appl. Physiol. 18*:619.

Astrand, P.-O., and B. Saltin. 1961*a*. Oxygen uptake during the first minutes of heavy exercise. *J. Appl. Physiol. 16*:971.

———. 1961*b*. Maximal oxygen uptake and heart rate in various kinds of muscular activity. *J. Appl. Physiol. 16*:977.

Balke, B. 1963. Physiological background for the assessment, evaluation and classification of physical fitness. *Proc. 1st Can. Fitness Seminar 1*:5.

———. 1965. Cardiac performance in relation to altitude. *Amer. J. Cardiol. 14*:796.

Bannister, R. G., and D. J. C. Cunningham. 1954. The effects on the respiration and performance during exercise of adding oxygen to the inspired air. *J. Physiol. 125*:118.

Bard, P. 1961. Blood supply of special regions. *In* P. Bard (ed.), *Medical Physiology*. C. V. Mosby Co., St. Louis.

Bendall, J. R. 1966. Muscle as a contractile machine, p. 7. *In* E. J. Briskey, R. G. Cassens, and J. C. Trautman (eds.), *The Physiology and Biochemistry of Muscle as a Food*. Univ. of Wis.

Bergstrom, J., L. Hermansen, E. Hultman, and B. Saltin. 1967. Diet, muscle glycogen and physical performance. *Acta Physiol. Scand. 71*:140.

Buller, A. J., J. C. Eccles, and R. M. Eccles. 1960. Differentiation of fast and slow muscles in the cat hind limb. *J. Physiol. 150*:399.

Chance, B., and G. R. Williams. 1956. The respiratory chain and oxidative phosphorylation. *Advance. Enzymol. 17*:65.

Cunningham, D. A., and J. A. Faulkner. 1969. The effect of training on aerobic and anaerobic metabolism during a short exhaustive run. *Med. & Sci. in Sports 1*:65.

Eberstein, A., and A. Sandow. 1961. Fatigue in phasic and tonic fibers of frog muscle. *Science 125*:383.

Edström, L., and E. Kugelberg. 1968. Histochemical composition, distribution of fibres and fatiguability of single motor units. Anterior tibial muscle of the rat. *J. Neurol. Neurosurg. Psychiat. 31*:424.

Ekblom, B., P.-O. Astrand, B. Saltin, J. Stenberg, and B. Wallstrom. 1968. Effect of training on circulatory response to exercise. *J. Appl. Physiol. 24*:518.

Ekelund, L. G., and A. Holmgren. 1967. Central hemodynamics during exercise, p. I-33. *In* C. B. Chapman (ed.), *Physiology of Muscular Exercise.* Amer. Heart Ass. Monograph No. 15.

Faulkner, J. A. 1968. New perspectives in training for maximum performance. *J. Am. Med. Ass. 205*:741.

Folkow, B., and H. D. Halicka. 1968. A comparison between "red" and "white" muscle with respect to blood supply, capillary surface area and oxygen uptake during rest and exercise. *Microvascular Res. 1*:1.

George, J. C. and R. M. Naik. 1957. Studies in the structures and physiology of the flight muscles of birds. *J. Anim. Morph. Physiol. 4*:23.

George, J. C., and R. V. Nene. 1965. Effect of exercise on the glycogen content of the red and white fibers of the pigeon pectoralis muscle. *J. Anim. Morphol. Physiol. 12*:246.

Gergely, S., D. Pragay, A. F. Scholz, J. C. Seidel, F. A. Sreter, and M. M. Thompson. 1965. Comparative studies on white and red muscle, p. 206. *In* S. Ebashi, F. Oosawa, T. Sekine, and Y. Tonomura (eds.), *Molecular Biology of Muscular Contraction.* Igaku Shoin, Tokyo.

Gray, S. D., E. Carlsson, and N. C. Staub. 1967. Site of increased vascular resistance during isometric muscle contraction. *Amer. J. Physiol. 213*:683.

Hanson, J. S., and B. S. Tabakin. 1965. Comparison of the circulatory response to upright exercise in 25 "normal" men and 9 distance runners. *Brit. Heart J. 27*:211.

Henneman, E., and C. B. Olson. 1965. Relations between structure and function in the design of skeletal muscles. *J. Neurophysiol. 28*:581.

Hermansen, L., E. Hultman, and B. Saltin. 1967. Muscle glycogen during prolonged severe exercise. *Acta Physiol. Scand. 71*:129.

Holloszy, J. O. 1967. Biochemical adaptations in muscle: effects of exercise on mitochondrial oxygen uptake and respiratory enzyme activity in skeletal muscle. *J. Biol. Chem. 242*:2278.

Holmgren, A., and H. Linderholm. 1958. Oxygen and carbon dioxide tensions of arterial blood during heavy and exhaustive exercise. *Acta Physiol. Scand. 44*:203.

Hultman, E., J. Bergstrom, and N. McL. Anderson. 1967. Breakdown and resynthesis of phosphorylcreatine and adenosine triphosphate in connection with muscular work in man. *Scand. J. Clin. Lab. Invest. 19*:56.

Jöbsis, F. F., and W. N. Stainsby. 1968. Oxidation of NADH during contractions of circulated mammalian skeletal muscle. *Respiration Physiol. 4*:292.

Karpatkin, S., E. Helmreich, and C. F. Cori. 1964. Regulation of glycolysis in muscle II. Effect of stimulation and epinephrine in isolated frog sartorius muscle. *J. Biol. Chem. 239*:3139.

Kugelberg, E., and L. Edström. 1968. Differential histochemical effects of muscle contractions on phosphorylase and glycogen in various types of fibres: relation to fatigue. *J. Neurol. Neurosurg. Psychiat. 31*:415.

Lawrie, R. A. 1953. The relation of energy-rich phosphate in muscle to myoglobin and to cytochrome oxidase activity. *Biochem. J. 55*:305.

Lind, A. R., and G. W. McNicol. 1967. Circulatory response to sustained hand-grip contractions performed during other exercise, both rhythmic and static. *J. Physiol. 192*:595.

Lloyd, B. B. 1967. World running records as maximal performances, p. I-218. *In* C. B. Chapman (ed.), *Physiology of Muscular Exercise.* Amer. Heart Ass. Monograph No. 15.

Lloyd, D. P. C. 1946. Facilitation and inhibition of spinal motoneurons. *J. Neurophysiol. 9*:421.

Magel, J. R., and J. A. Faulkner. 1967. Maximum oxygen uptakes of college swimmers. *J. Appl. Physiol. 22*:929.

Mager, M., W. F. Blatt, P. J. Natale, and C. M. Blatteis. 1968. Effect of high altitude on lactic dehydrogenase isozymes of neonatal and adult rats. *Amer. J. Physiol. 215*:8.

Margaria, R. 1967. Aerobic and anaerobic energy sources in muscular exercise, p. 15. *In* R. Margaria (ed.), *Exercise at Altitude.* Excerpta Medica Foundation, New York.

Margaria, R., R. D. Oliva, P. E. di Prampero, and P. Cerretelli. 1969. Energy utilization in intermittent exercise of supramaximal intensity. *J. Appl. Physiol. 26*:752.

Merton, P. A. 1956*a*. Voluntary strength and fatigue. *J. Physiol. 123*:553.

———. 1956*b*. Problems of muscular fatigue. *Brit. Med. Bull. 12*:219.

Mitchell, J. H., B. J. Sproule, and C. B. Chapman. 1958. The physiological meaning of the maximal oxygen intake test. *J. Clin. Invest. 37*:538.

Needham, D. M. 1960. Biochemistry of muscular action, p. 55. *In* G. H. Bourne (ed.), *The Structure and Function of Muscle,* Vol. 2. Academic Press, New York.

Olson, C. B., and C. P. A. Swett. 1966. Functional and histochemical characterization of motor units in a heterogeneous muscle (flexor digitorum longus) of the cat. *J. Comp. Neurol. 128*:475.

Pattengale, P. K., and J. O. Holloszy. 1967. Augmentation of skeletal muscle myoglobin by a program of treadmill running. *Amer. J. Physiol. 213*:783.

Paul, P., B. Issekutz, Jr., and H. Miller. 1966. Interrelationship of free fatty acids and glucose metabolism in the dog. *Amer. J. Physiol. 211*:1313.

Prewitt, M. A., and B. Salafsky. 1967. Effect of cross innervation on biochemical characteristics of skeletal muscles. *Amer. J. Physiol. 213*:295.

Rohmert, W. 1959. Die Armkrafte des Menschen im Stehen bei verschiedener Korperstellung. *Int. Z. Angew. Physiol. Einschl. Arbeitsphysiol. 18*:175.

Romanul, F. C. A. 1965. Capillary supply and metabolism of muscle fibers. *Arch. Neurol. 12*:497.

Rowell, L. B., H. G. Taylor, Y. Wang, and W. S. Carlson. 1964. Saturation of arterial blood with oxygen during maximal exercise. *J. Appl. Physiol. 19*:284.

Saltin, B., and J. Stenberg. 1964. Circulatory response to prolonged severe exercise. *J. Appl. Physiol. 19*:833.

Stainsby, W. N. 1965. Some critical oxygen tensions and their physiological significance, p. 29. *Proc. Int. Symp. Cardiovasc. Respir. Effects Hypoxia.* Hafner, New York.

Strydom, N. B., C. H. Wyndham, M. V. Radhenand, and C. G. Williams. 1962. Excess lactate turn-point in relation to maximum oxygen intake, p. 6. *Proc. Int. Congr. Physiol. Sci., 22nd.* Leiden.

Stubbs, S. St. George-, and M. C. Blanchaer. 1965. Glycogen phosphorylase and glycogen synthetase activity in red and white skeletal muscle of the guinea pig. *Can. J. Biochem. 43*:463.

Welch, H. G., and W. N. Stainsby. 1967. Oxygen debt in contracting dog skeletal muscle in situ. *Respiration Physiol. 3*:229.

Wilkie, D. R. 1950. The relation between force and velocity in human muscle. *J. Physiol. 110*:249.

# Chapter 28

# Summary and Discussion of Part 6

PANEL MEMBERS: E. E. GORDON, *Chairman*
E. ALLIN
R. E. DAVIS
A. HARPER
M. JUDGE
A. M. PEARSON

*Gordon:* The speakers have discussed muscle as related to development, growth, activity, and nutrition and have found many parallels in adaptation viewed from the biomolecular, biochemical, and histological levels. Thus a fetal isozyme of myosin adenosine triphosphatase (recalling the fate of fetal myo- and hemoglobin) progresses to the adult form. On the biochemical level increased creatine phosphokinase activity and the concomitant rise in contractile protein and intracellular constituents during early growth also point to increasing effectiveness of performance of the chemical-mechanical system of contraction. The $Ca^{++}$ transport capacity of the sarcoplasmic reticulum shows an upward trend corresponding to the period of greater activity of soluble enzymes. Histological and ultrastructural findings also coincide with maturation of the contractile unit in view of the greater number and area of myofibrils in muscle fibers and the increase in sarcomere length.

Dr. Perry rightly sees the rise in myosin adenosine triphosphatase and soluble enzyme activities consistent with adaptive changes to more effective energy-yielding systems, as the maturing animal exhibits increasingly rapid and frequent contraction-relaxation abilities. Whether physical activity is wholly or partially responsible for transition to adult structural, biomolecular, and biochemical patterns, it does suggest that one determinant of growth is exercise, and brings to mind the findings of increased myosin adenosine triphosphatase in acutely exercised adult rats (Iakovlev, 1957). Increased creatine phosphokinase has also been found in chronically trained rats in our laboratory (Gordon, unpublished observations). The interesting point, brought out by both Dr. Perry and Dr. Widdowson, is that young members of different species which are active early, or young animals encouraged to move about earlier than usual, sooner develop adaptive changes from fetal to adult patterns. Thus growth and physical activity are intimately related.

Nutrition holds a somewhat similar position to activity, for Dr. Widdowson has demonstrated that induced and spontaneous malnutrition in fetal and newborn pigs retards maturation. Undernutrition resembles inactivity in some respects: failure to accumulate myofibrillar protein, relative increase of sarcoplasm (Helander, 1961). But unlike inactivity, malnutrition in the young is devastating to every organ system, because it denies full maturation and capacity forever. This fact has large social implications.

One point is worth noting in Dr. Goldspink's data on myofibrillar growth in the young animal. Each small fiber contains about 500 myofibrils, and each large one 1,250; the relative areas of these myofibers are 3:8. Fiber to fiber there is 3.5 times the amount of myofibrillar protein in the large one compared to the smaller one. But a more realistic situation arising from analysis of muscle is the determination of the concentration per unit weight or mass. Consequently, the ratio of 3.5 needs elaboration. If we neglect the amounts of extracellular space between small and large fibers as a first approximation, at extreme conditions the ratio of small fibers to large ones *per unit weight* would be 8.3. Hence, the ratio of myofibrillar protein would be 5.3 to 5 $\left(\frac{500}{1{,}250} \times \frac{8}{3}\right)$. In ordinary muscle where there may be large and small fibers intermingled, the ratio in respect to large fiber would be less. We have found by biochemical methods a ratio of 4.5:5 between soleus and rectus femoris in young adult rats (Gordon, unpublished observations). Soleus consists of varied fiber sizes, but in rectus femoris fibers are uniformly large.

A second field of interest has been aroused by Dr. Goldspink's concept of interchangeability of small and large fibers under the proper circumstances of exercise and nutrition. Let me state first that this concept is not necessarily contradictory to accepted ideas, for we must remember that slow (red) and fast (white) muscle characteristics were initially ascribed to whole muscle, not to myofibers. It was on the basis of differential enzyme patterns that the concept of red and white fibers became identified with red and white (slow and fast) muscle. As a matter of fact, the so-called red soleus muscle presents a mosaic of fiber types in terms of enzyme patterns. Thus one can speak of transition from small to large fibers without violating the electrophysiologic associates of dichotomous innervation of fast and slow muscle.

Before one can accept Dr. Goldspink's proposal, however, its reconciliation with known facts need to be carried out. The prototype red or small fiber shows high succinic dehydrogenase and cytochrome oxidase which is quite low in the white or large fiber. We have found histochemically (Kowalski, Gordon, Martinez, and Adamek, 1969) that succinic dehydrogenase does indeed increase in the large fiber as well as in the small with chronic exercises without change in size. Histochemical as well as biochemical analysis (Gordon, unpublished observations) failed to indicate a change of activity for cytochrome oxidase in the large fiber. Moreover, histochemical determinations of phosphorylase activity—found to be characteristically low in red fiber by this technique—in some cases yielded significant tinctorial intensities in the small fibers after chronic training (Kowalski, Gordon, Martinez, and Adamek, 1969). Thus, if Dr. Goldspink is correct, mutability of enzyme profiles from one to the other fiber type would have to be demonstrated with appropriate change of fiber size following exercise, undernutrition, or other circumstances. Such changes have been demonstrated so far only in pathological or artificial situations (tenotomy, denervation, cross innervation). We have only partially documented mutability of enzyme patterns without changes in size as indicated above. More work along these lines is indicated with careful measurements of myofiber area.

While we have a good understanding of the details involving maturation of muscle tissue from the molecular to the structural level and its implications for adaptation, the determinants aside from physical activity and nutrition need further delineation. Acceptance of increase in cell number as well as size in early growth (Cheek, 1968) is challenged by Dr. Goldspink's direct counts of neonatal muscle. Even more intriguing is the relationship of red to white muscle fiber, a field which is attracting even greater attention but, as yet, has infrequently exploited physiologic stimuli to demonstrate adaptative changes.

## DISCUSSION

*Davies:* Are there significant changes in the size of the brain in malnutrition? Will these at all compare with the major changes that occur in muscle?

*Widdowson:* The brains of our malnourished pigs 1 year old are small when compared with well-nourished animals of the same age, but large when compared with well-nourished animals of the same body weight but only 4 weeks old. The total DNA in the brain was about the same in the undernourished pigs as in well-nourished 4-week-old animals of the same body weight; the weight of the brains was greater so the concentration of DNA in them was less. The total DNA was considerably less than it was in a well-nourished animal of the same age, but the concentration was about the same. Even after 1–3.5 years' rehabilitation the brain still contained significantly less DNA than it should (Dickerson, Dobbing, and McCance, 1967).

*Pearson:* Since the "runt" pig and the "starved" pig fail to develop normally even after being placed on adequate diet, could it be that the early dietary effects exert irreversible damage on the mechanism for protein synthesis? As you know there is evidence that protein levels in early life affect the ability of children to learn. Could the effects of inadequate nutrition influence not only muscle but also the central nervous system and the brain?

*Widdowson:* This may well be true, and the effects of malnutrition early in life on brain development and function both in children and experimental animals are being investigated by many groups of workers at the present time (Scrimshaw and Gordon, 1969).

*Weatherspoon:* You state that the "runt" pig was undernourished because of a poor blood supply. How was this poor blood supply determined or achieved?

*Widdowson:* A fetus that grows slowly and is small at birth has almost certainly received an insufficient supply of nutrients from its mother. This can only have occurred either because too little blood has reached it, or because the composition of the blood is faulty. All the fetuses in the litter in which the runt occurs receive blood of the same composition, and most of them grow normally, so composition cannot be the explanation. It has been shown in mice that runts always occur in the same position in the uterine horn, for example, at the ovarian end where the fetus has to share its blood supply with the ovary (McLaren and Michie, 1960).

Wigglesworth (1963) has produced the same effect experimentally in rats by partially occluding one of the uterine arteries. Runt piglets are almost always members of large litters, and those who remove piglets by Caesarian section say that runts tend to occupy a certain position in the uterus. It seems reasonable to assume, therefore, that they are small because in this position the blood supply is not as good as it is elsewhere. I must admit, however, that no quantitative measurements of blood flow have been made.

*Pearson:* If undernutrition decreases the number of fibers in a muscle, do these fibers attain a larger size upon restoration of adequate nutrition than is the case for normal fibers?

*Widdowson:* There is no clear evidence that undernutrition (calorie deficiency) decreases the number of fibers in a muscle. Montgomery, Dickerson, and McCance (1964) showed that it did not do so in the fowl. Montgomery (1962), however, believed that the number was decreased in children suffering from protein deficiency. We are at present investigating this in calorie-deficient and protein-deficient pigs, and in similar animals after rehabilitation.

*Judge:* How do pigs and ruminants compare with respect to the maturity of the muscle fibers at birth and to their capacity for compensatory muscle growth?

*Widdowson:* I do not know how pigs and ruminants compare with respect to the maturity of muscle fibers at birth, but I suspect that the muscle of ruminants would be more mature than that of pigs since ruminants are so much more mature in other ways.

*Allin:* Do the smallborn or undernourished piglets show true "catch-up" growth when given full nutrition? That is, do they show more rapid or more prolonged growth than normal piglets of their size or physiological age?

*Widdowson:* I am rearing some runt pigs along with their larger littermates. The runts show no sign of catch-up growth at present but they are still a long way from full size. I shall be able to answer your question more fully in about 2 years' time.

*Judge:* Is there any evidence that undernutrition *in utero* decreases the number of muscle fibers at birth? I.e., does the slowing of nuclear division you observed result in reduction of number of muscle fibers?

*Widdowson:* I cannot answer this question now, but should be able to do so very soon.

*Judge:* Does the failure of the muscle of your undernourished pigs to respond completely to rehabilitation indicate that hyperplasia was incomplete at the time of the restriction? Or is it likely that the suppression resulted from an interference with proliferation of nuclei during subsequent development of the cells?

*Widdowson:* Table 24.6 shows that the amount of DNA and hence the number of nuclei in the whole muscle was incomplete when the undernutrition was imposed, for there was a considerable increase during the first 2 weeks of rehabilitation.

*Davies:* Part of growth occurs because the sarcomeres themselves get longer; the I bands get longer although the filaments remain the same length so the number of the sarcomeres themselves must increase. Is it known how? It seems very unlikely that one can interpolate sarcomeres in a muscle that is bearing a load. If one thinks of a tug-of-war, how would you get the teams farther apart? You could easily put some more rope at the end and let them slide farther apart; similarly, you could add sarcomeres at the end where they are not bearing tension, around the tendons. Is there any evidence that this is the way the number of sarcomeres increases? Does it happen at the ends, the points that are clearly bearing less or no tension?

*Goldspink:* We have recently done some labeling experiments, and find that the newly formed myofibril proteins tend to be at the end of the muscle fibers. Mouse diaphragm makes a nice preparation because we are able to squash it and put the photographic film on the top. We did this at different ages after extracting the sarcoplasmic proteins and found that most of the label was at the end of the fibers. This would provide very strong evidence, if the results are correct, that the sarcomeres are added on to the end. The other piece of evidence is that the terminal sarcomeres tend to be nonfunctional in that when you stretch a young muscle fiber out the terminal sarcomeres do not pull out to the same extent as the middle ones. This is also evident in adult muscle, but in young muscle it is even more apparent. Your analogy of a rope and tug-of-war team is a good one as it is easy to visualize the new sarcomeres being knitted on to the end of the myofibrils beyond the point at which "the ropes" (myofibrils) are held.

*Pearson:* You stated that there are increases in the numbers and in the length of sarcomeres in a muscle during growth. Are there conditions during growth or adult life in which the numbers and length of the sacromeres decrease? If so, what appears to be responsible?

*Goldspink:* I cannot really answer this question except to say that the number of sarcomeres in series along the muscle may decrease in muscular dystrophy and other degenerative conditions. In dystrophy the fibers degenerate at different regions along their length, and this presumably means the number of sarcomeres will be reduced.

*Pearson:* Please discuss the sequence of structural changes that occur in the formation of new sarcomeres in active muscle.

*Goldspink:* Other than the information given in Chapter 25, we have no knowledge of the structural changes that occur in the formation of new sarcomeres. We are at present making a study of this subject and we hope to be in a position to be able to present some information shortly.

*Galloway:* Is there a relationship between the change in your proposed thick and thin filament overlap and normal bone growth, and what do you propose causes the change in the amount of overlap?

*Goldspink:* Yes, it is believed that the pulling of the sarcomeres takes place because the bone is elongating at a faster rate than the production of new sarcomeres. Most of the increase in sarcomere length takes place during the first few weeks after birth when the limb is lengthening very rapidly. Chapter 25 gives further information on this point.

*Allin:* If it is traction of skeletal growth that accounts for the increased sarcomere length during growth, then it might be expected that the change would be greatest during periods of swifter growth such as at adolescence. Also, once bone-lengthening ceased, addition of sarcomeres might catch up, resulting in reduced sarcomere length and increased filament overlap. Are either of these situations observed?

*Goldspink:* The greatest increase in bone length takes place just after birth, and it is during this time that the sarcomere length increases the most. The lengthening of the bone may cause a slight gap to occur between the end of the myofibrils and the tendon, which would enable additional sarcomeres to be assembled on the end. If this is the case, one would expect the addition of sarcomeres to stop soon after the bone had ceased to lengthen. This would mean that the resting length of the sarcomere that had already been pulled out would not be affected. According to our measurements neither of the situations you suggest occur in mouse muscle.

*Allin:* Why would ontogenetic reduction in filament overlap increase strength? I would expect just the reverse. I would expect reduced extensibility and increased shortening capacity as well. Also, if shorter sarcomeres are preferable for isometric action, adult postural muscles might be expected to show this feature. Is this the case?

*Goldspink:* A sarcomere length of 2.2 or 2.3 $\mu$ as found in very young mouse muscles means that the sarcomere will have a maximum functional overlap of cross bridges, and will thus be able to develop considerable isometric tension. However, these sarcomeres cannot shorten to any extent before the thin filaments overlap and the tension output would drop very considerably because of the distortion of the myofilamental lattice. Young muscle would therefore be expected to have a greater "isometric strength" but a lower "isotonic strength" than adult muscles. For length-tension plots please see Goldspink (1968). From the few measurements that we have carried out it seems that adult postural muscles have more or less the same sarcomere length as the phasic muscles. I think the postural muscles of the adult usually work over a more limited range of muscle lengths, however, as they are required to fix the position of the limb rather than to cause it to flex or extend. If one considers the soleus which fixes the foot in the fully extended (tiptoes) position, it would develop isometric tension with its sarcomeres almost fully shortened.

*Henrickson:* As a fiber develops, there is an increase in fibrils and sarcoplasm. What is the ratio between these growth parameters?

*Goldspink:* The amount of sarcoplasm cannot be accurately determined by electron microscopy.

*Dikeman:* If a sample was removed from the same muscle of different animals of approximately the same age, weight, breed, and sex in a "relaxed" or natural state in the animal, would you expect that there would be a difference in sarcomere length between these animals due to genetics or is the difference in sarcomere length between animals due only to changes caused by the biochemical changes occurring during rigor mortis?

*Goldspink:* A. R. Luff (1968) and I carried out some measurements on sarcomere length in different strains of mice. It was found that even in strains of mice which had different limb lengths the sarcomere length was more or less the same. In this work all the muscles were fixed, as soon after death as possible, with the limb in the fully extended position. The length that the limb attains appears to be closely connected to the number of sarcomeres produced in series along the length of the fibers, so that the sarcomere length in the adult animals is always about the same. Gross differences in sarcomere length will occur, of course, if the muscles set at different lengths in the carcass during rigor.

*Dutson:* You stated that the increase in fiber size appeared to be due to a splitting of the myofibrils. Do you have any evidence whether or not this process is reversible? Is there any coalescence of myofibrils in low nutrition or in disuse atrophy?

*Goldspink:* I think it is unlikely that coalescence of myofibrils occurs in low nutrition or disuse atrophy, because under these circumstances the myofibrils within the affected muscle fibers become smaller as well as less numerous.

*Greaser:* The theory that was presented to explain myofibril splitting depends on the assumption that the Z disk is a rigid structure. Is there any evidence that there is no change in size of the Z disk with changes in sarcomere length?

*Goldspink:* We visualize the Z disk as having a structure similar to that of a piece of loosely woven cloth. When tension is applied, the lattice spacing will increase to some extent until the tension applied to the disk becomes so great that the more central filaments will break. The rip thus started will extend across the disk with the direction of the weave. Work that we are presently engaged in indicates that the Z-disk lattice does change, both during the shortening of the sarcomere and during the growth of the myofibril. It appears, therefore, that the Z disk is not a completely rigid structure but that it is able to undergo a considerable amount of distortion before it rips.

*Reedy:* Since the splitting of fibrils would seem likely to leave the myofilament lattice highly parallel in the two daughter fibrils, one would expect to find numerous small families of neighboring fibrils in which A and Z lattices are parallel. Is the number of families equal to the number of fibrils in small fibers? Have you examined electron micrographs of transverse sections of large and small fibers in an effort to identify this?

*Goldspink:* Splitting may result in the A and Z lattice of adjacent myofibrils being highly parallel, initially. It is likely that this would be lost after one or two contractions, however. Certainly splitting is probably the reason why myofibrils tend to be straight-sided rather than spherical in shape and why they all have approximately the same sarcomere length. At the present time we are making an extensive study of the lattice in transverse-section electron micrographs of large and small fibers and we will certainly bear this point in mind.

*Dubowitz:* In Fig. 25.8, illustrating splitting of the fibril at the Z-band level, the A bands of the two "halves" seemed out of phase. Could there possibly be two closely approximated fibrils rather than one with a split?

*Goldspink:* Fig. 25.8 could well have been interpreted in the way you suggest. When we carried out these studies, however, we examined the split myofibrils along their entire length to the edge of the section or until they went out of the plane of the section. By examining myofibrils in this way it is possible to ascertain whether they are single myofibrils which have split to two adjacent myofibrils.

*Sanger:* You presented evidence that in "developing" mouse muscle, the sarcomeres elongate by gradually increasing the length of the I bands. This system would appear to present an ideal model for the roles of tropomyosin and troponin in the I bands. The use of fluorescent antibodies against the regulatory proteins may reveal changing patterns with increasing sarcomere length. Have you used this approach?

*Goldspink:* The length of the I band increases but the length of the I filament remains unchanged, so I am not sure what sort of information would be obtained by using fluorescent antibodies. This is certainly an interesting possibility.

*Allin:* There seems to be some confusion about the multiplication of muscle nuclei during the postnatal growth of striated muscle. Would Dr. Holtzer comment on this subject?

*Holtzer:* Some of the work that was referred to this morning (MacConnachie, Enesco, and Leblond, 1964) has been retracted. Leblond was kind enough to let me know that when they reexamined their material they in fact found no DNA synthesis *within* a muscle fiber. You will recall that there is a population of cells in muscle, first described by Mauro (1961) and more recently by Church, Noronka, and Allbrook (1966) and Ishikawa (1966), that are called "satellite cells." Satellite cells are mononucleated, they have not synthesized myosin or actin, and they are very readily confused with muscle nuclei under the light microscope. There is no increase in DNA in nuclei in cytoplasm in which myosin and actin are being synthesized. It is important to recognize the source of DNA synthesis or increment in DNA in developing muscle. Is it in satellite cells or connective tissue cells or fat cells, for example?

*Allin:* It should be clear from what Dr. Holtzer has said that it is misleading to speak of a "stage of hyperplasia" in muscle growth. Hyperplasia, by its usual definition (mitotic cell division), occurs throughout growth, with continuing incorporation of mononuclear myoblasts into the syncytia (multinucleate fibers). Nuclei within syncytia do not divide, nor do syncytia. Probably the dividing cells are the recently characterized satellite cells and possibly other mononuclear cells as yet unrecognized.

*Widdowson:* I agree with you that the phrase "stage of hyperplasia" is not a good one to use in respect of the development of striated muscle if there is no mitosis of the true muscle nuclei.

*Allin:* Cheek (1968) has chosen to consider the number of diploid nuclei to be the same as the number of muscle "cells," ignoring the existence of syncytia. This has operational advantages, but little else to recommend it. To say there is an increase in muscle nuclei or even syncytial nuclei is not to say there is an increase in the number of syncytia. Cheek's data say nothing at all about numbers of muscle fibers.

*Smith:* Are you suggesting that the increase in nuclei number per muscle fiber is due to mitosis of existing nuclei in the muscle fiber? Can this increase in nuclei be due to fusion of more myoblasts into the muscle fiber?

*Widdowson:* I did suggest that the increase in number of nuclei in the muscle fiber after birth is due to mitosis, on the basis of the paper by MacConnachie, Enesco, and Leblond (1964). Dr. Holtzer now tells us that these authors have retracted, and that the nuclei are now believed to increase in number by the incorporation of satellite cells into the muscle fiber.

*Hendrickson:* Each muscle fiber nucleus is thought to support a given area of the muscle fiber. Is this true?

*Widdowson:* I have no direct proof that each muscle fiber nucleus supports a given area of the muscle fiber, but it seems a reasonable assumption that there is an upper limit to the amount of cytoplasm which one nucleus can support. During early development the amount of cytoplasm to each nucleus is comparatively small. Later, the muscle fiber grows in size at a more rapid rate than the nuclei increase in number, so that the amount of cytoplasm to each nucleus rises. This is measured by the ratio of protein to DNA, and this reaches a ceiling which is similar in adult rat and man.

*Henrickson:* Would you expect all nuclei to have equal amounts of DNA? Is it not possible that nuclei will produce DNA based upon the amount needed for development?

*Widdowson:* It has been shown many times over that within any one species the diploid nucleus contains a constant amount of DNA. In the early stages of muscle development, when mitosis is still going on, synthesis of DNA occurs in the dividing nuclei, and the nuclei contain up to twice their resting amount of DNA according to the phase of mitosis of the nucleus at the particular moment when the measurement is made. After this stage in early fetal development is passed, and the myotubes are complete, it seems that there is no further mitosis, so the amount of DNA in each nucleus remains constant right through pre- and postnatal development into adult life.

*Davies:* You state that the increase in the number of muscle "cells" which occurs is 14-fold in boys and 10-fold in girls. Is there a similar increase in the number of cells in the pregnant uterus, or do the cells just increase and then decrease in size?

*Widdowson:* I know of no direct evidence about the cellular changes that take place in the pregnant uterus, but it is an interesting question, and one that could be answered easily. Administration of estrogens to immature rats caused an increase first in the extracellular phase and then in the cellular constituents (Talbot, Lowry, and Astwood, 1940), but whether the latter was due to hypertrophy was not investigated.

*Davies:* Dr. Goldspink said there is a constant number of muscle fibers after differentiation, but Dr. Widdowson stated that, in humans, there is a great increase.

*Widdowson:* I did not state that there is an increase in the number of muscle fibers after differentiation. I said that the number of nuclei in each fiber increases.

*Goldspink:* The number of fibers does not increase, although apparently the number of nuclei does increase during growth. The nuclei are believed to increase not by mitoses but by satellite cells which donate nuclei to growing fibers.

*Carpenter:* Could you suggest a measure of absolute numbers of muscle "cells" or fibers?

*Widdowson:* The only way I know of measuring the number of muscle fibers is by counting them in cross sections of whole muscles.

*Carpenter:* What evidence substantiates an increase in muscle fiber number in the rat?

*Widdowson:* The evidence comes from a paper by Morpurgo (1898). He stated that the number of fibers increases by 23% in m. radialis during the first 15 days after birth, and after this there is no further increase.

*Carpenter:* Could you explain the increase in $Mg^{++}$ from the 3-week piglet to the adult? Does this increase relate to the nuclei increase?

*Widdowson:* The increase in concentration of $Mg^{++}$ from 16 to 23 meq/kg of whole muscle from the 3-week piglet to the adult is related to the increase in cell mass at the expense of the extracellular phase. The small difference in the ratio intracellular protein:$Mg^{++}$ between the 3-week pig and the adult is not statistically significant.

*Judge:* How do nuclei increase in number during hypertrophy and hyperplasia?

*Widdowson:* Hypertrophy of muscle is generally taken to mean growth in the size of the fiber without an increase in the number of nuclei. Hyperplasia involves an increase in the number of nuclei. From what Dr. Holtzer says, it seems that this is brought about after birth, not by mitosis, but by the incorporation of satellite cells into the fiber.

*Pearson:* In view of the apparent increase in the number of "nuclei" or "satellite cells" as an animal grows, a biopsy using the protein-DNA ratio should provide an index of an animal's growth and composition. Would you please comment on this?

*Widdowson:* The protein-DNA ratio is what Cheek (1968) uses as a measure of the size of the muscle "cell." In male rats the ratio increases up to an age of about 14 weeks and thereafter remains approximately constant. In children, the ratio stabilizes at about 10.5 years in the female, but goes on increasing in the male at any rate up to 16 years. Cheek suggests that it may go on increasing up to 25 years.

*Perry:* I want to make a point of potential importance to the meat industry. There has always been a problem in carcass evaluation in estimating the amount of lean in relation to the amount of lipid. I think that these unusual amino acids might, when we know a little more about them, make a very nice quick method of deciding how much muscle you have in a carcass or a joint. This could be done by estimating methyl lysine or methyl histidine. We have to learn more about the distribution yet, but it could be quite a useful method.

*Maruyama:* What do you think is the role of granular $Mg^{++}$-activated adenosine triphosphatase activity in the developing embryo muscles? You showed that the $Ca^{++}$ uptake was rather ineffective in embryonic rabbit muscle cells—this was also true of embryonic chick muscle cells.

*Perry:* We believe that the $Mg^{++}$-activated adenosine triphosphatase of the sarcoplasmic reticular (granular) fraction is part of the $Ca^{++}$ transport system. In developing muscle only a small fraction is coupled into the transport system and hence can be readily measured in the absence of $Ca^{++}$ transport.

*Davies:* Is it possible that the $Ca^{++}$ transport system is mechanically less stable in the fetal rabbits and is more easily fragmented?

*Perry:* This of course is possible, though in our view unlikely. Careful electron microscope study has so far failed to show differences in the sarcoplasmic reticular preparations from adult and fetal muscle. The vesicles from both tissues are very similar in size distribution. We are currently investigating possible differences in the biochemical makeup of the two systems.

*Davies:* Have you tried to repeat the isolation procedures on both adult and fetal material in case the efficiency of $Ca^{++}$ transport really is high but gets damaged more easily? A second homogenization cycle might damage it even more.

*Perry:* No. In general, vesicles are more readily obtained from fetal muscle, which requires less mechanical treatment to liberate them.

*Harper:* How does Dr. Perry envisage the energy-requiring step fitting into the model of $Ca^{++}$ transport in muscle?

*Perry:* I have no personal scheme to fit that. These are facts we have and we are trying to find an explanation for them.

*Harper:* Is there any evidence of a link to the $Na^+$ pump, as there is for transport of glucose and amino acids, rather than direct energy requirement for the $Ca^{++}$ transport itself?

*Weber:* We have not found a specific effect of $Na^+$ or $K^+$ on $Ca^{++}$ uptake. In higher concentrations, however, cations are slightly inhibitory; the maximal steady state of $Ca^{++}$ filling (under physiological conditions, in the absence of $Ca^{++}$-precipitating agents) is lower in the presence of 100 mM KCl than in that of 50 mM.

*Duggan:* There is no evidence at this time that the $Na^+$ pump in muscle is associated with the $Ca^{++}$ pump of the sarcoplasmic reticulum. One major support for this view is the ineffectiveness of ouabain in inhibiting the $Ca^{++}$ pump, whereas the $Na^+$ component is very sensitive to this compound.

The uptake of $Ca^{++}$ by skeletal muscle microsomes is stimulated by $Na^+$ and more so by $K^+$ ions (Duggan, 1967, 1968), however. This stimulation is maximal at the physiological concentration of $K^+$ and, unlike that of the $Na^+$-$K^+$ transport adenosine triphosphatase there is no synergistic action of $Na^+$ and $K^+$ on $Ca^{++}$ transport. These ionic effects are more easily demonstrable when the microsomes are prepared in sucrose medium rather than one containing KCl, and may be the explanation of the many reports in the literature that $Ca^{++}$ uptake is either unaffected, or even inhibited, by $K^{+}$.

*Judge:* Do the large and small phase fibers you describe coincide with the other designations of fiber type, such as red and white, fast and slow?

*Goldspink:* The large phase fibers appear white and the small phase fibers appear red when unfixed, frozen sections are viewed with the light microscope. However, the large phase and the small phase fibers do not coincide with the fast and slow designation. That is to say, both the small fibers and the large fibers may be either fast or slow depending on what muscle they belong to. The conversion of fibers from the small to the large phase and vice versa is not accompanied by any change in the maximum velocity of contraction of the muscle.

*Gordon:* I think from a physiological point of view we may have to discard such terms as red, white, fast, slow, and use more specific designations.

*Jungk:* You mentioned that the change in size of an individual fiber only took about 1.5 days. How did you establish this?

*Goldspink:* One and a half days is believed to be the minimum time required for the jump in size of an individual muscle fiber. This period of time was calculated from data obtained by subjecting mice to alternating periods of starvation and normal feeding. During starvation the large phase fibers revert back to the small phase, and on refeeding some of the small phase fibers again develop into large phase fibers. During this experiment mice were serially sacrificed throughout the starvation and refeeding periods and the histograms of fiber size were examined. One and a half days was found to be the minimum time during which a change in the histograms was detectable. For further details refer to Goldspink (1965).

Work by Dreyfus, Kruh, and Schapira (1962) on the incorporation of labeled amino acids into muscle proteins showed that the incorporation into myosin reached a maximum within 2 days after administration of the isotope.

*Judge:* Is all the increase in girth of a muscle the result of development of large phase fibers? If so, how do the muscles that are constituted almost completely of small, red fibers increase in girth?

*Goldspink:* The muscles that consist entirely of small red fibers do not increase in girth to anything like the same extent as the bimodal muscles. Early in the life of the animal there is a slight increase in the diameter of the small fibers (see Fig. 25.4), until they reach the stable level at 20–25 μ. After this, these muscles in the mouse do not increase in girth very much, presumably because the intensity of the work load is not great enough to induce hypertrophy of any of their fibers. Incidentally, the muscles with small, red fibers are not affected to any extent when the animal is starved, whereas the bimodal muscles decrease in size very considerably (Rowe, 1968).

*Weatherspoon:* Is there a change in fiber type with muscle hypertrophy? If so, how is this affected by capillary density?

*Goldspink:* In general I am opposed to the typing of muscle fibers as this tends to suggest that muscle is a static tissue which, of course, is not true, as it possesses amazing adaptive ability. So many of the muscle fibers' characteristics can be changed by subjecting the animal to different conditions that a different classification would be required for each set of conditions used. I have described some of the changes associated with hypertrophy and their physiological significance. I have no wish to say whether they consititute a change in fiber type as I doubt the validity of the different classifications. However, I can say that the small fibers are red and the large fibers are white and that hypertrophy may change the muscle from red to white, apparently due to a dilution of the myoglobin and cytochromes. This may not be as straightforward as it appears because some types of exercise may induce an increase in myoglobin and cytochromes as well as hypertrophy of the fibers. It is very likely that capillary density is affected by hypertrophy because the increase in the fiber girth will mean that the capillary density per unit area will be decreased.

*Judge:* Why does the transition from small to large phase fibers apparently occur as a complete reversion in specific fibers rather than as a partial change in all fibers?

*Goldspink:* We do not know the answer to this question. I suspect that when the intensity of work load on a fiber becomes very great, protein synthesis is switched on and proceeds at a rapid rate until the fiber has reached a certain size at which it stops. In other words, once a threshold is superseded this triggers off the hypertrophy process. As the fibers contract in an all-or-nothing manner it is perhaps not surprising that they undergo hypertrophy in this manner. Of course, when the muscle is in normal use not all of the fibers are called upon to contract. If the work load is increased, this will bring more fibers into play until all of the fibers are involved. If it is increased still further, some of the fibers will be working much harder than they have done previously. It is probably only under these latter conditions that the "work-load threshold" in some of the fibers is exceeded and fibers are induced to undergo hypertrophy.

*Judge:* Assuming the large phase fibers you described are also pale, anaerobic-type fibers, why does the porcine longissimus dorsi, which seems to perform a "slow" function, develop predominantly pale fibers?

*Goldspink:* You cannot always equate redness with slowness. The relation between the speed of contraction and the color or type of metabolism of the fibers is an indirect one. Fast muscles are more efficient in shortening and doing work but very inefficient in maintaining tension. Some fast muscles are required to work very frequently, however (an example that has been mentioned is the ceratohyoid muscle of the bat, which is an extremely fast muscle, but it is also a muscle that is used very frequently when the bat is emitting the high-frequency sounds of its sonic detection mechanism). The relationship as I see it is as follows:

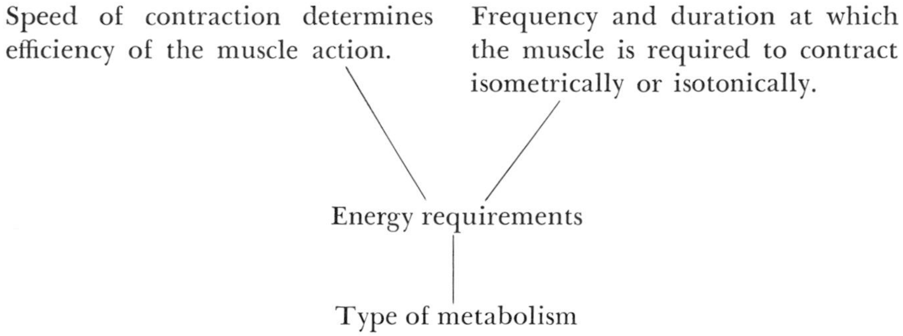

Speed of contraction determines efficiency of the muscle action.

Frequency and duration at which the muscle is required to contract isometrically or isotonically.

Energy requirements

Type of metabolism

The longissimus dorsi of the pig presumably is required to develop considerable tension, especially in present-day heavy hogs. Therefore, it may consist of hypertrophied slow fibers in which the mitochondria have been diluted by the additional myofibrillar material.

*Dubowitz:* If, as you suspect, the histochemically low succinic dehydrogenase is due to "dilution" of mitochondria by the myofibrils, do you find, as one would expect, a loss of the checkerboard pattern of fiber types, when the small phase fibers hypertrophy?

*Goldspink:* The checkerboard pattern of succinic dehydrogenase activity is changed but is not lost, because only a certain percentage of fibers undergoes hypertrophy. We have not in fact been able to devise a method of exercise which is intensive enough to stimulate all the small fibers to undergo hypertrophy.

*Allin:* Muscle fiber heterogeneity cannot be entirely accounted for by the growth-phase concept. Fiber size does not always correlate with fiber "type" differences. For example, biopsies of apparently normal muscle from healthy women may have smaller "white" fibers than "red," as Dr. Brooke had reported in Chapter 8. Some muscles in mice, if I am not mistaken, are histochemically mixed, yet homogeneous in fiber size. Also, in the rat gastrocnemius, small fibers and large fibers are both high in myofibrillar adenosine triphosphatase activity histochemically, while fibers of intermediate size are low in activity. Further, as Gauthier and Padykula (1966) have shown, mitochondria in different sorts and sizes of fibers differ not only in concentration but also in structure (e.g., number of cristae). I am not suggesting that fiber growth has nothing to do with metabolism. Large fibers do tend to be relatively poor in mitochondria and rich in glycolytic machinery. I doubt if this is to be explained, however, on the basis of dilution of mitochondria by myofibrillar material, with compensatory enhancement of anaerobic metabolism. It seems more likely that oxygen diffusion limitations favor anaerobic metabolism as fiber width increases. Another factor may be impulse frequency. Muscle fiber hypertrophy may be induced when contractile tension is increased (although tension itself may not be the stimulus). This entails an increase in impulse frequency, which in turn would promote glycolysis. Must hypertrophy and the shift toward anaerobiosis occur simultaneously, or may the latter precede the former? In mouse muscles which show progressive increase in proportion of large phase fibers, is there a matching increase in proportion of fibers which are histochemically "white" (type II, A, or whatever you prefer)?

*Goldspink:* First, let me say that I agree that muscle fiber heterogenity cannot be entirely accounted for by the nonsynchronous growth of the fibers. I do believe, however, that this is one of the main factors which determine the histochemical characteristics of the individual muscle fibers in the same muscle. Cytophotometric measurements of succinic dehydrogenase staining have been carried out on individual muscle fibers of different ages and sizes (Goldspink, 1969). These have shown that in three different muscles studied there was a linear relationship between fiber size and the concentration of the stain. This was found to be the case even for a unimodal muscle such as the mouse soleus although, of course, the range of fiber sizes was much less in this muscle than in the bimodal muscles. Even though the concentration of the stain was less in the large fibers the total amount per fiber was about the same irrespective of size. The fact that the total amount remains the same strongly suggests that the decrease in mitochondrial material as a result of hypertrophy is a dilution effect.

The shift from aerobic to anaerobic metabolism as the muscle fiber undergoes hypertrophy may be a combination of both the oxygen diffusion limitations and the decreased mitochondria:myofibril ratio. There is no evidence to suggest that the fibers become anaerobic before they undergo hypertrophy, although I do agree that tension may not be the direct stimulus which initiates hypertrophy. Qualitative descriptions of the mitochondria and other structures in fibers of different sizes are not evidence that these different sizes do not represent different stages of growth. As described by Drs. Green and Harris in Chapter 12, the cristae of mitochondria are capable of considerable conformational change according to the energy state at any particular time.

Regarding the last point you raised, there is an increase in the proportion of fibers which are histochemically white as the muscle matures. This is presumably why the metabolism of the muscle as a whole tends to be more anaerobic.

*Lewis:* How was fiber diameter determined?

*Goldspink:* The muscles were fixed on the bone with the limb in a standard position using Flemings without acetic acid or glutaraldehyde. Both of these fixatives (unlike many others) cause very little shrinkage of the muscle fibers. Usually 100 fibers were measured from each muscle and these were taken from transects drawn across the projected image of the section.

*Davies:* Does the extracellular compartment of muscle increase with growth as stated by you, or decrease as stated by Dr. Widdowson?

*Goldspink:* I believe that the extracellular compartments do increase during normal growth so that the percentage of extracellular fluid to fiber mass remains more or less constant. It is possible that the percentage of extracellular fluid decreases during the differentiation of the tissue but I do not think that this is the case during postembryonic growth. The percentage of water content of muscle drops only slightly during postembryonic growth, and this can be explained by the increase in the percentage of the insoluble myofibrillar material. The great problem with measuring the extracellular fluid volume during growth is that the membrane permeability of young muscle is different and the chloride space and inulin space mehods are not valid.

C. F. Hazlewood of the Texas Medical Center, Houston, has fed a lot of data concerning water content, membrane potentials, and distribution of ions into a computer and has reached the same conclusion as we did, i.e., that the percentage of extracellular space does not change much in growing muscles.

*Widdowson:* The extracellular compartment decreases with development as a percentage of the muscle, but of course the absolute amount in the muscle increases as the animal grows.

*Dubowitz:* Does exercise produce selective hypertrophy of the "large phase" fibers or of the "small phase"?

*Goldspink:* The histograms of fiber sizes show that the effect of exercise on a muscle that is normally bimodal is to increase the height of the large fiber peak without substantially changing the mode (the diameter at which the peak occurs). Therefore, it is concluded that as a result of exercise the large fibers do not get any larger but become more numerous. As the total number of fibers in the muscle is unchanged the additional large fibers must have arisen from small fibers which have undergone hypertrophy. This does not rule out the possibility that the large fibers are capable of further hypertrophy if the work load on the muscle is so great that all of the small fibers have been converted into large fibers. In conditions normally encountered in the life of the animal, it is probably safe to say that it is the small fibers that selectively undergo hypertrophy.

*Brekke:* In Chapter 25 you have implied that maximum strength and endurance of muscle could be changed, but not the rate of contraction. But with exercise, it seems that one's reflexes, i.e., the speed necessary for one to react, become sharpened. Is this then only the result of training or conditioning, or is the muscle really reacting faster to the nervous control to contract? If this is not due to physiological differences, why does it happen?

*Goldspink:* I think there is no doubt that changes can result from training that enable the neuromuscular system to respond faster with improved coordination. As far as I am aware, all the evidence suggests that these changes are associated with the nervous system and not with the muscle tissue itself.

*Peter:* Dr. Goldspink said he failed to find any changes in the twitch characteristics of muscles in which the sizes of fibers had been changed by exercise. To prove this he has to show that the new, larger fibers are actually contracting under his standard conditions of stimulation. Unless this has been demonstrated it is also possible that the "physiological" characteristics, as tested by electrically induced twitch, fail to change not because the enlarged fibers have not changed their electrical characteristics but because the enlarged fibers are not contracting and hence not contributing to the characteristics of the twitch.

*Goldspink:* In the fairly limited number of experiments we have carried out the large fibers must have been stimulated because of the much greater tension developed by the hypertrophied muscles. More rigorous studies than ours have been carried out by Binkhorst (1969). His work shows rather conclusively that the twitch contraction time is not changed as a result of subjecting the muscle to repeated exercise.

*Peter:* Dr. Goldspink demonstrated large, rapid increases in fiber diameter in his "exercised" mice. How much of this increase is due to movement of extracellular fluid into the muscle cells?

*Goldspink:* The percentage of water content of exercised muscles is slightly less than control muscles. Also, the evidence indicates that the percentage of extracellular fluid is less in exercised muscle. In effect, then, some of the extracellular water goes into the fibers. The hypertrophied fibers contain more water only because they are larger; their percentage water content is probably less.

*Davies:* Isometric contraction producing hypertrophy leaves the number of mitochondria constant whereas endurance exercise increases the number of mitochondria. Do you have any idea of the chemical nature of the stimuli which can increase myosin on the one hand and mitochondria on the other?

*Goldspink:* Unfortunately, I have no idea what the nature of the chemical link is between the mechanical events and the synthesis of myofibril or mitochondrial proteins. We intend to pursue this fascinating problem.

*Bocek:* You stated that there is an increase in fiber diameter and myofibrillar splitting within 1–2 days due to exercise. Is there any evidence of an increase in protein synthesis in these muscle fibers?

*Goldspink:* Recently we have been carrying out some radioisotope incorporation studies on growing muscle and have found that in transverse sections some of the fibers are heavily labeled while others show virtually no labeling. These findings tend to support the concept that muscle fibers grow in an all-or-nothing manner and that at any one time during the growth of the animal only a certain percentage of fibers are actively growing. We have not as yet carried out labeling experiments on exercised muscles. Such experiments would be most interesting.

*Edgerton:* High resistant exercise causes small phase fibers to increase in size; you have interpreted this as due to an increase in the number of myofibrils, and it accounts for an increase in strength. What then could be the advantage of a small phase fiber for endurance adaptation, if the two phases of muscles have the same number of mitochondria and they differ only in the dilution effect due to hypertrophy?

*Goldspink:* The small fibers have a higher mitochondria:myofibril ratio and, therefore, the ATP supply per myofibril would be expected to be better. Also the distance the ATP will have to move to get to the myofibril will be considerably less in the small fibers and this must be quite important. In conditions demanding endurance, it would be a disadvantage to be carrying the extra weight due to the additional myofibrils in the large or hypertrophied fibers.

*Barnard:* You stated that fiber hypertrophy was found in the soleus following tenotomy of a few days and inferred that this hypertrophy was similar to that observed in your long-term studies. Since the data available on myosin turnover indicate a life span of a least 20–30 days, it does not seem possible that short-term tenotomy can produce an increase in contractile proteins.

*Goldspink:* I cannot see any reason why the rate of buildup of contractile proteins during growth or hypertrophy is limited by the rate of turnover in normal mature muscle. The increase in contractile proteins during hypertrophy involves an increase in size of myofibrils and the formation of more myofibrils, whereas the turnover is merely the renewal of the existing myofibrils. Therefore, there is no reason to suppose that the turnover rate represents the muscle fibers' maximum capacity to synthesize contractile proteins.

*Bocek:* Is there an increased concentration of ribosomes at the ends of the fiber, where newly formed sarcomeres are assembled, in comparison to other areas of the fiber?

*Goldspink:* We have not examined the ends of the fibers to see if there is a greater ribosomal accumulation in this region. However, Paul Larson, who is at present working in the Muscular Dystrophy Laboratory in Newcastle, England, has shown me some of his electron micrographs of developing muscle which show ribosomes at the end of myofibrils; some of the ribosomes are in the form of a spiral around the terminal myosin filaments.

*Kang:* Did you mean more exercise results in greater body weight? If so, is this true for adults as well as growing animals?

*Goldspink:* Intensive exercise such as weightlifting tends to increase the animal's body weight because of the extra muscle mass which develops. I have observed that even when the exercised and control animals were supplied with the same amount of food the exercised mice put on more weight. Presumably intensive exercise in some way facilitates more protein retention. Apart from these few observations I have no other detailed knowledge on the subject.

*Pearson:* In Chapter 25 you classified the types of measurements of muscle activity into: (1) measurements of contractile strength; (2) endurance measurements; (3) measurements of rate of contraction; and (4) measurements of the efficiency of contraction. You then stated that contractile strength and endurance can be altered by physical training and exercise, whereas rate and efficiency of contraction are altered by growth, but not by training or exercise. What is the nature of the physiological controls that determine whether these parameters are influenced by growth or by exercise? How, if such controls are hormonal or chemically mediated, would it be possible to alter the effects of growth or exercise?

*Goldspink:* Bárány (1967) has shown that the rate of splitting of ATP by myosin adenosine triphosphatase is closely related to the speed of contraction of the muscle. Both Perry and Bárány have shown that the myosin adenosine triphosphatase changes during growth. If it is the myosin adenosine triphosphatase that governs the rate of contraction, it is quite feasible that once the muscle has received its adult form of myosin then no further change will take place in its maximum velocity shortening (i.e., of the unloaded muscle) even if it is exercised. Exercise will merely cause the muscle to produce more of the same type of myosin without changing its characteristics. The change in the speed of contraction during growth does in fact take place very early in the animal's life (Goldspink and Rowe, 1968), and may even be considered to be associated with differentiation and not with growth of the fibers.

Dr. Perry's suggestion that myosin adenosine triphosphatase may exist as different isozymes and that during growth there is a change from the "slow" fetal type of myosin adenosine triphosphatase to these different adult isozymes seems to me to be very reasonable. The type of myosin adenosine triphosphatase is apparently influenced by the nervous supply of the muscle as shown by Buller, Eccles, and Eccles (1960*b*) and thus is probably the main, if not the only, factor which determines what type of myosin adenosine triphosphatase the muscle should make.

*Pearson:* Dr. Perry has pointed out that the myosin of the fetus and that of the adult appear to be different. In light of Dr. Goldspink's finding that muscle fibers can change rapidly in size, would he comment on the rapidity of the change from fetal to adult myosin? How long is required?

*Perry:* Dr. Goldspink's experiments were carried out in the mouse, in which species the changes occur relatively rapidly. If similar myofibril changes occur in the rabbit I would expect them to take place more slowly. The time for the change from fetal to adult myosin as judged by the adenosine triphosphatase activity depends on the species (see Fig. 26.4). In the rabbit it is virtually complete 3 weeks after birth. In the guinea pig and fowl it is practically complete before birth.

*Pearson:* Is there any evidence that adult myosin can regress to fetal myosin in muscle degeneration or atrophy?

*Perry:* Yes, if we can assume that the extent of methylation of histidine is an index of the amount of the adult form of myosin. Mr. Lobley in my laboratory has recently shown that in Vitamin E dystrophy the 3-methyl histidine content of rabbit skeletal muscle myosin falls. This would be compatible with a reversion to the fetal type.

*Pearson:* Is it possible that adult and fetal myosin are merely isozymes of the same protein that vary in their proportion as a consequence of different functions as an animal becomes older?

*Perry:* Yes, this is the view that I hold.

*Allin:* Is there evidence suggesting that postnatal change in the character of rabbit myosin can be accelerated by inducing hyperactivity?

*Perry:* We have no evidence that the postnatal change in rabbit myosin can be accelerated by inducing hyperactivity. From analogy with other enzyme systems I would imagine that it can but presumably the change would be dependent on the type of stimulation given to the muscle.

*Allin:* Are there only two forms of myosin ("fetal" and "adult")? Might a single mature muscle fiber have a mixture of myosins?

*Perry:* The evidence is that there are differences in the preparations of myosin from fetal and adult tissue. It is suggested as the simplest hypothesis that there are two forms of the protein, and an attempt has been made on this basis to explain the differences observed in the low-molecular-weight components of the fetal and adult forms of myosin. It is possible that the observations could be explained by another hypothesis.

The second question is an interesting point to which we have given some thought but cannot give a definitive answer. It is possible that a particular myosin may be produced by a given cell type, but in my view it is more likely that both myosin types are produced by a given muscle cell, the proportion of the two types depending on the state of maturity and activity pattern of the cell.

*Allin:* What is the likely role of motoneuronal impulse pattern (frequency, train duration, etc.) and impulse total in the maturation and differentiation of muscle metabolism? If all muscle fibers in a heterogeneous muscle were given identical activation over a period of weeks or months would metabolic homogeneity be the eventual outcome?

*Perry:* I believe the motoneuronal impulse pattern is an important factor in maturation and differentiation of muscle metabolism, indeed the Buller, Eccles, and Eccles (1960*a*) type of experiment would suggest that this is so. I am not sure that identical activation of heterogeneous muscle would produce metabolic homogeneity, as I am not certain whether by this method alone you could produce complete interconversion between muscle fiber types.

*Allin:* I would like to ask Dr. Perry for his views on a question considered earlier by Dr. Mommaerts. Following nerve cross union of slow and fast muscles, would synthesis of new myosin occur, or modification of existing myosin?

*Perry:* If the lowering of specific enzyme activity of myosin after cross innervation is due to changes in the isozyme complement, I consider it will be produced by the synthesis of new myosin isozymes. Although the isozymes differ in extent of methylation of the histidine, I think it is unlikely that mechanisms exist for the demethylation in intact protein. This requires testing experimentally, however. There are also some suggestions that there are other differences in amino acid composition between the isozymes.

*Whitaker:* The difference in fetal and adult hemoglobin has important physiological significance. What do you believe the role of fetal myosin is?

*Perry:* Fetal myosin certainly has a lower specific adenosine triphosphatase activity, which is compatible with the slower contraction time of fetal muscle. As explained above, the enzymic activity of the myosin molecule has to be changed to alter the rate of ATP hydrolysis by the A filament, containing a fixed number of myosin molecules. A stricter analogy than hemoglobin to the adult and fetal forms of myosin would be the corresponding isozyme forms of creatine kinase, in which there are no very striking differences in enzymic properties.

*Scopes:* It seems possible that gene duplication and subsequent mutation in one of the genes could cause two molecular forms of the 20,000-mol-wt subunits of myosin. Does Dr. Perry have any information which would imply that the two (or more) subunits on each myosin molecule are different? The detection by electrophoresis of a number of molecular forms does not necessarily imply different functions for each form.

*Perry:* Agreed. The only evidence for difference in the small subunits of myosin is their different electrophoretic mobilities. At present we are attempting to isolate the subunits in sufficient quantities to compare them more rigorously.

*Judge:* What are the relative solubilities of the myosin isozymes?

*Perry:* Insofar as the myosin has not been separated into separate isozymes we have no direct information. I should expect their solubilities to be very similar. Certainly there is no obvious difference in the solubilities of myosin from mixed skeletal muscle of the fetus and white muscle of the adult rabbit.

*Pearson:* Your explanation of the effects of exercise on the young rabbit leads me to ask if you have compared the wild and domestic baby rabbit, since the former appears to develop earlier physiologically?

*Perry:* We have not compared the enzyme development in muscles of wild and domestic baby rabbits. It would be of interest to do so.

*Bailey:* Does heating to about 50° C change the various constituents of myosin? Would you anticipate any species specificity of these changes that might influence the binding capacity?

*Perry:* Locker (1956) showed that on heating myosin at about 50° C he was able to release material which probably corresponds to the small subunits. So far as I can remember these were fairly soluble, although I should expect the heavy-chain material to become completely insoluble in these conditions. I believe that there might be some species specificity in these changes but know of no evidence for it.

*Carpenter:* Could you suggest the regulatory mechanism of myosin methylation during "maturation" as implied in your presentation?

*Perry:* Not yet. It is of considerable interest that whereas in the rabbit the histidine residues are not methylated until after birth, lysine is methylated before birth. This suggests that the methylating system is present in both fetal and adult muscle as does the fact that actin from both fetal and adult muscle contains 3-methyl histidine. This implies that the availability of the appropriate myosin histidine changes during postnatal development. A simple explanation would be the replacement of one form of myosin by another whose histidine residues are more readily methylated. The situation may not be as simple as this, however, for cardiac myosin, which is presumably mainly of the fetal type as judged by the subfragments present, contains 3-methyl histidine but not as much as white skeletal muscle.

*Olcott:* What is known about the occurrence of methylated amino acids in the myofibrillar proteins of fish or other marine animals?

*Perry:* We have not yet carried out a general survey of the distribution of $\epsilon$-N-methyl lysine in myosins from different species. It is, however, present in rabbit skeletal and lobster tail muscle myosins. 3-methyl histidine is present in trout actin, and in lobster and crab myosins, although the samples of the latter two myosins so far analyzed were probably contaminated with actin.

*Sanger:* Your results suggest the speculation that the change in myosin adenosine triphosphatase in fetal versus adult rabbit myosin could be accounted for by a conformational change induced by the methylation of histidine in the headpiece component of the fetal myosin molecule.

*Perry:* On the face of it this would seem possible and a reasonable speculation. Photo-oxidation experiments, although not entirely conclusive, suggest that the 3-methyl histidine residues may not be involved in the biological activity of myosin. There is, however, a general correlation between specific enzymic activity and 3-methyl histidine content of myosin.

*Huszar:* In connection with Dr. Perry's paper I should like to mention some of our results (Huszar and Elzinga, 1969) concerning the $\epsilon$-N-methyl lysine (ML) content of myosin.

We first realized that myosin contains ML when we were analyzing some cyanogen-bromide fragments of rabbit white skeletal myosin. We consistently observed a shoulder after the lysine peak and when the basic column was extended to 40 cm we could resolve the shoulder from lysine. Its position was exactly that of an authentic sample of ε-N-mono-methyl lysine. The good resolution on this extended column enabled us to quantify the amount of this unique amino acid in myosin. After tryptic digestion of myosin all the ML was found in the heavy meromyosin and also in the subfragment-1. The amounts in the different fragments are as follows:

| | *per $10^5$* | *mole* | *per chain* |
|---|---|---|---|
| Myosin | 1.2 | 5.9 | 3.0 |
| HMM | 1.8 | 6.4 | 3.2 |
| Subfragment-1 | 2.3 | 2.6 | 2.6 |

Analysis of small alkali subunits and subfragment-2 shows that they contain no ML. Red myosin prepared from soleus and transversarius of rabbit also contains about 6 moles of ML per mole of myosin. The analysis of myosin prepared from newborn rabbits shows somewhat less ML than that found in adult myosin, but at least 4 moles per mole of myosin. The situation with respect to ε-N-methyl lysine is thus different from that concerning 3-methyl histidine: the latter, as Dr. Perry has shown, is not detectable in fetal or newborn myosin. Cardiac myosin from beef also contains 6 moles of ML per mole of myosin.

Dr. Perry has raised the question of possible histone contamination. Although myosin frequently contains a small amount of nucleic acid which might be associated with histones, we do not think it is likely that substantial amounts of histone could have been extracted under the mild conditions used for preparing myosin. As Dr. Perry pointed out, such histone contamination could not be expected to appear consistently in the same amount in a specific region of the myosin molecule; furthermore, we found the same amount of ML in reduced and S-β-carbamidomethylated myosin after 8 M urea treatment, and even in myosin exposed to 70% formic acid and subjected to repeated column chromatography.

Our results suggest that myosin contains six residues of ε-N-methyl lysines per mole of protein, although the possibility remains that myosin contains several partially methylated lysine residues. Experiments designed to resolve this question and further to localize ε-N-methyl lysine in myosin are in progress.

*Perry:* What do you think about our third radioactive peak? Can it be trimethyl lysine?

*Huszar:* We have not found di- or trimethyl lysine in our system yet. As far as our methyl lysine peak goes, it is always a single symmetrical peak. By proceeding with the work on the primary structure, I would anticipate that this peak can be identified. The only protein I know of that contains both mono- and trimethyl lysine is the wheat germ cytochrome C; in this case, in the analysis system developed by DeLange, Glazer, and Smith (1969), the monomethyl lysine is eluted before the di- and trimethyl lysine.

*Perry:* Have you had any evidence of this other component I mentioned, which I think might be trimethyl lysine? This is quite clearly in subfragment-1 and there are appreciable amounts of it. We are trying to identify it at the moment.

*Huszar:* We have done little work on it, but I hope to proceed by finding the primary structure of subfragment-1; we might then find the peptide which contains the residue and identify it.

*King:* Please complete the table Dr. Huszar started by estimating the methyl histidine content of actin and HMM (subfragment-1), also the methyl lysine content of actin.

Do these methyl amino acids have any functional role in adult actin or myosin?

*Perry:* The 3-methyl histidine of actin and heavy meromyosin subfragment-1 are shown in Table 26.1. Table 28.1 contains the figures for methyl lysine contents. Actin has no ε-N-monomethyl lysine or other methylated derivatives so far as we can determine. Dr. Huszar's figures for ε-N-monomethyl are much higher than ours for adult rabbit white muscle myosin.

In my view Huszar's figures represent the total amount of ε-N-methyl lysine and the as yet unidentified methylated amino acid, the presence of which we have reported (Hardy and Perry, 1969), possibly another

TABLE 28.1
*Methyl lysine content of myofibrillar proteins**

| Protein | N-methyl lysine (moles/mole) | "X"† |
|---|---|---|
| Myosin (adult rabbit) | 0.24 | 0.72 |
| | (0.16–0.33) | (0.59–0.82) |
| Myosin (fetal rabbit) | 0.10 | 0.58 |
| | (0.096–0.11) | (0.49–0.68) |
| Subfragment-1 (adult rabbit) | 0.59 | 1.75 |
| | (0.50–0.71) | (1.3–2.2) |
| Tropomyosin | 0 | 0 |
| Actin | 0 | 0 |

* Not presented during discussion. From M. Hardy, I. Harris, S. V. Perry, and D. Stone. 1969. "N-monomethyl-lysine and trimethyl lysine in myosin," *Proc. Biochem. Soc.*, London.

† Now confirmed as trimethyl lysine.

methylated derivative of lysine, which has not been resolved in his analytical system. As yet we cannot say with any certainty. Photo-oxidation studies suggest that the 3-methyl histidine is not concerned directly in the biological activity of myosin or actin. One would expect methyl of both histidines and lysines to increase the nonpolarity of the myosin molecule at specific points which would lead to localized conformational changes.

*Davies:* Dr. Perry, did you mean dimethyl—you said trimethyl?

*Perry:* I said trimethyl. I understand trimethyl lysine has recently been found in cytochrome C of wheat germ.

*Davies:* So you expect there is monomethyl and trimethyl?

*Perry:* We are sure of the monomethyl, but there is the other peak which clearly has an extra methyl group in it and the position is where one would expect the trimethyl lysine to be eluted.

*Davies:* But if it is tri then it has to be a terminal amino.

*Perry:* It is methylated on the C-nitrogen.

## *References*

Bárány, M. 1967. Adenosine triphosphatase activity of myosin correlated with speed of muscle shortening. *J. Gen. Physiol. 50* (suppl.) :197.

Binkhorst, R. A. 1969. The effect of training on some isometric contraction characteristics of fast muscle. *Pflügers Arch. 309:*193.

Buller, A. J., J. C. Eccles, and R. M. Eccles. 1960*a*. Differentiation of fast and slow muscles in the cat hind limb. *J. Physiol. 150:*399.

———. 1960*b*. Interactions between motoneurones and muscles in respect to the characteristic speeds of their responses. *J. Physiol. 150:*417.

Cheek, D. B. 1968. *Human Growth*. Lea and Febiger, Philadelphia.

Church, J. C. T., R. F. X. Noronka, and D. B. Allbrook. 1966. Satellite cells and skeletal muscle regeneration. *Brit. J. Surg. 53:*638.

DeLange, R. J., A. N. Glazer, E. L. Smith. 1969. Presence and location of an unusual amino acid, ε-N-trimethyllysine, in cytochrome C of wheat germ and neurospora. *J. Biol. Chem. 244:*1385.

Dickerson, J. W. T., J. Dobbing, and R. A. McCance. 1967. The effect of undernutrition on the postnatal development of the brain and cord in pigs. *Proc. Roy. Soc. B. 166:*396.

Dreyfus, J. C., J. Kruh, and G. Schapira. 1962. Muscular protein metabolism in normal and diseased states, p. 326. *In* F. Goss (ed.), *Protein Metabolism.* Springer-Verlag, Berlin.

Duggan, P. F. 1967. Potassium-activated adenosine triphosphatase and calcium uptake by sarcoplasmic reticulum. *Life Sci. 6:*561.

———. 1968. The monovalent cation stimulated calcium pump in frog skeletal muscle. *Life Sci.* *7*:913.

Gauthier, G. F. and H. Padykula. 1966. Cytological studies of fiber types in skeletal muscle. A comparative study of the mammalian diaphragm. *J. Cell Biol.* *28*:333.

Goldspink, G. 1965. Cytological basis of decrease in muscle strength during starvation. *Amer. J. Physiol.* *209*:100.

———. 1968. Sarcomere length during post-natal growth of mammalian muscle fibres. *J. Cell Sci.* *3*:539.

———. 1969. Succinic dehydrogenase content of individual muscle fibres at different ages and stages of growth. *Life Sci.* *8*:791.

Goldspink, G. and R. W. D. Rowe. 1968. Studies on postembryonic growth and development of skeletal muscle. *Proc. Roy. Irish Acad. Sci.* *66B*:85.

Hardy, M. F. and S. V. Perry. 1969. *In vitro* methylation of muscle proteins. *Nature* *223*:300.

Helander, E. 1961. Influence of exercise and restricted activity on the protein composition of skeletal muscle. *Biochem. J.* *78*:478.

Huszar, G. and M. Elzinga. 1969. ε-N-methyl lysine in myosin. *Nature* *223*:834.

Iakovlev, N. N. 1957. The effect of 2,4-dinitrophenol and of adrenaline on the carbohydrate-phosphoric metabolism of working muscles [in Russian]. *See Biol. Abstr.* #27541.

Ishikawa, H. 1966. Electron microscopic observations of satellite cells with special reference to the development of mammalian skeletal muscles. *Z. Anat. Entwicklungsgesch.* *125*:43.

Kowalski, K., E. E. Gordon, A. Martinez, and J. Adamek. 1969. Changes in enzyme activities of various muscle fiber types in rat induced by different exercise. *J. Histochem. Cytochem.* *17*:601.

Locker, R. H. 1956. The dissociation of myosin by heat coagulation. *Biochim. Biophys. Acta* *20*:514.

Luff, A. R. 1968. Physiological aspects of the post-natal development of skeletal muscle in the mouse. Ph.D. thesis, University of Hull, England.

MacConnachie, H. G., M. Enesco, and C. P. Leblond. 1964. The mode of increase in the number of skeletal muscle nuclei in the postnatal rat. *Amer. J. Anat.* *114*:245.

McLaren, A. and D. Michie. 1960. Control of pre-natal growth in mammals. *Nature* *187*:363.

Mauro, A. 1961. Satellite cell of skeletal muscle fibers. *J. Biophys. Biochem. Cytol.* *9*:493.

Montgomery, R. D. 1962. Muscle morphology in infantile protein malnutrition. *J. Clin. Path.* *15*:511.

Montgomery, R. D., J. W. T. Dickerson, and R. A. McCance. 1964. Severe undernutrition in growing and adult animals. *Brit. J. Nutr.* *18*:587.

Morpurgo, B. 1898. Ueber die post-embryonale Entwickelung der quergestreifetn Muskeln in weissen Ratten. *Anat. Anz.* *15*:200.

Rowe, R. W. D. 1968. Effect of low nutrition on size of striated muscle fibres in the mouse. *J. Exp. Zool. 167*:353.

Scrimshaw, N. S. and J. E. Gordon. 1969. *Malnutrition, Learning and Behavior.* MIT Press, Cambridge, Mass.

Talbot, N. B., O. H. Lowry, and E. B. Astwood. 1940. Influence of estrogen on the electrolyte pattern of the immature rat uterus. *J. Biol. Chem. 132*:1.

Wigglesworth, J. S. 1963. Experimental growth retardation in the foetal rat. *J. Path. Bact. 88*:1.

# PART 7

## Muscle Metabolism

# Chapter 29

# *Lipid as an Energy Source*

R. J. HAVEL

The transition of multicellular life from a sessile to a mobile state, coupled with intermittent availability of food, should favor development of mechanisms for efficient storage of energy. In sessile or sluggish creatures energy is stored as complex carbohydrate. Because deposition of glycogen is accompanied by increase in cell water, the caloric value of such stores is much lower than 4 cal/g. Fat, because of its lesser state of oxidation and insolubility in aqueous fluids, represents 9 cal/g, an impressive improvement in storage capacity. In organisms that need to store large quantities of reserve fuel, such as migrating insects, fish, and birds, virtually all readily available energy stores are triglycerides.

Striated muscles are the site of increased fuel consumption during muscular activity. Even at moderate rates of locomotion that are readily sustained for hours, they account for roughly 80% of energy use (Havel, 1970). Thus oxidative metabolism of striated muscle is of prime concern for metabolic homeostasis.

## EARLY STUDIES

More than 40 years ago, biochemists studying the energy sources for muscular contraction showed that glycogen is rapidly depleted and lactic acid is formed when certain muscles are stimulated and concluded that muscles burn mainly stored glycogen (or blood glucose). Almost 100 years ago, physiologists approached this question by measuring ventilatory exchanges of oxygen and carbon dioxide. Ventilatory RQ's ($CO_2/O_2$) near unity were often observed, consistent with use of carbohydrate. Evaluation of these early studies shows that: (1) The muscles used by biochemists were mainly specialized pale muscles which are poorly adapted for aerobic metabolism and they were often studied under hypoxic conditions. (2) High ventilatory RQ's at heavy work loads failed to reflect those of working muscles because accumulation of lactate led to metabolic acidosis with blowing off of additional $CO_2$ (Havel, 1969).

The question of use of fats was first raised from observations of ventilatory RQ's near 0.8 during exercise. In the 1930's, some reviewers concluded that both carbohydrate and fat ultimately are substrates for muscular contraction. The ready demonstration of carbohydrate use in muscle and failure to demonstrate use of fat directly led to the hypothesis that carbohydrate synthesized from fat at other sites was transported to the muscles.

Studies of the metabolism of fasting man, particularly the early study by Benedict (1915), showed clearly that stored fat is the primary metabolic fuel and that, with loss of glycogen stores after 2 to 3 days of fasting, the remainder is derived from protein. Even after an overnight fast, production of glucose from the liver was found to be sufficient for only about 30% of energy metabolism (Myers, 1950). Recent studies (Owen, Felig, Morgan, Wahren, and Cahill, 1969) have shown that production of glucose by liver and kidneys during starvation is equivalent to uptake of the gluconeogeneic precursors, glycerol, amino acids, and lactate (the latter does not provide for net synthesis of glucose since it represents reutilization via the Cori cycle). With prolonged fasting of obese subjects, the amount of glucose produced is not even sufficient for the observed metabolic requirements of the brain and of certain specialized tissues such as the mature erythrocyte, which can use only glucose as an energy source. In this unusual situation, acetoacetate and β-hydroxybutyrate become major fuels for the brain (Owen, Morgan, Kemp, Sullivan, Herrera, and Cahill, 1967). Thus fasting man at rest derives more than 80% of his energy from fat.

More direct evidence that skeletal and cardiac muscles burn fat was obtained when it was found that their uptake of glucose and other water-soluble metabolites in the postabsorptive state is grossly insufficient to account for consumption of oxygen. The careful studies of Zierler and coworkers in the early 1950's (Andres, Cader, and Zierler, 1956) showed this for the deep venous drainage of the human forearm. The RQ for this region was near 0.7, a value incompatible with substantial use of glucose. Thus, it was clear 15 years ago that, except for the brain and a few other tissues, fat is the main substrate of energy metabolism in man in the postabsorptive state. The chemical nature of this lipid remained elusive, since almost nothing was then known of the pathways for transport of endogenous lipids in the blood. It was known that ample lipid is present in blood plasma—in terms of fatty acid residues this amounts to about 10 g or 90 potential calories. As it turns out, most of this transport is accomplished by a single entity which accounts for only about 5% of these fatty acids.

THE FATTY ACID TRANSPORT SYSTEM (FIG. 29.1)

Dietary fat is absorbed in the upper small intenstine after hydrolysis to form partial glycerides and fatty acids which are dispersed in micelles by action of bile salts. The lipid residues diffuse into the mucosal epithetlial cells and are then reconverted to triglycerides and incorporated into large droplets covered by amphiphilic phospholipids and protein. These chylomicrons are secreted into the interstitial space, pass into lacteals, and are transported into the blood via the thoracic duct. Studies with chylomicrons obtained from thoracic duct lymph of animals fed $^{14}$C-palmitic acid showed that intravenously injected chylomicrontriglycerides are rapidly removed from the blood (Havel and Fredrickson, 1956). The removal process was accompanied by extensive hydrolysis of the triglycerides, since radioactivity initially present in the injected triglyceride fatty acids (TGFA) appeared rapidly in the small fraction of fatty acids that are not in ester linkage, now called free fatty acids (FFA). When $^{14}$C-palmitic acid as such was injected intravenously, it left the blood with remarkable rapidity; the mean life span in the circulation was about 3 min. These and other studies established the main mechanism for uptake of dietary fat in tissues (Havel, 1965). Hydrolysis occurs by action of a lipase (lipoprotein lipase) when the chylomicrons are adsorbed onto the surface of capillary endothelial cells. The FFA produced diffuse readily across the various plasma membranes to enter tissues where they can be stored (after reconversion to esterified lipid) or oxidized.

At the same time as the pathway for dietary fat was being defined, other studies showed that FFA are present in blood plasma at all times, independent of fat absorption, and that the concentration of these fatty acids is remarkably labile (Fredrickson and Gordon, 1958). On the basis of these studies and the rapid turnover rate mentioned above, it became apparent that FFA could account for transport of up to 2,000 cal of potential energy daily in man. Measurements of arteriovenous differences of FFA then showed that they enter the blood from regions rich in adipose tissue and leave the blood in many sites, particularly liver and cardiac muscle (Gordon, 1957). Uptake in skeletal muscle was not generally observed (subsequent tracer studies have shown that this results from simultaneous uptake of FFA into muscle and release from interspersed adipose tissue).

Lipid is stored in adipose tissue as triglycerides in large droplets. Entry of FFA into the blood is controlled by the activity of an intracellular

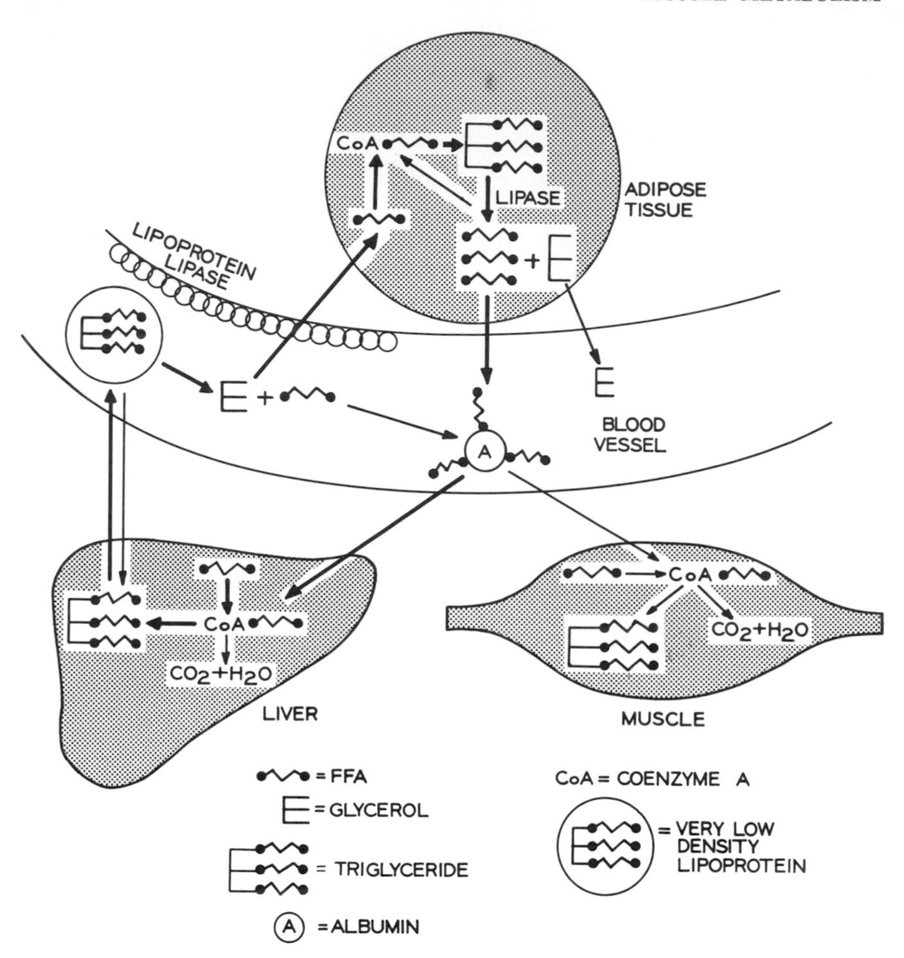


*Fig. 29.1.* The fatty acid transport system. Fatty acid residues in stored triglycerides in adipose tissue are mobilized after hydrolysis by an intracellular (hormone-sensitive) lipase and are transported in plasma in a complex with albumin. After entering various tissues, such as liver and muscle, they are either oxidized or stored temporarily after reesterification, mainly as triglycerides. From liver, triglycerides are secreted into the blood as very low density lipoproteins. Both endogenous triglycerides secreted from liver and exogenous triglycerides entering the blood as chylomicrons from the intestinal mucosa via the lymphatic system (not shown) are taken up in various tissues after the triglycerides have been hydrolyzed by a lipase (lipoprotein lipase) acting at the surface of capillaries.

lipase which hydrolyzes them to form FFA and glycerol (Havel, 1968). Both products diffuse readily into the blood, the latter in free solution, the former by interaction with binding sites on macromolecules; in blood plasma, virtually all FFA are bound to albumin. The activity of the lipase is regulated by the amount of cyclic 3′5′-AMP, which converts the enzyme to an active form. The formation of cyclic AMP, in turn, is regulated by the activity of the adenyl cyclase system which catalyzes its synthesis. As in other tissues, activity of the adenyl cyclase system is regulated by various hormones. Although important species differences exist, in most mammals the main activators of the hormone-sensitive lipase system in adipose tissue are the catecholamines, norepinephrine and epinephrine, and growth hormone (Havel, 1968). Epinephrine enters adipose tissue through the blood after secretion from the adrenal medulla, while norepinephrine is secreted locally at sympathetic nerve endings within the tissue. The principal demonstrated inhibitor of hormone-sensitive lipase is insulin. The high sensitivity of adipose tissue to catecholamines and insulin and their rate of action are such that availability of these hormones can account for minute-to-minute regulation of fat mobilization. Growth hormone, in contrast, has a delayed and sustained effect on lipolysis and its action appears to require *de novo* synthesis of protein.

In adipose tissue, there is an inverse relation between the activities of hormone-sensitive lipase and lipoprotein lipase (Vaughan and Steinberg, 1965). In the fed state, secretion of insulin both inhibits hormone-sensitive lipase and increases the activity of lipoprotein lipase, decreasing fat mobilization and promoting uptake of circulating triglycerides. In the fasting state, the reverse situation obtains because secretion of insulin decreases and possibly because that of growth hormone and norepinephrine increases. In this way, adipose tissue serves as a major site of caloric homeostasis directing appropriate amounts of FFA and triglycerides to tissues such as muscle.

Uptake of FFA from the blood into most tissues is regulated mainly by the plasma level and nutritional blood flow. In resting muscle, about half is removed with each circulation (extraction ratio equals 0.5) (Havel, Pernow, and Jones, 1967). When greater amounts enter than can be oxidized, they are stored temporarily as triglycerides, mainly in small fat droplets (Havel, 1970). This lipid provides a reserve store of energy in addition to glycogen.

Although it is recognized that oxygen consumption is determined mainly by the level of ADP, the mechanism by which oxidation of fatty acids is regulated is not certain. Potentially important control points have been identified (Newsholme and Gevers, 1967). Generally, these relate to reactions regulating metabolism at or near the branch point leading to oxidation or esterification (Fig. 29.2). Availability of α-glycerophosphate could regulate esterification, and its concentration does change in a manner consistent with a regulatory function in certain states. Oxidation of fatty acids in many tissues can be regulated by the reaction responsible for translocating fatty acyl residues from the cytosol

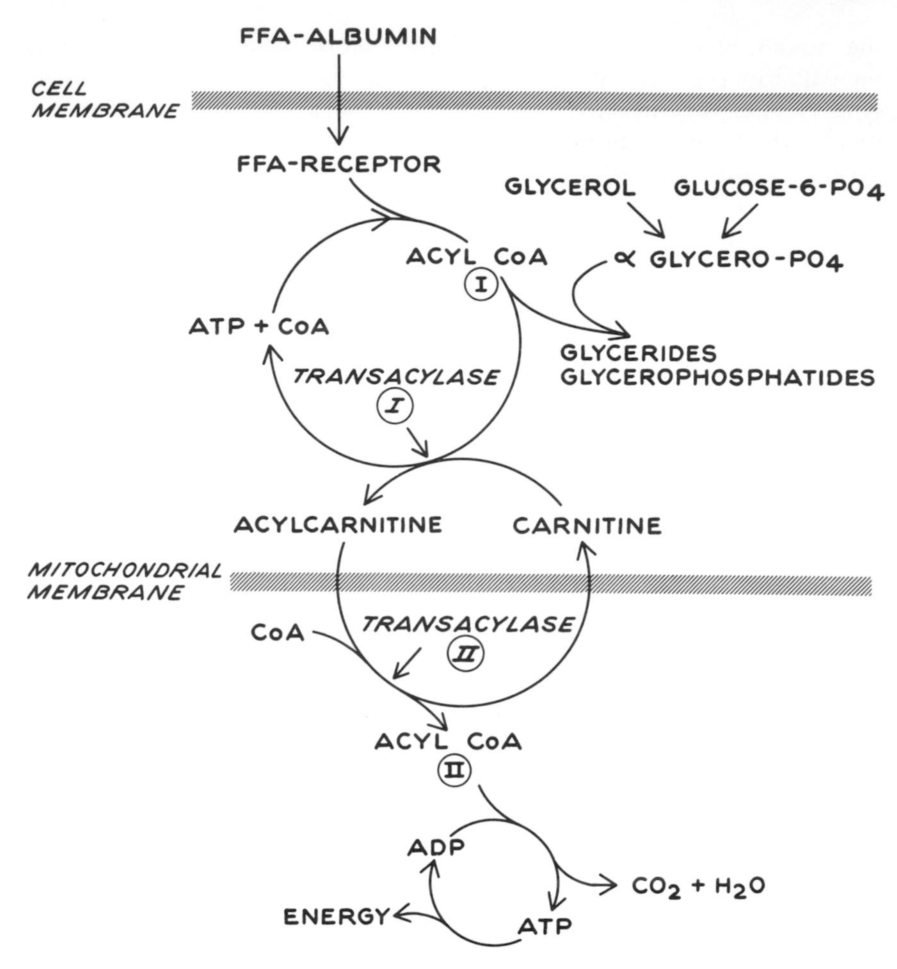


*Fig. 29.2.* Pathways for utilization of fatty acids. Fatty acids entering cells are activated as the coenzyme A derivative which can be converted in the extramitochondrial compartment to lipid esters for storage; oxidation of fatty acids requires transfer across the mitochondrial membrane, involving formation of the carnitine derivative, to provide access to the fatty acid oxidase system and the citric acid cycle.

into the mitochondria. This involves a transferase on a mitochondrial membrane which converts palmityl CoA to palmityl carnitine. The latter can enter mitochondria and be reconverted to palmityl CoA to enter the fatty acid oxidation spiral. The activity of this enzyme in muscle increases with fasting, but the nature of its regulation is unknown.

The pathways of oxidation and esterification are available in all tissues that metabolize FFA. The liver, as an important site of *de novo* synthesis of fatty acids from carbohydrate (lipogenesis) as well as of active uptake and oxidation of FFA, has an additional pathway available to it (Fig. 29.1). Excess quantities of fatty acids, after esterification to form triglycerides, can be packaged and secreted into the blood in a manner analogous to formation of chylomicrons in the intestinal mucosa. The lipid particles secreted from the liver are generally smaller and are referred to as very low density lipoproteins (VLDL). The endogenous triglycerides in these lipoproteins are always present in the blood and are metabolized by the same pathway described for exogenous chylomicron triglycerides (Havel, 1965). In addition to adipose tissue, they can be hydrolyzed by lipoprotein lipase in other capillary beds and thus contribute fatty acids for energy needs.

## USE OF LIPID IN DIFFERENT MUSCLES

It should now be clear that FFA derived from adipose tissue, together with exogenous TGFA from the intestine and endogenous TGFA from liver, can provide for energy needs of resting muscle. Since muscle also oxidizes carbohydrate from stored glycogen or blood glucose, the question arises of the function and relative importance of these sources of fuel in contracting muscle.

It is useful to consider this question in relation to the differing activities of muscles, particularly whether they serve continuous, intermittent but sustained, or sporadic functions. Cardiac muscle is the major example of the first of these. Such muscles ordinarily function completely aerobically and extract a large fraction of the oxygen from the blood. The heart is omnivorous in its selection of foodstuffs; the major metabolites extracted are FFA, glucose, and lactate (Opie, 1969). After a meal containing carbohydrate, glucose may be the most important, but in the postabsorptive state FFA are the major fuel. Lactate is extracted at all times, an unusual, and poorly understood, feature of this muscle, but one which could be related to its peculiar isozyme of lactic dehydrogenase. During heavy exercise, when large amounts are produced in skeletal muscle, lactate may become the major substrate of cardiac metabolism. This mechanism for modulating the metabolic acidosis has been little appreciated. Anaerobic glycolysis occurs in the heart only under conditions of severe hypoxia, as in ischemic disease or, presumably, during bursts of maximal exercise (Opie, 1968). It seems unlikely that appreciable glycogen is burned normally in the heart, even during fasting, since its concentration actually increases as the plasma level of FFA rises. This presumably reflects operation of the "glucose–fatty acid cycle" by which increased availability of FFA decreases the activity of the glycolytic pathway and the entry of pyruvate into the citric acid cycle. The "ketones," acetoacetate and β-hydroxybutyrate, are effectively burned in the heart, but low arterial concentration limits their contribution to oxidative metabolism except during prolonged starvation and in certain pathologic states. Use of triglycerides in chylomicrons and VLDL is expected from the high activity of lipoprotein lipase in heart muscle, and has been demonstrated in certain small mammals. In man, however, uptake of endogenous triglycerides under ordinary conditions is very small. That of chylomicrons has not been studied. In addition to glycogen, the heart contains triglycerides in small fat droplets between myofibrils and adjacent to mitochondria. These stores, like glycogen, increase during fasting and can be shown to be utilized in perfused hearts. Their contribution to energy metabolism under various conditions is unknown. There is some evidence that the heart contains a lipase which could hydrolyze these stores and that this enzyme is affected by hormonal stimuli.

Certain skeletal muscles such as the diaphragm are also continuously active, or, like the pectoral muscles of many birds, are capable of sustained activity for hours to days. These are typically "red" muscles and resemble the heart in high density of capillaries and in mitochondria with their contained machinery for oxidation of fatty acids. We have little quantitative information concerning the fuels of these muscles, but it is reasonable to assume that they resemble the heart, even perhaps in such unusual features as uptake of lactate, since the same isozyme of lactic dehydrogenase is present. Most quantitative data apply to mixed muscles studied under conditions of continuous, "steady-state" activity. These muscles contain fibers which resemble those of white or red muscle, or which are of intermediate structure. The white fibers are larger, surrounded by fewer capillaries, and contain less mitochondria and stored triglyceride than the red ones. Typical white muscles are not found in larger mammals. Where present (leg muscle of frogs, pectoralis of domestic fowls, "bottom" fish) they subserve intermittent, rapid movements rather than sustained activity. Their high content of glycogen and sparse capillaries and mitochondria indicate that they are primarily glycolytic tissues designed to contract in absence of oxygen and substrates requiring it for synthesis of ATP.

What then is the function of white fibers in mixed muscles? These muscles may differ from the heart because they require an ignition system. George and Berger (1966) observed that electrical stimulation of the pectoralis of certain birds rapidly depletes glycogen in white but not red fibers, and has suggested that the white fibers may contract separately at onset of flight while the red ones take over thereafter. This concept is supported by observations in individuals with genetically determined lack of myophosphorylase (McArdle's syndrome) (Pernow, Havel, and

Jennings, 1967). In this disorder, muscle glycogen cannot be converted to lactic acid. Typically, these individuals have difficulty initiating heavy muscular activity and may develop an electrically silent contracture with local swelling of the affected muscles. Presumably, this reflects depolarization and defective ion pumping consequent upon depletion of ATP. If activity is begun gradually and the work load progressively increased, they can usually perform normal activity and continue it for long periods. At times, the pain and associated swelling and cramping may gradually diminish and then disappear if they continue to exercise—the "second-wind" phenomenon. Ability to initiate exercise can be increased by measures which increase availability of substrate at onset of exercise, either by raising the arterial concentration of glucose or FFA or by increasing blood flow. Since muscles normally release substantial quantities of lactate virtually immediately with onset of activity while persistent activity is accompanied by little or none,* the functional disability in McArdle's syndrome seems to result from lack of an important component of the "starter" mechanism by which ATP is made available anaerobically until hemodynamic and metabolic adjustments are able to supply sufficient substrate and oxygen for this processs.

As already indicated, quantitative information concerning the metabolic mixture in active skeletal muscle is limited to steady state measurements for mixed (usually leg) muscles. Studies of arteriovenous differences of FFA require a radio-tracer technique in which $^{14}C$-labeled fatty acid bound to albumin is infused, either at a constant rate into a peripheral vein or into the artery supplying the exercising part. In this way uptake of FFA into active muscle can be distinguished from release from interspersed adipose tissue. Uptake of FFA in forearm muscles of fasting humans is sufficient to account for almost all of oxygen consumption at rest and during mild contractions associated with wrist movements (Zierler, Maseri, Klassen, Rabinowitz, and Burgess, 1968). With the increased rate of perfusion produced by muscular activity, uptake of FFA increases even though its extraction ratio falls (Havel, Pernow, and Jones, 1967). At rest, most FFA entering muscle are stored for substantial periods in ester form, but in the active state they are rapidly oxidized so that release of $^{14}CO_2$ from an exercising limb equals the uptake of $^{14}C$-fatty acid. Kinetic studies suggest that even in the active state some FFA are stored temporarily before oxidation, presumably in lipid esters (Havel, Ekelund, and Holmgren, 1967). In the postabsorptive state, leg exercise at moderate work load (about 400 kg m/min) is accomplished with a local RQ near 0.8, suggesting that fatty acids are the major energy

* Possibly lactate continues to be produced in white fibers and then diffuses into red fibers for oxidation.

source. Direct measurements indicate that about half of this is derived from FFA and very little more from endogenous triglycerides in plasma (Havel, Ekelund, and Holmgren, 1967). Thus, it appears that the storage pool of lipid esters (triglycerides) in the muscle is not only turning over but participates in net supply of fuel. Since measured uptake of glucose can supply only about 15% of energy needs, stored glycogen must also contribute to the metabolic mixture. Direct measurements (in specimens obtained by needle biopsy) of content of glycogen in the vastus lateralis during leg exercise have demonstrated this directly (Hultman, 1967). From the amount of lactate released it can be concluded that less than 1% of the energy needs are derived from glycolysis. Uptake of ketones also makes a relatively small contribution (Havel, Segel, and Balasse, 1969). From available evidence, two additional statements can be made about leg exercise in man. First, as the work load increases, so does use of glycogen. When glycogen stores are exhausted, heavy work loads cannot be sustained (Hultman, 1967). Second, with more prolonged exercise, moderate rates of work can continue in absence of glycogen. This presumably is possible because the rate of mobilization of FFA from adipose tissue increases progressively for several hours until it can serve as virtually the sole substrate (Young, Pelligra, and Adachi, 1966).

The rapid and sustained mobilization of fat from adipose tissue during exercise is an excellent example of the function of the hormone-sensitive lipase system in caloric homeostasis (Havel, 1970). At the onset of leg exercise in the postabsorptive state, the concentration of FFA in plasma rapidly falls, reflecting increased uptake into the active muscles. After a few minutes it begins to increase and rises rapidly for the next hour; thereafter, it increases slowly for the next 9 hours to reach a plateau at about 2mM, three to four times the basal level. During this period, the concentration of glucose gradually falls and also levels off after about 9 hours at 3–4 mM. Presumably, at this point, both liver and muscle glycogen stores are depleted and blood glucose is produced from such precursors as the glyceryl moiety of the triglycerides mobilized from adipose tissue, mainly for use by the central nervous system. The hormonal factors controlling this mobilization of fat are complex, but the increase during the first hour of exercise is probably mediated mainly by increased sympathetic nervous activity which releases norepinephrine near adipose tissue cells. In addition to greater lipolytic rate, increased perfusion of adipose tissue may play a role in this process. After a few minutes of such exercise, plasma levels of growth hormone rise abruptly and it can be assumed that this helps to sustain the increased lipolytic state. Secretion of growth hormone may also explain the slow return of fat mobilization to basal levels after exercise is discontinued, since its action is not only delayed in onset, but prolonged. Theoretically, inhibition of insulin secretion, by activation of sympathetic innervation of the $\beta$ cells of the pancreatic islets, could promote fat mobilization. This is not essential, however, since the process occurs rapidly in depancreatized dogs (Issekutz, Miller, and Rodahl, 1963) and in insuloprivic diabetes mellitus (Carlström, 1967).

Inhibition of fat mobilization during exercise has interesting consequences (Havel, Segel, and Balasse, 1969). This can be achieved by administering $\beta$-adrenergic blocking agents or drugs, such as nicotinic acid, which prevent activation of hormone-sensitive lipase. Results of such blockade include increased RQ of the active muscles (and, presumably, of other tissues as well), increased uptake of glucose and breakdown of glycogen and, frequently, hypoglycemia. These changes reflect greater use of carbohydrate in muscle upon release of the "brake" normally applied by oxidation of fatty acids, and represent an additional example of the glucose-fatty acid cycle.

## CONCLUSION

Like most nonnervous tissues, striated muscle preferentially burns fatty acids. The broad outlines of the pathways involved and the means by which fatty acids are made available to muscle, particularly for the increased demands of activity, are known. In mixed skeletal muscles, glycogenolysis has the important function of providing extra ATP at onset of activity, but it is required for sustained activity only at high work loads when energy needs greatly exceed the rate of uptake of fatty acids and other blood-borne substrates. More information is needed about the regulation of fatty acid oxidation and use of stored triglycerides in muscle. Additional studies of the comparative inter- and intraspecific biochemistry and physiology of muscle should provide much useful information about the various ways in which the widely varying metabolic requirements of this tissue are met.

## ACKNOWLEDGMENTS

The research for this paper was supported by USPHS grant HE–06285.

## *References*

Andres, R., G. Cader, and K. L. Zierler. 1956. The quantitatively minor role of carbohydrate in oxidative metabolism by skeletal muscle in intact man in the basal state. Measurements of oxygen and glucose uptake and carbon dioxide and lactate production in the forearm. *J. Clin. Invest. 35*:671.

Benedict, F. G. 1915. *A Study of Prolonged Fasting*. Carnegie Inst. Wash. Publ. No. 203.

Carlström, S. 1967. Studies on fatty acid metabolism in diabetics during exercise I. Plasma free fatty acid concentration in juvenile, newly diagnosed diabetics during exercise. *Acta Med. Scand. 181*:609.

Frederickson, D. S., and R. S. Gordon, Jr. 1958. Transport of fatty acids. *Physiol. Rev. 38*:585.

George, J. C., and A. J. Berger. 1966. *Avian Myology*. Academic Press, New York.

Gordon, R. S., Jr. 1957. Unesterified fatty acid in human blood plasma II. The transport function of unesterified fatty acid. *J. Clin. Invest. 36*:810.

Havel, R. J. 1965. Metabolism of lipids in chylomicrons and very low density lipoproteins, p. 499. *In Handbook of Physiology; Section 5, Adipose Tissue.* Amer. Physiol. Soc., Washington.

———. 1968. The autonomic nervous system and intermediary carbohydrate and fat metabolism. *Anesthesiology 29*:702.

———. 1970. The fuels for muscular exercise. *In* E. Buskirk (ed.), *Science and Medicine of Exercise and Sports*. Harper & Row. (In press.)

Havel, R. J., L.-G. Ekelund, and A. Holmgren. 1967. Kinetic analysis of the oxidation of palmitate-1-$C^{14}$ in man during prolonged heavy muscular exercise. *J. Lipid Res. 8*:366.

Havel, R. J., and D. S. Fredrickson. 1956. The metabolism of chylomicra I. The removal of palmitic acid-1-$C^{14}$ labeled chylomicra from dog plasma. *J. Clin. Invest. 35*:1025.

Havel, R. J., B. Pernow, and N. L. Jones. 1967. Uptake and release of free fatty acids and other metabolites in the legs of exercising men. *J. Appl. Physiol. 23*:90.

Havel, R. J., N. Segel, and E. O. Balasse. 1969. Effect of 5-methylpyrazole-3-carboxylic acid (MPCA) on fat mobilization, ketogenesis and glucose metabolism during exercise in man. *In* R. Paoletti (ed.), *Advances in Experimental Medicine and Biology*. Plenum, New York. (In press.).

Hultman, E. 1967. Studies on muscle metabolism of glycogen and active phosphate in man with special inference to exercise and diet. *Scand. J. Clin. Lab. Invest. 19:* (suppl. 94).

Issekutz, B., Jr., H. I. Miller, and K. Rodahl. 1963. Effect of exercise on FFA metabolism of pancreatectomized dogs. *Amer. J. Physiol. 205*:645.

Myers, J. D. 1950. Net splanchnic glucose production in normal man and in various disease states. *J. Clin. Invest. 29*:1421.

Newsholme, E. A., and W. Gevers. 1967. Control of glycolysis and gluconeogenesis in liver and kidney cortex. *Vitamins and Hormones 25*:1.

Opie, L. H. 1968. Metabolism of the heart in health and disease I. *Amer. Heart J. 76*:685.

———. 1969. Metabolism of the heart in health and disease II. *Amer. Heart J.* *77*:100.

Owen, O. E., P. Felig, A. P. Morgan, J. Wahren, and G. F. Cahill, Jr. 1969. Liver and kidney metabolism during prolonged starvation. *J. Clin. Invest.* *48*:574.

Owen, O. E., A. P. Morgan, H. G. Kemp, J. M. Sullivan, M. G. Herrera, and G. F. Cahill, Jr. 1967. Brain metabolism during fasting. *J. Clin. Invest.* *46*:1589.

Pernow, B., R. J. Havel, and D. Jennings. 1967. The second wind phenomenon in McArdle's syndrome. *Acta Med. Scand.* *472* (Suppl.) :294.

Vaughan, M., and D. Steinberg. 1965. Glyceride synthesis, glyceride breakdown and glycogen breakdown in adipose tissue: mechanisms and regulation, p. 239. *In Handbook of Physiology: Section 5: Adipose Tissue.* Amer. Physiol. Soc., Washington.

Young, D. R., R. Pelligra, and R. R. Adachi. 1966. Serum glucose and free fatty acids in man during prolonged exercise. *J. Appl. Physiol.* *21*:1047.

Zierler, K., A. Maseri, G. Klassen, D. Rabinowitz, and J. Burgess. 1968. Muscle metabolism during exercise in man. *Trans. Ass. Amer. Physicians* *81*:266.

# Chapter 30

# *Interrelationships of Free Fatty Acids, Lactic Acid, and Glucose in Muscle Metabolism*

B. ISSEKUTZ, JR.

After the work of A. V. Hill and the discovery of the Embden-Meyerhof pathway it was generally accepted that carbohydrate represented the only energy source for the contracting muscle. In contrast to this, it is now recognized that under physiological conditions both carbohydrate and fat serve as direct fuels during physical exercise. In other words, liver glycogen, adipose tissue, and muscle glycogen have to be accepted as energy sources, and some consideration must be given to the possible role of fat stored within or near the muscle cells.

There is, however, some controversy as to the relative contributions of these sources to the highly elevated energy metabolism of the working muscles. This controversy seems to be due, at least partly, to the great variety of experimental conditions chosen by the various investigators to attack the problem. The approaches include studies on faradically stimulated muscle groups of anesthetized animals, and experiments carried out on exercising men and dogs. Conclusions were based on results of chemical analyses of the working muscle and on measurements of blood flow and arteriovenous differences across the muscle or across the splanchnic area (to obtain the hepatic sugar output in exercising men). A considerable body of information also came from experiments using $^{14}C$ tracer techniques, radiopalmitate or radioglucose, in the form of a single injection or constant rate infusion. All these approaches have their merits, but they also have their limitations. Each approach has technical difficulties, as for instance, the rapid breakdown of glycogen in a small biopsy sample taken from a tissue, the metabolic rate of which is elevated 20 to 30 times, or the determination and evaluation of small arteriovenous differences obtained at a highly increased blood flow. In connection with the isotope dilution technique such problems arise as the recycling of the label between the product and the substrate, compartmentalization, and equilibrium between the $^{14}CO_2$ and the body's bicarbonate pool. These require special kinetic analyses and mathematical handling of the data. One has to point out, however, that, owing to the accelerated turnover in the various compartments, a steady state exercise tends to reduce the error connected with the isotope dilution technique, while it may enhance the inherent problems of other approaches. Furthermore, the results may be influenced by such factors as anesthesia (by inhibiting the participation of extramuscular energy depots), the duration and severity of exercise, the preceding diet, etc.

Controversies arise if we try to compare data obtained on stimulated muscle groups with those derived from exercising animals or men, or if we try to put short heavy work and prolonged moderate and light work in the same category, assuming that there are only quantitative differences; but the relative contributions of the various energy sources remain the same under all circumstances. It must also be emphasized that the term "heavy work" is a very relative one. It greatly depends on the "physical fitness" of the subject or, in more physiological terms, it depends on his ability to adjust his circulatory and respiratory system to the $O_2$ demand of the working muscles. This makes it difficult to compare data obtained at different work loads on different subjects, and it is even more difficult to apply results derived from dogs running on a treadmill directly to bicycling men, where there is more isometric component in the work.

We are perhaps on safer ground when we are investigating those factors which may affect the participation of various energy sources. Here we may assume that the basic principles of regulation of metabolism are similar in all mammalian organisms. Therefore, we shall restrict our discussion to the interaction of these factors rather than put the emphasis on the greatly variable absolute values.

### EFFECTS ON THE ADIPOSE TISSUE AS AN ENERGY SOURCE

A large body of evidence of various types is now available to prove that plasma free fatty acids (FFA) can serve as a direct fuel for the oxidative metabolism of working muscles (Fritz, Davis, Hultrop, and Dundee, 1958; Fritz, 1960; Friedberg and Estes, 1962; Havel, Pernow, and Jones, 1967). When an exercise begins at a level of energy expenditure 6–7 times the basal metabolic rate, the rate of removal of FFA from the plasma approximately doubles, as shown with radiopalmitate on men (Carlson and Pernow, 1961; Friedberg, Harlan, Trout, and Estes, 1960; Havel, Naimark, and Borchgrevink, 1963), on dogs (Issekutz, Miller, Paul, and Rodahl, 1964; Paul and Issekutz, 1967), as well as on horses (Carlson, Fröberg, and Persson, 1965). This is probably due to the increased blood flow and to the enlarged capillary surface in the working muscle, simply because the FFA become available to more muscle cells. In addition, possibly the elevated metabolic activity plays a role, since Fritz (1958) showed that electrical stimulation increased the FFA uptake *in vitro*. Whether the plasma FFA level will decrease, increase, or remain unchanged depends on the response of the adipose tissue. As the work continues the FFA level will rise and, after 4–5 hours of exercise, it is usually in the range of 2–2.5 $\mu$eq/ml, that is, 3–5 times the resting value. This response is very much the same in man (Basu, Passmore, and Strong, 1960; Rodahl, Miller, and Issekutz, 1964) as in the dog (Paul and Issekutz, 1967). With the rise of plasma FFA the rate of removal of FFA, presumably by the working muscle, increases, and after about 3–4 hours of work some 70–75% of the dog's $CO_2$ output arises from the oxidation of plasma FFA (Issekutz, Paul, and Miller, 1967). There is a straight line correlation between the rate of oxidation of FFA and the plasma FFA level (Paul and Issekutz, 1967). This response certainly cannot be seen in experiments carried out on Nembutal-anesthetized animals when a muscle group is electrically stimulated because: (a) only a small fraction of the musculature is working, and (b) Nembutal markedly inhibits the FFA turnover (Armstrong, Steele, Altszuler, Dunn, Bishop, and DeBodo, 1961). As to the stimulus for this striking mobilization of FFA, one has to give consideration to the interplay of at least three hormones. It was shown that in man exercise increased the plasma concentration of norepinephrine (Vendsalu, 1960), and caused a marked increase in growth hormone level (Glick, Roth, Yalow, and Berson, 1965). The effects of these potent lipolytic hormones are opposed by insulin, the concentration of which markedly decreases in the plasma. The question of whether this decrease, which can be found in man (Devlin, 1963) as well as in the dog (Issekutz, Paul, and Miller, 1967), is due to the increased release of norepinephrine (Porte and Williams, 1966), or to the decline of plasma glucose level, requires further investigation. In the absence of insulin in the pancreatectomized dog, the FFA turnover is in the same range at rest as in the normal dog after about a 3-hour run. It increases further during exercise, and it is far in excess of the energy expenditure, close to 90% of which is now covered by the oxidation of plasma FFA alone (Issekutz, Paul, and Miller, 1967).

Probably the interaction of insulin and growth hormone plays a role in the response of adipose tissue to exercise in nonfasting subjects. Havel, Naimark, and Borchgrevink (1963) showed that in a 2-hour very light exercise (wrestlers walking on a horizontal treadmill) 41–49% of the $CO_2$ output derived from the oxidation of FFA (plasma FFA: 1.1–1.4 $\mu$eq/ml) if the subjects were fasting, and only 10% if they were not fasting (plasma FFA: 0.2 $\mu$eq/ml). In one of our earlier studies on men, the plasma FFA was at first elevated by a 3-hour exercise (alternating treadmill and bicycle work) to values of 1.2–1.6 $\mu$eq/ml, and then in a 15-min resting period a carbohydrate lunch was given. The plasma glucose rose from about 65 mg% to 80–82 mg% and the FFA decreased to 0.9–1.0 $\mu$eq/ml (Rodahl, Miller, and Issekutz, 1964). The effect lasted for about 1 hour and continuation of exercise raised the plasma FFA to values of 1.6–2.2 $\mu$eq/ml. In another study, carried out with radiopalmitate on dogs, glucose was infused at a constant rate of about four times the hepatic sugar output during the preexercise resting period (105 min), and during the subsequent treadmill run (3 hours). It took about 1.5 hours of exercise before the predominantly carbohydrate metabolism (characterized by high RQ, elevated plasma insulin level, and low FFA turnover) slowly changed to fat metabolism (decrease in RQ, decrease of plasma insulin level, and rising rate of turnover and oxidation of FFA). The adipose tissue covered more than half of the energy expenditure only in the third hour (Issekutz, Paul, and Miller, 1967).

These moderate or light exercises are characterized by a relatively constant and low blood lactate level (less than 20 mg%).

In "heavy work" we are facing an entirely different situation. Due to the insufficient $O_2$ supply the blood lactic acid concentration rises, and the heavier the work for a given individual, the higher the lactate level. This seems to have a profound effect on the participation of the adipose tissues as an energy source. Cobb and Johnson (1963) reported that, at about the same energy output, subjects with sedentary occupations showed a greater decrease in plasma FFA than physically trained individuals. In the former group the lactate level rose to an average value of 59 mg%, while in the latter it reached only 11 mg%. Experiments on dogs with radiopalmitate as a tracer revealed that heavy exercise could actually inhibit the release of FFA to the extent that the FFA turnover became less than it was at rest (Issekutz, Miller, Paul, and Rodahl, 1965*a*; Issekutz, Miller, and Rodahl, 1966). There was an inverse correlation between the changes in FFA release and the increase of blood lactate measured at the end of a 30-min exercise. Further studies on resting dogs showed that lactate infusion decreased the plasma FFA level (Issekutz and Miller, 1962), and markedly inhibited the rate of entry of FFA (Issekutz, Miller, Paul, and Rodahl, 1965*b*), and that this effect, unlike the similar effect of glucose, did not require the presence of insulin. On the contrary, the lactate inhibition of FFA release proved to be even more dramatic on the highly elevated FFA turnover of pancreatectomized dogs (Miller, Issekutz, Paul, and Rodahl, 1964; Issekutz, Miller, and Rodahl, 1966). Björntorp (1965) then showed that lactate inhibited the activity of lipase in the isolated epididymal fat pad.

In our recent experiments* we infused L(+)-lactic acid (or sodium lactate) into running dogs, the FFA turnover of which was elevated by a 1-hour exercise. The lactate infusion simulated the response to a heavy exercise, but without inadequate $O_2$ supply. As Fig. 30.1 shows, lactate infusion drastically depresses the rate of release, and parallel with this decreases also the rate of oxidation of FFA. The contribution of adipose tissue to the $CO_2$ output drops from 60% to 20%. On cessation of lactate infusion, blood lactate quickly drops, and the FFA turnover rapidly rises. A summary of these experiments is presented in Fig. 30.2. When the logarithms of FFA entry (release) were plotted against the logarithms of blood lactate level, a straight line correlation was obtained ($P < 0.001$). This indicates that at a lactate concentration of about 50 mg %, which is generally considered to be a moderate level, and at which the plasma FFA level is around 0.5 $\mu$eq/ml, the turnover rate of FFA, even if it were completely oxidized (1 meq = 2.5 cal), could not cover more than 30% of an energy expenditure of 4.7 cal/$m^2$ min, at which our dogs are usually working.

## HEPATIC SUGAR OUTPUT

Earlier literature contains ample evidence to show that muscle contraction increases the glucose uptake *in vitro* as well as *in vivo*. If the plasma glucose concentration is to stay at a reasonable level, which is a prerequisite for the continuation of exercise, then the rate of hepatic glucose output has to match the elevated peripheral utilization. Elevated sugar output was clearly demonstrated on men by measurements of blood flow and arteriovenous glucose differences across the splanchnic area (Rowell, Masoro, and Spencer, 1965), and by the isotope dilution radioglucose tracer techniques in men (Reichard, Issekutz, Kimbel, Putnam, Hochella, and Weinhouse, 1961) as well as in dogs.

In athletes working at an $O_2$ uptake of 2–2.5 liters/min, Rowell, Masoro, and Spencer (1965) found values ranging from 107 to 689 mg/min. The average was 295 mg/min, an increase of about threefold over the resting value. Bergström and Hultman (1967) measured values of 200–300 mg/min at the end of a 20–25 min work at 400 kpm/min on the bicycle ergometer. These values also represented a two- to threefold increase above the preexercise glucose output. A similar increase was obtained on running normal dogs with the constant rate in-

* Experiments carried out with A. Issekutz and D. Nash. These studies were supported by a grant from the Medical Research Council of Canada.

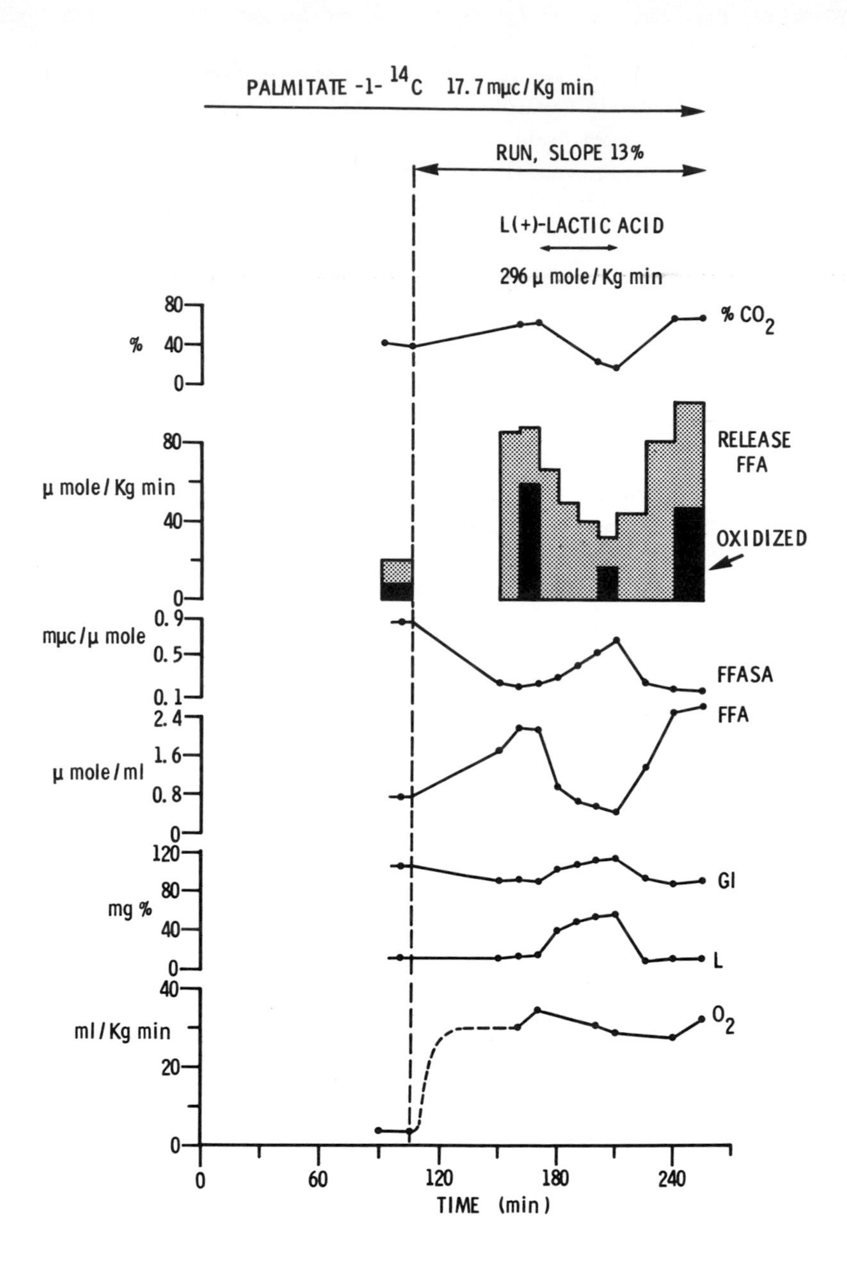


*Fig. 30.1.* Effect of L(+)-lactic acid infusion on the rate of release and oxidation of FFA in an exercising dog. Exercise started at t = 105 and lactic acid was infused from t = 170 to t = 210.

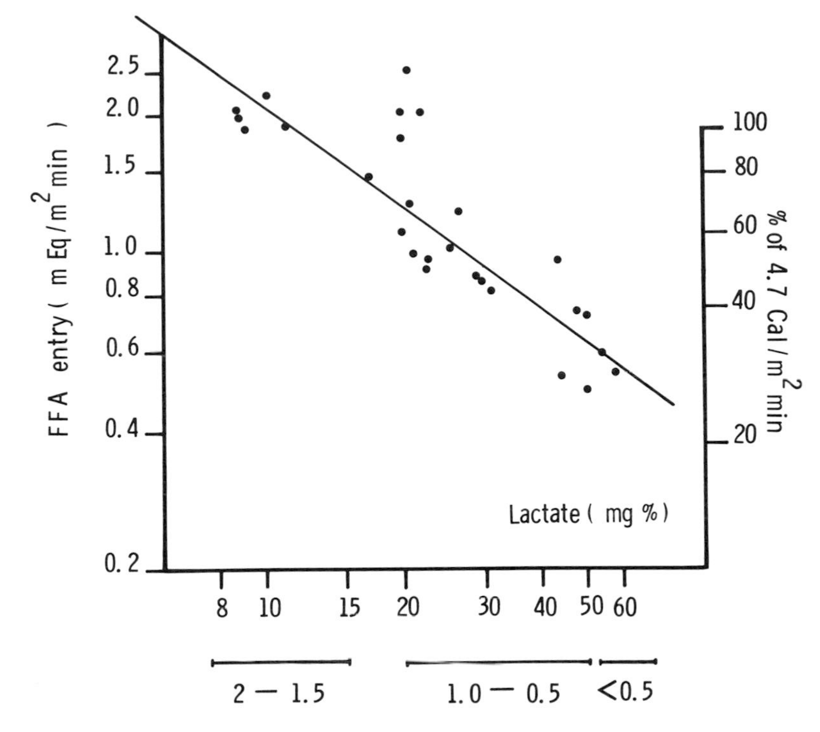


*Fig. 30.2.* Correlation between the logarithms of FFA release and the logarithms of blood lactate level following L (+) -lactate infusions into exercising dogs. The scale on the right shows the possible contribution of plasma FFA to an energy expenditure of 4.7 cal/m² min, assuming 100% oxidation of the FFA turnover.

Bottom lines give the approximate range of plasma FFA level (μeq/ml) at various concentrations of blood lactate.

fusion technique using uniformly labeled radioglucose as tracer (Issekutz, Paul, and Miller, 1967; Paul and Issekutz, 1967). The average value of Rowell, Masoro, and Spencer (1965) obtained by hepatic vein catheterization if calculated as mmole/m² min (approx. 0.9 mmole/m² min) agrees remarkably well with the value obtained in running dogs (0.89–0.91 mmole/m² min).

If, during exercise, the entire glucose output is oxidized, it can account for only 10–11% of the $O_2$ uptake. Experiments on dogs showed that, at an energy expenditure of 4.7 kcal/m² min (six times the resting value), the entire hepatic output was converted to $CO_2$ and 10–15% of $CO_2$ output derived from plasma glucose. This, of course, does not say that all the glucose taken up by the muscles is oxidized in the muscle. A part of it can be converted to lactate and then oxidized in the liver. In this respect, measurements of the $^{14}CO_2$ output of the intact animal cannot give any information. It should, however, be mentioned that Havel, Pernow, and Jones (1967) measured the rate of oxidation of FFA (palmitate-1-$^{14}C$ as a tracer), the RQ, the glucose uptake, and the lactate production of the working legs of bicycling men. From the arteriovenous differences they calculated that the glucose uptake could account for 16% (range from 6–26%) of the $CO_2$ output of the leg, and that 36% (range from 8–84%) of the glucose carbon was released in the form of lactate. A common feature of hepatic response to exercise in all these studies is that the hepatic glucose output increases rather slowly. It takes at least a half-hour or more to reach the new elevated level. Once this is reached it often remains constant, at least in the dog, for several hours, while the rising FFA take an increasing part in the energy expenditure (Issekutz, Paul, and Miller, 1967). The hepatic glucose output declines only towards the end of the exercise, which is followed by a decrease in the plasma glucose level, and eventually by cessation of work.

When unlabeled glucose was infused at a constant rate of about four times the resting hepatic glucose output, exercise further increased the glucose turnover. Yet in the first hour only 27% of the energy expenditure could be covered by the plasma glucose and about 48% from plasma FFA. The remaining 25% had to come from other sources. Only after 3 hours of exercise, when the FFA turnover was sufficiently elevated, could the muscle obtain its full energy need from the plasma.

In pancreatectomized dogs the hepatic glucose output, even at rest, is of the same order of magnitude as in the exercising normal animals. Work increases it slightly, but the contribution of the 300–400 mg% plasma glucose to the $CO_2$ output remains very low and, on the average, only about 6% of the energy expenditure is covered by the oxidation of plasma glucose. Apparently, physical work does not increase the glucose permeability of muscle cells enough to make use of the high plasma glucose level, and a normal hormonal balance is needed. It is also possible, however, that the extremely high FFA level in these animals inhibits the utilization of glucose, since it was shown (Randle, Newsholme, and Garland, 1964) that palmitate decreased the glucose uptake of rat hemidiaphragm.

The common feature of both extramuscular energy depots is the sluggishness of their response to exercise. At the beginning of work there is at least a 30-min period during which intramuscular energy sources have to supply 50% or more of the energy, and in short, heavy exercise the adipose tissue and the liver together cannot cover more than 40% of the requirement.

### INTRAMUSCULAR ENERGY DEPOTS: LIPIDS

The idea that lipids stored in or near the muscle cells could serve as energy sources is not new. In 1913, Greene suggested that the king salmon uses intramuscular lipids while migrating upstream during the spawning season. Neptune, Sudduth, and Foreman (1959) showed that the lipid content of rat diaphragm decreases if it is incubated in a substrate-free medium. The glycogen, which disappeared from the muscle and was not found as lactate, could account for only 10% of the $O_2$ uptake (Neptune and Foreman, 1959). It was also demonstrated that radiopalmitate entered a larger pool, which was then subsequently oxidized to $^{14}CO_2$ in a substrate-free medium (Neptune, Sudduth, Foreman, and Fash, 1959).

The need to assume the participation of intramuscular fat in exercise metabolism arose when the oxidation of plasma FFA proved to be much too low to explain the RQ measured on exercising dogs (Issekutz, Miller, Paul, and Rodahl, 1964). It was estimated that in a 30-min heavy exercise about half of the fat oxidation arose from intramuscular lipids. This assumption was further strengthened by experiments in which the intramuscular fat was at first depleted by a 30-min heavy run, and then radiopalmitate was infused over a 3-hour recovery period. After cessation of the infusion the radioactivity in the plasma decreased very rapidly, due to the very high FFA turnover. Then the test run started, and showed that between the 10th and 40th min of exercise approximately 120 times more radioactviity was exhaled as $^{14}CO_2$ than disappeared from the blood.

Havel, Ekelund, and Holmgren (1967), using radiopalmitate and $NaH^{14}CO_3$ in the form of a single injection (pulse labeling), carried out a detailed kinetic analysis of the oxidation of FFA in exercising men. The conclusion was reached that only about half of the FFA uptake was directly oxidized, the other half entering larger compartments. This latter part reentered the oxidative pathways within 30 min. In the dog, as in man, the RQ indicated that the oxidation of plasma FFA could account only for about half of the total fat oxidation. The role of intramuscular lipids as an energy source was challenged by Masoro, Rowell, McDonald, and Steiert (1966). They found no difference in the phospholipid and triglyceride content between the quiescent and electrically stimulated (1 or 3 twitches/sec for 5 hours) gastrocnemius or soleus of anesthetized pigtail monkeys. In view of the findings made on rats, on dogs, and on men, it is difficult to ascribe these negative results to species differences. It may be that the pentobarbital depleted the usable intramuscular fat to such an extent that contractile activity could not cause any further decrease, but some consideration must be given also to the possibility that the undoubtedly very high lactate concentration in the maximally working muscle could inhibit the intramuscular lipase.

Muscle glycogen has always been looked upon as the classical energy source of contractile activity. We now know much more about the participation of adipose tissue and the liver in energy metabolism, but in view of their slow response to exercise, the role of muscle glycogen should not be underestimated. First of all, whenever the $O_2$ supply is insufficient (and this is certainly the case in the first 2 or 3 min of any physical work), glucose-6-phosphate is the only source which can prevent the complete depletion of the creatine phosphate $\rightarrow$ ATP system in the muscle. Secondly, the inadequate $O_2$ supply during heavy work and the accompanying rise of blood lactate may not only depress the response of the adipose tissue but also reduce the availability of fatty acids from intramuscular lipids. This may be because the locally accumulating lactate could inhibit the lipase (Björntorp, 1965), and, because inadequate $O_2$ supply also increases the concentration of $\alpha$-glycerophosphate (Peterson, Gaudin, Bocek, and Beatty, 1964), which in turn can act as an FFA acceptor. Consequently, whenever the $O_2$ requirement exceeds the supply, the oxidative metabolism is expected to be shifted towards carbohydrate utilization. However, carbohydrate is a rather inefficient source of energy. The oxidation of 1 mmole of fatty acid (=2.5 kcal) yields about 3.6 times more energy than that of glucose (=0.69 kcal), and more than 50 times as much as the breakdown of glucose to lactate (=0.048 kcal). Under these circumstances, therefore, one may expect a rapid breakdown of large quantities of muscle glycogen proceeding at a very high rate at the beginning of exercise, before the onset of the Pasteur effect (inhibition of anaerobic glycolysis by $O_2$) and then at a somewhat lower rate as the turnovers of both plasma glucose and FFA rise. In a series of investigations carried out with needle biopsy techniques Bergström and Hultman (1967) showed that heavy exercise on the bicycle ergometer decreased the glycogen content of the working quadriceps femoris from 1.38 g/100 g to 0.76 g/100 g in the first 15 min. After a resting period of the same length, the subsequent exercise periods decreased the muscle glycogen to 0.42, 0.17, and 0.08 g/100 g. Complete depletion of glycogen was followed within a few minutes by cessation of the work. In another study, Bergström, Hermansen, Hultman, and Saltin (1967) have found a good correlation between the initial glycogen content of the working muscle and the duration of heavy exercise until exhaustion. They used three different diets—a carbohydrate-rich, a mixed, and a fat-rich diet. Before exercise the average glycogen content of the leg muscle was more than five times higher on the carbohydrate diet (3.31 g/100 g) than on the fat diet (0.63 g/100 g). It was concluded that the initial glycogen content is a determinant for the capacity to perform long-term heavy exercise. Upon closer examination of the published tables, this conclusion does not seem to be completely warranted; after the high carbohydrate diet the muscle glycogen was, at the point of exhaustion, on the average only 0.2 g/100 g lower (and in four out of nine cases actually higher) than in the same individual on a fat diet before the start of exercise. One has to give some consideration also to the possibility that the liver glycogen content changes with the diet, and the hepatic sugar output can also be a limiting factor. During such heavy exercise, which increases the blood lactate four- to fivefold, carbohydrate is undoubtedly the major energy source because lactate may greatly limit the mobilization of extra- and intramuscular lipid sources, but it is difficult to estimate the participation of these various energy sources. Under these conditions the use of RQ is hazardous because "excess" nonmetabolic $CO_2$ is produced from the bicarbonate-lactic acid reaction (Issekutz and Rodahl, 1961) and this may lead to a gross overestimation of carbohydrate oxidation. Since we have no clue concerning the magnitude of the elevated lactate turnover, it is not possible to estimate what portion of the glycogen was oxidized and how much was involved in the low-energy-yielding breakdown to lactate. On the basis of our studies on exercising dogs with lactate infusion, it seems to be a safe guess that, in heavy exercise, the FFA cannot supply more than 20–30% of the energy. This could explain the favorable effect of a carbohydrate diet on the duration of heavy physical work, as was shown earlier also by Christensen and Hansen (1939).

In an attempt to study the participation of intramuscular sources in the energy expenditure when the rate of release of FFA is suppressed, normal and pancreatectomized dogs were pretreated with Na-nicotinate (Issekutz and Paul, 1968). This drug is a powerful inhibitor of lipase activity in the adipose tissue (Carlson, 1963), where it rapidly accumulates (Carlson and Hanngren, 1964). In exercising men, it prevented the increase of the FFA turnover and reduced its contribution to the $CO_2$ output (Carlson, Havel, Ekelund, and Holmgren, 1963). When dogs were given about 100 mg/kg Na-nicotinate i.v. or i.a. prior to a steady exercise, the release of FFA from the adipose tissue was greatly limited for about 2 hours. Under possibly identical conditions two experiments were carried out on each dog, one with radioglucose and another one with radiopalmitate as a tracer. The contribution of plasma glucose and that of plasma FFA and to the $CO_2$ output was subtracted from the total $CO_2$ output. Similarly, the amount of $O_2$ required for the oxidation of plasma glucose and that of FFA was subtracted from the total $O_2$ uptake. The remaining $CO_2$ output and $O_2$ uptake provided an RQ which excluded these substrates carried by the circulation to the working muscles, and which presumably represented the participation of intramuscular energy sources in the oxidative metabolism.

Fig. 30.3 summarizes the results of these studies carried out on normal and diabetic dogs running on a treadmill (slope 15%, speed 100 m/min) at a rather steady energy expenditure of about 4.7 cal/m$^2$ min with and without sodium nicotinate. In the control animals without nicotinate, after 60 minutes' run, FFA supplies about 50% of the energy and plasma glucose covers an additional 12%. As was mentioned above, in the 3rd hour some 85% of the expenditure is covered by extramuscular sources (adipose tissue plus hepatic glucose output). In the pancreatectomized dogs there was no difference between the 2nd and 3rd hours of exercise. About 85% of the energy was derived from FFA and about 6% from plasma glucose. When normal dogs received a continuous infusion of glucose at a rate of about 3.5–4 times the resting hepatic glucose output, the usual decline of plasma glucose was prevented and the glucose level remained at 110–115 mg % throughout the exercise. The result was that in the 2nd hour some 74% and in the 3rd hour more than 90% of the energy requirement was covered by the plasma.

Sodium nicotinate administered prior to the exercise greatly delayed the rise of plasma FFA which, in the first 2 hours, could supply no more than about 37% of the energy expenditure. Since only 9.5% of the energy was derived from the oxidation of plasma glucose, more than 50% of the requirement had to come from energy depots stored in or near the muscle. The calculated intramuscular RQ (0.87) indicated that some 31% of the total energy expenditure arose from the oxidation of glycogen and 21% from the intramuscular lipids. The assumption that muscle lipids can serve as energy sources was strengthened considerably by similar experiments carried out on pancreatectomized dogs. In these animals, after nicotinate, during the 2 hours' run, only about 35% of the energy was supplied via blood circulation, and an intramuscular RQ of 0.77 was obtained. This suggests that some 47% of the 4.7 cal/m$^2$ min derived from muscle lipids and 18% from glycogen. Assuming that in the exercising dog the working muscles represent about 40% of the body weight, one could estimate how much of the intramuscular sources were used by 1 kg muscle in 2 hours, when the FFA turnover was depressed but the $O_2$ supply was adequate. One arrives at values which show that in the normal dog about 4.9 g glycogen and 1.3 g fatty acids were oxidized per kg muscle from its endogenous stores. The corresponding values for

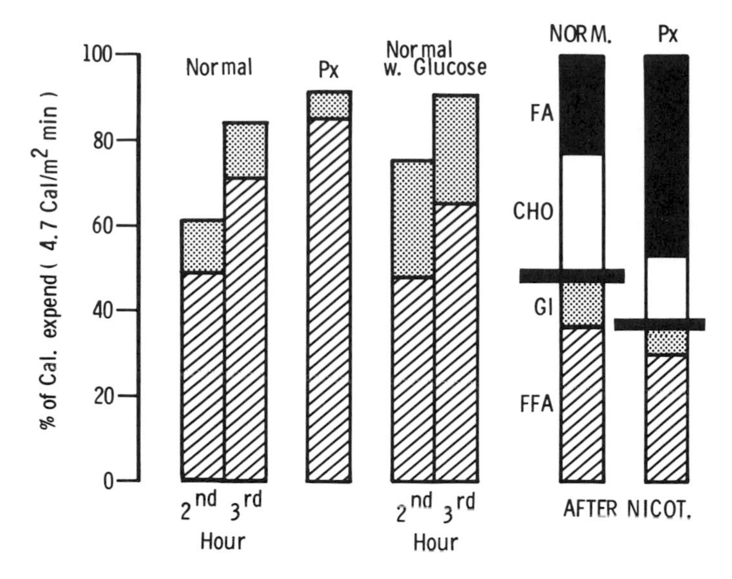


*Fig. 30.3.* Summary of the results obtained on exercising dogs. *Hatched columns,* contribution of plasma FFA; *dotted columns,* plasma glucose; *white,* intramuscular carbohydrate; *black,* intramuscular lipids. *Px,* pancreatectomized dogs.

pancreatectomized dogs were 2.8 g glycogen and 2.9 g fatty acid. The amount of fatty acids used from intramuscular lipids of normal dogs is small and may fall within the standard deviation of chemical analysis, especially if these fatty acids come from a variety of glyceride esters, as may well be the case.

This type of calculation, by us and others, certainly underestimates the breakdown of muscle glycogen because it is based on the RQ and excludes the anaerobic utilization of muscle glycogen (formation of liver glycogen from muscle glycogen via three carbon units).

Recently, we attempted to work out a new approach which (see footnote, p. 627) permits us to follow the utilization of muscle glycogen in the intact exercising dog without needle biopsy. The assumption on which the method is based is that the skeletal muscle is essentially a nongluconeogenic tissue (Scrutton and Utter, 1968) and, during exercise, it does not form phosphoenol-pyruvate from pyruvic acid via oxalacetate. Therefore, if glucose-$^{14}C$ is infused at a constant rate as it enters the glucose-6-phosphate pool in the muscle, it is diluted by the breakdown of muscle glycogen, and the specific activity of lactate, being equal to that of pyruvate (from which it is formed), can be a measure of this dilution.

Since the glucose uptake (presumably by the working muscle) is known, one can calculate the breakdown of glycogen according to the formula

$$G = g_u \left( \frac{GLSA}{2LSA} - 1 \right)$$

where $G$ represents the rate of breakdown of muscle glycogen expressed as $\mu$mole glucose per kg min, $g_u$, the glucose uptake ($\mu$mole/kg min), *GLSA* and *LSA* are the specific activities of plasma glucose and lactate, respectively, both in m$\mu$c/$\mu$mole.

The crucial procedure in this approach is the measurement of the low LSA. Sufficient accuracy could be achieved by concentrating the lactic acid on an anion exchange resin and then eluting it with a small volume of 0.2 M NaCl.

Figs. 30.4 and 30.5 show the effect of sodium-nicotinate on the breakdown of muscle glycogen in two experiments. When the nicotinate was infused during the resting period prior to exercise, the rise of FFA was prevented for about 1 hour. In the first 30 min of exercise there was a rapid breakdown of glycogen (137 $\mu$mole/kg min), but the rate declined as the hepatic glucose output slowly increased from 40 $\mu$mole/kg min to 65 $\mu$mole/kg min. This rate of glucose production, however, could not be maintained and, as the blood sugar decreased, the utilization of muscle glycogen rose again. After 103 min of run at a low FFA (0.4 $\mu$eq/ml) and low blood sugar level (56 mg %) the animal collapsed, obviously as a result of the sudden decrease of the hepatic glucose output. One can estimate that the total amount of muscle glycogen broken down during the 103 min was about 6 g/kg muscle.

Fig. 30.5 shows the effect of nicotinate during exercise. In this experiment the dog ran on an 18% slope for 150 min. The hepatic sugar output was rather constant at 46 $\mu$mole/kg min and an additional 34 $\mu$mole/kg min was supplied by the muscle glycogen. Then the release of FFA was depressed by nicotinate, and the plasma FFA fell from 1.2 $\mu$eq/ml to 0.2 $\mu$eq/ml. Both the hepatic glucose output and the breakdown of glycogen rose. The liver became exhausted rather quickly and, after 120 min of exercise, the glucose specific activity rose sharply, indicating a rapid fall of glucose production with a resulting decrease of plasma glucose level. As the hepatic glucose output decreased, the utilization of muscle glycogen rose. Again the duration of exercise was limited by the rate of glucose production and consequently by the glucose level. The increased utilization of muscle glycogen after nicotinate was recently shown also on exercising men with the biopsy technique (Bergström, Hultman, Jorfeldt, Pernow, and Wahren, 1969).

In summary, we would like to conclude that the working muscle can use fatty acids just as well as carbohydrate from both extramuscular and

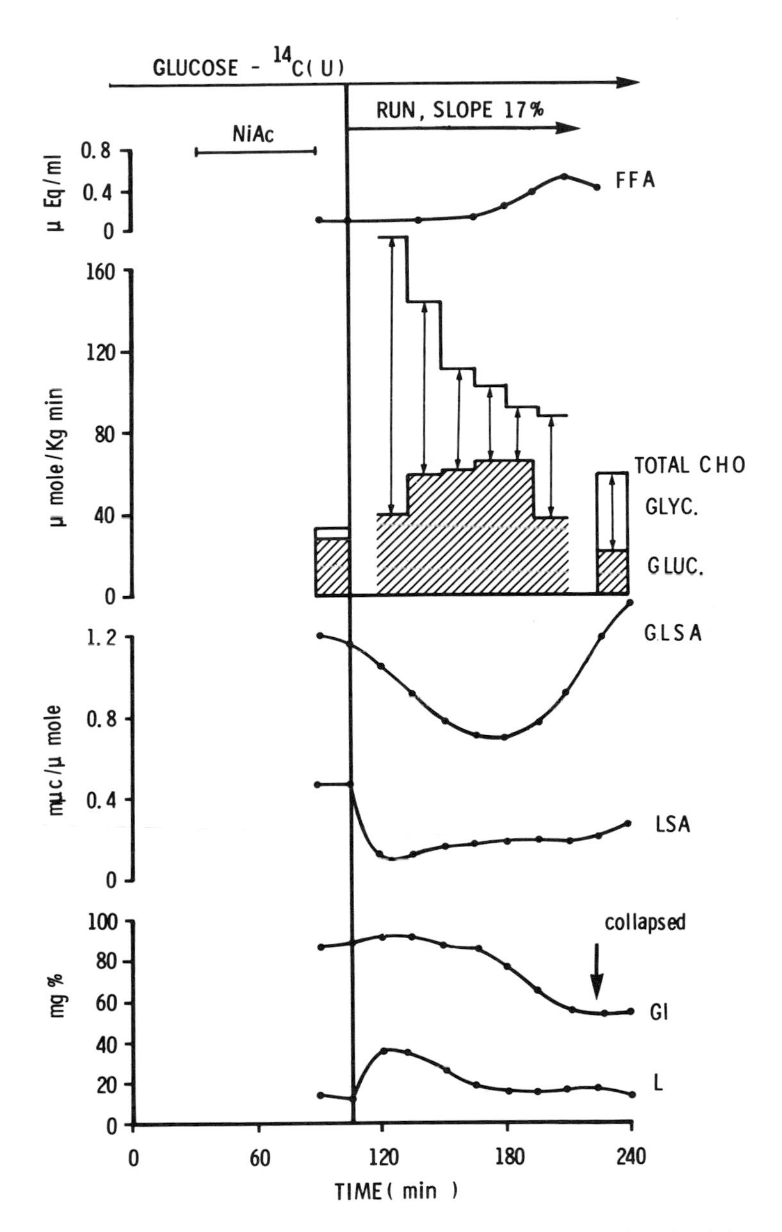


*Fig. 30.4.* The effect of sodium nicotinate (1.8 mg/kg min for 60 min) in an experiment with glucose-$^{14}C$ as a tracer (primed, constant infusion). Treadmill run on a slope of 17%, speed 100 m/min. *Hatched columns,* uptake of plasma glucose; *white,* breakdown of muscle glycogen calculated from glucose specific activity (GLSA) and lactate specific activity (LSA).

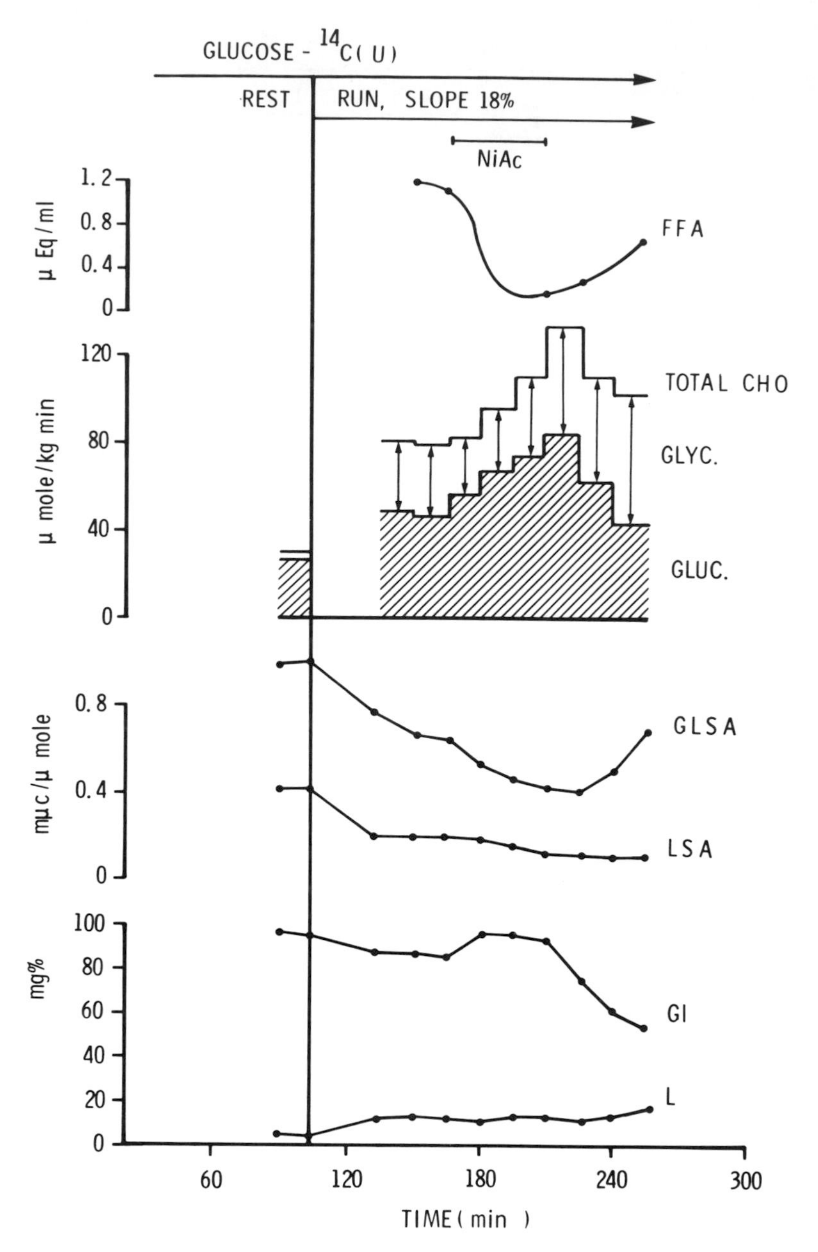


*Fig. 30.5.* Effect of sodium nicotinate on the carbohydrate metabolism during exercise. NiAc infusion (1.35 mg/kg min for 45 min) started in the 60th min of exercise (slope 18%, speed 100 m/min). *Hatched columns,* uptake of plasma glucose; *white,* breakdown of muscle glycogen calculated from GLSA and LSA.

from intramuscular stores. Whether fat or carbohydrate is the major source of energy seems to depend entirely on the experimental conditions, leading to controversies as soon as we try to draw a general conclusion from any particular type of study. For instance, fat covers more than 80% of the energy requirement for a dog running on the treadmill on a 10% slope for 5 hours (Paul and Issekutz, 1967), or for an obese subject working after 25 days of complete starvation for 4 hours at an RQ of 0.71 (Issekutz, Bork, Miller, and Wroldsen, 1967), or for the exercising pancreatectomized dog (Issekutz, Paul and Miller, 1967). Carbohydrate is the major fuel when a person performs a heavy exercise on the bicycle ergometer, after being on a carbohydrate-rich diet for several days (Bergström, Hermansen, Hultman, and Saltin, 1967; Christensen and Hansen, 1939).

The relative severity and the type of the exercise also seem to play important roles in the fuel supply, because the increased lactate production depresses the response of adipose tissue and perhaps also the utilization of intramuscular lipids. Working on the bicycle against high friction calls for the development of high tension in the exercising muscles. This can lead to a less sufficient $O_2$ supply during cycling than during a treadmill run, and hence to a greater utilization of carbohydrate, that is, to a more rapid breakdown of muscle glycogen in the former than in the latter type of exercise.

Finally, it should be pointed out that the regulation of the metabolism during exercise seems to be aimed at maintaining the blood sugar as long as possible at a level acceptable to the central nervous system. During exercise the muscles and the brain are in strong competition for the blood glucose; the decrease of plasma insulin and the rise of both growth hormone and noradrenalin production make possible the mobilization of FFA as the most effective fuel for exercise. The utilization of intramuscular energy sources, as well as the usual 15–20 mg % decrease in the plasma glucose level, seem to keep the glucose uptake and the contribution of the hepatic glucose at a possibly low level (10–15% of the energy expenditure). From this point of view, the rising FFA concentration may serve a double role. It is not only a fuel for the working muscle; it may also enhance the formation of liver glycogen from three carbon units (lactate, pyruvate, glycerol) (Williamson, 1967). Such an effect certainly would explain the observation that blocking the release of FFA from the adipose tissue by nicotinate at first increases the hepatic glucose output, then often leads to a catastrophic decrease and to the eventual collapse of the animal (Figs. 30.4–5). This would also explain the nicotinate-induced decrease of hepatic sugar output in both resting (Paul, Issekutz, and Miller, 1966) and exercising pancreatectomized dogs (Issekutz and

Paul, 1968). Furthermore, the breakdown of muscle glycogen may be considered more as a device for maintaining the liver glycogen rather than for providing the small amount of energy which can be released via anaerobic glycolysis, which becomes the major energy source only during insufficient $O_2$ supply. That hepatic gluconeogenesis is going on during exercise was indicated by the work of Rowell, Kraning, Evans, Kennedy, Blackman, and Kusumi (1966) who, in exercisng men, measured the removal of lactate and the RQ across the splanchnic area. They found extremely low RQ values (down to zero).

At present, one cannot estimate the magnitude and the importance of gluconeogenesis. This would require measurements of lactate turnover during exercise, because even at low blood lactate the turnover rate may be very high. This fascinating chapter of exercise metabolism has hardly been touched as yet.

## *References*

Armstrong, D. T., R. Steele, N. Altszuler, A. Dunn, J. S. Bishop, and R. C. DeBodo. 1961. Regulation of plasma free fatty acid turnover. *Amer. J. Physiol. 201*:9.

Basu, A., R. Passmore, and J. A. Strong. 1960. The effect of exercise on the level of NEFA in the blood. *Quart. J. Expl. Physiol. 45*:312.

Bergström, J., L. Hermansen, E. Hultman, and B. Saltin. 1967. Diet, muscle glycogen and physical performance. *Acta Physiol. Scand. 71*:140.

Bergström, J., and E. Hultman. 1967. A study of the glycogen metabolism during exercise in man. *J. Clin. Lab. Invest. 19*:218.

Bergström, J., E. Hultman, L. Jorfeldt, B. Pernow, and J. Wahren. 1969. Effect of nicotinic acid on physical working capacity and on metabolism of muscle glycogen in man. *J. Appl. Physiol. 26*:170.

Björntorp, P. 1965. Effect of lactic acid on adipose tissue metabolism in vitro. *Acta Med. Scand. 178*:253.

Carlson, L. A. 1963. Studies on the effect of nicotinic acid on catecholamine-stimulated lipolysis in adipose tissues in vitro. *Acta Med. Scand. 173*:719.

Carlson, L. A., S. Fröberg, and S. Persson. 1965. Concentration and turnover of free fatty acids of plasma and concentration of blood glucose during exercise in horses. *Acta Phys. Scand. 63*:434.

Carlson, L. A. and Å. Hanngren. 1964. Initial distribution in mice of $^3$H-labeled nicotinic acid studied with autoradiography. *Life Sci. 3*:867.

Carlson, L. A., R. J. Havel, L.-G. Ekelund, and A. Holmgren. 1963. Effect of nicotinic acid on the turnover rate and oxidation of the free fatty acids of plasma in man during exercise. *Metabolism 12*:837.

Carlson, A., and B. Pernow. 1961. Studies on lipids during exercise II. The arterial plasma free fatty acid concentration during and after exercise and its regulation. *Lab. Clin. Med. 58:*673.

Christensen, E. H., and O. Hansen. 1939. Arbeitsfähigkeit und Ernährung. *Skand. Arch. Physiol. 81:*160.

Cobb, L., and W. Johnson. 1963. Hemodynamic relationships of anaerobic metabolism and plasma free fatty acids during prolonged strenuous exercise in trained and untrained subjects. *J. Clin. Invest. 42:*800.

Devlin, J. G. 1963. The effect of training and acute physical exercise on plasma insulin-like activity. *Jr. J. Med. Sci.* 423.

Friedberg, S. J., and H. Estes, Jr. 1962. Direct evidence for the oxidation of free fatty acids by peripheral tissues. *J. Clin. Invest. 41:*677.

Friedberg, S. J., W. R. Harland, Jr., D. L. Trout, and E. H. Estes, Jr. 1960. The effect of exercise on the concentration and turnover of plasma nonesterified fatty acids. *J. Clin. Invest. 39:*215.

Fritz, I. B. 1960. Effects of insulin on glucose and palmitate metabolism by resting and stimulated rat diaphragms. *Amer. J. Physiol. 198:*807.

Fritz, I. B., D. G. Davis, R. H. Holtrop, and H. Dundee. 1958. Fatty acid oxidation by skeletal muscle during rest and activity. *Amer. J. Physiol. 194:*379.

Glick, S. M., J. Roth, R. S. Yalow, and S. A. Berson. 1965. The regulation of growth hormone secretion. *Recent Progr. Hormone Res. 21:*241.

Greene, Ch. W. 1913. The storage of fat in the muscular tissue of the King salmon and its resorption during the fast of spawning migration. *Bull. Bur. Fisheries 23:*73.

Havel, R. J., L.-G. Ekelund, and A. Holmgren. 1967. Kinetic analysis of the oxidation of palmitate-1-$C^{14}$ in man during prolonged heavy muscular exercise. *J. Lipid Res. 8:*366.

Havel, R. J., A. Naimark, and Ch. F. Borchgrevink. 1963. Turnover rate and oxidation of free fatty acids of blood plasma in man during exercise: studies during continuous infusion of palmitate-1-$C^{14}$. *J. Clin. Invest. 42:*1054.

Havel, R. J., B. Pernow, and N. L. Jones. 1967. Uptake and release of free fatty acids and other metabolites in the legs of exercising men. *J. Appl. Physiol. 23:*90.

Issekutz, B., Jr., W. M. Bork, H. I. Miller, and A. Wroldsen. 1967. Plasma free fatty acid response to exercise in obese humans. *Metabolism 16:*492.

Issekutz, B., Jr., and H. Miller. 1962. Plasma FFAs during exercise and the effect of lactic acid. *Proc. Soc. Exp. Biol. Med. 110:*237.

Issekutz, B., Jr., H. I. Miller, P. Paul, and K. Rodahl. 1964. Source of fat oxidation in exercising dogs. *Amer. J. Physiol. 207:*583.

———. 1965*a*. Aerobic work capacity and plasma FFA turnover. *J. Appl. Physiol. 20:*293.

———. 1965*b*. Effect of lactic acid on free fatty acids and glucose oxidation in dogs. *Amer. J. Physiol. 209:*1137.

Issekutz, B., Jr., H. I. Miller, and K. Rodahl. 1966. Lipid and carbohydrate metabolism during exercise. *Fed. Proc. 25:*1415.

Issekutz, B., Jr., and P. Paul. 1968. Intramuscular energy sources in exercising normal and pancreatectomized dogs. *Amer. J. Physiol. 215:*197.

Issekutz, B., Jr., P. Paul, and H. I. Miller. 1967. Metabolism in normal and pancreatectomized dogs during steady state exercise. *Amer. J. Physiol. 213:*857.

Issekutz, B., Jr., and K. Rodahl. 1961. Respiratory quotient during exercise. *J. Appl. Physiol. 16:*606.

Masoro, E. J., L. B. Rowell, R. M. McDonald, and B. Steiert. 1966. Skeletal muscle lipids II. Non-utilization of intracellular lipid esters as an energy source for contractile activity. *J. Biol. Chem. 241:*2626.

Miller, H. I., B. Issekutz, Jr., P. Paul, and K. Rodahl. 1964. Effect of lactic acid on plasma free fatty acids in pancreatectomized dogs. *Amer. J. Physiol. 207:*1226.

Neptune, E. M., Jr., and D. R. Foreman. 1959. Endogenous glycogen of rat diaphragm and its theoretical capacity to support respiration. *J. Biol. Chem. 234:*1942.

Neptune, E. M., Jr., H. C. Sudduth, and D. R. Foreman. 1959. Labile fatty acids of rat diaphragm muscles and their possible role as the major endogenous substrate for maintenance of respiration. *J. Biol. Chem. 234:*1659.

Neptune, E. M., Jr., H. C. Sudduth, D. R. Foreman, and F. J. Fash. 1959. Phospholipid and triglyceride metabolism of exercised rat diaphragm and the role of these lipids in fatty acid uptake and oxidation. *J. Lipid Res. 1:*229.

Paul, P., and B. Issekutz, Jr. 1967. Role of extra-muscular energy sources in the metabolism of the exercising dog. *J. Appl. Physiol. 22:*615.

Paul, P., B. Issekutz, Jr., and H. I. Miller. 1966. Interrelationship of free fatty acids and glucose metabolism in the dog. *Amer. J. Physiol. 211:*1313.

Peterson, R. D., D. Gaudin, R. M. Bocek, and C. H. Beatty. 1964. α-Glycerophosphate metabolism in muscle under aerobic and hypoxic conditions. *Amer. J. Physiol. 206:*599.

Porte, D., Jr., and R. H. Williams. 1966. Inhibition of insulin release by norepinephrine in man. *Science 152:*1248.

Randle, P. J., E. A. Newsholme, and P. B. Garland. 1964. Regulation of glucose uptake by muscle 8. Effects of fatty acids, ketone bodies and pyruvate, and of alloxan-diabetes, and starvation, on the uptake and metabolic fate of glucose in rat heart and diaphragm muscles. *Biochem. J. 93:*652.

Reichard, G. A., B. Issekutz, Jr., P. Kimbel, R. C. Putnam, N. J. Hochella, and S. Weinhouse. 1961. Blood glucose metabolism in man during muscular work. *J. Appl. Physiol. 16:*1001.

Rodahl, K., H. I. Miller, and B. Issekutz, Jr. 1964. Plasma free fatty acids in exercise. *J. Appl. Physiol. 19:*489.

Rowell, L. R., K. K. Kraning, T. O. Evans, J. W. Kennedy, J. R. Blackman, and F. Kusumi. 1966. Splanchnic removal of lactate and pyruvate during prolonged exercise in man. *J. Appl. Physiol. 21:*1773.

Rowell, L. R., E. J. Masoro, and M. J. Spencer. 1965. Splanchnic metabolism in exercising man. *J. Appl. Physiol. 20:*1032.

Scrutton, M. C., and M. F. Utter. 1968. The regulation of glycolysis and neoglycogenesis in animal tissues. *Ann. Rev. Biochem. 37*:249.

Vendsalu, A. 1960. Studies on adrenaline and noradrenaline in human plasma. *Acta Physiol. Scand. 49* (suppl. 173) :1.

Williamson, J. R. 1967. Effects of fatty acids, glucagon and antiinsulin serum on the control of gluconeogenesis and ketogenesis in rat liver, p. 229. *In* G. Weber (ed.), *Advances in Enzyme Regulation,* Vol. 5. Pergamon, New York.

# Chapter 31

# Compartmentation of Pathways of Carbohydrate Metabolism in Muscle Cells

B. R. LANDAU

For almost 20 years now investigators have presented data that suggest the existence in a muscle cell of more than one pool of glucose-6-phosphate, or alternatively, of a pathway for glycogen formation from glucose not involving glucose-6-phosphate. It is my purpose to examine the evidence for these hypotheses.

Beloff-Chain, Chain, Bovet, Pocchiari, Cantanzaro, and Longinotti (1953) reported that hexose phosphates could enter cells and that glycogen formation by rat diaphragm incubated in a medium containing glucose was nearly the same as glycogen formation in a medium containing glucose-1-phosphate when no insulin was present. When insulin was added, glycogen formation from glucose, but not from glucose-1-phosphate, was increased. Glucose-6-phosphate was isomerized to fructose-6-phosphate by phosphohexoisomerase leached from the diaphragm into the medium; from this mixture of phosphates, the diaphragm did not synthesize glycogen. The observation that glucose-1-phosphate, but not glucose-6-phosphate, was converted to glycogen was concluded to be inconsistent with the generally accepted theories of glucose metabolism in which glucose-6-phosphate is always the first intermediate.

Beloff-Chain, Cantanzaro, Chain, Masi, Pocchiari, and Rossi (1955) reported that at the beginning of incubation of diaphragm with glucose-$^{14}C$, insulin markedly stimulated the incorporation of $^{14}C$ into glycogen and oligosaccharides of diaphragm, but that there was no significant stimulation of $^{14}C$ incorporation into hexose monophosphates (including glucose-1-phosphate, glucose-6-phosphate, and fructose-6-phosphate assayed together), $CO_2$, and lactate. Incorporation into the hexose phosphates increased with time (Fig. 31.1). Incorporation into $CO_2$ was consistently unresponsive to insulin, while incorporation into lactate was generally so. A phosphate-buffered medium was usually used but results

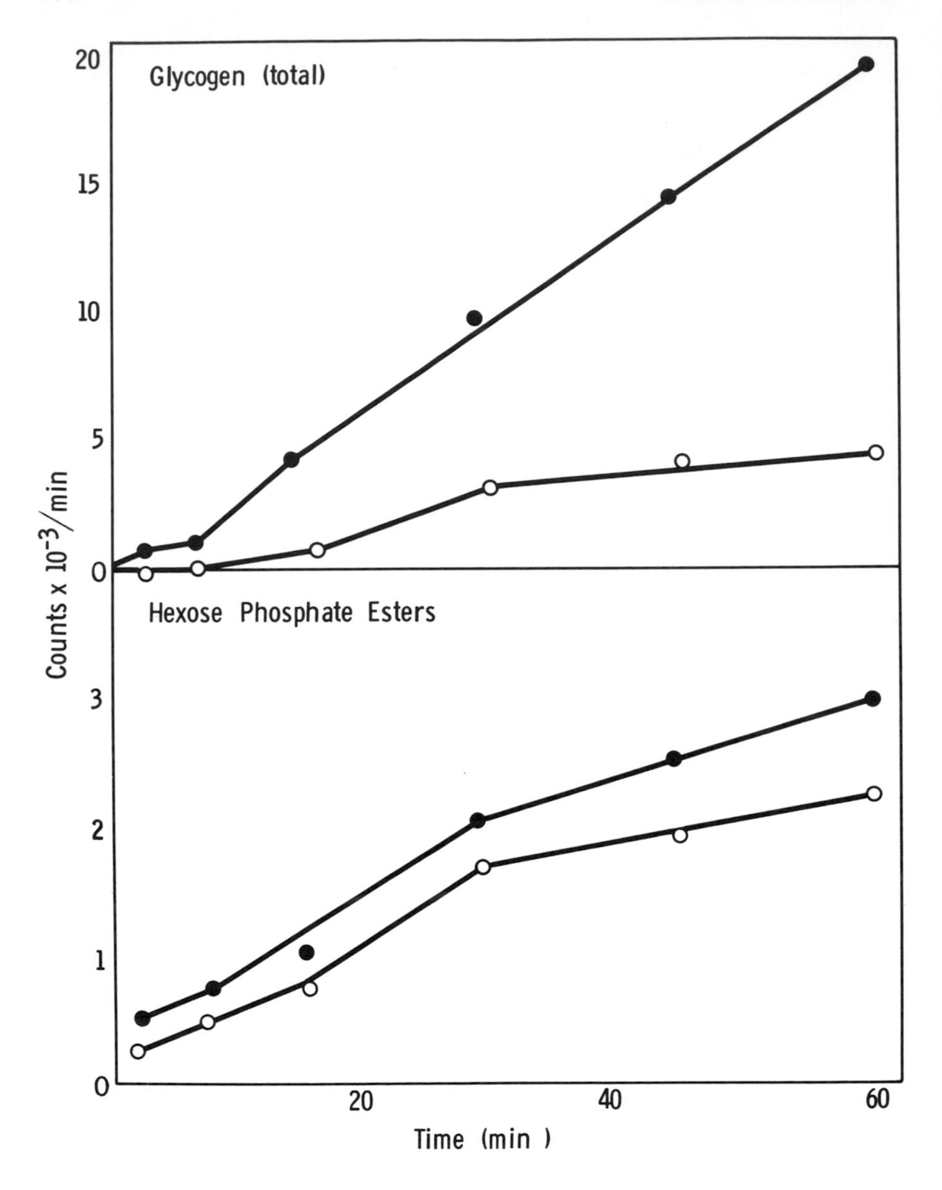


*Fig. 31.1.* Incorporation of $^{14}C$ from glucose-$^{14}C$ into glycogen and hexose-6-phosphate in rat diaphragm incubated in the absence (*open circle*) and presence (*solid circle*) of insulin. From Fig. 9 of Beloff-Chain, Cantanzaro, Chain, Masi, Pocchiari, and Rossi (1955).

were similar using a bicarbonate-buffered medium. These results were concluded to be compatible with the assumption that the simple hexose monophosphates do not represent the main intermediates in the conversion of glucose to glycogen, i.e., glucose → glucose-6-phosphate → glucose-1-phosphate → glycogen, but are breakdown products of glycogen metabolism. A significant increase in incorporation into glucose-1-phosphate may have occurred, but may have been masked by incorporation into the other hexose phosphates and the study of Beloff-Chain, Chain, Bovet, Pocchiari, Cantanzaro, and Longinotti (1953) seems to suggest that glucose-1-phosphate but not glucose-6-phosphate is a precursor of glycogen. The data are also compatible with the existence of two pools of hexose phosphates, one of which is in the pathway to glycogen; while rapidly turning over, it is small in comparison to the second pool, which is intermediate in the formation of $CO_2$ and lactate.

Shaw and Stadie (1957) reported observations on the effect of insulin on glucose metabolism by rat diaphragm incubated in a phosphate-saline medium. They confirmed the finding of Beloff-Chain, Cantanzaro, Chain, Masi, Pocchiari, and Rossi (1955) that insulin did not increase the synthesis from glucose of lactate, but did increase that of glycogen. Glucose-6-phosphate and glucose-1-phosphate isolated from diaphragm incubated with glucose-$^{14}C$ were both labeled; insulin increased the incorporation of $^{14}C$ into both. These observations were interpreted to indicate that the glucose was metabolized within the diaphragm to glucose-6-phosphate, then to glucose-1-phosphate, and then to glycogen, and that this pathway was insulin-responsive. There was no incorporation into fructose-1,6-diphosphate but this was explained by the report (Shaw and Stadie, 1959) that phosphofructokinase but not phosphohexoisomerase was inactive in the absence of bicarbonate. Because of this, glucose was presumed not to be metabolized to lactate in the insulin-sensitive pathway, i.e.:

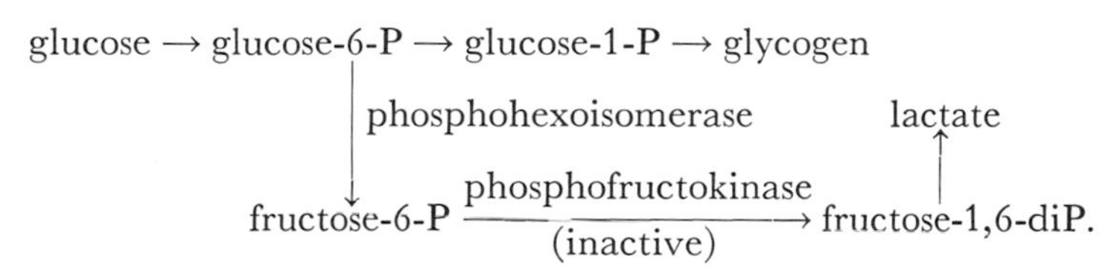


Shaw and Stadie (1957) considered that the lactate formed in the diaphragm incubated in the phosphate-saline medium arose from a second Embden-Meyerhof pathway. When glucose-$^{14}C$ was incubated in the medium with unlabeled glucose-6-phosphate and fructose-1,6-diphosphate, lactate formation was enhanced by the presence of the esters and the specific activity of the lactate diluted. This was taken to indicate that the esters were intermediates in the conversion of the glucose to lactate. This was coupled with the fact that when the esters were isolated from the medium at the end of the incubation they were labeled with $^{14}C$. When unlabeled glucose-1-phosphate was incubated with glucose-$^{14}C$, the glucose-1-phosphate in the medium did not become labeled. There was thus no exchange between the esters in the tissue and in the medium since, as we have noted, under the conditions of these experiments glucose-1-phosphate in the tissue was labeled and fructose-1,6-diphosphate was not. Shaw and Stadie therefore concluded the existence of a second pathway, glucose → glucose-6-P → fructose-1,6-diP → lactate. Since at the end of the incubation the enzymic activities for this pathway were not found in the medium, they concluded that the reactions did not occur in the medium due to leaching of enzymes from the tissue. There was no mention of the reason that lactate formation in the phosphate medium by this pathway was not prevented through inactivity of phosphofructokinase.

Shaw and Stadie (1959) assumed that the insulin-nonresponsive pathway was structurally localized within the cells of the diaphragm or on their surface. Since glycogen is intracellularly located the insulin-sensitive pathway was concluded to be intracellular, and the pathway for glucose conversion to lactate was tentatively assumed to be a "cell surface enzyme system." Their scheme of separate Embden-Meyerhof pathways is depicted in Fig. 31.2.

Subsequently, Shaw and Stadie (1959) reported the results of experiments similar to those they had done in the phosphate-saline medium, but now in a bicarbonate-phosphate medium. The results were somewhat different. Insulin increased the conversion of glucose-$^{14}C$ to lactate-$^{14}C$, and fructose-1,6-diphosphate in the tissue became labeled. This would be in accord with the view that the insulin-sensitive pathway forms glycogen as well as lactate, i.e., that phosphofructokinase is active in the presence of bicarbonate. Villee and Hastings (1949) had previously reported that in bicarbonate buffer insulin increased conversion of glucose to lactate by diaphragm.

Some experiments that Shaw and Stadie (1957) had done in the phosphate-saline medium were not repeated in the bicarbonate-containing medium. Thus, exchange between glucose-1-phosphate in the medium and tissue was not examined, though it was shown that when labeled glucose and unlabeled fructose-1,6-diphosphate were incubated, the fructose-1,6-diphosphate became labeled. Indeed, from just the data obtained using the bicarbonate-saline medium one could conclude that there exists a single, insulin-sensitive Embden-Meyerhof pathway in diaphragm, one in which both glycogen and lactate are formed and in which fructose-1,6-diphosphate in the medium can exchange, perhaps through cut edges, with fructose-1,6-diphosphate in the tissue. The possibility must also be entertained, since muscle is bathed *in vivo* in bicarbonate-buffered fluid, that experiments using bicarbonate buffer are more likely

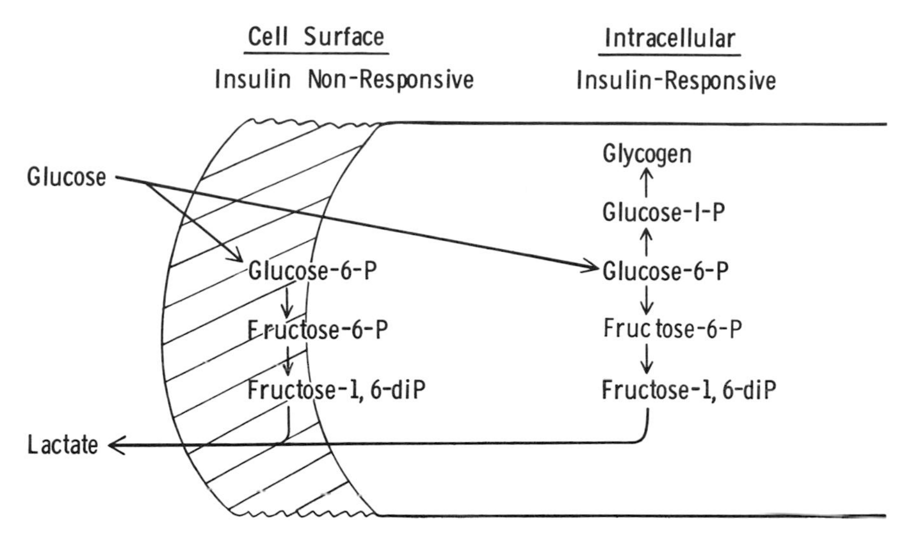


*Fig. 31.2.* Pathways in rat diaphragm as assumed by Shaw and Stadie (1957, 1959). The conversion of fructose-6-phosphate to lactate by the intracellular pathway was assumed to be inoperative in the absence of bicarbonate.

to give results reflecting physiological circumstances. If so, it must be assumed that the data using the phosphate-saline medium do indicate two pathways, but that in some manner one is an artifact of the conditions.

If there is a single Embden-Meyerhof pathway in a cell, the pathway followed by glucose once it is converted to glucose-6-phosphate should be the same as that followed by glucose-6-phosphate itself (Fig. 31.3). Thus, the proportion of glucose converted to lactate to that converted to glycogen should be the same as for glucose-6-phosphate. If, however, there are two pathways, as postulated by Shaw and Stadie, and exogenous glucose-6-phosphate has access only to the pool of the non-glycogen-forming pathway, these proportions should be different (Fig. 31.4). Thus, the fact that glucose-6-phosphate is not converted to glycogen while glucose is converted means that the ratio of glucose to glucose-6-phosphate conversion to glycogen would be larger than that for lactate or for $CO_2$. Since the lactate and $CO_2$ would in this scheme be derived from both pools, their ratios would not be equal unless the proportion of lactate to $CO_2$ formed from each pool was the same.

Shaw and Stadie (1959) reported two experiments in which they incubated glucose-6-phosphate, $^{14}C$ with diaphragm in the bicarbonate-containing medium and measured incorporation of $^{14}C$ into glycogen and lactate. They concluded that "insignificant amounts of glucose-6-phosphate were incorporated into intracellular glycogen," and used this as

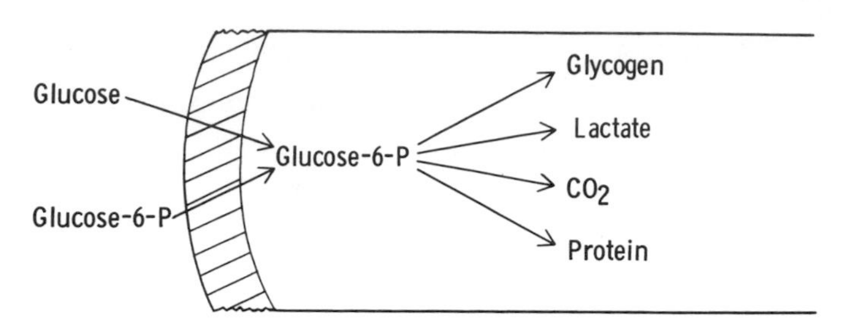


$$\frac{\text{Glucose} \longrightarrow \text{Glycogen}}{\text{Glucose-6-P} \longrightarrow \text{Glycogen}} = \frac{\text{Glucose} \longrightarrow \text{Lactate}}{\text{Glucose-6-P} \longrightarrow \text{Lactate}} = \frac{\text{Glucose} \longrightarrow CO_2}{\text{Glucose-6-P} \longrightarrow CO_2} = \text{etc.}$$

*Fig. 31.3.* Reactions in one compartment. Carbons from glucose going to the products should be in the same proportions as carbons from glucose-6-phosphate.

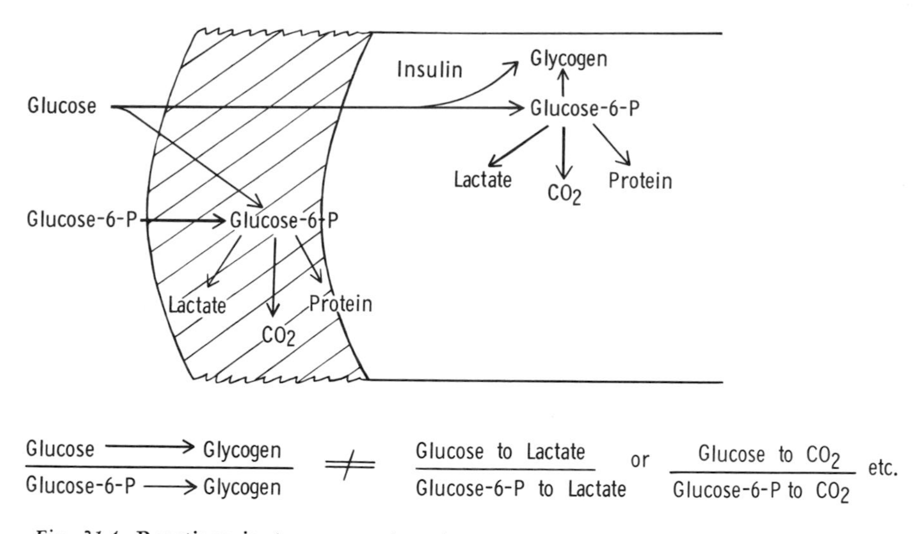


*Fig. 31.4.* Reactions in two compartments.

evidence for the existence of two Embden-Meyerhof pathways in muscle. But the amounts reported were not insignificant. Table 13.1 shows the ratios between the amounts of glucose and glucose-6-phosphate incorporated into glycogen and lactate. The amount incorporated into glycogen from glucose-$^{14}C$ was 66 to 177 times more than the amount incorporated from glucose-6-phosphate,$^{14}C$ but with one exception (177:62 in Experiment 1) for each experiment, in the absence and presence of insulin, there was a similarly greater incorporation into lactate from glucose-$^{14}C$

than from glucose-6-phosphate,$^{14}$C. This would suggest a single Embden-Meyerhof pathway in muscle rather than two pathways.

The design of the experiments for Table 31.1 is not completely adequate for the comparisons just made, since the glucose-$^{14}$C was not incubated in the presence of unlabeled glucose-6-phosphate nor the glucose-6-phosphate,$^{14}$C in the presence of unlabeled glucose. One cannot, therefore, compare the patterns of incorporation in the certain knowledge that results were derived from flasks whose contents were identical in composition except for the substrate labeled. Further, it was not shown definitively that the incorporations were from glucose-6-phosphate,$^{14}$C; they could have been from small quantities of glucose-$^{14}$C present as a contaminant or formed by hydrolysis of the phosphate during the course of incubation.

Landau and Sims (1967) performed experiments with both substrates present in each incubation and alternately labeled. Incubation was for 2 hours. The results are presented in Table 31.2. Glucose was converted preferentially to glycogen as compared to glucose-6-phosphate, relative to their conversion to $CO_2$ and lactate. The ratios for $CO_2$ and lactate are similar, suggesting that they, but not glycogen, were derived from a single pool of glucose-6-phosphate. The preferential conversion of glucose to glycogen means that we cannot accept this experiment as evidence for the existence of a single Embden-Meyerhof pathway in the muscle cell. The failure to observe identical ratios does not, however, prove the existence of two pathways or of a pathway for glucose to glycogen not involving glucose-6-phosphate; several other possibilities exist to explain the ratios. Sims and Landau (1966) and Landau and Sims (1967) attempted to eliminate them but were not able to do so completely.

TABLE 31.1

*Comparison of incorporation by rat diaphragm of $^{14}$C from glucose-$^{14}$C and glucose-6-phosphate, $^{14}$C into glycogen and lactate*

| Ratio | Expt. | Insulin Absent | Insulin Present |
|---|---|---|---|
| Glucose ——→ Glycogen / Glucose-6-P → Glycogen | 1 / 2 | 177 / 66 | 125 / 69 |
| Glucose ——→ Lactate / Glucose-6-P → Lactate | 1 / 2 | 62 / 62 | 116 / 87 |

From Shaw and Stadie (1959); see Table VIII of Landau and Sims (1967).

TABLE 31.2

*Comparison of incorporation by rat diaphragm of $^{14}C$ from glucose-$^{14}C$ and glucose-6-phosphate, $^{14}C$ into various products*

| Ratio | No. of Observations | Insulin Absent | Insulin Present |
|---|---|---|---|
| Glucose ⟶ Glycogen / Glucose-6-P → Glycogen | 5 | 37 (24–49) | 70 (41–117) |
| Glucose ⟶ Lactate / Glucose-6-P → Lactate | 3 | 7.4 (5.8–9.2) | 8.3 (6.8–10.7) |
| Glucose ⟶ $CO_2$ / Glucose-6-P → $CO_2$ | 5 | 8.8 (6.1–11.4) | 9.1 (8.3–10.3) |
| Glucose ⟶ Protein / Glucose-6-P → Protein | 2 | 17 (16–17) | 9 (5–13) |

From Landau and Sims (1967); mean and ranges are recorded.

First, the possibility existed that the glucose-6-phosphate was metabolized in the medium, i.e., that the second pathway was in the medium and as such it was an artifact of the system. We showed that glucose-6-phosphate was not metabolized in a medium which had been incubated with diaphragm and from which the diaphragm was then removed. Possibly metabolism occurred in the extracellular space of the tissue, where enzymes leaching from the cell would be more concentrated and where perhaps other conditions were more suitable for metabolism than in the medium.

Second, since the diaphragms were cut in their removal from the rats, the metabolism of the glucose-6-phosphate might have occurred in the area of the cut edge to which glucose-6-phosphate might have had access; in this area, because of damage to the tissue, glycogen might not have been synthesized. In contrast, glucose would have had access to the whole tissue. This possibility was eliminated by determining the incorporation of $^{14}C$ from glucose-$^{14}C$ and glucose-6-phosphate,$^{14}C$ into glycogen and protein in the cut edge as well as in the central and medial portions of the diaphragm. In all these portions glucose preferentially was incorporated into glycogen as compared to glucose-6-phosphate relative to their incorporation into protein (incorporation into protein rather than lactate was measured since the protein was presumably fixed in location while lactate could have diffused). An intact diaphragm preparation (Kipnis and Cori, 1957), i.e., one in which the diaphragm is attached to the rib cage, was also incubated, and again glucose was preferentially incorporated into glycogen, providing further evidence that cut edges were not responsible for the apparent existence of two pathways.

One of the pathways could possibly have been on the surface of the diaphragm, perhaps in the mesothelial cells covering the muscle cells of the diaphragm. To eliminate this possibility the diaphragm was sectioned in depth. In each section $^{14}C$ from both glucose and glucose-6-phosphate,$^{14}C$ was incorporated, and again glucose was preferentially incorporated into glycogen relative to incorporation into protein. The possibility cannot be eliminated that the glucose-6-phosphate was metabolized in other than muscle cells, e.g., in connective tissue cells distributed throughout the diaphragm.

It was also possible that the appearance of two pathways was created through changes in the tissue during the course of incubation. Thus, if the cell membranes of the diaphragm had been permeable at the beginning of incubation to glucose but not to glucose-6-phosphate, glycogen and lactate would have been formed only from glucose. Then, if toward the end of the incubation, the diaphragm deteriorated and permeability to glucose-6-phosphate developed but glycogen formation declined, lactate would be preferentially formed from the glucose-6-phosphate. This possibility was eliminated by the fact that the ratios of incorporation of $^{14}C$ after 2 hours of incubation were similar to those after 40 and 80 min of incubation.

Landau and Sims (1967), therefore, interpreted their data as evidence for the existence in muscle either of (1) two pools of glucose-6-phosphate, one of which is preferentially converted to glycogen relative to $CO_2$, protein, and lactate, or possibly of (2) a pathway from glucose to glycogen not having glucose-6-phosphate as an intermediate.

Beloff-Chain, Betto, Cantanzaro, Chain, Longinotti, Masi, and Pocchiari (1964) extended their observations of 1953 by incubating diaphragm with $^{14}C$-labeled glucose, glucose-1-phosphate, and glucose-6-phosphate, and measuring incorporation into glycogen, lactate, and $CO_2$ (Table 31.3). A phosphate buffer was used. Glucose and glucose-6-phosphate, each at a concentration in glucose equivalents of 10 mg/ml, were converted to similar extents to lactate. Two to three times as much glucose-6-phosphate as glucose was converted to $CO_2$, but four times as much glucose as glucose-6-phosphate was converted to glycogen in the absence and nine times as much in the presence of insulin. Glucose-1-phosphate was converted to glycogen to about twice the extent of glucose in the absence of insulin and to an equal extent in its presence. It was oxidized to $CO_2$ in twice the amount and converted to lactate at about two-thirds the amount. The much greater conversion of glucose-1-phos-

TABLE 31.3

*Conversion of glucose, glucose-1-phosphate and glucose-6-phosphate to products by rat diaphragm*

| Substrate | Glucose | | Glucose-1-P | | Glucose-6-P | |
|---|---|---|---|---|---|---|
| Insulin | − | + | − | + | − | + |
| To glycogen and oligosaccharides | 236 | 461 | 411 | 413 | 61 | 54 |
| To lactate | 865 | 708 | 559 | 527 | 841 | 843 |
| To $CO_2$ | 125 | 132 | 217 | 220 | 291 | 324 |

From Beloff-Chain, Betto, Cantanzaro, Chain, Longinotti, Masi, and Pocchiari, 1964; results in $\mu$m moles converted by 50 mg of tissue after 90 min incubation.

phate than glucose-6-phosphate to glycogen, relative to their conversion to lactate and $CO_2$, led to the conclusion that this was additional evidence that glucose-6-phosphate was not converted to glucose-1-phosphate and is not on the pathway of glycogen synthesis from glucose. Bicarbonate buffer was not used in these experiments, the substrates were not incubated together alternately labeled, and their concentrations were relatively high.

Smith, Taylor, and Whelan (1967) theorized that glucose-1-phosphate might be the first intermediate in glucose conversion to glycogen through the reaction: glucose + glucose-1,6-diphosphate → 2 glucose-1-phosphate, where glucose-1,6-diphosphate would be formed by the reaction: glucose-1-phosphate + ATP → glucose-1,6-diphosphate + ADP. The sum of these reactions would then be: glucose + ATP → glucose-1-phosphate + ADP.

Pocchiari (1968) reviewed and extended the studies of Beloff-Chain and her collaborators. In Pocchiari (1968) and D'Agnolo, Baroncelli, Betto, Cantanzaro, Longinotti, and Pocchiari (1969) they reported patterns of metabolism similar to those reported by Landau and Sims (1967) on incubating diaphragm with glucose and glucose-6-phosphate alternately labeled and at two of three different concentrations. Thus, under these conditions, while there was preferential incorporation of glucose into glycogen, there was not preferential oxidation of glucose-6-phosphate to $CO_2$ relative to lactate. At the highest concentration—66 mM—of glucose-6-phosphate there was no preferential incorporation of glucose into glycogen. This result is consistent, for this concentration, with a single pool of glucose-6-phosphate, or with access of the exogenous glucose-6-phosphate to both pools. But hydrolysis of glucose-6-phosphate to glucose to even a small percentage, or a small contamination of the glucose-6-phosphate, $^{14}C$ with glucose-$^{14}C$, might be the explanation.

Dully, Bocek, and Beatty (1969) incubated glucose-$^{14}C$ and glucose-6-phosphate,$^{14}C$ with skeletal muscle from monkeys under conditions similar to those of Landau and Sims (1967). Their results and interpretations were essentially the same as had been made for diaphragm. They observed, however, that insulin resulted in a small but significant stimulation of incorporation of $^{14}C$ from glucose-$^{14}C$, but not of incorporation from glucose-6-phosphate, $^{14}C$ into $CO_2$ and lactate. Small increases in incorporation from glucose-$^{14}C$ observed by Landau and Sims (1967) were not statistically significant. Dully, Bocek, and Beatty (1969) also observed that one-fifth to one-eighth the quantity of $^{14}C$ from glucose-1-phosphate,$^{14}C$ as from glucose-$^{14}C$ was incorporated into glycogen. This is in contrast to the results of Beloff-Chain, Betto, Cantanzaro, Chain, Longinotti, Masi, and Pocchiari (1964) using rat diaphragm in which, as already noted, incorporation from glucose-1-phosphate,$^{14}C$ was at least as much as from glucose-$^{14}C$ and much better than from glucose-6-phosphate,$^{14}C$. The latter authors compared glucose with the hexose phosphates at the same concentration in a phosphate buffer, while Dully, Bocek, and Beatty (1969) incubated with glucose at five times the concentration of the hexose phosphates and in a bicarbonate buffer.

Before further examination of the data for muscle, an account of the evidence for a pathway in liver for the conversion of glucose to glycogen, not involving glucose-6-phosphate, is relevant.

Niemeyer and his coworkers in 1956 showed that in liver slices glucose is a better precursor of glycogen than either glucose-6-phosphate or glucose-1-phosphate. They considered (Figueroa, Niemeyer, and Gonzalez, 1956) that this might indicate that glucose did not have these phosphates as intermediates, but they recognized that these observations could simply indicate a greater permeability of the slices to glucose than to the phosphates. Figueroa, Pfeifer, and Niemeyer (1962), using a liver homogenate to eliminate permeability as a factor, found that glucose-6-phosphate did not inhibit, nor the addition of hexokinase stimulate, the incorporation of $^{14}C$ of glucose-$^{14}C$ into glycogen. The glucose-6-phosphate did decrease the formation of $^{14}CO_2$. They considered that, if the pathway of metabolism were glucose $\rightarrow$ glucose-6-P $\begin{smallmatrix} \nearrow \text{glycogen} \\ \searrow CO_2 \end{smallmatrix}$, then the addition of unlabeled glucose-6-phosphate, taking into consideration the concentration at which it activates glycogen synthetase, should have so diluted the pool of $^{14}C$-labeled glucose-6-phosphate formed from glucose-$^{14}C$, as to decrease incorporation into $CO_2$ and glycogen. They therefore tentatively concluded that glucose $\rightarrow$ glucose-6-phosphate was not a necessary reaction in glycogen formation.

Hers, Larner, and Brown (1964) discussed experiments similar to those of Figueroa, Pfeifer, and Niemeyer (1962); they considered, however, that the results were due to an exchange reaction catalyzed by amylo-1,6-glucosidase: glucose-$^{14}C$ + glycogen → glucose + glycogen-$^{14}C$. Hue and Hers (1969) confirmed and extended these observations, while Brown (Hers, Larner, and Brown, 1964) noted that in addition to the exchange there might also be a net synthesis of glycogen through the formation of new 1,6 branch points.

Figueroa and Pfeifer (1964) subsequently reported that while liver slices formed $^{14}CO_2$ in greater quantity from glucose-6-phosphate-$^{14}C$ than from glucose-$^{14}C$, glycogen-$^{14}C$ was formed in smaller quantity. These results are similar to those just described for muscle (Beloff-Chain, Betto, Cantanzaro, Chain, Longinotti, Masi, and Pocchiari, 1964).

London (1966) set up a mathematical model of hepatic glycogen metabolism and on inserting experimentally determined values for the concentrations of the intermediates and the kinetic constants of the enzymes, concluded that the assumption that the reactions occur in a homogeneous phase without compartments is not a good one. He suggested the possibility of two pools of glucose-6-phosphate, a suggestion also made by Segal and Lopez (1963).

Threlfall (1966) reported the results of experiments in which he injected glucose-$^{14}C$ into fed rats and then rapidly removed their livers into liquid nitrogen. The specific activities of glucose-6-phosphate and UDP-glucose in the livers were then determined. The specific activity of the UDP-glucose was greater than that of glucose-6-phosphate, leading to the conclusion that glucose-6-phosphate was not an obligatory intermediate in glycogen formation. The possibility of two pools of glucose-6-phosphate as an alternative was considered. The specific activities in the liver of glycerol phosphate and lactate were found to be only slightly lower than that of glucose-6-phosphate, so the pool of glucose-6-phosphate from which the UDP-glucose was derived would have had to be small relative to the pool from which the glycerol phosphate and lactate were derived. Livers from fasted rats and samples of livers excised from anesthetized rats were examined to test the possibility that glycogen breakdown during the course of killing the rats and removing their livers lowered the specific activity of the glucose-6-phosphate. Specific activities were similar to those observed in the fed rats.

Threlfall and Heath (1968) studied the specific activities of glucose, UDP-glucose, glycerol phosphate, and glucose-6-phosphate during a 35-min period following the administration of fructose-$^{14}C$ to rats. From kinetic analysis of the data they concluded that glycolysis and gluconeogenesis are compartmentalized in rat liver. A major observation upon which this conclusion is based is that the specific activity of glucose-6-phosphate was less than that of glucose during the 20–35-min period (glucose-6-phosphate: glucose was from 0.47–0.51). Muntz and Vanko (1962) made a similar observation and concluded from it that glucose-6-phosphate is not a necessary intermediate in the conversion of fructose to glucose. They had, however, isolated the glucose-6-phosphate after washing the liver briefly with saline and blotting and weighing it before homogenization; during this time glycogen breakdown could have occurred. In the experiments of Threlfall and Heath (1968) considerable effort was made to eliminate glycogen breakdown as an explanation for the observation.

Threlfall and Heath (1968) did not consider whether or not their observations represented metabolism in a single cell type, since several types exist in liver, or metabolism in the overall liver. Even if the compartmentation were in a single cell type, the cells of this type could be metabolically nonhomogeneous, e.g., through exposure to differing environments, so that glycolysis could have proceeded in one cell simultaneously with gluconeogenesis in another. Hepatic cells first presented with the blood from the portal vein and hepatic artery are exposed to a higher oxygen content than the hepatic cells in the region of the central vein. Differences in glycogen content and response of hepatic cells to fasting have been reported to be a function of the position of the cells in the liver lobule (Maximow and Bloom, 1949). Thus the data of Threlfall and Heath (1968) do not allow the assignment of compartmentation to the hepatic cell.

Hue and Hers (1969) examined each step in the conversion of glucose to glycogen as it occurs in mouse liver homogenate. They could find no evidence for formation of glucose-1-phosphate other than via glucose-6-phosphate, and they demonstrated that glucose-6-phosphate was a necessary intermediate in the conversion of glucose to UDP-glucose.

Antony, Srinivasan, Williams, and Landau (1969) employed another approach to determine the existence of either two pools of glucose-6-phosphate or a pathway from glucose to glycogen not involving glucose-6-phosphate. Liver slices were incubated with pyruvate 1 $^{14}C$ and glucose-6-$^{14}C$. Slices were used so that the liver cells were exposed to a relatively uniform environment and could be frozen in liquid nitrogen within 5 sec. Pyruvate-1-$^{14}C$ in its conversion to glucose-6-phosphate and the glucose unit of glycogen should label carbons 3 and 4 of these compounds (Fig. 31.5). Glucose-6-$^{14}C$ should of course label carbon 6. The glucose from glycogen and glucose-6-phosphate was isolated from the slices and the activities in their individual carbons determined. Relative to the activity in carbon 4, there was as much activity in carbon 6 of glucose-6-

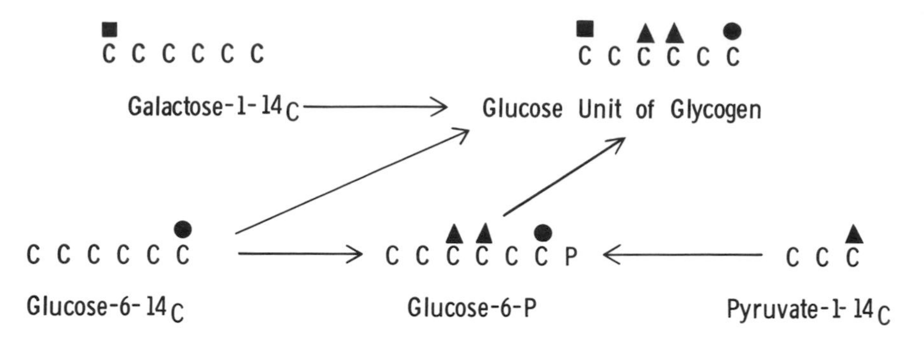


*Fig. 31.5.* Labeling pattern in glucose-6-phosphate and the glucose unit of glycogen formed from galactose-1-$^{14}$C, glucose-6-$^{14}$C, and pyruvate-1-$^{14}$C.

phosphate as there was in glycogen (Table 31.4, ratios 0.97 and 0.78). Galactose-1-$^{14}$C was actually incubated simultaneously with the glucose-6-$^{14}$C and pyruvate-1-$^{14}$C (Fig. 31.5). Relative to the activity in carbon 4 there was greater incorporation of galactose carbon, measured by incorporation into carbon 1, into glucose of glycogen than into glucose-6-phosphate (Table 31.4, ratios 2.39 and 1.75). Since galactose incorporation into glycogen proceeds via UDP-glucose and not glucose-6-phosphate, this shows that under the conditions of the experiment a pathway to glycogen in which glucose-6-phosphate is not an intermediate could be detected. Assuming glucose-6-phosphate is an obligatory intermediate in the conversion of pyruvate to glycogen, there was then no preferential incorporation of glucose carbon into glycogen compared to glucose-6-phosphate carbon. This is good evidence that glucose conversion into glycogen has glucose-6-phosphate as an intermediate to at least the same extent as pyruvate. Therefore, there is no evidence for two pools of glucose-6-phosphate in liver from these experiments, and the evidence is against a pathway not involving glucose-6-phosphate for glucose conversion to glycogen.

In contrast to the results with liver slices, similar incubation of diaphragm resulted in a preferential incorporation of glucose carbon relative to pyruvate carbon into glycogen, compared to glucose-6-phosphate (Table 31.4, ratios 1.60 and 1.68). It was possible that more glucose carbon than pyruvate was deposited in glycogen at an early period in the incubation, because pyruvate carbon must traverse a larger number of intermediates than glucose carbon in its conversion to glycogen. The distribution in glycogen might, therefore, represent deposition over the entire incubation period, while the distribution in glucose-6-phosphate was that in a single pool at the time the incubation was terminated.

Incubations of 30 and 60 min duration were done in three separate experiments. In two, the preferential incorporation of glucose into glycogen occurred through the 30–60-min period, but in one it did not.

Possibly the preferential incorporation into glycogen occurred as suggested for the experiments of Figueroa, Pfeifer, and Niemeyer (1962), through an exchange reaction between glucose and glycogen catalyzed by amylo-1,6-glucosidase. This appears unlikely: first, in the presence of insulin as employed in these experiments the quantity of glucose-$^{14}$C incorporated into glycogen is relatively large, so that the exchange reaction would have to be very active; second, preferential incorporation did not occur in liver slices, where the enzyme would also be expected to be active (though its activity in liver and muscle might be different as a function of the conditions the effect in liver, but not in muscle, being obscured). With regard to this second point, the concentration of free glucose available to participate in the exchange in muscle might be expected to be lower than in liver (Williams, Exton, Park, and Regan, 1968). These data, then, while not conclusive, further suggest the existence of two pools of glucose-6-phosphate in muscle or a pathway from glucose to glycogen not involving glucose-6-phosphate.

When muscle is incubated with high concentrations of glucose in the absence of insulin, so that the utilization of glucose is the same as it would be with a lower glucose concentration in the presence of insulin,

TABLE 31.4

*Ratio of incorporation of $^{14}$C of glucose-$^{14}$C and galactose-$^{14}$C into glycogen to that into glucose-6-phosphate, relative to the incorporation of $^{14}$C of pyruvate-1-$^{14}$C into glycogen and glucose-6-phosphate*

| Tissue preparation | Incubation time (min) | Glucose-$^{14}$C | Galactose-$^{14}$C |
|---|---|---|---|
| Liver slices | 45 | 0.97 (0.94–0.99) | 2.39 (1.90–2.87) |
| | 90 | 0.78 (0.52–0.95) | 1.75 (1.34–2.49) |
| Diaphragm | 30 | 1.60 (1.17–2.22) | |
| | 60 | 1.68 (1.66–1.71) | |

From Anthony, Srinivasan, Williams, and Landau, 1969; mean and range are recorded.

the glucose in the absence of insulin is converted primarily to lactate; in the presence of insulin it is converted to glycogen (Larner, Villar Plasi, and Richman, 1959). If insulin simply increased glucose entrance into muscle (Levine, 1966), this difference should not exist. It would be explicable if, in the absence of insulin, glucose is utilized primarily by an insulin-nonresponsive pathway in which glucose is converted to lactate, and in the presence of insulin by the glycogen-forming insulin-responsive pathway. It also seems reasonable that when, during feeding, blood glucose concentration increases and insulin is released from the pancreas, the glucose should be directed to storage rather than to glycolysis. In exercise, glucose, perhaps through an exercise factor (Havivi and Wertheimer, 1964) would be expected to be preferentially directed to the glycolytic pathway.

There is reason to believe that glycogen is formed in the area of the sarcotubules, since on fractionation of muscle, glycogen synthetase is found in large degree in a fraction containing the sarcotubules (Andersson-Cedergren and Muscatello, 1963; P. Bailey, unpublished observations). Adenyl cyclase is also associated with this fraction (Rabinowitz, Desalles, Meisler, and Lorand, 1965; L. Mandel, unpublished observations) as are the hexokinase enzymes that have been designated types I and II (Katzen, 1967, and unpublished observations). Hexokinase II as well as glycogen synthetase are less than normal in the sarcotubular fraction isolated from muscle of rats with streptozotocin-induced diabetes mellitus (P. Bailey and H. M. Katzen, unpublished observations).

Sims and Landau (1966) speculated that insulin may act on the sarcotubular membrane and not on the remaining membrane of the cell (Fig. 31.6). Insulin might affect the T system in a manner which could increase the access of glucose to the sarcoplasmic reticulum. Such a scheme seems in accord with all observations made thus far. Its most important aspect is that insulin would increase glucose transport into the environment of the cell, where its conversion to glycogen would be favored. $CO_2$ and lactate would be derived from a single pool, in keeping with the similar ratios for lactate and $CO_2$ observed by Dully, Bocek, and Beatty (1969), Pocchiari (1968), and Landau and Sims (1967) in their experiments (see Table 31.2). The small increases produced by insulin in the conversion of glucose to lactate that have been observed would be explainable either through the breakdown of some of the glycogen with subsequent conversion to lactate or by some mixing of the glucose-6-phosphate pools. Under conditions where glycogen breakdown is accelerated, as in epinephrine action, conversion to lactate would be increased through this scheme. The mechanism by which insulin acts, whether on

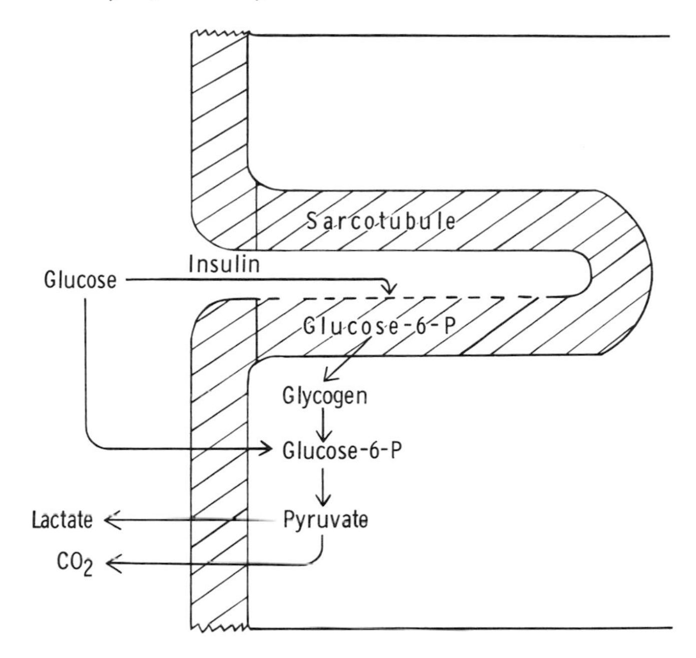


*Fig. 31.6.* Proposed model of glucose metabolism in rat diaphragm.

adenyl cyclase, a carrier, etc., would then be an extension of these speculations.

An alternative to the scheme proposed in Fig. 31.6 would show the conversion of glucose to glycogen without the involvement of glucose-6-phosphate as an intermediate. Insulin would then in some manner both have to increase the transport of glucose into the cell and also favor the conversion of the glucose into glycogen.

Thus, many investigators have made efforts to establish whether or not in the muscle cell (as well as the liver cell) there is compartmentation of the pathways of carbohydrate metabolism or a pathway from glucose to glycogen not requiring glucose-6-phosphate as an intermediate. Thus far none of the results allows differentiation between these two possibilities nor conclusively establishes the existence of at least one of them. The evidence suggests that either or both may exist in muscle; if this is the case, it could have considerable importance for our understanding of the mechanism by which alterations in carbohydrate metabolism in muscle occur as a consequence of changes in hormonal, dietary, and other conditions.

## *References*

Andersson-Cedergren, E., and U. Muscatello. 1963. The participation of the sarcotubular system in glycogen metabolism. *Ultrastructure Res. 8:*391.

Antony, G. J., I. Srinivasan, H. R. Williams, and B. R. Landau. 1969. Studies on the existence of a pathway in liver and muscle for the conversion of glucose into glycogen without glucose-6-phosphate as an intermediate. *Biochem. J. 111:*453.

Beloff-Chain, A., P. Betto, R. Cantanzaro, E. B. Chain, L. Longinotti, I. Masi, and F. Pocchiari. 1964. The metabolism of glucose-1-phosphate and glucose-6-phosphate and their influence on the metabolism of glucose in rat diaphragm muscle. *Biochem. J. 91:*620.

Beloff-Chain, A., R. Cantanzaro, E. B. Chain, I. Masi, F. Pocchiari, and C. Rossi. 1955. The influence of insulin on carbohydrate metabolism in the isolated diaphragm muscle of normal- and alloxan-diabetic rats. *Proc. Roy. Soc. B. 143:*481.

Beloff-Chain, A., E. B. Chain, D. Bovet, F. Pocchiari, R. Cantanzaro, and L. Longinotti. 1953. Metabolism of hexose phosphate esters. *Biochem. J. 54:*529.

D'Agnolo, G., V. Baroncelli, P. Betto, R. Catanzaro, L. Longinotti, and F. Pocchiari. 1969. Glucose-6-phosphate pools in isolated rat diaphragm. *Experientia 25:*697.

Dully, C. C., R. M. Bocek, and C. H. Beatty. 1969. The presence of two or more glucose-6-phosphate pools in voluntary skeletal muscle and their sensitivity to insulin. *Endocrinology 84:*855.

Figueroa, E., H. Niemeyer, and C. Gonzalez. 1956. Influence of DL-glyceraldehyde and of sorbose-1-phosphate on glycogen synthesis from glucose by rat liver slices. *Acta Physiol. Latino-Amer. 6:*112.

Figueroa, E., and A. Pfeifer. 1964. Incorporation of $^{14}C$-glucose and $^{14}C$-glucose-6-phosphate into glycogen and $CO_2$ by rat liver slices. *Nature 204:*576.

Figueroa, E., A. Pfeifer, and H. Niemeyer. 1962. Incorporation of $^{14}C$-glucose into glycogen by whole homogenate of liver. *Nature 193:*382.

Havivi, E., and H. E. Wertheimer. 1964. A muscle activity factor increasing sugar uptake by rat diaphragms *in vitro. J. Physiol. 172:*342.

Hers, H. G., J. Larner, and J. Brown. 1964. Discussion, p. 81. *In* W. J. Whelan and M. P. Cameron (eds.), *Ciba Foundation Symposium: Control of Glycogen Metabolism.* Little, Brown, Boston.

Hue, L., and H. G. Hers. 1969. A reevaluation of the pathway by which glucose is converted into glycogen in a liver homogenate. *FEBS Letters 3:*41.

Katzen, H. M. 1967. The multiple forms of mammalian hexokinase and their significance to the action of insulin, p. 335. *In* G. Weber (ed.), *Advances in Enzyme Regulation,* Vol. 5. Pergamon, New York.

Kipnis, D. M., and C. F. Cori. 1957. Studies of tissue permeability III. The effect of insulin on pentose uptake by the diaphragm. *J. Biol. Chem. 224:*681.

Landau, B. R., and E. A. H. Sims. 1967. On the existence of two separate pools of glucose-6-phosphate in rat diaphragm. *J. Biol. Chem. 242:*163.

Larner, J., C. Villar Plasi, and D. J. Richman. 1959. Insulin-stimulated glycogen formation in rat diaphragm. *Ann. N. Y. Acad. Sci. 82*:345.

Levine, R. 1966. The action of insulin at the cell membrane. *Amer. J. Med. 40*:691.

London, W. P. 1966. A theoretical study of hepatic glycogen metabolism. *J. Biol. Chem. 241*:3008.

Maximow, A., and W. Bloom. 1949. *A Textbook of Histology* (5th ed.), p. 430. W. B. Saunders Co., Philadelphia.

Muntz, J. A., and M. Vanko. 1962. The metabolism of intraportally injected fructose in rat liver *in vitro*. *J. Biol. Chem. 237*:3582.

Pocchiari, F. 1968. Initial stages in the conversion of glucose into glycogen, p. 129. *In* W. J. Whelan (ed.), *Control of Glycogen Metabolism*. Academic Press, New York.

Rabinowitz, M., L. Desalles, J. Meisler, and L. Lorand. 1965. Distribution of adenyl-cyclase activity in rabbit muscle fractions. *Biochim. Biophys. Acta 97*:29.

Segal, H. L., and C. G. Lopez. 1963. Early effects of glucocorticoids on precursor incorporation into glycogen. *Nature 200*:143.

Shaw, W. N., and W. C. Stadie. 1957. Coexistence of insulin-responsive and insulin-non-responsive glycolytic systems in rat diaphragm. *J. Biol. Chem. 227*:115.

— —. 1959. Two identical Embden-Meyerhof enzyme systems in normal rat diaphragms differing in cytological location and response to insulin. *J. Biol. Chem. 234*:2491.

Sims, E. A. H., and B. R. Landau. 1966. Insulin responsive and non-responsive pools of glucose-6-phosphate in diaphragmatic muscle. *Fed. Proc. 25*:835.

Smith, E. E., P. M. Taylor, and W. J. Whelan. 1967. Hypothesis on the mode of conversion of glucose into α-glucose-l-phosphate. *Nature 213*:733.

Threlfall, C. J. 1966. Role of glucose-6-phosphate in diaphragmatic muscle. *Nature 211*:1192.

Threlfall, C. J., and D. F. Heath. 1968. Compartmentation between glycolysis and gluconeogenesis in rat liver. *Biochem. J. 110*:303.

Villee, C. A., and A. B. Hastings. 1949. The metabolism of $C^{14}$-labeled glucose by the rat diaphragm *in vitro*. *J. Biol. Chem. 179*:673.

Williams, T. F., J. H. Exton, C. R. Park, and D. M. Regan. 1968. Stereospecific transport of glucose in the perfused rat liver. *Amer. J. Physiol. 215*:1200.

# Chapter 32

# Summary and Discussion of Part 7

PANEL MEMBERS: H. A. LARDY, *Chairman*
E. ALLEN
R. G. KAUFFMAN
J. W. PORTER
H. J. SALLACH

*Lardy:* The energy support systems of muscle cells influence the color, texture, and composition of this tissue. They have, therefore, a bearing on the practical aspects of meat production as well as on the fundamental questions concerning regulation of biochemical processes in contractile cells.

Both Dr. Havel and Dr. Issekutz described the transition from early concepts emphasizing the glycogen-lactate system as an energy source for muscle contraction to the realization that fat contributes much more of the energy under ordinary circumstances and nearly all of the energy in fasting and diabetic subjects. Dr. Havel described the processes that provide fatty acids to muscle and the regulation of fatty acid release from adipose tissue by hormones such as the catechol amines and somatotropin through their "second messenger" 3′,5′-cyclic AMP. Dr. Issekutz made mention of these phenomena and also emphasized the importance of insulin in controlling circulating free fatty acid concentrations.

Dr. Havel emphasized that within muscle cells there are again elaborate mechanisms for regulating fatty acid oxidation. These involve reesterification to neutral fat which may be temporarily stored intracellularly, activation of fatty acids by conversion to the corresponding acyl CoA in the cytosol, transacylation with carnitine—the fatty acid carrier across the mitochondrial membrane—and transfer back to CoA in the mitochondria. The ultimate regulation of oxidation rate in the mitochondria is mediated by the energy demand on the cell through its influence on the ratio ATP:ADP + $P_i$.

Dr. Havel pointed out that muscles that work on a sustained basis have been demonstrated to be omnivorous. Their selection of materials for fuel depends on the many physiological factors that regulate the availability of glucose, fat, amino acids, lactate, etc. Both Havel and Issekutz indicated that skeletal muscles that may work rapidly, and for relatively brief periods, depend to a considerably greater extent on stored glycogen as a source of energy. The lactate produced by such muscles may provide fuel for the muscles working on a more sustained basis, as well as supplying a carbon source for gluconeogenesis in liver and kidney.

In general, the regulatory mechanisms seem to adapt to dietary changes that vary the intake of different classes of foodstuffs. Dr. Issekutz pointed out "that the regulation of the metabolism during exercise seems to be aimed at maintaining the blood sugar as long as possible at a level acceptable to the central nervous system."

Dr. Landau addressed himself to the frequently raised possibility of more than one pool of glucose-6-phosphate in muscle cells or the possibility that glycogen may be formed from glucose by a path not involving glucose-6-phosphate. He carefully examined the relevant experiments in the literature and disclosed some errors of interpretation of the data by the collectors. The sarcotubular system of muscle provides somewhat greater opportunity for compartmentation than is the case with many other mammalian tissues. After all these considerations, Landau concluded that a choice cannot yet be made between the existence of two pools of glucose-6-phosphate in muscle or the existence of a pathway from glucose to glycogen not involving the Robison ester.

## DISCUSSION

*Scopes:* Is there any real evidence that glucose can enter a muscle cell without getting phosphorylated? Having got in (if it can do so), can it then become phosphorylated at all readily?

*Landau:* Without added insulin and at glucose concentrations generally encountered in the blood, the muscle cell membrane is usually the major rate-limiting step in glucose uptake, and free glucose will not accumulate intracellularly. At higher concentrations phosphorylation can become limiting and intracellular free glucose can accumulate (see Morgan, Henderson, Regen, and Park, 1961). Specifically in regard to our observation, there is evidence for free glucose accumulating in diaphragm incubated at very high glucose concentrations, as 2,000 mg/100 ml, in the presence of insulin (see Park, Bornstein, and Post, 1955). At a concentration of 500 mg/100 ml the free glucose content of muscle was low and not measurably changed by insulin addition. In our experiments diaphragms were incubated with glucose at a concentration of 180 mg/100 ml.

*Wood:* Am I right in saying that, if Dr. Landau's explanation is correct, this differential effect between glucose and glucose-6-phosphate should be abolished if a muscle homogenate is used? Has he carried out any experiments on these lines?

*Landau:* Since in our explanation access is assumed of exogenous glucose-6-phosphate to one pool of glucose-6-phosphate and not another (the second associated with glycogen formation and speculatively with the sarcotubular system), the differential effect would not be expected, if in the homogenate glucose-6-phosphate had access to both these pools. I would expect that you are correct and this would be the case in an homogenate. We have not carried out experiments with muscle homogenates. I believe the study would be complicated by the little (if any) glycogen synthesis or effect of insulin added *in vitro* likely to occur using an homogenate, and the hydrolysis of the glucose-6-phosphate occurring due to the presence of nonspecific phosphatases. If an effect were observed I think it could be attributed, with greater justification than for the intact tissue, to exchange reactions such as that catalyzed by the amylo-1,6-glucosidase.

*Sallach:* I have a question about the experiments that utilize either cold glucose and hot glucose-6-phosphate or vice versa. I do not know what the transport or uptake of the phosphorylated esters may be, but in the case of the labeled glucose, if cold glucose-6-phosphate were taken up, would this inhibit the hexokinase sufficiently so that you would "force" labeled glucose into glycogen via the 1–6 glucosidase exchange referred to by Hers, Larner, and Brown (1964)?

*Landau:* When one has this paired flask situation, hexokinases in both flasks have to be at the same activity. When you are measuring whether the $^{14}C$ is from glucose or from glucose-6-phosphate, you are going to have the same size pool of glucose-6-phosphate though it may be of different specific activity. So far as utilization is concerned, the quantity of glucose-6-phosphate that is utilized is about 10% of the material added. We do not know the intracellular concentration of glucose-6-phosphate.

*Sallach:* My point is that if you were inhibiting the hexokinase then more $^{14}C$ glucose, *per se,* would have accumulated inside the cell to undergo the exchange. It would not be phosphorylated by the hexokinase which might be inhibited; therefore there might be more to undergo the exchange and this might give you an abnormal incorporation into glycogen from the free nonphosphorylated glucose-$^{14}C$. Have there been any experiments in that connection? I realize that you degraded glycogen to get individual carbon labeling, but have there been any enzymic degradations of glycogens that have been isolated that would, perhaps, indicate whether you were dealing with a 1–6 or 1–4 link?

*Landau:* We have not answered the question of a 1–4 or 1–6 link in our experiments. I think the glucosidase reaction is certainly a possibility for the preferential incorporation of glucose into glycogen. The arguments against it in muscle are as follows. First of all, if this were so it would mean a tremendously large exchange, because of the quantity of glucose that is going to glycogen preferentially. It represents at least 25% of the glucose being taken up and a good quantity of glucose is utilized by diaphragm in the presence of insulin. The second thing that can be said is that the quantity of glucose that is usually found free in diaphragm, or within muscle in general, is relatively small as compared with that in liver, which is freely permeable. At the same time one does not find this preferential incorporation in liver. Now if one were to say that the conditions were different so that the 1–6 glucosidase was behaving differently in the two tissues, and this is why you are getting exchange in one and not in the other, this would still remain a possibility.

*Sallach:* Yes, you could be dealing with a glucokinase in liver. I do not know the nutritional status of the animals that were used in the liver experiments. In the interest of keeping metabolism simpler, I would prefer to have fewer pathways.

*Landau:* I would much prefer that too. Also, we have made quantitations of pathways in muscle (Green and Landau, 1965), which become very difficult to evaluate when there are two Embden-Meyerhof pathways.

*Porter:* What information is available on the activation and inhibition of partially purified adipose tissue lipase?

*Havel:* The enzyme has been difficult to purify, but partially purified preparations are activated by cyclic-AMP. Whether all hormones which rapidly activate or inhibit the lipase operate through the adenyl cyclase system is uncertain.

*Sallach:* Is control by hormones and substrate specificity of cardiac lipase similar to adipose lipase?

*Havel:* This is an important but unresolved question. In perfused hearts, release of glycerol (presumably from stored triglyceride) increases with fasting and in presence of diabetes mellitus, suggesting that the two enzymes may be activated under similar circumstances.

*Allen:* In red-meat animals there is a large variation between animals as to the quantity of intramuscular lipid in any particular muscle. For example, it would not be unusual to find a 10–15-fold difference in intramuscular lipid content of the longissimus dorsi muscle of two bovine animals with a similar nutritional history. Would you suspect that some of this variation may be due to differences in the activities of lipoprotein lipase or hormone-sensitive lipase in the muscle?

*Havel:* Enzymic differences could certainly be a factor.

*Kauffman:* The statement was made that fatty acids were transported across the membrane in a passive fashion. How would you explain this in terms of the role of insulin and other substances that may suggest an active transport?

*Havel:* The evidence for facilitated diffusion is based upon studies of tissues *in vitro* and relates to effects of temperature, etc. There is no direct effect of insulin on uptake of FFA by any tissue that we know anything about. The extraction fraction of FFA *in vivo* in the splanchnic region and heart is not affected by insulin. Insulin inhibits the hormone-sensitive lipase by a mechanism which is independent of its effect on glucose transport. This is clearly defined and requires very low levels of the hormones.

*Kauffman:* How do you explain the quite large differences in the uptake and deposition of FFA in different types of muscle?

*Havel:* I think I indicated in my presentation that there may be a flow limitation. If you are referring to effects such as Carlson, Liljedahl, and Wirsén (1965) produced when they increased fat mobilization and found greater triglyceride deposition in red than in white fibers, there could, of course, be enzymic differences which could be a factor. In addition, as we heard earlier this week, the blood flow to red fibers even at rest would be expected to be much more effective since the fibers often are much smaller and are more completely surrounded by capillaries.

*Sallach:* Is muscle lipase different from that of adipose lipase in that the muscle enzyme is apparently not inhibited by nicotinate?

*Issekutz:* Carlson and Hanngren (1964) have shown that nicotinate accumulated in the adipose tissue, and that the triglyceride content of the red fibers of rats was decreased by nicotinate (Carlson, Fröberg, and Nye, 1966). The authors concluded that local lipid stores play an important part in oxidative metabolism when the flux of FFA is inhibited. The lack of a lipase-inhibitory effect in the muscle may be due to the elective distribution of nicotinate in the body.

*Allen:* Since most of the lipids in muscle are esterified, I would like to know how sodium nicotinate would distinguish between one lipase and another. In other words, what is the comparison between mobilization of intramuscular lipid versus that outside the muscle?

*Issekutz:* Carlson and Hanngren (1964) were the first to discover the effect of nicotinate on the adipose tissue lipase by showing the distribution of tritium-labeled nicotinate in the mouse. He found a great accumulation in the adipose tissue, but almost none elsewhere. Apparently it does not enter the muscle so fast, and that may be the explanation.

*Kauffman:* Is there any information about the utilization of fatty acids (present in muscle) as a source of contractile energy, that would shed light on lipid transport into and out of muscle?

*Issekutz:* I have to refer to the part of Chapter 30 dealing with the available evidence related to the possible utilization of intramuscular lipids. I am not aware of any detailed study describing the mechanism by which FFA are transported across the membrane of the skeletal muscle, and I do not know of any evidence that under physiological conditions lipids would be released from the muscle into the circulation.

*Allen:* Do you have any information on whether or not muscles that are trained for extended exercise show an increase in intramuscular lipid during the training period?

*Issekutz:* The only information I know of comes from the observation made by Greene (1913) on the King salmon, the "dark" muscle of which can store large quantities of fat at the beginning of spawning migration. The possible effect of training on intramuscular lipid is certainly an interesting problem worth studying.

*Allen:* Could you offer any reason why the liver and adipose tissue respond in such a sluggish manner to short, heavy exercise? Is there an appreciable increase in synthesis of certain enzymes such as the hormone-sensitive lipase during this period of time?

*Issekutz:* It is, of course, possible that there is an increase in the synthesis of certain enzymes as exercise proceeds, but it is equally possible that it requires some time to bring about the necessary changes in hormonal balance, norepinephrine, growth hormone, etc.

*Barnard:* You stated that glucose permeability did not change in muscle during exercise. Might it be that your work load was not severe enough to need blood glucose as a fuel?

*Issekutz:* What I meant was that in the pancreatectomized dog exercise did not increase the glucose permeability of the muscle enough to utilize a blood sugar of about 400 mg %. Despite this high concentration, the contribution of plasma glucose to the $CO_2$ output was less than in the normal dog exercising at a blood glucose level of 80–90 mg % and at the same energy expenditure.

*In vivo,* in the running dog, one cannot increase the metabolic rate of the muscle to the same extent as under *in vitro* conditions by electrical stimulation. In this latter case, activity had an insulin-like effect (Sacks and Smith, 1958). In addition, in the pancreatectomized dog the highly elevated FFA may inhibit the glucose transfer across the muscle membrane (Randle, Newsholme, and Garland, 1964).

*Porter:* You mentioned your results on the decrease in insulin of plasma during exercise. Do you know what happens to the insulin? What are the relative rates of oxidation of glucose and palmitic acid by *in vitro* preparations from muscle?

*Issekutz:* The decrease of plasma insulin level during exercise shows only that the rate of release is less than the rate of elimination of the insulin. Only a turnover study with labeled insulin could answer your first question. As was pointed out, there are two factors which may indicate that the rate of release of insulin is depressed during exercise: (a) despite the elevated hepatic glucose output, the plasma glucose level decreases, and (b) physical work increases the norepinephrine concentration in the plasma (Vendsalu, 1960), which in turn may decrease the plasma insulin level (Porte and Williams, 1966). Concerning your second question, I have to point out that the relative rates of *in vitro* oxidation of glucose and palmitate depend on the composition of the medium. There is now rather convincing evidence showing that if both FFA and glucose are present in the medium, the heart and the hemidiaphragms prefer the former as a fuel (Shipp, Opie, and Challoner, 1961), and palmitate in physiological concentration actually inhibits the glucose uptake (Randle, Newsholme, and Garland, 1964). During prolonged exercise, the three- to-fourfold-elevated FFA level may be responsible for the predominantly fat metabolism.

*Cassens:* We (Cassens, Bocek, and Beatty, 1969) have studied the effect of glucose–fatty acid interaction in *in vitro* preparations of striated skeletal muscle of the rhesus monkey. Octanoate had no effect on glucose uptake but caused a marked increase in oxygen consumption and $CO_2$ production. Incubation of muscle fibers in the presence of octanoate-1-$^{14}C$ demonstrated that increasing the octanoate concentration increased both uptake and oxidation of this fatty acid.

*Allen:* Do you have any information relating to the halflife of a particular fatty acid in the muscles of dogs or other experimental animals during periods of rest and exercise?

*Issekutz:* No, I do not have any information in this respect. Dr. Havel carried out detailed kinetic analyses with the "pulse-labeling" technique on man. Perhaps he could tell more.

*Havel:* Our studies in intact humans and rats indicate that a large fraction of FFA leaving the blood in working muscles is oxidized in a minute or so; this fraction is much smaller at rest, and a correspondingly larger fraction enters pools of lipid esters which turn over very much more slowly.

*Allen:* Does physical training cause an increase in the total capillary network of the muscle?

*Issekutz:* It is generally believed that training increases the capillarization of the muscle and this may explain the fact that during exercise at about the same $O_2$ uptake, the blood lactate level rises much less in trained individuals than in untrained persons (Cobb and Johnson, 1963).

*Allen:* Is there a greater uptake of fatty acids into active muscle fibers than into inactive fibers in the same muscle? Could uptake of labeled palmitate be used to identify active fibers radio-autographically?

*Issekutz:* I do not think that such a comparison between inactive and active fibers was ever made. I agree that autoradiographic investigations carried out on small animals with sufficient amounts of radiopalmitate could produce very interesting findings.

*Waldman:* Is there evidence which indicates a selective use by cardiac or skeletal muscle of fatty acids in regard to chain length or level of unsaturation?

*Havel:* Various investigators are not in complete agreement, but studies with labeled fatty acids have failed to show substantial differences in these respects.

*Barnard:* I noticed that your subject with McArdle's syndrome developed a fairly large $O_2$ debt. Since lactate does not accumulate, what do you think causes the debt?

*Havel:* The $O_2$ debt obviously cannot be related to accumulation of lactate; it could reflect regeneration of high-energy phosphate stores or persistence of certain synthetic processes activated during exercise, such as hepatic gluconeogenesis.

*Barnard:* Have you measured glucose output by the liver in this subject? It might indicate a high rate of gluconeogenesis.

*Havel:* No, we have not. Uptake of blood glucose into tissues of the exercising leg was similar to that of control subjects, however.

*Romans:* Can excess FFA be taken up by muscle and reesterified to storage triglyceride after FFA mobilization has caused elevated plasma levels of FFA in excess of those needed for uptake and energy production by the muscle?

*Havel:* This is known to occur. During fasting, myocardial triglyceride content increases and catecholamine-induced fat mobilization is accompanied by increased deposition of triglycerides in many tissues, including skeletal muscle.

*Kauffman:* What is the role of proteins as an energy source during catabolism (starvation)?

*Havel:* Proteins serve as the major source of carbon (in addition to glyceride glycerol from adipose tissue) for hepatic and renal gluconeogenesis.

*Kauffman:* Why do interfascicular lipids accumulate in muscle in abnormally high quantities during rehabilitation after starvation?

*Havel:* I have no information on this. Possibly it would be related to the associated "adaptive hyperlipogenesis."

*Allen:* Are there any known examples of fat depots that lack the enzymes necessary to synthesize fatty acids or to mobilize fatty acids for energy requirements? In other words, are there any fat depots known that are not dynamic but are strictly "storage vats?"

*Havel:* Probably not in most mammals, although some depots, which may have other functions, tend to be mobilized last (e.g., retro-orbital fat). In synthesis of fatty acids, there are important species differences. For example, lipogenesis in birds occurs primarily in liver, so that most stored lipids in adipose tissue are derived from plasma triglycerides.

*McLoughlin:* In view of the role of fat in muscle metabolism, is it likely that dietary factors may play a part in the determination of muscle type, analogous, for example, to the effects of innervation, particularly in an evolutionary context? Ruminants, with a high dietary fat intake, have predominantly red muscles with a high capacity for oxidative metabolism, whereas the musculature of the monogastric pig has a high content of anaerobically metabolizing fibers. This generalization may not be true, but is such an effect of diet a possibility?

*Issekutz:* This is an interesting thought and I am not aware of any study along these lines. An extensive, comparative histological study could perhaps give support to this suggestion.

*Tipton:* What is the mechanism by which lactic acid is inhibitory to your fatty acids?

*Issekutz:* Björntorp (1965) showed that lactic acid added *in vitro* to adipose tissue of the rat inhibited the lipase, decreased the release of FFA and decreased also the release of glycerol. Before his work, I assumed that lactic acid increased the formation of $\alpha$-glycerophosphate providing two hydrogens for the dihydroxy acetone phosphate.

*Sallach:* Has lactate inhibition of adipose lipase been studied at the enzyme level? How does lactate inhibit oxidation of fatty acids?

*Issekutz:* As I mentioned, Björntorp (1965) has shown that lactate inhibited the lipase activity in the isolated epididymal fat pad. *In vivo,* lactate depresses the oxidation of FFA, but this can well be due to the inhibition of the release of FFA from the adipose tissue and to the marked decrease of the FFA concentration in the plasma.

*Porter:* There appears to be some discrepancy between your data on the relative importance of carbohydrate and fatty acids as energy sources for muscle during exercise. Would you care to comment on this point?

*Havel:* I do not believe that there are any discrepancies that cannot be attributed to nutritional state, intensity of exercise, or possibly to species differences. In general, studies in both humans and dogs show that fatty acids are the chief, but not the sole, lipid substrate and that use of carbohydrate increases with rising work loads.

*Issekutz:* In the introduction to Chapter 30, I discussed a number of reasons which may lead to controversies in this field if one tries to draw general conclusions, disregarding differences in the severity and duration of the work, in the type of exercise (bicycling or running), in preceding diet, and in the response of blood lactate. All these factors greatly influence the rise of plasma FFA, on which seems to depend the contribution of fatty acids to the oxidative metabolism during exercise. In addition, I should point out that at a high $O_2$ uptake (50 ml/kg per min), some 80–90% of the FFA turnover is immediately oxidized, while at more moderate work (25–30 ml/kg per min) only about 60–65% of it is converted to $CO_2$. If we base our comparisons on the plasma FFA level and on the $O_2$ uptake, then most of the apparent discrepancies concerning the contribution of fatty acids will disappear.

*Allen:* Is there any evidence available to show that muscles other than cardiac muscle utilize any appreciable quantity of lactic acid as an energy source?

*Havel:* I am not aware of this, although, as I indicated, it seems reasonable for red muscles with similar enzymic machinery.

*Allen:* Is all adipose tissue (including intramuscular) innervated by the sympathetic nerve endings?

*Havel:* The answer is yes, probably. For white adipose tissue, however, there is controversy whether the autonomic nerves innervate fat cells directly or only small blood vessels.

*Allen:* You mentioned that the white fibers of muscle have less stored triglyceride than red ones. Unlike glycogen, is not most of the stored triglyceride in muscle found between muscle bundles rather than within the fibers?

*Havel:* This is true in most mammals used for meat. It is less true in some laboratory species. The important point is that some muscles such as cardiac muscle do contain substantial triglyceride stores close to mitochondria within the fibers.

*Kauffman:* Is there any reason to believe that lipids deposited interfascicularly, because of their anatomical proximity to muscle fibers, play a greater role as a source of energy?

*Issekutz:* It is conceivable that interfascicular fat can play a role as an energy source. It should be mentioned, however, that Carlson, Liljedahl, and Wirsén (1965) found Sudan-stainable material in the red fibers in rows between myofibers. After 24 hours of norepinephrine infusion (highly elevated turnover rate of FFA) the fat droplets filled the sarcoplasma.

*Allen:* Have you studied lipolysis or fatty acid synthesis as it relates to increases in lipid of dystrophic chicken muscle?

*Cosmos:* We have not studied lypolysis or fatty acid synthesis in the dystrophic muscle as yet, although this is a very important problem in the disease. It is apparent from our data that the dystrophic animal is capable of protein synthesis in the immature stages; this ability is greatly depressed in the sexually mature muscle. Since in the latter there can be no turnover of protein, the muscle tissue is lost and lipids infiltrate those areas once occupied by muscle cells. Whether the adipocytes seen in the adult tissue result from a stimulation of adipogenesis or develop from dystrophic cells unable to utilize fat is still unknown.

## References

Björntorp, P. 1965. Effect of lactic acid on adipose tissue metabolism *in vitro*. *Acta Med. Scand. 178*:253.

Carlson, L. A., S. O. Fröberg, and E. R. Nye. 1966. Acute effects of nicotinic acid on plasma, liver, heart and muscle lipids. Nicotinic acid in the rat. II. *Acta Med. Scand. 180*:571.

Carlson, L. A. and Å. Hanngren. 1964. Initial distribution in mice of $^3$H-labeled nicotinic acid studied with autoradiography. *Life Sci. 3*:867.

Carlson, L. A., S. Liljedahl and C. Wirsén. 1965. Blood and tissue changes in the dog during and after excessive free fatty acid mobilization. A biochemical and morphological study. *Acta Med. Scand. 178*:81.

Cassens, R. G., R. M. Bocek, and C. H. Beatty. 1969. The effect of octanoate on carbohydrate metabolism in red and white muscle of the rhesus monkey. *Amer. J. Physiol. 217*:715.

Cobb, L. and W. Johnson. 1963. Hemodynamic relationships of anaerobic metabolism and plasma free fatty acids during prolonged strenuous exercise in trained and untrained subjects. *J. Clin. Invest. 42*:800.

Green, M. R. and B. R. Landau. 1965. Contribution of the pentose cycle to glucose metabolism in muscle. *Arch. Biochem. Biophys. 111*:569.

Greene, C. W. 1913. The storage of fat in the muscular tissue of the King salmon and its resorption during the fast of spawning migration. *Bull. Bur. Fisheries 23*:73.

Hers, H. G., J. Larner, and J. Brown. 1964. Discussion, p. 81. In W. J. Whelan and M. P. Cameron (eds.), *Ciba Foundation Symposium: Control of Glycogen Metabolism*. Little, Brown, Boston.

Morgan, H. E., M. J. Henderson, D. M. Regen, and C. R. Park. 1961. Regulation of glucose uptake in muscle I. The effects of insulin and anoxia on glucose transport and phosphorylation in the isolated, perfused heart of normal rats. *J. Biol. Chem. 236*:253.

Park, C. R., J. L. Bornstein, and R. L. Post. 1955. Effect of insulin on free glucose content of rat diaphragm *in vitro. Amer. J. Physiol. 182*:12.

Porte, D., Jr., and R. H. Williams. 1966. Inhibition of insulin release by norepinephrine in man. *Science 152*:1248.

Randle, P. J., E. A. Newsholme, and P. B. Garland. 1964. Regulation of glucose uptake by muscle 8. Effects of fatty acids, ketone bodies and pyruvate, and of alloxan-diabetes, and starvation, on the uptake and metabolic fate of glucose in rat heart and diaphragm muscles. *Biochem. J. 93*:652.

Sacks, J. and J. F. Smith. 1958. Effects of insulin and activity on pentose transport into muscle. *Amer. J. Physiol. 192*:287.

Shipp, J. C., L. H. Opie, and D. Challoner. 1961. Fatty acid and glucose metabolism in the perfused heart. *Nature 189*:1018.

Vendsalu, A. 1960. Studies on adrenaline and noradrenaline in human plasma. *Acta Physiol. Scand. 49* (suppl. 173) :1.

# PART 8

## Biology of Muscle as a Food

# Chapter 33

# *Morphology of Muscle as a Food*

R. G. CASSENS

Morphology is an essential, integral part of any concentrated effort to study muscle as a food. In this chapter I want to show how useful, important, and necessary it is, not by a great review of the many existing morphological studies, but rather by using a limited number of illustrations that show direct application to the use of muscle as a food.

The express purpose of morphological investigations is to visualize, to reconstruct an arrangement, to interpret, and finally to understand how the cell or tissue works. Today, we simply say that we comprehend the structure of a tissue from molecular to gross levels and understand how the systems operate and are driven biochemically and physiologically to produce the function. Muscle is a beautiful example—everyone has an image of the muscle fibrillar protein molecules, visualizes their spatial arrangement and relation to other muscle cell components, and finally understands how the whole of the cell functions to produce contraction and subsequent motion.

Let me emphasize here that I believe morphology is a prerequisite to a coordinated and complete approach to the study of muscle as a food; but histological studies are often rather weak to stand alone. It is only in combination and permutation with biochemistry and physiology that the full benefit of morphology is reaped in the gathering of new knowledge about structure and function.

The pioneering efforts to relate muscle to meat were simple and economic. It was a quantitative anatomical approach to the question of how much protein was present. Later, concern for quality developed. It was quickly recognized that all muscles are not similar, nor do they have similar properties for human consumption. Let us consider texture as an example. The cut surface of various muscles has quite different textures, spoken of as coarse or fine; the rule of thumb is, the finer the texture the more desirable the muscle for use as a food. Why? It is true that the bundle arrangement and connective tissue characteristics affect the visual texture. Finally Locker (1960) related muscle contraction state to meat tenderness; this one simple morphological observation led to a number of other studies (Herring, Cassens, and Briskey, 1965; Locker and Hagyard, 1963; Marsh and Leet, 1966), which substantiated the real importance of this observation that the fundamental function of muscle contraction was of great significance to the economic consideration of meat tenderness. Some of these observations were just being made in 1965, at the time of the first symposium on the physiology and biochemistry of muscle as a food. The implications of contraction state in the area of muscle as a food were extensively discussed at that symposium.

The light optics microscope and phase contrast microscope are rather ordinary tools in our sophisticated scientific society, but they have a potential as yet unexploited, residing in the ability of the user to transpose the artifacts produced by preparation and staining to the structure, function, and life of the tissue. Only the user can recognize the unusual or deviate, and visualize the perspective from live animal to cooked meat.

Then too, it is often difficult to quantitate morphological results. Communication of one's interpretation of a picture is so impossible that simple linear measurements are often reported. Let us again illustrate with one of the earliest problems presented to the study of muscle as a food. Muscle fiber size was a suitable parameter for simple quantitative assessment and it was of great interest as the following questions were asked: First, can the postnatal growth of muscle be accounted for exclusively by increases in fiber size, or is there also an increase in number of fibers? Second, are the extremes in fiber size significant in any way to the properties of the muscle for use as food? Simple questions, but at least two of the chapters in this symposium report very fundamental research that is applicable. Goldspink (Chapter 25) has described increases in length and girth of muscle in relation to physiological parameters, but that does not preclude extension of his results to answer the questions posed above; application of his techniques and data on myofibril proliferation to other applied problems seems most logical. Or, consider the remarks of Holtzer and Bischoff (Chapter 3). They speak of cell differentiation as being a historical process, yet they are vitally interested in a most fundamental investigation of proliferative and quantal mitosis—the implications here to muscle differentiation, growth, and ultimate properties are clearly evident. We now know that a high correlation exists between fiber size and contraction state (Herring, Cassens, Suess, Brungardt, and Briskey, 1967); therefore the original thought that larger fibers are associated with tougher meat is correct, but it has only recently been shown that fiber size is greatly influenced by contraction state. This knowledge in turn calls for more fundamental morphological and biochemical studies on muscle contraction in order to understand the real basis of meat tenderness. The contraction state of the muscle can also be judged morphologically by the form of isolated fibers. In highly contracted muscle, a number of fibers are passively shortened by their actively contracting neighbors, and these passively shortened fibers are folded or kinked to fit the shortened form. The straight, contracted fibers and passively shortened, kinky, or wavy fibers can be easily visualized in post-rigor teased or macerated preparations. Cooper, Breidenstein, Cassens, Evans, and Bray (1969) have demonstrated a good association between extent of passively contracted fibers and sarcomere length in beef muscle. A study of simple morphology of the muscle cell can be of great value in estimating its value for muscle as a food.

It is impossible to mention all the problems of meat science that have common ground with muscle biology, so I am forced to draw on only a few examples. The muscle from domestic animals is unique in some subtle ways, and we think this uniqueness stems from the selection and environment imposed on these animals in order that they produce the greatest amount of edible protein at the most efficient cost. This uniqueness about which I speak leads to problems when the muscle is converted to meat. I would like now to list five such situations and illustrate each with an attempt at morphological study of the problem.

## MUSCLE LIPID

The United States Department of Agriculture standards for beef grading contain a provision for visual fat content (marbling) of the muscle, with higher fat content (within certain limits) indicating a more palatable and desirable meat. This is an important issue, in view of current investigation of the value of saturated animal fats for human consumption. Havel (Chapter 29) and Issekutz (Chapter 30) both discuss the use of muscle lipids as an energy source. This is a most active, controversial, and pertinent area of investigation, and one that holds in sway the future of muscle for human consumption. What has been done morphologically? The histology textbooks describe the fat cell and its morphogenesis, and a recent electron microscopic description of fat cell morphogenesis has been presented by Sheldon (1969). Moody and Cassens (1968*a*) have studied fat cell morphology in bovine muscle with results that I mention here as worth future investigation. Fig. 33.1 illustrates typical fat cell masses that contain different numbers of cells. The strikingly obvious feature is that the fat cell size increases as the number of fat cells per mass increases. Muscle with three levels of visual fat was studied, and

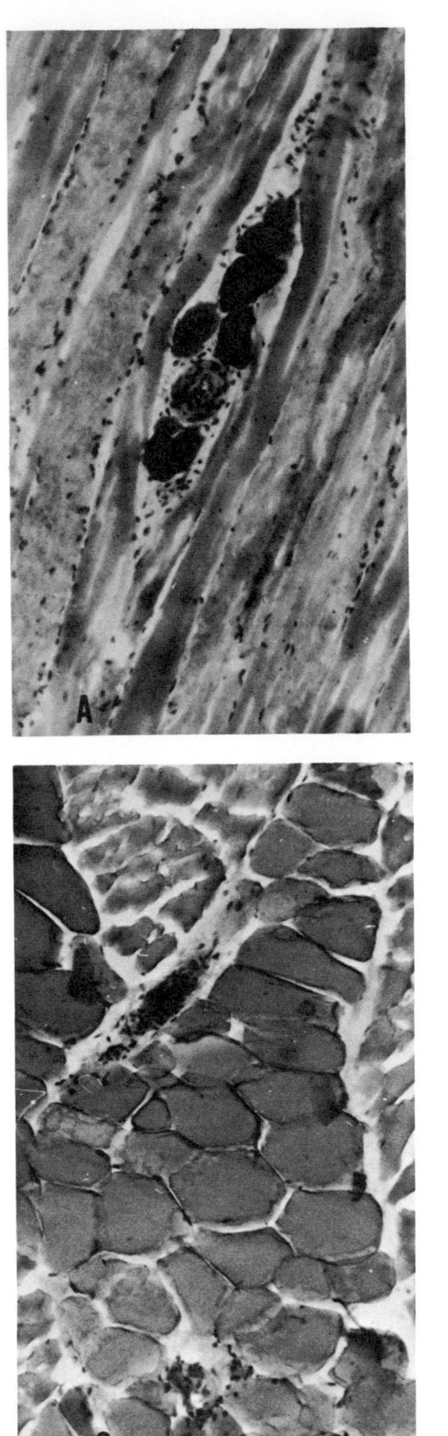


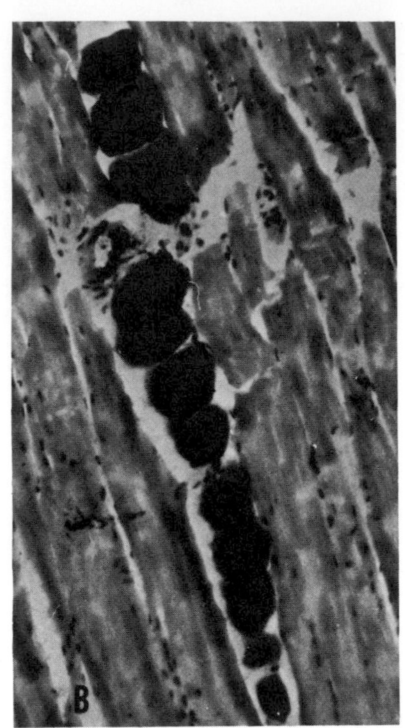


*Fig. 33.1.* Photomicrograph showing typical fat cell mass classifications. *A,* 5–10 cells; *B,* 11–20 cells; *C,* above 20 cells. Oil Red 0, Hematoxylin. ×55. From Moody and Cassens (1968).

as the level of visual fat increased the average fat cell size also increased. Figures for both of these observations are given in Table 33.1.

It was suggested that in order for a fat cell to attain a large size, it had to be present in a group of cells rather than as an isolated one or few cells, and also, that once a muscle is committed to increase fat, both the size and number of fat cells apparently increase.

### ANIMAL PHYSIOLOGY

The emphasis now diverts more to the role of morphological studies to animal physiology. Lister, in Chapter 34, discusses animal physiology and the importance of the animal to the conversion of muscle to meat. The supply and removal of blood to muscle is critical, particularly in view of the stress susceptibility syndrome reported for porcine animals. The blood supply is normally under rather stringent control in skeletal muscle, and appears to be associated closely with metabolic activity. Cooper, Cassens, and Briskey (1969) have studied capillary distribution in pig muscle with the alkaline phosphatase reaction. This procedure reportedly describes active capillaries, and the reaction product occurs at the site of the endothelial cell nuclei. Fig. 33.2 illustrates the results. Most of the capillaries were located axially within the fasciculi with only a few at the periphery. There were also segments of capillary rings around some fibers located peripherally in the bundles. This was interpreted to indicate that when the microcirculatory system entered the bundle, it traversed the outer fibers until it reached the axial portion of the bundle where it paralleled the fiber. When serial sections were incubated for the NADH-TR reaction, it was seen that the number of capillaries

TABLE 33.1

*Relative size of fat cells from three marbling groups of bovine longissimus*

| Fat cell masses | Marbling groups | | |
|---|---|---|---|
| | Traces | Small | Moderate |
| 5–10 cells | 7.0 ± 1.4 | 8.3 ± 2.3 | 9.9 ± 2.6 |
| 11–20 cells | 9.3 ± 3.6 | 11.4 ± 2.2 | 11.7 ± 2.1 |
| >20 cells | 12.4 ± 4.5 | 17.6 ± 4.7 | 20.5 ± 5.1 |
| Mean | 9.9 ± 2.2 | 12.4 ± 2.2 | 14.0 ± 3.0 |

Figures are means ± SD. Means represent the number of grid squares required to cover the area of one fat cell. Each grid square covered an area of 0.004 mm². Any two means not underscored by the same line are significantly different ($P < 0.05$). Means were determined from 15 cells.

From Moody and Cassens (1968*a*).

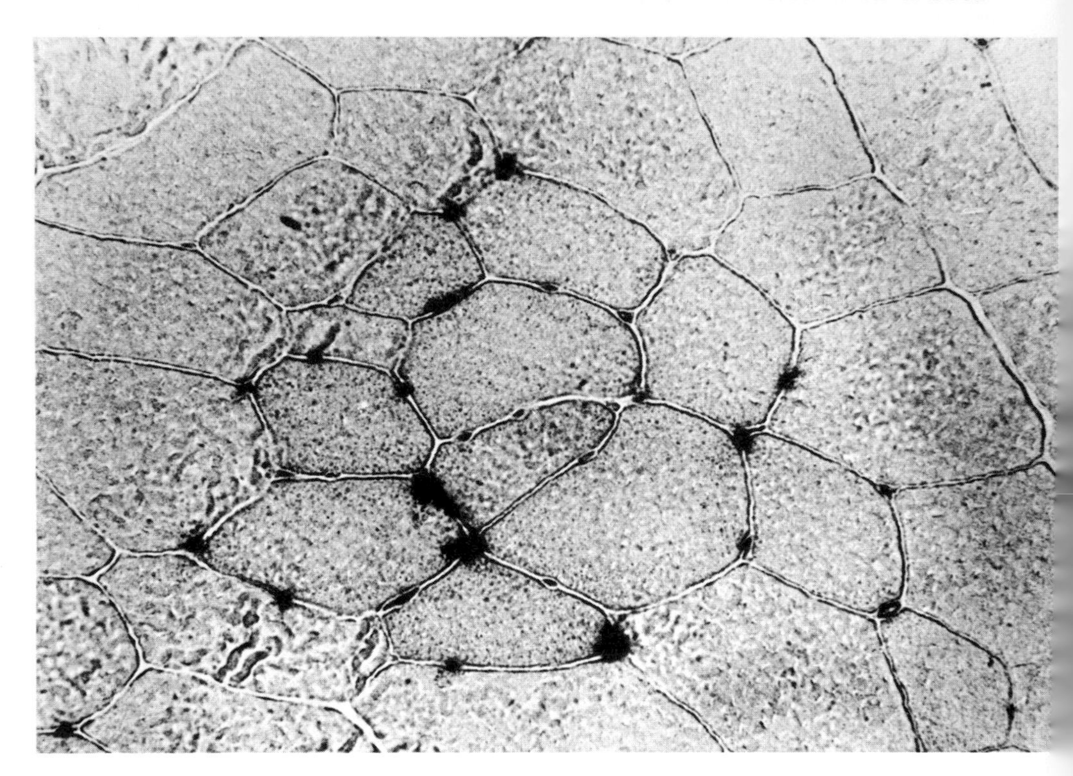

*Fig. 33.2.* Fresh frozen section of porcine longissimus muscle reacted for alkaline phosphatase. Capillaries are seen as dark reaction deposits. 10 $\mu$ thick section. $\times$ 212. From Cooper, Cassens, and Briskey (1969).

surrounding each fiber or fiber group appeared in proportion to its redness or its oxidative metabolic activity. In many cases, the only capillaries associated with white fibers appeared to be shared with adjacent red fibers. There are rather large areas of totally white fibers in pig muscle, and it was difficult to believe that these areas had such a poor or absent capillary supply. These findings have, however, been substantiated recently by Morita, Cassens, and Briskey (1969) in studies on myoglobin localization in pig muscle. The histochemical myoglobin reaction is also positive for trapped red blood cells, and it shows a pattern almost identical to that of the alkaline phosphatase reaction. C. R. Ashmore (pers. comm., 1969) has expressed the opinion that selection of domestic animals for muscle production has led us to the point of near abnormality in some muscle, or at least to muscle that cannot respond to severely stressful situations. An example would be pig muscle (with low capillary supply to white fibers) that develops an unusually high lactic acid content in certain stressful situations.

## FIBER TYPES AND ABNORMALITIES

The usefulness of the light microscope is greatly expanded with the use of histochemical techniques. The previous description of capillary distribution is an example of histochemical study; I would now like to focus attention on an active area of muscle histochemistry, the study of red and white fibers. The literature produced during the past 10 years is tremendous, and the most recent findings are discussed in the present symposium. I will again limit myself to a very few morphological experiments that were designed to answer a question about the use of muscle as a food.

Fig. 33.3 shows that porcine muscle is composed of more than one fiber type; the gross characteristics, physiological properties, and biochemical machinery of muscle are all closely associated with the fiber-type composition of the muscle. The minimum number of fiber types is two, with the red fiber being rich in oxidative enzymes and poor in phosphorylase while the white fiber is poor in oxidative enzymes but rich in phosphorylase. There is a range in properties from red to white, however, as there is with most biological phenomena, and fibers intermediate in some property between red and white are recognized. This figure shows a unique property of porcine muscle. The red fibers are found in discrete groups or clumps of about five to seven fibers toward the center of the bundle (Moody and Cassens, 1968*b*) and are surrounded by white fibers. This unusual distribution holds for a number of muscles of the pig that have been examined. In redder muscles, almost the entire center of the bundle is composed of one large group of red fibers surrounded by one or two layers of white fibers at the periphery. The distribution within a clump of red fibers appears to be with the most positively reacting fibers in the interior surrounded by less and less positive fibers. This arrangement is very unlike the checkerboard or scattered arrangement of red fibers that is known for other mammalian muscle, and that is found in bovine muscle (Fig. 33.4). The fiber distribution pattern of pig muscle is reminiscent of the small groups or islands of similar fiber types seen in cross-innervated or reinnervated muscle. The real significance of this observation in pig muscle is unknown but it certainly merits further investigation.

A morphological study has been conducted at the University of Wisconsin Muscle Biology Laboratory over the past few years on the fiber type composition of pig muscle. Certain pigs are stress-susceptible; their muscle undergoes a very rapid glycolysis post mortem and is ultimately undesirable for use as a food (i.e., pale, soft, and exudative, or PSE). Some members of our laboratory studied the physiology and biochemistry of this abnormality, and we thought it necessary to establish the histochemistry of the muscle in order that it might be correlated with the

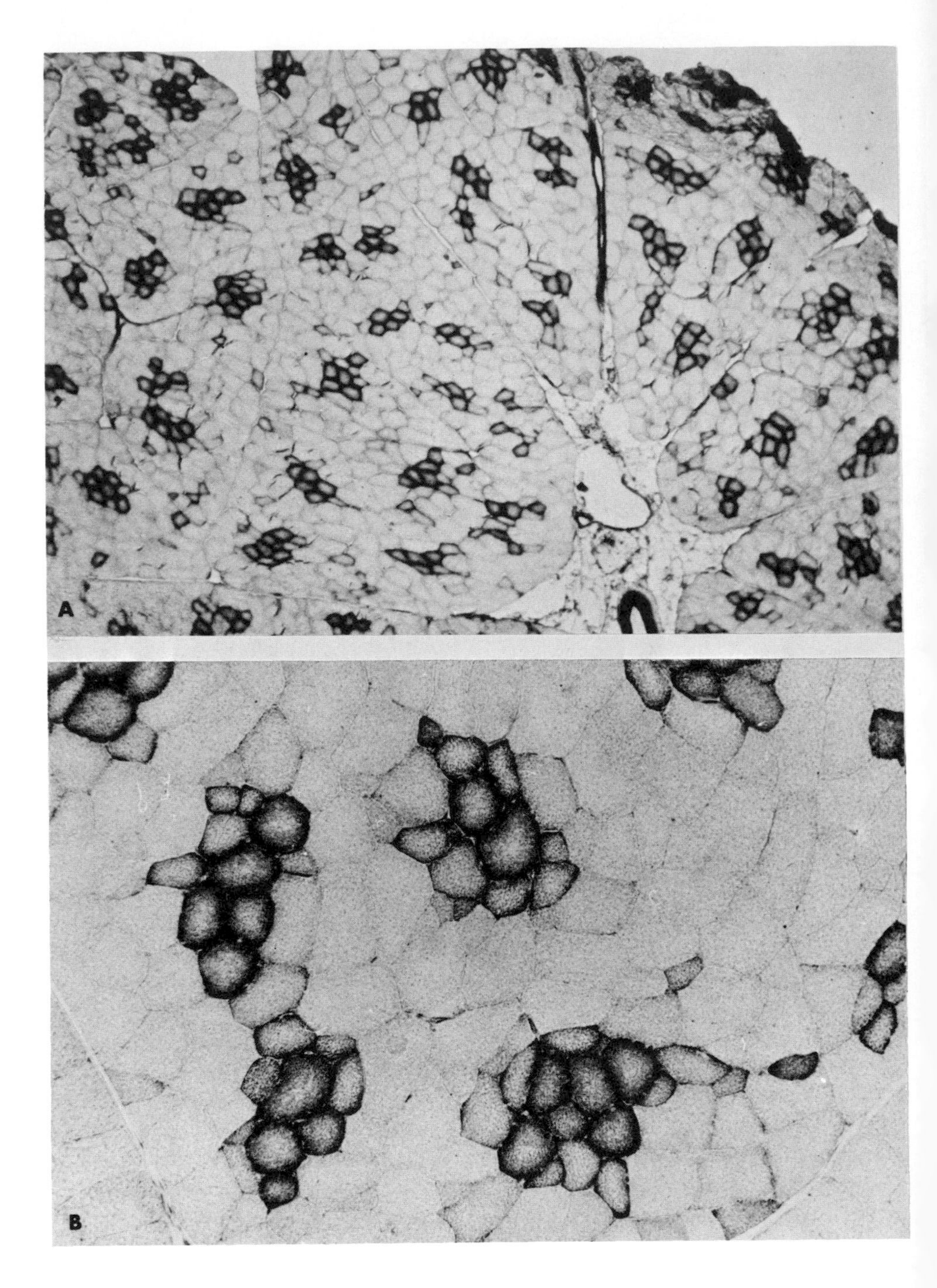


*Fig. 33.3.* Porcine muscle, reacted for NADH-TR, that shows unusual grouping pattern of red fibers. *A*, × 16, and *B*, × 100.

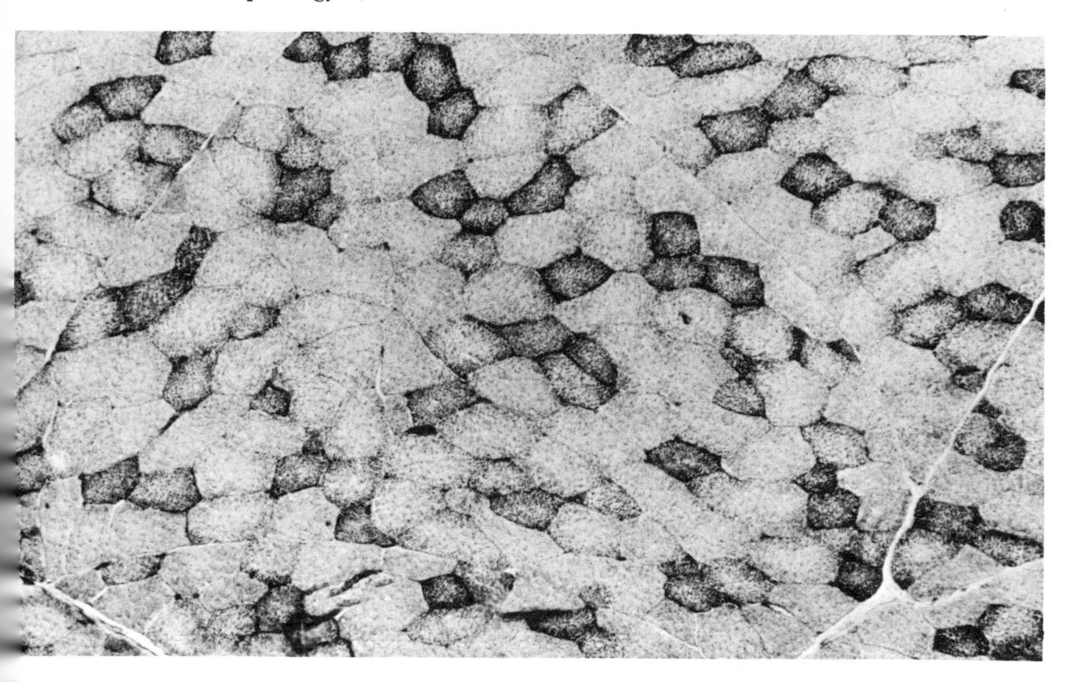

'ig. 33.4. Bovine muscle reacted for NADH-TR that shows typical checkerboard distribution pat-ern. × 100. Unpublished data from S. Morita, University of Wisconsin.

physiological and biochemical properties of the muscle. Lister, Sair, Will, Schmidt, Cassens, Hoekstra, and Briskey (1970) studied fiber characteristics of PSE and normal pig muscle with Sudan black B, cytochrome oxidase, and succinic dehydrogenase techniques. They concluded that PSE muscle fibers were larger than fibers from normal muscle and that there was a larger area of red fibers per bundle in the PSE muscle; it is important to note that a differentiation of intermediate fibers was not made. These results then were difficult to align with the biochemical work. Kastenschmidt, Hoekstra, and Briskey (1968) found that the metabolic intermediate patterns were consistent with the concept that phosphorylase was the primary control site for post mortem glycolysis and that phosphorylase activity was higher in the muscle from stress-susceptible animals that were destined to become PSE. Also, Quass and Briskey (1968) found that biochemically determined myosin–adenosine-triphosphatase activity was higher in the muscle destined to become PSE. Theoretically, red fibers should have a lower phosphorylase and adenosine triphosphatase than white fibers. Cooper, Cassens, and Briskey (1969) reexamined the problem in muscle destined to become PSE and found that, in fact, a great proportion of the red fibers were not true red fibers. These fibers, which were called intermediate, were not only positive to the oxidative enzyme reactions used earlier but were also positive for phosphorylase and adenosine triphosphatase. The essence of the problem then resides in the finding that certain intermediate fibers in muscle from stress-susceptible porcine animals have a substantial phosphorylase and adenosine triphosphatase, even though they are positive for NADH-TR, and therefore the muscle is apparently quite well equipped to glycolyze better than muscle from normal pig.

Typical intermediate fibers are illustrated in Fig. 33.5. The NADH-TR reaction shows a typical response of a group of positive fibers. However, only two of the fibers are uniformly dense while the others are slightly less reactive, and have a subsarcolemmal deposition of diformazan with lighter centers; this pattern is characteristic for intermediate fibers. The reaction for amylophosphorylase reveals that indeed only the two true red fibers are negative for amylophosphorylase. The very dense uneven phosphorylase reaction is also typical of muscle destined to be PSE (the reaction color is very dark blue).

These findings led to a study of muscle fiber differentiation during growth and development of the animal. We thought that a knowledge of the fiber-type differentiation process would be useful in order to know when the stage was being set for the high intermediate fiber content and ultimately for the PSE condition. Cassens, Cooper, Moody, and Briskey (1969) used the Sudan black B technique to visualize red and white fibers in pig longissimus muscle at the following developmental stages: 8–9 weeks' gestation, 9.5–11 weeks' gestation, 12–13 weeks' gestation, 1 day postnatal, 13-day-old, 180-day-old, and 24-month-old. There was no differentiation of fiber types in any of the fetal stages or at 1 day; the fibers were all Sudan black B positive and therefore taken as type I or red. Fig. 33.6 shows the process from 1 day onwards. At 1 day of age there was no apparent differentiation, the fibers were rounded rather than polygonal and there was much loose connective tissue present. The differentiation of fiber types was clear by the time the animal had reached 13 days of age, and about 60% of the fibers were Sudan black B negative or white. Red fibers composed only a small portion of the total fibers (about 15%) by the time the animal had reached 200 days of age, and there appeared to be little further change as the pig matured to 24 months of age. This work established the general pattern of fiber differentiation in pig muscle, but Cooper, Cassens, Kastenschmidt, and Briskey (1970) have studied the process in greater detail and with different techniques. They found with the myosin–adenosine-triphosphatase reaction that a slight differentiation was apparent at 1 day of age, but that

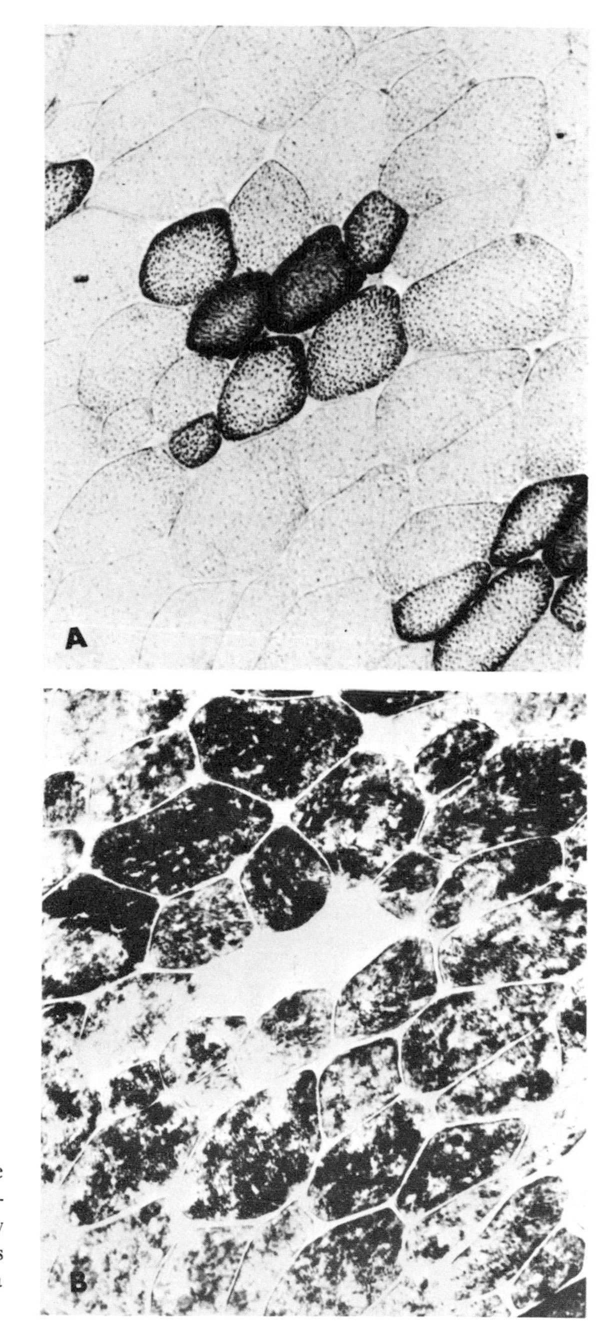


ig. 33.5. Serial sections of porcine muscle ·acted for NADH-TR (*A*) and amylophos-horylase (*B*). *A*, two red fibers (uniformly ositive) surrounded by intermediate fibers ighter reaction). × 200. Unpublished data om R. Sair, University of Wisconsin.

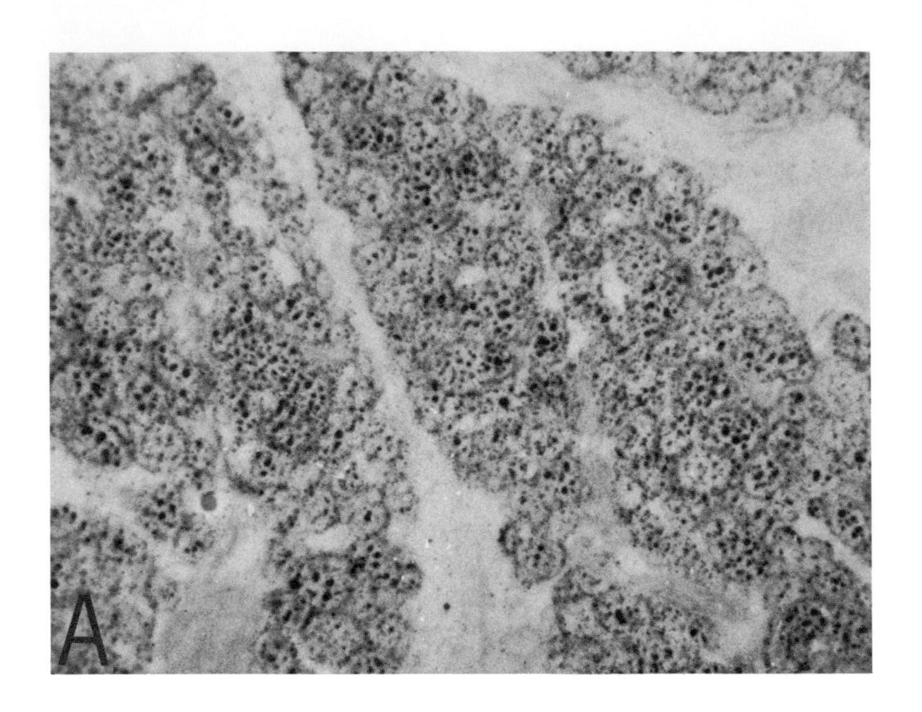
A

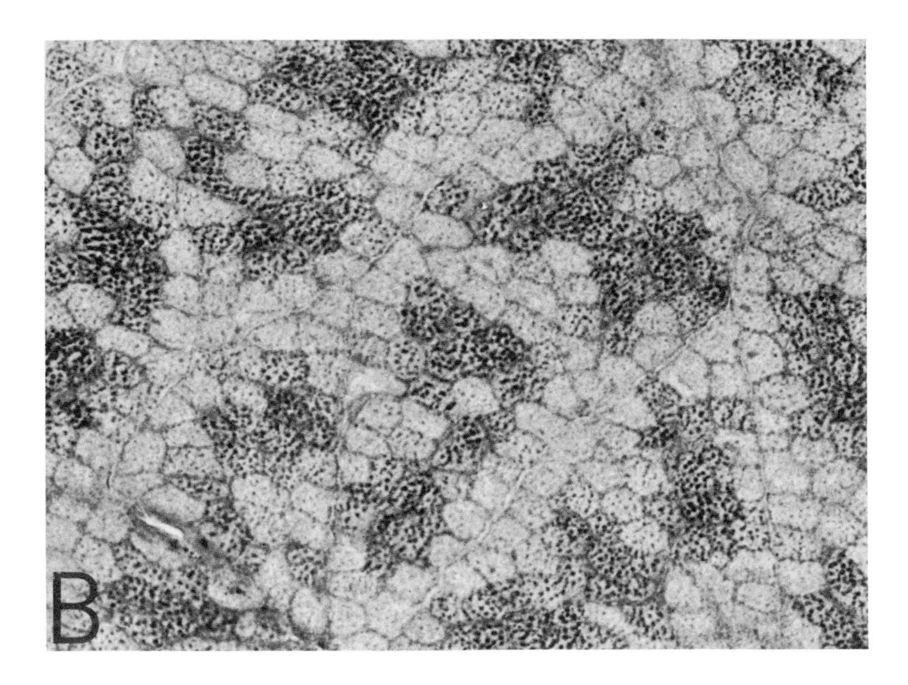
B

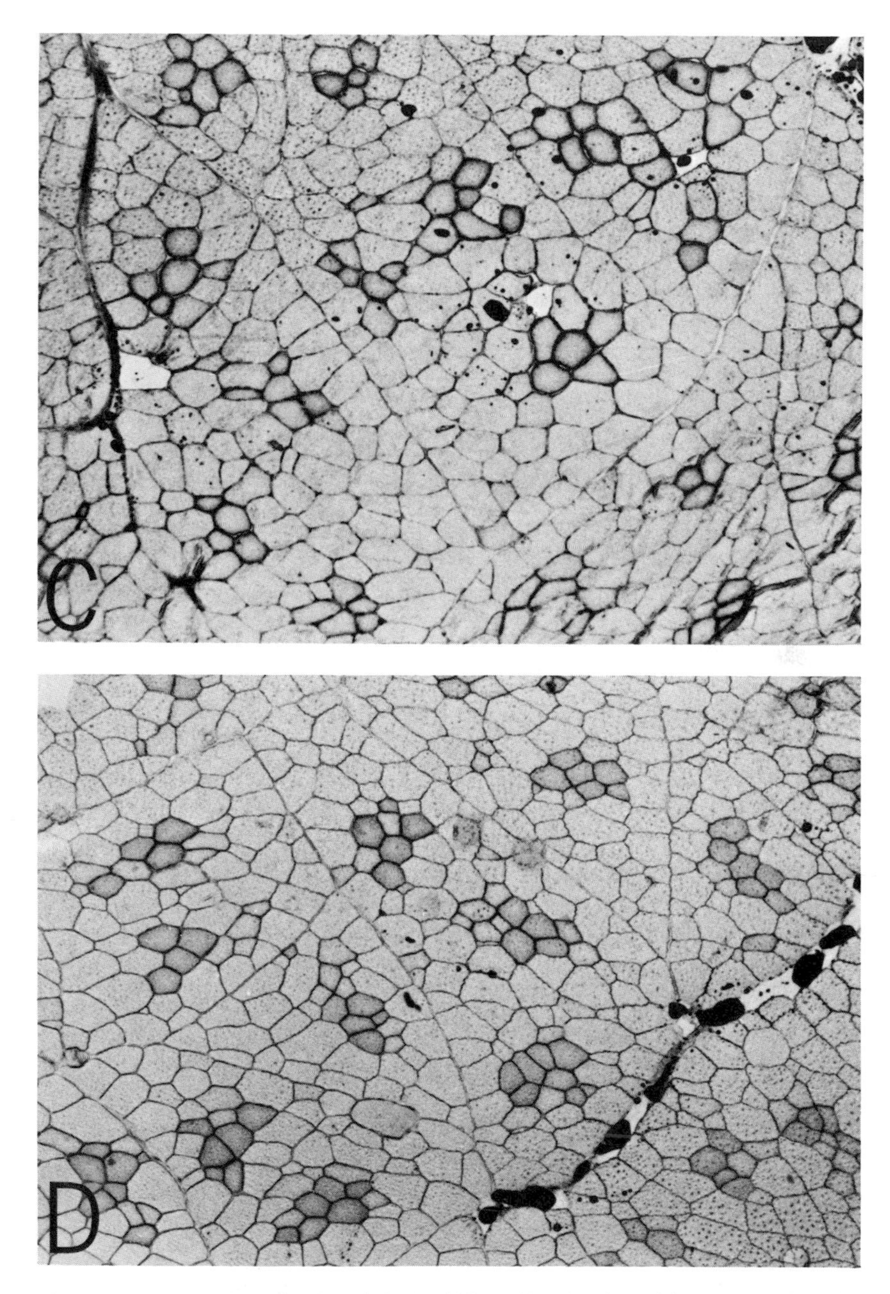


*Fig. 33.6. A,* 1-day-old animal. × 443. *B,* 13-day-old animal. × 279. *C,* approximately 180-day-old animal. × 44. *D,* approximately 24-month-old animal. × 44. All are frozen sections 10 μ thick from longissimus muscle and are stained with Sudan black B. From Cassens, Cooper, Moody, and Briskey (1969) .

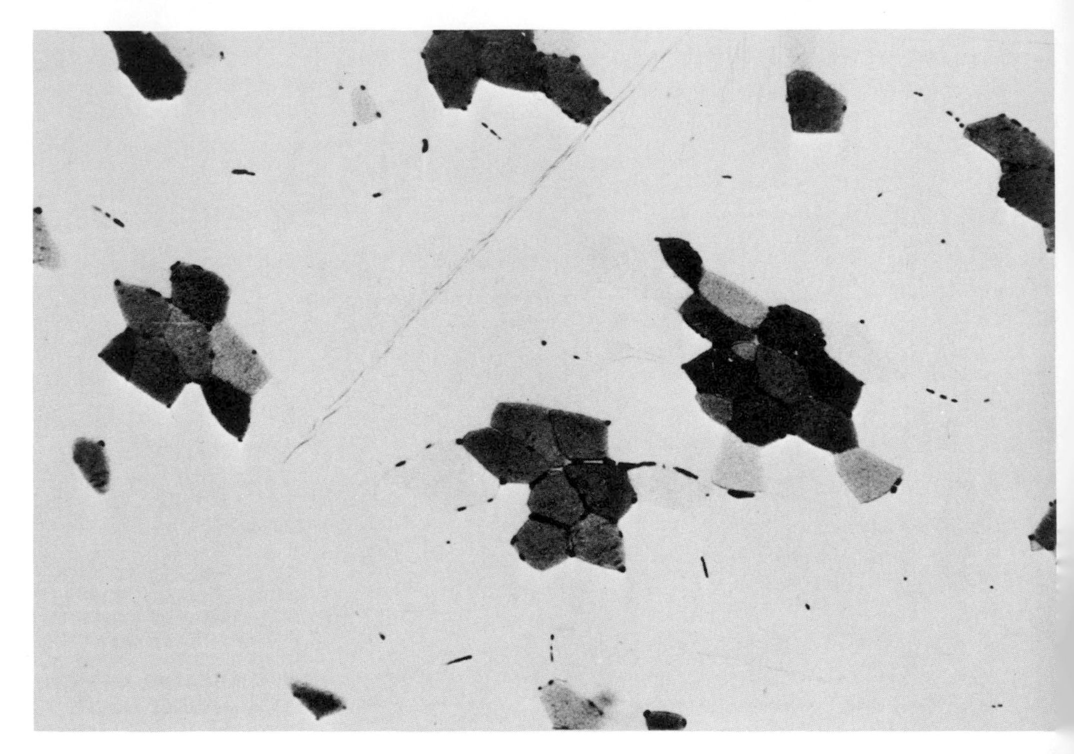

*Fig. 33.7.* Cross section of porcine striated muscle showing the distribution of myoglobin. *Above,* longissimus, *right,* trapezius. The longissimus shows positive intermediate and negative fibers,

generally the fibers all appeared red; at 1 week of age a differentiation was clearly evident. Type II and intermediate fibers could be separately classified at 4 weeks of age. The authors concluded that type II fibers increased over the period 4 weeks to 6 months at the expense of intermediate fibers, whereas the type I fibers decreased only slightly.

Morita, Cooper, Cassens, Kastenschmidt, and Briskey (1970) applied their histochemical myoglobin technique to the longissimus muscle of developing pigs. They found an essentially negative myoglobin reaction in 1-day-old pigs, but the typical adult pattern was evident by 3 weeks of age. Fig. 33.7 shows the typical myoglobin reaction pattern for longissimus muscle of adult pig. Three distinguishable gradations of reaction intensity are apparent: strong, intermediate, and negative. Serial sections reacted for other histochemical tests showed that myoglobin-positive and myoglobin-negative staining occurred in red and white fibers, respectively. Morita, Cassens, and Briskey (1970) have disclosed further interesting differences in the myoglobin staining pattern among pig, beef, and

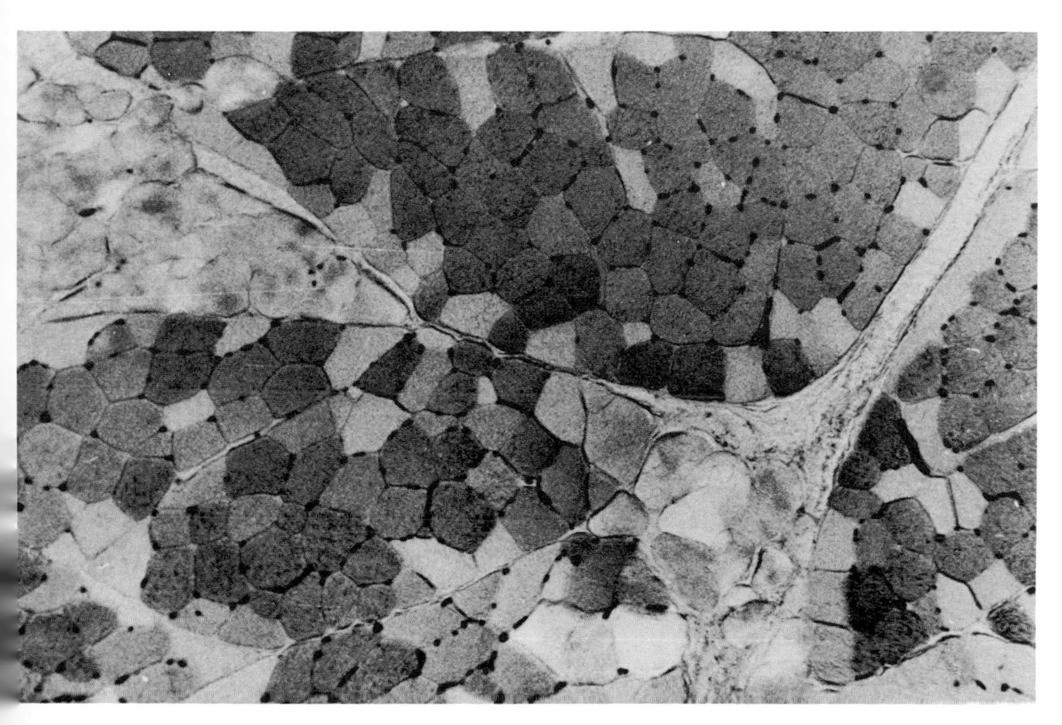

while the trapezius shows more uniformity of reaction, but red and white fibers are still differentiated. Hemoglobin in erythrocytes trapped in capillaries is strongly stained. Benzidine reaction, 30 μ thick section. × 100. From Morita, Cassens, and Briskey (1969*a*).

rabbit muscle: the results align well with biochemical analysis but give additional information about localization.

Our histochemical studies have also revealed an unusual, large fiber in pig muscle (Cassens, Cooper, and Briskey, 1969). Fig. 33.8 shows the general morphological features and histochemical reaction pattern of the giant fiber. Typical giant fibers are round and are usually larger than surrounding fibers. They often appear at the periphery of a fascicle and have apparently compressed the surrounding fibers to conform to the circular shape. They display a variable reaction for NADH-TR, a negative reaction for phosphorylase, and a positive reaction for adenosine triphosphatase. Giant fibers therefore do not correspond to either of the classical red or white fiber reaction patterns. White fibers react positively for both amylophosphorylase and adenosine triphosphatase, red fibers negatively; giant fibers react negatively for phosphorylase and positively for adenosine triphosphatase. The negative reaction for amylophosphorylase is interesting in view of the fact that it is a soluble sarcoplasmic

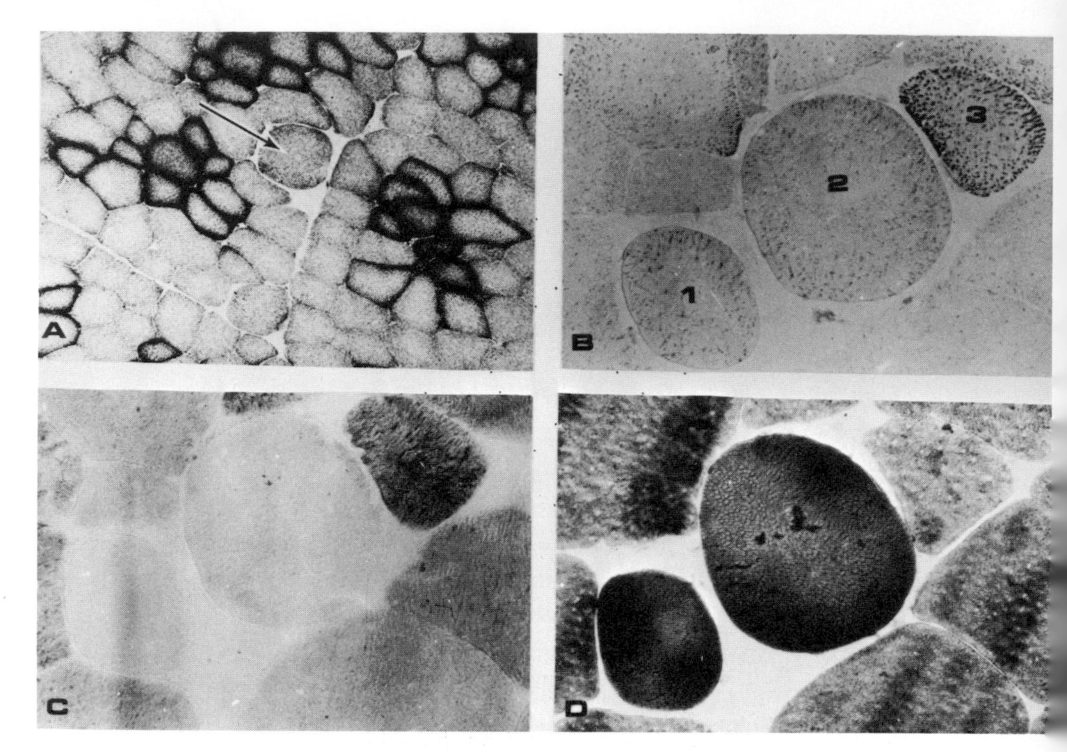


*Fig. 33.8. A,* photomicrograph of porcine longissimus muscle. Arrow indicates giant fiber. Fresh frozen section 10 μ thick reacted for NADH-TR. × 142; *B–D,* photomicrographs of porcine longissimus muscle which show giant fibers. Fiber 2 is a typical giant fiber. Serial sections reacted for NADH-TR (*B*), amylophosphorylase (*C*) and adenosine triphosphatase (*D*). 10 μ thick. × 360. From Cassens, Cooper, and Briskey (1969).

enzyme that could be lost from the cell, if cell membrane permeability was altered. We have found giant fibers in longissimus, psoas major, semitendinosus, and biceps muscles of both adult and growing animals. They comprise less than 1% of the total fiber population, but our observation is that they occur frequently in muscle of stress-susceptible pigs but rarely in muscle of normal or stress-resistant pigs.

## MYOFIBRILLAR CHANGES

I will pass now from the light optics microscope to the next order of resolution, offered by the electron microscope. As the resolution is increased one approaches the level of molecular arrangement, and therefore very exciting ground, but one also moves further away from the living cell through the preparative methodology required; interpretation must be carefully tendered. Great advances have been made in describing the fine structure of the muscle cell components, but I shall again call on only a few publications to illustrate the use of the electron microscope for consideration of the morphology of muscle as a food.

It is safe to say that everyone is familiar with the sliding filament theory and with the location of certain proteins within the myofibril (see Huxley [1969] for recent review). This is classical work of the most fundamental nature, but it still has application to the use of muscle as a food. We discussed earlier the significance of contraction state to tenderness, but a more obvious question is, are there structural changes incurred by the myofibril during post mortem conversion, and if so, are such changes related to applied properties of the meat such as color and water-binding capacity?

Cassens, Briskey, and Hoekstra (1963*a*) were the first to apply the electron microscope to a study of structural changes in muscle post mortem. The objective was to investigate any structural changes incurred during conversion post mortem that might be related to ultimate properties of the muscle as food (i.e., water-binding, color, tenderness).

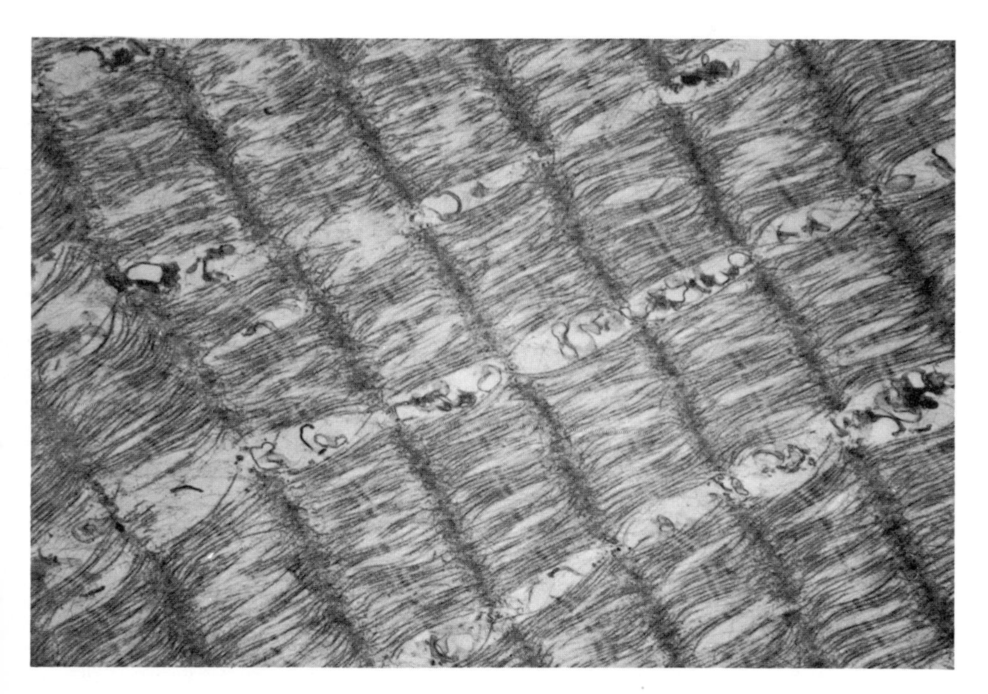

*Fig. 33.9.* Myofibril fraction from normal muscle isolated immediately after death. × 25,000. From Greaser, Cassens, Briskey, and Hoekstra (1969*b*).

Muscle that underwent a normal change post mortem exhibited a gradual disruption of sarcoplasmic components with little if any change in the myofibrils. Muscle that underwent a very rapid change post mortem revealed a very rapid disruption of sarcoplasmic components and some apparent disruption of the myofilaments. The authors thought these changes were reflected in decreased water-binding capacity of such rapidly glycolyzing muscle. During the same period, Cassens, Briskey, and Hoekstra (1963*b*) described the ultrastructure of contraction bands; these can be produced by thaw rigor and cause the muscle to be undesirable for use as a food.

Greaser, Cassens, Briskey, and Hoekstra (1969*b*) carried this type of investigation one step further by isolating myofibrillar, mitochondrial, heavy sarcoplasmic reticulum, and light sarcoplasmic reticulum fractions from homogenates of normal and rapidly glycolyzing (PSE) muscle at 0 hour and 24 hours post mortem. No differences were observed between myofibrils isolated at death from muscle that was subsequently normal or PSE, but there were different types of myofibrils recognizable in all samples. Fig. 33.9 shows the most common structure. The myofibrils were highly contracted with no I band present and with a dense band at the Z line. This supercontractee conformation probably resulted from the disruption during homogenization that released divalent cations and that

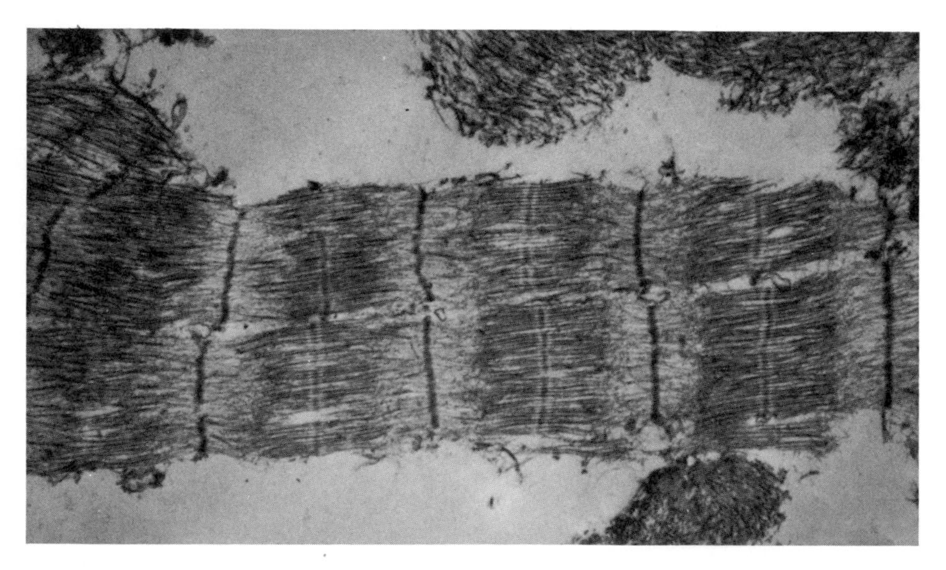

*Fig. 33.10.* Myofibril fraction from normal muscle isolated immediately after death. Note relaxed myofibrils. × 25,000. From Greaser, Cassens, Briskey, and Hoekstra (1969*b*).

in turn stimulated contraction in the presence of the residual muscle ATP. The next most common pattern is shown in Fig. 33.10. Such relaxed myofibrils composed less than 10% of the total. The Z lines were well preserved and displayed the typical zigzag appearance. The major apparent difference from intact muscle was that the filaments at the A-I border were somewhat tangled and disoriented, causing this boundary to be less clearly defined than in most intact muscle preparations. The third myofibrillar pattern is shown in Fig. 33.11. The A-band material appeared to be partly extracted, with many of the thick filaments missing or broken. There was also tangling and disorganization of the thin filaments, but the extra material on the center of the thick filaments which gives rise to the M line could still be observed. Micrographs of 24-hour post mortem myofibril preparations from normal and PSE muscle are shown in Figs. 33. 12–13, respectively. The supercontracted and extracted types of myofibrils were not seen in 24-hour preparations. The 24-hour myofibrils appeared to be very similar to that observed with intact fresh muscle, with clearly defined A and I bands, Z lines, and M lines. Some breakage and loss of material from Z lines and adjacent thin filaments was seen. The myofibrils from 24-hour PSE muscle showed two major differences from normal. The first was the marked increase in width of Z lines. The zigzag structure was obscured and it appeared that there was a precipitation of material of unknown origin in this region. It could be precipitated sarco-

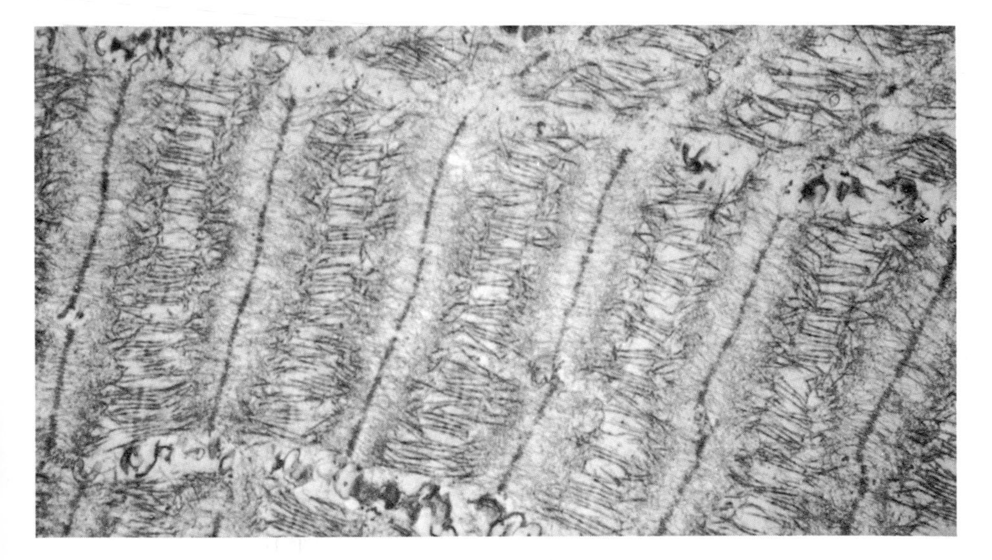

*Fig. 33.11.* Myofibril fraction from normal muscle isolated immediately after death. × 25,000. From Greaser, Cassens, Briskey, and Hoekstra (1969*b*).

*Fig. 33.12.* Myofibril fraction from normal muscle isolated at 24 hours post mortem. × 25,000. From Greaser, Cassens, Briskey, and Hoekstra (1969*b*).

*Fig. 33.13.* Myofibril fraction from PSE muscle isolated at 24 hours post mortem. × 25,000. From Greaser, Cassens, Briskey, and Hoekstra (1969*b*).

plasmic protein, solubilized and reprecipitated myofibrillar protein, or disorganized Z-line proteins. The second difference was the distinct granular appearance of the myofilaments, most likely due either to precipitated sarcoplasmic proteins or to changes in the conformation and properties of the myofibrillar proteins themselves. Some great changes in conformation of mitochondrial membranes were seen from PSE muscle and also as muscle was aged 24 hours. The sarcoplasmic reticulum membranes showed little change as they aged 24 hours in muscle, however, even though there was a great loss of biochemical activity (Greaser, Cassens, and Hoekstra, 1967; Greaser, Cassens, Briskey, and Hoekstra, 1969*a*).

Peachey (Chapter 13) has reviewed the form of the sarcoplasmic reticulum and T system in striated muscle. There has been extensive biochemical work (stimulated by Marsh's original observations [1952] on the relaxing factor) on the function of the sarcoplasmic reticulum, and we know that it serves as a $Ca^{++}$ pump and is the intermediary for excitation-contraction coupling. Realizing then that I may be accused of redundancy, I will say that the sarcoplasmic reticulum is a beautiful model for study in association with problems related to muscle as a food.

Greaser, Cassens, Hoekstra, and Briskey (1969*a, b, c*) have studied the membrane fine structure of sarcoplasmic reticulum. These studies were prompted by the observation that sarcoplasmic reticulum loses its biochemical activity as a function of post mortem time (Greaser, Cassens, and Hoekstra, 1967; Greaser, Cassens, Hoekstra, Briskey, Schmidt, Carr, and Galloway, 1969), and the rate of loss is closely associated with the

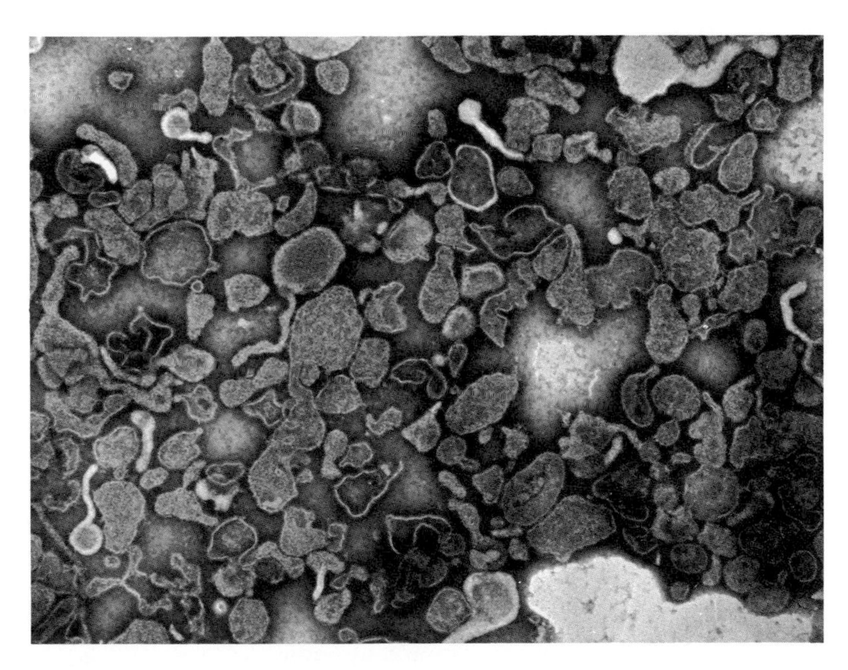

*Fig. 33.14.* Sarcoplasmic reticulum fragments negatively stained with 1% PTA. × 80,000. From Greaser, Cassens, Hoekstra, and Briskey (1969*b*).

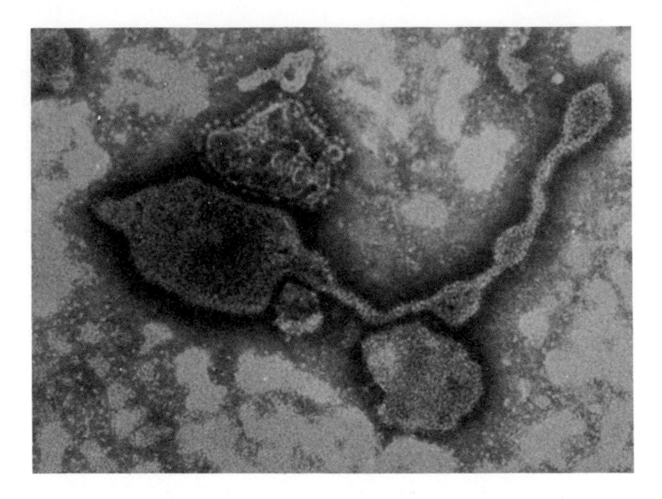

*Fig. 33.15.* Membrane fragments negatively stained with 1% PTA. × 160,000. From Greaser, Cassens, Hoekstra, and Briskey (1969*b*).

*p*H decline curve. The technique of negative staining was used to visualize the membrane fine structure.

Fig. 33.14 shows a typical sarcoplasmic reticulum fraction. Most of the vesicles are roughly spherical, with some having tails of variable length. Both a mitochondrial and a sarcoplasmic reticulum fragment are shown in Fig. 33.15. The surface of the sarcoplasmic reticulum fragment had a fine granular appearance when viewed face-on, while the mitochondrial fragment had a rough, more irregular surface appearance. This latter membrane was lined with the well-known mitochondrial subunits of approximately 90 Å in diameter. The sarcoplasmic reticulum membranes also showed subunits but they were about 35 Å in diameter with a 50–60-Å center-to-center spacing. These sarcoplasmic reticulum subunits have been reported previously by Ikemoto, Sreter, and Gergely (1966), Engel and Tice (1966), and Martonosi (1968). In the studies with pig muscle sarcoplasmic reticulum it was surprising that the subunit structure of 24-hour post mortem and biochemically inactive preparations was indistinguishable from 0-hour preparations. Thus, the subunit structure did not appear to be the critical property of the membranes for $Ca^{++}$ accumulation. Fig. 33.16 demonstrates the appearance of calcium-oxalate-loaded membranes. These irregular deposits were seen after incubation of the preparations in a medium containing ATP, $MgCl_2$, $K_2C_2O_4$ and $CaCl_2$. They were never observed if one or more of the essential ingredients were omitted from the incubation medium. The shape of the

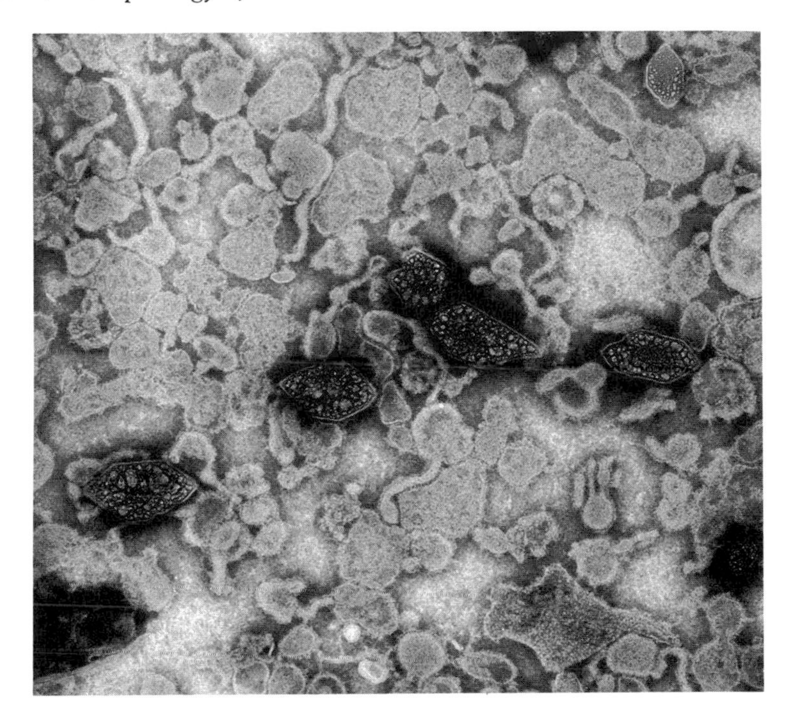

*Fig. 33.16.* Sarcoplasmic reticulum fragments after $Ca^{++}$ accumulation. Negatively stained with 1% PTA. Calcium oxalate deposits were found in approximately 20% of the vesicles. The dense stain layer next to the loaded vesicles suggested that they were not as flattened as nonloaded ones. × 80,000. From Greaser, Cassens, Hoekstra, and Briskey (1969*b*).

calcium-oxalate–loaded vesicles was markedly altered, the loaded membrane fragments having a strikingly angular appearance compared to normal vesicles.

## THE FUTURE

The future for the study of morphology of muscle as a food holds great promise. On the one hand, consider the techniques available that have not yet been applied with great authority to muscle as a food—scanning electron microscopy, electron microscope histochemistry, microincineration, electron microprobe analysis, polarizing microscopy, interference microscopy, autoradiography, and others. On the other, consider the problems of muscle as a food that lend themselves to the available techniques—characteristic differences of muscles and species, muscle abnormalities, muscle growth, and differentiation and others. The tools and problems are available.

ACKNOWLEDGMENTS

This contribution was prepared during the term of support by grant no. FD–00107–11 (USPHS). The author thanks E. J. Briskey for his advice and encouragement.

## *References*

Cassens, R. G., E. J. Briskey, and W. G. Hoekstra. 1963*a*. Electron microscopy of post-mortem changes in porcine muscle. *J. Food Sci. 28*:680.

———. 1963*b*. Similarity in the contracture bands occurring in thaw rigor of muscle and other violent treatments. *Biodynamica 9*:165.

Cassens, R. G., C. C. Cooper, and E. J. Briskey. 1969. Giant fibers in skeletal muscle of adult animals. *Acta Neuropath. 12*:300.

Cassens, R. G., C. C. Cooper, W. G. Moody, and E. J. Briskey. 1969. Histochemical differentiation of fiber types in developing porcine muscle. *J. Anim. Morph. Physiol. 15*:135.

Cooper, C. C., B. B. Breidenstein, R. G. Cassens, G. Evans, and R. W. Bray. 1969. The influence of marbling and maturity on the palatability of beef muscle II. Histological considerations. *J. Anim. Sci. 27*:1542.

Cooper, C. C., R. G. Cassens, and E. J. Briskey. 1969. Capillary distribution and fiber characteristics in skeletal muscle of stress-susceptible animals. *J. Food Sci. 34*:299.

Cooper, C. C., R. G. Cassens, L. L. Kastenschmidt, and E. J. Briskey. 1970. Histochemical characterization of muscle differentiation. (Submitted.)

Engel, A. G., and L. W. Tice. 1966. Cytochemistry of phosphatases of the sarcoplasmic reticulum I. Biochemical studies. *J. Cell Biol. 31*:473.

Greaser, M. L., R. G. Cassens, E. J. Briskey, and W. G. Hoekstra. 1969*a*. Postmortem changes in subcellular fractions from normal and pale, soft exudative porcine muscle 1. Calcium accumulation and adenosine triphosphatase activities. *J. Food Sci. 34*:120.

———. 1969*b*. Post-mortem changes in subcellular fractions from normal and pale, soft, exudative porcine muscle 2. Electron microscopy. *J. Food Sci. 34*:125.

Greaser, M. L., R. G. Cassens, and W. G. Hoekstra. 1967. Changes in oxalate-stimulated calcium accumulation in particulate fractions from post-mortem muscle. *J. Agr. Food Chem. 15*:1112.

Greaser, M. L., R. G. Cassens, W. G. Hoekstra, and E. J. Briskey. 1969*a*. Effect of *p*H-temperature treatments on the calcium accumulating ability of purified sarcoplasmic reticulum. *J. Food Sci. 34*:633.

———. 1969*b*. Purification and ultrastructural properties of the calcium accumulating membranes in isolated sarcoplasmic reticulum preparations from skeletal muscle. *J. Cell. Physiol. 74*:37.

———. 1969*c*. Effects of diethyl ether and thymol on the ultrastructural and biochemical properties of purified sarcoplasmic reticulum fragments from skeletal muscle. *Biochim. Biophys. Acta 193*:73.

Greaser, M. L., R. G. Cassens, W. G. Hoekstra, E. J. Briskey, G. R. Schmidt, S. D. Carr, and D. E. Galloway. 1969. Calcium accumulating ability and compositional differences between sarcoplasmic reticulum fractions from normal and pale, soft, exudative porcine muscle. *J. Anim. Sci. 28*:589.

Herring, H. K., R. G. Cassens, and E. J. Briskey. 1965. Sarcomere length of free and restrained bovine muscles at low temperature as related to tenderness. *J. Sci. Food Agr. 16*:379.

Herring, H. K., R. G. Cassens, G. G. Suess, V. H. Brungardt, and E. J. Briskey. 1967. Tenderness and associated characteristics of stretched and contracted bovine muscles. *J. Food Sci. 32*:317.

Huxley, H. E. 1969. The mechanism of muscular contraction. *Science 164*:1356.

Ikemoto, N., F. A. Sreter, and J. Gergely. 1966. Localization of Ca-uptake and ATPase activity in fragments of sarcoplasmic reticulum. *Fed. Proc. 25*:465.

Kastenschmidt, L. L., W. G. Hoekstra, and E. J. Briskey. 1968. Glycolytic intermediates and co-factors in "fast" and "slow-glycolyzing" muscles of the pig. *J. Food Sci. 33*:151.

Lister, D., R. A. Sair, J. A. Will, G. R. Schmidt, R. G. Cassens, W. G. Hoekstra, and E. J. Briskey. 1970. Metabolism of striated muscle of "stress-susceptible" pigs breathing oxygen or nitrogen. *Amer. J. Physiol. 218*:102.

Locker, R. H. 1960. Degree of muscular contraction as a factor in tenderness of beef. *Food Res. 25*:304.

Locker, R. H., and C. J. Hagyard. 1963. A cold shortening effect in beef muscles. *J. Sci. Food Agr. 14*:787.

Marsh, B. B. 1952. The effects of ATP on the fibre volume of a muscle homogenate. *Biochim. Biophys. Acta 9*:247.

Marsh, B. B., and N. G. Leet. 1966. Studies in meat tenderness III. The effects of cold shortening on tenderness. *J. Food Sci. 31*:450.

Martonosi, A. 1968. Sarcoplasmic reticulum V. The structure of sarcoplasmic reticulum membranes. *Biochim. Biophys. Acta 150*:694.

Moody, W. G., and R. G. Cassens. 1968*a*. A quantitative and morphological study of bovine longissimus fat cells. *J. Food Sci. 33*:47.

———. 1968*b*. Histochemical differentiation of red and white muscle fibers. *J. Anim. Sci. 27*:961.

Morita, S., R. G. Cassens, and E. J. Briskey. 1969. Localization of myoglobin in striated muscle of the domestic pig: benzidine and $NADH_2$-TR reactions. *Stain Tech. 44*:283.

———. 1970. Histochemical localization of myoglobin in skeletal muscle of rabbit, pig, and ox. *J. Histochem. Cytochem.* (In press.)

Morita, S., C. C. Cooper, R. G. Cassens, L. L. Kastenschmidt, and E. J. Briskey. 1970. A histochemical study of myoglobin in developing muscle of the pig. (Submitted.)

Quass, D. W., and E. J. Briskey. 1968. A study of certain properties of myosin from skeletal muscle. *J. Food Sci.* *33*:180.

Sheldon, H. 1969. Morphology and growth of adipose tissue. *Meat Industry Res. Conf. Proc.* (Chicago, Ill.).

# Chapter 34

# *The Physiology of Animals and the Use of Their Muscle for Food*

D. LISTER

It is not always appreciated that the meat we eat was once a part of a living system. The animal from which it came may have been a few days, months, or even years old, and may have been wild, or domesticated and intensively housed. It may have suffered a long and agonizing death, or it may have been slaughtered in a quick and humane fashion. The manner of death is known to be of considerable consequence to the quality of meat produced, but only recently have the implications of methods of rearing and of genotype environment interactions generally been considered to any degree. In the following pages I want to consider the anatomical and physiological characteristics of meat-producing animals both in the light of current breeding and husbandry practices and of the consequences of slaughter.

## SOMATIC DEVELOPMENT

### *General considerations*

Animals which are reared primarily for meat are slaughtered, usually, at a relatively early stage of development. The extent of postnatal life may be only as long as the time spent *in utero*. For example, pigs and lambs are commonly slaughtered before they become sexually mature; beef cattle usually are mature, but they will still be much younger than the age at which the epiphyses of the long bones fuse (Crichton, Aitken, and Boyne, 1960).

In animal-breeding programs selection pressure is applied to a considerable extent on the propensity to grow rapidly, to deposit lean tissue preferentially and to do this with the maximum efficiency of conversion of food to body tissue. How far the hormonal status of growing animals is affected by such procedures is debatable; how the animals will respond to different types of husbandry, transport, and slaughter, is even more so.

*Anatomy and physiology*

The overwhelming importance of growth in relationship to every function of the body was recognized by Edwards (1824): "Dans la jeunesse tout parait tendre au développement et à l'acroissement du corps." For example, 90% of the protein nitrogen in the food of the newborn piglet is incorporated into the tissues and never presents itself for excretion by the kidney as urea (McCance and Widdowson, 1956), and yet the piglet is still fully capable of maintaining blood urea and creatine within normal limits provided growth is taking place as it should. Growth is as important as the lungs, kidneys, and other integrated organs in maintaining homeostasis in the young animal.

Adequate nutrition is of prime importance to form and functional changes, but even when food is restricted to an extent which prevents the normal increase in the body weight of young animals, some anatomical and physiological reorganization takes place (Hogan, 1928; Jackson, 1932; McCance, 1960). Such differential development (Huxley, 1932) is thought to provide for the functional needs of animals, often irrespective of their rate of growth (Elsley, McDonald, and Fowler, 1964; Fowler, 1968). Individual tissues develop in the order nervous, skeletal, muscular, and adipose, and this order has been shown to be common to farm animals (Hammond, 1932; Pàlsson, 1939, 1940; McMeekan, 1940*a, b, c*), birds (Wilson, 1952, 1954*a*), and to man (Jackson, 1909; Scammon, 1930).

The consequences of developmental changes in the relative proportions of the tissues and organs of growing animals are reflected in their metabolism. Table 34.1, for example, accounts for the resting heat production or basal metabolic rate of an adult man in terms both of the contribution a tissue or organ makes in the proportion it represents of the total body, and its contribution on a unit weight basis. It will be obvious that changes in the proportions of kidney or of muscle will have quite different effects on basal metabolic rates, and this is borne out empirically. Thus, an adult man with approximately 43% muscle (Wilmer, 1940) has a lower basal metabolism than a newborn baby, which contains about 25% muscle. This difference is largely accounted for by the greater proportion of brain, kidneys, liver, and heart in babies, all of which contribute appreciably to resting heat production (Column 3, Table 34.1).

Fowler (1968) has attempted to account for intake and requirements for metabolizable energy (which are, of course, related to resting metabo-

TABLE 34.1

*Resting heat production of a 35-year-old man, weight 75 kg, height 170 cm*

| Organ | cal/min | Basal metabolism (%) | cal/g/min |
|---|---|---|---|
| Whole body | 1200 | 100.0 | 0.016 |
| Muscle | 456 | 38.0 | 0.018 |
| Liver | 149 | 12.4 | 0.057 |
| Intestinal tract | 91 | 7.6 | 0.032 |
| Kidneys | 90 | 7.5 | 0.171 |
| Spleen | 76 | 6.3 | 0.032 |
| Heart | 53 | 4.4 | 0.083 |
| Brain | 36 | 3.0 | 0.036 |
| Pancreas | 16 | 1.3 | 0.069 |
| Blood | 13 | 1.1 | 0.002 |
| Salivary glands | 8 | 0.7 | 0.056 |

Reproduced from Lewis (1962), by permission of the publishers.

lism) in terms of the major tissues of the growing pig (Fig. 34.1). It is clear that the intake of metabolizable energy increases as the animal increases in size. Voluntary food intake stabilizes, however, when an animal approaches about 50% of its mature weight (Oslage and Fliegel, 1965; Kay, 1967) and this may be a response to the accumulation of fat (Graham, 1969). The need for an increased intake of energy is accounted for, in part, by the higher maintenance requirement of an increased overall size. In addition, the efficiency of utilization of metabolizable energy reflects the difference in the energy cost of the deposition of protein and fat. The cost of protein deposition is only approximately 7 kcal of metabolizable energy per 1 g of protein, as opposed to almost 12 kcal for each 1 g of fat (Kielanowski, 1965), which is deposited preferentially in adults. The efficiency of conversion of food, i.e., the amount of food required to produce unit gain in live weight, will fall, therefore, as an animal increases in size, and the fall will be greater if the animal shows a tendency to lay down excessive amounts of fat. The practical benefits, in terms of food consumption and utilization, to be derived from a reduced ability to deposit fat, or, alternatively, a propensity to deposit muscle, are clear.

*The effect of hormones*

The preceding paragraphs have indicated that as animals age their rate of increase in overall size is matched, relatively speaking, by some tissues and organs; some exceed it, whilst others fall behind. This implies some degree of internal control which, in the main, is exercised by the endocrine system.

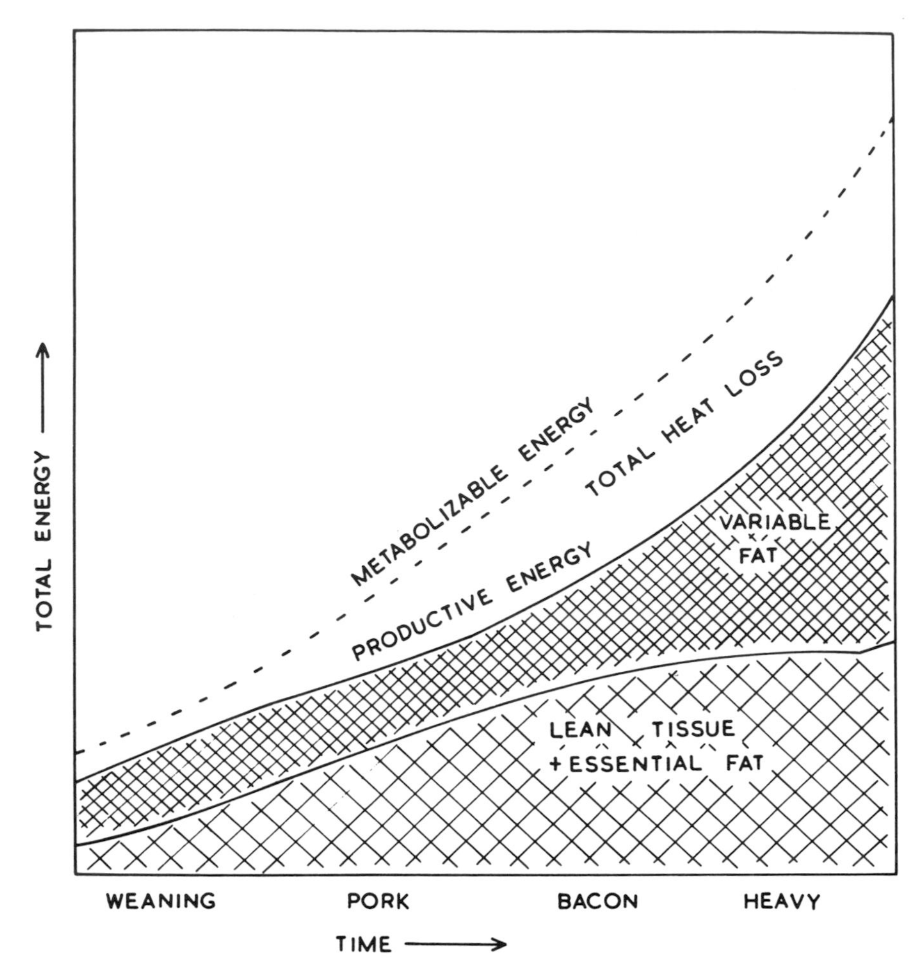


*Fig. 34.1.* Partition of metabolizable energy in the growing pig on unrestricted feeding (Fowler, 1968) .

*A. Thyroid.* Absence or atrophy of the thyroid is incompatible with satisfactory growth and development (Horsley, 1886) . In humans, cretinism results, some of its symptoms being reduced skeletal and somatic growth, arrested sexual development, and subnormal size of liver, kidney, and heart. There may be, however, an increased retention of nitrogen despite the slower growth of hypothyroid animals (see Table 34.2) and the slower release of animo acids from peripheral tissues (Bondy, 1949) might suggest increased protein synthesis. The rate of basal metabolism

TABLE 34.2

*Influence of thyroidectomy for 12 months on adult male rats*

| Organ and treatment | Organ wt (g) | $H_2O$ (%) | Protein (g) Total | Protein (g) % dry |
|---|---|---|---|---|
| Liver | | | | |
| Thyroid X | 11.0 | 67.3 | 1.6 | 46.2 |
| Control | 14.9 | 68.0 | 2.5 | 52.2 |
| Kidney | | | | |
| Thyroid X | 2.1 | 72.5 | 0.3 | 57.3 |
| Control | 3.1 | 73.9 | 0.5 | 56.3 |

| | Body wt (g) | N balance* (g/kg/day) |
|---|---|---|
| Thyroid X | 354–483 | +0.33 |
| Control | 356–565 | +0.25 |

* Nitrogen balance prior to autopsy.

Reproduced from Leathem (1964), by permission of the publishers.

may fall to less than 50% of normal and there may be increased susceptibility to infection.

Less dramatic effects of the thyroid are associated with normal growth and development, and the implications, as far as farm animals are concerned, have recently been reviewed in detail by Falconer and Draper (1968) and need not be extensively considered here. It is important to note the general decline in thyroid activity and basal metabolic rate with increasing age (Post and Mixner, 1961). Thyroid activity considerably above the normal rate may also lead to slower rates of growth (Fig. 34.2). The slow rate of growth in hyperthyroidism may result from the reduced efficiency of utilization of food at the higher basal metabolic rate and an increased voluntary food intake may not be sufficient to meet the requirements for maintaining normal rates of growth. There appears to be decreased protein synthesis (Leathem, 1964) and an association between thyroxine and an uncoupling effect on oxidative phosporylation has been suggested (Hoch, 1962; Smith and Hoijer, 1962).

*B. Hypothalamus.* The regulation of energy balance is also exercised by the activities of other endocrine glands, and the topic has recently been reviewed by Hervey (1969). There is appreciable evidence that an important part of the controlling mechanisms is centered in the ventromedial nuclei of the hypothalamus (Hetherington and Ranson, 1939; Brobeck, 1946; Kennedy, 1950); lesions produced in these sites in rats will induce gross obesity, and fatty livers and full stomachs will be

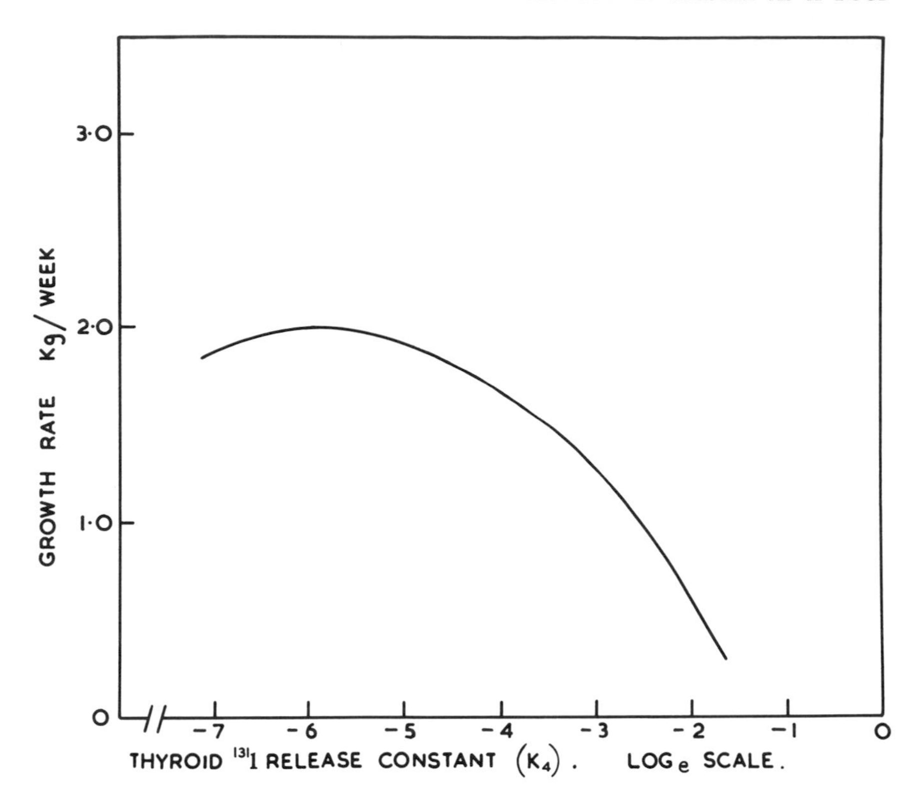


*Fig. 34.2.* Correlation between the growth rate and thyroid activity of lambs kept in a controlled environment and fed *ad lib* a standard diet. Highly significant curvilinear correlation; $P < 0.001$, variance ratio = 17.3. Falconer and Draper (1968).

revealed when the animals are autopsied. The obesity is known to result from a loss of appetite control and consistently reduced activity. The association can be seen in the opposite sense, on a more moderate scale, in normal female rats during estrus when activity is increased and food intake depressed (Wang, 1923; Slonaker, 1925; Kennedy and Mitra, 1963). The consequence of this is a loss in body weight (Brobeck, Wheatland, and Strominger, 1947), but the loss is made good during diestrus when activity declines and intake rises.

Various hypotheses have been put forward to identify an internal regulatory system for energy balance. These are: (1) Brobeck's (1948) "thermo regulatory" theory aimed at the maintenance of body temperature; (2) Mayer's (1955) "glucostatic" theory which suggests a response to differences between arterial and venous plasma concentrations of glucose; and (3) Kennedy's (1953) suggestion that the long-term modification of food intake and bodily activity is a response to changes in the amount of fat in the body, fat acting as an "energy-memory." Hervey (1969) considers the "lipostatic" theory most plausible and extends it by suggesting a possible sensory mechanism for the level of fatness in the body. He considers that different amounts of body fat would provide different dilutions of a suitable hormone to which the central nervous system is sensitive, the degree of dilution acting as an internal indicator of obesity or leanness. Progesterone has been implicated, but its effect would be primarily in females and in, say, early pregnancy when the storage of fat is to be initiated (Hytten, 1963). Adrenal glucocorticoids seem nearer to fulfilling the requirement and, under conditions similar to those when the effect of progesterone has been studied, adrenalectomy leads to accumulation of lean in the carcass (Hervey, pers. comm., 1966).

*C. Adrenals.* The effects of adrenal steroids on somatic development do not appear to be quite so dramatic as are their effects on electrolyte balance and during stress, which will be considered below. In many ways adrenocortical hormones are similar in their action to thyroxine in that they may be anabolic or catabolic depending on the output and tissue. On the whole, they create a less favorable nitrogen balance in the body, perhaps as a consequence of their capacity to mobilize stores of protein. Some tissues, e.g., the liver, can accumulate protein by the incorporation of amino acids simultaneously derived from the breakdown of carcass protein under the influence of corticosteroids (Silber and Porter, 1953; Wilmer and Foster, 1960). Several processes may contribute to the effect of corticoids on muscle: the reduced rate of incorporation of amino acids, (Clark, 1953; Fritz, 1956), the increased breakdown of protein, or the decreased synthesis of protein.

*D. Pituitary.* Hypophysectomy restricts growth and even induces a loss of body weight. It limits the retention of nitrogen and the incorporation of amino acids into liver and muscle (Kostyo and Knobil, 1959; Manchester and Young, 1959). Replacement therapy, however, does not require complete pituitary extracts, but growth hormone is essential for a uniform distribution of protein (Scow, 1959; Donovan and Jacobsohn, 1960), and it may be that it is this mechanism which is elicited by thyroid administration to induce the latter's growth-promoting effect. Growth hormone would, therefore, seem to be the most important of the pituitary hormones in the present context, although the exact function of the hormone has not been elucidated (Hunter, 1968).

Apart from its effects on nitrogen metabolism (Leathem, 1964) growth hormone has, perhaps, a more established function in increasing lipolysis to maintain fuel supplies for tissues; the trigger mechanism for this operation may be related to the level of blood glucose (Hunter, 1968).

*E. Gonads.* Androgens and estrogens are known to stimulate growth and nitrogen retention to an extent which varies from one species to another, and androgens would seem to be more effective than ovarian steroids (Leathem, 1964; Kochakian, 1966; Bradfield, 1968). The androgen-induced nitrogen retention is largely due to a protein anabolic action on skeletal muscle. Not all muscles respond uniformly, however; the acromiotrapezius of the rat, for example, shows a high response to androgens (Korner and Young, 1955), whereas there is a distinct lack of response in the thigh muscles (Scow and Hagan, 1957). Castration decreases the rate of growth of animals and the response found in entire males and females to improved planes of nutrition (Prescott, 1963; Robinson, 1964; Lister, 1965).

The integration of the activities of the various endocrine glands and, in particular, the central role of the adenohypophysis has been extensively covered by McCann, Dhariwal, and Porter (1968) and indicates the enormous strides which have been made in the field in recent years. For the present, it is sufficient to consider in outline the regulation of adrenal and thyroid activity by adrenocorticotrophic hormone (ACTH), and thyrotropin (TSH), both of which are produced by the pituitary, and the release of growth hormone.

The part played by the hypothalamus in increasing the rate of secretion of ACTH through its corticotropin-releasing factor (CRF) seems to be well established (Porter, Dhariwal, and McCann, 1967). Lesions produced in the supposed active sites of release of CRF lead to the secretion by the adrenals of corticosterone at very low rates in response to stress. Thyroxine will stimulate the release of ACTH (Bohus, Lissak, and Meyer, 1965–66) although extensive stress will diminish it, and the further synthesis and secretion of ACTH is inhibited by its own concentration in blood (Vernikos-Danellis and Trigg, 1967).

The hypothalamus is also responsible for the control of TSH secretion (Guillemin, Yamazaki, Jutisz, and Sakiz, 1962), and TSH-releasing factor (TRF) has been purified from hypothalamic tissue taken from the pig, sheep, ox, and from man (Guillemin, Burgus, Sakiz, and Ward, 1966; Schally, Bowers, and Redding, 1966*a, b;* Schally, Bowers, Redding, and Barrett, 1966). The secretion rate of TSH is affected by thyroid hormones and TRF, which tend to reduce it; Sinha and Meites (1965–66) have shown that TSH-releasing activity is much greater in hypothalamic tissue derived from thyroidectomized rats. Rees (1966) has shown that thyroxine and triiodothyronine lowered the TSH content of the pituitary when they were given in high doses and it is now felt that the concentration of TRF in the fluid surrounding the pituitary determines the secretory rate of TSH (Greer, Matsuda, and Scott, 1966).

The relationship between thyroid hormone levels and growth hormone secretion, as pointed out earlier, has long been known, thyroidectomy resulting in a decrease in growth hormone secretion and cessation of growth (Schooley, Friedkin, and Evans, 1966). The increased rate of release of growth hormone during fasting is well documented (Roth, Glick, Yallow, and Berson, 1964) but this effect is reduced in thyrotoxicosis and when adrenal corticoids are administered (Frantz and Rabkin, 1964; Burgess, Smith, and Merimee, 1966). The involvement of the hypothalamus as the source of production of growth-hormone-releasing factor is now established, and its physiological role has been demonstrated in the response, for instance, to hypoglycemia (Krulich and McCann, 1966) or after hypophysectomy (Müller, Arimura, Saito, and Schally, 1967).

It follows from the way in which patterns of growth are established in animals that the delicate and intricate interrelationships of various hormones which bring about developmental changes must, perforce, be altered if selection pressures are applied for particular characteristics, e.g., to bring about differences in the proportions and distribution of lean tissue. As a primary consideration, selection for increased nitrogen deposition alone must, to some extent, become a selection against the protein catabolic actions of, for example, excessive thyroid and adrenal secretion. But it will be obvious that the problem cannot be so simply resolved, and the whole basis of the endocrinological control of energy balance must be involved.

## LIVING MUSCLE TO FOOD

The conversion of muscle to the component tissue of a cut of meat can be summarized as the effects of the degradation of ATP in the period from death to post rigor (Bendall, 1966*a*). It is becoming increasingly obvious that it is the factors which affect the rate and extent of the changes brought about by the disappearance of ATP that confer on meat its excellence or unacceptability. It is true that commercial handling practices after slaughter can influence the subsequent quality of meat, but they can only do this within limits set by the physiological and biochemical characteristics of an animal before and at the time of slaughter.

The importance of ATP in life is in the provision of energy and, in this context, for the musculature in particular. Bendall (1966*b*, 1969) has summarized the function of ATP in muscle and it is sufficient here to state its three main uses: to drive the $Na^{+}$-$K^{+}$ pump in the plasmalemma;

to drive the $Ca^{++}$ pump in the longitudinal elements of the sarcoplasmic reticulum; and as the immediate source of contractile energy. The energy needed for each of the three processes is of the order 1:100:1,000 respectively, the energy required for contraction being overwhelmingly the most expensive of the three, with that required for the $Ca^{++}$ pump a relatively close second. After death the minimal rate of breakdown of ATP is probably that required by the $Ca^{++}$ pump (Bendall, 1969), for it is known that myofibrillar adenosine triphosphatase activity is strongly inhibited by falling *p*H in the range found during rigor onset (Perry, 1956; Bendall, 1960; Wismer-Pedersen, 1966).

It is clear that in order to maintain an appropriate energy supply extremely efficient resynthetic mechanisms are required for the working muscle, and it is the extent of interference with the resynthetic mechanisms by preslaughter handling and the slaughter processes which concern us most.

Even at rest the contribution which muscle makes to body heat production is about 38% of the total (see above), and most of this is contributed by the respiratory muscles (Lewis, 1962). The utilization of ATP is almost minimal, and resynthesis by the system of oxidative rephosphorylation sited mainly in the mitochondria is entirely adequate. During vigorous exercise, or under any physiological circumstances when energy production is called for and the oxygen supply is limiting, the inefficient anaerobic resynthesis of ATP by the glycolytic system of the sarcoplasm is utilized and muscle glycogen is converted to lactic acid.

To ensure a slow rate of decline in the *p*H of muscle after death, it is necessary to have high levels of ATP and creatine phosphate (CP) present in the muscle initially (Bate-Smith and Bendall, 1956). This is also an essential requirement for the production of meat of optimal quality (for extensive review see Lawrie, 1966). If these criteria are to be met, it is clear that, both before and after death, either the rate of breakdown of ATP in muscle must be kept very low and minimal, or the rate of resynthesis of ATP must be adequate for current needs and to maintain an appreciable balance of CP. In practice, any condition which increases the rate of breakdown of ATP, e.g., the increase of myofibrillar adenosine triphosphatase activity when muscle is stimulated, or which reduces the rate of resynthesis, must bring about a reduction in *p*H (Needham, 1960). In dying muscle, for instance, even though adenosine triphosphatase activity is minimal, anaerobic resynthesis will not be sufficient to maintain normal levels of ATP indefinitely, and consequently the *p*H falls.

The relationship between myofibrillar and sarcoplasmic adenosine triphosphatase activity and *p*H is by no means clear. It is usually accepted that the initial fall in *p*H from the *in vivo* value to that found immediately after death results from myofibrillar adenosine triphosphatase activity and the subsequent changes from nonmyofibrillar adenosine triphosphatase activity at lower *p*H (Bendall, 1960). It is possible, however, to increase the rate of *p*H fall in muscle throughout its entire time course by eliciting tetanic contractions which, in themselves, do not reduce the initial *p*H of the muscle significantly (Hallund and Bendall, 1965; Bendall, 1966*a*). Some animals, in which the muscle *p*H shortly after death is relatively low and in which a rapid rate of change of *p*H is observed subsequently until rigor is complete, have been shown to have higher myosin adenosine triphosphatase activity (Quass and Briskey, 1968). A higher rate of sarcoplasmic adenosine triphosphatase activity, which might have been expected, has proved difficult to demonstrate in similar animals (Bendall and C. C. Ketteridge, unpublished observations, 1969; Greaser, Cassens, Hoekstra, Briskey, Schmidt, Carr, and Galloway, 1969).

The animal which is to be slaughtered provides additional problems, for the removal of blood precludes the possibility of recovery which, in life, would relieve the consequences of lactate formation (Cori, 1931); the nutritional status of the animal also has some bearing, for the reserves of glycogen in the muscle will determine the extent of the change in its *p*H post mortem (Bernard, 1877; Bate-Smith and Bendall, 1956).

*The physiological implications of slaughter practices*

Animals are normally taken to a central plant for slaughter and this might involve their traveling for long or short distances, with or without a commensurate period of fasting. They may be held in small or large groups, and may fight or remain relatively quiescent, until presented for stunning by one of a variety of methods which include poleaxing, electrical, carbon dioxide, or captive-bolt stunning. Then, perhaps even with no preslaughter treatment, they are bled by severance of the carotid artery and the jugular vein, or the vena cava. There would seem to be little point, however, in comparing all the possible combinations of preslaughter handling and stunning techniques. Perhaps more could be gained by a consideration of the general effects of long- and short-term nervous and muscular stimulation and the ways in which they affect the subsequent changes in muscle post mortem.

The act of sticking and draining the blood from the body of an animal removes the means whereby the aerobic resynthesis of ATP and ADP can be maintained in the conversion of muscle glycogen or fatty acids to carbon dioxide and water. Additionally it also prevents the transport to other parts of the body of microorganisms which may leak through the gut wall into the blood stream at death, especially in fatigued or starved animals, and provide a source of taint in the carcass (Ficker, 1905; Haines, 1937; Robinson, Ingram, Case, and Benstead, 1953). Efficient bleeding is therefore considered to be essential for the production of meat with a long shelf-life. The value of this would seem questionable, however, for as much as 50% of the total blood volume may be retained in the whole body (Thornton, 1949) and even the psoas and longissimus dorsi may contain appreciable amounts (Lawrie, 1950). However, it may well be that the prevention of spoilage is the argument of most consequence for draining the blood from an animal, for it has been shown that the maintenance of high blood $pO_2$ is not essential to the production of good-quality meat provided the rate of breakdown of ATP in muscle is minimal (Lister, Sair, Will, Schmidt, Cassens, Hoekstra, and Briskey, 1970). In any case, the normal biochemical changes in the production of meat from living tissues are essentially anaerobic and more extensive and prolonged than may occur *in vivo*. It would seem that the rate and extent of anaerobic glycolysis, and therefore of *p*H, would be of greater importance in determining the characteristics of muscle after death, and the ways in which these may be affected are considered next.

It is quite clear that, under commercial conditions, animals which are to be slaughtered are subjected to various degrees of stimulation of one kind or another. One argument put forward to explain the variation in the quality of meat is that it is brought about as a consequence of differences in the metabolism of muscles of animals induced, to a greater or lesser extent, when different types of animals are subjected to different stressors. It is commonly observed, for example, that the quality of muscle produced from animals stunned electrically by an inexperienced man is frequently poorer than that from animals stunned by one who is experienced. There seems no reason to suppose that different animals forced into a dark pen to be anesthetized with carbon dioxide, for example, should not respond in different ways both physiologically and metabolically. The degree of response would seem to be one which varies from one animal, or group of animals, to another and ultimately leads to the concept of stress susceptibility and nonsusceptibility in animals (Briskey and Lister, 1969).

The effects of stimulation of muscle on meat quality were first examined by Bendall (1966*a*). In his experiments, Bendall curarized Large White pigs and, by applying artificial respiration, maintained them in this condition, under which no nervous impulses could reach the muscles for a period before sticking. This treatment produced an extremely high *p*H in the longissimus dorsi muscle after bleeding and a slow rate of *p*H fall subsequently when comparisons were made with muscles from electrically stunned control animals (Fig. 34.3a). When Sair, Lister, Moody, Cassens, Hoekstra, and Briskey (1970) repeated the experiment with Chester White and Poland China pigs, curarization induced higher levels of *p*H, ATP, and CP in the muscle, but the effect was not nearly so marked as Bendall had found earlier with Large Whites. The rates of biochemical change in muscle from curarized Poland China pigs were still appreciably higher than were found in that of conventionally slaughtered Large White pigs. The metabolism of muscle from Chester Whites, on the other hand, resembled that of Large White pigs. It was also found (Lister, Sair, Will, Schmidt, Cassens, Hoekstra, and Briskey, 1970) that Poland China pigs accumulate appreciable amounts of lactic acid in the longissimus dorsi muscle when they are made anoxic and that Chester Whites do not do so to the same extent; nor, presumably, do Large Whites. This cannot be the whole story, for Pietrain pigs (Fig. 34.3b), which are comparable with Poland China animals (Lister, Scopes, and

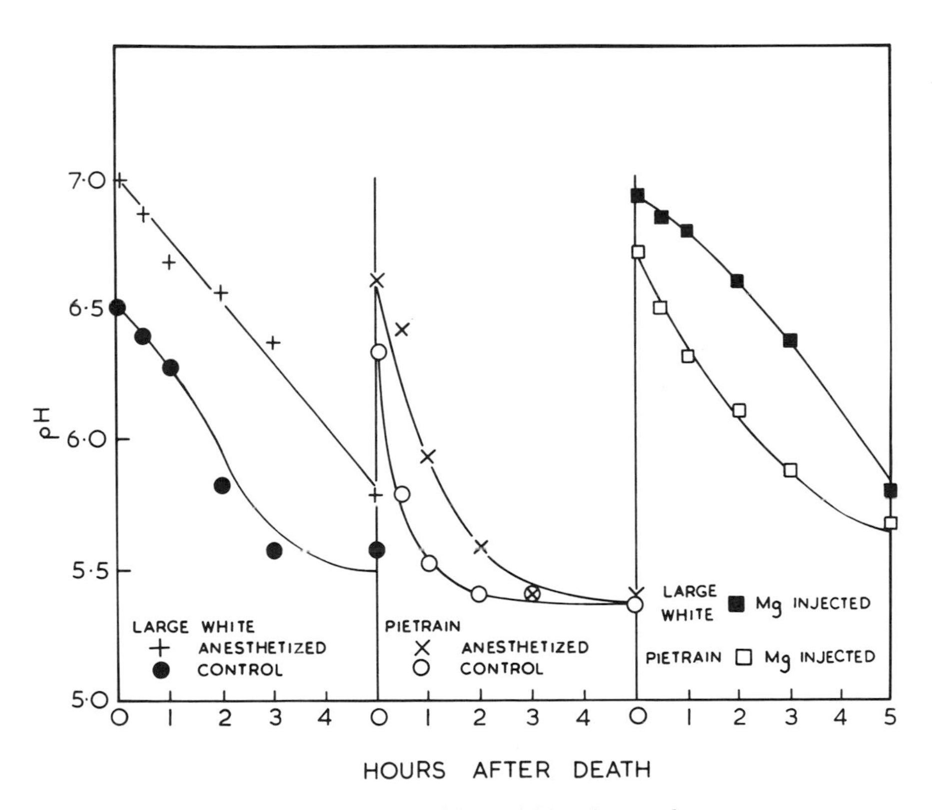


*Fig. 34.3.* *p*H changes in Large White and Pietrain muscle post mortem.

Bendall, 1969), when injected with magnesium sulphate solution to block adenosine triphosphatase activity, show moderately high muscle *p*H at death and slower rates of *p*H change afterwards (Fig. 34.3c). However, the blood drained from the animals at sticking may be almost black in color with a correspondingly low $pO_2$ (Lister, unpublished observation, 1969). The indication emerges that struggle at the time of slaughter can only result in a lowering of the *p*H of a sample of muscle taken from an animal shortly after sticking, and the *p*H changes in the muscle subsequently will be more or less rapid. That some animals will suffer an anoxic response is also clear, and these animals (Poland China, Pietrain) may be classified as stress-susceptible; but the basic cause of the undesirable changes in *p*H appears to be the mechanisms which are responsible for regulating the rate of ATP turnover in muscle (Lister, Scopes, and Bendall, 1969).

*Hormones and muscle metabolism*

It was suggested earlier that selection programs to improve the carcasses of meat-producing animals were also likely to bring about changes in their hormonal balance. If this were so, the question then arises as to how this might, in turn, influence the metabolism of muscle to produce adverse changes in meat.

It has been common practice to account for the paradoxical increase in amount, and decrease in quality, of the meat of contemporary animals in terms of Selye's (1951) "General Adaptation Syndrome." That there is an apparent decrease in the response of the adrenals of some animals to stressful situations is abundantly clear (Topel, 1969*b*), and the reduction in overall size and especially the degeneration of the adrenal cortex are also well documented (Henry and Billon, 1959; Cassens, Judge, and Briskey, 1965; Judge, Briskey, Cassens, Forrest, and Meyer, 1968). The association between these characteristics and the quality of muscle is not completely clear, however. In stress-susceptible animals there is an extremely rapid accumulation of lactic acid in the longissimus dorsi muscle at death. It is not certain whether this arises from a stagnant anoxia induced by circulatory disturbance in adrenal insufficiency (Ingle, 1950; Sprague, 1951), or from the poor clearance of lactate from muscle (Topel, 1968; Lister, Sair, Will, Schmidt, Cassens, Hoekstra, and Briskey, 1970), or from a combination of the two. Whatever the mechanism involved, the accumulation of lactic acid can be reduced, to a limited extent, by treating susceptible animals with $\alpha$-blocking or $\beta$-stimulating adrenergic agents (Lister, Sair, Will, Schmidt, Cassens, Hoekstra, and Briskey, 1970). This treatment might be expected to reduce the consequences of the excessive release of constrictor substances which occurs in adrenal insufficiency (Sheehan, 1948). However, treatments of this kind or with corticosteroids will not prevent the vascular collapse and death of animals which are subject to excessive sympathetic bombardment (Overman and Wang, 1947; Wiggers, Ingraham, Roemhilk, and Goldberg, 1948; Topel, Bicknell, Preston, Christian, and Matsushima, 1968). This latter syndrome is most commonly found to be the cause of death in animals of the "meaty" breeds.

The action of the adrenal catecholamines and glucocorticoids on the microcirculation and the way in which the type of muscle fiber, i.e., red or white, is involved, has been reviewed recently by Topel (1968) and Merkel (1968). Both authors concluded that the picture was by no means clear and that some of the evidence was conflicting. In addition, whilst most of the evidence showed that there were disturbances of the circulation in a tissue, there was no indication how adrenal activity *per se* affected the rate and extent of glycolysis. It may well be that the answer to this problem lies in the adrenal's role in the regulation of electrolyte balance, of cell membrane function and sensitivity and, in turn, on the various adenosine triphosphatases of cell systems (Rothstein, 1968).

The relationship of thyroid activity, muscle metabolism, and meat quality is, again, very vaguely understood, and views on the importance of thyroid function are conflicting (Ludvigsen, 1960; Topel and Merkel, 1966; Judge, Briskey, Cassens, Forrest, and Meyer, 1968; Wismer-Pedersen, 1969). Where measurements of thyroid function, in terms of Protein Bound Iodine (PBI) in serum, have been made (Judge, Briskey, Cassens, Forrest, and Meyer, 1968), the results suggest that animals whose muscle becomes pale, soft, and exudative after death show signs of reduced thyroid activity, i.e., high levels of PBI and low $^{131}I$ uptake. The feeding of goitrogens to pigs for some time before slaughter is known to increase, but not consistently (Topel and Merkel, 1966), the incidence of this condition (Briskey, 1961).

It is difficult to imagine, however, a condition where the thyroid and adrenal glands do not act in concert, and there is ample evidence to show that low activity of one is accompanied by low activity of the other and vice versa. This is especially well documented in the case of the response of animals to cold or heat (Burton and Edholm, 1955; Kirmiz, 1962), and the experiments of Judge, Briskey, Cassens, Forrest, and Meyer (1968) clearly demonstrate the thyroid and adrenal relationship. Apart from any central effects which low thyroid activity may have on thermoregulation in the pig (Ingram and Slebodzinski, 1965), undoubtedly the adequate dissipation of heat by way of the peripheral vasculature (Mount, 1966), as in the hyperthermia of fever, is prevented by modified adrenal activity.

*Changes in muscle and meat quality*

The relationship between the rate of change of *p*H in muscle and the quality of meat in terms of the amount of drip produced, its toughness, and to a lesser extent its color, appears to be a function of the temperature of the muscle when a particular *p*H value is reached. It would seem from recent work (Lister, unpublished observations, 1969) that the amount of drip released from a sample of muscle can be predicted with some accuracy if the temperature is known at which a *p*H of about 6.0 is reached. This current work extends the findings of Wismer-Pedersen and Briskey (1961). It will be appreciated, moreover, that the more rapid the change in *p*H after death, the higher the muscle temperature will be at the lower *p*H values. The resultant denaturation and precipitation of sarcoplasmic proteins on the myofibrils (Bendall and Wismer-Pedersen, 1962; McLoughlin, 1963; Scopes and Lawrie, 1963), is responsible for the production of pale, soft, and exudative pork, the etiology and practical and commercial consequences of which have been well described by Briskey (1964) and Bendall and Lawrie (1964) (see also Topel, 1969*a*).

The phenomenon may not be restricted to pig carcasses alone, for it will be evident that even though the decline in muscle *p*H may be slow, as it is in beef (Marsh, 1954), inadequate refrigeration of a large bulk of carcass may not bring about sufficient reduction in muscle temperature to prevent extensive denaturation of tissue if its *p*H has dropped below 6.0.

The alleged toughness of pale, soft, and exudative meat (Hegarty, Bratzler, and Pearson, 1963) may be associated with the degree of shortening, i.e., reduction of sarcomere length, to be expected when the *p*H of muscle falls rapidly and its temperature is still above 19° C (Locker and Hagyard, 1963). Toughening will also result, and for the same reason, if the rate of *p*H fall is slow and the muscle temperature is lower than 14° C. This is demonstrated in the "cold shortening" effect described by Locker and Hagyard (1963).

The long-term effects of excessive muscular and nervous stimulation have been observed for some time both in pigs and cattle and the mechanism of the effects has been clearly established. Dark-cutting beef was described as long ago as 1774 (Kidwell, 1952) and subsequently associated with Claude Bernard's (1877) phenomenon of "alkaline rigor" (Bate-Smith and Bendall, 1949). A comparable phenomenon became apparent in pigs in Northern Ireland in 1933 when, in order to improve hygienic, sanitary, and humanitarian conditions of slaughter, legislation was introduced to have centralized slaughtering arrangements for meat animals whereas, before, "on the farm" slaughtering had been practiced. There was an immediate and disastrous rise in the incidence of tainting in hams which was eventually traced to the higher ultimate *p*H of the musculature (see Fig. 34.4) arising from the excessive depletion of glycogen brought about during movement, either motorized or ambulatory, from the farm to the slaughterhouse. The taint organisms responsible were shown to flourish at the prevailing higher *p*H produced from glycogen-depleted muscle (Callow, 1937; Ingram, 1962). Glazy bacon (Callow, 1935) is another consequence of high ultimate *p*H, and is usually held to be undesirable because of its fiery appearance and stickiness. Dark-cutting meat has, however, been declared by some authors to

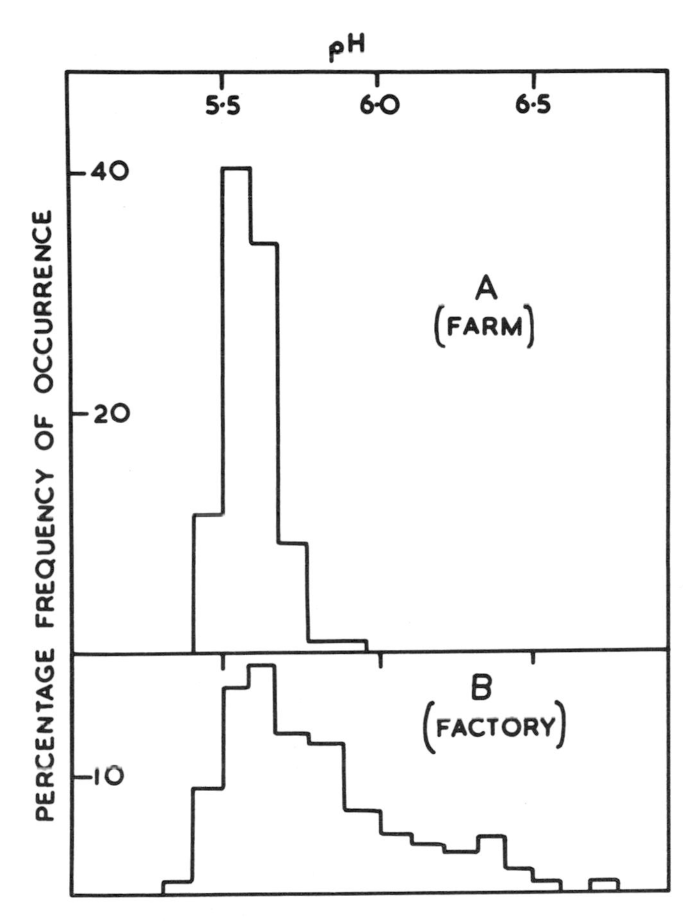


*Fig. 34.4.* Frequency histograms showing the range of acidity in the psoas major muscle of carcasses from (*A*) rested (farm-killed) and (*B*) fatigued (factory-killed) pigs. Unpublished observations, E. H. Callow, quoted by Ingram (1962).

be organoleptically advantageous because of its increased tenderness, and certainly the commercial problems associated with it are now not quite so marked (Wismer-Pedersen, 1969). There is said to be some reduction in shrink on curing, and loss when cooked (Kauffman, Carpenter, Bray, and Hoekstra, 1964). Meat of high ultimate *p*H is also known to produce a superior product when it is freeze-dried (Penny, Voyle, and Lawrie, 1964), the desired high *p*H resulting from the injection of adrenaline before slaughter (Lawrie, 1958; Hedrick, Boillot, Brady, and Naumann, 1959). Other workers have employed this and other techniques with some success, to produce meat and meat products of high ultimate *p*H (Hall, 1950; Brissey, 1952; Rongey, Kahlenberg, and Naumann, 1959).

It is generally acknowledged, however, that meat with a high ultimate *p*H is not completely acceptable commercially because of its reduced appeal when displayed, its dryness, alleged toughness, and reduced penetration to curing salts. The latter characteristic may result from the close texture and increased water-binding capacity associated with meat of high *p*H (Hamm, 1960).

Despite the obvious nature of the condition, it has occasionally proven difficult to produce meat of high ultimate *p*H, experimentally, either by fasting cattle for prolonged periods (Howard and Lawrie, 1956) or by transporting pigs over prolonged distances (Cuthbertson and Pomeroy, 1969). Yet the condition is known to be present commercially and likely to increase (Harrington, 1969). High ultimate *p*H has been observed by the author in the longissimus dorsi muscles of commercial pigs which have been held overnight without feeding at temperatures around 0° C (Lister, unpublished observations, 1969). Briskey, Bray, Hoekstra, Grummer, and Philips (1959) showed, experimentally, that pork with a high ultimate *p*H could be produced under similar environmental conditions. Pigs of a susceptible or nonsusceptible type, which, respectively, have fast or slow rates of glycolysis in muscle after death, may when stressed for a number of hours, lose, retain, or restore muscle glycogen, according to the duration of treatment. Stress-susceptible animals become progressively depleted of glycogen with time; resistant animals lose glycogen initially, but restore it as time passes (Judge, Cassens, and Briskey, 1967). The present tendency toward increases in the incidence of dark-cutting beef may be a preliminary indication of increased rates of glycolysis in cattle.

## THEORETICAL AND PRACTICAL CONSIDERATIONS

It is commonly acknowledged that most problems of meat quality which are associated with the physiology of animals are most obviously seen in pigs and their carcasses. I shall conclude this paper, therefore, with reference mainly to pigs, but there are obviously points of overlap with other species.

It has often been suggested that problems of meat quality have arisen because of the changes in management, housing, and breeding, which have been brought about in recent years to produce pigs with better growth rates and superior carcasses. Nowhere has this been more consistently achieved nor documented than in the Danish Progeny Testing Stations which, on average, in the period 1926–63, have recorded reductions of about 8 days in the time required by pigs to grow from 20 to 90 kg, and in the total amount of fat by about one-fifth, the latter being replaced by lean tissue. There have been concomitant reductions in both the total amount of food consumed on test and in the amount required for unit gain in live weight (Clausen, Nørtoft Thomsen, and Pedersen, 1964). Kielanowski (1966) used these data to show that the change in the average daily deposition of protein was the most important factor in the progress observed in Danish pig production over the years and this was responsible for 77% of the increased efficiency of food utilization. The contributions made by decreased intake of food and lower maintenance requirement to improving food conversion efficiency were 11% and 12% respectively. Kielanowski concludes that the increased daily deposition of protein was brought about by selection for leanness of carcasses and economy of food conversion. There seems to be no reason to suppose that the dramatic changes which have been brought about in the composition of the carcasses of selected lines of pigs at the Iowa Agricultural Experiment Station (Christian, 1969; Lasley, 1969) have not been achieved with the same effects on their metabolism.

It might, therefore, be worth considering the ways in which increased rates of protein deposition may be brought about in the living animal. There is no doubt that voracious feeding in young animals leads to increased live weight gains, but it also leads to excessive deposition of fat, and the efficiency of food utilization is impaired. However it is clear from the work reported above that the increased levels of protein deposition did not come from increased food consumption, because restricted feeding is practiced in Danish Progeny Testing Stations. It is also known that excessively lean breeds of pig tend to have smaller appetites when compared on the basis of live weight (Bichard, 1968). Food restriction will improve carcass grading and to some extent the efficiency of food utilization, but the overall rate of growth will be reduced (McMeekan, 1940*a*, *b*, *c*), and excessive restriction reduces the daily deposition of protein and ultimately may reverse it (Pomeroy, 1941). It seems clear, therefore, that whilst the pattern of feeding employed in the Danish Testing Stations was likely to achieve acceptable amounts of muscle in carcasses, it is unlikely to have contributed significantly to a long-term improvement in performance and production.

Intensively housed pigs are kept, usually, at a temperature which is below but fairly close to the zone of thermal neutrality (Mount, 1968*a*, *b*) and confined in a way which limits the amount of exercise that may be taken. The latter, itself, may contribute to impaired meat quality (Rülcker, 1967; Wismer-Pedersen, 1969). Limitation of activity may also help to minimize the maintenance requirement for, with animals so ill-equipped to dissipate heat as pigs are, relatively mild muscular activity may cause a rise in body temperature and an associated rise in metabolic rate. Maintenance requirements under optimal housing conditions are likely to be minimal, and this presumably explains the relatively small contribution (12%) which reductions in maintenance requirements have made to improved economy of food conversion (Kielanowski, 1966).

To explain the increased rates of protein deposition in some contemporary pig populations it would seem necessary to examine the fundamental control of nitrogen balance in the growing animal. The control of protein anabolism appears to rest in the combined effects of the thyroid, pituitary, adrenals, and (to a lesser extent) the gonads. The improvement in performance and productivity of pigs has not been confined to intact males and females alone and it would appear, therefore, that the effects of androgens and estrogens contribute only slightly, through significantly (Rhodes, 1969), to improved protein deposition. It is worth noting, however, that problems of meat quality were found, in a Danish study (Wismer-Pedersen, 1959), more often in intact females than in hogs, which might be expected to grow more slowly.

The striking increase in leanness and susceptibility to stress of pigs in the Iowa selection program, the associated adrenal insufficiency (Topel, 1969*b*), and the implication of reduced thyroid function, lead one to suppose that increases in the incidence of stress susceptibility and the pale, soft, and exudative syndrome of porcine muscle (Clausen and Nørtoft Thomsen, 1962; Topel, Merkel, and Wismer-Pedersen, 1967; Briskey and Lister, 1969), are a natural consequence of long-term selection against the protein catabolic actions of excessive rate of secretion of thyroid and adrenal hormones. It is, however, highly unlikely that the modification of energy balance is brought about by these two hormones acting in isolation; the whole system of endocrine control, and, in particular, the hypothalamus and pituitary, must inevitably be integrated to achieve this. The small mature size, for example, often found in animals which demonstrate excessive leanness during growth (Bichard, 1968) might, to some extent, be explained by an induced deficiency of growth hormone which is not elaborated and produced by the pituitary in sufficient amounts when thyroid hormone is lacking (Eartly and Leblond, 1954).

The preceding account pinpointed pale, soft, and exudative muscle as a consequence of the metabolic response to stress of relatively short duration before slaughter. It will be appreciated that stress over a long period of time may elicit the production of "dark-cutting meat" as the consequence of depletion of the reserves of muscle glycogen, as it does in susceptible pigs (Judge, Cassens, and Briskey, 1967). Moreover, it seems unlikely that pigs will be the only species to suffer an increased susceptibility to stress as selection for leanness continues, for broiler chickens, with a generation interval of only 6 months, lend themselves well to genetic improvement over a relatively short period of time, and even cattle are subjected to appreciable selection pressure. Already, cattle are beginning to show rates of change in the *p*H of muscle after death which exceed the rate normally to be expected (Hamm, pers. comm., 1969). Chicken muscle may be prevented from suffering excessive water loss only because the cooling practices adopted in commercial processing, and also the relatively small size of the carcass, reduce muscle temperature before its *p*H has dropped to a point where excessive denaturation of the sarcoplasmic proteins can take place.

It is clear that many of the important adverse characteristics of meat, e.g., excessive drip, toughness, cooking loss, etc., are secondary consequences of the rate of *p*H change in muscle and are related, primarily, to temperature change during *p*H fall. There are practical and economic limits to what can be done by commercial refrigeration practices, and we must now consider what can be done to restore more moderate rates of *p*H change in muscle post mortem.

Animals, like ourselves, may be "fearfully and wonderfully made" (Psalm 139, v. 14), but they are, apparently, neither sufficiently wonderfully nor fearfully made to withstand all the unceasing demands of producers, processors, and consumers without responding adversely in some way or another.

## *References*

Bate-Smith, E. C., and J. R. Bendall. 1949. Factors determining the time course of rigor mortis. *J. Physiol. 110*:47.

———. 1956. Changes in muscle after death. *Brit. Med. Bull. 12*:230.

Bendall, J. R. 1960. Post-mortem changes in muscle, p. 227. *In* G. H. Bourne (ed.), *Structure and Function of Muscle,* Vol. 3. Academic Press, New York.

———. 1966*a*. The effect of pre-treatment of pigs with curare on the post-mortem rate of *p*H fall and onset of rigor mortis in the musculature. *J. Sci. Food Agric. 17*:333.

———. 1966*b*. Muscle as a contractile machine, p. 7. *In* E. J. Briskey, R. G. Cassens, and J. C. Trautman (eds.), *The Physiology and Biochemistry of Muscle as a Food.* Univ. of Wis.

———. 1969. *Muscles, Molecules and Movement.* Heinemann Educational Books, London.

Bendall, J. R., and R. A. Lawrie. 1964. Watery pork. *Anim. Breed. Abs. 32*:1.

Bendall, J. R., and J. Wismer-Pedersen. 1962. Some properties of the fibrillar proteins of normal and watery pork muscle. *J. Food Sci. 27*:144.

Bernard, C. 1877. *Leçons sur le diabète et la glycogenèse animale.* Baillière, Paris.

Bichard, M. 1968. Genetic aspects of growth and development in the pig, p. 309. *In* G. A. Lodge and G. E. Lamming (eds.), *Growth and Development of Mammals.* Butterworth Scientific Publications, London.

Bohus, B., K. Lissak, and B. Mezei. 1965–66. Effect of thyroxine implantation in the hypothalamus and the anterior pituitary on pituitary-adrenal function in rats. *Neuroendocrinology 1*:15.

Bondy, P. K. 1949. Effect of the adrenal and thyroid glands upon the rise of plasma amino acids in the eviscerated rat. *Endocrinology 45*:605.

Bradfield, G. E. 1968. Sex differences in the growth of sheep, p. 92. *In* G. A. Lodge and G. E. Lamming (eds.), *Growth and Development of Mammals.* Butterworth Scientific Publications, London.

Briskey, E. J. 1961. Relationship of feeding to meat composition and properties. *Proc. 13th Res. Conf., Amer. Meat Inst. Foundation, Univ. Chi.*

———. 1964. Etiological status and associated studies of pale, soft, exudative porcine musculature. *Advance. Food Res. 13*:89.

Briskey, E. J., R. W. Bray, W. G. Hoekstra, R. H. Grummer, and P. H. Philips. 1959. Effect of various levels of exercise in altering the chemical and physiological characteristics of certain pork ham muscles. *J. Anim. Sci. 18*:153.

Briskey, E. J., and D. Lister. 1969. Influence of stress syndrome on chemical and physical characteristics of muscle post mortem, p. 177. *In* D. G. Topel (ed.), *The Pork Industry: Problems and Progress.* Iowa State.

Brissey, G. E. 1952. U.S. Patent No. 2596067.

Brobeck, J. R. 1946. Mechanism of the development of obesity in animals with hypothalamic lesions. *Physiol. Rev. 26*:541.

———. 1948. Food intake as a mechanism of temperature regulation. *Yale J. Biol. Med. 20*:545.

Brobeck, J. R., M. Wheatland, and J. L. Strominger. 1947. Variations in regulation of energy exchange associated with estrus, diestrus and pseudo-pregnancy in rats. *Endocrinology 40*:65.

Burgess, J. A., B. R. Smith, and T. J. Merimee. 1966. Growth hormone in thyrrotoxicosis: Effect of insulin-induced hypoglycemia. *J. Clin. Endocrinol. 26*:1257.

Burton, A. C., and O. G. Edholm. 1955. *Man in a Cold Environment*. E. Arnold, London.

Callow, E. H. 1935. Electrical resistance of muscular tissue in relation to curing, p. 57. *Ann. Rept. Food Invest. Bd.* (London).

———. 1937. The "ultimate *p*H" of muscular tissue, pp. 34, 49. *Ann. Rept. Food Invest. Bd.* (London).

Cassens, R. G., M. D. Judge, and E. J. Briskey. 1965. Porcine adrenocortical lipids in relation to post mortem striated muscle characteristics. *Proc. Soc. Exp. Biol. Med. 120*:854.

Christian, L. L. 1969. Limits for rapidity of genetic improvement for fat, muscle, and quantitative traits, p. 154. *In* D. G. Topel (ed.), *The Pork Industry: Problems and Progress.* Iowa State.

Clark, I. 1953. Effect of cortisone upon protein synthesis. *J. Biol. Chem. 200*:69.

Clausen, H., and R. Nørtoft Thomsen. 1962. 50th report on comparative tests with pigs from state recognised breeding centres. Rept. 331, Beretn. Forsøgslab, Copenhagen.

Clausen, H., R. Nørtoft Thomsen, and O. K. Pedersen. 1964. 52nd report on comparative tests with pigs from state recognised breeding centres, 1962–63. Rept. 344, Beretn. Forsøgslab, Copenhagen.

Cori, C. F. 1931. Mammalian carbohydrate metabolism. *Physiol. Rev. 11*:143.

Crichton, J. A., J. N. Aitken, and A. W. Boyne. 1960. Effect of plane of nutrition during rearing on growth, production, reproduction and health of dairy cattle. *Anim. Prod. 2*:45.

Cuthbertson, A., and R. W. Pomeroy. 1969. The effect of length of journey by road to abattoir and resting and feeding before slaughter of bacon pigs. *Anim. Prod. 12*:37.

Donovan, B. T., and D. Jacobsohn. 1960. Growth responses of mammary glands and other tissues in hypophysectomized female rats treated with thyroxine, insulin, cortisone, and pregnant mare serum gonadotropin. *Acta Endocrinol. 33*:197.

Eartly, H., and C. P. Leblond. 1954. Identification of the effects of thyroxine mediated by the hypophysis. *Endocrinology 54*:249.

Edwards, W. F. 1824. *De l'influence des agents physiques sur la vie.* Paris.

Elsley, F. W. H., I. McDonald, and V. R. Fowler. 1964. Effect of plane of nutrition on the carcasses of pigs and lambs when variations in fat content are excluded. *Anim. Prod. 6*:141.

Falconer, I. R., and S. A. Draper. 1968. Thyroid activity and growth, p. 109. *In* G. A. Lodge and G. E. Lamming (eds.), *Growth and Development of Mammals.* Butterworth Scientific Publications, London.

Ficker, M. 1905. Über die Keimdichte der normalen Fehleimant des intestinal Traktus. *Arch. Hyg., Berl. 54*:354.

Fowler, V. R. 1968. Body development and some problems of its evaluation, p. 195. *In* G. A. Lodge and G. E. Lamming (eds.), *Growth and Development of Mammals.* Butterworth Scientific Publications, London.

Frantz, A. G., and M. T. Rabkin. 1964. Human growth hormone: clinical measurement, response to hypoglycemia and suppression by corticosteroids. *New Engl. J. Med. 271:*1375.

Fritz, I. B. 1956. Effects of adrenal cortex on the release of nitrogen-15 from tissue proteins of rats previously fed glycine-$N^{15}$. *Endocrinology 58:*484.

Graham, N. McC. 1969. The influence of body weight (fatness) on the energetic efficiency of adult sheep. *Aust. J. Agric. Res. 20:*375.

Greaser, M. L., R. G. Cassens, W. C. Hoekstra, E. J. Briskey, G. R. Schmidt, S. D. Carr, and D. E. Galloway. 1969. Calcium accumulating ability and compositional differences between sarcoplasmic reticulum fractions from normal and pale, soft, exudative porcine muscle. *J. Anim. Sci. 28:*589.

Greer, M. A., K. Matsuda, and A. K. Scott. 1966. Maintenance of the ability of rat pituitary homotransplants to secrete TSH by transplantation under the hypothalamic median eminence. *Endocrinology 78:*389.

Guillemin, R., R. Burgus, E. Sakiz, and D. N. Ward. 1966. Purification of the hypothalamic hormone TSH-releasing factor (TRF). *Compt. Rend. Acad. Sci. 262:*2278.

Guillemin, R., E. Yamazaki, M. Jutisz, and E. Sakiz. 1962. Présence dans l'extrait de tissues hypothalamiques d'une substance stimulant la secretion de l'hormone hypophysaire thyréotrope (TSH). Première purification par filtration sur Gel-Sephadex. *Compt. Rend. Acad. Sci. 255:*1018.

Haines, R. B. 1937. Microbiology in the preservation of animal tissues. *Spec. Rept. Food Invest. Bd.* (London) No. 45.

Hall, G. O. 1950. U.S. Patent No. 2513094.

Hallund, O., and J. R. Bendall. 1965. The long-term effect of electrical stimulation on the post-mortem fall of *p*H in the muscles of Landrace pigs. *J. Food Sci. 30:*296.

Hamm, R. 1960. Biochemistry of meat hydration. *Advance. Food Res. 10:*356.

Hammond, J. 1932. *Growth and Development of Mutton Qualities in the Sheep.* Oliver and Boyd, Edinburgh and London.

Harrington, G. 1969. Carcass classification in other countries, p. 2. *Inst. Meat. Bull.* No. 64.

Hedrick, H. B., J. B. Boillot, D. E. Brady, and H. D. Naumann. 1959. Etiology of dark-cutting beef. *Univ. Missouri Res. Bull.* No. 717.

Hegarty, G. R., L. J. Bratzler, and A. M. Pearson. 1963. Relationship of some intracellular protein characteristics to beef muscle tenderness. *J. Food Sci. 28:*525.

Henry, M., and J. Billon. 1959. Nouvelles observations sur l'influence des aggressions non spécifiques sur la qualité de la viande de porc. *5th Europe. Meeting of Meat Res. Workers* (Paris).

Hervey, G. R. 1969. Regulation of energy balance. *Nature 222:*629.

Hetherington, A. W., and S. W. Ranson. 1939. Experimental hypothalamico-hypophyseal obesity in the rat. *Proc. Soc. Exp. Biol. 41*:465.

Hoch, F. L. 1962. Biochemical actions of thyroid hormones. *Physiol. Rev. 42*:605.

Hogan, A. G. 1928. Some relations between growth and nutrition, p. 67. *In Growth.* Yale Univ.

Horsley, V. 1886. Further researches into the function of the thyroid gland and into the pathological state produced by removal of the same. *Proc. Roy. Soc. 40*:6.

Howard, A., and R. A. Lawrie, 1956. Studies on beef quality II. Physiological and biochemical effects of various pre-slaughter treatments. *Spec. Rept. Food Invest. Bd.* (London) No. 63.

Hunter, W. M. 1968. A diminished role for growth hormone in the regulation of growth, p. 71. *In* G. A. Lodge and G. E. Lamming (eds.), *Growth and Development of Mammals.* Butterworth Scientific Publications, London.

Huxley, J. S. 1932. *Problems of Relative Growth.* Methuen, London.

Hytten, F. E. 1963. Nutritional aspects of foetal growth, p. 50. *Proc. 6th Int. Nutr. Cong.* (Edinburgh).

Ingle, D. J. 1950. Biological properties of cortisone. *J. Clin. Endocrinol. 10*:1312.

Ingram, D. L., and A. Slebodzinski. 1965. Oxygen consumption and thyroid gland activity during adaptation to high ambient temperatures in young pigs. *Res. Vet. Sci. 6*:522.

Ingram, M. 1962. The importance of *p*H in the microbiology of meat. *Medlemsbl. Norske Vetforen.* No. 8 (suppl.).

Jackson, C. M. 1909. On the prenatal growth of the human body and the relative growth of the various organs and parts. *Amer. J. Anat. 9*:119.

———. 1932. Structural changes when growth is suppressed by undernourishment in the albino rat. *Amer. J. Anat. 51*:347.

Judge, M. D., E. J. Briskey, R. G. Cassens, J. C. Forrest, and R. K. Meyer. 1968. Adrenal and thyroid function in "stress-susceptible" pigs. *Amer. J. Physiol. 214*:146.

Judge, M. D., R. G. Cassens, and E. J. Briskey. 1967. Muscle properties of physically restrained stressor-susceptible and stressor-resistant porcine animals. *J. Food Sci. 32*:565.

Kauffman, R. G., Z. L. Carpenter, R. W. Bray, and W. G. Hoekstra. 1964. Biochemical properties of pork and their relationship to quality I. *p*H of chilled, aged and cooked muscle tissue. *J. Food Sci. 29*:65.

Kay, M. 1967. Quoted by Blaxter, K. L., p. 339. *In* G. A. Lodge and G. E. Lamming (eds.), *Growth and Development of Mammals.* Butterworth Scientific Publications, London.

Kennedy, G. C. 1950. Hypothalamic control of food intake in rats. *Proc. Roy. Soc. B. 137*:535.

———. 1953. Role of depot fat in the hypothalamic control of food intake in the rat. *Proc. Roy. Soc. B. 140*:578.

Kennedy, G. C., and J. Mitra. 1963. Hypothalamic control of energy balance and the reproductive cycle in the rat. *J. Physiol. 166*:395.

Kidwell, J. F. 1952. Muscular hypertrophy and "Black Cutter" beef. *J. Hered. 43*:157.

Kielanowski, J. 1965. Estimates of the energy cost of protein deposition in growing animals, p. 13. *In* K. L. Blaxter (ed.), *Energy Metabolism*. Academic Press, London and New York.

———. 1966. Conversion of energy and the chemical composition of gain in bacon pigs. *Anim. Prod. 8*:121.

Kirmiz, J. P. 1962. *Adaption to Desert Environment*. Butterworths, London.

Kochakian, C. D. 1966. Regulation of muscle growth by androgens, p. 81. *In* E. J. Briskey, R. G. Cassens, and J. C. Trautman (eds.), *Physiology and Biochemistry of Muscle as a Food*. Univ. of Wis.

Korner, A., and F. G. Young. 1955. Effect of methylandrostenediol on the weight and protein content of muscles and organs of the rat. *J. Endocrinol. 13*:84.

Kostyo, J. L., and E. Knobil. 1959. Stimulation of leucine-2-$C^{14}$ incorporation into the protein of isolated rat diaphragm by simian growth hormone added into vitro. *Endocrinology 65*:395.

Krulich, L., and S. M. McCann. 1966. Effect of alterations in blood sugar on pituitary growth hormone content in the rat. *Endocrinology 78*:759.

Lasley, J. F. 1969. Relationship of breeding and reproduction to carcass quality and quantity characteristics, p. 145. *In* D. G. Topel (ed.), *The Pork Industry: Problems and Progress*. Iowa State.

Lawrie, R. A. 1950. Some observations on factors affecting myoglobin concentrations in muscle. *J. Agr. Sci. 40*:356.

———. 1958. Physiological stress in relation to dark-cutting beef. *J. Sci. Food Agr. 9*:721.

———. 1966. *Meat Science*. Pergamon, Oxford.

Leathem, J. H. 1964. Some aspects of hormone and protein metabolic interrelationships, p. 343. *In* H. N. Munro and J. B. Allison (eds.), *Mammalian Protein Metabolism*. Academic Press, New York and London.

Lewis, H. E. 1962. Regulation of the body temperature, p. 755. *In* H. Davson and M. G. Eggleton (eds.), *Principles of Human Physiology*. J. and A. Churchill Ltd., London.

Lister, D. 1965. Effect of nutrition before and after birth on adult size and structure. Ph.D. thesis, University of Cambridge.

Lister, D., R. A. Sair, J. A. Will, G. R. Schmidt, R. G. Cassens, W. G. Hoekstra, and E. J. Briskey. 1970. Metabolism of striated muscle of "stress-susceptible" pigs breathing oxygen or nitrogen. *Amer. J. Physiol. 218*:102.

Lister, D., R. K. Scopes, and J. R. Bendall. 1969. Some properties of the muscle of Pietrain pigs. *Anim. Prod. 11*:288.

Locker, R. H., and C. J. Hagyard. 1963. A cold shortening effect in beef muscles. *J. Sci. Food Agr. 14*:787.

Ludvigsen J. 1960. Maladaptation syndromes in pigs. *Proc. 2d Int. Anim. Nutr. Conf.* (Madrid, Spain).

McCance, R. A. 1960. Severe undernutrition in growing and adult animals I. Production and general effects. *Brit. J. Nutr. 14*:59.

McCance, R. A., and E. M. Widdowson. 1956. Metabolism, growth and renal function of piglets in the first days of life. *J. Physiol. 133*:373.

McCann, S. M., A. P. S. Dhariwal, and J. C. Porter. 1968. Regulation of the adenohypophysis. *Ann. Rev. Physiol. 30*:589.

McLoughlin, J. V. 1963. Effect of rapid post mortem *p*H fall on the extraction of the sarcoplasmic and myofibrillar proteins of pig muscle. *Proc. 9th Conf. Europe. Meat Res. Workers* (Budapest) Paper 33.

McMeekan, C. P. 1940*a*. Growth and development in the pig, with special reference to carcass quality characters I. *J. Agr. Sci. 30*:276.

———. 1940*b*. Growth and development in the pig, with special reference to carcass quality characters II. Influence of the plane of nutrition on growth and development. *J. Agr. Sci. 30*:387.

———. 1940*c*. Growth and development in the pig, wtih special reference to carcass quality characters III. Effect of the plane of nutrition on the form and composition of the bacon pig. *J. Agr. Sci. 30*:511.

Manchester, K. L., and F. G. Young. 1959. Hormones and protein biosynthesis in isolated rat diaphragm. *J. Endocrinol. 18*:381.

Marsh, B. B. 1954. Rigor mortis in beef. *J. Sci. Food Agr. 5*:70.

Mayer, J. 1955. Regulation of energy intake and the body weight: the glucostatis theory and the lipostatis hypothesis. *Ann N.Y. Acad. Sci. 63*:15.

Merkel, R. A. 1968. Implication of the circulatory system of skeletal muscle to meat quality, p. 204. *Proc. 21st Reciprocal Meat Conf. Amer. Meat Sci. Ass.*

Mixner, J. P., K. T. Szabo, and R. F. Mather. 1966. Relation of thyroxine secretion rate to body weight in growing female Holstein-Friesian cattle. *J. Dairy Sci. 49*:199.

Mount, L. E. 1966. Basis of heat regulation in homoeotherms. *Brit. Med. Bull. 22*:84.

———. 1968*a*. *The Climatic Physiology of the Pig.* E. Arnold Ltd., London.

———. 1968*b*. Adaptation in swine, p. 277. *In* E. S. E. Hafez (ed.), *Adaption of Domestic Animals.* Lea & Febiger, Philadelphia.

Müller, E. E., A. Arimura, T. Saito, and A. V. Schally. 1967. Growth hormone-releasing activity in plasma of normal and hypophysectomized rats. *Endocrinology 80*:77.

Needham, D. M. 1960. Biochemistry of muscular action, p. 55. *In* G. H. Bourne (ed.), *Structure and Function of Muscle,* Vol. 2. Academic Press, New York.

Oslage, H. J., and H. Fliegel. 1965. Nitrogen and energy metabolism of growing-fattening pigs with an approximately maximal feed intake, p. 297. *In* K. L. Blaxter (ed.), *Energy Metabolism.* Academic Press, London and New York.

Overman, R. R., and S. C. Wang. 1947. Contributory role of the different nervous factors in experimental shock: sublethal hemorrhage and sciatic nerve stimulation. *Amer. J. Physiol. 148*:289.

Pàlsson, H. 1939. Meat qualities in the sheep with special reference to Scottish breeds and crosses I. J. *Agr. Sci. 29*:544.

———. 1940. Meat qualities in the sheep with special reference to Scottish breeds and crosses II. *J. Agr. Sci. 30*:1.

Penny, I. F., C. A. Voyle, and R. A. Lawrie. 1964. Some properties of freeze-dried pork muscles of high or low ultimate *p*H. *J. Sci. Food Agr. 15*:559.

Perry, S. V. 1956. Nucleotide metabolism and intracellular organization in skeletal muscle, p. 364. *Proc. 3rd Int. Cong. Biochem.* (Brussels) .

Pomeroy, R. W. 1941. Effect of a submaintenance diet on the composition of the pig. *J. Agr. Sci. 31*:50.

Porter, J. C., A. P. S. Dhariwal, and S. M. McCann. 1967. Response of the anterior pituitary-adrenocortical axis to purified CRF. *Endocrinology 80*:679.

Post, T. B., and J. P. Mixner. 1961. Thyroxine turnover methods for determining thyroid secretion rates in dairy cattle. *J. Dairy Sci. 44*:2265.

Prescott, J. H. D. 1963. Influence of sex and certain steroids on growth and carcass composition in pigs. Ph.D. dissertation, University of Nottingham.

Quass, D. W., and E. J. Briskey. 1968. A study of certain properties of myosin from skeletal muscle. *J. Food Sci. 33*:180.

Rees, C. P. van 1966. The effect of triiodothyronine and thyroxine on thyrotrophin levels in the anterior pituitary gland and blood serum of thyroidectomized rats. *Acta Endocrinol. 51*:619.

Rhodes, D. N. 1969. *Meat from Entire Male Animals.* J. and A. Churchill, London.

Robinson, D. W. 1964. Plane of nutrition and compensatory growth in pigs. *Anim. Prod. 6*:227.

Robinson, R. H. M., M. Ingram, R. A. M. Case, and J. G. Benstead. 1953. Whalemeat: bacteriology and hygiene. *Spec. Rept. Food Invest. Bd.* (London) No. 59.

Rongey, E. H., O. J. Kahlenberg, and H. D. Naumann. 1959. Some factors influencing the uniformity and stability of colour in cured hams. *Food Tech. 13*:640.

Roth, J., S. M. Glick, R. S. Yallow, and S. A. Berson. 1964. Influence of blood glucose on the plasma concentration of growth hormone. *Diabetes 13*:355.

Rothstein, A. 1968. Membrane phenomena. *Ann. Rev. Physiol. 30*:15.

Rülcker, C. 1967. Influence of physical training and pre-slaughter exercise on fluid retention, colour and adenosine triphosphate of pork muscle. *Acta Vet. Scand. 8*:189.

Sair, R. A., D. Lister, W. G. Moody, R. G. Cassens, W. G. Hoekstra, and E. J. Briskey. 1969. Action of curare and magnesium on striated muscle of stress-susceptible pigs. *Amer. J. Physiol.* (In press.)

Scammon, R. E. 1930. The measurement of the body in childhood. *In* J. A. Harris, C. M. Jackson, D. G. Paterson and R. E. Scammon (eds.), *The Measurement of Man.* Univ. of Minn.

Schally, A. V., C. Y. Bowers, and T. W. Redding. 1966*a*. Presence of thyrotropic hormone-releasing factor in porcine hypothalamus. *Proc. Soc. Exp. Biol. Med. 121*:718.

———. 1966*b*. Purification of thyrotropic hormone-releasing factor from bovine hypothalamus. *Endocrinology 78*:726.

Schally, A. V., C. Y. Bowers, T. W. Redding, and J. F. Barrett. 1966. Isolation of thyrotropin releasing factor (TRF) from porcine hypothalamus. *Biochem. Biophys. Res. Comm. 25*:165.

Schooley, R. A., S. Friedkin, and E. S. Evans. 1966. Re-examination of the discrepancy between acidophil numbers and growth hormone concentration in the anterior pituitary gland following thyroidectomy. *Endocrinology 79*:1053.

Scopes, R. K., and R. A. Lawrie. 1963. Post mortem lability of skeletal muscle proteins. *Nature 197*:1202.

Scow, R. O. 1959. Effect of growth hormone and thyroxine on growth and chemical composition of muscle, bone and other tissues in thyroidectomized-hypophysectomized rats. *Amer. J. Physiol. 196*:859.

Scow, R. O., and S. N. Hagan. 1957. Effect of testosterone propionate on myosin, collagen and other protein fractions in striated muscle of gonadectomized rats. *Endocrinology 60*:273.

Selye, H. 1951. *1st Annual Report on Stress.* Acta Inc., Montreal.

Sheehan, H. L. 1948. Shock in obstetrics. *Lancet 254*:1.

Silber, R. H., and C. C. Porter. 1953. Nitrogen balance, liver protein repletion and body composition of cortisone-treated rats. *Endocrinology 52*:518.

Sinha, D., and J. Meites. 1965–66. Effects of thyroidectomy and thyroxine on hypothalamic concentration of "thyrotropin releasing factor" and pituitary content of thyrotropin in rats. *Neuroendocrinology 1*:4.

Slonaker, J. R. 1925. The effect of copulation, pregnancy, pseudopregnancy and lactation on the voluntary activity and food consumption of the albino rat. *Amer. J. Physiol. 71*:362.

Smith, R. E., and D. J. Hoijer. 1962. Metabolism and cellular function in cold acclimation. *Physiol. Rev. 42*:60.

Sprague, R. G. 1951. Cortisone and ACTH. A review of certain physiologic effects and their clinical implications. *Amer. J. Med. 10*:567.

Thornton, H. 1949. *Textbook of Meat Inspection.* Bailliere, Tindall and Cox, London.

Topel, D. G. 1968. Endocrine influence on the micro-circulatory system, p. 194. *Proc. 21st Reciprocal Meat Conf., Amer. Meat Sci. Ass.*

———. 1969*a*. *The Pork Industry: Problems and Progress.* Iowa State.

———. 1969*b*. Relation of plasma glucocorticoid levels to some physical and chemical properties of porcine muscle and the porcine stress syndrome, p. 91. *Proc. Int. Symp. on Meat Quality* (Zeist, May, 1969).

Topel, D. G., E. J. Bicknell, K. S. Preston, L. L. Christian, and C. Y. Matsushima. 1968. Sudden death in swine and its relationship to blood and muscle changes. *Mod. Vet. Prac. 49*:40.

Topel, D. G., and R. A. Merkel. 1966. Effect of exogenous goitrogens upon some physical and biochemical properties of porcine muscle and adrenal gland. *J. Anim. Sci. 25*:1154.

Topel, D. G., R. A. Merkel, and J. Wismer-Pedersen. 1967. Relationship of plasma 17-hydroxycorticosteroid levels to some physical and biochemical properties of porcine muscle. *J. Anim. Sci. 26*:311.

Vernikos-Danellis, J., and D. N. Trigg. 1967. Feedback mechanisms regulating pituitary ACTH secretion in rats bearing transplantable pituitary tumors. *Endocrinology 80*:345.

Wang, G. H. 1923. *The Relation between 'Spontaneous' Activity and Oestrus Cycle in the White Rat.* Comparative Psychology Monographs, Vol. 2, Ser. 6. Williams and Wilkins, Baltimore.

Wiggers, H. C., R. C. Ingraham, F. Roemhilk, and H. Goldberg. 1948. Vasoconstriction and the development of irreversible hemorrhagic shock. *Amer. J. Physiol. 153*:511.

Wilmer, H. A. 1940. Changes in structural components of human body from six lunar months to maturity. *Proc. Soc. Exp. Biol. Med. 43*:545.

Wilmer, J. S., and T. S. Foster. 1960. Influence of adrenalectomy and individual steroid hormones upon the metabolism of acetate-1-$C^{14}$ by rat liver slices I. *Can. J. Biochem. Physiol. 38*:1387.

Wilson, P. N. 1952. Growth analysis of the domestic fowl I. Effect of plane of nutrition and sex on live-weights and external measurements. *J. Agr. Sci. 42*:369.

———. 1954*a*. Growth analysis of the domestic fowl II. Effect of plane of nutrition on carcass composition. *J. Agr. Sci. 44*:67.

———. 1954*b*. Growth analysis of the domestic fowl III. Effect of plane of nutrition on carcass composition of cockerels and egg yield of pullets. *J. Agr. Sci. 45*:110.

Wismer-Pedersen, J. 1959. Quality of pork in relation to rate of *p*H change post mortem. *J. Food Sci. 24*:711.

———. 1966. Effects of post-mortem changes on meat quality. *Ziets, Tierz. Zucht. 82*:308.

———. 1969. Modern production practices and their influence on stress conditions, p. 163. *In* D. G. Topel (ed.), *The Pork Industry: Problems and Progress.* Iowa State.

Wismer-Pedersen, J., and E. J. Briskey. 1961. Relationship of post-mortem acidity and temperature. *Food Technol. 15*:232.

# Chapter 35

# *The Metabolism of Muscle as a Food*

L. L. KASTENSCHMIDT

It is now quite generally accepted that the rate of post mortem metabolism in muscle is a variable quantity which is profoundly affected by many factors, some of which have been discussed in other chapters of this book. It is perhaps equally well accepted that this variable rate of post mortem metabolism has important implications in the ultimate usefulness of muscle as a food (Briskey, 1964; Briskey, Kastenschmidt, Forrest, Beecher, Judge, Cassens, and Hoekstra, 1966).

In the pages that follow, I shall attempt to present an integrated view of our current understanding of some of the biochemical principles which may account for the variability in the rate of post mortem metabolism in porcine muscle.

The regulation of energy-yielding processes in muscle was discussed at the previous symposium (Lardy, 1966) and it is evident that these processes in muscle are subject to multisite control by a large number of intermediates such as $P_i$, ATP, ADP, AMP, glucose-6-phosphate, and citrate. Fig. 35.1 presents an integrated view of the metabolism of skeletal muscle. When a muscle becomes anoxic under physiological conditions, the glycolytic rate is greatly accelerated (Williamson, 1966). The rate-limiting reactions under these conditions are thought to be those catalyzed by phosphorylase and phosphofructokinase (PFK) (Helmreich and Cori, 1965). Furthermore, some additional control is probably exerted at the pyruvate kinase step (Lardy, 1966).

Space does not permit a complete discussion of the regulation of the glycolytic system here. Let it suffice to make a simple statement about each of the regulatory metabolites. Inorganic phosphate serves as a substrate for the phosphorylase and glyceraldehyde-3-phosphate dehydrogenase steps, as an allosteric inhibitor of aldolase, and as an allosteric activator of PFK and hexokinase. ATP is a substrate in the hexokinase

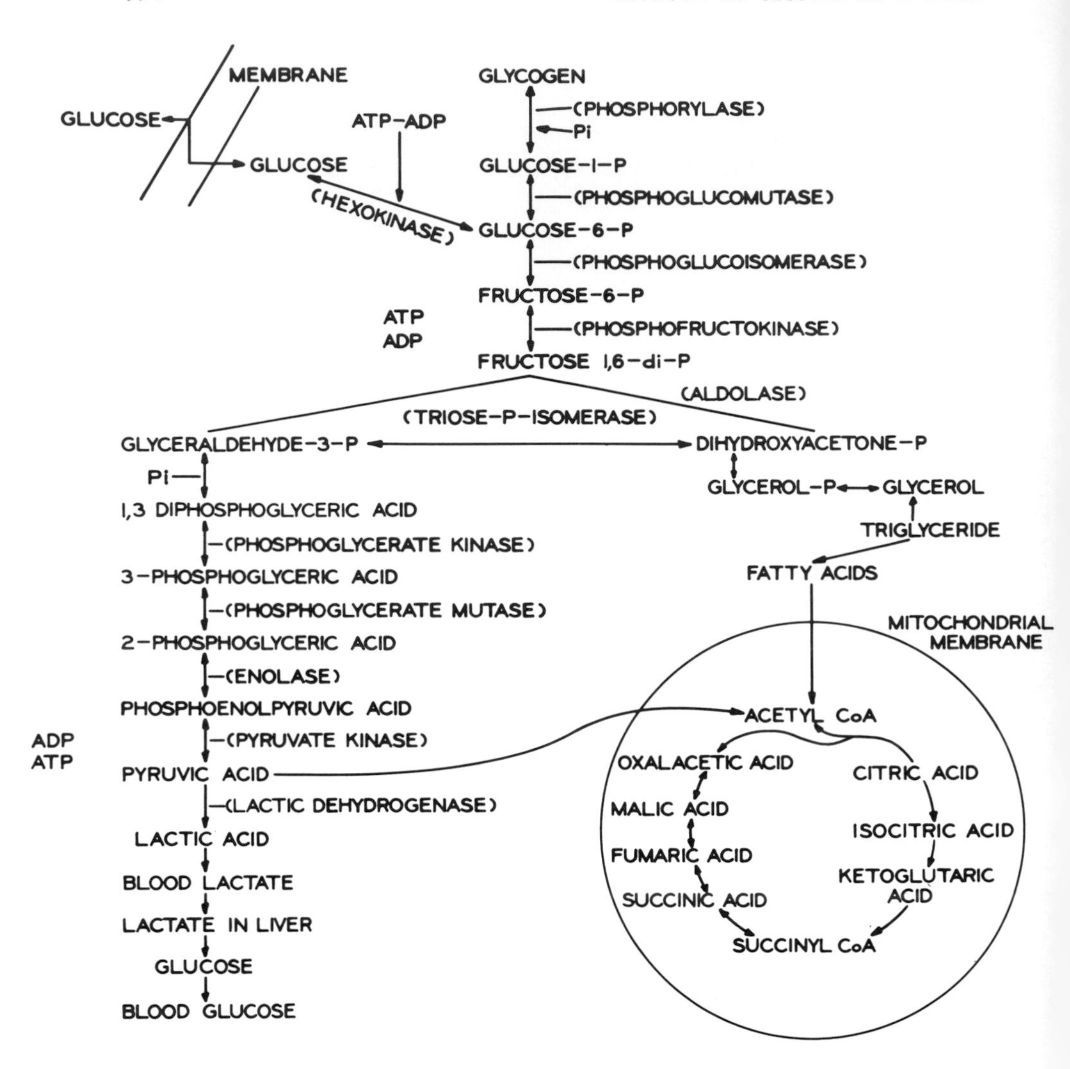


*Fig. 35.1.* A simplified overview of muscle metabolism.

and PFK reactions and inhibits the phosphorylase, phosphoglycerokinase, PFK, and pyruvate kinase reactions. Adenosine monophosphate (5′-AMP) is an important activator of phosphorylase b and serves together with fructose-6-phosphate, 3′5′-cyclic AMP, fructose diphosphate, citrate, $K^+$, and $NH_4^+$ as allosteric activators of PFK.

Glucose-6-phosphate is a powerful inhibitor of hexokinase. This inhibition is counteracted by $P_i$.

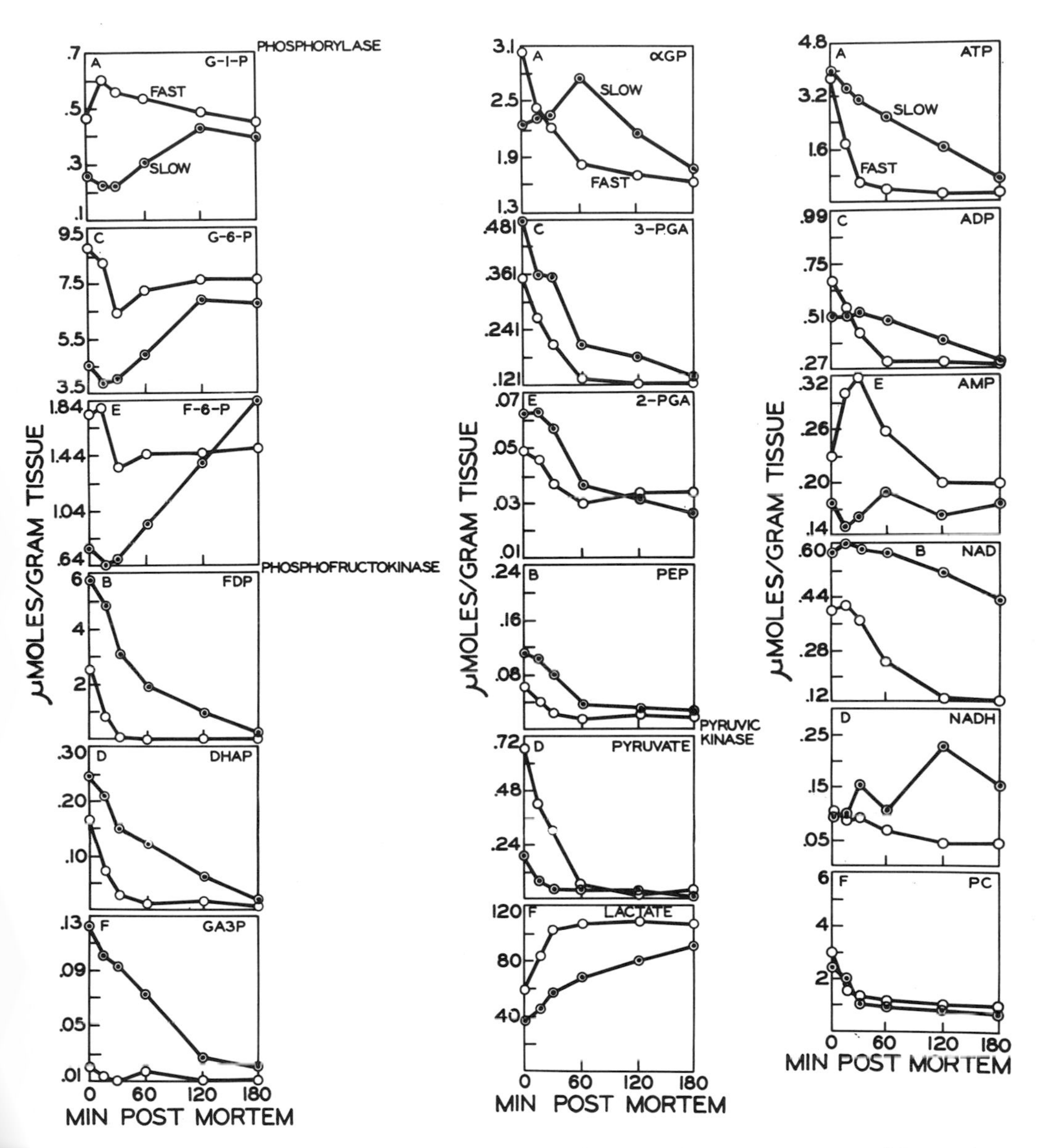


*Fig. 35.2.* Changes in glycolytic intermediate levels as a function of time post mortem. Mean values from 6 fast-glycolyzing muscles and 3 slow-glycolyzing muscles obtained from Poland China pigs are presented. *Solid circles,* mean values from fast-glycolyzing muscles; *open circles,* mean values from slow-glycolyzing muscles. From Kastenschmidt (1966).

## POST MORTEM GLYCOLYTIC CONTROL

Our first approach was to compare glycolytic intermediate and cofactor levels in porcine muscles having markedly different rates of post mortem glycolysis. These studies were conducted almost exclusively on the longissimus muscle. For convenience in speaking about post mortem glycolytic rate, we have adopted the following conventions (Kastenschmidt, Hoekstra, and Briskey, 1966): (1) "Fast-glycolyzing" muscles are those having a *p*H of 5.5 or less at 30 min post mortem; (2) "slow-glycolyzing" muscles have a *p*H of 6.0 or higher at 60 min post mortem; (3) "stress-resistant" animals are those which can withstand ante mortem stress and whose muscles after death are usually slow-glycolyzing; (4) "stress-susceptible" animals are those which cannot tolerate ante mortem stress. They usually have fast-glycolyzing muscles or expire before they can be exsanguinated.

Samples were excised at various times post mortem and rapidly frozen with liquid nitrogen. Analyses were conducted to determine the levels of most of the intermediates in the glycolytic pathway and for nucleotides, phosphocreatine, $P_i$, and glucose. This type of experiment was conducted on samples obtained from about 40 animals having "fast" or "slow" rates of post mortem glycolysis. A complete presentation of the experimental details has been published (Kastenschmidt, Hoekstra, and Briskey, 1966, 1968; Kastenschmidt, 1966).

Fig. 35.2 illustrates the type of data obtained and the sampling times used. It is evident that fast-glycolyzing muscles are operating at higher levels of hexose monophosphate compounds and at lower levels of the intermediates from fructose diphosphate through phosphoenolpyruvate than slow-glycolyzing muscles.

Another means of clearly illustrating this point is to express the data from fast-glycolyzing muscles relative to the values from slow-glycolyzing muscles. This treatment is analagous to the "crossover plot" of Chance and Williams (1956). The glycolytic intermediate data obtained from samples taken 15 min post mortem are expressed in crossover plot format in Fig. 35.3. At this time the glycolytic flux is about 2.0 and 0.6 $\mu$M lactate/g per min for fast- and slow-glycolyzing muscles, respectively. Apparently both the phosphorylase and PFK enzymes were activated in fast glycolyzing muscle, but the latter tended to become rate-limiting under these conditions.

In attempting to explain the apparent activation of the glycolytic system in fast-glycolyzing muscle, the concentrations of several activators of glycolysis were measured. As shown in Table 35.1, the level of ATP was markedly decreased while the levels of AMP and $P_i$ were increased.

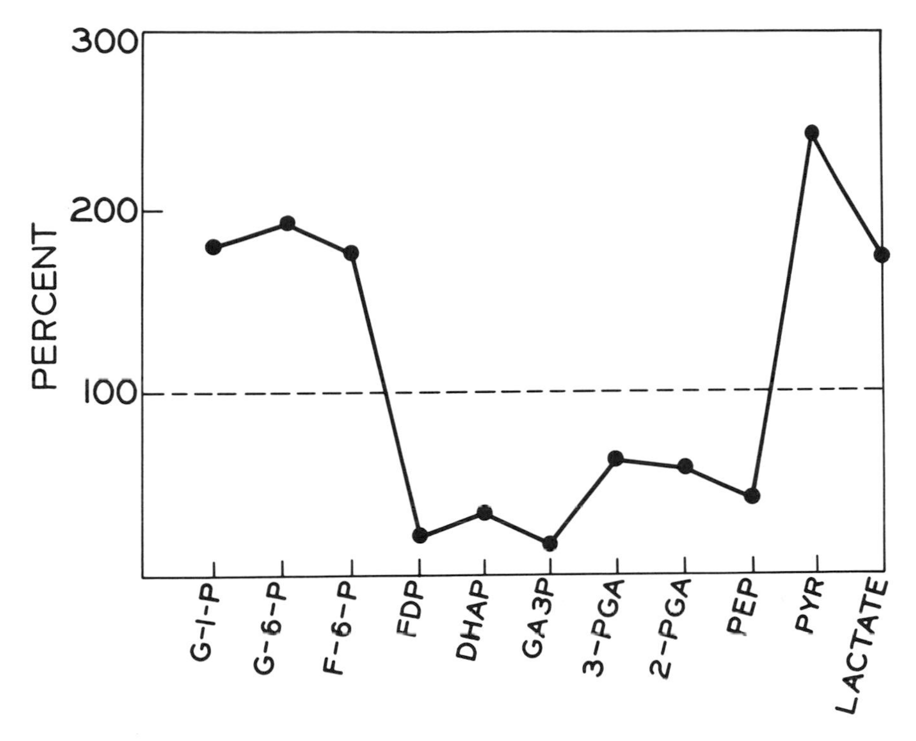


*Fig. 35.3.* Crossover plot of glycolytic intermediate levels observed in fast- and slow-glycolyzing muscles at 15 min post mortem. Calculated from mean values of 22 longissimus muscles with "slow" post mortem glycolysis and 15 muscles with "fast" post mortem glycolysis. Values found in slow-glycolyzing muscles are considered to be 100%.

These metabolite changes would tend to activate both the phosphorylase and PFK enzymes (Williamson, 1966). Since, in the case of phosphorylase, it has been shown that the enzyme is primarily in the *b* form at 10 min post mortem (Sayre, Briskey, and Hoekstra, 1963), one might ask if the observed changes in ATP and AMP are large enough to account for the presumed changes in enzyme activity. Another factor which may partially answer this question is that the divalent metal ions $Mg^{++}$ and $Ca^{++}$ can markedly tighten (Table 35.2) the binding of AMP to phosphorylase (Kastenschmidt, Kastenschmidt, and Helmreich, 1968*a, b*). Liberation of these divalent metal ions post mortem might help override the inhibitory effects of glucose-6-phosphate on phosphorylase. In addition, Ebashi and Endo have already demonstrated a role for $Ca^{++}$ in the activation of phosphorylase *b* kinase. Finally, it should be pointed

TABLE 35.1

*Levels of regulatory intermediates 15 min post mortem*

| Glycolytic rate | Breed* | ATP | ADP | AMP | $P_i$ |
|---|---|---|---|---|---|
| | PC | 3.76†,‡ | 0.505†,‡ | 0.148†,‡ | — |
| Slow | CW | 5.04 | 0.911 | 0.111 | 0.60†,‡ |
| | Hamp | 5.51 | 0.733 | 0.094 | 0.36 |
| Fast | PC | 1.77 | 0.552 | 0.299 | 6.50 |

* PC = Poland China; CW = Chester White; Hamp = Hampshire.
† Expressed as $\mu$moles/g, wet wt.
‡ From Kastenschmidt (1966).

out that not all the glucose-6-phosphate present may be in contact with the enzyme (Helmreich and Cori, 1965). Further studies to clarify these points are under way in our laboratory.

## RECONSTITUTED GLYCOLYTIC SYSTEMS

The studies outlined above do not rule out the possibility that there may be inherent differences in the control properties of the glycolytic enzymes in fast-glycolyzing muscles. We have attempted to test this hypothesis by studying glycolytic control properties *in vitro* (Kastenschmidt, unpublished data, 1967). Sarcoplasmic extracts were prepared from fast-glycolyzing and slow-glycolyzing muscles. Endogenous substrates were removed by passage of the extract through a Sephadex column. Known concentrations of substrates and cofactors were then added. Fig. 35.4 compares the lactate production of sarcoplasmic extracts from fast-glycolyzing

TABLE 35.2

*The effect of divalent cations on the binding of 5′-AMP to phosphorylase b*

| Additions (mM) | $K_{diss}$ 5′-AMP ($M \times 10^4$) |
|---|---|
| None | 3.7 |
| $SrCl_2$ (20) | 1.6 |
| $MgCl_2$ (10) | 2.2 |
| $MgCl_2$ (20) | 1.2 |
| $MgCl_2$ (30) | 1.3 |
| $MnCl_2$ (20) | 0.8 |
| $CaCl_2$ (20) | 0.5 |
| Glucose-1-P (50) | 0.8 |
| Glucose-1-P (50) + $MgCl_2$ (20) | 0.8 (0.3) |

From Kastenschmidt, Kastenschmidt, and Helmreich (1968*b*).

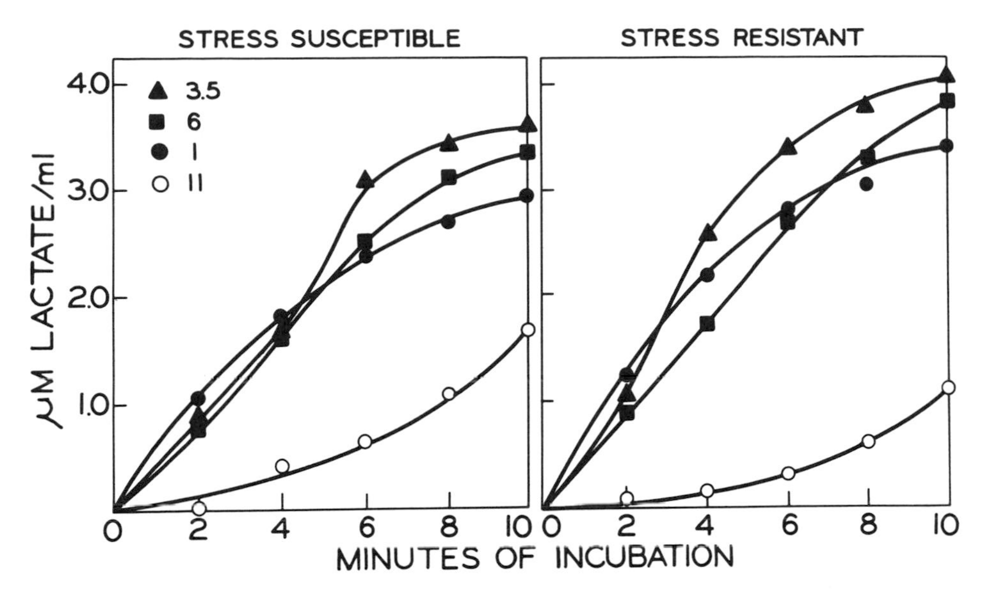


*Fig. 35.4.* Lactate production *in vitro* at varying levels of ATP at 37° C. Sarcoplasmic proteins were extracted with 50 mM imidazole, 0.3 M sucrose, *p*H 7.4, from a muscle homogenate prepared from frozen muscle stored in liquid nitrogen. The incubation medium contained: sarcoplasmic protein, 0.63 mg/ml; imidazole, 100 mM; $P_i$, 10 mM; NAD, 0.2 mM; cysteine, 0.55 mM; creatine, 20 mM; $MgSO_4$, 4.4 mM; glycogen, 0.4% (w/v); and ATP at the levels indicated. The reaction was started by adding glycogen to the incubation medium. Aliquots were removed from the medium at the times indicated, deproteinized, and assayed for lactate by enzymatic techniques.

muscles at a constant level of $P_i$ (10 mM). Concentrations of ATP in excess of 3.5 mM inhibited both systems in a similar manner. Fig. 35.5 illustrates the converse experiment where $P_i$ was varied at a constant level of 6 mM ATP. The shapes of the velocity vs. $P_i$ concentration curves are also quite similar. On the basis of this evidence, it appears unlikely that the control properties of the two systems are grossly different.

## CESSATION OF POST MORTEM GLYCOLYSIS

It has frequently been observed that the ultimate *p*H in muscle can vary. In certain instances, even when adequate glycogen is present, glycolysis ceases when the muscle *p*H is appreciably higher than the usual 5.4–5.5 (Newbold and Lee, 1965). Glycolytic intermediate data (Kastenschmidt, 1966) were consistent with the view that in porcine muscle under the usual conditions, inhibition at the PFK step was the primary cause for cessation of post mortem glycolysis. Conversely, Newbold and Lee (1965) provided evidence that in diluted muscle minces cessation of

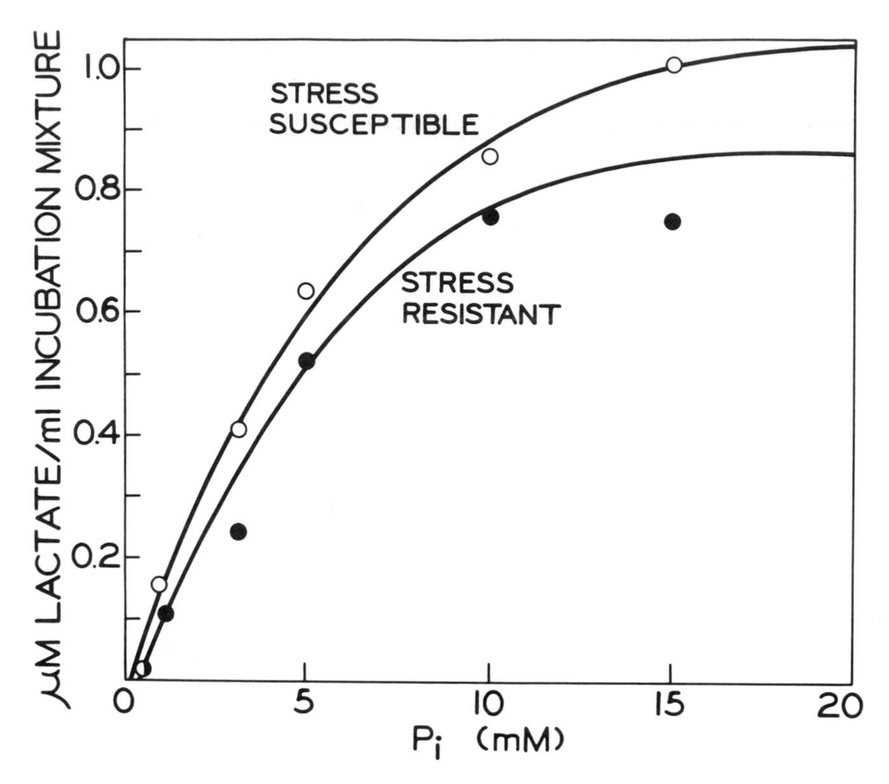


*Fig. 35.5.* Maximal rates of lactate production at varying levels of inorganic phosphate in the incubation medium. Rates were calculated from experiments analagous to those in Fig 35.4 and are expressed as μM lactate produced per ml reaction mix per min. Incubation conditions identical to those described in legend to Fig. 35.4 except that ATP was held constant at 6 mM and $MgSO_4$ was 6 mM.

glycolysis occurred as a result of inactivation of phosphorylase. This was particularly evident when minces were diluted with 1 volume of 0.16 M KCl. A plausible explanation for the KCl effect might be that KCl facilitated the PFK step, since it is known to be an activator of this enzyme (Lardy, 1966). Unfortunately, the KCl mince experiments have not yet been carried out on porcine muscle.

### IN VIVO OXYGEN UPTAKE AND POST MORTEM GLYCOLYSIS

These early studies (Kastenschmidt, Hoekstra, and Briskey, 1966, 1968) pointed to the possibility that the fast-glycolyzing muscles were in an oxygen-deficient state prior to the time of death. In an effort to delineate the effect of anoxia *per se,* Lister, Sair, Will, Schmidt, Cassens, Hoekstra, and Briskey (1970) studied the effects of administering oxygen

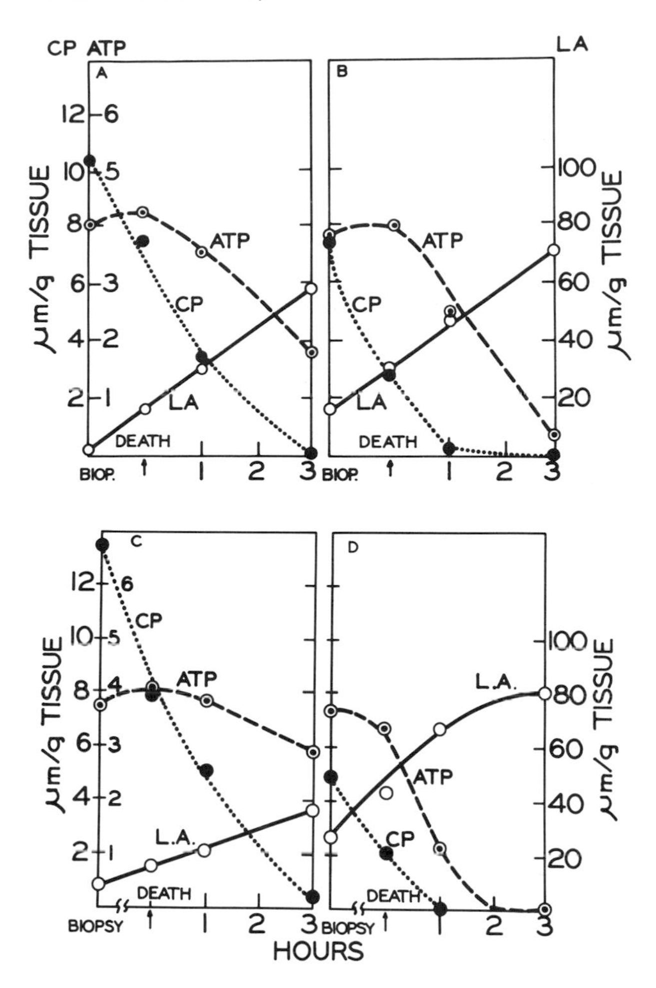


*Fig. 35.6.* Effect of breathing oxygen or nitrogen on adenosine triphosphate, creatine phosphate, and lactic acid values in skeletal muscle. *A,* stress-resistant pigs (oxygen) ; *B,* stress-susceptible pigs (oxygen) ; *C,* stress-resistant pigs (nitrogen) ; *D,* stress-susceptible pigs (nitrogen) . Biopsy samples were taken immediately before exsanguination. From Lister, Sair, Will, Schmidt, Cassens, Hoekstra, and Briskey (1970) .

or nitrogen immediately before death to anesthetized stress-resistant and stress-susceptible animals. They found that the stress-susceptible animals either were made anoxic more easily or responded to anoxia to a greater extent. An interesting finding was that there was a very rapid production of lactic acid during the first 2 min after exsanguination (Fig. 35.6) , as

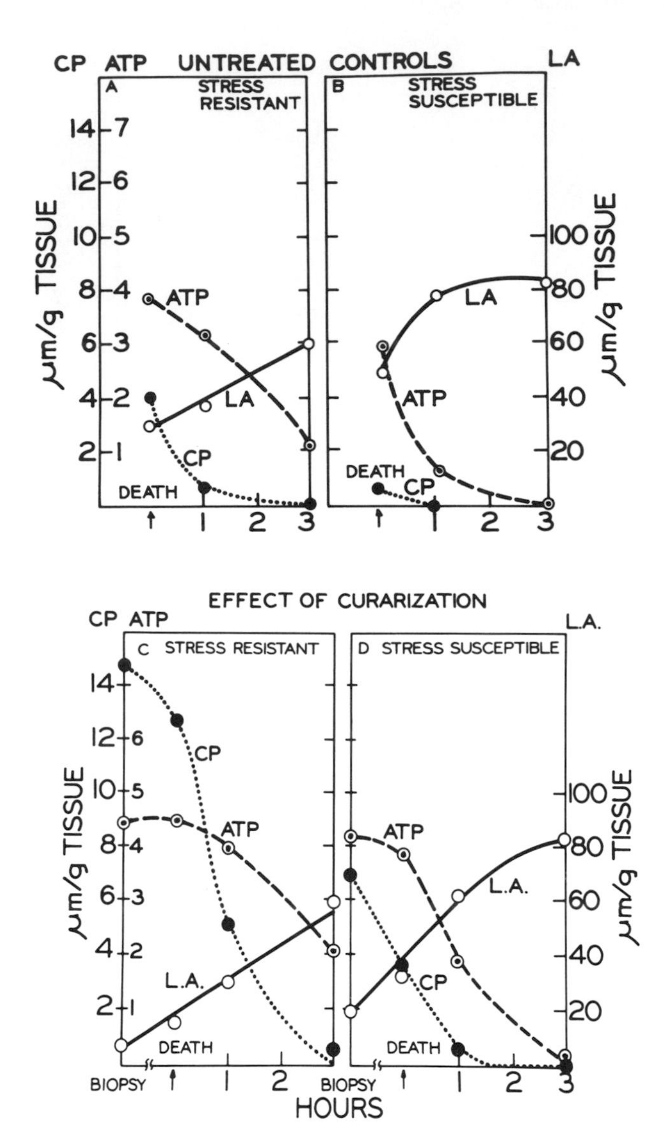


*Fig. 35.7.* Effect of curarization on adenosine triphosphate, creatine phosphate, and lactic acid in skeletal muscle. *A,* stress-resistant pigs (control) ; *B,* stress-susceptible pigs (control) ; *C,* stress-resistant pigs (curare) ; *D,* stress-susceptible pigs (curare) . From Lister, Sair, Will, Schmidt, Cassens, Hoekstra, and Briskey (1970) .

the biopsy samples were removed immediately before exsanguination. This rapid lactic acid formation could partially explain the high levels of lactic acid in fast-glycolyzing muscles that had been observed previously. Vaso-active agents were partially successful in reducing the lactic acid levels found in the samples taken 2 min post mortem.

### EXPERIMENTAL ALTERATION OF ATP AND CREATINE PHOSPHATE LEVELS

Sair, Lister, Moody, Cassens, and Hoekstra (1970) were particularly interested in the levels of high-energy intermediates such as ATP and creatine phosphate in muscle at the time of death. In another study, they found that by preventing muscle stimulation during exsanguination by the injection of d-tubocurarine chloride, the lactic acid accumulation could be reduced. Using this treatment, they found about 15 $\mu$M/g creatine phosphate in biopsy samples of muscles from stress-resistant animals. Only about 7 $\mu$M/g of creatine phosphate was found in similar samples from stress-susceptible animals (Fig. 35.7). The ATP levels in muscles of both groups were slightly over 4 $\mu$M/g. The most successful agent to date by which the levels of ATP and creatine phosphate can be elevated is $MgSO_4$ in anesthetizing doses. This compound, when infused in 50% solution at the rate of 0.2–0.25 ml/kg over a 20-min period, led to a marked elevation in creatine phosphate levels in the muscles of stress-susceptible pigs to the levels found in stress-resistant pigs (Sair, Lister, Moody, Cassens, and Hoekstra, 1970) (Fig. 35.8). Furthermore, it completely prevented the rapid post mortem changes and lowered initial levels of lactic acid in the muscles. $Mg^{++}$ is known to have anesthetic and vaso-dilatory properties (Maxwell, Elliot, and Burnell, 1965), but whether this is the complete explanation for its dramatic effect on post mortem glycolysis is not known at present.

### IN VITRO STUDIES ON THE OXYGEN UPTAKE OF PIG MUSCLE FIBERS

In view of the marked production of lactic acid upon exposure to anoxia, Sair (1970) designed experiments to determine whether the rapid post mortem changes in muscles from stress-susceptible animals are due to lower oxygen uptake, i.e., fewer functioning mitochondria leading to low levels of ATP and creatine phosphate immediately after death, or whether these muscles utilize ATP faster than muscles from stress-resistant animals given equal levels of ATP and creatine phosphate.

Muscle fiber groups from stress-susceptible pigs had a lower oxygen consumption than those from stress-resistant animals. Further investigation showed that when both types of animals were given $MgSO_4$ similar oxygen consumption values were observed in the isolated fiber groups, as shown in Table 35.3.

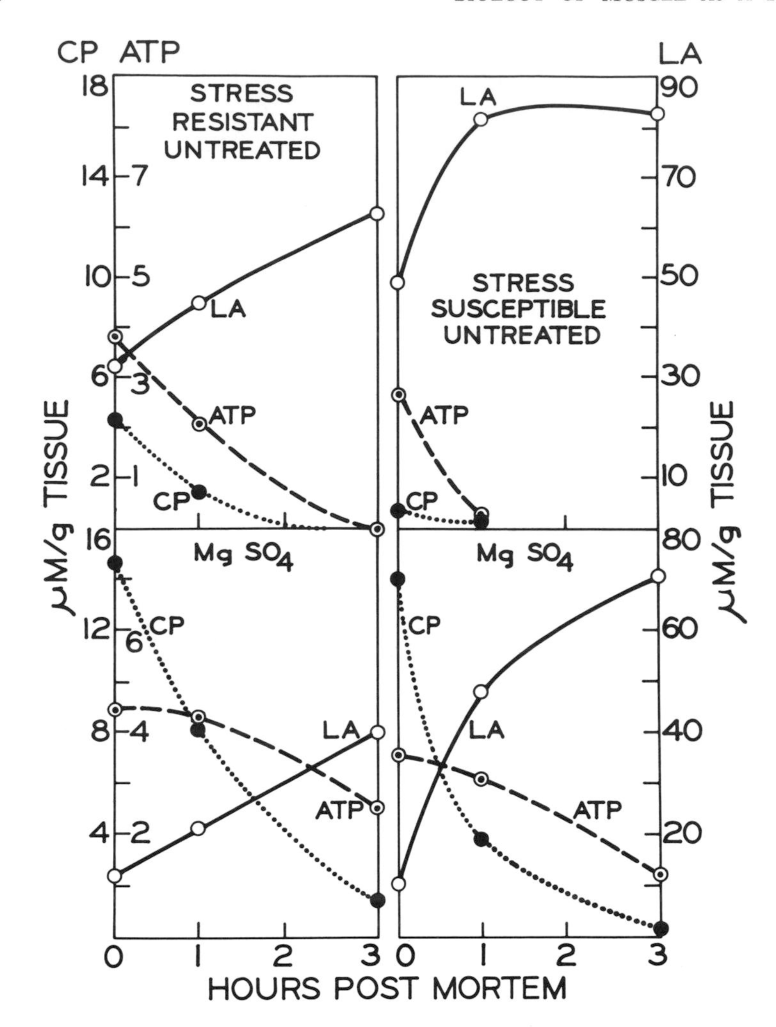


*Fig. 35.8.* Effect of $MgSO_4$ treatment on adenosine triphosphate, creatine phosphate, and lactic acid levels in stress-resistant and stress-susceptible pigs. Animals were injected intravenously with 0.20–0.25 ml/kg of body weight of a 50% solution of magnesium sulfate. From Sair, Lister, Moody, Cassens, and Hoekstra (1970).

When muscle fiber groups from stress-resistant and stress-susceptible animals were incubated under anoxic conditions, a similar rate of lactate production was observed during the period from 15 to 120 min. Nevertheless, the transition from an oxygenated to an anaerobic environment led to a marked increase in the lactate production in the case of the stress-susceptible muscle fibers. These data are illustrated in Fig. 35.9. They are consistent with the second possibility, namely that fiber groups

TABLE 35.3

*Effect of $MgSO_4$ treatment on oxygen consumption of muscle fiber groups from stress-susceptible and stress-resistant pigs**

| Fiber group | μM oxygen/g tissue per 2 hr | |
|---|---|---|
| Untreated | | |
| Stress-susceptible | 8.30 ± 0.65 | |
| | | $P < 0.05$ |
| Stress-resistant | 13.43 ± 0.84 | |
| Magnesium treatment | | |
| Stress-susceptible | 13.01 ± 0.74 | |
| | | NS |
| Stress-resistant | 12.00 ± 0.14 | |

* Incubation media contained: KCl, 2.46 mM; NaCl, 123.2 mM; $CaCl_2$, 9.70 mM; $KH_2PO_4$, 1.23 mM; $MgSO_4 \cdot 7\ H_2O$, 0.62 mM, glycylglycine, 25.5 mM and glucose, 5.55 mM. Approximately 250 mg of muscle fiber groups were included in each flask. Incubation was at 37° C.

From Sair (1970).

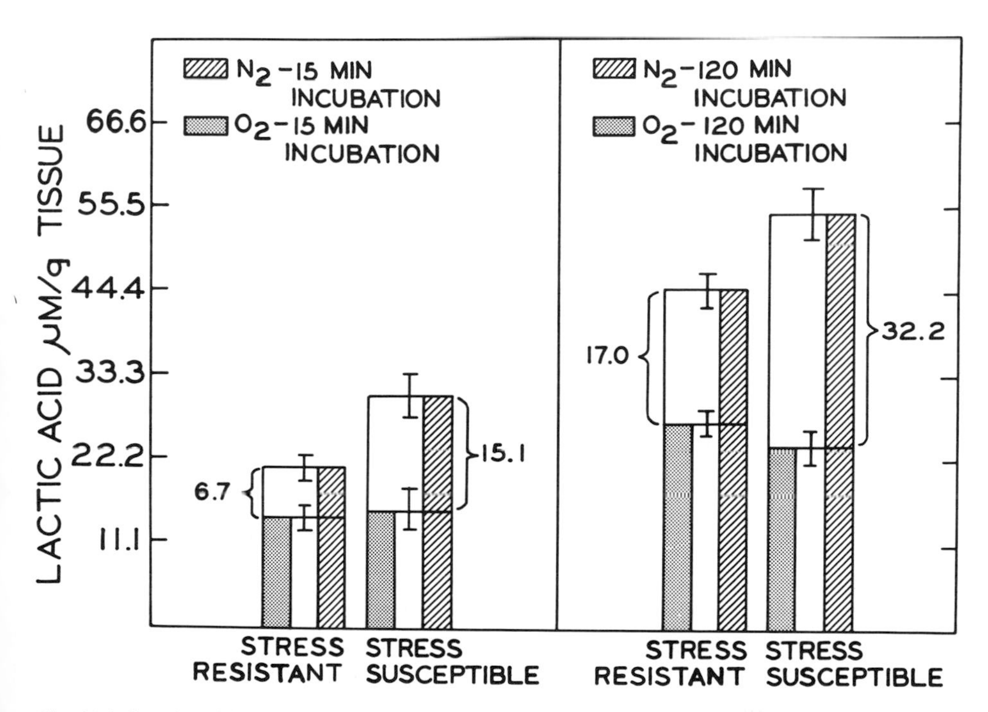


*Fig. 35.9.* Lactic acid production by muscle fiber groups incubated *in vitro* in the presence or absence of oxygen. One standard error of the mean is indicated by the bars. From Sair (1970).

TABLE 35.4

*$Ca^{++}$ accumulation* by the heavy sarcoplasmic reticulum fraction from longissimus muscle from stress-resistant and stress-susceptible pigs*

| Fiber group | % initial extensibility | | | |
|---|---|---|---|---|
| | 100† | 75 | 50 | 0 |
| Stress-resistant | 4.19 ±0.47 | 3.01 ±0.71 | 2.10 ±0.36 | 1.39 ±0.30 |
| Stress-susceptible | 4.00 ±0.41 | 2.67 ±0.47 | 2.24 ±0.52 | 1.63 ±0.43 |
| Significance | NS | NS | NS | NS |

Strips were incubated in a water-saturated nitrogen atmosphere at 37° C. Values are mean values ± SE, in $\mu$ moles $Ca^{++}$/mg protein. The incubation medium contained 100 mM KCl, 5 mM histidine, 5 mM $MgCl_2$, 5 mM ATP, and 0.2 mM $CaCl_2$ (containing 0.1 $\mu$c of $^{45}Ca^{++}$). Incubation was for 15 min at *p*H 7.2 and 23° C.

* Experiment 1.

† 100% extensibility was taken at 30 min post mortem.

from stress-susceptible animals utilize their ATP more rapidly, especially during the aerobic to anaerobic transitional period.

## THE RELATIONSHIP BETWEEN POST MORTEM METABOLISM AND THE INTEGRITY OF THE SARCOPLASMIC RETICULUM

Greaser (1968) investigated the relationship between the $Ca^{++}$-binding ability of the isolated sarcoplasmic reticulum and the rate of post mortem glycolysis. A loss of ability to accumulate $Ca^{++}$ would result in an increase in the concentration of $Ca^{++}$ in the sarcoplasm and elevated myofibrillar adenosine triphosphatases. Muscles which have the most rapid loss of $Ca^{++}$-binding ability were found to have poor muscle quality. More recently, Schmidt (1969) investigated the relationship between $Ca^{++}$-binding ability and rigor mortis in muscle under carefully controlled conditions. In agreement with the work of Greaser (1968), a marked loss in $Ca^{++}$-binding ability with time post mortem was observed. However, at equal points of rigor development the $Ca^{++}$-binding abilities of fast-glycolyzing and slow-glycolyzing muscle were similar. These data are shown in Table 35.4. While they argue against physical damage to the sarcoplasmic reticulum as a causative factor for rapid post mortem glycolysis, they do not preclude impairment of the $Ca^{++}$-binding ability of the sarcoplasmic reticulum by more physiological means such as lowered membrane potential, *p*H, etc.

TABLE 35.5

*erum calcium, magnesium, sodium, potassium, and inorganic phosphorus concentrations*

| | $Ca^{++}$ (mg/100 ml) | $Mg^{++}$ (mg/100 ml) | $\frac{Ca^{++}}{Mg^{++}}$ | $Na^{+}$ (meq/1) | $K^{+}$ (meq/1) | $P_i$ (mg/100 ml) |
|---|---|---|---|---|---|---|
| ress-resistant pigs (n = 10) | 7.59 ±0.27 | 2.32 ±0.12 | 3.35 ±0.19 | 131.5 ±5.8 | 6.00 ±0.36 | 6.8 ±0.1 |
| ress-susceptible pigs (n = 10) | 8.98 ±0.59 | 2.28 ±0.13 | 4.02 ±0.33 | 143.4 ±6.1 | 6.18 ±0.48 | 7.3 ±0.1 |
| gnificance | $P < 0.01$ | NS | $P < 0.01$ | $P < 0.05$ | NS | $P < 0.01$ |

Mean value ±SE.
From Schmidt (1969).

BLE 35.6

*um glucose, total protein, alkaline phosphatase, lactic dehydrogenase, glutamic oxalacetic transaminase, and blood urea rogen concentrations*

| | Glucose (mg/100 ml) | Total protein (g/100 ml) | Alkaline phosphatase (mU/ml) | Lactic dehydrogenase (mU/ml) | Serum glutamic oxalacetic transaminase (mU/ml) | Blood urea nitrogen (mg/100 ml) |
|---|---|---|---|---|---|---|
| ss-resistant igs (n = 10) | 96 ± 3 | 7.43 ±0.16 | 48.1 ± 3.7 | 332 ± 27 | 54 ± 3 | 15 ± 1 |
| ss-susceptible gs (n = 10) | 96 ± 3 | 7.43 ±0.12 | 80.5 ± 5.4 | 422 ± 38 | 69 ± 9 | 19 ± 1 |
| ificance | NS | NS | $P < 0.01$ | $P < 0.01$ | $P < 0.01$ | $P < 0.01$ |

ean value ±SE.
om Schmidt (1969).

## BLOOD METABOLITE PROFILE AND POST MORTEM GLYCOLYSIS

In an effort to find means of identifying stress-susceptible animals prior to death, Schmidt (1969) studied several serum metabolites which can be measured by a Technicon 1260 Autoanalyzer. The following data were obtained from 10 animals whose carcasses had poor-quality muscle and 10 others which had good-quality muscle. Table 35.5 shows that the stress-susceptible animals had higher levels of serum $Ca^{++}$, $Na^{+}$, and $P_i$. Table 35.6 shows that these same animals also had higher levels of lactic dehydrogenase, alkaline phosphatase, glutamate-oxaloacetate transaminase, and blood urea nitrogen. While this experiment demonstrates that there are definite enzymic and electrolyte differences in the serum obtained from pigs, these differences are rather small. Further studies are in progress.

It should be noted that Addis and Kallweit (1969) have reported higher levels of muscle-type lactic dehydrogenase isozymes in serum from stress-susceptible pigs. We have partially confirmed these results using assays for total lactic dehydrogenase (LDH), creatine phosphokinase (CPK), and both LDH and CPK isozymes. Large daily variations in enzyme levels and isozyme patterns were observed. We are currently trying to elucidate the possible reasons for the large daily variations.

### ELECTRICAL ACTIVITY AFTER DEATH AND POST MORTEM GLYCOLYSIS

Other workers (see Bendall, 1966) have hypothesized that the amount of excitement and struggling produced before and during slaughter was to a large extent responsible for the variability in the rates of post mortem glycolysis. In an effort to substantiate this hypothesis, Schmidt (1969) in our laboratory studied the electrical activity in the longissimus dorsi muscle of pigs before, during, and after stunning using two concentric needle electrodes coupled to an integrated electromyograph. This technique should be a good measure of the amount of muscular contraction during the period studied. Fig. 35.10 shows the EMG pattern of stress-resistant animals. Notice the large number of spikes at the time of stunning. Fig. 35.11 shows the EMG pattern for stress-susceptible animals. Note the fewer spikes, which indicate that there is less muscular contraction in the stress-susceptible animals. These results argue against the idea that the struggle at death is in large part responsible for the variability in post mortem glycolysis.

### CONCLUDING STATEMENTS

The evidence presented here and elsewhere suggests that much of the variability in post mortem glycolytic rates can be explained on the basis of altered regulatory cofactor levels. There is good evidence that stress-susceptible muscles utilize their energy supplies much more rapidly than do stress-resistant muscles. The reason that this is so still eludes us, but progress is being made.

The role of the muscle membrane appears to be a fruitful area to investigate. In this connection, the observation by Schmidt and coworkers that the stress-susceptible animal has decreased electrical activity during the first 2–3 min after death is extremely interesting. Whether this implies a lowered membrane potential and increased release of $Ca^{++}$ into the sarcoplasm is an area of very active investigation. Already there are some data to suggest this may be the case (Forrest and Briskey, 1967).

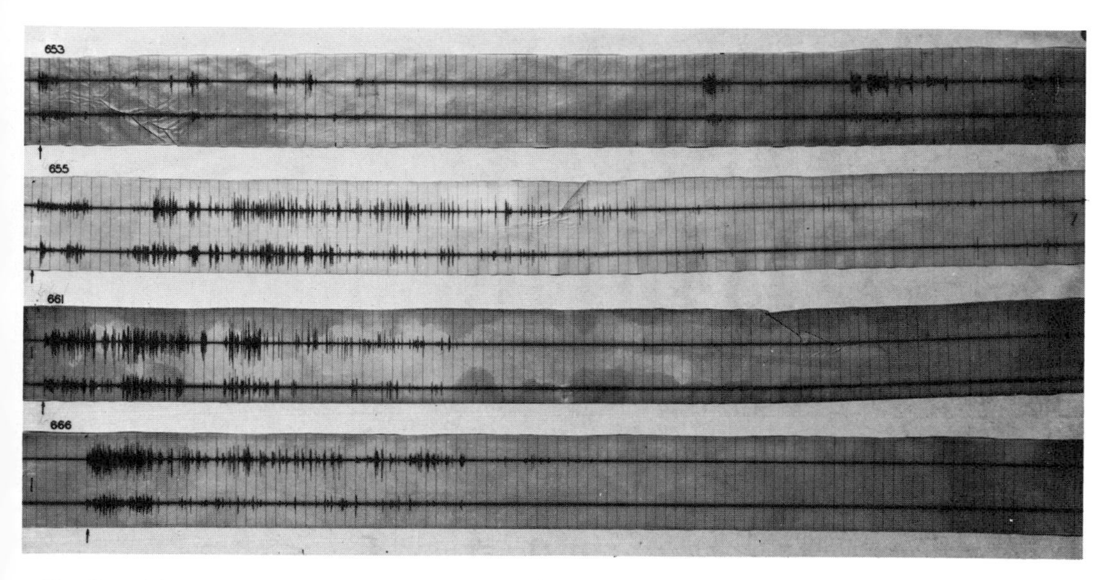


*Fig. 35.10.* Electromyographic recordings from longissimus muscles of stress-resistant Chester White pigs. The point of stunning is indicated by arrow. Each vertical line on the strips represents 1 sec. From Schmidt (1969).

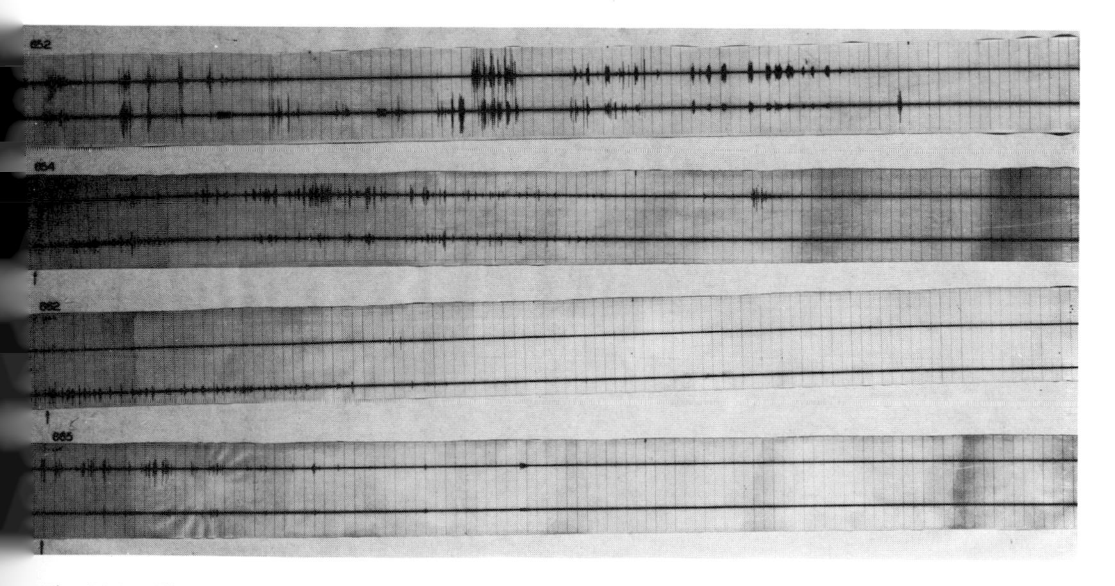


*Fig. 35.11.* Electromyographic recordings from longissimus muscles of stress-susceptible pigs. See Fig. 35.10 for other details. From Schmidt (1969).

Another factor which must be put in proper perspective is the role of adrenal and thyroid hormones in post mortem muscle metabolism. Are all these hormonal effects explainable in terms of regulating the flow of blood in the muscle? These and other unsolved problems undoubtedly will occupy muscle biologists for a long time to come.

## ACKNOWLEDGMENTS

I am deeply indebted to my colleagues at the Muscle Biology Laboratory in Madison for providing a good deal of the information used in this chapter and for making their manuscripts available to me before publication.

## *References*

Addis, P. B., and E. Kallweit. 1969. Relationships between the isoenzymes of lactodehydrogenase (LDH) in blood serum and meat property characteristics in the pig. *Die Fleischwirtschaft 49*:218.

Bendall, J. R. 1966. The effect of pretreatment of pigs with curare on the post-mortem rate of *p*H fall and onset of rigor mortis in the musculature. *J. Sci. Food Agr. 17*:333.

Briskey, E. J. 1964. Etiological status and associated studies on pale, soft, exudative porcine musculature. *Advance. Food Res. 13*:89.

Briskey, E. J., L. L. Kastenschmidt, J. C. Forrest, G. R. Beecher, M. D. Judge, R. G. Cassens, and W. G. Hoekstra. 1966. Biochemical aspects of post-mortem changes in porcine muscle. *J. Agr. Food Chem. 14*:201.

Chance, B., and G. R. Williams. 1956. The respiratory chain and oxidative phosphorylation. *Advance. Enzymol. 17*:65.

Ebashi, S., and M. Endo. 1968. Calcium ion and muscle contraction. *Progr. Biophys. Mol. Biol. 18*:123.

Forrest, J. C., and E. J. Briskey. 1967. Response of muscle to electrical stimulation. *J. Food Sci. 32*:1.

Greaser, M. L. 1968. Ultrastructural and biochemical properties of isolated sarcoplasmic reticulum from skeletal muscle. Ph.D. thesis, Univ. of Wisconsin, Madison, Wis.

Helmreich, E., and C. F. Cori. 1965. Regulation of glycolysis in muscle, p. 91. *In* G. Weber (ed.), *Advances in Enzyme Regulation,* Vol. 3.

Kastenschmidt, L. L. 1966. Regulation of post-mortem glycolysis in striated muscle. Ph.D. thesis. Univ. of Wisconsin, Madison, Wis.

Kastenschmidt, L. L., W. G. Hoekstra, and E. J. Briskey. 1966. Metabolic intermediates in skeletal muscles with fast and slow rates of post-mortem glycolysis. *Nature 212*:288.

———. 1968. Glycolytic intermediates and co-factors in "fast-" and "slow-glycolyzing" muscles of the pig. *J. Food Sci. 33*:151.

Kastenschmidt, L. L., J. Kastenschmidt, and E. Helmreich. 1968*a*. Subunit interactions and their relationship to the allosteric properties of rabbit skeletal muscle phosphorylase *b*. *Biochemistry* 7:3590.

———. 1968*b*. The effect of temperature on the allosteric transitions of rabbit skeletal muscle phosphorylase *b*. *Biochemistry* 7:4543.

Lardy, H. A. 1966. Regulation of energy-yielding processes in muscle, p. 31. *In* E. J. Briskey, R. G. Cassens, and J. C. Trautman (eds.), *The Physiology and Biochemistry of Muscle as a Food*. Univ. of Wis.

Lister, D., R. A. Sair, J. A. Will, G. R. Schmidt, R. G. Cassens, W. G. Hoekstra, and E. J. Briskey. 1970. Metabolism of striated muscle of "stress-susceptible" pigs breathing oxygen or nitrogen. *Amer. J. Physiol. 218*:102.

Maxwell, G. M., R. G. Elliot, and R. H. Burnell. 1965. Effects of hypermagnesemia on general and coronary hemodynamics of the dog. *Amer. J. Physiol. 208*:158.

Newbold, R. P., and C. A. Lee. 1965. Post-mortem glycolysis in skeletal muscle. The extent of glycolysis in diluted preparations of mammalian muscle. *Biochem. J. 97*:1.

Sair, R. A. 1970. Physiological and biochemical studies on pale, soft, exudative porcine muscle. Ph.D. thesis. Univ. of Wisconsin, Madison, Wis.

Sair, R. A., D. Lister, W. G. Moody, R. G. Cassens, and W. G. Hoekstra. 1970. The action of curare and magnesium on the striated muscle of "stress-susceptible" pigs. *Amer. J. Physiol. 218*:108.

Sayre, R. N., E. J. Briskey, and W. G. Hoekstra. 1963. Porcine muscle glycogen structure and its association with other muscle properties. *Proc. Soc. Exp. Biol. Med. 112*:164.

Schmidt, G. R. 1969. Physical and metabolic changes during the development of rigor mortis in porcine muscle. Ph.D. thesis, Univ. of Wisconsin, Madison, Wis.

Williamson, J. R. 1966. Glycolytic control mechanisms II. Kinetics of intermediate changes during aerobic-anoxic transition in perfused rat heart. *J. Biol. Chem. 241*:5026.

# Chapter 36

# Chemistry of Muscle Proteins as a Food

D. E. GOLL, N. ARAKAWA, M. H. STROMER, W. A. BUSCH, AND
R. M. ROBSON

Although a number of topics could be discussed under the above heading, we have chosen to deal only with those changes that occur in muscle proteins during the onset and resolution of rigor mortis. Furthermore, some of the post mortem changes in sarcoplasmic and stroma proteins will be considered only briefly, and principal emphasis will be on post mortem changes in the myofibrillar proteins and some possible causes for these changes. This choice is made for several reasons:

(1) Changes in muscle proteins during heating, freezing, emulsification, etc., are not well understood in molecular terms, and meaningful discussion is therefore difficult.

(2) Many of the attributes which determine the desirability of muscle as a food are established during the first 24–48 hours after death; hence, it is important to understand the nature of the changes that occur in muscle proteins during this period.

(3) The myofibrillar proteins constitute over 50% of the total protein in muscle and are also directly implicated in the water-holding capacity, emulsification properties, and tenderness of muscle; thus, it seems appropriate to place principal emphasis on the myofibrillar proteins when examining post mortem changes in muscle and their relation to the use of muscle as a food.

(4) Post mortem alterations in the myofibrillar proteins have received much recent attention and several new kinds of changes have just been discovered; consequently, it should be instructive at this point to review these changes and their possible meaning in the use of muscle as a food.

In this chapter, we will first summarize a few of the studies on post mortem changes in the sarcoplasmic and stroma protein fractions and then discuss some of the recent work on post mortem changes in myofibrillar proteins. We will then consider some of the evidence against any extensive post mortem proteolysis of muscle proteins, and finally describe, in terms of alterations of specific proteins, some possible causes for the onset and resolution of rigor mortis.

## DEFINITION OF RIGOR MORTIS

The need for a clear definition of the term "rigor" at the outset of any discussion of post mortem muscle is evident from the confusion and controversy that exist in the literature. For example, most muscle biologists use the term "rigor" to refer to any muscle that has had its ATP* removed (Elliott, Lowy, and Worthington, 1963; Huxley, 1968; Huxley and Brown, 1967; Reedy, Holmes, and Tregear, 1965). Ordinarily, this ATP depletion is accomplished by removing the muscle from the cadaver, separating it into fiber bundles, and immersing the bundles in buffered 50% glycerol for 24–48 hours at 2°. When the term "rigor" is used in this sense, it follows that the contractile proteins are also in "rigor" in biochemical preparations such as myofibrillar suspensions, natural actomyosin, or reconstituted actomyosin, since the procedures used to make these preparations would cause almost complete ATP removal from the contractile complex. This would be true whether such contractile preparations were made from muscle immediately after death or from post mortem muscle that was left *in situ* until 24–48 hours after death.

On the other hand, many medical and agricultural scientists use the term, rigor mortis, to refer to the stiffening or hardening that occurs in a cadaver during the first 24 hours after death. This definition is couched in terms of time post mortem and is obviously related to muscle properties *in situ*. Moreover, there is not any direct reference to ATP depletion in this concept of rigor mortis. ATP depletion, however, does occur in muscle left *in situ*, and the time at which ATP is depleted often corresponds closely to the onset of hardening or stiffening. This apparent relationship between ATP depletion and the onset of hardening has probably caused much of the confusion regarding use of the term "rigor mortis." It has become evident, however, during the past few years, that rigor mortis has two closely related but quite different aspects (Davies, 1966): (1) a shortening or contraction of the post mortem muscle fiber; and (2) a loss of muscle extensibility. The present evidence indicates

* Less common abbreviations used in this presentation are: EGTA, 1,2-bis-(2-dicarboxymethylaminoethoxy)-ethane; ITP, inosine 5′-triphosphate; Γ/2, ionic strength calculated on a molarity basis. All temperatures are given in ° C.

that loss of muscle extensibility is directly related to ATP depletion and that post mortem muscle loses its extensibility only after ATP concentration has fallen to less than 10–20% of its initial level. On the other hand, post mortem shortening requires ATP as an energy source (Bendall, 1951) ; shortening must, therefore, precede total loss of extensibility in post mortem muscle. Moreover, it now appears that, although loss of extensibility is a necessary condition for maintenance of the rigidity or stiffness developed in rigor, it is not itself the primary cause of rigor stiffening (Briskey, 1964; Goll, 1968; Newbold, 1966) . Rather, it seems probable that rigidity in rigor is a direct result of post mortem muscle shortening, since attempted shortening by muscles on opposing sides of the same bone would produce a stiffness or hardness similar to that observed in an isometric contraction (Goll, 1968; Goll and Robson, 1967) . Inextensibility is essential to this hypothesis since this condition keeps post mortem muscle in its shortened state after its ATP supply is exhausted and it can no longer actively generate tension. If the sarcomeres in post mortem muscle were not inextensible, they would be free to return to their rest length after the ATP supply had been exhausted, and this would result in loss of the stiffness or rigidity caused by the shortening.

In this chapter we will use only the shortening-rigidity definition of rigor, both because it seems more congruous with the literal meaning of the words "rigor mortis," ("stiffness of death") and because the shortening-rigidity concept can be more directly related to post mortem changes in muscle as a food. Since the shortening-rigidity concept of rigor is based on the premise that attempted shortening by muscles on the opposite sides of the same bone causes stiffness or rigidity in a cadaver, the onset and resolution of rigor mortis is most easily monitored in a quantitative fashion by measuring the isometric tension development of post mortem muscle strips. An example of the results obtained by this type of measurement is shown in Fig. 36.1. In this measurement, the onset of tension development corresponds to the onset of rigor mortis, and the gradual loss in ability to maintain isometric tension can be related to the resolution of rigor (Goll, 1968; Goll and Robson, 1967) . Although the measurements shown in Fig. 36.1 are for bovine semitendinosus muscle, extensive experimentation in our laboratory has shown that similar patterns, differing only in time course and extent of tension development, can be obtained from bovine, rabbit, and porcine muscle at storage temperatures of 2°, 16°, 25°, or 37°. Therefore, post mortem tension development and release is a widespread phenomenon and is not limited to bovine muscle at 2°, as seemed to be indicated by our earlier experiments on this subject (Busch, Parrish, and Goll, 1967; Jungk, Snyder, Goll, and McConnell, 1967) .

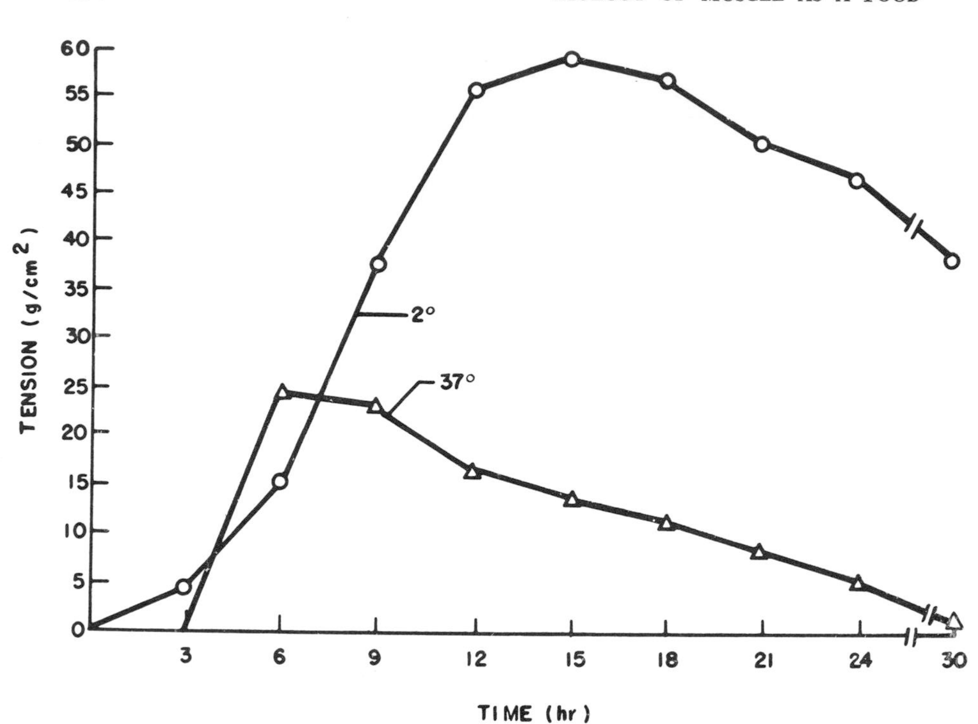


*Fig. 36.1.* Post mortem changes in isometric tension development of bovine semitendinosus muscle at 2° C and 37° C. Tension was measured on muscle strips immersed in 50 mM KCl, 60 mM K-phosphate, *p*H 7.2, 1 mM $MgCl_2$ and 1 mM $NaN_3$ by using an E & M Physiograph and isometric myograph transducers (cf. Goll, 1968).

The exact relationship between the isometric tension measurement of rigor mortis and the widely used extensibility measurement of rigor mortis is not yet clear, although Newbold (1966) has shown that the time course of these two measurements is clearly different in bovine muscle at 2°. It is obvious that some loss of extensibility must accompany the shortening process itself, since shortening is caused by an interaction between the thick and thin filaments in muscle. However, Huxley and Brown (1967) have shown that in contracting or shortening muscle only about 20% of the cross bridges, which extend outward from the surface of the myosin or thick filament, are actually linked to the actin or thin filament at any one time. On the other hand, almost 100% of the myosin cross bridges are bound to the thin filament in muscle, which has been depleted of ATP and which is therefore inextensible (Huxley, 1968; Huxley and Brown, 1967). Thus, post mortem muscle, in its shortening phase, should be approximately five times more extensible than post mortem muscle that has lost its ATP and has therefore completed its extensibility phase. These considerations indicate that the shortening and extensibility concepts of rigor can be clearly distinguished from one another, both in time and in mechanism. Furthermore, although the extensibility concept suggests that rigor mortis is simply the result of ATP depletion in post mortem muscle, the shortening-rigidity concept portrays rigor mortis as a more complex phenomenon in which events presently obscure act first to initiate tension development and then to cause gradual loss of the muscle's ability to maintain tension. Later we will describe some evidence, recently obtained in our laboratory, which indicates that $Ca^{++}$ may be involved in both initiation and loss of isometric tension development in post mortem muscle.

## POST MORTEM CHANGES IN SARCOPLASMIC PROTEINS

The sarcoplasmic protein, or "myogen" fraction of muscle, is usually defined in operational terms as that fraction of muscle protein which is soluble at low ionic strength—defined for our purposes here as any ionic strength below 0.2. Sarcoplasmic protein makes up approximately 30–34% of total protein in skeletal muscle of mature mammals, although this proportion will vary according to age of animal and among smooth, cardiac, and skeletal musculature.

Post mortem changes in the sarcoplasmic protein fraction have long interested medical researchers and those concerned with muscle as a food, but it has been very difficult to obtain any definitive data regarding these changes, largely because of the tremendous complexity of the sarcoplasmic protein fraction. This fraction contains most of the glycolytic and biosynthetic enzymes together with other proteins such as myoglobin. Consequently, the sarcoplasmic protein fraction contains at least 50–100 different proteins, and at the present level of sophistication in protein chemistry, it is plainly an impossible task to separate and study the effect of post mortem aging on all of them simultaneously. Moreover, Neelin and Ecobichon (1966) have reported that the protein composition of sarcoplasmic extracts depends on the method and severity of homogenization used and whether extraction is done with a hypotonic salt solution or a hypertonic sucrose solution (0.44 M sucrose). Thus, it was observed that starch gel electrophoresis patterns of a "sarcoplasmic protein" fraction extracted from minced chicken muscle by vigorous homogenization in a veronal buffer ($\Gamma/2 = 0.08$, $pH = 8.0$) did not exhibit any detectable, consistent change during post mortem storage at 2° for 48 hours, although an additional electrophoretic component appeared slowly with increasing time of post mortem storage in extracts of red chicken muscle prepared by the same procedure. However, starch gel electrophoresis patterns of "sarcoplasmic protein" fractions extracted by gentle homogenization in sucrose lacked many of the components seen in fractions prepared by veronal buffer extraction of muscle. Moreover, it was observed that during post mortem aging, these missing components gradually appeared in the electrophoretic patterns of sarcoplasmic proteins extracted by sucrose (Neelin and Rose, 1964). Subsequently it was found (Neelin and Ecobichon, 1966) that one of the components lacking in gel electrophoresis patterns of the sucrose-extracted sarcoplasmic protein fraction was lactic dehydrogenase. These results led Neelin and Ecobichon (1966) to suggest that many of the so-called "sarcoplasmic proteins" are, in fact, associated *in vivo* with cytoplasmic structures such as the sarcolemma or the sarcoplasmic reticulum. This suggestion has been supported by Hultin and Westort's recent finding (1969) that lactic dehydrogenase appears to be closely associated with the membranes of a purified sarcolemmal preparation. Severe homogenization, particularly in hypotonic salt solutions, may cause disintegration of these cytoplasmic structures and release the associated proteins into the sarcoplasmic protein fraction. Gentle homogenization in hypertonic sucrose solutions, on the other hand, probably preserves these cytoplasmic structures, and they and their associated proteins are removed from the sarcoplasmic protein fraction by centrifugation. This hypothesis accounts for the lack of certain electrophoretic components in sucrose-extracted sarcoplasmic protein fractions, since these missing components are presumably associated with the intact cytoplasmic organelles. Neelin and Ecobichon (1966) further suggest that the gradual reappearance of these missing components in gel electrophoresis patterns of post mortem muscle indicates that post mortem storage causes disruption of the cytoplasmic structures and release of their associated proteins, even in the absence of severe homogenization. Several recent studies have provided strong support for these suggestions. Thus, Greaser, Cassens, and Hoekstra (1967), Greaser, Cassens, Briskey, and Hoekstra (1969*a, b*) and Eason (1969) have found that during post mortem storage sarcoplasmic reticular membranes very quickly lose their ability to sequester $Ca^{++}$, even though their adenosine triphosphatase activity and ultrastructure do not undergo large post mortem changes. Furthermore, Osner (1966) and Reed, Houston, and Todd (1966) have shown that the sarcolemma is a very labile system and may change rapidly after death. Neelin and Ecobichon's hypothesis should therefore be considered in studying post mortem changes in the sarcoplasmic proteins.

Many studies have attempted to determine the gross effects of post mortem storage on sarcoplasmic proteins by measuring changes in sarcoplasmic protein solubility. Because of the many different homogenization conditions and extracting solvents used, it is difficult to compare accurately the results of many of these studies. Moreover, many investigators have used water to extract sarcoplasmic proteins from both at-death and post mortem muscle. Since the *p*H of post mortem muscle is usually below 6.0, whereas the *p*H of at-death muscle is near 7.0, use of unbuffered water for extraction of sarcoplasmic proteins results in extraction of the at-death fraction at a *p*H value near 7.0 while the post mortem fractions are extracted at some varying *p*H value, usually below 6.0. This difference in extraction *p*H alone may cause substantial differences in the type of protein fractions obtained, regardless of any post mortem alterations in the sarcoplasmic proteins themselves. A low ionic strength buffer should be used to extract sarcoplasmic proteins, particularly in those studies involving extraction of post mortem muscle. In spite of these methodological complications, it is possible to conclude from the studies reported thus far that sarcoplasmic protein solubility is highest immediately after death and either remains unchanged or decreases by some variable amount during post mortem storage. The rate and severity of any post mortem decrease in solubility depend critically on both rate of *p*H decline in the post mortem tissue and temperature of post mortem storage. If muscle *p*H decreases below 6.0 while muscle temperature is still 35° or higher, sarcoplasmic protein solubility is decreased substantially (Sayre and Briskey, 1963; Scopes, 1964; Scopes and Lawrie, 1963). Consequently, post mortem storage of muscle at temperatures above 30° results in a large decrease in sarcoplasmic protein solubility (Chaudhry, Parrish, and Goll, 1969). On the other hand, rapid cooling of at-death muscle to temperatures of 0–4° and subsequent storage at these temperatures causes sarcoplasmic protein solubility to remain high, near the at-death level (Borchert and Briskey, 1965; Goll, Henderson, and Kline, 1964). Conditions of post mortem storage intermediate between these two extremes result in a slight loss of sarcoplasmic protein solubility; these losses usually vary between 10–30% of the original solubility.

Recent studies using starch gel electrophoresis or analytical ultracentrifugation to fractionate the sarcoplasmic protein fraction have indicated that the decreased solubility of sarcoplasmic proteins after post mortem storage at 37° is due, at least in part, to denaturation and consequent insolubility of creatine kinase (Borchert, Powrie, and Briskey, 1969; Scopes, 1964; Scopes and Lawrie, 1963), and probably also to denaturation of phosphoglucomutase (Borchert, Powrie, and Briskey, 1969), lactic dehydrogenase (Kronman and Winterbottom, 1960), phosphorylase *b* (Kronman and Winterbottom, 1960), and triosephosphate isomerase and F-protein (Borchert, Powrie, and Briskey, 1969). F-protein is a basic sarcoplasmic protein first described by Scopes (1966). Numerous other investigations (Aberle and Merkel, 1966; Borchert, Powrie, and Briskey,

1969; Lawrie, Penny, Scopes, and Voyle, 1963; Maier and Fischer, 1966; Neelin and Rose, 1964; Scopes, 1964), however, have shown that if post mortem muscle is stored at temperatures of 5° or lower, gel electrophoresis patterns of sarcoplasmic proteins extracted from at-death muscle are very similar to those of sarcoplasmic proteins extracted from post mortem muscle, even after 336 hours post mortem (Aberle and Merkel, 1966). This similarity between at-death and post mortem gel electrophoresis patterns extends even to naturally occurring pale, soft, exudative porcine tissue, which has 50% lower sarcoplasmic protein solubility than at-death muscle (Borchert, Powrie, and Briskey, 1969).

Although studies with cellulose ion-exchange chromatography have shown that the elution profile of a sarcoplasmic protein fraction extracted from at-death muscle differs from the elution profile of sarcoplasmic proteins extracted from post mortem muscle stored at 2° (Fujimaki and Deatherage, 1964; Rampton, Anglemier, and Montgomery, 1965; Thompson, Davidson, Montgomery, and Anglemier, 1969), the results of these studies are very difficult to interpret because the 14–20 fractions resolved by cellulose ion-exchange chromatography almost certainly do not represent homogeneous proteins and because individual proteins appear to be spread among as many as eight of the 14–20 different fractions. Any further work on fractionation of sarcoplasmic proteins by cellulose ion-exchange chromatography should probably consider a preliminary ammonium sulfate fractionation of the crude sarcoplasmic protein extracts in order to reduce the complexity of the fractions applied to the columns. At the present time, starch or polyacrylamide gel electrophoresis would appear to be the technique of choice for monitoring post mortem qualitative changes in the sarcoplasmic proteins; larger numbers of samples can be conveniently tested simultaneously, smaller amounts of time are required per run, and resolving power is greater than with most other techniques available for protein fractionation. However, although it is possible to obtain as many as 35 different fractions from a single vertical starch gel electrophoresis run (Scopes, 1964; Scopes and Lawrie, 1963), even gel electrophoresis does not separate sarcoplasmic proteins into a series of homogeneous fractions, and considerable care must also be exercised in attempting to interpret changes in starch gel electrophoresis patterns in terms of quantitative changes in specific proteins.

In summary, the recent starch gel electrophoresis results show that sarcoplasmic proteins do not undergo large changes in composition during post mortem storage at temperatures of 5° or lower, even in circumstances where sarcoplasmic protein solubility may be lowered. This indicates that the sarcoplasmic proteins do not experience extensive post mortem proteolysis, and, furthermore, suggests that the myofibrillar or stroma proteins are not proteolytically degraded during post mortem storage, since such degradation should cause both an increase in the amount of sarcoplasmic protein extracted and the appearance of new protein components in the sarcoplasmic protein fraction. Because of this, it is presently difficult to relate changes in the sarcoplasmic protein fraction *per se* to changes in the properties of muscle used as a food. It seems obvious, however, that because of their role in post mortem glycolysis, the sarcoplasmic proteins may produce muscle conditions that have important effects on the myofibrillar and stroma protein fractions of muscle. Further studies on post mortem changes in the sarcoplasmic proteins should use measurements of specific properties of individual proteins, such as enzymic activity or ultracentrifugal behavior, to characterize more completely the nature of any possible alterations induced by post mortem storage.

## POST MORTEM CHANGES IN STROMA PROTEINS

The stroma protein fraction may be defined as that fraction of muscle protein which is insoluble in neutral aqueous solvents. Stroma proteins make up 10–15% of total muscle protein, but because they are insoluble in their native state, it has been difficult to determine both the nature and the number of different proteins contained in this fraction. It seems probable, however, that collagen, elastin, and reticulin make up the largest proportion of the stroma protein fraction, and this discussion will be limited to a brief review of the post mortem changes in connective tissue proteins. It is now evident that connective tissue has an important role in the use of muscle as a food, since it is probably responsible for the "background" toughness of muscle as opposed to the "actomyosin" toughness due to the myofibrillar proteins themselves (Herring, Cassens, and Briskey, 1967). "Total muscle toughness," however, includes both "background" and "actomyosin" toughness, so connective tissue content is not always highly related to toughness of muscle, particularly in those muscles that have low connective tissue content and whose toughness is therefore due primarily to "actomyosin" toughness.

As indicated previously (Goll, 1965), very little is known about post mortem changes in connective tissue. The first report on the effect of post mortem aging of connective tissue was that of Prudent (1947) who found that collagen and elastin content of bovine muscles, as measured by alkali-insoluble nitrogen of cooked and uncooked samples, did not change during post mortem storage for 1, 2, 5, 10, 20, or 30 days at 4°. The techniques used by Prudent, however, would be relatively insensitive for detection of a small amount of connective tissue solubilization caused by post mortem storage, and would probably be unable to detect any post mortem changes in structure of the connective tissue proteins that might render them more susceptible to solubilization by heating. In fact, Winegarden (1950) later reported that the force required to shear through strips of collagenous connective tissue and ligamentum nuchae (a tissue rich in elastin) was slightly smaller after 35 days of post mortem storage than after 10 days of post mortem storage. About this same time, Husaini, Deatherage, and Kunkle (1950) and Hershberger, Deans, Kunkle, Gerlaugh, and Deatherage (1951) found that alkali-insoluble protein of bovine muscle was decreased by 10–30% after 15 days of post mortem storage at 4°. A large amount of animal variation was noted in these studies, however, and subsequently Wierbicki, Kunkle, Cahill, and Deatherage (1954) and Wierbicki, Cahill, Kunkle, Klosterman, and Deatherage (1955) reported that alkali-insoluble protein content of bovine muscles did not change between 3 and 15 days of post mortem storage. This finding has been confirmed recently in poultry muscle by Khan and van den Berg (1964) and Sayre (1968), who found that the amount of alkali-insoluble protein in chicken muscle remained constant during post mortem storage at 0° for 2, 4, 24, or 48 hours, or even for periods as long as 5 months.

These studies show that post mortem storage does not cause any large change in amount of connective tissue present in muscle, but they do not eliminate the possibility that post mortem storage may cause subtle changes in structure of the fibrous connective tissue proteins, thereby rendering them more susceptible to solubilization by heating. Unfortunately, only a few studies have been reported which have attempted to detect subtle changes in connective tissue proteins caused by post mortem aging, and consequently, the exact effects of post mortem aging on connective tissue are still not clear. In one of the earliest attempts to detect changes in connective tissue proteins as a result of post mortem storage, Sharp (1963) found that hydroxyproline extractability did not change during aseptic post mortem storage of bovine muscle for 172 days at 37°, and provided strong evidence against any conversion of insoluble collagen into soluble collagen during post mortem aging. Herring, Cassens, and Briskey (1967), however, have reported recently that the percentage of total muscle hydroxyproline that was solubilized by heating bovine semitendinosus muscles at 77° for 10 min increased significantly after post mortem storage at 4° for 10 days. Both these results suggest that during post mortem storage muscle collagen is altered in a subtle way that does not result in its solubilization at temperatures of 37° or less, but does cause it to be more readily solubilized by temperatures of 50–80°. This conclusion is supported by the finding of McClain, Mullins,

Hansard, Fox, and Boulware (1965) that total soluble collagen did not change between 30 min and 7 days of post mortem storage at 4° but that there was a gradual conversion of neutral soluble collagen into acid-soluble collagen during this period. Recently, McClain, Pearson, Brunner, and Crevasse (1969) have shown that connective tissue from pale, soft, exudative porcine muscle has a higher salt-soluble collagen content, and is more readily solubilized by heating in 0.9% saline at 65° for 10 min than connective tissue from normal porcine muscle. Since pale, soft, exudative muscle is usually caused by a rapid post mortem decline in *p*H, so that *p*H values of 5.5 or less are reached while muscle tissue temperature is still above 35°, McClain's results suggest that post mortem storage can alter the number and strength of cross linkages between and within collagen molecules. Clayson, Beesley, and Blood (1966) have suggested that post mortem changes in the cross linkages between collagen molecules are probably not due to proteolytic enzymes but rather may be caused by traces of oxidized nitrogen in a highly active form, such as nitroxyl radicals.

Although the studies just described indicate that post mortem aging may cause some weakening or rupture of cross linkages between collagen molecules, and that this weakening or rupture increases the solubility of the collagen molecule at temperatures of 50–80°, de Fremery and Streeter (1969) have reported direct evidence that 24 hours of post mortem storage at 2° does not cause collagen from poultry muscle to become more susceptible to solubilization by heating. They showed that 24 hours of post mortem storage at 2° has no effect on the amount of alkali-insoluble protein in muscle that had been heated under conditions that resulted in solubilization of 80% of total muscle hydroxyproline. They also found, in confirmation of earlier results, that the amount of alkali-insoluble protein in chicken muscle does not change during post mortem aging at 2° for periods up to 8 days.

While it seems evident that post mortem aging has no large effects on connective tissue solubility at temperatures of 37° or lower, it is not yet clear whether post mortem storage may alter connective tissue so as to make it more labile at temperatures of 50–80°. De Fremery and Streeter's (1969) finding that connective tissue in chicken muscle is not made more heat-labile by post mortem aging may have been due to the fact that their heat treatment was so severe that it concealed any differences which originally existed. Indeed, it is possible that there were differences in de Fremery and Streeter's experiments between the rates at which the hydroxyproline was solubilized from the at-death and the 24-hour samples. However, definite proof of this conjecture must await further experimental results, and at the present time it can only be indicated that any effects of post mortem aging on connective tissue are very subtle and probably limited to changes in the number or strength of cross bridges between connective tissue proteins.

## POST MORTEM CHANGES IN MYOFIBRILLAR PROTEINS

The myofibrillar proteins make up 50–55% of total muscle protein and may be defined, in general terms, as those muscle proteins that are not extracted at ionic strengths of 0.1 to 0.2 but are extracted at ionic strengths of 0.4 to 1.5. It is now known, however, that some of the recently discovered myofibrillar proteins can be extracted at very low ionic strengths, below 0.01, and that once they have been extracted, many of the myofibrillar proteins are soluble at all ionic strengths from 0 to 1.5. Consequently, it may be most appropriate to define the myofibrillar proteins simply as those proteins that constitute the myofibril.

Although the number of known myofibrillar proteins has increased rapidly during the past few years, there are not as many different myofibrillar proteins in muscle as there are sarcoplasmic proteins. There are at present six myofibrillar proteins sufficiently well characterized to be classified as separate protein components of the myofibril: actin, myosin, tropomyosin, troponin, α-actinin, and β-actinin. A seventh myofibrillar protein, termed M substance because of its supposed location in the M line, has recently been described by Ebashi and coworkers (Masaki, Takaiti, and Ebashi, 1968). The specific arrangement of these proteins to form the interdigitating thick and thin filament structure of muscle has already been described in this symposium (see Chapters 16–19) and will not be discussed further here. The same is true for many of the chemical and biological properties of the myofibrillar proteins, and here we will only point out that actin and myosin together are both sufficient and necessary for a contractile response. The other four myofibrillar proteins apparently have the function of modifying or regulating the actin-myosin interaction in such a way that contraction can be rapidly initiated or stopped in the constant presence of ATP. Thus, the tropomyosin-troponin complex causes the actin-myosin interaction to require $Ca^{++}$; removal of $Ca^{++}$ in the presence of the tropomyosin-troponin complex results in cessation of contraction and dissociation of the actin-myosin complex. On the other hand, α-actinin has been shown to have the ability to accelerate or strengthen the contractile response of actomyosin suspensions in *in vitro* systems.

Because of their obvious importance in muscle contraction and because it was long suspected that they were primarily responsible for many of the physical changes observed in post mortem muscle, the myofibrillar proteins have been the subject of many careful studies. Until recently, however, most investigations of post mortem alterations in the myofibrillar proteins have been limited to measurements of myofibrillar protein solubility in different solvent systems as a criterion of post mortem alteration. Although the relatively small number of different myofibrillar proteins would seem to offer the possibility that post mortem changes in myofibrillar protein solubility could be interpreted in terms of alterations in specific proteins, myofibrillar protein solubility is complicated by the many strong and highly specific interactions which occur among the myofibrillar proteins, and by the fact that certain substances such as ATP or Kl have marked effects on myofibrillar protein solubility. Consequently, these early studies did not produce any clear insight into the effect of post mortem storage on the structure of myofibrillar proteins. Indeed, the early work of Bate-Smith, Bendall, and their associates at Cambridge provided some evidence for the idea that physical changes observed in post mortem muscle were due primarily to the loss of ATP during post mortem storage, and that post mortem muscle was in fact quite similar to at-death muscle that had its ATP removed. In terms of this concept, there are no post mortem changes in the myofibrillar proteins other than those caused by bacterial action. Recent ultrastructural and biochemical studies, primarily by Marsh and his associates in New Zealand; Fujimaki, Arakawa, and their coworkers in Tokyo; Yasui, Fukazawa and their coworkers in Hokkaido; and Stromer, Robson, Chaudhry, and Henderson in our laboratory, have shown clearly that myofibrillar proteins undergo at least two kinds of specific alterations during post mortem storage: (1) a loss of Z-line structure and, in some cases, complete removal of both the Z line and the M line; and (2) modification of the actin-myosin interaction, this modification resulting in a 20–80% increase in the $Mg^{++}$-modified actomyosin adenosine triphosphatase activity, and at the same time, causing a greater sensitivity to dissociation by ATP. We certainly do not exclude the possibility that myofibrillar proteins may experience other types of post mortem alterations in addition to the two that we have mentioned, but recent evidence indicates that these two do indeed occur in post mortem muscle, and it may be instructive at this point to review this evidence and to discuss some possible causes for these alterations.

### *Loss of Z-line structure in post mortem muscle*

The most direct evidence for the loss of Z-line structure in post mortem muscle comes from the ultrastructural investigations of Stromer (Stromer and Goll, 1967*a, b;* Stromer, Goll, and Roth, 1967) and Henderson (1968) ; Yasui and coworkers (Fukazawa and Yasui, 1967; Takahashi, Fukazawa, and Yasui, 1967) ; and Davey and Gilbert (1967, 1969) (for a summary of these investigations see Chapter 33). Loss of Z-line structure in post mortem muscle has also been indicated in a biochemical fashion by studies showing that the bonds holding actin filaments to the Z line are weakened and perhaps even disrupted by post mortem storage (Chaudhry, Parrish, and Goll, 1969; Davey and Gilbert, 1968*a, b;* Penny, 1968).

Before discussing these recent biochemical studies, it will be necessary to review briefly a few features of myofibrillar protein solubility. It is now known that myosin can be solubilized in 30–60 min by extraction with 0.4–0.6 M KCl solutions at *p*H 7.0, but that actin is more difficult to solubilize, and extraction times of 5–10 hours are normally required to extract appreciable amounts of actin from the myofibril. Since Haga, Yamamoto, Maruyama, and Noda (1966) have shown that actin extraction is preceded by a break between the actin filaments and the Z line, this relative inextractability of actin presumably originates from the strength of the bonds binding actin filaments to the Z line. Moreover, since actin has a strong affinity for myosin, the relative inextractability of actin may also have marked effects on the extractability of myosin. Thus, in the absence of any agent to weaken the actin-myosin interaction, actin will bind myosin and prevent its complete solubilization. In these circumstances, the rate of myofibrillar protein extraction will be controlled primarily by the rate of actin extraction.

It is also known that certain polyanions, such as ATP and pyrophosphate, have the ability to dissociate the actin-myosin complex. Therefore, if the bonds between actin and the Z line are intact, it is possible, by use of the ATP- or pyrophosphate-containing solvents, to extract myosin almost completely from the myofibril without solubilizing any actin. Myosin has a solubilizing effect on actin (Haga, Yamamoto, Maruyama, and Noda, 1966), so complete extraction of myosin without any extraction of actin is possible only when pyrophosphate or ATP is present to prevent any interaction between the myosin in solution and the actin bound to the residue. If the bonds holding actin to the Z line have been weakened or disrupted, however, some actin will be extracted even by ATP- or pyrophosphate-containing solutions, and this actin will then appear together with myosin in high-ionic-strength pyrophosphate or ATP extracts of myofibrils.

With this short review of a few factors affecting myofibrillar protein solubility, we wish to describe briefly the recent biochemical evidence that suggests that the I-Z bonds are weakened or disrupted by post mortem storage. Davey and Gilbert (1968*a*) reported that after 17 days of post mortem storage, extraction of myofibrillar protein by a 0.6 M KCl solution at *p*H 9.2 was completed within 10 min, whereas 60 min were required for completion of protein extraction from myofibrils stored for only 2 days post mortem. Even though the rate of myofibrillar protein extraction increased with increasing time post mortem, the total amount of protein that could be solubilized by 0.6 M KCl at *p*H 9.2 remained constant during post mortem storage at 75–87% of total myofibrillar protein. Since the rate of actin solubilization would be the limiting factor in rate of myofibrillar protein extraction in Davey and Gilbert's experiments, these results suggest that 17 days of post mortem storage has increased the rate at which actin is solubilized. The most probable cause of this increase is weakening of the bonds holding actin filaments to the Z line. By using a 0.6 M KCl solution containing 1 mM $Mg^{++}$ and 10 mM pyrophosphate, Davey and Gilbert (1968*a*) also found that the amount of protein extractable from bovine myofibrils increases from 54% of the total myofibrillar protein in myofibrils stored for 1 day to 75% of total myofibrillar protein in myofibrils stored for 17 days post mortem. Since myosin makes up 50–55% of total myofibrillar protein, it appears that pyrophosphate-containing solutions will extract myosin almost quantitatively from 1-day myofibrils, but that they extract considerable amounts of protein in addition to myosin from 17-day myofibrils.

In a second paper, Davey and Gilbert (1968*b*) further characterized the protein solubilized from myofibrils by pyrophosphate-containing solutions, showing that only myosin was extracted from at-death myofibrils, but that extracts of post mortem myofibrils contained actin and a second water-soluble protein in addition to myosin. The nature of the water-soluble protein was not clear: Davey and Gilbert showed that it was not tropomyosin and suggested that it might be $\alpha$- or $\beta$-actinin, or possibly some form of denatured actin. The results clearly indicate that, after several days of post mortem storage, the I-Z bonds are weakened to the extent that they can no longer resist pyrophosphate extraction, and actin begins to appear in pyrophosphate extracts of myofibrils.

Independently, and almost simultaneously with Davey and Gilbert's reports, Penny (1968), working with rabbit muscle, and Chaudhry (Chaudhry, Parrish, and Goll, 1969), working with both rabbit and bovine muscle, reported similar findings on post mortem changes in myofibrillar protein solubility. Penny used $Ca^{++}$ -modified adenosine triphosphatase activity to determine the amount of myosin extracted and found, in agreement with Davey and Gilbert, that pyrophosphate-containing solutions extract myosin almost quantitatively from at-death myofibrils, but that pyrophosphate extraction of myofibrils stored for 4 days at 15–18° or for 14 days at 4° solubilizes additional protein which is not myosin and which Penny termed actin plus tropomyosin. Moreover, both Penny (1968) and Chaudhry (Chaudhry, Parrish, and Goll, 1969) found that with increasing time of post mortem storage, 0.5 to 1.0 M KCl solutions extract increasing amounts of protein from myofibrils, again suggesting that the rate of actin solubilization is substantially increased by post mortem storage.

Since precautions were taken in these four studies to prevent bacterial growth during post mortem storage, these results on myofibrillar protein solubility, together with the ultrastructural observations on Z-line degradation and removal, offer clear and conclusive evidence that the Z line and perhaps the I-Z bonds undergo substantial disruption and degradation during post mortem storage.

*Alteration of actin-myosin interaction in post mortem muscle*

The second kind of biochemical alteration that occurs in post mortem myofibrils is a modification of the actin-myosin interaction. The exact nature of this modification remains unclear, and it can presently be detected only because it results in a 20–80% increase in actomyosin adenosine triphosphatase activity, and at the same time, increases the sensitivity of the actin-myosin complex to dissociation by ATP.

The increased adenosine triphosphatase activity of actomyosin from post mortem muscle was first noted by Fujimaki, Arakawa, and coworkers (Fujimaki, Arakawa, Okitani, and Takagi, 1965*b;* Fujimaki, Okitani, and Arakawa, 1965), who discovered that both the $Mg^{++}$- and $Ca^{++}$-modified adenosine triphosphatase activities of myosin B extracted from rabbit muscle after 2 days of post mortem storage at 4° were 15–25% higher than the same activities of myosin B extracted from at-death muscle. Robson (Goll and Robson, 1967; Robson, Goll, and Main, 1967) has used both myofibrils and myosin B from bovine muscle in an extensive investigation on the effects of post mortem storage on the nucleosidetriphosphatase activity of actomyosin (see Table 36.1). The $Ca^{++}$- and $Mg^{++}$-modified adenosine triphosphatase activities and the $Mg^{++}$-modified inosine triphosphatase activity of myofibrils prepared from bovine muscle after 24 hours of post mortem storage at either 2° or 16° are 20–60% higher than the corresponding activities of myofibrils prepared from at-death muscle. Moreover, $Mg^{++}$-modified adenosine triphosphatase activity decreases to the at-death level after 312 hours of post mortem storage, whereas the $Ca^{++}$-modified adenosine triphosphatase and the $Mg^{++}$-modified inosine triphosphatase activities remain high, even after 312 hours of post mortem storage at 2° or 16°. This post mortem increase in actomyosin nucleosidetriphosphatase activity is quite remarkable, particuarly since it has been found (Fujimaki, Arakawa, Okitani, and Takagi, 1965*b;* Fujimaki, Okitani, and Arakawa, 1965) that myosin B extracted from post mortem rabbit muscle contains a higher proportion of actin,

TABLE 36.1

*The effect of post mortem storage on nucleosidetriphosphatase activity of bovine myofibrils**

| Substrate and activator | Time and temperature of post mortem storage | | | | |
|---|---|---|---|---|---|
| | 0 hr | 24 hr, 2° C | 24 hr, 16° C | 312 hr, 2° C | 312 hr, 16° C |
| TP, 5 mM $Ca^{++}$ | 0.106 ± .008† | 0.133 ± .005 | 0.132 ± .008 | 0.187 ± .014 | 0.168 ± .009 |
| TP, 5 mM $Mg^{++}$ | 0.051 ± .002 | 0.081 ± .007 | 0.074 ± .003 | 0.047 ± .004 | 0.054 ± .007 |
| TP, 5 mM $Ca^{++}$ | 0.022 ± .002 | 0.015 ± .002 | 0.018 ± .002 | 0.026 ± .002 | 0.028 ± .002 |
| TP, 5 mM $Mg^{++}$ | 0.046 ± .004 | 0.066 ± .007 | 0.062 ± .008 | 0.080 ± .006 | 0.075 ± .007 |

* Conditions of assay: 120 mM KCl, 24 mM Tris, *p*H 7.6, 5 mM ATP or ITP, 0.75–0.90 mg myofibrils/ml, total ionic strength assay medium = 0.18.

† Figures are μmoles $P_i$/mg protein per min expressed as means ±SE of 16 determinations.

which is not enzymically active, and a lower proportion of myosin, which is enzymically active, than does myosin B extracted from at-death muscle. Indeed, in view of the results of Davey and Gilbert (1968*a, b*), Penny (1968), and Chaudhry, Parrish, and Goll (1969), indicating that actin becomes more extractable with increasing time of post mortem storage, it would be expected that high ionic strength extracts of post mortem muscle would contain higher proportions of actin than similar extracts of at-death muscle. Consequently, the post mortem increase in actomyosin nucleosidetriphosphatase activity cannot be explained by a higher proportion of myosin in myofibrils or myosin B prepared from post mortem muscle. Okitani, Takagi, and Fujimaki (1967) observed that storage of myosin B in 0.6 M KCl at *p*H 6.0 and 25° results in a 50–100% increase in $Mg^{++}$-modified adenosine triphosphatase activity, so it is possible that the post mortem increase in myofibrillar nucleosidetriphosphatase activity is caused by oxidation of some sulfhydryl groups on myosin or simply by exposure of myosin to *p*H values of 6.0 or less for prolonged periods of time. In spite of Okitani's observations, however, the nature of the changes causing increased nucleosidetriphosphatase activity in post mortem muscle remains unknown, although this increased enzymic activity has been observed also by Herring (1968) and by Greaser, Cassens, Briskey, and Hoekstra (1969*a*).

The increased sensitivity of myosin B prepared from post mortem muscle to dissociation by ATP was also first observed by Fujimaki, Arakawa, Okitani, and Takagi (1965*a, b*) (see Fig. 36.2). Myosin B prepared from at-death muscle and dissolved in 0.5 M KCl at *p*H 7.0 is not fully dissociated until the ATP concentration is raised to 0.6 mM, whereas under the same conditions, 0.2 mM ATP will dissociate myosin B

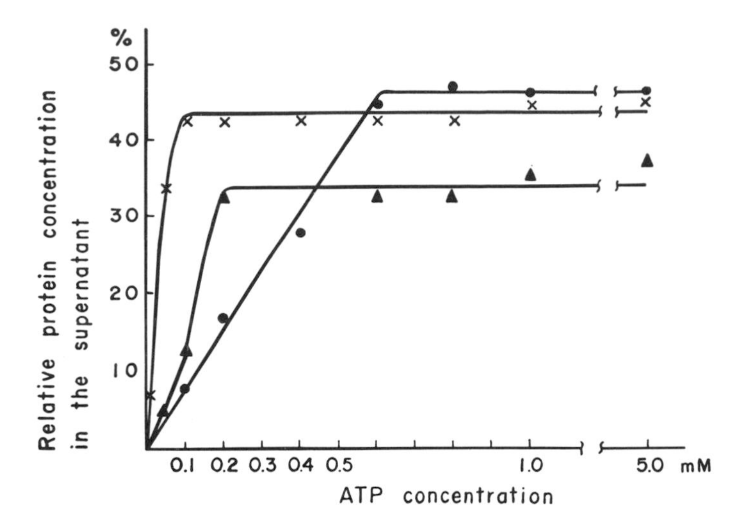


*Fig. 36.2.* The effect of post mortem storage on the dissociation of myosin B by ATP. Relative protein concentration in the supernatant (ordinate) refers to the percentage of total protein remaining in the supernatant of a myosin B solution after the addition of ATP as indicated on the abscissa and centrifugation at 100,000 × *g* for 180 minutes at 5° C. The myosin B solution contained 0.6 M KCl, 1 mM $MgCl_2$, 20 mM Tris-maleate, *p*H 7.0, plus varying levels of ATP. *Solid circle,* myosin B prepared from rabbit muscle immediately after death; *Solid triangle,* myosin B prepared from rabbit muscle after 2 days of post mortem storage at 4° C; *X,* myosin B prepared from rabbit muscles after 7 days of postmortem storage at 4° C. From Fujimaki, Arakawa, Okitani, and Takagi, 1965*a,* by permission of the authors and of the Agricultural Chemical Society of Japan.

prepared from muscle after 2 days of post mortem storage at 4°, and only 0.1 mM ATP is required to dissociate myosin B prepared from muscle after 7 days of post mortem storage at 4°. This increased sensitivity to dissociation by ATP may be interpreted most easily and directly as indicating a weakening of the actin-myosin interaction during post mortem storage. Such weakening has also been suggested by Penny (1968), who observed that 1 M KCl will extract larger amounts of protein from myofibrils that have been stored for 7 days at 4° than from freshly prepared myofibrils, and that this increased extractability is due almost entirely to increased solubilization of myosin. Since 1 M KCl solutions do not cause dissociation of the actin-myosin complex, this increase in myosin extractability is most easily interpreted as indicating a weakening of the actin-myosin interaction in stored myofibrils, so that actin can no longer bind myosin strongly enough to prevent its solubilization. Okitani, Takagi, and Fujimaki (1967) have discovered that myosin B prepared from at-death muscle and stored at *p*H values of 5.7 or lower requires less ATP for dissociation than the same myosin B before storage. so it is possible that the languid actin-myosin interaction in post mortem muscle is also caused simply by prolonged exposure of the actin-myosin complex to *p*H values of 6.0 or less.

Although biochemical studies have thus far offered the most direct evidence that post mortem muscle conditions cause weakening of the actin-myosin interaction, some recent ultrastructural observations also suggest the same thing. Ultrastructurally, this weakening is seen as a tendency of rigor-shortened sarcomeres to actually lengthen between 12 and 300 hours post mortem, even in the absence of ATP. This lengthening or "relaxation" was first observed independently and almost simultaneously by Gothard, Mullins, Boulware, and Hansard (1966) and by Stromer, Goll, and Roth (1967), working with bovine muscle, and Takahashi, Fukazawa, and Yasui (1967), working with chicken muscle. The electron micrographs of Stromer, Goll, and Roth (1967) clearly show that both rigor shortening and postrigor lengthening occur by a sliding of interdigitating filaments past one another, a circumstance that could only obtain if some "slippage" was occurring at the points of interaction between the myosin cross bridges and the actin filaments. Both ultrastructural and biochemical studies, therefore, indicate plainly that the actin-myosin interaction is weakened by post mortem storage, although the exact nature of the changes in myofibrillar architecture that cause this weakening remains unknown.

*Post mortem changes in regulatory proteins*

Tropomyosin, troponin, and α-actinin all have the ability to modify the actin-myosin interaction in *in vitro* suspensions. The tropomyosin-tropinin complex can apparently prevent the actin-myosin interaction in the presence of MgATP and the absence of $Ca^{++}$, whereas α-actinin appears to strengthen or promote the actin-myosin interaction in the presence of $Mg^{++}$ and ATP, regardless of $Ca^{++}$ concentration. Consequently, it occurred to us that post mortem alterations in α-actinin or the tropomyosin-tropinin complex may account for weakening of the actin-myosin interaction in post mortem muscle. We wish to report some results of our recent studies on post mortem changes in α-actinin and tropomyosin-troponin-containing preparations. Our studies were initiated by the observation that superprecipitation of myosin B extracted from post mortem muscle occurs more rapidly than superprecipitation of myosin B extracted from muscle immediately after death (Fig. 36.3).

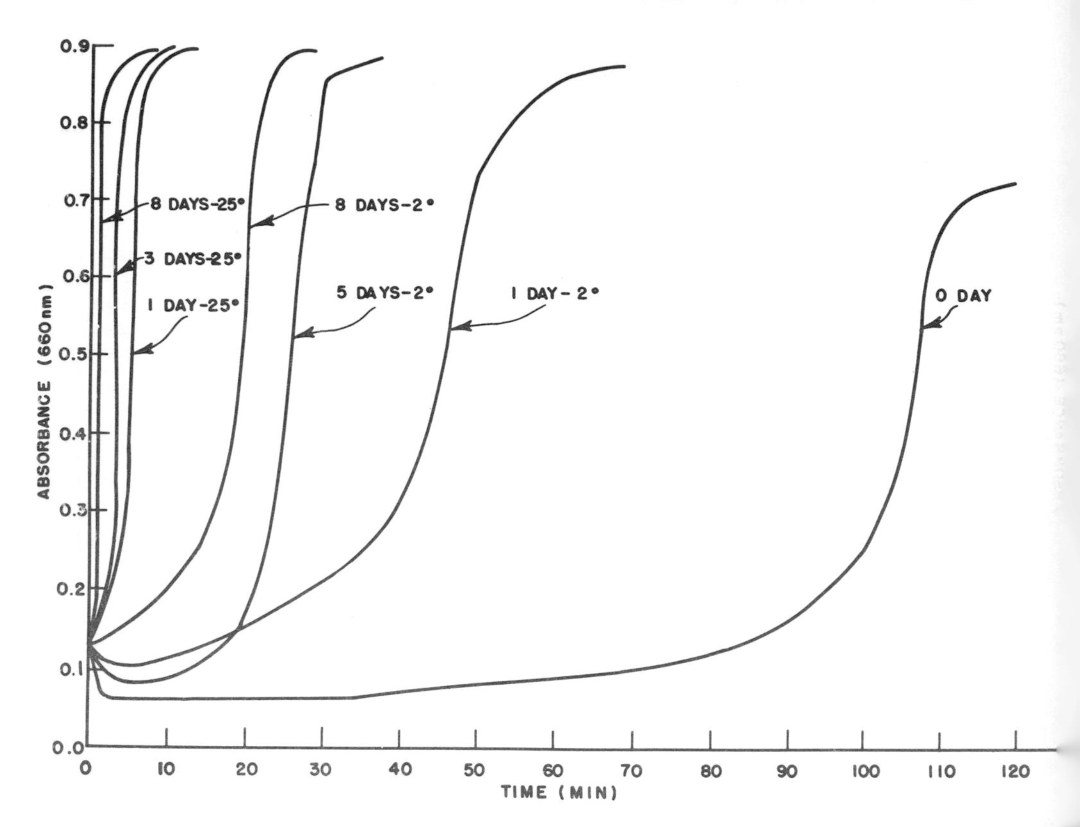


*Fig. 36.3.* Effect of post mortem storage on the superprecipitation of rabbit myosin B in the absence of added $Ca^{++}$. Conditions of superprecipitation assay: 1 mM $MgCl_2$, 1 mM ATP, 100 mM KCl, 10 mM Tris-acetate, *p*H 7.0, 0.4 mg myosin B/ml, 27° C. Time and temperature of post mortem storage are indicated.

Superprecipitation studies in which $Mg^{++}$ is the only added modifier are actually done, however, in the presence of some unknown level of $Ca^{++}$ due to the existence of contaminating $Ca^{++}$ in reagents and water. Consequently, we have examined the effects of post mortem storage on the rate of myosin B superprecipitation in the presence of EGTA, a $Ca^{++}$-chelator. The results of these tests, done at very low $Ca^{++}$ concentrations (Fig. 36.4), clearly show that $Ca^{++}$ sensitivity of myosin B decreases with increasing time of post mortem storage, and that this decrease in $Ca^{++}$ sensitivity occurs more rapidly during storage at 25° than at 2°. In fact, after 3 days of post mortem storage at 25°, the rate of myosin B superprecipitation at very low $Ca^{++}$ concentrations is very similar to the rate of superprecipitation of at-death myosin B in the presence of 0.05 mM $Ca^{++}$.

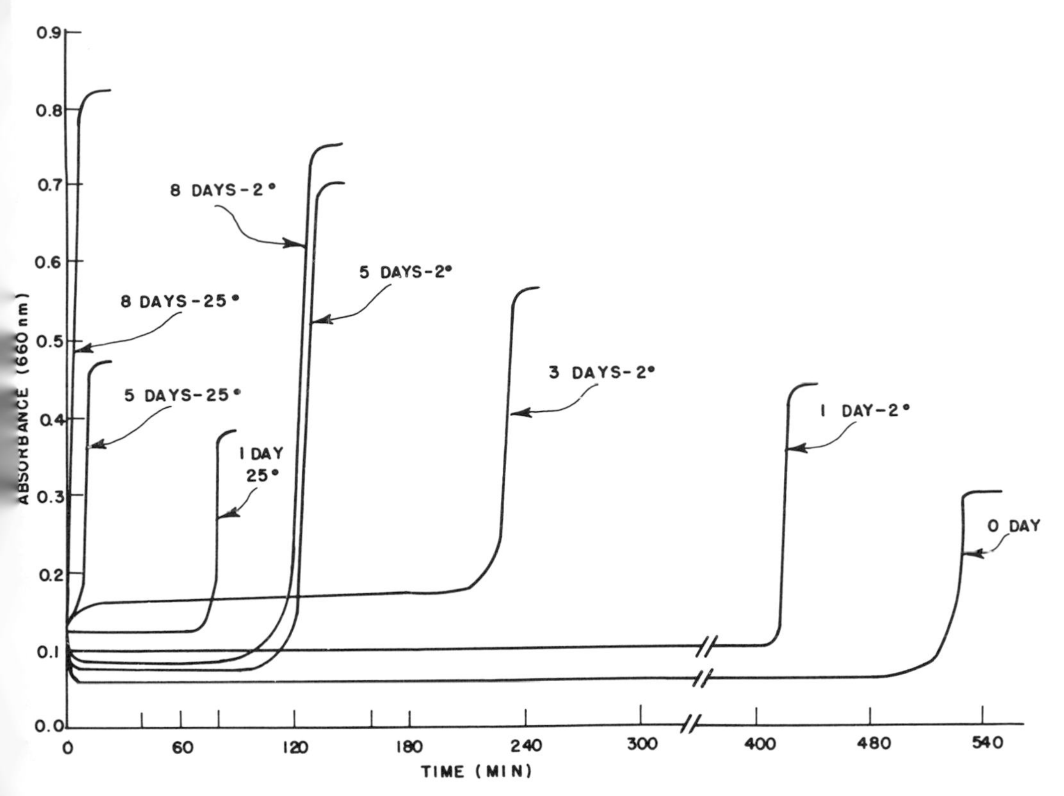


*Fig. 36.4*. Effect of post mortem storage on the superprecipitation of rabbit myosin B at low $Ca^{++}$ concentrations. Conditions of superprecipitation assay: 1 mM $MgCl_2$, 1 mM ATP, 0.1 mM EGTA, 100 mM KCl, 10 mM Tris-acetate, *p*H 7.0, 0.4 mg myosin B/ml, 27° C. Time and temperature of post mortem storage indicated.

These results suggest that myosin B extracted from muscle after 3 days of post mortem storage at 25° either does not contain the tropomyosin-troponin complex, or it contains this complex in a form that is unable to confer $Ca^{++}$ sensitivity on the actin-myosin interaction.

Although superprecipitation of myosin B in the presence of EGTA suggests that the tropomyosin-troponin complex is altered during post mortem storage, subsequent experiments in the presence of 0.05 mM $Ca^{++}$ showed that the post mortem increase in rate of myosin B superprecipitation is not due entirely to post mortem alterations in the troponin-tropomyosin complex. Thus, it was observed that even in the presence of added $Ca^{++}$, the rate of myosin B superprecpitation increases with increasing time of post mortem storage, and that this increase occurs more rapidly during storage at 25° than at 2° (Fig. 36.5). Even after post

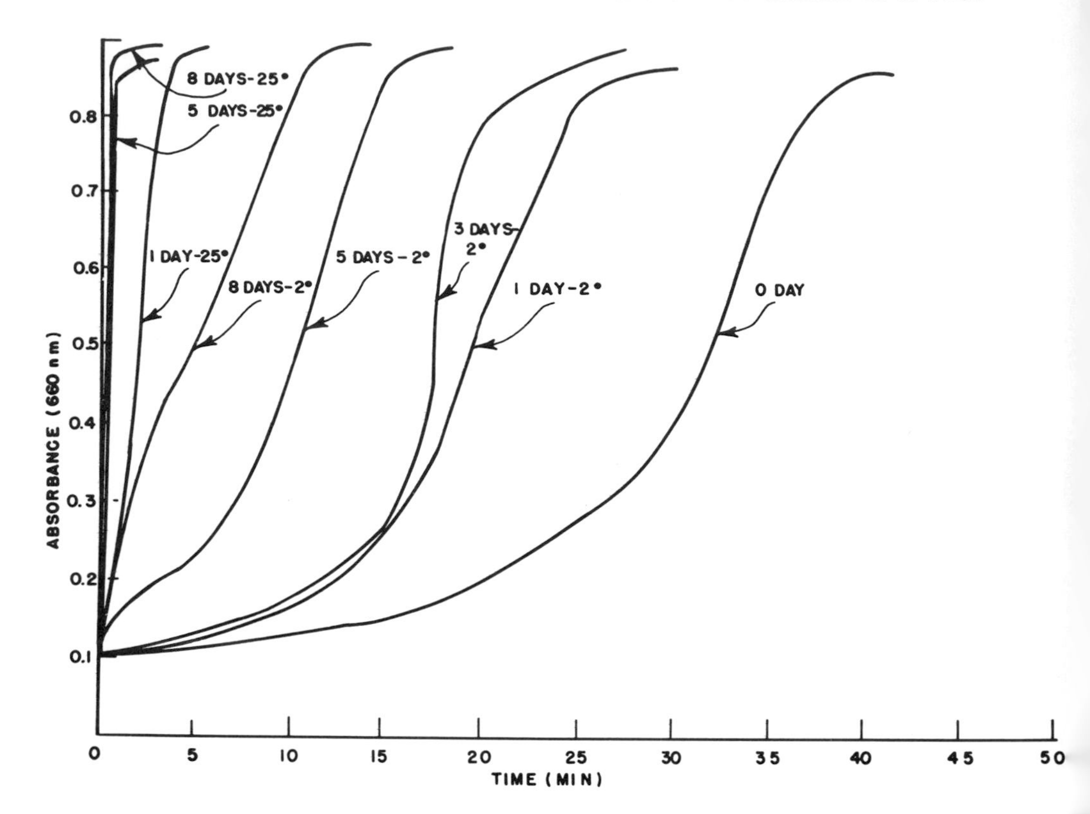


*Fig. 36.5.* Effect of post mortem storage on the superprecipitation of rabbit myosin B in the presence of added $Ca^{++}$. Conditions of superprecipitation assay: 1 mM $MgCl_2$, 1 mM ATP, 0.05 mM $CaCl_2$, 100 mM KCl, 10 mM Tris-acetate, *p*H 7.0, 0.4 mg myosin B/ml, 27° C. Time and temperature of post mortem storage indicated.

mortem storage for 8 days at 25°, however, myosin B superprecipitation always occurs more rapidly in the presence of 0.05 mM $Ca^{++}$ than in the presence of 0.1 mM EGTA, i.e., at very low $Ca^{++}$ concentration. Since α-actinin has the ability to increase the rate of actomyosin superprecipitation, these results might suggest either that myosin B extracted from post mortem muscle contains more α-actinin than myosin B extracted from at-death muscle, or that the ability of α-actinin to accelerate actomyosin superprecipitation is increased by post mortem storage. The first suggestion is supported by the fact that α-actinin is apparently located in or near the Z line (Goll, Mommaerts, and Seraydarian, 1967; Goll, Mommaerts, Reedy, and Seraydarian, 1969; Masaki, Endo, and Ebashi, 1967) and, as we have already indicated, post mortem storage causes disruption of the Z line and a more rapid extraction of actin. Hence, it is entirely possible that myosin B prepared from post mortem muscle contains larger amounts of α-actinin than myosin B prepared from at-death muscle.

Since the myosin B superprecipiattion tests indicated that both α-actinin and the tropomyosin-troponin complex may be altered by post mortem storage, our subsequent studies were directed toward assay of the

TABLE 36.2

*Flow sheet showing procedures for extraction of α-actinin and the tropomyosin-troponin complex from rabbit myofibrils at 2° C*

| Supernatant | Sediment |
|---|---|
| | I. Myofibrils, washed three times, suspended in 150 mM KCl<br>(a) Centrifuge at 1000 × *g* for 10 min |
| Supernatant (discard) | II. Sediment<br>(a) Resuspend in 5 vol (original muscle weight) 1 mM EDTA, *p*H 7.6<br>(b) Centrifuge at 1000 × *g* for 10 min |
| Supernatant (discard) | III. Sediment<br>(a) Resuspend in 5 vol $H_2O$<br>(b) Centrifuge at 2000 × *g* for 15 min |
| Supernatant (discard) | IV. Sediment<br>(a) Resuspend in 5 vol $H_2O$<br>(b) Centrifuge at 2000 × *g* for 30 min |
| Supernatant (discard) | V. Sediment<br>(a) Resuspend in $H_2O$ to total volume 12-fold the initial muscle weight<br>(b) Centrifuge at 14,000 × *g* for 20 min |
| Supernatant (discard) | VI. Sediment<br>(a) Resuspend in $H_2O$ to total volume 12-fold the initial muscle weight<br>(b) Centrifuge at 14,000 × *g* for 30 min |
| Supernatant (discard) | VII. Sediment<br>(a) Adjust *p*H to 8.5 with solid Tris<br>(b) Add 2-mercaptoethanol to a final concentration of 7.13 mM<br>(c) Store at 2° C for 64–72 hr<br>(d) Add 1 M Tris · HCl, *p*H 8.5, to make final Tris concentration 10–11 mM<br>(e) Centrifuge at 14,000 × *g* for 60 min |
| Supernatant (Extract A) | VIII. Sediment<br>(a) Resuspend in same volume of 10 mM Tris · HCl, *p*H 8.5, as obtained in Extract A<br>(b) Centrifuge at 14,000 × *g* for 60 min |
| Supernatant (Extract B) | Sediment (discard) |

effects of α-actinin or tropomyosin-troponin preparations made from post mortem muscle on the adenosine triphosphatase activity and superprecipitation rate of reconstituted actomyosin suspensions. In these studies, intact rabbit muscle was stored for varying periods of time post mortem at 2° or 25° and then α-actinin or the tropomyosin-troponin complex was extracted by using the procedure developed by Robson (1969) in his studies on purification of α-actinin (Table 36.2). Briefly, this procedure consists of preparation of myofibrils, followed by washing the myofibrils with water to lower the ionic strength, and then extraction of the swollen myofibrils by 1–2 mM Tris at pH 8.5 and 2° for 64–72 hours. These crude extracts were then fractionated between 0% and 30% and between 30% and 75% ammonium sulfate saturation to produce a crude α-actinin extract (0–30%) and a crude tropomyosin-troponin extract (30–75%). Hereafter in this presentation, we will refer to these fractions as the $P_{0-30}$ and the $P_{30-75}$ fractions. The evidence we have already discussed in this presentation indicates that post mortem storage results in some disruption of myofibrillar structure and that this disruption might cause increased extractability of α-actinin and the tropomyosin-troponin complex; therefore the water-washes of the myofibrils were saved and also fractionated with ammonium sulfate to produce a $P_{0-30}$ and a $P_{30-75}$ wash fraction. The effect of post mortem storage on the yields of these four preparations is shown in Table 36.3. There is no obvious trend toward either more or less extraction of α-actinin and the tropomyosin-troponin complex with increasing time of post mortem storage to 8 days post mortem at 25°. Neither does post mortem storage cause any evident increase in the amount of α-actinin and tropomyosin-troponin extracted in the water washes.

The effect of post mortem aging on the ability of the $P_{0-30}$ fraction to accelerate the rate of superprecipitation of reconstituted actomyosin sus-

TABLE 36.3

*Effect of post mortem storage on amount of protein extracted from rabbit myofibrils by water washing or a 3-day, pH 8.5 extraction*

| Time of post mortem storage (hrs) | mg protein fraction/g of muscle | | | |
|---|---|---|---|---|
| | $P_{0-30}$ (3-day) | $P_{30-75}$ (3-day) | $P_{0-30}$ (wash) | $P_{30-75}$ (wash) |
| 0 | 1.14 | 1.65 | 0.81 | 3.24 |
| 72 | 0.88 | 2.94 | 0.79 | 3.69 |
| 96 | 0.79 | 1.95 | 0.78 | 3.53 |
| 120 | 1.23 | 2.09 | 0.74 | 3.56 |
| 168 | 0.62 | 1.31 | 0.52 | 2.02 |
| 192 | 1.76 | 2.54 | 1.18 | 2.96 |

pensions is shown in Fig. 36.6. The activity of the $P_{0-30}$ fraction decreased with increasing time of post mortem storage, but even after 8 days at 25° it still clearly possessed the ability to accelerate the turbidity response of actomyosin suspensions. Hence, the increased rate of superprecipitation for myosin B prepared from post mortem muscle is evidently not due to any post mortem increase in ability of α-actinin to accelerate the turbidity response of reconstituted actomyosin.

The activity of the $P_{40-75}$ fraction also decreased slightly during post mortem storage (Fig. 36.7), but even after 7 days at 25°, it still clearly possessed the ability to make superprecipitation of reconstituted actomyosin sensitive to $Ca^{++}$. The superprecipitation tests shown in Figs. 36.6–7 were confirmed by adenosine triphosphatase assays (not shown here), which also indicated that post mortem storage causes some decrease, but not complete loss, of activity of the $P_{0-30}$ and $P_{30-75}$ fractions.

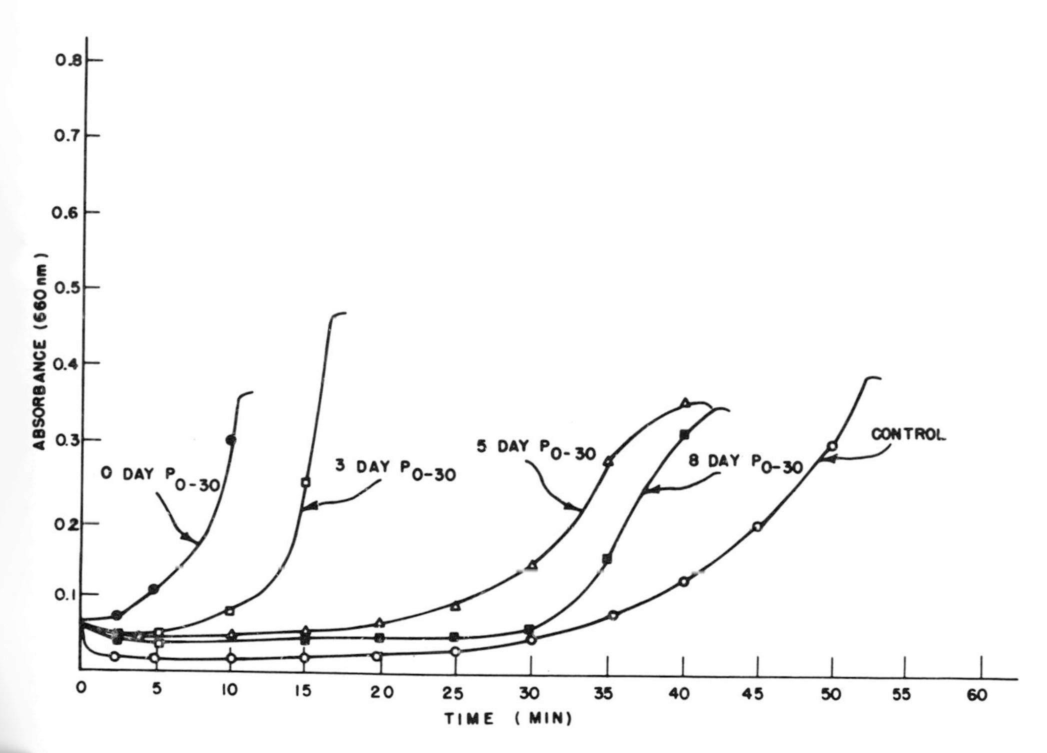


Fig. 36.6. Effect of the $P_{0-30}$ fraction prepared from post mortem rabbit muscle on superprecipitation of reconstituted actomyosin. Conditions of superprecipitation assay: 1 mM $MgCl_2$, 1 mM ATP, 0.05 mM $CaCl_2$, 100 mM KCl, 20 mM Tris-acetate, *p*H 7.0, 0.4 mg actomyosin/ml, 0.08 mg $P_{0-30}$/ml when added, 25° C. Muscle was stored at 25° C for indicated time post mortem.

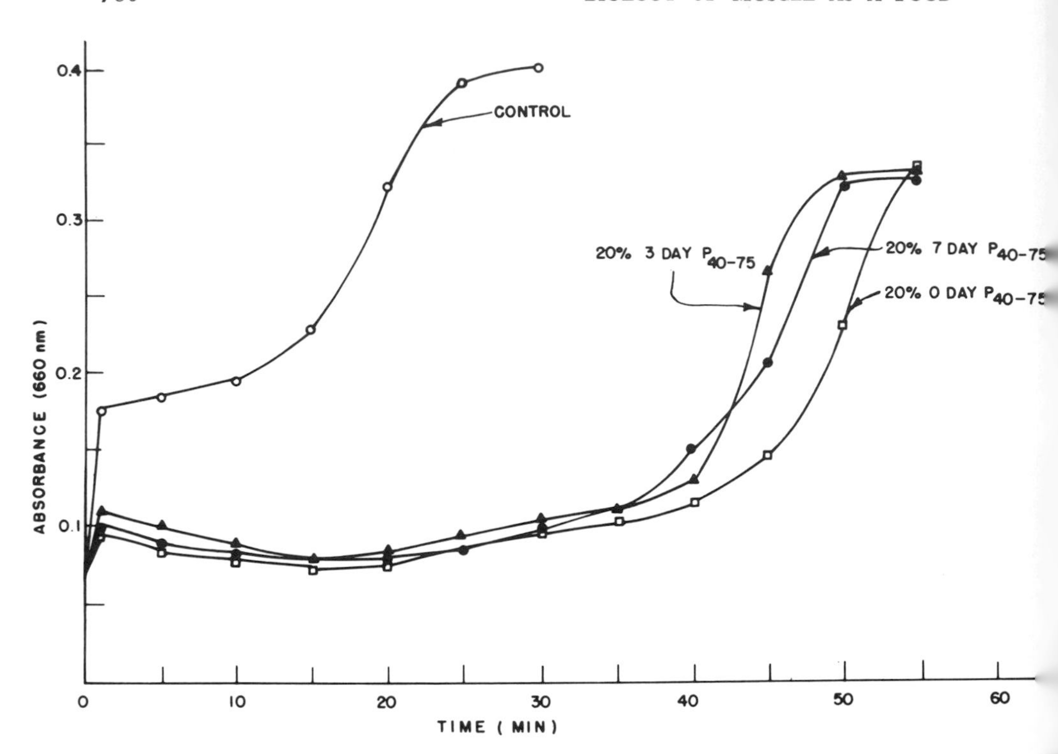


*Fig. 36.7.* Effect of a $P_{40-75}$ fraction prepared from post mortem rabbit muscle on superprecipitation of reconstituted actomyosin. Conditions of superprecipitation assay: 1 mM $MgCl_2$, 1 mM ATP, 0.05 mM EGTA, 67 mM KCl, 20 mM Tris-acetate, *p*H 7.0, 0.4 mg actomyosin/ml, $P_{40-75}$ indicated as percentage of actomyosin present, 27° C. Muscle stored at 25° C for indicated times post mortem.

Analytical ultracentrifugal diagrams of the $P_{0-30}$ and $P_{30-75}$ fractions after various times of post mortem storage at 25° (Figs. 36.8 and 36.9) confirm the conclusion that post mortem storage does not cause extensive degradation of α-actinin or the tropomyosin-troponin complex. Even after 8 days at 25°, the $P_{0-30}$ fraction consisted primarily of a protein species sedimenting with an observed sedimentation coefficient of 6.1 S and was similar ultracentrifugally to the $P_{0-30}$ fraction prepared from at-death muscle. Similarly, the sedimentation pattern of the $P_{30-75}$ fraction after 8 days of post mortem storage at 25° was nearly identical to the sedimentation pattern of the $P_{30-75}$ fraction prepared from at-death muscle (Fig. 36.9). Both fractions consisted primarily of a protein sedimenting with an observed sedimentation coefficient of 4.5 S, characteristic of the tropomyosin-troponin complex. Occasionally, we observed that the $P_{30-75}$ wash preparations contained, in addition to the 4.5 S tropomyosin-troponin complex, large amounts of a more slowly sedimenting species resembling uncomplexed tropomyosin or troponin (cf. Fig. 36.9, 8-day wash). Evidently, such preparations contained an excess of tropomyosin or troponin (probably the former) together with the tropomyosin-troponin complex. This situation was observed most frequently in the wash solutions and did not appear to be related in any way to length of post mortem storage.

The results of the studies just described clearly indicate that the post mortem changes in rate of myosin B superprecipitation are probably not due to changes in α-actinin or the tropomyosin-troponin complex alone. We therefore examined the possibility that the increased rate of superprecipitation for myosin B from post mortem muscle might originate from loss of the ability of myosin B to interact with α-actinin or the tropomyosin-troponin complex. These experiments were done by testing the effects of the $P_{0-30}$ and $P_{30-75}$ fractions on myosin B prepared from muscle after varying times of post mortem storage. The results show that the $P_{15-25}$ fraction (an α-actinin fraction) had very little effect on myosin B prepared from either at-death or post mortem muscle (Fig. 36.10). This is opposite to the effect that would be expected if an increased interaction with α-actinin were responsible for the faster rate of superprecipitation observed in myosin B from post mortem muscle. Moreover, as shown in Table 36.4, the $P_{40-75}$ fraction prepared from either at-death or from 3-day, 25° muscle is able to cause a substantial increase in $Ca^{++}$ sensitivity of myosin B prepared from muscle after 8 days' post mortem storage at 16°. The increased adenosine triphosphatase activity of actomyosin from post mortem muscle is also illustrated in Table 36.4, since the $Mg^{++}$- plus $Ca^{++}$-modified adenosine triphosphatase activity increases from a specific activity of 0.234 μmoles $P_i$/mg protein per min for at-death myosin B to 0.300 μmoles $P_i$/mg protein per min for 8-day myosin. It is probable that this increased adenosine triphosphatase activity is related to the increased rate of superprecipitation observed from myosin B prepared from post mortem muscle, although the cause for either of these two phenomena remains unknown.

Our studies on post mortem changes in α-actinin and the tropomyosin-troponin complex suggest, therefore, that neither of these protein fractions is extensively degraded during post mortem storage but that both retain at least part of their activity, even after 8 days post mortem at 25°. Consequently, it appears that the increased rate of superprecipitation observed in myosin B prepared from post mortem muscle is not due

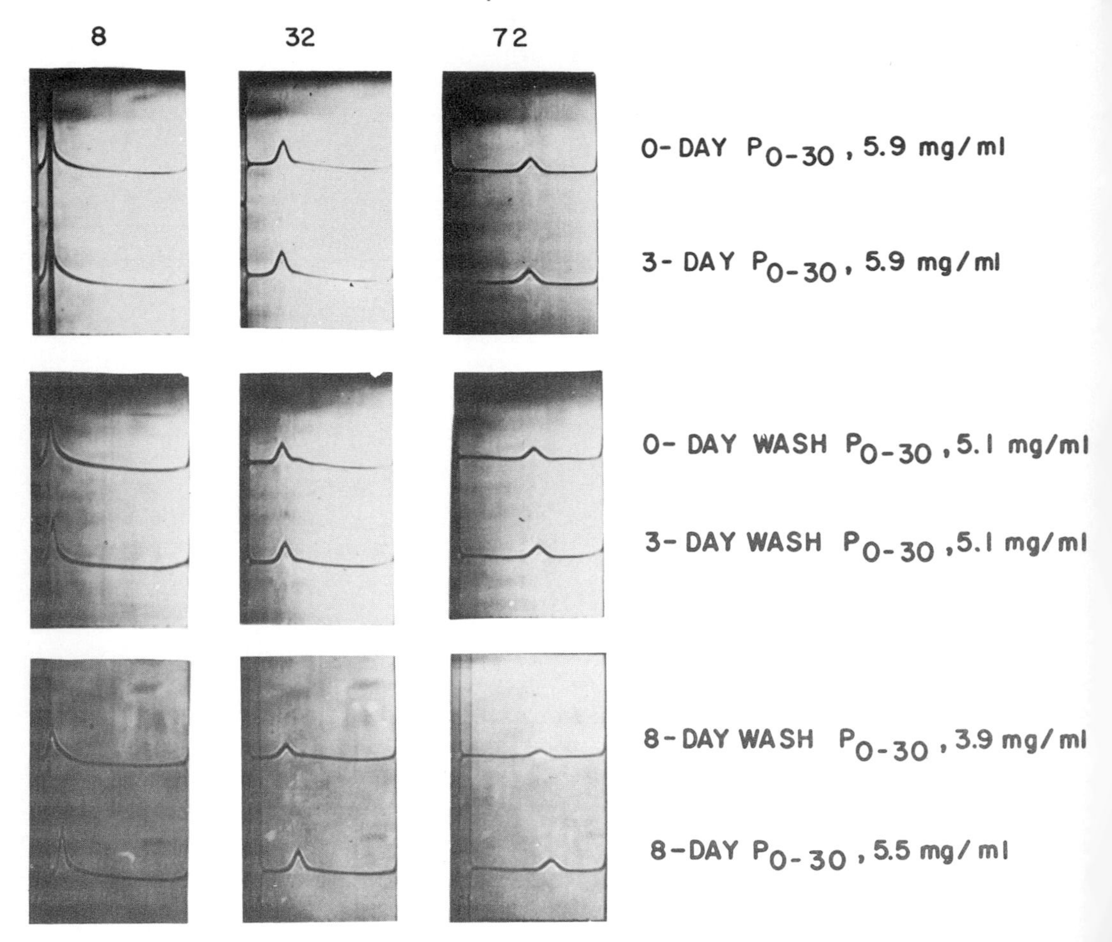


*Fig. 36.8.* Analytical ultracentrifuge diagrams of the $P_{0\text{-}30}$ fraction prepared from post mortem rabbit muscle. All samples were dissolved in 100 mM KCl, 20 mM Tris-acetate, *p*H 7.5, at the concentrations indicated. Temperature of run = 20.0° C, phase plate angle = 65°. Muscle was stored at 25° C for the indicated times post mortem.

primarily to changes in α-actinin or the tropomyosin-troponin complex but probably originates principally from alterations in actin and myosin themselves. Since α-actinin is apparently located in or next to the Z line (Goll, Mommaerts, and Seraydarian, 1967; Goll, Mommaerts, Reedy, and Seraydarian, 1969; Masaki, Endo, and Ebashi, 1967), it is presently difficult to reconcile Henderson's (1968) observation that the Z line in rabbit muscle is extensively disrupted and usually totally removed after 24 hours of post mortem storage at 25° with our finding that α-actinin is

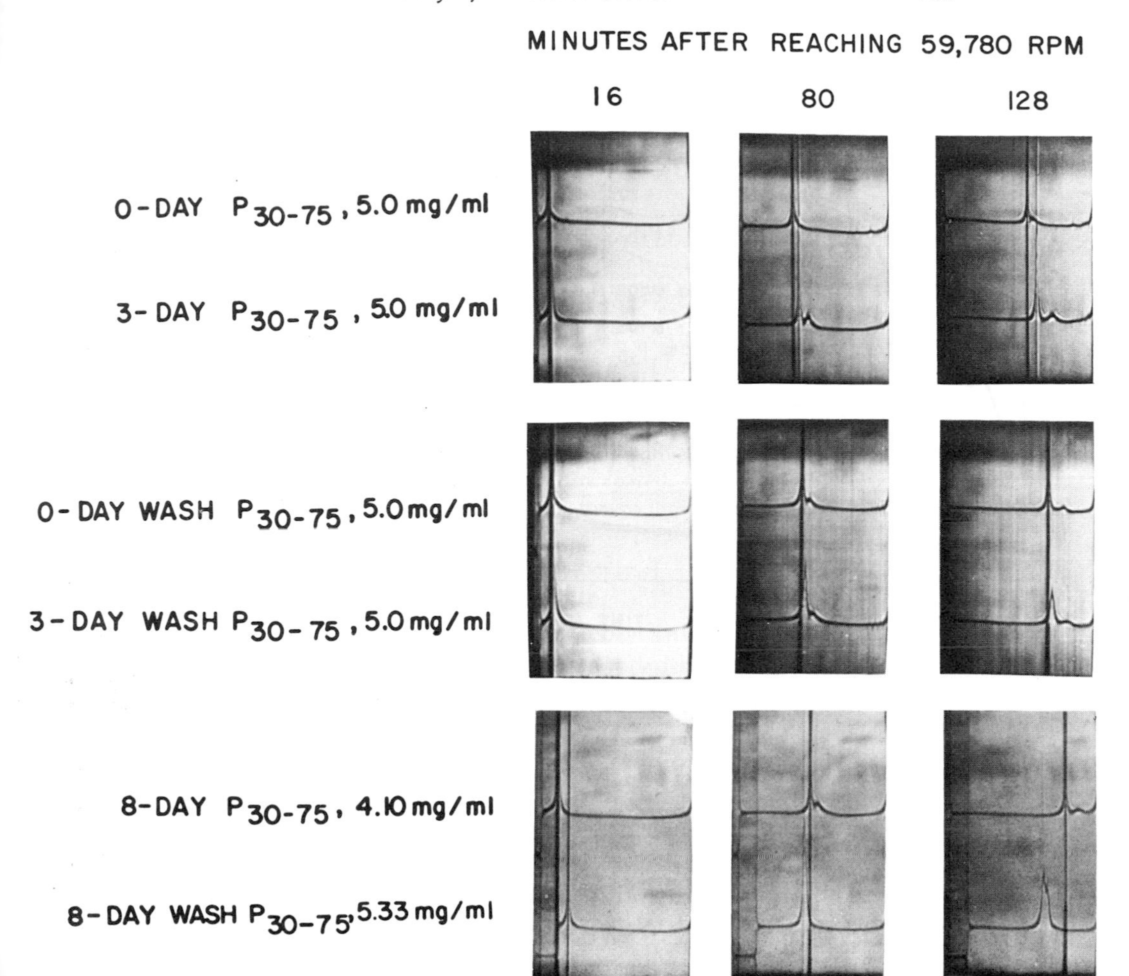


*Fig. 36.9.* Analytical ultracentrifuge diagrams of the $P_{30-75}$ fraction prepared from post mortem rabbit muscle. All samples were dissolved in 98 mM KCl, 20 mM Tris-acetate, *p*H 7.5, at the concentrations indicated. Temperature of run = 20.0° C, phase plate angle = 65°. Muscle was stored at 25° C for the indicated times post mortem.

not extensively degraded by post mortem storage. Further studies on this dichotomy may provide information on the role of α-actinin in muscle.

## THE EVIDENCE AGAINST PROTEOLYSIS IN POST MORTEM MUSCLE

Ever since Hoagland's early report (Hoagland, McBryde, and Powick, 1917) that proteolysis was an important factor contributing to post mortem changes in muscle proteins, the extent and role of proteolysis in post mortem muscle have been shrouded in controversy and confusion.

meromyosin, by either trypsin, chymotrypsin, or subtilisin. Both Bodwell and Pearson (1964) and Martins and Whitaker (1968) have shown that neither myosin nor actomyosin is affected by catheptic enzymes which have been partially purified from muscle, and, moreover, Goll and Robson (1967) and Robson, Goll, and Main (1967) have shown by using nucleosidetriphosphatase activity that there is no proteolytic cleavage of myosin *in situ* during periods of post mortem storage for as long as 13 days at either 2° or 16°. Furthermore, Okitani, Arakawa, and Fujimaki (1965) have shown that the sedimentation profile of myosin B does not change during post mortem storage. Together, these three lines of evidence strongly suggest that myosin itself is not degraded proteolytically during post mortem storage.

The second proteolytically vulnerable site in the myofibril is the tropomyosin-troponin complex, which is rapidly degraded by trypsin into small peptides (Ebashi and Kodama, 1966). The evidence discussed earlier shows first that the yields of α-actinin and the tropomyosin-troponin fraction are not changed by post mortem storage, and secondly, that it is possible to prepare functionally active α-actinin and tropomyosin-troponin fractions which exhibit normal sedimentation patterns from muscle stored for eight days at 25°. These two findings show clearly that there is no proteolytic destruction of either α-actinin, tropomyosin, or troponin during post mortem storage.

The third possible site for post mortem proteolysis of myofibrils is at or near the Z line. Stromer, Goll, and Roth (1967) have shown that trypsin very quickly removes the Z line from myofibrils, and we have already summarized the evidence showing that post mortem storage causes extensive degradation of the Z line. Moreover, both Busch (1969) and Penny (1968) have found that Z-line degradation occurs only during post mortem storage of intact muscle, which would contain all catheptic enzymes found *in situ* in either blood or muscle, and not during storage of myofibrils prepared from at-death muscle; such myofibrils would not contain catheptic enzymes. Although this evidence would suggest that proteolysis is responsible for degradation of Z lines in post mortem muscle, both Fukazawa and Yasui (1967) and de Lumen (pers. comm., 1968) have shown that incubation of at-death myofibrils having intact Z lines with partially purified cathepsins or lysosomal preparations, made from either muscle or blood, does not cause any detectable loss of Z-line structure. Moreover, Busch (1969) has recently discovered that Z-line degradation in post mortem muscle may occur by an entirely different mechanism which does not involve proteolytic enzymes; this mechanism will be discussed in the next section of this chapter. These results suggest that Z-line degradation in post mortem muscle is not due to proteolysis.

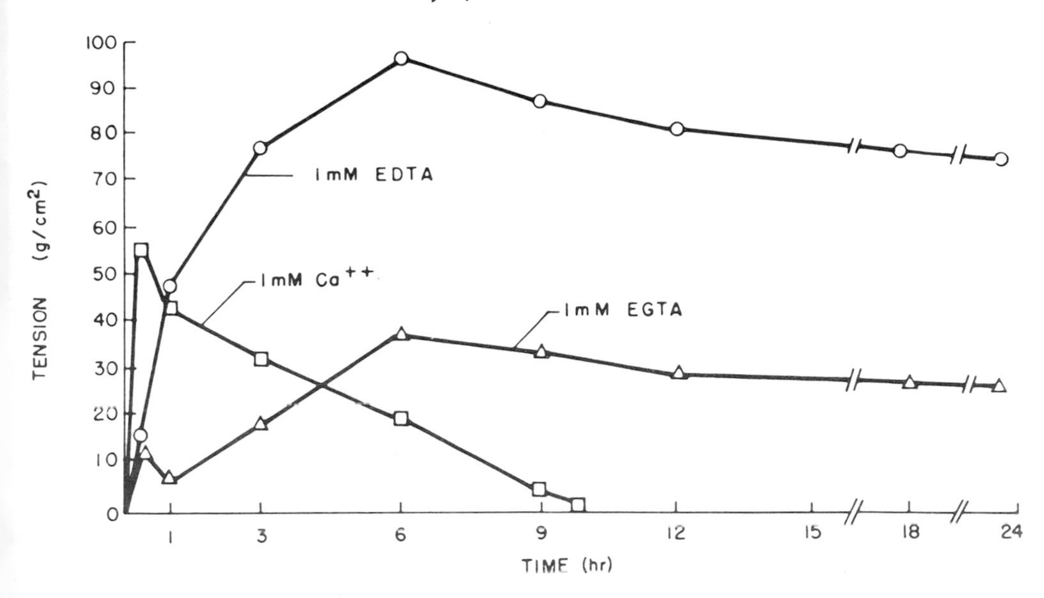


*Fig. 36.11.* Effect of $Ca^{++}$, EDTA, and EGTA on post mortem isometric tension development of rabbit psoas muscle at 25° C. Muscle strips were immersed in 80 mM KCl, 60 mM Tris-acetate, *p*H 7.1, 1 mM $NaN_3$, 1 mM deoxycholate containing either 1 mM $Ca^{++}$ plus 5 mM $MgCl_2$, 1 mM EGTA plus 5 mM $MgCl_2$, or 1 mM EDTA. Isometric tension was measured by use of an E & M Physiograph and isometric myograph transducers.

These results *in toto* clearly indicate that post mortem muscle does not undergo any extensive proteolysis and that proteolysis is probably not an important factor contributing to post mortem changes in muscle proteins. Considerably more emphasis should, therefore, be placed on study of the effects of *p*H, temperature, and ATP and $Ca^{++}$ levels on muscle proteins.

### POSSIBLE CAUSE OF POST MORTEM SHORTENING AND Z-LINE DEGRADATION IN POST MORTEM MUSCLE

Although the ultrastructural and biochemical evidence described in the preceding sections shows clearly that myofibrillar proteins undergo at least two kinds of biochemical changes during post mortem storage, the cause of these alterations remains unknown. Some recent results, however, have suggested that $Ca^{++}$ may have an important role in post mortem alteration of the myofibrillar proteins. The possible importance of $Ca^{++}$ was first suggested by the finding (Greaser, Cassens, and Hoekstra, 1967; Greaser, Cassens, Briskey, and Hoekstra, 1969*a, b*) that, within several hours after death, the sarcoplasmic reticulum in porcine muscle begins to lose its ability to sequester $Ca^{++}$. These findings have been confirmed recently in rabbit and bovine muscle by Eason (1969), in our laboratory. Moreover, Eason found that the onset of tension development or shortening in post mortem muscle strips is closely correlated to the time at which the sarcoplasmic reticulum begins to lose its $Ca^{++}$-sequestering ability. That this relationship exists in both rabbit and bovine muscle at post mortem storage temperatures of 2° or 37° suggests that release of $Ca^{++}$ from sarcoplasmic reticular membranes is the event that initiates shortening or isometric tension development in post mortem muscle.

As a result of Eason's findings, Busch (1969) initiated a careful study on the effects of added $Ca^{++}$ or EGTA (a $Ca^{++}$-chelator) on post mortem isometric tension development of muscle strips immersed in a saline solution (Fig. 36.11). As expected, addition of $Ca^{++}$ to the solution bathing the muscle strips caused almost immediate tension development, thus supporting Eason's conclusion. But muscle strips also developed isometric tension when 1 mM EGTA, or even 1 mM EDTA, was added to the bathing solution. This result must indicate that the sarcolemmal membrane is impermeable to EGTA or EDTA during the first few hours post mortem. The impermeability persists even when 1 mM deoxycholate is added to the bathing solutions. The most significant part of the

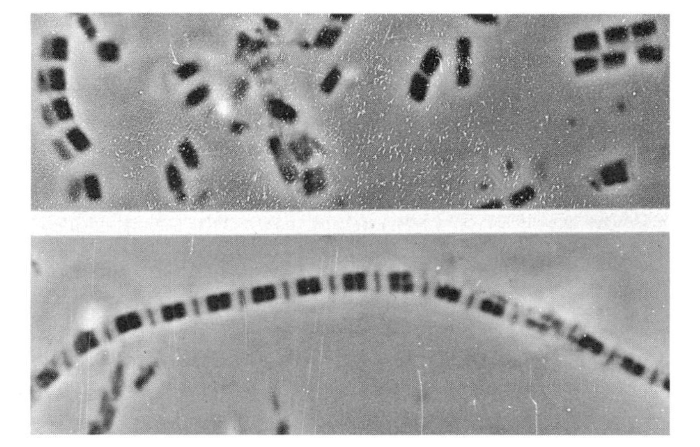

*Fig. 36.12.* Phase micrographs of myofibrils prepared from rabbit muscle strips that had been suspended isometrically for 24 hours at 37° C. *Above,* after suspension in 80 mM KCl, 60 mM Tris-acetate, *p*H 7.1, 5 mM $MgCl_2$, 1 mM $CaCl_2$, 1 mM $NaN_3$, and 1 mM deoxycholate. × 2000. *Below,* after suspension in 80 mM KCl, 60 mM Tris-acetate, *p*H 7.1, 1 mM EDTA, 1 mM $NaN_3$, and 1 mM deoxycholate. × 2000.

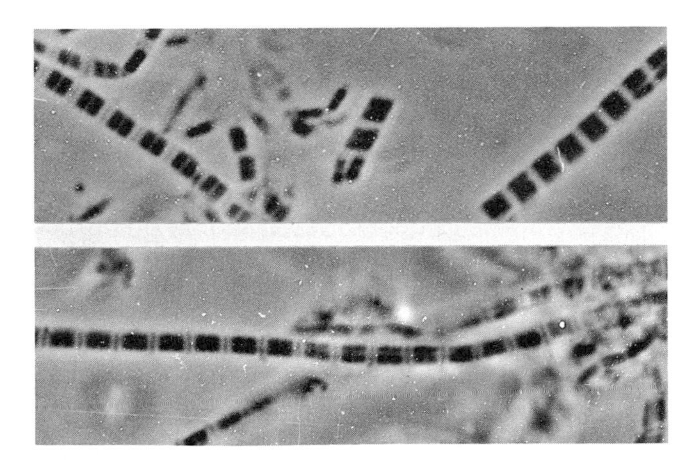

*Fig. 36.13.* Phase micrographs of myofibrils prepared from rabbit muscle strips that had been suspended isometrically for 24 hours at 37° C in the absence of deoxycholate. *Above,* after suspension in 80 mM KCl, 60 mM Tris-acetate, *p*H 7.1, 5 mM $MgCl_2$, 1 mM $CaCl_2$, and 1 mM $NaN_3$. × 2000. *Below,* after suspension in 80 mM KCl, 60 mM Tris-acetate, *p*H 7.1, 1 mM EDTA, and 1 mM $NaN_3$. × 2000.

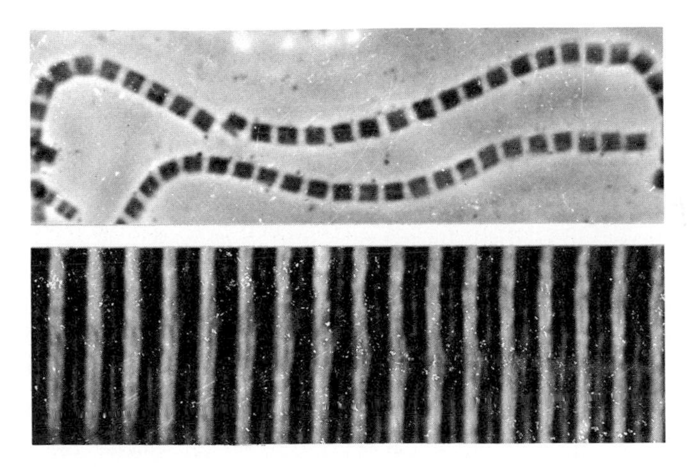

*Fig. 36.14.* Phase micrographs of fibrils prepared by teasing of rabbit muscle strips that had been suspended isometrically for 20 hours at 37° C. *Above,* after suspension in 80 mM KCl, 60 mM Tris-acetate, *p*H 7.1, 5 mM $MgCl_2$, 1 mM $CaCl_2$, 1 mM $NaN_3$, and 1 mM deoxycholate. × 1720. *Below,* after suspension in 80 mM KCl, 60 mM Tris-acetate, *p*H 7.1, 1 mM EDTA, 1 mM $NaN_3$, and 1 mM deoxycholate. × 2000.

isometric tension patterns shown in Fig. 36.11, however, is the fact that in the presence of EGTA or EDTA, isometric tension development does not decrease with increasing time of post mortem storage, but remains nearly unchanged for 24 hours. On the other hand, in the presence of added $Ca^{++}$ isometric tension development very quickly and sharply decreases back to zero within 9 hours post mortem.

Phase microscopic examination of myofibrils prepared by homogenization of muscle fibers after they had been bathed for 24 hours in solutions containing 1 mM $Ca^{++}$, 1 mM EDTA, or 1 mM EGTA showed that exposure to 1 mM $Ca^{++}$ for 24 hours at 37° causes marked structural alterations in myofibrils (Figs. 36.12–14). In those experiments where 1 mM deoxycholate was added to the bathing solution to increase sarcolemmal permeability, myofibrils prepared from strips bathed in 1 mM $Ca^{++}$ were fragmented extensively by the homogenization procedure used in their preparation and most of them were only 2–3 sarcomeres in length (Fig. 36.12). Moreover, these $Ca^{++}$-treated myofibrils did not possess any Z lines. On the other hand, myofibrils from strips bathed in solutions containing 1 mM EDTA exhibited a normal appearance in the phase microscope, being 8–10 sarcomeres or more in length and having intact Z lines (Fig. 36.12, lower). Even when deoxycholate was omitted from the bathing solution and the sarcolemmal membrane restricted the availability of $Ca^{++}$ to the myofibril, bathing in 1 mM $Ca^{++}$-containing solutions still caused substantial fragmentation and loss of Z lines in approximately one-half the sarcomeres examined (Fig. 36.13). In order to determine whether the homogenization and washing procedures used for myofibril preparation were removing degraded Z lines in the $Ca^{++}$-treated fibers or whether the $Ca^{++}$ treatment itself was causing Z-line removal, some experiments were done in which the bathed muscle strips were carefully teased into fibrils by use of a fine forceps. Examination of these "teased" fibrils in the phase microscope showed that exposure to 1 mM $Ca^{++}$ caused Z-line removal, even in intact muscle strips before homogenization (Fig. 36.14). It was also observed during these experiments that bathing in 1 mM $Ca^{++}$ decreased the strength of the lateral attachments

*Fig. 36.15 (above)*. Electron micrograph of rabbit muscle immediately after death. Note the fibrillar structure of the Z line and the wide I band. × 29,631.

*Fig. 36.16 (below)*. Electron micrograph of rabbit muscle after isometric suspension for 9 hours at 37° C in 80 mM KCl, 60 mM Tris-acetate, *p*H 7.1, 5 mM $MgCl_2$, 1 mM $CaCl_2$, 1 mM $NaN_3$, and 1 mM deoxycholate. Remnants of Z-line material remain attached to thin filaments from opposing sarcomeres, but the Z line is now split down its center, leaving a less dense area. This area usually contains some filaments, but in a few instances appears almost devoid of filaments. × 43,326.

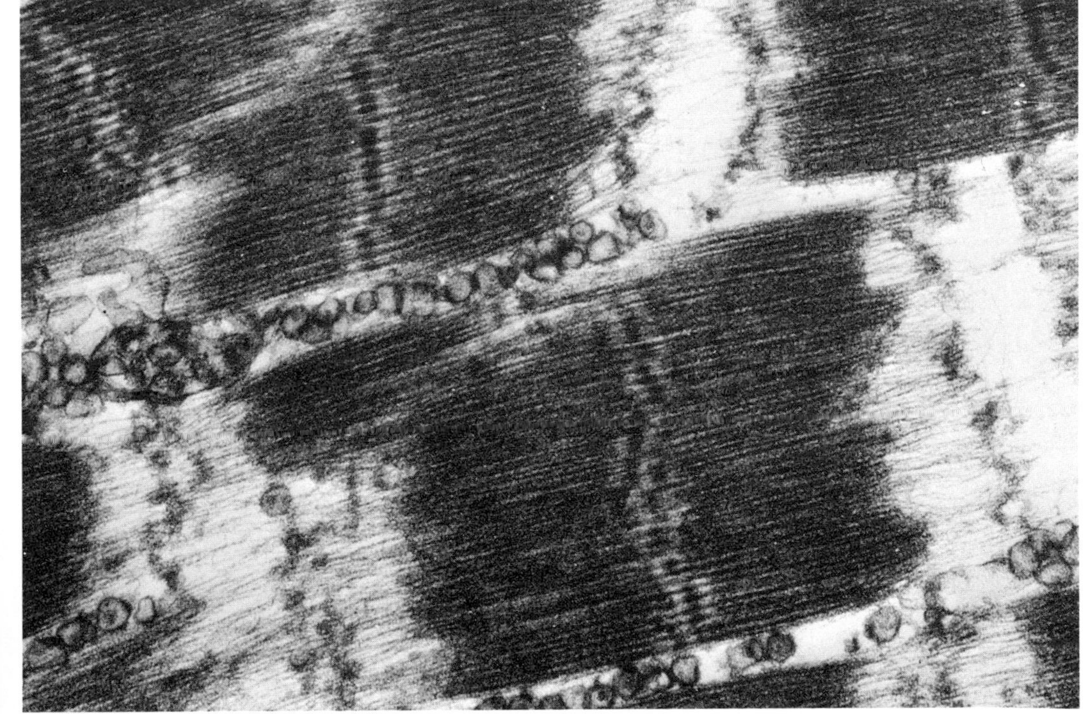

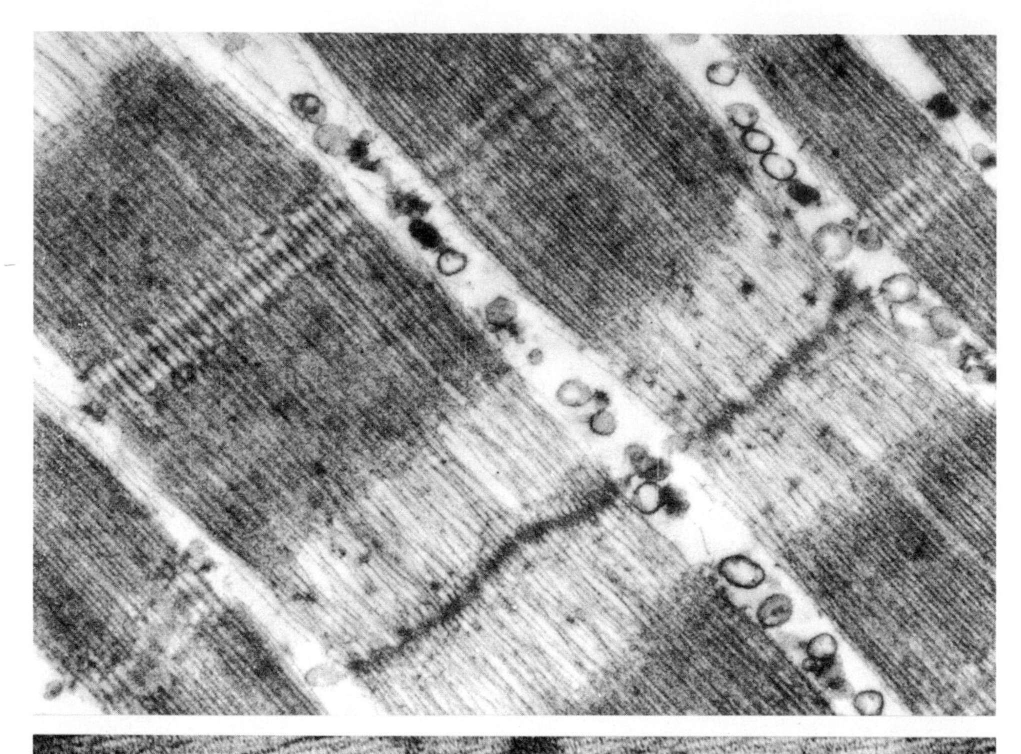

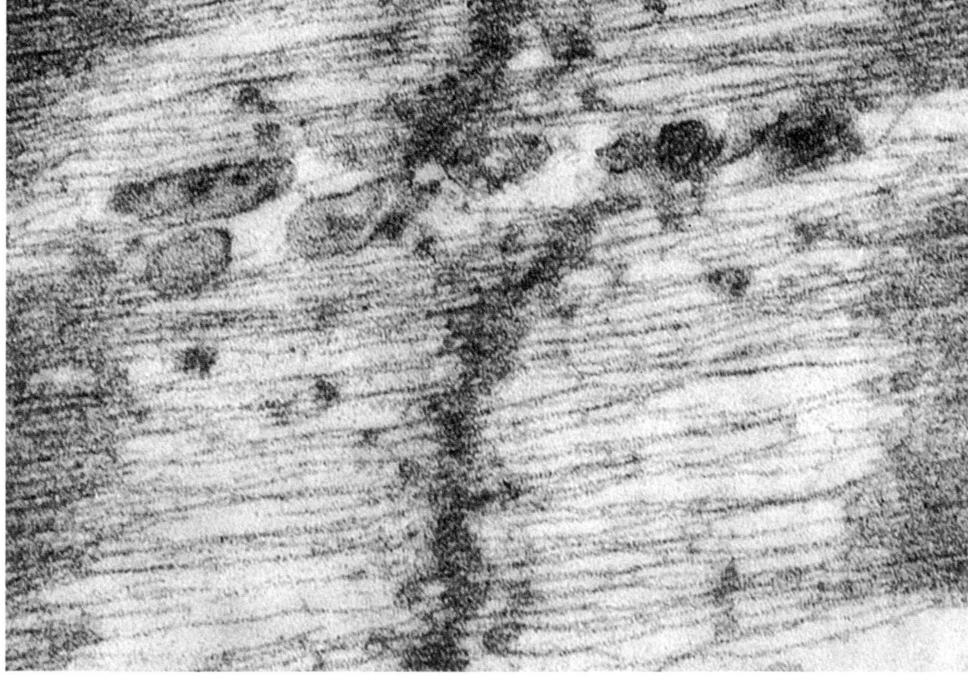

among adjacent fibers, so that the $Ca^{++}$-treated strips could be teased into individual fibers much more easily than the EDTA-treated strips.

Since the phase microscopic studies indicate that bathing in 1 mM $Ca^{++}$-containing solutions has marked effects on myofibrillar structure, Busch (1969) has extended these observations by a careful electron microscopic study on the effects of $Ca^{++}$ on the ultrastructure of myofibrils (see Figs. 36.15–19). For comparative purposes, the ultrastructure of muscle sampled immediately after death is shown in Fig. 36.15. This micrograph shows the characteristic features of at-death muscle, with wide I bands and distinct Z lines which have a fibrillar appearance. After 9 hours of exposure to 1 mM $Ca^{++}$ (about the point that isometric tension development returns to zero), the Z line appears to have been split down the center (Fig. 36.16). Remnants of what is apparently Z-line material remain attached to the ends of the thin filaments from apposing sarcomeres. This appearance suggests that the effect of $Ca^{++}$ is on the Z line itself and not on the I-Z junction. Approximately 25% of the sarcomeres in fibrils treated with $Ca^{++}$ for 9 hours still possessed intact Z lines, although these Z lines were always very broad and amorphous in appearance. Another 25% of the sarcomeres in 9-hour, $Ca^{++}$-treated fibrils had their Z lines totally removed. On the other hand, after 9 hours' exposure to 1 mM EDTA- or 1 mM EGTA-containing solutions, myofibrils still exhibited intact Z lines (Figs. 36.17–18). It was noted consistently during this study that Z lines from EDTA-treated fibers were often broad and amorphous in appearance (Fig. 36.18), although they remained intact for as long as 48 hours post mortem at 37°. Z lines from EGTA-treated fibers, on the other hand, exhibited a distinct fibrillar structure, very similar to that seen in at-death muscle (*cf.* Figs. 36.15 and 36.17). This observation suggests that $Mg^{++}$, which is chelated by EDTA but not by EGTA, is essential for maintenance of normal Z-line structure, although the absence of $Mg^{++}$ does not cause Z-line removal. Twenty-four hours' exposure to 1 mM $Ca^{++}$ caused complete removal of Z lines, and open spaces could be seen between thin filaments extending from adjacent sarcomeres (Fig. 36.19).

*Fig. 36.17* (*above*). Electron micrograph of rabbit muscle after isometric suspension for 9 hours at 37° C in 80 mM KCl, 60 mM Tris-acetate, *p*H 7.1, 5 mM $MgCl_2$, 1 mM EGTA, 1 mM $NaN_3$, and 1 mM deoxycholate. The fibrillar appearance of the Z line in this muscle is very similar to that seen in at-death muscle (cf. Fig. 36.15). × 38,399.

*Fig. 36.18* (*below*). Electron micrograph of rabbit muscle after isometric suspension for 9 hours at 37° C in 80 mM KCl, 60 mM Tris-acetate, *p*H 7.1, 1 mM EDTA, 1 mM $NaN_3$, and 1 mM deoxycholate. Although the Z line can still be seen as an electron-dense structure, it has lost its fibrillar appearance and appears amorphous and structureless. × 92,936.

*Fig. 36.19.* Electron micrograph of rabbit muscle after isometric suspension for 24 hours at 37° C in 80 mM KCl, 60 mM Tris-acetate, *p*H 7.1, 5 mM $MgCl_2$, 1 mM $CaCl_2$, 1 mM $NaN_3$, and 1 mM deoxycholate. Z lines are completely absent in this muscle and open spaces without filaments can often be seen between thin filaments from adjacent sarcomeres. × 31,566.

These ultrastructural observations, together with the earlier biochemical studies showing that post mortem isometric tension development begins approximately at the same time that sarcoplasmic reticular membranes lose their ability to sequester $Ca^{++}$, clearly indicate that the release of $Ca^{++}$ from sarcoplasmic reticular membranes causes two distinct effects in post mortem muscle: (1) initiation of post mortem shortening or isometric tension development within 10–30 min after the sarcoplasmic reticulum begins to lose its $Ca^{++}$-sequestering ability; and (2) degradation and the eventual complete removal of the Z line; this degradation requires several hours before its effects can be detected by ultrastructural examination. Furthermore, since the rapid decline in isometric tension development of $Ca^{++}$-treated strips can be related easily and directly to loss of Z lines in these strips, it is now evident that loss of the ability of post mortem muscle strips to maintain isometric tension development is due primarily to loss of Z line integrity in these strips.

## CONCLUSIONS

It is now clear that post mortem storage causes at least two kinds of biochemical changes in the myofibrillar apparatus: (1) disruption and degradation of the Z line, and (2) weakening of the actin-myosin interaction. Post mortem degradation of the Z line is detected most easily by direct ultrastructural examination of stored muscle, but it is also indicated in a biochemical fashion by a post mortem increase in the rate of actin extractability. Post mortem weakening of the actin-myosin interaction, on the other hand, is indicated most clearly biochemically by an increased susceptibility of the actin-myosin complex to dissociation by ATP, and possibly also by the increased nucleosidetriphosphatase activity of actomyosin preparations made from post mortem muscle. Ultrastructurally, post mortem weakening of the actin-myosin interaction is seen as a lengthening or relaxation of rigor-shortened sarcomeres in the absence of ATP. Our recent results indicate that post mortem Z-line degradation is caused by $Ca^{++}$, which is released when the sarcoplasmic reticular membranes in post mortem muscle lose the ability to accumulate $Ca^{++}$ against a concentration gradient. On the other hand, the cause of post mortem weakening of the actin-myosin interaction is not yet clear. It is not due primarily to post mortem changes in $\alpha$-actinin or the tropomyosin-troponin complex, but probably originates from post mortem changes in actin and myosin themselves. It is obvious that post mortem disruption and eventual removal of Z lines would cause loss of the ability of post mortem muscle strips to maintain isometric tension and would also have important effects on the use of muscle as a food. Future studies of the effects of $Mg^{++}$, $Ca^{++}$, ATP, and *p*H changes on the myofibrillar proteins may be expected to clarify our understanding of post mortem changes in the molecular architecture of muscle.

## ACKNOWLEDGMENTS

We wish to express our sincere appreciation to Joanne Temple, Charlene Johnson, and Karen Schwarz for their collaboration, and to Bruce Eason and D. W. Henderson for some of the data cited in this review.

The research results reported from our laboratory were supported in part by USPHS grant GM–12488 and by Iowa Agriculture Experiment Station Project No. 1549. During this investigation, Nobuhiko Arakawa was the recipient of a visiting professorship from Training Grant UI 01043–03, National Center for

Urban and Industrial Health (USPHS), and R. M. Robson held an NIH predoctoral fellowship, 5–FI–GM–28, O41.

## References

Aberle, E. D., and R. A. Merkel. 1966. Solubility and electrophoretic behavior of some proteins of post-mortem aged bovine muscle. *J. Food Sci. 31*:151.

Bendall, J. R. 1951. The shortening of rabbit muscles during rigor mortis: its relation to the breakdown of adenosine triphosphate and creatine phosphate and to muscular contraction. *J. Physiol. 114*:71.

Bodwell, C. E., and A. M. Pearson. 1964. The activity of partially purified bovine catheptic enzymes on various natural and synthetic substrates. *J. Food Sci. 29*:602.

Borchert, L. L., and E. J. Briskey. 1965. Protein solubility and associated properties of porcine muscle as influenced by partial freezing with liquid nitrogen. *J. Food Sci. 30*:138.

Borchert, L. L., W. D. Powrie, and E. J. Briskey. 1969. A study of the sarcoplasmic proteins of porcine muscle by starch gel electrophoresis. *J. Food Sci. 34*:148.

Briskey, E. J. 1964. Etiological status and associated studies of pale, soft, exudative porcine musculature. *Advance. Food Res. 13*:89.

Busch, W. A. 1969. Subcellular events associated with the development and release of isometric tension in post-mortem bovine, rabbit, and porcine skeletal muscle. Ph.D. thesis, Iowa State University, Ames, Iowa.

Busch, W. A., F. C. Parrish, and D. E. Goll. 1967. Molecular properties of post-mortem muscle. 4. Effect of temperature on adenosine triphosphatase degradation, tension parameters, and shear resistance of bovine muscle. *J. Food Sci. 28*:680.

Chaudhry, H. M., F. C. Parrish, and D. E. Goll. 1969. Molecular properties of post-mortem muscle 6. Effect of temperature on protein solubility of rabbit and bovine muscle. *J. Food Sci. 34*:183.

Clayson, D. H. F., J. A. Beesley, and R. M. Blood. 1966. The chemistry of the natural tenderization or maturation of meat. *J. Sci. Food Agr. 17*:220.

Davey, C. L., and K. V. Gilbert. 1966. Studies in meat tenderness II. Proteolysis and the aging of beef. *J. Food Sci. 31*:135.

———. 1967. Structural changes in meat during aging. *J. Food Technol. 2*:57.

———. 1968*a*. Studies in meat tenderness 6. The nature of myofibrillar proteins extracted from meat during aging. *J. Food Sci. 33*:343.

———. 1968*b*. Studies in meat tenderness 4. Changes in the extractability of myofibrillar proteins during meat aging. *J. Food Sci. 33*:2.

———. 1969. Studies in meat tenderness 7. Changes in the fine structure of meat during aging. *J. Food Sci. 34*:69.

Davies, R. E. 1966. Recent theories on the mechanism of muscle contraction and rigor mortis, p. 39. *Proc. 3d Meat Industry Res. Conf., Amer. Meat Inst. Found.* (Chicago, Ill.).

De Fremery, D., and I. V. Streeter. 1969. Tenderization of chicken muscle: the stability of alkali-insoluble connective tissue during post-mortem aging. *J. Food Sci. 34*:176.

Eason, B. A. 1969. Purification and properties of skeletal muscle microsomes. Ph.D. thesis, Iowa State University, Ames, Iowa.

Ebashi, S., and A. Kodama. 1966. Native tropomyosin-like action of troponin on trypsin-treated myosin B. *J. Biochem. 60*:733.

Elliott, G. F., J. Lowy, and C. R. Worthington. 1963. An X-ray and light-diffraction study of the filament lattice of striated muscle in the living state and in rigor. *J. Mol. Biol. 6*:295.

Fujimaki, M., N. Arakawa, A. Okitani, and O. Takagi. 1965*a*. The dissociation of the "myosin B" from the stored rabbit muscle into myosin A and actin and its interaction with ATP. *Agr. Biol. Chem. 29*:700.

———. 1965*b*. The changes of "myosin B" ("actomyosin") during storage of rabbit muscle II. The dissociation of "myosin B" into myosin A and actin, and its interaction with ATP. *J. Food Sci. 30*:937.

Fujimaki, M., and F. E. Deatherage. 1964. Chromatographic fractionation of sarcoplasmic proteins of beef skeletal muscle on ion-exchange cellulose. *J. Food Sci. 29*:316.

Fujimaki, M., A. Okitani, and N. Arakawa. 1965. The changes of "myosin B" during storage of rabbit muscle I. Physico-chemical studies on "myosin B." *Agr. Biol. Chem. 29*:581.

Fukazawa, T., and T. Yasui. 1967. The change in zigzag configuration of the Z-line of myofibrils. *Biochim. Biophys. Acta 140*:534.

Goll, D. E. 1965. Post-mortem changes in connective tissue, p. 161. *Proc. 18th Reciprocal Meat Conf.* (Chicago, Ill.).

———. 1968. The resolution of rigor mortis, p. 16. *Proc. 21st Reciprocal Meat Conf.* (Chicago, Ill.).

Goll. D. E., D. W. Henderson, and E. A. Kline. 1964. Post-mortem changes in physical and chemical properties of bovine muscle. *J. Food Sci. 29*:590.

Goll, Darrel E., W. F. H. M. Mommaerts, M. K. Reedy, and K. Seraydarian. 1969. Studies on α-actinin-like proteins liberated by tryptic digestion of α-actinin and of myofibrils. *Biochim. Biophys. Acta 175*:174.

Goll, Darrel E., W. F. H. M. Mommaerts, and K. Seraydarian. 1967. Is α-actinin a constituent of the Z-band of the muscle fibril? *Fed. Proc. 26*:499.

Goll, D. E., and R. M. Robson. 1967. Molecular properties of post-mortem muscle 1. Myofibrillar nucleosidetriphosphatase activity of bovine muscle. *J. Food Sci. 32*:323.

Gothard, R. H., A. M. Mullins, R. F. Boulware, and S. L. Hansard. 1966. Histological studies of post-mortem changes in sarcomere length as related to bovine muscle tenderness. *J. Food Sci. 31*:825.

Greaser, M. L., R. G. Cassens, E. J. Briskey, and W. G. Hoekstra. 1969*a*. Post-mortem changes in subcellular fractions from normal and pale, soft, exudative porcine muscle 1. Calcium accumulation and adenosine triphosphatase activities. *J. Food Sci. 34*:120.

———. 1969*b*. Post-mortem changes in subcellular fractions from normal and pale, soft, exudative porcine muscle 2. Electron microscopy. *J. Food Sci. 34*:125.

Greaser, M. L., R. G. Cassens, and W. G. Hoekstra. 1967. Changes in oxalate-stimulated calcium accumulation in particulate fractions from post-mortem muscle. *J. Agr. Food Chem. 15*:1112.

Haga, T., M. Yamamoto, K. Maruyama, and H. Noda. 1966. The effect of myosin and calcium on the solubilization of F-actin from muscle mince. *Biochim. Biophys. Acta 127*:128.

Henderson, D. W. 1968. Effect of temperature on post-mortem structural changes in rabbit, bovine and porcine skeletal muscle. Ph.D. thesis, Iowa State University, Ames, Iowa.

Herring, H. K. 1968. Muscle contraction and tenderness, p. 47. *Proc. 21st Reciprocal Meat Conf.* (Chicago, Ill.).

Herring, H. K., R. G. Cassens, and E. J. Briskey. 1967. Factors affecting collagen solubility in bovine muscles. *J. Food Sci. 32*:534.

Hershberger, T., R. Deans, L. E. Kunkle, P. Gerlaugh, and F. E. Deatherage. 1951. Studies on meat III. The biochemistry and quality of meat in relation to certain feeding management practices. *Food Technol. 5*:523.

Hoagland, R., C. N. McBryde, and W. C. Powick. 1917. Changes in fresh beef during storage above freezing. *U.S. Dept. Agr. Bull.* 433.

Hultin., H. O., and C. Westort. 1969. Sarcolemmae from chicken skeletal muscle 2. Properties. *J. Food Sci. 34*:172.

Husaini, S. A., F. E. Deatherage, and L. E. Kunkle. 1950. Studies on meat II. Observations on relation of biochemical factors to changes in tenderness. *Food Technol. 4*:366.

Huxley, H. E. 1968. Structural difference between resting and rigor muscle; evidence from intensity changes in the low-angle equatorial X-ray diagram. *J. Mol. Biol. 37*:507.

Huxley, H. E., and W. Brown. 1967. The low-angle X-ray diagram of vertebrate striated muscle and its behavior during contraction and rigor. *J. Mol. Biol. 30*:383.

Jungk, R. A., H. E. Snyder, D. E. Goll, and K. G. McConnell. 1967. Isometric tension changes and shortening in muscle strips during post-mortem aging. *J. Food Sci. 32*:158.

Khan, A. W., and L. van den Berg. 1964. Changes in chicken muscle proteins during aseptic storage at above-freezing temperatures. *J. Food Sci. 29*:49.

Kronman, M. J., and R. J. Winterbottom. 1960. Post-mortem changes in the water-soluble proteins of bovine skeletal muscle during aging and freezing. *J. Agr. Food Chem. 8*:67.

Lawrie, R. A., I. F. Penny, R. K. Scopes, and C. A. Voyle. 1963. Sarcoplasmic proteins in pale, exudative pig muscles. *Nature 200*:673.

Locker, R. H. 1960. Proteolysis in the storage of beef. *J. Sci. Food Agr. 11*:520.

McClain, P. E., A. M. Mullins, S. L. Hansard, J. D. Fox, and R. F. Boulware. 1965. Acid and salt-soluble collagen in bovine muscle. *Proc. Soc. Exp. Biol. Med. 119*:492.

McClain, P. E., A. M. Pearson, J. R. Brunner, and G. A. Crevasse. 1969. Connective tissues from normal and PSE porcine muscle 1. Chemical characterization. *J. Food Sci. 34*:115.

Maier, G. E., and R. L. Fischer. 1966. Acrylamide gel disc electrophoretic patterns and extractability of chicken breast muscle proteins during post-mortem aging. *J. Food Sci. 31*:482.

Martins, C. B., and J. R. Whitaker. 1968. Catheptic enzymes and meat tenderization 1. Purification of cathepsin D and its action on actomyosin. *J. Food Sci. 33*:59.

Masaki, T., M. Endo, and S. Ebashi. 1967. Localization of 6 S component of α-actinin at Z-band. *J. Biochem. 62*:630.

Masaki, T., O. Takaiti, and S. Ebashi. 1968. "M-substance," a new protein constituting the M-line of myofibrils. *J. Biochem. 64*:909.

Neelin, J. M., and D. J. Ecobichon. 1966. Enzymic activities in pre-rigor and post-rigor sarcoplasmic extracts of chicken pectoral muscle. *Can. J. Biochem. 44*:735.

Neelin, J. M., and D. Rose. 1964. Progressive changes in starch gel electrophoretic patterns of chicken muscle proteins during "aging" post-mortem. *J. Food Sci. 29*:544.

Newbold, R. P. 1966. Changes associated with rigor mortis, p. 213. *In* E. J. Briskey, R. G. Cassens, and J. C. Trautman (eds), *The Physiology and Biochemistry of Muscle as a Food*. Univ. of Wis.

Okitani, A., N. Arakawa, and M. Fujimaki. 1965. The changes of "myosin B" during storage of rabbit muscle III. Sedimentation studies on "myosin B." *Agr. Food Chem. 29*:971.

Okitani, A., O. Takagi, and M. Fujimaki. 1967. The changes of "myosin B" during storage of rabbit muscle IV. Effect of temperature, *p*H, and ionic strength on denaturation of "myosin B" solution. *Agr. Biol. Chem. 31*:939.

Osner, R. C. 1966. Influence of processing procedures on the post-mortem permeability of chicken muscle sarcolemmas to protein. *J. Food Sci. 31*:832.

Parrish, F. C., Jr., D. E. Goll, W. J. Newcomb, B. O. de Lumen, H. M. Chaudhry, and E. A. Kline. 1969. Molecular properties of post-mortem muscle 7. Changes in nonprotein nitrogen and free amino acids of bovine muscle. *J. Food Sci. 34*:196.

Penny, I. F. 1968. Effect of aging on the properties of myofibrils of rabbit muscle. *J. Sci. Food Agr. 19*:518.

Prudent, I. 1947. Collagen and elastin content of four beef muscles aged varying periods of time. Ph.D. thesis, Iowa State University, Ames, Iowa.

Rampton, J. H., A. F. Anglemier, and M. W. Montgomery. 1965. Fractionation of bovine sarcoplasmic proteins by DEAE-cellulose chromatography. *J. Food Sci. 30*:636.

Reed, R., T. W. Houston, and P. M. Todd. 1966. Structure and function of the sarcolemma of skeletal muscle. *Nature 211*:534.

Reedy, M. K., K. C. Holmes, and R. T. Tregear. 1965. Induced changes in orientation of the cross-bridges of glycerinated insect flight muscle. *Nature 207*:1276.

Robson, R. M. 1969. Purification and properties of α-actinin from rabbit striated muscle. Ph.D. thesis, Iowa State University, Ames, Iowa.

Robson, R. M., D. E. Goll, and M. J. Main. 1967. Molecular properties of post-mortem muscle 5. Nucleoside triphosphatase activity of bovine myosin B. *J. Food Sci. 32*:544.

Sayre, R. N. 1968. Post-mortem changes in extractability of myofibrillar protein from chicken pectoralis. *J. Food Sci. 33*:609.

Sayre, R. N., and E. J. Briskey. 1963. Protein solubility as influenced by physiological conditions in the muscle. *J. Food Sci. 28*:674.

Scopes, R. K. 1964. The influence of post-mortem conditions on the solubilities of muscle protein. *Biochem. J. 91*:201.

———. 1966. Isolation and properties of a basic protein from skeletal muscle sarcoplasm. *Biochem. J. 98*:193.

Scopes, R. K., and R. A. Lawrie. 1963. Post mortem lability of skeletal muscle proteins. *Nature 197*:1202.

Sharp, J. G. 1963. Aseptic autolysis in rabbit and bovine muscle during storage at 37° C. *J. Sci. Food Agr. 7*:468.

Stromer, M. H., and D. E. Goll. 1967*a*. Molecular properties of post-mortem muscle II Phase microscopy of myofibrils from bovine muscle. *J. Food Sci. 32*:329.

———. 1967*b*. Molecular properties of post-mortem muscle 3. Electron microscopy of myofibrils. *J. Food Sci. 32*:386.

Stromer, M. H., D. E. Goll, and L. E. Roth. 1967. Morphology of rigor-shortened bovine muscle and the effect of trypsin on pre- and post-rigor myofibrils. *J. Cell Biol. 34*:431.

Takahashi, K., T. Fukazawa, and T. Yasui. 1967. Formation of myofibrillar fragments and reversible contraction of sarcomeres in chicken pectoral muscle. *J. Food Sci. 32*:409.

Thompson, G. B., W. D. Davidson, M. W. Montgomery, and A. F. Anglemier. 1968. Alterations of bovine sarcoplasmic proteins as influenced by high temperature aging. *J. Food Sci. 33*:68.

Wierbicki, E., V. R. Cahill, L. E. Kunkle, E. W. Klosterman, and F. E. Deatherage. 1955. Effect of castration on biochemistry and quality of beef. *J. Agr. Food Chem. 3*:244.

Wierbicki, E., L. E. Kunkle, V. R. Cahill, and F. E. Deatherage. 1954. The relation of tenderness to protein alterations during post-mortem aging. *Food Technol. 8*:506.

Winegarden, M. W. 1950. Physical and histological changes of three connective tissues of beef during heating. Ph.D. thesis, Iowa State University, Ames, Iowa.

# Chapter 37

# *Summary and Discussion of Part 8*

PANEL MEMBERS: R. A. LAWRIE, *Chairman*
A. EDGAR
R. HAMM
R. HENRICKSON
J. JAY
R. SLEETH
G. WELLINGTON

*Lawrie:* The three words "as a food" occur in each of the four chapters which constitute the final part of this symposium. In earlier chapters detailed consideration has been given to the most recent fundamental knowledge on how our machine—to continue Dr. Marsh's apt analogy—is constructed, of what it is constructed and how it works. In the concluding chapters, consideration has been given to what happens when we are ready to eat it. Such terms as beef, lamb, and pork have now been used in referring to the material of the machine. These words are more familiar in the factory and at home. But it is important to emphasize that the material is the same whatever we call it. The names which various specialists (in one or another discipline) employ to describe meat are not meant to confuse but to permit, more readily, unequivocal identification of what is a complex system.

A major advantage of the morphological approach to muscle, especially when combined with histochemical staining techniques, is the possibility of visualizing in some detail how its chemically defined constituents are arranged. Such information clarifies how muscle develops in normal and abnormal circumstances and it is becoming increasingly evident that histological data correlate usefully with the organoleptic attributes of meat which are detected at the macroscopic level. In elaborating this theme in Chapter 33, Dr. Cassens mentioned the connection established by New Zealand workers between the degree of sarcomere shortening, as observed histologically, and toughness as determined by the palate. He and his colleagues have recently further elucidated this finding by showing that the degree of contraction is directly related to the size of the muscle fibers. Since large fiber size has long been associated with toughness, this observation is particularly interesting. Another feature of great importance for an understanding of meat quality is muscle differentiation. The histological evidence which Dr. Cassens presented drew attention to the dangers of oversimplifying meat; it cannot be regarded as derived from either "red" or "white" muscles. Quite apart from other considerations, there are at least three types of fiber—one of these being of very large cross-sectional area. Moreover, it seems that the manner of their distribution within a given muscle cannot fail to affect the attributes of the meat derived from it. Thus some of the peculiarities of pork, possibly including its tendency to become PSE in character, may well reflect the fact that such "red" fibers as pig muscles possess cluster in the central region of the fiber bundles. In other mammalian muscle the "red" fibers have no preferred location. When muscle morphology is studied at the next order of discrimination, further features, which can be related to meat quality, are revealed. Electron micrographs show minute, but significant, differences in sarcomere patterns and in the form of the sarcotubular system, which correspond with the relative tendency to develop PSE character.

In Chapter 34 Dr. Lister carried the theme of the inevitability of the connection between muscle biochemistry and meat quality from the tissue level to that of the intact animal. He emphasized, however, that the nature of meat depends not only on muscle differentiation but also on the reaction of the animal to conditions during growth and development, in the immediate preslaughter period, at death, and during the immediate post mortem period. The nutrition of the young, actively growing animal substantially determines the relative proportions of its tissues and organs; these, in turn, determine its metabolic requirements and its feed conversion efficiency. The author also stressed the paramount effect of the body's hormones in determining its precise nature in any individual; only a beginning has yet been made to elucidate their involvement with the quantity and quality of muscular tissue in meat animals. Much of the effect on meat quality of the circumstances at the death of the animal are expressed in the rate and extent of post mortem glycolysis—and by the concomitant fall of *p*H and of ATP. Dr. Lister reminded us that among pigs there are breed differences in the characteristic rates of post mortem glycolysis which can be related to the likelihood that the various breeds will develop PSE musculature. Such undesirable conditions in meat may well reflect an overintensive selection for feed conversion efficiency. He speculates that further aspects of stress susceptibility in the meat from domestic animals are likely to arise before fundamental knowledge of hormonal action is sufficiently precise to permit effective control.

After acknowledging the significance of post mortem glycolysis in determining meat quality, Dr. Kastenschmidt gave detailed consideration, in Chapter 35, to the biochemical conditions which could account for its variability. He drew attention to the possibility that such hormones as those of the adrenals and thyroid, which were clearly implicated, may exert their influence by controlling the blood flow through muscles. It was thus interesting to note his conclusion that the degree of struggling at death is *not* mainly responsible for the rate of post mortem glycolysis; and that the muscles of "stress-susceptible" pigs are more readily made anoxic than those of "normal" animals. The observed efficacy of $MgSO_4$ in conserving the *in vivo* levels of ATP and of creatine phosphate may be due to its vasodilatory action; and its preslaughter administration brought about a normal biochemical pattern in "stress-susceptible" animals. Although there are promising possibilities for using biochemical parameters in live animals to predict stress susceptibility, these are not yet sufficiently well defined to permit effective prophylactic action.

In Chapter 36, Dr. Goll showed how investigations of muscle proteins were elucidating the changes occurring when meat is conditioned by chill storage for some days after death. It has long been appreciated that meat becomes more tender under these circumstances but the biochemical reasons for this desirable change have been a source of controversy. The work of Sharp (1959) seemed to show conclusively that the breakdown of connective tissue proteins was not involved; and the findings of Davey and Gilbert (1968) suggested that cleavage of actin rods from their attachment to Z-line material was implicated. Dr. Goll provided data which substantiated that loss of Z-line material occurred when meat become more tender; but he also presented important detailed evidence indicating that other subtle changes occur at this time. Careful study has suggested that while the $\alpha$-actinin and tropomyosin-troponin complex both become more readily extractable there is no evidence for any chemical alteration in these species. On the other hand there is some change in the manner of linkage between actin and myosin. It also seems that the degradation of the Z-line material is initiated by $Ca^{++}$ released from the sarcotubular system post mortem. Such studies clearly illustrate how far into basic areas one may be required to proceed in order to understand "classical" meat phenomena.

It is worth reiterating that the biology of muscle as a food is an integrated study necessarily involving concepts and knowledge which start at conception of the animal—and proceed through growth and maturity, to slaughter, processing and preservation, cooking and tasting, to assimilation; and beyond to the areas of nutrition and other branches of medicine. At each stage in this long chain there is knowledge to be gained which will help our understanding of other stages in the chain. Fundamental work on muscle is most important for an understanding of meat; the problems of meat in industry and in our home are most important in guiding fundamental studies on muscle. The machine is edible; and its edibility represents an essential mechanism, controlling and modifying all that goes before.

DISCUSSION

*Maruyama:* The effect of $Ca^{++}$ on the disappearance of the Z lines of muscle fiber might be related to the observations that actin can be more easily solubilized from myosin-removed myofibrils in $Ca^{++}$-containing medium (Haga, Yamamoto, Maruyama, and Noda, 1966).

*Goll:* Yes, I think it is. We have carefully studied the paper by Haga and his coworkers and are in agreement with the results presented there. Our experiments at this stage do not eliminate some other possibilities for the effects of $Ca^{++}$, however, and we at present cannot clearly differentiate between a direct effect of $Ca^{++}$ on the Z line and some type of indirect effect due to an influence of $Ca^{++}$ on other intracellular components, such as activation of a muscle cathepsin.

*Wellington:* To what extent will post mortem disruption and eventual removal of Z lines affect the use of muscle as a food?

*Goll:* We do not yet have explicit information on the effects of loss of Z line, but some indirect evidence obtained in our experiments together with some deductive reasoning would suggest that disruption or loss of Z line causes considerable improvement in tenderness. I am not sure whether loss of Z lines would have any effect on water-holding capacity, keeping qualities, or emulsifying properties of meat.

*Jay:* Does the removal of the Z-line material bring about any measurable changes in muscle texture or tenderization?

*Goll:* We have not attempted to measure these directly, but it is easily observable that our $Ca^{++}$-treated muscle strips (which do not contain Z lines) are much more fragile and easily broken into small pieces than are our EDTA- or EGTA-treated strips, which have intact Z lines. Moreover, we have noticed that $Ca^{++}$-treated strips can be "teased" into individual fibers much more easily than EDTA- or EGTA-treated strips; this suggests that $Ca^{++}$ is also acting to weaken or disrupt interfibrillar attachments as well as intrafibrillar Z lines.

*Olcott:* Do you have information on the rates of post mortem disappearance of Z line from "white" as compared to "red" fibers?

*Goll:* We do not have any explicit information on this point, but our observations lead us to believe that Z lines in red muscle are less labile than those in white muscle. If so, this raises some intriguing possibilities concerning Z-line composition.

*Greaser:* Would you comment on our observations that the Z lines of myofibrils from rapid post mortem glycolyzing pig muscle were apparently intact at 24 hours post mortem in spite of the fact that the sarcoplasmic reticulum $Ca^{++}$ was presumably released earlier in these muscles than in normal ones?

*Goll:* I cannot offer any explanation for your results other than to suggest that temperature may be a factor. Our experiments were done with muscle strips incubated at 37° C, whereas I presume your experiments were done with muscle that was being chilled to 2–4° C during the first 24 hours post mortem. We have observed that Z-line removal, even in the presence of $Ca^{++}$, requires much longer at temperatures below 25° C than at temperatures of 25° C or higher (about 3–5 days for disruption of Z-line in porcine muscle at 2° C). Consequently, the effects of $Ca^{++}$ on Z-line structure may not be ultrastructurally observable within the first 24 hours post mortem. AT 37° C in our experiments, Z lines were completely absent after 24 hours in the presence of 1 mM $Ca^{++}$.

*Jay:* Does citrate act like EGTA relative to Z-line stability post mortem?

*Goll:* We have not tested the effect of citrate on post mortem Z-line stability, but since citrate can chelate $Ca^{++}$, it may be expected that its presence may stabilize Z lines.

*Jay:* Have you determined the effect of urea and/or guanidine HCl on Z line?

*Goll:* Not in any specific fashion. At high concentrations (4 M or more) both these substances dissolve myofibrils, including the Z line, almost completely. At lower concentrations (1–3 M), urea appears to dissolve Z lines faster than the rest of the myofibril (Rash, Shag, and Biesele, 1968).

*Jay:* If it is true that Z-line removal can be affected by exposure to as little as 1 mM $Ca^{++}$, what is the fate of the Z-line material?

*Goll:* We do not know, although we would very much like to.

*King:* Have you tried to relate the loss of activity of the sarcoplasmic reticulum, post mortem, to release or activation of lysosomal enzymes?

*Cassens:* We have not made a specific attempt to study the influence of lysosomal enzymes. This area was discussed in Chapter 17 of *The Physiology and Biochemistry of Muscle as a Food* (1966). The problem with work on change post mortem in the sarcoplasmic reticulum is the great difficulty encountered in trying to separate the cause-effect relationship: does change in the sarcoplasmic reticulum promote *p*H fall or does *p*H fall elicit a change in the sarcoplasmic reticulum properties? Greaser and coworkers have accumulated data on this problem for both *in vivo* (Greaser, Cassens, Hoekstra, Briskey, Schmidt, Carr, and Galloway,

1969) and *in vitro* (Greaser, Cassens, Briskey, and Hoekstra, 1970) systems.

*Hamm:* We have studied the release or nonrelease of mitochondrial enzyme during the development of rigor mortis. An efflux does occur and we think that a certain damage occurs to the mitochondria, but the membranes do not show a great disintegration. What damage might be observable to the membranes?

*Cassens:* There might be some change in the membranes that is not detectable with electron microscopy.

*Ratcliff:* Both Dr. Cassens and Dr. Kastenschmidt hinted at the role that may be played by sarcoplasmic reticulum in relation to meat quality parameters, mention being made of muscle contraction and therefore tenderness. The role of reticulum has been shown by Krzywicki and Ratcliff (1967) to be wider than this; they studied the phospholipid content of cell fractions of muscle, and showed that rate of glycolysis in post mortem pork longissimus dorsi is related to the amount of reticulum in the muscle. This has a bearing on the incidence of PSE condition of the meat. Subsequent work has indicated that myosin adenosine triphosphatase activity and rate of $Ca^{++}$ uptake by the post mortem muscle are similarly related to *p*H of the muscle (in pork) and, therefore, to the amount of reticulum.

*Hamm:* I understand that we do not yet know whether the microsomal adenosine triphosphatase or the myofibrillar adenosine triphosphatase or the combined effect of several adenosine triphosphatases is responsible for the breakdown of ATP in muscle post mortem. Bendall (1960) has postulated that the majority of the ATP breakdown after death is carried out by a sarcoplasmic adenosine triphosphatase, which is now known to consist mainly of fragments of the sarcoplasmic reticulum. Greaser, Cassens, Briskey, and Hoekstra (1969) have recently shown that myofibrillar adenosine triphosphatase may play a more important role than was previously suggested. From Table 2 in this paper of Greaser and others, however, we can see that the activity of the adenosine triphosphatase in the mitochondrial fraction is of the same magnitude as the adenosine triphosphatase of the myofibrillar fraction, at least during the first 3 hours post mortem. I wonder why the authors did not discuss the possible importance of mitochondrial adenosine triphosphatase but only the role of microsomal and myofibrillar adenosine triphosphatase.

Is there any way to differentiate between the different adenosine triphosphatases in muscle tissue *in situ* by specific inhibition of each? In contrast to the myofibrillar and sarcotubular forms, for instance, the mitochondrial adenosine triphosphatase should not be inhibited by PCMB.

*Lister:* I think this is quite a useful point and one which, so far as I am aware, has not been considered in detail. I have nothing to add on this except that rapid rates of *p*H fall can be found in some pigeon breast muscle in which the role of mitochondrial adenosine triphosphatase might be important.

*Henrickson:* When tissue loses its ability to accumulate $Ca^{++}$, has the sarcoplasmic reticulum been damaged? Is the reason for loss due to the inability of the reticulum to function?

*Kastenschmidt:* We believe the reticulum fails to accumulate $Ca^{++}$ due to a physiological mechanism rather than simply due to physical damage to the reticulum.

*Jay:* When an animal dies, essentially all of the post mortem changes that take place in the tissues may be viewed as leading towards the ultimate destruction of the tissue, with the organic components being converted to inorganic compounds by microorganisms as the ultimate act. Since the muscles of PSE animals undergo post mortem changes more rapidly than do non-PSE muscles, might this fact not suggest that the animal producing PSE pork represents the more natural "wild-type" animal as opposed to the non-PSE type which might reflect the success of selection methods for meat production?

*Cassens:* I take the view that the PSE-producing animal is the result of selection. This very muscular animal may have muscle fibers that cannot cope with their environment (Cooper, Cassens, and Briskey, 1969). The animals have a high number of intermediate fibers, and the present postulate is that these intermediate fibers can glycolyze very well post mortem. One must also mention the stress-susceptibility problem (Lister, Sair, Will, Schmidt, Cassens, Hoekstra, and Briskey, 1970; Sair, Lister, Moody, Cassens, Hoekstra, and Briskey, 1970) that often appears to be associated with the very muscular animals. So I feel that selection for the most desirable meat-producing animal has led to the verge of muscle abnormality that is now revealing itself in problems such as PSE.

*Briskey:* Lohse (1966) compared the post mortem quality of "wild" and domestic pigs and found the former to be quite superior.

*Jay:* In your opinion, have we reached a point in the selection of meat animals where we should perhaps sacrifice some of our meat quality attributes in favor of a more durable, more stress-resistant animal?

*Lister:* I am sure that animal breeders have not placed enough emphasis on the quality of the increased amounts of leaner carcasses they have produced or, perhaps, we would not have the problems we have now. Advances in meat technology will help to alleviate some of the problems of watery meat, but in the long term we must reappraise the criteria by which meat animals are selected.

*Briskey:* It is not a matter of sacrificing meat quality for stress-resistance; in fact the stress-resistant animal has the highest quality in its muscle post mortem.

*Jay:* How do the physiological parameters of our slaughter animals compare with those of their prototypes in the wilds?

*Lister:* The modern way of life brings many problems to farm animals, and pigs in particular, as it does to some of the human population. Ready access to virtually unlimited supplies of food and lack of exercise produce the well-documented obesity and vascular disorders in animals as well as man. I would have thought that these aspects of environment appreciably influence the physiology of present day farm animals to make them different from their "prototypes." Selection for fashionable, and sometimes arbitrary, characteristics of breeding stock can and does, as we have seen, result in major physiological upheavals.

*Jay:* What is known about the genetics of stress-susceptible and stress-resistant slaughter animals; are these phenomena known to be genetically determined and if so, what is the nature of expression of the respective traits in terms of dominance and recessiveness?

*Kastenschmidt:* Certainly there are both strain and breed differences with respect to this trait. See Allen, Forrest, Chapman, First, Bray, and Briskey (1966) for the pertinent literature. It is difficult to determine the dominance and recessiveness of these traits as one cannot clearly identify the stress-susceptible animal prior to severe treatment. Even after the carcass is available, there are differences in the degree of expression of the consequences of the stress-susceptibility syndrome.

*Sleeth:* The emphasis you place on feeding practices and rate of growth is intriguing. What management practices, such as breed, rate of gain, etc., would you recommend to result in superior carcasses?

*Lister:* If the only task of animal breeders and feeders is to produce stock which yield large amounts of lean meat in the most economical fashion, then they appear to be doing an excellent job and I can suggest no alternatives. As yet, research and technology have to find the means of selecting against, or nullifying, the deleterious effects on meat quality of some of the characteristics being exploited by stockmen.

*Henrickson:* Is it possible or feasible to cause muscle deposition preferentially? For example, could one cause the beef animal to deposit protein preferentially in the fibers of the semitendinosus instead of the biceps or semimembranosus?

*Lister:* I would have thought it wellnigh impossible, under commercial conditions, to cause the preferential development of one muscle rather than another. Sophisticated exercise techniques are a possibility for certain muscles but it would seem more reasonable to look for mutant types of animal which demonstrate the characteristic one is interested in and then hope to establish that characteristic by selective breeding.

*Henrickson:* Is it feasible to have synthesis of myofibrillar protein without sarcoplasmic protein synthesis?

*Lister:* Yes, to some extent; the proportions vary in normally developing muscle.

*Briskey:* Exercise can also result in the preferential synthesis of myofibrillar protein.

*Henrickson:* Does it seem feasible to increase the quantity of muscle in an animal without an adequate supporting hormonal system?

*Lister:* No, the limits for protein deposition must be largely under hormonal control. Problems arise when animals are selected for particular characteristics, e.g., leanness, for it seems that the hormonal requirements for this are not compatible with those for stress resistance.

*Sleeth:* Assume that selection of domestic animals for muscle production has resulted in abnormalities whereby the muscle cannot respond to certain stress conditions, the result being lowered ultimate product quality. What guidelines would you recommend breeders and producers use in their future breeding and selection programs to ensure optimum quality of muscle for food?

*Cassens:* I feel that selection should be based more on muscle characteristics than on total muscle. For example, one might expect in the future that selection should include some provision for fiber type. Cooper, Cassens, Kastenschmidt, and Briskey (1970) have shown that fiber-type differentiation occurs rather early in the life of the animal. The information could perhaps be obtained from biopsy.

*Jay:* I would like to underscore an apparent dichotomy which exists in our meat animal practices. Dr. Lister called attention to the importance of the overall health of the animal to the proper functioning of its endocrine glands and the importance of this on subsequent meat quality. When it comes to the quartering of slaughter animals, we subject them to a set of rather unnatural conditions, e.g., we tend to discourage exercising beyond a minimum, we subject them to special diets so that they deposit lean and fat tissue in a place and manner that we regard as being desirable, etc. In brief, our slaughter animals are deliberately subjected to a set of unnatural conditions for our selfish desires. It would appear, therefore, that we should expect abnormality in the overall physiology of our slaughter animal populations relative to the corresponding wild and nonslaughter types.

*Fukazawa:* We have found that the α-actinin fraction, especially the 6 S component, is more extractable at 48 hours post mortem than at 0 hour post mortem. Your result, however, seems to differ from ours. Could this difference be due to the preparation procedure?

How do you get the α-actinin fraction from rabbit skeletal muscle, and do you think the Z line has some backbone material but not α-actinin? From your presentation, the Z line seems to be separatable into two parts. Do you have any further opinion about this separation?

*Goll:* I agree that differences in our preparative procedures are a likely cause for the difference between our results. In view of the degradation which is easily observable in Z lines in post mortem rabbit muscle, we had expected that α-actinin would become more extractable with increasing time of post mortem storage. Even after repeated efforts, however, we have been unable to observe any trend in either total yield or extractability of the 6 S α-actinin species with increasing time of post mortem storage. It should also be pointed out here that we have estimated the proportion of 6 S species in our preparations by measuring peak area in schlieren diagrams. Post mortem storage did not have any consistent effect on the proportion of 6 S species in our preparations. Consequently, our results are not due to the possibility that our preparations from at-death muscle contained less 6 S species and more contaminating protein than preparations from post mortem muscle. In our opinion, many of the α-actinin preparations described in the literature contain such low proportions of the 6 S α-actinin species (perhaps only 5–10% of total protein) that it would be very difficult to estimate yields of α-actinin by using these preparations.

We extracted our α-actinin-containing preparations from myofibrils by simply letting them stand at very low ionic strength and *p*H 8.5 for 60–72 hours. Details of our procedure are described in Chapter 36.

We are of the opinion that the Z line does contain considerable material in addition to α-actinin, and there is no reason to believe that this material is composed of only two different components. Our belief rests on the simple fact that Huxley and Hanson (1960) estimated that the Z line makes up 4–6% of the total material in the myofibril and that we have been unable to obtain α-actinin in yields larger than 0.6–0.8% of the total protein in the myofibril. Thus, 75–80% of the Z-line material is not accounted for by α-actinin. We at present do not have any information on the nature of the other substances in the Z line.

*King:* Do you have any ideas concerning the nature of the postrigor changes in actin or myosin molecules? For example, did you look at your superprecipitation system in the analytical ultracentrifuge?

*Goll:* We have examined our myosin B preparations in the analytical ultracentrifuge, but not in a very thorough or systematic fashion. Consequently, the only statement I can make about post mortem changes in ultracentrifugal behavior of rabbit myosin B is that myosin B from muscle after 2–3 days of post mortem storage appears to contain more of the "gel component" (as defined by Johnson and Rowe, 1964). The only other suggestion I can offer concerning the nature of the post mortem alterations in actin and myosin is that these changes may involve sulfhydryl (-SH) groups on either one or both of these molecules.

*Sayre:* Could the decreased time required for superprecipitation be due to increased extractability of α-actinin into the myosin B preparation even though the "specific activity" of the α-actinin is decreased with aging?

*Goll:* Yes, it could. We have expended considerable effort attempting to obtain an unequivocal answer to this possibility but with only qualified success. We cannot extract any more α-actinin from myosin B prepared from post mortem muscle than from myosin B prepared from at-death muscle, but these experiments are technically complex, and it is always difficult to be certain that all of the α-actinin is being extracted. We have also added at-death α-actinin ($P_{0-30}$ fraction) to at-death and to post mortem myosin B preparations and have observed that added α-actinin usually has very little effect on either the adenosine triphosphatase activity or the rate of turbidity development of either at-death or post mortem myosin B preparations. Occasionally, however, added α-actinin will accelerate the rate of turbidity development of post mortem myosin B preparations. Since post mortem α-actinin has a lower specific activity than at-death α-actinin, post mortem myosin B would have to contain considerably more α-actinin than at-death myosin B to account for the differences in rate of turbidity development that we observe between these preparations. Consequently, our two lines of evidence suggest rather strongly that the increased rate of turbidity development for myosin B from post mortem muscle is not due to the presence of larger amounts of α-actinin in these preparations, although the complexity of the system prevents an unequivocal answer on this point.

*King:* Have you studied the effect of heat or the effect of freezing muscle on retention or loss of α-actinin, troponin, or tropomyosin as a function of postrigor storage of the muscle?

*Goll:* No, although we know from *in vitro* studies that α-actinin and the tropomyosin-troponin complex are quite stable in the temperature range of 0–45° C at *p*H 7.0.

*King:* Was β-actinin in one of your fractions? If so, do you have any comments on its stability?

*Goll:* We do not have any information on the possible presence of β-actinin in our fractions, nor on its post mortem stability, since we have not attempted to assay for it. This would be an interesting study.

*Maruyama:* Your finding that actomyosin becomes more easily superprecipitable post mortem is very interesting. Have you studied the effect of increasing concentration of KCl on the superprecipitation of actomyosin from muscle stored for 8 days at 25° C? The reason is that the KCl concentration range of onset of superprecipitation is shifted to a higher concentration. If the clearing response would take place at a higher KCl (say 0.18–0.20 M) then I would suspect that some of the -SH groups of myosin were oxidized; PCMB-treated myosin + actin shows such a behavior (Maruyama and Ishikawa, 1964*a, b*). Unfortunately this -SH-blocking effect is irreversible. Could you store post mortem muscle by rinsing with an -SH-reagent like dithiothreitol (DTT) (I think it would penetrate into muscle fibers after death)? I would expect this to protect the -SH of myosin from oxidation.

*Goll:* Yes, we have done our superprecipitation assays at several different KCl concentrations, and have observed that the range of onset of superprecipitation is shifted to a higher KCl concentration in post mortem muscle. I agree with Dr. Maruyama's suggestion that this behavior may reflect some oxidation of -SH groups during post mortem storage, but we have not yet had the opportunity to do the experiment he suggests with DTT.

*Dubowitz:* I was interested in the clusters of type I fibers in the porcine muscle. Are there any data on the nerve terminals and innervation of pig muscle?

*Swatland:* I am at the moment in the process of examining the morphology of the motor innervation of pig muscle by staining with methylene blue, cutting thick frozen sections, and examining only complete and clearly visible terminal axons. Since I have only collected information from 225 terminal axons at the present time I am only able to offer the following impression: 13% of the terminal axons supply double end plates, 2% supply triple end plates, and 24% of the terminal axons show some degree of sprouting. The fine axonal branches lying within the sarcolemmal folds of the motor end plates form two distinct patterns, either a conventional aborization or a distinct circular form. (This may be likened to the convergence and fusion of the two main limbs of a tree-like form.) Intermediate types are found. In the animals examined (live weight, 180–280 lbs) the maximum diameter presented by the neural elements of the motor end plate has ranged from 15$\mu$ to 50$\mu$. Axons are sometimes observed in a distended or vesiculated state or with a beaded appearance resembling that of a degenerating axon. The completed study will compare breeds and parts of the muscle composed of predominantly white or red fibers, as well as some description of the ultrastructure.

*Wierbicki:* You have shown that muscle fibers of pig and cattle are subdivided into dark, semidark, and light, and that the color differences are caused by uneven myoglobin distribution among the fibers. Since myoglobin is the bearer of oxygen, it seems that only some myofibrils are able to take part in aerobic ATP restoration and aerobic glycolysis.

*Cassens:* Morita, Cassens, and Briskey (1970*b*) have shown that essentially all fibers in beef longissimus show some degree of positive reaction for myoglobin, but in pig longissimus muscle it appears that a large number of fibers show very little or no reaction. This corresponds with the general thought that red fibers depend mainly on aerobic metabolism while white fibers depend mainly on anaerobic metabolism. Even though one must interpret histochemical results with caution, the above finding seems very interesting in view of the problems encountered in pig muscle as regards glycolysis in post mortem muscle (see Morita, Cassens, and Briskey, 1969).

*Topel:* You indicated that fiber type is highly related to development of PSE. What is the correlation between fiber type and color of post mortem longissimus muscle of the pig at 24 hours post mortem?

*Cassens:* We think more in terms of fiber type making the muscle vulnerable to a faster change post mortem (Cooper, Cassens, and Briskey, 1969). We have not done statistical correlations in this work, but recently Morita, Cassens, Kauffman, and Briskey (1970*a*) have approached the problem by studying myoglobin distribution in porcine longissimus muscle. They found a correlation of 0.23 between color score and area percentage of myoglobin positive fibers.

*Henrickson:* Why are some fibers large and adjacent fibers small? This may be true regardless of whether they are red or white.

*Cassens:* As muscle is sectioned, the plane of the section may pass anywhere from the belly of the fiber to the tip of the tapering end. This can, of course, result in different fiber sizes in adjacent fibers. Then too, the stage of growth and the processes of hypertrophy or atrophy should be considered.

*Henrickson:* Would you expect the composition of the myofibrillar proteins to differ among individual muscle fibers?

*Goll:* Yes. There is abundant evidence showing that myosin in "red" fibers is structurally different from myosin in "white" fibers, and it is possible that the other myofibrillar proteins in "red" fibers are also different from those in "white" fibers. Moreover, the proportion of myofibrillar to sarcoplasmic proteins varies slightly among different kinds of muscles and may therefore be expected to vary among individual muscle fibers.

*Sleeth:* If you were a meat processor, what ante and post mortem management practices—holdover feeding, stunning techniques, refrigeration—would you have in effect to ensure optimum product quality? How extensive can quality factors (tenderness, flavor, juiciness) be affected by ante and post mortem handling practices?

*Lister:* Refrigeration is, in my opinion, the most important management practice to ensure optimum product quality, and should be used to an extent defined by economic and commercial considerations. The exact limits will be decided by criteria that you know more about than I do.

Present techniques for stunning animals leave much to be desired, but unfortunately I see no likelihood of radical innovations being introduced in the near future. There are many interesting new methods of producing rapid and relaxed anesthesia in animals but there are difficulties in their use under commercial conditions and the fact that there may be detectable, and potentially hazardous, residues left in the carcass.

We sometimes forget that whatever may be desirable from the physiological point of view cannot always be justified in economic terms. If I were a meat processor my main concern would be profitability and it would be up to me to assess which physiologically desirable practices gave the best financial return for me. These would obviously vary from packer to packer depending upon, for example, the particular program for procuring animals.

*Briskey:* While I agree with some of the points which have been made by Dr. Lister, there is little reported information that would justify holdover feeding; stunning should be accomplished under conditions where there is a minimum of excitability and oxygen deprivation. Temperature control holding pens and slaughter floors should be considered.

*Edgar:* If it is true that differences in stunning technique affect meat quality, would you suggest that we study more thoroughly not only technique but stunning methods?

*Lister:* I think it is necessary to have a complete reappraisal of stunning methods. There is relatively little scientific merit in any of the procedures used at present. But this is true not only of stunning methods but of the whole of the slaughter process from the transport of animals to the time carcasses enter the cooler and, in some cases, after this.

*Jay:* Are our present meat quality attributes inconsistent with sound animal physiology?

*Lister:* It does seem as though some commercial requirements for meat are incompatible with the increased rates of protein deposition and economy of food utilization such as are found in some contemporary animals. This may be a result of not using appropriate indices to select animals for breeding, for not all lean animals die before they are intended to, nor do they produce meat which is PSE.

*Sleeth:* What is the possibility of finding a means to identify stress-susceptible animals ante mortem?

*Kastenschmidt:* We are currently spending a considerable amount of time on this problem. We have studied serum enzyme levels in response to a brief heat stress and have tried to relate this to ultimate muscle quality. Generally speaking, animals which eventually have PSE muscles show larger increases in serum CPK and lactic dehydrogenase than do stress-resistant controls. I believe that we eventually will be able to identify those animals which are stress-susceptible.

*Hamm:* I would like to make a comment on your measurements of LDH activity in the serum. We found that the variation in the total LDH activity of the serum is much greater in PSE pigs than in normal animals. Nevertheless, the serum of PSE animals showed a significantly higher activity of the isozymes LDH V, LDH IV, LDH III, and LDH II. On the other hand, the LDH V and LDH IV activity of the longissimus dorsi muscle post mortem was significantly lower in PSE animals than in normal pigs. Therefore, I suppose that in PSE animals the subunit M-isozymes penetrated the cell membranes and went into the blood.

*Kastenschmidt:* Animals which exhibit PSE muscles in their carcasses show greater elevations in total serum LDH and CPK than animals which have normal muscles when both groups are subjected to a mild thermal stress, which suggests that the muscle membrane may in fact be more permeable.

*Jay:* Is there any correlation between stress-susceptibility or stress-resistance and sex of animal?

*Kastenschmidt:* It is not clear whether such a relationship exists.

*Edgar:* Would the measurement of electrical activity in muscle of pigs immediately after death be a possible means of separating stress-susceptible animals?

*Kastenschmidt:* I think this is a possibility worth exploring.

*Sleeth:* Does the use of $MgSO_4$ have any practical or commercial significance in controlling PSE?

*Kastenschmidt:* I believe it has great potential importance; however, there are some problems to solve. First, the method of administration—we have routinely injected it into the ear vein over a 20-min period. A too-rapid injection results in the premature death of the animal and a too-slow injection results in ineffective anesthesia. This method of injection is perhaps too cumbersome to be of practical importance. Other possibilities which might be useful are to administer it intraperitoneally or as an enema. I seriously doubt if feeding high levels would be effective since much of its effectiveness is probably dependent on its rather peculiar anesthetic properties.

*Henrickson:* When $MgSO_4$ is used in anesthetizing doses, should one expect the sarcoplasmic and/or myofibrillar proteins to be altered?

*Kastenschmidt:* One would expect certain $Mg^{++}$-activated or inhibited enzymes to be affected by $Mg^{++}$, notably myofibrillar adenosine triphosphatase. It has not yet been conclusively demonstrated that the $Mg^{++}$ concentration in muscle is increased by anesthetizing doses of $MgSO_4$.

*Edgar:* Do you have any suggestions on preslaughter treatment that could assure higher levels of ATP?

*Lister:* In most animals the elimination of struggle at death would assure higher levels of ATP. Animals of a stress-susceptible nature are a continuing headache. While various medicaments prove more or less beneficial, there is no treatment which can be guaranteed to raise levels of CP and ATP to a point where a sufficiently slow decline in *p*H post mortem can be assured.

*Edgar:* Would you expect the rate of glycolysis to be affected by administering to unanesthetized pigs moderately high levels of oxygen before exsanguination?

*Kastenschmidt:* I doubt whether any long-term control over post mortem glycolysis would be achieved. In the experience of Lister, Sair, Will, Schmidt, Cassens, Hoekstra, and Briskey (1970) in our laboratory, stress-susceptible pigs react much more strongly to anoxia and show markedly accelerated rates of post mortem glycolysis.

*Solberg:* Can dark-cutting meat be obtained through adrenaline injection just prior to slaughter?

*Lister:* Yes. A large dose of adrenaline given to some stress-prone pigs within, say, 10 min of slaughter will lead to darkening of the musculature. Normally, however, it takes appreciably longer for the glycogen reserves to become sufficiently depleted by adrenaline treatment to produce dark-cutting meat.

*Greaser:* How do you reconcile the fact that $Mg^{++}$ inhibits the *p*H decline post mortem (supposedly due to reduced ATP breakdown) while it activates both myofibrillar and sarcoplasmic reticulum adenosine triphosphatases?

*Lister:* I used the phrase "block adenosine triphosphatase activity" in its general sense, i.e., to bring about the minimal breakdown of ATP in muscle. Whether this is achieved at the motor end plate or by a direct effect on some adenosine triphosphatase is not known; perhaps both mechanisms are involved. That such an effect reduces the rate of *p*H fall post mortem is not doubted.

There is some debate whether the so called $Mg^{++}$-activated adenosine triphosphatase activity is as important as is often thought, or whether such an effect is really caused by contaminating $Ca^{++}$ (Bendall, 1969). Whether this is so or not is hair-splitting, for the concentration of $Mg^{++}$ used in the experiment was, in any case, completely unphysiological as are, of course, studies of $Mg^{++}$ activation *in vitro*.

*Solberg:* It should be pointed out that in addition to ATP, creatine phosphate, extensibility, and *p*H changes in conversion of living tissue to meat, there is also the conversion from an aerobic system to an anaerobic system, a temperature change from 37–39° C to 0–5° C, and possibly a change in cell permeability with a subsequent modification in osmotic pressure.

The osmotic equivalent of physiological saline is well known as representing *in vivo* muscle. Is there any knowledge pertaining to the alteration of this value during the conversion of living tissue into meat and, if so, what values have been determined? If possible can this be discussed through consideration of the sarcoplasmic material separate from myofibrillar material?

*Lister:* I have no information on this, but the biochemical changes in cells during the rigor process might be expected to induce some changes in osmotic pressure. Of course, by the time the rigor process is complete there would be appreciable damage to and disruption of cell membranes and osmotic forces would not then be of particular consequence.

*McLoughlin:* You suggested that absence of evidence of muscular contraction in carcasses of stress-susceptible pigs indicated that contraction was not an important factor. However, if the muscle *p*H was already below 6.8 to 6.7 in these carcasses the muscle would not be capable of contraction and would not respond to electrical stimulation. We find muscle which becomes PSE has lost its capacity to respond to electrical stimulation very soon (3–5 min) after death because of the low initial *p*H (as low as 6.0 at 5 min). To me, absence of ability to contract merely indicates that glycolysis has already proceeded appreciably—possibly even because of muscle contraction just at death.

*Kastenschmidt:* Dr. McLoughlin has alluded to a subtle point, namely the difference between an action-potential-induced contraction and an inability to relax. We believe that the data presented argue against an action-potential-induced contraction and make the second possibility even more likely. While the *p*H of the muscle certainly has an effect on the ability of a muscle to contract, it is likely that the situation is more complicated. We have been able to stimulate muscles electrically to contract even when the *p*H on the surface had reached 6.3, which is considerably below the value Dr. McLoughlin suggested.

*Scopes:* Dr. Kastenschmidt is aware that Dr. Newbold and I have obtained results with beef muscle which differ considerably with regard to levels of glycolytic metabolites and cofactors from those that he has presented. I have also observed similar differences when using pig muscle. This difference may be due to our different methods of making extracts. Rather than ask a question about these, I should like to query the level of an intermediate which I have not attempted to measure. This is the value for reduced NAD, which I see has decreased since the earlier publications, but is still, I feel, a factor of 100 or so too high. If you calculate what the NADH level should be, assuming that LDH maintains equilibrium between its substrates (Dr. Kastenschmidt's figures for pyruvate, lactate, and oxidized NAD I do not dispute), then you must conclude that either LDH is restricted in activity by some inhibitor, or that most of the NADH is in a separate compartment.

*Kastenschmidt:* The values for reduced NAD are consistent with our earlier publication (Kastenschmidt, Hoekstra, and Briskey, 1968). I do not mean to imply, as Dr. Scopes suggests, that this is only cytoplasmic reduced NAD. Rather, the values given are those obtained from extracting a muscle powder, prepared at liquid nitrogen temperatures, by the method of Klingenberg (1963). When one considers the known fact that there is some reduced NAD in the mitochondria, which we are also measuring in other methods, then the comparison of expected NADH from LDH equilibrium and measured total tissue levels becomes a meaningless exercise.

*Sybesma:* The control mechanism of glycolysis in muscle could be challenged much more in those muscles which have a less developed mitochondria system (Gauthier, 1969) than in others. Inactivation of the mitochondria then would occur more rapidly under certain unfavorable intramuscular circumstances. Would you think that this is one of the possibilities of a rapid post mortem glycolysis?

*Kastenschmidt:* This certainly is a possibility. Nevertheless, there are several examples in the literature where muscles especially adapted for anaerobic lactate production have very effective glycolytic control mechanisms. This is very well illustrated for the case of frog sartorius muscle. It is difficult to design an experiment to demonstrate unequivocally that stress-susceptible animals have indeed fewer mitochondria or that they are inactivated more rapidly during unfavorable intramuscular circumstances.

*Jay:* Aside from the more obvious effects of rapid *p*H fall, what are some of the less obvious effects of this phenomenon in PSE pork on muscle proteins? For example, is protein denaturation significantly different in stress-susceptible animals from that in stress-resistant animals?

*Kastenschmidt:* It is quite clear that both sarcoplasmic and myofibrillar proteins are being denatured in PSE muscle (Sayre and Briskey, 1963). Studies are now under way to clarify which of the major myofibrillar proteins are affected by the combination of low *p*H and high temperature.

*Henrickson:* What role does the nervous system play in the metabolism of muscle?

*Kastenschmidt:* I assume you mean in the metabolism of muscle post mortem. The electromyographic evidence of Schmidt, Goldspink, Roberts, Kastenschmidt, Cassens, and Briskey (1970) from our laboratory demonstrated that all electrical activity in terms of action potentials ceases within the first 5 min or less after stunning the animal. Hence, one must assume that the role of the nervous system must be exerted prior to or immediately after the time of death. At this time, of course, the nervous system can play an important role in regulation of blood flow to muscles and in adrenal responses as well as through muscular contraction.

*Henrickson:* The tissue from stress-susceptible animals is usually light-colored. We assume this is due to less myoglobin. Has the iron of the heme molecule been destroyed or has the animal failed to manufacture more myoglobin?

*Kastenschmidt:* We have been unable to demonstrate significantly less myoglobin in PSE muscles so I really cannot speculate about the second part of this question.

*Solberg:* What is the possibility that the paleness observed in some tissue is related to the drip through the leaching out of the highly water-soluble myoglobin?

*Lister:* Undoubtedly some myoglobin will be lost in the exudate from muscle, but most studies have shown that the amounts of myoglobin are similar in PSE and normal muscle. It is now generally accepted that the paleness arises from a loss of opacity of the surface layers of tissue caused by denaturation of the proteins.

*Edgar:* Has any complete study been made of the rate of glycolysis of porcine muscle in relation to temperature?

*Kastenschmidt:* No.

*Edgar:* It is interesting to note that the rapid drop in *p*H after death is accompanied by a high muscle temperature. Do you have any data to indicate that pigs with high preslaughter temperatures might have more rapid *p*H changes after death?

*Lister:* I have no data on this, but Dr. Sybesma has. There appears to be an association between high temperatures in muscle at death and rate of *p*H fall. How important this is, is not certain.

*Jay:* Your treatment of the topic of post mortem changes showing a lack of proteolysis is quite convincing. While you have been concerned with aseptic muscle, we have been unable consistently to show myofibrillar protein breakdown during low-temperature spoilage of muscle held for several weeks (Jay, 1966). In view of the foregoing as well as your data on sarcoplasmic and stroma proteins post mortem, what is the mechanism by which meat tenderization is achieved through post mortem aging?

*Goll:* I do not know, but a very interesting hypothesis, based on our data, might be that the release of $Ca^{++}$ from the sarcoplasmic reticulum (which in turn causes a $Ca^{++}$-induced degradation of the Z line) is the event principally responsible for the post mortem increase in tenderness. We have observed that both rate and extent of Z-line degradation are much more extensive at higher temperatures (above 16° C), so this Z-line hypothesis is also consistent with the observations showing that high-temperature aging improves meat tenderness.

*Passbach:* You commented on the possible role of the adrenal gland on $Na^{+}$ and $K^{+}$ and the effect on membrane permeability as related to PSE. Would you comment on the possible role of mineralocorticoids on the glycolytic rate of PSE muscle?

*Lister:* I have no information on the possible role of these, apart from your own observations (Passbach, Mullins, Wipf, and Paul, 1969). Presumably the glycolytic rate of PSE muscle reflects the rate of breakdown of ATP and it would be interesting to know how adenosine triphosphatase activity in muscle is affected by adrenal hormones.

*Jay:* What is the relative contribution of the various morphologic muscle entities to the overall edible and desirable quality of meat?

*Cassens:* I know of no figures to quote but will say that a number of morphologic factors do contribute to one's general opinion of a sample of meat. These include fat cells, connective tissue, and myofibrils; their state and properties are altered by change post mortem and by cooking. The entire issue is extremely involved and is just the reason that a conference such as this is conducted—so that a better understanding of the biochemistry, physiology, and morphology of muscle and its components can be made, and then the pieces reconstructed to comprehend meat quality more fully.

*Sleeth:* You suggested that toughness of pale, soft, exudative meat may be associated with degree of shortening of the sarcomere. Could tenderness be ensured by preventing muscle shortening during rigor mortis?

*Lister:* Probably not. Herring (1968) has recently reviewed the literature pertaining to this problem and there would seem to be little doubt that one of many factors contributing to toughness in meat is related to the extent of overlap of actin and myosin filaments. The amount of connective tissue and the nature of the cross linkages between and within the structural units also probably contribute to toughness. The contribution made by these is unlikely to be appreciably affected by the contractile state of the muscle.

*Edgar:* You have observed that in highly contracted muscle the passively shortened fibers are of a different configuration. Can you account for this in its relationship to meat tenderness?

*Cassens:* Cooper, Breidenstein, Cassens, Evans, and Bray (1968) showed a greater content of kinked fibers (passively shortened) in E maturity muscle than in A and B maturities. The E maturity muscle was less tender.

*Henrickson:* In muscle contraction, some fibers contract to various degrees. For example, some fibers within a bundle may be straight, some buckled or kinky, and some severely twisted. Why this difference?

*Cassens:* First, one should consider the factor of active or passive contraction. As a fiber contracts, it may pull adjacent fibers along with it. Then, in a fixed sample, one sees actively contracted (straight) and passively contracted (kinked) fibers. The more basic question is what causes only certain fibers in a bundle to contract. A motor unit may be distributed over a number of bundles and innervate only certain fibers in a bundle. A stimulus would therefore elicit contraction in only a certain number of fibers in the bundle.

*Sleeth:* You suggested a high correlation between fiber size and muscle contraction and their relation to tenderness. Can fiber size and contraction be controlled or managed to result in optimum tenderness?

*Cassens:* We have shown (Herring, Cassens, Suess, Brungardt, and Briskey, 1967) very clearly that a decrease in muscle length produces an increase in fiber diameter. Dr. Marsh can answer your question about controlling such factors to assure tenderness.

*Marsh:* To some extent the degree of shortening, and hence of toughening, can be controlled by appropriate treatment after slaughter. Cold shortening, caused by the early post mortem exposure of muscle to low temperatures, can be prevented by avoiding this temperature range during rigor onset. Appreciable shortening and consequent toughening occur when lamb carcasses are cooked rapidly shortly after slaughter, and excised bovine muscles display the same effect. It has not yet been established, however, that cold shortening occurs to a significant extent when a beef carcass or side is cooked before rigor onset; perhaps its size (and consequent slow rate of cooling) may prevent the effect.

*Sleeth:* What evidence do you have that the finer the texture the more desirable the muscle for use as a food?

*Cassens:* Cooper, Breidenstein, Cassens, Evans, and Bray (1968) reported that bundle size is positively related to shear force ($r = 0.39$) and negatively related to tenderness score ($r = -0.41$). Visual appraisal of texture actually is a rather good estimate of bundle size; this indicates that a unit of muscle microstructure is being satisfactorily judged and is therefore of consequence in view of the above correlations.

*Kropf:* Are the various structural areas of the sarcomere more vulnerable to heat effects in the relaxed or hypercontracted state than in the less severely contracted muscle?

*Goll:* Your question is a good one since many studies that attempt to explain meat tenderness apparently fail to recognize that tenderness is a property of cooked meat, at least for most *Homo sapiens.* Unfortunately, we are also guilty of this omission, and we have not done any ultrastructural studies comparing the effects of heat on relaxed or contracted muscle. The studies we have done thus far on the effects of heat on muscle show (rather surprisingly) that the Z line is remarkably stable to heat and that the most noticeable effect of heat is some fragmentation in the I and A bands. Even this fragmentation occurs only at temperatures above 80° C, a temperature where meat is rather well done. These results are still rather preliminary, but based on them I would suggest that degradation of the Z line before cooking would produce rather large beneficial effects on meat tenderness and that these effects would be carried over to the cooked product without much change.

*Kropf:* Why is contracted muscle progressively less tender than relaxed muscle? What structural or chemical difference explains the difference in the above cases for cooked muscles?

*Goll:* I do not have unequivocal answers for any of these questions, but can offer my ideas concerning these effects. I suspect that one major reason that contracted muscle is less tender than relaxed muscle is that contracted muscle has more filaments per unit of muscle cross section than relaxed muscle. This occurs because the amount of thick-thin filament overlap increases as a muscle shortens, and if shortening proceeds far enough, thin filaments also begin to overlap in the center of the A band. Our preliminary results suggest that this shortening carries over into the cooked state without much change. If shortening is very severe (sarcomere lengths of 1.5 μ or less), then the ends of the thick filaments begin to push against the Z line, causing considerable disruption of the Z line and perhaps also of the thick filaments themselves. The severity of this Z-line disruption can be seen clearly in Fig. 2 of Stromer, Goll, and Roth (1967), where the Z line can be seen only as a dark amorphous mass. As I suggested before, I suspect that the beneficial effects of Z-line degradation on tenderness pass through without much change to the cooked product. As is probably evident from my answer, I think at present that the state of the Z line has a major role in what is often called "actomyosin toughness." Interestingly, we have observed on several occasions that the Z line in older (more mature) animals is less labile than the Z line in younger animals; it may also account for part of the tenderness differences observed between young and old animals.

In addition to the structural changes I have discussed, there may be some chemical changes also contributing to tenderness differences between contracted and "super-contracted" (sarcomere lengths less than 1.5 μ) muscle, but the nature of any such chemical differences is entirely unknown at the present time.

*Edgar:* One of the reasons for toughness in meat is excessive loss of moisture in cooking. Would you expect that in addition to the increased solubility of connective tissue on aging, enzyme-induced tenderness could reduce the ability to contract and force out water?

*Goll:* If by enzyme-induced tenderness you are referring to catheptic enzymes, then I should point out that I do not believe that cathepsins play any major role in post mortem tenderization. I think it is intuitively obvious that loss of Z lines would prevent muscle shortening because in the absence of Z lines, individual sarcomeres would shorten, but since they would not be connected to one another, there would not be any shortening of the entire fiber. Thus, loss of Z lines would reduce the amount of water lost due to fiber contraction or shortening. I should also add, however, that I think the improvement in tenderness due to additional water retained in muscle without Z lines would be very small compared to the tenderness improvement due directly to fragmentation of fibers without Z lines.

*Jay:* Would you elaborate more on the possible mechanism of the increased adenosine triphosphatase activity of post mortem muscle components? Is it conceivable that this is due at least in part to increased hydration capacity of these proteins, so that more reactive sites facilitate their enzymic activities?

*Goll:* It is possible that increased hydration is involved in the higher specific adenosine triphosphatase activity of actomyosin from post mortem muscle, but at present I think that a more probable explanation is partial oxidation of -SH groups on actin or myosin or both, similar to that suggested by Dr. Maruyama in this discussion as a possible cause for the increased rate of superprecipitation in post mortem actomyosin.

*Hamm:* The increase in tenderness during storage of postrigor muscles is due to a certain disintegration of the myofibrillar system, as has been demonstrated by several authors. The changes of the rheological properties of muscle post mortem may support the conception of a loosening of the actomyosin system *in situ*. Beef muscle was intensively comminuted and homogenized in distilled water at different times post mortem. $T_0$ and $T_A$ were measured in a rotating viscosimeter. $T_0$ describes the firmness and plasticity of the homogenate; it is defined as the pushing force at which the substance just starts to flow. $T_A$ is the force which is needed for shearing the system.

One hour post mortem the rheological data $T_0$ and $T_A$ are low; this means that the homogenate is a soft paste with a low viscosity. This is certainly due to the dissociation of actomyosin in the presence of ATP. The particles are highly flexible and follow the laminary flowing of the homogenate. Rigor mortis, however, causes a remarkable increase in the rheological values, i.e., an increase of the rigidity of the particles. The myofibrils are now less flexible; they are hindering each other and, consequently, the internal friction of the system is increased. Homogenates of postrigor muscles show significantly lower rheological values than those of rigor muscles. This reduced viscosity must be due to a decrease in the rigidity of the fibrils which indicates a certain disintegration of the actomyosin system. These changes are so strong that the rheological values after 3 days are about the same as the values immediately after slaughter.

*Karmas:* How can the large water-binding of muscle tissue be explained in morphological terms?

*Cassens:* Morphology may provide an answer when more aspects of molecular arrangement are discovered.

*Edgar:* The muscles of different species of meat animals vary widely in water-binding (Hamm, 1960). Furthermore, we find a wide variation within a species such as beef, and in fact, there even appears to be an association with age and quality of the beef carcass. The higher quality grades such as Good and Choice have higher water-binding properties than the lower quality grades of Canner and Cutter. What do we know about the difference in morphology of the muscle of old and young beef animals? Is the water-binding property of muscle primarily related to its chemical status or to its chemical morphology? Further, how might morphology *per se* influence water-binding?

*Cassens:* That is a very involved question and I will give only a few thoughts. First, the amount of lipid might influence results, depending on the method used. Cooper, Breidenstein, Cassens, Evans, and Bray (1968) have shown a shorter sarcomere length in older, more mature carcasses—this might affect water-binding. Then too, one would have to consider the ante mortem treatment of the animal and the post mortem change the muscle has undergone. Chemical status is probably the factor responsible for different water-binding capacities but these changes are quite conceivably detectable in morphology.

*Hamm:* Water-holding capacity is an important feature of meat quality, and there is a drastic drop in water-holding capacity during the development of rigor mortis. There is a regrettable lack of knowledge about the morphology of this problem. Could the electron microscope be used?

*Cassens:* Samples for electron microscopy are dehydrated and subjected to a number of other severe treatments. I think that use of electron microscopy to gain insight into the above problem would be rather indirect. It could provide useful information in regard to the integrity of protein structures and also provide dimensional or spacing data.

*Jay:* In view of the striking glycolytic differences between "stress-susceptible" and "stress-resistant" muscles, what is the difference between the water-holding capacity of the respective muscle proteins, and how much of this difference, if any, do you feel is due to the glycolytic effects?

*Kastenschmidt:* Muscle proteins from PSE muscles are considerably lower in water-binding capacity (Briskey, 1964). A large part of this decrease, we feel, is due to the low *p*H and high temperature conditions at the onset of rigor mortis. It is difficult to separate glycolytic (*p*H) from temperature effects; however, it is clear (Borchert and Briskey, 1964) that lowering the temperature by partial freezing with liquid nitrogen leads to retention of acceptable water-binding capacity.

*Edgar:* When prerigor muscle is minced with NaCl, the water-binding ability is improved. Can this be a result of the weakening of the actin-myosin interaction?

*Goll:* Do you mean that the water-holding capacity is improved relative to the water-holding capacity of prerigor muscle not minced in NaCl? If so, it is possible that part of this increase is due to "weakening" of the actin-myosin interaction since, *in vitro,* the actin-myosin interaction is weakened by elevating the ionic strength in the presence of ATP (prerigor muscle should contain ATP). However, any increase in water-holding capacity due to weakening of the actin-myosin interaction should disappear as the ATP level falls to zero in post mortem muscle. Therefore it seems likely that the added NaCl is acting to bind some additional water itself, and that in the presence of ATP, NaCl may cause some "loosening" of the thick filament structure (depending on ATP concentration at the time of the NaCl addition and the amount of NaCl added) and that this "loosening" of the thick filament structure makes more sites on myosin available for water-binding. Such "loosening" of the thick filaments would probably not be reversible upon the disappearance of ATP.

*Henrickson:* Is there evidence as to how the myosin cross-bridges to actin when muscle is in rigor?

*Goll:* Not in terms of molecular details, although Huxley's recent X-ray diffraction results (Huxley, 1968; Huxley and Brown, 1967) suggest that almost all the myosin cross bridges are attached to actin in muscle without ATP, whereas only part of them are attached to actin at any given instant in contracting muscle. The possibility remains that the actin-myosin interaction in post mortem muscle is different from that in living muscle, but there is not any conclusive evidence on this point.

*Wellington:* How can the scanning electron microscope by used in morphological studies of muscle as a food?

*Cassens:* I think it can be used to further such aspects as connective tissue, fiber type, and contraction state research. These morphological features are related to the use of muscle as a food and the scanning microscope might well provide some additional useful information.

*Sleeth:* What is the relation of number and size of fat cells to quality; is there a species difference relative to number and size, and do you feel the present USDA standards for beef grading relative to visual fat content (marbling) adequately reflect quality?

*Cassens:* I do not know if there is a species difference in fat cell size but there is a large difference for different sites within an animal. For example, Moody and Cassens (1968) showed in bovine muscle that subcutaneous fat has larger fat cells than intermuscular fat, and that both subcutaneous and intermuscular fat cells are much larger than those of intramuscular fat. Histological estimates of fat follow generally with chemical fat determinations, but the accuracy is rather poor (Moody and Cassens, 1968; Cooper, Breidenstein, Cassens, Evans, and Bray, 1968). Breidenstein, Cooper, Cassens, Evans, and Bray (1968) showed that the effect of marbling was greater for juiciness, flavor, and general opinion than for tenderness. Increased marbling generally improved palatability although the differences were not significant in some cases, and a rather wide range in marbling had to be considered to give significant results.

*Henrickson:* What causes a connective tissue cell to fill with fat?

*Cassens:* We do not know, but assume it must be due to increase or decrease in an enzyme or hormone system.

*Henrickson:* To what extent do connective tissue (collagen, elastin, reticulin) proteins undergo a form of rigor mortis? Do collagen fibers change after death to become less tender? Has any change in the fiber been histologically shown?

*Goll:* If rigor mortis is defined in terms of either ATP loss or stiffness, it seems unlikely that connective tissue *per se* undergoes rigor mortis. I do not think that collagen fibers change after death to become less tender (become more highly cross-linked). Rather, it seems more likely that collagen may change during post mortem storage to become "more tender" (less highly cross-linked, due to acidity development in post mortem muscle). I do not know of any extensive histological study demonstrating post mortem changes in connective tissue fibers.

*Henrickson:* What is known about the sarcolemma? For example, what is the thickness of the sarcolemma? Does it vary in thickness among fiber types? Do large fibers have a different sarcolemma?

*Goll:* It depends on how the sarcolemma is defined. The muscle cell is immediately surrounded by a lipoprotein membrane which is 75–100 Å in thickness. This membrane may be called the "plasmalemma," although some authors prefer to use the term sarcolemma to refer to only this membrane. Immediately exterior to the plasmalemma is an amorphous area 300–500 Å in width. Outside this amorphous area is a layer about 300 Å in width. This layer appears to consist of small collagen fibrils. Exterior to the collagen-containing area is a zone 100 Å in width that appears to consist of a mesh of fine filaments. It is probably a combination of these outer three zones that Wang (1956) earlier referred to as the "fibrous envelope." The exact relationship of these four zones to the sarcolemma and the muscle fiber remains a matter of some controversy. Also, the chemical composition of the four zones is still not clear; for example, it has not been settled whether the sarcolemma contains collagen or not. In view of these uncertainties, it is difficult to ascertain whether the sarcolemma varies among different muscle fiber types or among different sizes of muscle fibers. We have recently obtained some indirect evidence that the sarcolemma associated with "red" muscle fibers differs from that associated with "white" muscle fibers, but the nature of this difference is entirely unknown. In short, the sarcolemma is badly in need of further characterization.

*Jay:* What is the relative susceptibility of PSE muscle to microbial spoilage?

*Cassens:* It has generally been thought that microbial spoilage occurs more slowly in PSE muscle due to the lower *p*H. The excessive moisture that forms in packages may be a confounding factor, however.

*Jay:* Have there been any physiologic studies conducted on the Kobe beef animals in Japan? If so, how do they compare with the more conventional slaughter cattle of comparable breed?

*Lister:* I have no knowledge of any work done on Kobe beef.

*Hamm:* We have observed some cases of an extremely high rate of post mortem glycolysis and of breakdown of ATP in longissimus dorsi muscles of beef animals. Within 1 hour after slaughter, all ATP was transformed into IMP. The rate of these changes corresponded to that in PSE muscles of pigs but it did not necessarily lead to wateriness or paleness of meat. Nevertheless, in such muscles, the low water-holding capacity of rigor muscle is reached within 1 or 2 hours instead of 10 to 24 hours as in normal muscles. Therefore, presalting the hot meat for production of frankfurter type sausages does not improve the water-holding capacity of such meat. It would be of interest to know whether such a high rate of ATP breakdown and glycolysis of bovine muscles was found also in other laboratories and whether a relation of ante mortem treatment to such post mortem conditions in beef has been observed.

## *References*

Allen, E., J. C. Forrest, A. B. Chapman, N. First, R. W. Bray, and E. J. Briskey. 1966. Phenotypic and genetic associations between porcine muscle properties. *J. Anim. Sci. 25*:962.

Bendall, J. R. 1960. Post-mortem changes in muscle, p. 227. *In* G. H. Bourne (ed.), *The Structure and Function of Muscle,* Vol. 3. Academic Press, New York.

———. 1969. *Muscles, Molecules and Movement.* Heinemann Educational Books, London.

Borchert, L. L. and E. J. Briskey. 1964. Prevention of pale, soft, exudative porcine muscle through partial freezing with liquid nitrogen post-mortem. *J. Food Sci. 29*:203.

Breidenstein, B. B., C. C. Cooper, R. G. Cassens, G. Evans, and R. W. Bray. 1968. The influence of marbling and maturity on the palatability of beef muscle I. Organoleptic and biochemical considerations. *J. Anim. Sci. 27*:1532.

Briskey, E. J. 1964. Etiological status and associated studies of pale, soft, exudative porcine musculature. *Advance. Food Res. 13*:89.

Cooper, C. C., B. B. Breidenstein, R. G. Cassens, G. Evans, and R. W. Bray. 1968. The influence of marbling and maturity on the palatability of beef muscle II. Histological considerations. *J. Anim. Sci. 27*:1542.

Cooper, C. C., R. G. Cassens, and E. J. Briskey. 1969. Capillary distribution and fiber characteristics in skeletal muscle of stress-susceptible animals. *J. Food Sci. 34*:299.

Cooper, C. C., R. G. Cassens, L. L. Kastenschmidt, and E. J. Briskey. 1970. Enzyme changes during muscle differentiation. (Submitted.)

Davey, C. L., and K. V. Gilbert. 1968. Studies in meat tendencies. Changes in the extractability of myofibrillar proteins during meat aging. *J. Food Sci. 33*:2.

Gauthier, G. F. 1969. Ultrastructure of three fiber types in mammalian skeletal muscle. *This volume,* p. 103.

Greaser, M. L., R. G. Cassens, E. J. Briskey, and W. G. Hoekstra. 1969. Post-mortem changes in subcellular fractions from normal and pale, soft, exudative porcine muscle 1. Calcium accumulation and adenosine triphosphatase activities. *J. Food Sci. 34*:120.

Greaser, M. L., R. G. Cassens, W. G. Hoekstra, and E. J. Briskey. 1970. Effect of *p*H-temperature treatments on the calcium accumulating ability of purified sarcoplasmic reticulum. *J. Food Sci. 34*:633.

Greaser, M. L., R. G. Cassens, W. G. Hoekstra, E. J. Briskey, G. R. Schmidt, S. D. Carr, and D. E. Galloway. 1969. Calcium accumulating ability and compositional differences between sarcoplasmic reticulum fractions from normal and pale, soft, exudative porcine muscle. *J. Anim. Sci. 28*:589.

Haga, T., M. Yamamoto, K. Maruyama, and H. Noda. 1966. The effect of myosin and calcium on the solubilization of F-actin from muscle mince. *Biochim. Biophys. Acta 127*:128.

Hamm, R. 1960. Water-binding of muscle tissue. *Advance. Food Res. 10.*

Herring, H. K. 1968. Muscle contraction and tenderness, p. 47. *Proc. 21st Reciprocal Meat Conf.* (Chicago, Ill.).

Herring, H. K., R. G. Cassens, G. G. Suess, V. H. Brungardt, and E. J. Briskey. 1967. Tenderness and associated characteristics of stretched and contracted bovine muscles. *J. Food Sci. 32*:317.

Huxley, H. E. 1968. Structural difference between resting and rigor muscle; evidence from intensity changes in the low-angle equatorial X-ray diagram. *J. Mol. Biol. 37*:507.

Huxley, H. E. and W. Brown. 1967. The low-angle X-ray diagram of vertebrate striated muscle and its behavior during contraction and rigor. *J. Mol. Biol. 30*:383.

Huxley, H. E. and J. Hanson. 1960. The molecular basis of contraction in cross-striated muscle, p. 183. *In* G. H. Bourne (ed.), *The Structure and Function of Muscle,* Vol. 1. Academic Press, New York.

Jay, J. 1966. Influence of post-mortem conditions on muscle microbiology, p. 390. *In* E. J. Briskey, R. G. Cassens, and J. C. Trautman (eds.), *The Physiology and Biochemistry of Muscle as a Food.* Univ. of Wis.

Johnson, P. and A. J. Rowe. 1964. An ultracentrifuge study of the actin-myosin interaction, p. 279. *In* J. Gergely (ed.), *Biochemistry of Muscle Contraction.* Little, Brown, Boston.

Kastenschmidt, L. L., W. G. Hoekstra, and E. J. Briskey. 1968. Glycolytic intermediates and co-factors in "fast-" and "slow-glycolyzing" muscles of the pig. *J. Food Sci. 33*:151.

Klingenberg, M. 1963. Diphosphopyridine nucleotide, p. 528. *In* H. U. Bergmeyer (ed.), *Methods of Enzymatic Analyses.* Academic Press, New York.

Krzywicki, K. and P. W. Ratcliff. 1967. The phospholipids of pork muscle and their relation to the post-mortem rate of glycolysis. *J. Sci. Food Agr. 18*:252.

Lister, D., R. A. Sair, J. A. Will, G. R. Schmidt, R. G. Cassens, W. G. Hoekstra, and E. J. Briskey. 1970. Metabolism of striated muscle of "stress-susceptible" pigs breathing oxygen or nitrogen. *Amer. J. Physiol. 218*:102.

Lohse, B. 1966. Demonstration on meat quality, p. 201. *In Meat for Food,* Sonderdruck aus *Z. Tierzuchtung Zuchtungsbiologie,* Mariensee, Germany.

Maruyama, K. and Y. Ishikawa. 1964*a*. Effects of some sulfhydryl compounds on the magnesium enhanced adenosine triphosphatase activity of myosin B. *J. Biochem. 56*:372.

———. 1964*b*. Effect of chloromercuribenzoate on the Mg-enhanced ATPase activity of actomyosin. *Annot. Zool. Jap. 37*:134.

Moody, W. G. and R. G. Cassens. 1968. A quantitative and morphological study of bovine longissimus fat cells. *J. Food Sci. 33*:47.

Morita, S., R. G. Cassens, and E. J. Briskey. 1969. Localization of myoglobin in striated muscle of the domestic pig: Benzidine and $NADH_2$-TR reactions. *Stain Tech. 44*:283.

———. 1970. Myoglobin localization in rabbit, porcine and bovine muscle. (Submitted.)

Morita, S., R. G. Cassens, R. G. Kauffman, and E. J. Briskey. 1970. Myoglobin localization and PSE muscle. (Submitted.)

Passbach, F. L., A. M. Mullins, V. K. Wipf, and B. A. Paul. 1969. Influence of aldosterone on pale, soft, exudative (PSE) porcine muscle, p. 64. *In Proc. 15th Europe. Meeting of Meat Res. Workers* (Helsinki).

Rash, J. E., J. W. Shag, and J. J. Biesele. 1968. Urea extraction of Z-bands, intercolated disks, and desmosomes. *J. Ultrastructure Res. 24*:181.

Sair, R., A. D. Lister, W. G. Moody, R. G. Cassens, W. G. Hoekstra, and E. J. Briskey. 1970. Action of curare and magnesium on striated muscle of stress-susceptible pigs. *Amer. J. Physiol. 218*:108.

Sayre, R. M. and E. J. Briskey. 1963. Protein solubility as influenced by physiological conditions in the muscle. *J. Food Sci. 28*:674.

Schmidt, G. R., G. Goldspink, T. Roberts, L. L. Kastenschmidt, R. G. Cassens, and E. J. Briskey. 1970. Electromyography and membrane potential in skeletal muscle of "stress-susceptible" and "stress-resistant" pigs. (Submitted.)

Sharp, J. G. 1959. Observation on aseptic autolyses in muscle. *Proc. 5th Europe. Meat Res. Conf.* (Paris, France).

Stromer, M. H., D. E. Goll, and L. E. Roth. 1967. Morphology of rigor-shortened bovine muscle and the effect of trypsin on pre- and post-rigor myofibrils. *J. Cell Biol. 34*:431.

Wang, H. 1956. The sarcolemma and fibrous envelope of striated muscles in beef. *Exp. Cell Res. 11*:452.

# Index